OCT 20 '09

DATE DUE

	MAY 0 9 2011		

Demco, Inc. 38-293

The American West at Risk

The American West at Risk

Science, Myths, and Politics
of Land Abuse and Recovery

Howard G. Wilshire, Jane E. Nielson,
Richard W. Hazlett

UNIVERSITY PRESS
2008

OXFORD
UNIVERSITY PRESS

Oxford University Press, Inc., publishes works that further
Oxford University's objective of excellence
in research, scholarship, and education.

Oxford New York
Auckland Cape Town Dar es Salaam Hong Kong Karachi
Kuala Lumpur Madrid Melbourne Mexico City Nairobi
New Delhi Shanghai Taipei Toronto

With offices in
Argentina Austria Brazil Chile Czech Republic France Greece
Guatemala Hungary Italy Japan Poland Portugal Singapore
South Korea Switzerland Thailand Turkey Ukraine Vietnam

Copyright © 2008 by Oxford University Press, Inc.

Published by Oxford University Press, Inc.
198 Madison Avenue, New York, New York 10016

www.oup.com

Oxford is a registered trademark of Oxford University Press

Library of Congress Cataloging-in-Publication Data
Wilshire, Howard Gordon, 1926-
The American west at risk : science, myths, and politics of land abuse and
recovery / Howard G. Wilshire, Jane E. Nielson, Richard W. Hazlett.
     p. cm.
 Includes bibliographical references and index.
 ISBN 978-0-19-514205-1
1. Nature conservation—West (U.S.) 2. Conservation of natural resources—
West (U.S.) 3. Nature—Effect of human beings on—West (U.S.) 4. Land use—
Environmental aspects—West (U.S.) 5. West (U.S.)—Environmental conditions.
I. Nielson, Jane E. II. Hazlett, Richard W. III. Title.
QH76.5.W34W57 2008
333.720978—dc22   2007024318

9 8 7 6 5 4 3 2 1
Printed in the United States of America
on acid-free paper

COVER STORY

The spectacular cover photograph was taken by Sam Chase, an oil-industry employee, on a work-related flight from Sacramento to Bakersfield, California, the morning of December 20, 1977. In the photo, an enormous plume of dust sweeps out of the Tehachapi Mountains and rises 5,000 feet above the southern San Joaquin Valley. It vividly illustrates how natural forces extend human disturbances on arid lands, resulting in severe erosion. In this case an extreme wind storm was stripping overgrazed lands, unprotected farmlands, urban developments, and dirt roads and tracks. All are especially vulnerable to wind erosion.

When the small plane took off from Sacramento, Bakersfield Airport's pre-recorded weather report spoke only of morning fog and light winds. The storm's violence had kept airport workers at home, unable to change the recording. Approaching Bakersfield, everyone on the plane could see the futility of trying to land there. Sam took the photograph after the plane turned east to Tehachapi—apparently he and the pilot were the only ones who hadn't lost their stomachs.

The incredible record left by this storm is unparalleled in the geologic literature, and we are grateful to Sam for making this unique photograph available.

We dedicate this work to our family—Ruth and Dave, David, Paul, and Collette, Ben, Owen, and Scé, and our grandchildren David Hiroshi and family, Ayako, Amanda, Ann, Jacob, and Daniel Miyahara; and Madalyn and Nathaniel Wilshire— and to their futures. It is also dedicated to our students, who give us hope that the western United States and the world can support the communities of living things that support us all.

ACKNOWLEDGMENTS

We could not have completed this book without the assistance and support of many generous friends and colleagues. Most notably, we benefited enormously from the critical technical reviews of specialists and generalists among our past and present colleagues in the U.S. Geological Survey and from many other experts whose work we encountered in researching the book. We are very grateful for the technical reviews of chapters and appendices from Dr. Fred Swanson, U.S. Forest Service, Pacific Northwest Research Station; Dr. Kristiina Vogt, College of Forest Resources, University of Washington; Dr. Colin Campbell of the Association for the Study of Peak Oil & Gas, Oil Depletion Analysis Centre, Stockholm, Sweden; Dr. Warren Hamilton, international expert on earth and planetary sciences, Geophysics Department, Colorado School of Mines; Dr. David Pimentel, ecologist with very broad experience in maintenance of Earth's life support systems, Department of Entomology, Systematics and Ecology, Cornell University; Dr. Charles Francis, Director of the Center for Sustainable Agriculture Systems, University of Nebraska; Vernon Brechin, indefatigable nuclear weapons production and testing researcher; Dr. William Schlesinger, President, Institute of Ecosystem Studies, Millbrook, New York, with wide experience of desert and forest ecosystem responses to global climate change; Dr. Eric Grosfils, Professor of Geology, Pomona College; Allison Jones, ecological grazing impacts researcher, Wild Utah Project; Dr. Douglas Markle, Professor of Fisheries, Oregon State University; Dr. Michelle Marvier, Assistant Professor of Biology, Santa Clara University; Claire Todd, University of Washington; and Rue Furch, Sonoma County, California, Planning Commissioner.

We also owe very special thanks to Dr. D. D. Trent, former Geology Professor at Citrus Community College, Glendora, California, and our good friend, whose broad geologic experience and very excellent textbook on environmental geology have informed and guided us. In addition, Dr. Trent reviewed nearly every chapter of this book. We have had especially helpful discussions on a wide range of subjects with a number of past and present U.S. Geological Survey researchers, including Dr. Jack Barbash on historical and pesticide issues, who also critiqued an early draft of our conclusions; Dr. Jayne Belnap on soil crusts; Dr. Michael Bogan on southwestern U.S. environmental problems; Dr. Robert Webb, a dear friend and former co-author, on water supply and the impacts of dams; Dr. Grecia Matos on U.S. industrial materials consumption; and Gail P. Thelin on the distribution of western U.S. cropland.

Discussions with J. A. (Tony) Fallin, including issues of energy, water, and pollution have been invaluable, along with his wonderful letters—complete with detailed and artistic maps and diagrams. We also consulted with Dr. Norman Ellstrand, University of California–Riverside, on genetically modified organisms; Mark Jorgensen, on disease transmission in mountain sheep, Anza Borrego State Park, California; Patrick Diehl and Tori Woodard on Utah grazing issues; Daniel Patterson on many issues affecting desert wildlife; Linda McElver, San Luis Obispo, California, on human health issues; Leda Beth Gray, Dr. David Drake, and Dr. Jeff Lovitch on wildlife issues; and Dr. Art Lachenbruch on earth science models. Another very special thank you goes to Dr. Angela Jayko for access to U.S. Geological Survey tape recordings of 1998 debris flow and landslide scenes, with survivor stories.

Thanks especially to Lenny Siegel of the Pacific Studies Center, Mountain View, California, for access to a large library of government literature on military issues, and to Anna Tellez, Susie Bravos, and Michael Moore, our valued friends and always-helpful U.S. Geological Survey Library staff in Menlo Park, California. We have benefited also from the information, references, and answers to our questions supplied by Professor John Reganold, Washington State University; Professor Steven Wing, University of North Carolina–Chapel Hill; Preston Sullivan, ATTRA (National Sustainable Agriculture Information Service); Adina Merelander, University of California–Berkeley; Ralph Nobles, retired nuclear physicist; Diane D'Arrigo, Nuclear Information and Resource Service; Professor Emeritus John Letey, University of California–Riverside; Brett D. Lieberman, Newhouse News Service; and Jodi Frediani, environmental forestry consultant. In 1998, Geirr Fagnastøl gave Howard Wilshire and Jane Nielson a tour of his mountainside farm near Voss, Norway, including its shared støl (summer pasture), and explained the system of sharing grazing rights handed down from Viking days. He also helped Jane research her Nielson forebears, who had farmed the very same land until 1857.

Our intent in preparing this book was to open the scientific knowledge of environmental issues to nonscientists, especially to citizens concerned about environmental problems in their own communities, and to lawyers and reporters trying to make sense of competing legal and political claims. Taking a leaf from the popular-science guides of our beloved former U.S. Geological Survey colleague N. King Huber, we sought out chapter critiques from many nonscientists of our acquaintance. Their influence on the writing has been profound. We owe very special thanks to Dee Cope, a skilled technical editor and dear friend, who critiqued all chapters, including the most difficult subjects. Among the many others who shared their time and thoughts on how to explain technical information to nonspecialists, we especially thank independent environmental and science film makers Doug Prose and Diane LaMacchia; citizen-activists Ernest Goitein, Claire Feder, Olive and Henry Mayer, and H. R. Downs; ecologist-activists Edie and Jim Harmon;

marketing specialist Garry Lambert; environmentalist Karen Schambach; Hanford downwinder June Stark Casey; Nevada Test Site downwinder and Webmaster Wayne Simister; and concerned citizens Ursula and Erich Baur-Senn.

For sharing knowledge on many subjects, from politics to legal issues, we owe many thanks to attorney Jeff Ruch, Executive Director, Public Employees for Environmental Responsibility, staunch defender of government employees who try to do their jobs; scientist-attorney Brendan Cummings of the Center for Biological Diversity, who understands both biology and ecology and how to defend the natural world; our dear friend Lyn Sims, for alerting us to the continued presence of aerially sprayed "simulant" *Serratia marcescens* in San Francisco's population; and Rita DeSouza, for her experience and many inspirational no-holds-barred, no-quarter-given thoughts and insights.

We are deeply grateful for Molly Eckler's expert production of electronic artwork, not to mention her wonderfully accurate renderings of all Laguna de Santa Rosa fauna plus much of its flora and her staunch defense of it all; for Byron Kesler's splendid hand-drawing of the wind energy map; and to the Pomona College Department of Geology for generously funding the illustration work. In addition, we are indebted to our friend and former colleague Dr. John Dohrenwend for generously sharing his own photographs and his gorgeous scenes from satellite images; to James Catlin of the Wild Utah Project for access to his detailed maps of off-road vehicle incursions into Wilderness Study Areas; and to Gail Hoskisson, Southern Utah Wilderness Alliance, for photographs of energy exploration damage in Utah Wilderness Study Areas. We are also indebted to Lighthawk and pilot Steve Parker, also pilot Zach Miller, of Pomona College, for extended photography overflights. We must also give special recognition to Sam Chase, who many times has allowed Howard Wilshire to use his spectacular photograph of the December 1977 southern San Joaquin Valley windstorm, shown on the book's cover, for educational purposes.

We are grateful to the many people who eased the creative and communications process, especially Lori Keala of the Pomona College Geology Department, who always makes our Pomona visits delightful and helped keep the authors in touch. Sue Baur provided critical help in the face of computer rebellions. We also owe many thanks to our family: Paul Wilshire helped with figures and photographs and kept our Macintosh computers and peripherals operational and upgraded, while recycling outdated equipment; David Wilshire provided the big screen for better editing, probably saving Howard's eyes and neck, and led a photograph-the-authors project. The professional skills of Ben Pike created our Web site (www.losingthewest.com), making it both artistic and user friendly. Owen Pike donated critical computer drafting consultations, and Owen and Scé Pike provided an electric neck massager to keep us going. Thanks also to Ruth Miyahara, and all our grandchildren and great-grandchildren, for their patience with our distracted states of mind and short visits over the past nine years.

Not least, we thank our Oxford editors—first Cliff Mills, whose invaluable comments helped us shape and focus a nearly overwhelming wealth of information. Many thanks also go to Cliff's successor, Peter Prescott, to Peter's fabulous assistants Alycia Somers and Tisse Takagi, to our skilled and patient production editor, Sara Needles, and our skilled copyeditor Trish Watson, who have held our hands through the submission and production process, faithfully responding to every inquiry, great and small.

Above all, we salute the many devoted public servants at all levels of government, who try earnestly to make sure that federal, state, and local policies are based on valid scientific data and analyses, against political pressure and on slim funding. We also salute the land managers in the field, who continually must contend with angry citizens expressing many different viewpoints about how they should be doing their jobs. We once toiled among you, and we recognize your desire to do what is best for the environment, while facing the demands of a growing population and the need to feed yourselves and your families.

CONTENTS

The American West at Risk

Introduction   Obeying Nature

> *Science...does not compromise....[It] forces ideas to compete in a dynamic process. This competition refines or replaces old hypotheses, gradually approaching a more perfect representation of the truth....The natural process of a bureaucracy...tends to compromise competing ideas. The bureaucracy then adopts the compromise as truth and incorporates it into its being.*
> John M. Barry, *Rising Tide*

This book focuses on the human-caused environmental woes of America's 11 contiguous western states, its mostly arid western continental frontier. In the nineteenth century, penny pamphlets and dime novels mythologized the American west, making icons of its prospectors, "cowboys," northwestern loggers, and wide open spaces. The west was free of encroaching neighbors and government controls, open to fresh starts. As Robert Penn Warren wrote, in *All the King's Men*, "West...is where you go when the land gives out and the old-field pines encroach...when you are told that you are a bubble on the tide of empire...when you hear that thar's gold in them-thar hills...." But the "West" was more than gold and oil bonanzas—it was also a land of rich soils, bountiful fisheries, immense, dense forests, desert wonders, and sparkling streams. It is no myth that the western states were America's treasure house.

The romantic myths related to "winning" the west tend to obscure both its basic objective of resource exploitation and the huge public expenditures that supported every aspect, bestowing fortunes on a few. Western resources supported U.S. industrial growth and affluent lifestyle, but now they are highly depleted or largely gone, and the region is in danger of losing the ability to sustain an even moderately comfortable future. Much of what we have done to these magnificent lands opened them to devastating erosion and pollution. Today, whole mountains are being dismantled to produce metals from barely mineralized zones. Entire regions may be devastated in the attempt to extract the last possible drops of petroleum. We soon could cut down the last remnants of ancient western forests, along with the possibility of ever again seeing their like. Large-scale farming has opened vulnerable western soils to erosion by water and wind, perhaps inviting another dust bowl era. Irrigating vast crop acreages has converted many of them to salt farms, perhaps resembling the conditions that spelled doom for the ancient Babylonian Empire.

The how and why of these risks—the past and impending losses—are the theme of this book, along with proposals, strategies, hopes, and even *fantasies* about how

3

to salvage what is left and rebuild western lifelines. Most chapters describe land uses that degrade and deplete this slow-healing, mostly arid region—and especially its public lands. Each one explains how natural forces spread the negative impacts of forest clearings, farming and grazing, mining, roads and pipelines, all aspects of military training and weapons manufacturing and testing, urbanizing sprawl and excessive water developments, recreation, waste disposal, and energy extraction and use—resulting in severe erosion and flooding, reduced and compromised water supplies, and degraded air quality. We address manufacturing impacts under most of these headings. The final chapter explains the basic natural processes themselves, and a glossary defines scientific terms and concepts.

The authors expect that environmental science and analysis teachers and students may use this book, but also intend it for a guide for lawyers, journalists, other researchers, and people with environmental problems in their own neighborhoods. To help our audience understand the technical issues, we have endeavored to present them in terms that the general reader can understand. Clearly, no single book could possibly cover all the issues, case studies, and implications of these topics—some have whole libraries devoted to them. Each chapter is supported by detailed references for deeper exploration, which support the text and provide resources for the reader's class projects or local issues. Additional references, and an alphabetical bibliography, can be found on the Web site www.losingthewest.com.

The scientific information we rely on is backed by the best and most highly validated research and analyses that we could find, free from the influence of special interests. Most of it was performed by government or academic scientists at public expense. Many of the studies are published in peer-reviewed journals, but a very large component exists in academic and governmental reports that most citizens barely or never hear about, let alone see. The prose in these limited-distribution reports often features intractable jargon that would defeat nonscientist readers. Many chapters lean heavily on our own observations or detailed geologic research in nearly every western state—in particular, lead author Howard Wilshire's many published studies on human surface impacts.

Seeing the television images of Neil Armstrong's footprint on the soft lunar surface in 1969—so much like dry desert soils—Howard wondered how long that print would last in the Moon's airless and rainless environment. A colleague's calculations suggested that it will be recognizable more than a million years from now, setting Howard to examining his own tracks on desert streambeds and dark pebble-mosaic surfaces. Immediately, he realized they were no different from the marks of the burgeoning and abrasively noisy off-road motorcycles and dune buggies that were shattering his prized desert silence. He then began investigating the fate and durability of human imprints on the Earth under the eroding, transporting, and depositing processes of gravity, rain, and wind.

Since then, Howard has studied the impacts of motorized off-highway recreation (chapter 11), coal mining and mine reclamation projects (chapter 4), mountaintop wind farms (chapter 12), roads and pipelines (chapter 5), and the recovery of old mining ghost towns and military camps, including still-visible tank tracks and trampled base camps from World War II troop training (chapter 6). With colleagues from the U.S. Geological Survey, he also critiqued the site-selection procedures for a proposed nuclear waste dump (chapter 10). Along with Howard, Jane

Nielson and Richard Hazlett have performed geologic studies for evaluating the mineral resources of proposed wilderness areas and have performed numerous environmental reviews of proposed developments, required by state and federal environmental laws. Jane and Rick also bring to this book strong backgrounds in forestry, groundwater, energy, agriculture and grazing, and environmental toxicology issues. All the authors have helped journalists, lawyers, and concerned citizens comprehend research bearing on a multitude of environmental issues, and all have taught university-level courses in addition to performing research for the U.S. Geological Survey.

The host of valid scientific investigations that we cite, and many others, strongly indicate that preserving as much of the natural world as possible is essential to sustaining human health and safety, as well as the future food supply. Like many or most Americans, we once uncritically accepted the inevitability of human "progress" and ignored its Earth-degrading aspects. We expected to find that past and current human land uses do not preclude a prosperous future, but finally have accepted that most data point the other way. As Shakespeare's contemporary Sir Francis Bacon reputedly put it, "Nature, to be commanded, must be obeyed." That being the case, an understanding of Earth processes, and the significance of their biological connections, is a critical frame for societal and governmental attitudes toward the Earth and natural processes—and, in particular, for framing national land-use policies. These connections inevitably lead science into the murkier realms of bureaucratic stasis and political conflicts.

Land uses always have political implications. The federal government funded nineteenth-century western exploration, including many scientific investigations. The government also supplied the conquering armies that secured the west for Euro-American settlement and exploitation and gave away lands and support for settlers in all directions. Many federal policies have contributed and still do contribute to degrading the west and depleting its resources. Sadly, some government science has been done very badly, to the detriment of all Americans and at the hazard of many lives. In the following chapters, we discuss many past and current political conflicts over western land uses and the environmental consequences, and try to expose the misguided science that supported some very bad outcomes. The issues do not necessarily pit environmentalists against landowners or named political parties against each other, nor do they invoke right versus left splits. Some draw all these strands together. The policies and laws that we espouse are ones that good science and practical experience indicate can help us obey nature and better protect us from nature's worst ravages.

In 1968, University of California Professor Garrett Hardin published "The Tragedy of the Commons,"[1] a now-famous essay on the tendency of people to overuse and destroy any finite resource that anyone can access without restriction, especially under growing population pressure. Hardin likened such resources to the New England town commons, which were open grazing lands for many generations. It is easy to compare those commons to America's overused public lands, but Hardin pointed out that the atmosphere and water bodies, which nobody owns—including rivers, lakes, groundwater, and oceans—also are commons. Their ability to absorb all our waste is finite in terms of human life spans. Continuing to use them as dumping grounds is what destroys these life support systems for us as well as for wildlife. The multiple threats from human-accelerated global climate warming, only now becoming obvious, vividly illustrate the severe consequences of using the environment for a dumping ground.

Hardin and others have pointed out that overusing these common resources has allowed businesses to keep profits high. By not having to pay royalties for using up resources, or for cleanup or disposal, industries can raise their bottom lines. Misuses of public lands and private crop lands represent the "externalized" costs of doing business. But someone always pays, and in the United States, taxpayers suffer from polluted air and water at the same time that they subsidize private farmers, public lands graziers, and the private companies that log and mine the public's lands. More recently, Paul Hawken and colleagues[2] have noted that natural resources—and the natural processes that create those resources—are the support for industrial economies. The functions of nature have substantial financial values that are ignored at great peril. They should be considered "natural capital" and factored into industrial balance sheets.

More than a century ago, a rising public consciousness began to recognize the value of preserving and protecting natural lands. That conservation movement gave rise to national parks, national forest preserves, and other land-conserving political movements. It also created land management roles and conservation and preservation mandates (figure I.1) for a number of federal agencies, including the U.S. Department of the Interior's Bureau of Land Management, Fish and Wildlife Service, Bureau of Indian Affairs, Bureau of Reclamation, and National Park Service; the U.S. Department of Agriculture's U.S. Forest Service; and even the U.S. Department of Defense. Many states created management agencies for state lands in the same time frame; however, broader agency mandates include such diverse and sometimes irreconcilable agendas as national security, economic enterprise, and local resident interests.

Since the early 1960s, national environmental laws—notably the Wilderness Act, the National Environmental Policy Act, the Endangered Species Act, the Clean Water Act, and the Clean Air Act—have reached beyond public lands to preserve or restore a clean and safe environment *virtually everywhere* in the United States. Again, many western states followed suit. These noble and ambitious laws are based on valid scientific principles, drawing especially on the fields of chemistry, hydrology, and ecology. They have taken America a long distance toward clearing our air and water natural capital sectors, and protecting native species and natural lands for the economic and other benefits that they provide to humankind. It would be hard to overestimate how much worse off our environment would be without them.

National and state bureaucracies do not always implement or rigorously enforce the landmark environmental laws and regulations as the framers had envisioned, however. There is still a long way to go before America's rivers and streams are "swimmable and drinkable" as envisioned by the Clean Water Act, for example. And certainly many good-seeming laws have unintended negative consequences.

Since 1980, the very idea of regulating land uses, and air and water quality, has come under attack—along with attacks on the very existence of public lands and the credibility and relative neutrality of science itself. Rejecting regulation, the anti-public lands movement has warped Garrett Hardin's concerns to suggest that all lands should be in private hands. Public relations "sound science" campaigns tout alternative information or trumped-up "controversies" that support a favored project, and question valid data that could obstruct a dump, resort, or expanded

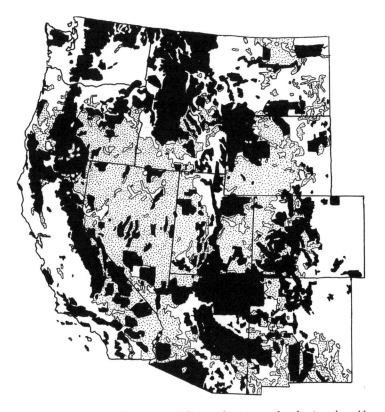

Figure 1.1 Map showing public and private lands in the 11 western states. The U.S. Department of the Interior Bureau of Land Management oversees the largest public acreage by far (stippled), and the rest (shaded black) is managed by the Department of the Interior's National Park Service, Bureau of Indian Affairs, Bureau of Reclamation, and U.S. Fish and Wildlife Service; by the U.S. Department of Agriculture Forest Service; and by the U.S. Department of Defense. Private lands are shown in white.

bombing range "only" to protect fish, birds, insects, lizards, or the habitats that support them. Many scientists are poor at public relations and have had a difficult time defending their research from false charges of poorly framed studies, of overly alarmist conclusions, or, ironically, of bias. And some scientists fear seeming strident or less than objective, even in defense of natural systems, preventing them from speaking out for better policies at federal and state levels. Tragically, some of the best government scientists have come under more and more political control and have seen the hearts cut out of their data-based technical reports in order to support the policies of one federal administration or another.

Western U.S. public lands, about 47% of the region, are this nation's patrimony—the bulk of its remaining natural capital. As Garrett Hardin pointed out, allowing free use of them will mean utter destruction. There is still a lot to save, and much of the damage can be reversed. But utter destruction is the risk that the nation is taking, and what we authors hope to avoid. Theoretically, the west's public lands belong to every American citizen, and we all have a stake in how the public lands are managed and how our food is grown, especially since so much public money goes into it. Environmental laws give the public the most say about land uses on both public and private lands, and our democratic form of government also gives the public a say about how the public purse is managed.

Citizen engagement is the first step on the trail toward working with, instead of against, nature. The lack of an ongoing discourse between scientists and the public obscures many critical issues, however, and most are not easily generalized for public consumption. Print journalists still produce excellent investigative reports on scores of environmental hazards and their effects on our lives, but apparently fewer people are reading them. The broadcast media is less patient and less focused, so it cannot explore the scientific nuances. Television and radio reports often highlight apparent controversies instead of investigating what may lie behind conflicting presentations, even when one side has an obvious bias.

But we authors find that Americans overwhelmingly want to protect their environment. More than half the public understand viscerally that wildlife and wild lands are important to their well-being and want to know more about how these connections are made. Although national park use may be down, Americans cherish the existence of parks and public forest lands and understand the need for wilderness protection. They cherish and even identify with their local natural areas. It is the authors' hope that this book will help to increase and broaden public awareness of the risks from overuse and misuse of America's western lands, and help citizens to better address national and local land use issues.

Not surprisingly, we have failed to come up with many easy fixes. Most of the ideas are practiced and practical but take a long time to bear fruit. Some of our fantasy solutions, or approaches to solutions, may be our readers' nightmares. If you don't like our solutions, please invent some of your own—just be sure that they obey nature.

1   Once and Future Trees

Forêts précèdent les peuples, et les déserts les suivent. (Forests precede civilizations and deserts follow them.)
Chateaubriand (1768–1848), quoted in Aldous Huxley,
The Human Situation

I am trying to save the knowledge that the forests and this planet are alive, to give it back to you who have lost this understanding.
Paulinho Paiakan, Kayapo Leader, Brazil, 1990

Along the Colorado Plateau's high-standing Mogollon Rim in northern Arizona's Coconino National Forest stands a small patch of big trees that matured well before Europeans came to North America. Massive ponderosa pines, and even pinyon pines and western junipers, tower above the forest floor, shutting out all but the most shade-tolerant competitors. Few places like this one still exist anywhere in the United States, even on national forest lands. A tourist hoping to see all the diversity that earliest European arrivals found commonplace in the western landscape must seek out a wide scattering of isolated enclaves across the region.

Western forests no longer contain the grand glades and lush thickets that our forerunners encountered because most woodlands, especially those owned by the public, largely serve a wide variety of human purposes, as campsites or home sites, board-feet of lumber, potential jobs, recreational playgrounds, and even temples of the spirit. We also rely on forests to maintain habitat for endangered species and seed banks for restoring depleted biodiversity—and to provide us with clean air and water, stable hillside soils, and flood control in wet years. Forests must perform these roles while being consumed, fragmented by roads, and heavily eroded. But there is no guarantee that these most beloved and iconic of natural resources can sustain such a burden.

Federal, state, and local government agencies oversee and regulate western U.S. forest lands and their uses, trying to manage the complex and only partly understood biological interactions of forest ecology to serve public needs. But after nine decades of variable goals, and five decades of encroaching development, western woodlands are far from healthy. Urban pollution and exotic tree diseases, some brought by humans, are killing pines, firs, and oaks. Loggers have more than decimated the oldest mountainside forests—most valuable for habitat and lumber alike—with clearcutting practices that induce severe soil erosion. Illegal clearings for marijuana farms are increasing. Drought, following a long history of too much fire prevention, promotes widespread, devastating fires. Salvage logging follows the fires, promoting more erosion and habitat losses.

As these stresses converge toward a crisis, rapid climate warming is reducing the survival potential of many tree species, if not of entire interdependent plant and animal communities (ecosystems). If the climate warms too fast and droughts stretch out, many of our highly logged and trampled and driven-over western forests could perish, depriving us of all their critical services.

Preserving Forests for the Trees

Trees have served humanity's economic and spiritual needs and wants as long as people have lived on the Earth. Before and after the rise of civilizations, people cleared woods for farming, cut trees for building shelters, and burned them to cook food and keep warm.[1] Over the last 10,000 years, extensive and pervasive human uses, for firing pottery and smelting metals; making tools, paper, and other equipment; and building communities, boats, roads, and vehicles have radically transformed the world's forests. While ancient Mycenean and Greek cultures venerated trees and preserved many sacred groves for religious rituals,[2] their unregulated forest cutting also denuded hills and mountains. Severe erosion followed, stripping upland forest soils and flooding huge sediment loads into rivers. The eroded forest soils choked harbors and pushed coastlines seaward from important coastal cities such as Ephesus, Troy (Illios), and Mytelene,[3] which ended up landlocked or buried under sediment.

Like the ancients, European civilizations long depended on forests for their primary energy source—as critical then as petroleum and coal are to us now (see chapter 12). Ships made of wood were essential to navigation and economic expansion until the latter nineteenth century. By the seventeenth century, farmland clearings and massive construction projects, including whole navies, had severely reduced the size and quality of northern Europe's timber resources—a classic illustration of Garrett Hardin's 1968 "Tragedy of the Commons." A worldwide search for ship masts ensued, dictating both foreign policy and military actions from the Baltic to the New World.[4] Once global explorers discovered vast Asian and American timberlands, Europe's rulers established colonies to exploit them, among them the founding settlements of these United States.

The European settlers moved into forests of the temperate zone, between the Tropic of Cancer and the Arctic Circle. These forests covered about a billion acres of what is now the contiguous United States. Native Americans had made impacts upon the woods, yet mature stands of large to giant trees from several hundred to several thousand years old were abundant and widespread. To avoid words laden with scientific controversy and legal confrontation, we refer to these pre-Columbian North American woodlands as "heritage forests."

Although mostly arid, the west harbors dense, undulating stands of trees hugging the northwestern coast—still the world's most impressive stretch of temperate rain forest. Interior forests grow mostly in the mountains, the islands of moisture punctuating dry prairies and deserts. The largest, tallest, and oldest trees on

Earth—the great groves of coast redwoods and stately Douglas firs; monumental giant sequoia; gnarled, weather-beaten timberline bristlecone pines; and dark, coastal fog-moistened spruce and yew—all grow in the west.

Few of the heritage forests are left today. The remaining scraps cover less than 63 million acres, around 6% of the original.[5] North America's earliest settlers saw no reason to preserve forests that could harbor hostile Indians and predatory animals. Wrote one, "Upon first glance, the woods gave...the impression of a 'wild and savage hue....'"[6] For many generations, most Americans accepted clearing forests as virtuous activity. Except in a few early forest reserve areas, they indulged in unchecked cutting and logging until the early twentieth century. Forests representing almost all major ecosystems have been logged over at least once, or converted to myriad human uses.[7]

By 1907, agricultural clearings, mostly in the eastern states, had shrunk U.S. woodlands some 28%,[8] raising the possibility that rapacious lumber companies and railroad interests could denude America of its forests. Leading foresters and politicians of the day seemed to understand that forest processes were important "natural capital"—predicting that the nation could be facing not only a timber famine, but also the loss of precious topsoils similar to what happened in Mycenean Greece, along with depleted groundwater supplies and degraded water quality. They also feared that loss of evapotranspiration from trees—the process of drawing water from the ground through plant roots, upward through stems or trunks, and releasing it through leaves or needles—might even dry the climate.[9]

As it turned out, formally classified U.S. forest land declined only 2% after 1907 even though the global demand for wood products increased fivefold between 1900 and 2000. This was due to a combination of social forces and enlightened government regulations, which helped the United States avoid timber famine and complete forest denudation. Fossil fuels replaced wood (see chapter 12), and plastics manufacturing from petroleum (see appendix 3) began shifting the world toward substitutes for many wood products. Meanwhile, the federal government enacted nationwide forest management policies that allowed American forests to recover, also preserving many heritage forest remnants in national parks and forest reserves. Millions of farmers moved from low-quality or degraded farm lands to find work in cities (see chapter 2), and forests regrew on those lands, as well. Also, after World War II, the United States imported foreign lumber in tremendous quantities.

Forests of predominately native tree species still occupy about 70% of the 1620-era woodlands nationwide.[10] But the overwhelming majority of American forests now mostly contain small-diameter trees, no more than 30 inches across, from the second or third regrowth cycle. Only about 7% of American forests are protected in national and state parks, wilderness areas, and other conservation reserves.[11] And large forest regions are in the process of conversion to other uses as you read these pages.

Most western forests are on federal public lands, variously under the supervision of the U.S. Department of Agriculture's Forest Service (USFS) and two U.S. Department of the Interior agencies, the National Park Service (NPS) and Bureau of Land Management (BLM). State and county preserves protect some heritage forest stands, especially in California, and environmental groups have purchased a number of others.

The national forests, under USFS management, were born in the early twentieth century after the most productive and accessible forests, generally on lowlands, had come under private ownership. The remaining timber stands mostly were in steep remote terrain, which private landholders had spurned as less accessible and generally less productive. Up to WWII, the abundance of private timber, and the difficulty and expense of logging mountainsides far from any road, protected the trees in national forests. But when demand for forest products surged after the war, logging severely depleted the private forests, leaving national forests the nation's largest timber resource.

Although the remaining groves of great old trees are only a few percent of the nation's remaining harvestable timber acreage, they contain greater volumes of relatively unblemished wood than equivalent acreages of regrown timber. The old forests also are principal habitat for dwindling native North American wildlife. Campaigns for preserving them are intended largely to save the life support of threatened and endangered species. Nature worshipers and recreating urbanites—a vocal and active segment of the citizenry—also campaign to preserve heritage groves as monuments of the spirit. Some of the preservation campaigns have turned into mystical experiences. Perhaps the apotheosis of modern tree worship is Julia Butterfly Hill, who lived high in the branches of a coast redwood for more than two years, developing a close personal relationship with the tree named "Luna."

Julia Hill's tree-sitting campaign succeeded in saving just one heritage forest remnant, while commercial companies still get permits to cut in and around thousands more.[12] Ironically, many people hope to preserve heritage forests *and* continue consuming high-quality lumber for their homes, boats, and decks. The conflict between these disparate goals, both within individuals and across the culture, continues to inflame tensions over western U.S. forests.

Most people do not realize that there are even better reasons for saving forests—that trees' natural functions are just as important to human lives as to wildlife. *We need to save forests for the services that trees provide.* Trees are Earth's main source of airborne oxygen, absorbing carbon dioxide (CO_2) and releasing oxygen for people and all other animals to breathe. This same process of forest respiration also reduces the atmospheric greenhouse gas concentrations that drive rapid climatic warming. Trees also help purify surface and underground water. The clean water that forests provide is nearly as important for human survival and health as oxygen—and clean water is a critically declining resource (see chapter 9).

Former USFS chief Mike Dombeck has

> worried that we may, as a society, lose our appreciation of what the land does for us; why open space is important....The fact that a single tree sequesters about 13 pounds of carbon each year. That a single tree produces enough oxygen for a family of four to breathe. The water filtration functions of the vegetation on the landscape. It's important for people to appreciate and connect to the land.[13]

Since extensive tree cutting on hill slopes commonly results in severe soil erosion and siltation, Dombeck might have added that trees add essential nutrients to the soils in which they grow, hold soil on slopes, and help prevent catastrophic floods (see chapters 5, 8, 9, 13). These are the top six reasons for humans to preserve forests and worship trees.

The Nature of Forests

USFS and BLM aspire to take over forest management from nature. So far, however, they have not proved that people can manage heritage forests to preserve natural ecosystems while extensively and intensively producing lumber, paper, and other commodities. Some conservationists clamor for the agencies to protect the old woods in their presettlement state or, slightly more realistically, return them to that state. To reach either goal requires better knowledge of the actual state of both the presettlement and today's forests, and also expertise in how to stabilize small remnant glades that are surrounded by clearings and second- or later growth woodland. The most difficult challenge is preserving the whole diversity of forest ecosystems, particularly in heritage forest remnants outside of parks and preserves. The first, most critical step is finding out how forests grew and thrived before intense human occupation.[14]

Woodlands are shaped by climate and the soils that they grow in. The old temperate forests of North America matured in temperate-zone soils, which resist degradation and erosion better than do soils in the tropics, where rainfall levels are extremely high. The temperate zone's lower rainfall and cooler climate make temperate soils good nutrient storehouses.[15] But understanding how temperate woodlands grow, what makes them thrive, and how best to preserve a forest or any other ecosystem, whether stressed by humans or by natural forces, remains the subject of considerable research efforts.

Biogeographer F. E. Clements[16] proposed that forests develop through broadly predictable plant succession processes, with intimate links between all the plant and animal species. Widespread-seeding, fast-growing "pioneer" trees easily colonize an unforested area, increasing shade and moisture as they mature. This critical first stage helps longer lived "successor" species spread their seeds, grow, and eventually take over. Clements demonstrated that a natural forest generally transitions through one or more successional stages (or "seres") over 100–500 years, eventually reaching a "climax" mix of forest species that may not change for millennia. The climax community's CO_2 absorption is more stable and sluggish than that of younger forests because the older forest promotes less new growth. A climax forest must age several centuries before achieving the full maturity of old growth (table 1.1).[17] The total soil nutrient and water use of old growth forests generally equal the combined weight of vegetation and animal life (biomass) in the ecosystem.

Plant succession processes create much of the biological diversity (biodiversity) in natural ecosystems. A typical succession pattern in western temperate forests starts after one or more of the natural destructive forces—high wind, natural fire, flood, drought plus beetle infestations, disease, landslide, volcanic eruption, and even meteorite impacts—open a clearing or a series of clearings. Pioneer grasses, wildflowers, and shrubs sprout very quickly in the clearing, and then successor

Table 1.1 Forest Successional Stages

Stage	Age Range (Years)
Young forest	1–70
Mature forest	70–150
Early old growth	150–250
Mid old growth	250–500
Late old growth	500–750

"Young," "mature" nomenclature based on figure 4 in:
J. F. Franklin and T. A. Spies. Composition, Function,
and Structure of Old-Growth Douglas-Fir Forests. In
L. F. Ruggiero et al., eds. Wildlife and Vegetation of
Un-managed Douglas-fir Forests. *USDA Forest Service*
General Technical Report PNW-GTR-285, 1991;71-80.
Available: www.humboldt.edu/~storage/pdfmill/Batch%205/
unmanaged.pdf. Ends of ranges are not to be taken
literally.

aspens, willows, and lodgepole pines grow up to replace them. If nothing else changes, these successors yield in turn to climax-stage firs, which dominate the forest until they die from old age, or until another disturbance clears them away and resets the ecological clock. The strongest support for Clements's succession model is the broad geographical extent of uniform forest communities across many mountain ranges in the American west, and even at common elevations and latitudes worldwide.

Ecologist A. S. Waitt's forest dynamics concept modified Clements's long-term successional model by accounting for the effects of short-term changes. Waitt showed that frequent events, such as fires and windfalls, down to and including the natural aging and death of individual trees, continually open up gaps in natural forests and create an ever-changing "gap-mosaic" architecture (figure 1.1).

> Following the death of a large tree and its fall, a canopy gap forms. The area below this gap becomes the site of increased regeneration and survival of trees. Trees grow, the forest builds, the canopy closes, and the gap disappears. Eventually, the mature forest in the vicinity of the former gap suffers the mortality of a large tree and the new gap is formed and the cycle is repeated.[18]

The gaps offer a large variety of habitats, maximizing forest biodiversity.

H. A. Gleason's contrasting "individualistic community" concept[19] proposes that a forest is simply a collection of individual trees, plants, and animals without significant codependent relationships, only coincidentally requiring similar ecologic conditions. The well-documented interdependence of plant and animal communities favor Clements's successional model, but this individualistic community concept is useful for understanding exotic plant invasions (see chapter 3), which can radically alter the survival prospects of native species.[20]

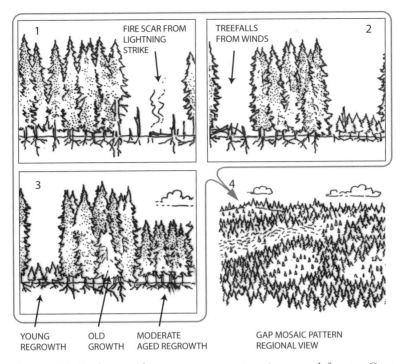

Figure 1.1 Evolution of gap mosaic structure in natural forests. Gaps periodically open from lightning strikes (1) and strong winds (2), as well as insect infestations, landslides, flooding, and trees dying of old age. Successor plants colonize gaps as they form (3), leading to a mosaic of differently aged trees and successional ecosystems (4).

Cataclysms

Climatic forces, along with other natural forces driven by gravity, the sun, and Earth's own internal heat engine, modify the land surface in many ways (see chapter 13). When a tree falls, a volcano erupts, or a vehicle impacts a hill slope or stream, natural forces start transforming the site through erosion, siltation, biological decomposition, and the like. All the forces operate continuously and interrelate complexly. Fossil tree leaves and branches, found in some sedimentary rocks—including thick coal seams (see chapter 12)—testify that natural forces periodically devastated forests before humans or their axes appeared on Earth.

One immense cataclysm, the Chicxulub meteorite that slammed into what is now Mexico's Yucatán Peninsula some 65 million years ago, destroyed vast tracts of North American forest. Natural processes reestablished plant cover and added animal species over about the next 10 million years. The resulting ecosystems were entirely different communities than before the meteorite impact.[21] Natural climate changes are much more common than meteorite impacts, or even volcanic eruptions or landslides, and pose the most serious long-range challenge to the survival of temperate-climate forests worldwide.

Climate variations have kept North America's forests in a state of continuous flux for at least five million years. Over the past two and a half million years, western North America was mostly moister and cooler than it is now. A number of times, glaciers—sheets of ice, some as much as two miles thick—emerged from far northern latitudes and blanketed the North American landscape as far south as Kansas. Glaciers also emerged from high mountain ranges, carving sloping river canyons into steep-walled valleys with broad, marshy floors, thus creating the topography of Yosemite, Glacier, and Rocky Mountain National Parks. During ice ages, warm intervals like the present occurred less than 10% of the time. Today's rapid warming, aided by human burning of fossil fuels, is much higher than would have been achieved by natural climate changes at this point in Earth history.

North American forests started adjusting to the warmer conditions when the most recent ice age ended, about 10,000 years ago. As the glaciers retreated, forests began to spread northward into areas that had been ice-covered for a hundred thousand years. This northward reforestation is still taking place. All this means that America's heritage forests developed in a relatively short period of climate and geologic stability within a much longer history of harsh, erratically changing conditions. At present, the basic mix of species in forest communities is changing on thousand-year time scales in many places—and in a few locations, the adjustments are happening much faster.[22]

Human factors

During ice ages, glaciers tied up such large volumes of water that ocean levels declined by hundreds of feet, turning shallow ocean floor into dry land bridges between continents and connecting areas that now appear as islands. As the last ice age ended, North America's Indian forerunners came from Siberia on a land bridge that now lies under the Bering Sea, together with large mammals such as the grizzly, moose, buffalo, and elk. Eventually, most of America's pre-Columbian human population lived in forests and cut trees to make lodges or tipis, canoes, and much more. Geographer Thomas Vale explained, "Pre-European peoples humanized areas on the North American continent, including parts of the American west...." But these are minor human imprints that do not fundamentally modify the natural world.[23]

Some of the highly civilized native cultures built permanent communities, even cities, and did significantly degrade the local ecology with irrigated agriculture (see chapters 2, 9). By one estimate, all forest within six to nine miles of the prehistoric metropolis of Cahokia, near modern-day St. Louis, was cut down to make way for farming. Tree cutting for fuel and construction lumber extended much farther.[24] These effects were ecologically significant to a much lesser degree than those of later settlers or the present.[25]

So European colonists moved into forest-mosaics shaped both by nonhuman and by human forces. Deep forest areas were not uniformly dense and did not all consist of giant trees. Instead, groves of sparse mature trees, accompanied by ground-level grasses, ferns, and small shrubs, alternated with sunny openings of meadow grass punctuated by seedling trees and bushes, or with zones of densely packed younger trees and understory shrubs. At irregular intervals, human- or lightning-sparked fires burned the forests, prairies and other grasslands, and even wetlands.

Native Americans were sophisticated land managers. They manipulated plant species to sustain food and fiber supplies, often deliberately setting forest fires to promote growth of their favorite food plants and clear away brush and litter (see chapter 3).[26] The fires they set rarely were severe—they improved forage and habitat for game animals, opened hunter pathways, and made hunting easier, particularly of deer and other game that prefer feeding in open forest understory. The fires also enhanced biodiversity and opened up niches for saplings and healthy young trees.

Forest Health

Forest managers need a way to assess the health of forest ecosystems to guide their management practices. Otherwise, they—and we—will not know whether forests are thriving now or if they can thrive in the future under our relentless pressures. First and foremost, sustaining forests requires preserving stable slopes and soils. All human clearings destabilize slopes and soils, which in turn increases floods and lowers water quality, oxygen production, and biodiversity. Clearings also reduce the forests' CO_2 absorption.[27] The number and size of human clearings have vastly expanded over the past 50 years, breaching forest integrity, soils, slopes, and streams on scales that natural processes rarely accomplish, short of cataclysms.

In the western United States, 64% of exploitable timber grows on public lands, most of it in the Pacific Northwest states under USFS jurisdiction, or in the arid Great Basin and southwestern states under BLM management.[28] Other significant unprotected forest tracts are state owned, or privately held and governed by state forestry regulations.[29] Regulations and guidelines vary by agency and category of land ownership.

Effectively, the USFS is the largest timber supplier in the United States. It plans and prepares "timber sales"—actually, sales of permits for logging timber on federal lands—offers the sales for bidding, awards the contracts, and administers the eventual harvest of trees belonging to all U.S. citizens. In addition, agriculture and grazing (chapters 2, 3), mining (chapter 4), reservoirs (chapter 9), military training (chapter 6), utility corridors and roads for oil and gas exploration and production (chapters 5, 12), plus urban–suburban developments (chapter 8) all encroach on western regrown and heritage woodlands. Large-scale recreational developments on national forest lands (chapter 11) and increased suburban settlements at the forests' edges tend to influence national forest management policies.

Federal and state land managers and elected officials have a duty to evaluate the sustainability of logging practices and lumber yields and to enforce laws protecting endangered species. At the same time, they must also respond to appeals from other commercial forest users and address the public's sometimes conflicting demands for access and fire suppression. Both USFS and BLM must manage federal lands to reconcile the often incompatible goals of "multiple use"—fostering economic and recreational uses while trying to preserve natural resources. The agencies tend to allow extensive clearing in and around western U.S. forests for myriad consumptive human uses.

Sustaining harvests

More than 90% of all the world's wood products come from natural forests, primarily in the northwestern United States, Russia, and rainy equatorial countries.[30] Logging takes place in eight of the 11 western states, principally the Pacific Northwest states plus California, Idaho, Montana, Wyoming, Colorado, and parts of northern Arizona and New Mexico. The timber mostly supplies construction lumber, and wood for paper and biomass fuel. All by itself, the United States uses 20% of all the world's produced wood—nearly 50 cubic feet per year for every American. This is more than three times the global average, and double the lumber consumption of most other industrialized countries.[31]

Many of the people who call for endangered species protections also demand abundant forest products, with little idea of the paradox or environmental consequence. Globally, the consumption of wood products is rising, driven both by world population growth and by many developing countries' goals for economic expansion. By 1999, fuel wood and charcoal consumption grew to just over half of all wood cut.[32] Paper accounts for another 20% of the global annual cut. Worldwide wood consumption is likely to reach more than 77 billion cubic feet per year by 2010—roughly equivalent to cutting down 77 million trees of 50–65 foot height. Only about a third of wood products are recycled.

An agreeable climate and healthy soil are the most vital factors supporting the long-term sustainability of a forest. Under ideal natural conditions, changes from human exploitation may do little ecological damage to a forest that grew and will regrow, but where conditions have severely deteriorated, the same changes can force ecosystems decline or extinction. Given optimal soil and climate combinations, plant and animal associations can resist substantial stresses from drought, insect infestation, logging, or grazing. An ecosystem is much more vulnerable under less than optimal conditions.[33] But even under optimal conditions, frequent repeat harvests ultimately render a forest incapable of recovering from logging impacts. At each cut, the soil loses nutrients, its layered structure degrades, and the overall erosion potential increases (see chapter 13).

There are appropriate and inappropriate ways to cut a forest, depending upon the landscape and forest type. To avoid overcutting and devastating soil losses, timber-dependent economies must harvest timber under sustained-yield principles, continuously producing wood from the same areas for indefinite periods. Selective cutting can be the least damaging approach to logging. To achieve sustainability, the rate of forest cutting must never exceed the growth rate of harvestable trees. Roads must be correctly placed, preferably high on a slope, and care taken to limit the surface disturbances from cutting and hauling. For example, highly conserving Switzerland has long restricted timber cutting in its mountain cantons to selected trees scattered through a forest tract. This practice of selective cutting preserves the stability of wooded slopes, prevents floods from increasing in number and height, and protects vital ecological services, to the benefit of villagers and graziers living in the valleys below. Low Swiss population pressures keep timber demands well within sustainable limits.

In practice, foresters cut forest patches on rotation, allowing adequate intervals for regrowth depending on the type of tree. A stand of Australian eucalyptus may yield marketable products once every ten years, but plantation pine, a major source of pulpwood and paper, commonly needs to grow for at least 20 years between harvests. To sustain the yield from a pine forest, only 5% of the timber can be cut each year. On the other hand, eastern U.S. oak and beech forests require 200-year rotation periods, which partly explains high oak and beech lumber prices.

Unkind cuts

The USFS records estimate that more than 60% of western U.S. logging is selective,[34] but the large per capita U.S. demand for wood products makes loggers prefer clearcutting. Unlike selective cutting, clearcuts utterly eradicate a patch of forest (figure 1.2). Heavy machinery fells and extracts only the larger trees while crushing smaller trees and shrubs and overturning soil layers that typically take thousands of years to develop (see chapter 5, box 5.1; and chapter 13). Clearcutting allows logging companies to extract large amounts of timber in short periods of

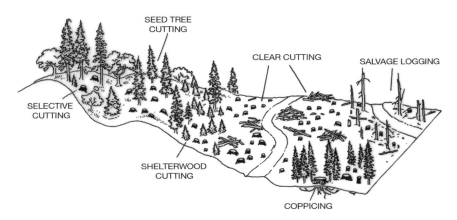

Figure 1.2 Forest harvesting techniques. *Selective cutting* removes relatively few carefully selected trees of top value and leaves forest structure mostly undisturbed. *Shelterwood cutting* removes large valuable trees piecemeal in two to three cuts over a period of 10 years and largely preserves forest diversity and habitat. *Seed tree cutting* leaves a few large trees to provide seed stock and shelter for regrowth. *Clearcutting* removes all commercially valuable timber and mostly destroys forest structure and habitat. *Coppicing* cuts trees able to regrow through root sprouts, with no need for reseeding. *Salvage logging* removes valuable timber from forests killed by fire, insects, or other natural processes.

time, and to replant more easily. It is the simplest and, in terms of profit margins, most efficient way to harvest a forest on a rotational basis.

Individual clearcuts are variably sized and shaped, ranging from a few to several hundred acres. Removing logs from clearcut areas requires abundant roads and skid trails for removing the cut trees, which are even more damaging than the cuts themselves. Road reclamation is feasible and desirable but rarely happens on western public lands (see chapter 5). The roads open forests to off-road vehicles, which exacerbate the damage (see chapter 11).

Seed tree harvesting, a less abrasive form of clearcutting, typically leaves 5–12 mature trees standing per acre to help reseed logged areas. Other practices that encourage rapid forest regeneration include shelter wood harvesting, which removes only mature trees, and coppicing—a highly productive alternative to clearcutting, which leaves stumps behind to sprout new trunks and regenerate another crop (figure 1.2). Coppicing is best for growing tree varieties with short rotation intervals (5–10 years), but unfortunately, it requires particularly intensive herbicide, insecticide, and synthetic fertilizer applications to suppress undesired plants and give the sun exposure to preferred tree species.[35]

Various skidding alternatives depend upon costs and the slope of cut-over land. Dragging the logs out with motorized equipment, such as tractors, significantly disturbs the soil on low slopes of less than about 57% (30° angle). For dragging logs to loading points across steeper hillsides, loggers set poles and string metal cables between the poles and remaining trees—but even this technique can critically disturb soil. On steeply sloping land, some loggers use helicopters, and even balloons, to remove high value logs without damaging soils. In recent years, some environmentally concerned loggers have reverted to hauling logs with mules and horses, because they correctly perceive that animal haul paths are narrower and cause less soil disturbance than do machine-made trails.[36]

Tree roots hold the soil on slopes. As the Myceneans discovered, tree roots help keep the rainwater runoff to a minimum, thereby minimizing erosion and soil losses that can choke streams, prevent fish from spawning, and bury farmlands in flood times (see chapter 2). Conversely, cutting trees severely diminishes the effect of living roots that hold soil on a slope against the pull of gravity. Wholesale root decay after intensive logging may take 5–10 years, depending upon climate, soil type, and the kinds of trees. But even if new plants can establish seedlings and restore some root systems in that time, the slope's strength still tends to decline.

Slopes held in place only by dead, weakening roots are open to severe mass wasting processes—everything from gullying to large-scale landsliding—that deliver loads of sediment into streams (see chapter 13). Soil creep, the slow, gravity-induced, downslope movement of soil masses on steep slopes, is a very widespread phenomenon on both cleared and uncleared forest slopes (chapter 13). The sediment shed from clearcuts can far exceed what soil creep delivers in uncut forests, however. When trees are suddenly removed, their evapotranspiration stops, letting soil moisture and shallow groundwater levels suddenly rise toward the surface, which further increases the slope's failure potential. Denuded slopes become particularly vulnerable to debris flows, which are thick slurries of mud and debris that surge downslope, picking up trees and other rubble and flattening everything in their paths (chapter 13).[37]

Shallow landsliding increases dramatically on steep clearcut slopes, as much as two to three times normal rates.[38] Landslide inventories in Oregon's Cascade Range recorded significantly higher soil mass movement rates in clearcuts compared to surrounding forests.[39] In Siuslaw National Forest, Oregon, about three-quarters of all rain-triggered landslides in one 1975 storm came from clearcuts.[40] A 1999 study on a part of the Humboldt County, California, Headwaters Forest reported that the highest proportion of landslides occurred in areas that had been clearcut over the previous 15 years.[41] Landslides move sediment and logging slash, plugging up road culverts and diverting streams and their tributaries. In some landscapes, the water diversions led to severe gully erosion in forests (see chapter 13).[42] In wet years, the increased rainwater coming off clearcuts increases eroded sediment loads. Carried into fast-running streams, the sediments aid in lowering stream channels and undercutting stream banks, potentially inducing more landslides and reactivating old ones.[43]

Many loggers formerly removed clearcutting residues and slash to reduce fires, leaving mountainsides bare and highly prone to erosion. The presently accepted best practice is to leave some deadwood and forest litter on the ground to trap moisture and provide nutrients for forest regrowth. But erosion levels still are high, especially on slopes. After cutting, many loggers burn the piles of slash left in clearcuts and even have dropped napalm on them from low-flying aircraft. The high heat of napalm sterilizes the remaining soil to depths of several centimeters, eradicating any surviving organic layers and baking the surface to a hard crust that cannot absorb rainwater. Compared to undisturbed forested hillsides on similar slopes with equivalent landslide potential, stormwater runoff is higher from the compacted, torn, mixed, and baked soils, yielding 3–30 times more eroded sediment.[44]

Plant losses take away shade and increase evaporation, which dries out upper soil layers and reduces the internal cohesion between soil particles. Groundwater rising at deeper levels from evapotranspiration losses may not balance these soil drying effects. On slopes steeper than 10% (6° angle), noncohesive dry soils release streams of loosened mineral bits and other soil matter, termed "dry ravel." In the first 24 hours after a 1980s fire in the Oregon Coast Range, clearcut areas on slopes steeper than 60% (31° angle) lost between 2 and 16 dump truck loads of eroded sediment per acre, more than half in the form of dry ravel.[45] Moister areas are much less likely to produce dry ravel. Even in southern California, the less sunny north-facing slopes tend to retain more soil moisture than do south-facing ones, so burn scars in chaparral vegetation on north-facing slopes release 10 times less dry ravel than do the south-facing slopes.[46]

Hillside logging requires roads, but roads on slopes are highly destructive. National forests contain nearly 450,000 miles of roads, mostly in the western states, built at taxpayer expense for timber cutting and fire control (see chapter 5).[47] Even accounting for clearcuts, erosion from roads is 10 times greater than the erosion directly associated with logging in the western United States.[48] The study of a steeply sloping California clearcut showed a shocking increase of more than 200% of eroded sediment in the first six years after the operation had constructed roads low on slopes or near watercourses, and used tractors for skidding logs downslope to haul-trails built in stream channels.[49] Clearcuts in northern Idaho yielded 770 times more eroded sediment over the six-year period following logging road construction compared to equivalent unlogged forest slopes. Landslides produced 70% of the measured sediment.[50] More than half of all landslides in northern Idaho forests come from roads carved into slopes, and a further 30% originated in areas combining roads, fire burns, and logging scars.[51] An air photo study of Vancouver Island indicated that more than a third of all landslides in logged areas were somehow related to roads.[52]

Forest–water connections

The stream and spring water in natural forests is remarkably pure, and the streams coming from them transport very small sediment loads even in steep terrain. These filtering services are provided by forest soils and leaf litter—the natural filters that remove suspended solids and dissolved compounds from rain runoff (figure 1.3). The forest understory of branches and smaller undergrowth, plus litter on the ground, also cushions soils against pounding raindrops. Where water collects on the mat of forest litter, various chemical processes and animal activities create openings that let the water rapidly seep underground, replenishing the groundwater that many cities and rural dwellers depend upon (see chapter 13). Cutting forests and stripping or paving over woodlands and other natural habitats limit the areas where groundwater may be naturally replenished. Not surprisingly, groundwater supplies are steadily diminishing all across the naturally arid western United States (see chapter 9).

Flat lands are obviously less vulnerable to slope and soil stability problems than is hilly country, although flat lands bear the flood burdens of extensive silt deposits and water pollution. But forested watersheds protect developed areas downstream from severe flooding in all but the worst storms. The leaves, branches, soil, and

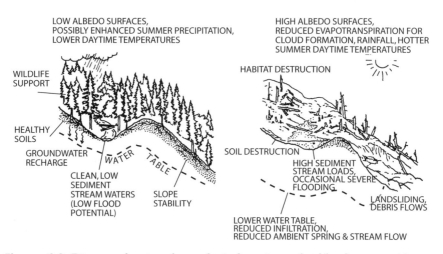

Figure 1.3 Diagram showing the ecological services a healthy forest provides on slopes, contrasted with the damage from excessive clearcutting and severe fires.

roots that obstruct and retain running water in a forest also capture and divert much of the precipitation falling on forests, and take up large amounts in plant evapotranspiration processes. These diversions lower the amount of rain that runs off and slow flow rates, particularly on slopes (see chapter 13). During a downpour, the water flows coming out of forests are much less intense than the drainage from poorly vegetated or unvegetated areas. Streams from forested areas also take a much longer time to reach flood levels and fall back again and so produce less damaging floods than do streams from less vegetated and developed areas (see chapter 8, figure 8.7).

Soils eroded from partly logged to denuded slopes pollute water supplies, clog filtration systems, and promote heavy silt accumulations in downstream waterways. Removing that sediment is one of the most expensive treatments required for purifying municipal water supplies, for both reducing wear on pumps and, even more important, removing toxic chemicals attached to sediment particles.[53] Water purification costs are high for municipalities and their consumers, but forests do it for free. New York City's former mayor David Dinkins understood these facts when he advanced a plan in July 1993 to upgrade management and reinforce protection for 2,000 square miles of forests and farmlands around the city's 19 reservoirs. The protected areas, as far away as the Catskill Mountains, have provided New York with water since 1842. Dinkins's plan for improving and maintaining the system cost $720 million—much cheaper than the billions it would cost to build new water treatment and purification systems, let alone find ways to dispose of treatment plant wastes.[54]

Gases for life

The normal respiration of trees, taking up CO_2 and releasing oxygen to the lower atmosphere, is a critical forest ecosystem function that supports human life and helps regulate greenhouse gases. The oxygen that trees release is the main source of the oxygen that people—and all other animals—have to breathe. Trees retain CO_2 in their wood and leaves, making forests a critical factor in regulating the global carbon cycle and worldwide atmospheric temperatures. Cutting forests eliminates their role in removing CO_2 from the atmosphere and so helps to warm the climate. Once the forest has been cut, soils degrade and release substantial amounts of CO_2 to the atmosphere, further adding to climate warming.

Sustaining diversity

The concept of biodiversity describes the wide variety of plants and animals in ecosystems, which maintain competitive or collaborative interactions of many kinds, providing food and shelter and performing myriad other services mutually necessary for all species' survival—including humans. This diversity of plants and animals in ecosystems provides natural "factories" that break down wastes of all sorts, clean up polluted groundwater, and support complex food chains. A few of the life forms in any ecosystem perform vital, keystone functions, meaning that many other animal or plant species could not survive without them.[55] The catastrophic biodiversity reductions worldwide since the end of WWII largely correspond to destroyed or degraded forests and tropical marine ecosystems.

Human clearings and roads subdivide habitat and reduce diversity. Such clearings, especially logging clearcuts, are becoming more abundant and pervasive in forests than are the gaps produced by natural fires and other natural clearing processes. And natural gaps also continue to open. The clearings constitute barriers, subdividing forests into many small segments and making some too small to function as ecosystems (see chapter 5). The effects of forest fragmentation have been especially devastating to habitat.[56] Unfortunately, current government policies are promoting roads and vehicular recreation in unfragmented lands with wilderness values, while allowing many more clearings for development in national forests.

Forest fragmentation markedly increases the amount of edge habitat, transitional to clearings, at the expense of denser forest interior zones. This change reduces life support for many songbirds and other animals of the forest interior, exposing them to edge habitat predators, parasites, and disease, and so decreasing their survivability. A few small forest clearings in a large forest stand may affect biodiversity very little, while the same-size clearings could be devastating to ecosystems in a small forest. Adding numerous human clearings to natural openings—the deep canyons, sand dunes, and mesas or buttes—that already subdivide western forests can disrupt plant succession and disproportionately reduce biodiversity even in large forests.

Large predators and migratory herbivores, such as the red wolf, elk, and grizzly bear, need to roam across hundreds of square miles of unfragmented inner forest. As interior habitats decrease in size, the numbers of animal and plant species they support also decline: the species-area effect. A 90% reduction in habitat area can cut biodiversity by half.[57] Reduced biodiversity impedes future forest regrowth, especially from the loss of pollinating birds and insects, while obstructing the maintenance of balanced, healthy forest ecosystems. Heavy logging, with heavy soil losses, also opens a woodland to bioinvasion, letting nonnative species take over and preventing native ecosystems from reestablishing themselves (see chapter 3).

Western forest fragmentation already has contributed to the disappearance of predatory animals that need extensive unfragmented forest habitat, while encouraging cowbirds, mockingbirds, hawks, ravens, and rats to expand into human residential areas. All these species can easily adapt to rapidly changing conditions and thrive in unstable environments, supported by garbage dumps and open neighborhood dumpsters. Ravens have become the dominant bird species in woodlands adjoining many California and Arizona population centers.

Burning Issues

What foresters regarded as a "healthy forest" in 1891 is far different from our understanding of healthy forests today. The federal mission to serve multiple public interests in managing public woodlands has become progressively more difficult to work out, given both the increase in population pressure and a much deeper scientific understanding of our negative impacts on forest ecosystems. Scientific studies show that healthy forests are diverse ecological systems, ever-changing on extended time lines that greatly exceed ordinary human economic interest. To support diversity, and what may be regarded as a truly healthy forest condition, federal and state land managers must look beyond timber harvests and factor in the natural changes that modify forests, especially fires.[58]

Ecological studies reveal that fires are natural and common in most ecosystems. Over millions of burning and regrowth cycles, most plants have adapted to fire, some developing a thick bark, rich in fire-retarding tannin, some storing food in roots to sustain themselves after a burn, and others resprouting vigorously. A number of fire-adapted trees, such as lodgepole pine, produce cones that can release seeds only after intense heating. Heat and flame return some nutrients to the soil even while depleting others, so that surviving plants and seedlings—especially pioneers—often thrive on burned ground.

Fire-prone plant communities are "self-reinforcing: The type of plants that grow on a burn determines the nature of the fuel complex...[which] determines the intensity and frequency of [the next] fire and its future biological effects."[59] Thus, changing numbers and locations of plant communities influences future fire patterns, which influence future landscape development, the whole grand cycle operating in a time frame of thousands of years.

Experience also shows that human land uses force such rapid changes on natural ecosystems that they cannot recover easily, and sometimes not at all. Natural processes never stop, and always amplify ecological disruptions. If human disruptions exceed natural thresholds, the ecological response can be devastating. These are serious challenges to such agencies as the USFS and BLM and are forcing significant changes to forest management.[60] Fire control is an area where natural thresholds have been crossed. On no issue of forest management have attitudes changed more.

Crowning blows

Woodlands burn in three different ways:

- Ground fires burn root structures and oxygen-rich soil.
- Surface fires burn in underbrush and downed dead logs and branches, called deadfall.
- Crown fires burn the forest canopy, its upper tree branches and leaves.

Although unspectacular and slow moving, from the human point of view ground fires are among the most destructive because they kill the roots of trees and are very difficult to control. Surface fires commonly move slowly enough for firefighters to contain them at the fire front, but may spread into forest canopy as they advance. Ground and surface fire are called "cool" fires because maximum temperatures remain between 500°F and 1,000°F. High summer temperatures, wind, and rising air drafts from the heat of a ground fire, can whip flames upward into the canopy, generating crown fires.[61]

Crown fires move more or less independently of the surface but can set surface fires as they spread, moving as much as 10 miles per hour under gusty conditions. Temperatures may be as high as 2,000°F, enough to melt glass and soften steel beams so much that they bend like a pretzel. Far-spreading hot crown fires do more to destroy valuable stands of timber than do smaller burns, and usually defy firefighters' attempts to contain them until the weather changes to less windy or higher humidity conditions.

Prior to Lewis and Clark's western expedition, fires were relatively cool and did not often develop into crown fires that burned the larger trees. The fires were relatively frequent, cleaning out leaf litter and deadfall and reducing the fuel available for subsequent burns. Frequent cool fires reduced the number of crown fires. Intervals between crown fires in presettlement times sometimes can be deduced from studying the number and appearance of growth rings in a tree's wood, which show its age.[62] Prehistoric fire-charring in some Oregon tree ring records suggest that large crown fires might have swept a typical patch of western Oregon Douglas Fir forest an average of once every 200–400 years.[63] Yellowstone National Park's major 1988 fire supports this conclusion, since the most recent previous conflagration of similar size occurred in the 1700s, an interval of about 200 years. Outside Yellowstone, several crown fires have occurred each decade since 1970—a disturbing trend.

Until the later twentieth century, the U.S. government policy was to fight forest fires. The policy started with the "Peshtigo Blow Up" fire of October 1871, the most destructive forest fire in North American history.[64] Ignoring the contribution of clearcut slash that the loggers had left on the ground, an influential 1899 U.S. Geological Survey report quite unscientifically concluded that wildfires are "always evil, without a single redeeming feature."[65] Propping up the "evil fires" notion, the 1910 Great Idaho fire, also partly fueled by clearcut slash, destroyed three million acres of pine woods in the northern Rocky Mountains, killed more than 80 people, and carried wood ash as far north as Greenland, about 3,000 miles away.

Protected from burning for decades, forest litter and undergrowth began accumulating both in logged and regrown woodlands and in unlogged heritage forests. Cattle grazing on USFS and BLM meadows and clearings eliminated grasses and wildflowers, permitting the growth of small, weak trees in dense ("doghair") thickets (see chapter 3). Historical photographs, especially in the substantial photo coverage of California's Sierra Nevada range, show general increases in forest density and cover throughout the west since the mid-nineteenth century. Between 1866 and 1961, forest cover on the floor of Yosemite Valley increased approximately 20%. Photos of Lake Tahoe's Emerald Bay area show even greater increases between 1873 and 1994.[66]

Post-1910 forest fire suppression may have helped reduce human-set fire frequencies for a time. But over the succeeding decades, thick undergrowth and deadfall accumulations began to alarm fire specialists. As the densely crowded trees matured, insect infestations and plant diseases flourished. Weakened and dying timber stands covered thousands of acres, and some whole woodlands suddenly perished, greatly increasing the load of highly flammable dead wood. In 1970, the NPS faced rising firefighting costs, increasing numbers of hot crown fires, and rising losses of private property and human life in forest-fringe suburbs. To address these problems, the NPS began allowing some natural fires to burn freely in remote parklands and inaugurated controlled burning to remove deadfall and increase biodiversity.[67] Eventually the six-million-acre 1988 Yellowstone fire, a literal backcountry firestorm in the national park's densely overgrown lodgepole pine forest, drew public attention to the problem. That fire underscored the new policy's wisdom and added controlled burning to national forest management tools.

But the NPS's burn policies came too late and accomplished too little. NPS and BLM had to continue fighting major fires on national forest lands to protect property interests. Meanwhile, studies of controlled burn areas showed that they can increase erosion,[68] and both environmental groups and forest-margin residents have challenged them for impairing air quality. Additional concerns attach to the use of fire retardants and foams, both to contain controlled burns and to fight fires in remote areas. The chemicals are toxic to both terrestrial and aquatic native animals and plants, and encourage weed invasions, thereby reducing species diversity.[69]

Restoration or salvage

Trying to understand the density of presettlement forests, Arizona forestry researchers mapped the 1890 distribution of trees in northern Arizona's Ponderosa pine forests.[70] Their "restored forest" model is the expected mosaic of predominantly large, old pine trees and dead snags, punctuated by meadows and aspen groves. It suggests that the region's presettlement forests had supported 10–50 trees per acre (averaging 23 per acre), interspersed with patches of ground-level grasses and shrubs. By contrast, northern Arizona's human-managed forests today average 851 trees per acre. To determine whether the model forest can promote cooler fires, the researchers cut 50–90% of trees in overgrown experimental forests[71] and later burned the same areas to promote grassy growth. They planned to observe the experimental area over a decade or more to evaluate the restoration model.[72]

Before restoration theories could be fully tested, extensive hot fires burned huge western woodland areas in the summers of 2000–2003, fulfilling decades-old fears. Excepting the costly 2000 Los Alamos, New Mexico, fire, which started as a controlled burn, the fires swept across lands where prescribed fires had been blocked for years. The year 2000 fires utterly denuded some forest lands, wiping out their capacity to "recover" through natural processes—meaning produce harvestable timber—for more than a century to come. The 2002 wildfires covered more than 300,000 acres, and the 2003 fires consumed 400,000 acres of scrub and forest. In southern California alone, the 2003 fires took 1,500 houses and 13 lives.

The fires attracted political attention that promoted a version of the yet-unproved "forest restoration" model. Largely throwing science out the window, the untested idea of removing "excess fuel through controlled burning and thinning" suddenly became the favored approach to managing forests.

In 1995, President Bill Clinton had anticipated something like forest restoration when he opened recently burned national forest areas to two years of "emergency salvage" logging. Legislation promising thousands of new jobs and increased timber industry profits allowed commercial loggers to remove burned, diseased, and insect-infested trees, as well as trees downed by wind and other causes, ostensibly to prevent future fires. For convenience, the loggers also could remove some living trees. In the northern Rocky Mountains alone, 318 woodlands, including 25 heritage forests, were proposed for salvage logging.[73] The provisions required salvage projects to observe all applicable environmental laws but suppressed public reviews and input.

Since 2001, the George W. Bush administration has conflated "salvage logging" with "restoration" as "hazardous fuels reduction" measures, meant to mitigate future severe forest fires by removing "excess" dead, dry wood from regrowing forests. In 2003, the USFS supervised the Red Star Restoration Project to salvage still-usable trees from a fresh 2,000-acre burn 15 miles west of Lake Tahoe. Both the Clinton and Bush salvage logging operations failed to live up to promises that they would create jobs and produce greater financial return than the normal timber sales process. Meanwhile, they opened burned forests and heritage forest areas to logging of viable standing trees.

Paradoxically, salvage logging does not necessarily reduce fire hazards, and may actually *increase* them. Fires often begin when lightning strikes dead snags, but the main fuel that feeds a conflagration is the finer material on or near the forest floor, including brush, pine needles, small trees, and other debris of no value to logging operations—especially when partly burned. Removing larger snags and logs alters local conditions: wind, sun exposure, and evaporation all increase, along with the fire hazard. Large moisture-retaining logs are not left on the ground to slow advancing fires. Disturbances caused by salvage logging also reduce seedling regeneration by 71%.[74]

The law of healthy forests

Following the 2002 fires, Congress declared that "throwing billions at fire fighting is fruitless without sweeping restoration" and adopted the Healthy Forests Initiative (HFI), loosely based on the ideas behind forest restoration. Voted into law and signed by President George W. Bush in December 2003, HFI mandates selective tree cutting and prescribed or carefully controlled low-level burning, and deadfall log and underbrush removal. These measures aim to reduce the density of trees and tree debris in national forests to "pre-twentieth century status," ignoring both the lack of information on the pre-twentieth century status of any western forest ecosystems other than ponderosa pines in northern Arizona,[75] and the too-short time lapse for evaluating the restoration experiments. At the same time, the government proposed substantial changes to such mainstay environmental laws as the National Forest Management Act, the National Environmental Policy Act (NEPA), and the Appeals Reform Act.[76]

HFI emphasizes cutting instead of burning for fire control and would sanction forest cutting projects, including salvage logging, for "fuels reduction." At the same time, HFI allows the projects "categorical exclusions" from formal environmental impact studies under NEPA to assess potential adverse impacts. Not surprisingly, the amounts of proposed "restoration" cutting appear extreme in many cases.[77] HFI also reduces the time required for issuing forest clearing permits, in some cases by trimming or relaxing existing environmental regulations. In essence, HFI puts the benefits of cutting trees for fire reduction far above threats to soil stability and endangered species.

The industry-friendly Bush administration previously had tried to streamline environmental reviews of logging proposals, easing the way for "approval to thin underbrush and small noncommercial trees as well as conduct some 'thinning out of commercial grade wood' [i.e., large trees] in areas at high risk of fire." All the proposed regulations increase the difficulty of mounting citizen appeals against abusive logging plans, and extend the assault on forests into roadless natural areas.

Many of the roadless areas include heritage forest stands, which potentially could obtain wilderness status.[78] Environmental groups understand the need to thin forests for emergency fire control but tend to view the HFI proposals as a rerun of the 1995 fire-salvage project, aimed at opening national forests and critical endangered species habitats to unregulated commercial cutting.[79] A proposal to apply HFI to Alaska's lush Tongass National Forest, too wet ever to have sustained a significant wildfire, only increased suspicions of government deceit and duplicity. The suspicions seemed to be confirmed when the USFS produced a brochure promoting HFI, which featured a 1909 photograph purporting to show sparse large trees in an unlogged part of California's Sierra Nevada. Environmental groups quickly discovered that the actual scene was a freshly logged forest in Montana.[80]

Cutting Costs

The USFS often has appeared willing to sell timber at any price, even in its normal timber harvest process. Salvage logging programs may be even less remunerative.[81] Criticisms of the agency's financial performance generally focus on below-cost timber sales, common in Rocky Mountains states, which yield revenues lower than the costs of preparing and administering them.[82] The yearly distribution depends on market strengths or weakness, and also varies with accounting methods. According to the U.S. General Accounting Office, the 1992–1994 timber harvests eventually returned to the Treasury less than a quarter of the funds Congress had appropriated for the USFS to use in preparing timber sales and administering harvests. The timber sale program often has required annual congressional supplements to break even, using taxpayer subsidies to cover losses from below-cost sales. Clearly, below-cost timber sales are a drain on the public purse.

Environmentalist critics charge that the USFS's Emergency Salvage Logging Program actually costs the U.S. Treasury more than $50 million,[83] representing yet another below-cost logging operation on federal lands. Revenue from the Red Star salvage contract, supposed to help sustain the commercial logging industry, proved insufficient even to support the USFS's management operations or help pay for cleaning up the debris, which remained on the ground in highly flammable piles.[84] Part of the revenue shortfall has been blamed on the late 1990s drop in lumber prices, which sent the logging industry into decline—ostensibly from Canadian competition, imported lumber, and lumber mill closures in the western United States.[85] But western lumber industry organization data indicate that the west's lumber oversupply from the late 1990s into 2003 was due more to increased plant efficiencies and the competitive globalized economy than to strident conservationist lawsuits and the much maligned Endangered Species Act.[86]

Timber sale proceeds tend to get kicked back to states and into USFS programs, rather than to the U.S. Treasury. The U.S. General Accounting Office has reported that the Treasury received only 10% of total USFS timber sale receipts for fiscal years 1992–1994; the rest went to special payments and accounts.[87] For example, a 1908 law directed the USFS to return 25% of receipts to the states for building county roads and schools in national forest areas, because they do not generate state tax revenues. These payments exceed cash timber sale receipts for many forests, making them extreme cases of below-cost sales.[88] To keep the revenue-sharing payments high, many western counties have become timber sale advocates, often ignoring potential adverse environmental impacts and negative economics.

Timber sales receipts feed so many special internal USFS accounts and trust funds that critics charge USFS managers can keep themselves and their staffs employed with timber sales funds. The sale revenues can be used for projects to mitigate logging damages to wildlife habitat, potentially providing some wildlife managers an incentive to support damaging logging projects.[89] One example is the so-called "K-V Fund,"[90] which can receive any amount from timber sale receipts for reforestation and timber stand improvement projects, as well as for timber sale activities. The economic calculus of salvage logging is similarly complicated by mandated payments to states and deposits to special funds. Salvage profits are added to the Timber Salvage Sale Fund, another special account, instead of returning to the U.S. Treasury. But the Timber Salvage Sale Fund does not cover total costs of salvage logging, which must come from additional congressional appropriations (taxpayer money). The USFS does not audit its own accounts, and the actual financial costs or benefits of lumber sales on national forest lands are unknown.[91]

Other financial fiascos are outright frauds, including timber theft and abusive logging practices that contribute to deteriorating forest health. Fraudulent bidding and related practices are long-term problems, particularly in the Pacific Northwest.[92] To ensure profitability, some bidders have pushed the plethora of USFS standards to the limits, or even beyond. The USFS has altered bidding practices to curb the worst of the abuses, but the public cannot scrutinize completed contracts, and the USFS does not compare actual harvest returns with estimated sale revenues, so the effectiveness of those alterations is unassessed and unknown.[93]

Real costs

Salvage logging is supposed to improve forest health,[94] but like other logging, it actually exacts a steep environmental price. The machines and process of skidding overturn and destroy charred and vulnerable forest soils, remove nutrient-rich organic matter, and destroy ecological habitats that had survived earlier fires or insect infestations. Following a large fire, the salvaged forests are slower to regrow because the soils recover more slowly than in unlogged burns. And both the economically marginal 1995 and Red Star salvage projects attracted bids by allowing loggers to extract large living trees. USFS forest salvage proposals for watersheds that provide urban drinking water supplies to cities, such as Phoenix, Arizona, and Durango, Colorado, have raised concerns over the likelihood of increased erosion with increased sediment loading and turbidity in streams.[95]

In addition to paying many costs of cutting forests, taxpayers bear the "externalized" costs of deteriorating forests. Possibly because loggers cut mostly the large pines in western forests, firs—and especially Douglas firs—have become dominant in the cut-over areas.[96] Firs are more easily damaged by drought, insects, and diseases and contribute disproportionately to catastrophic wildfire risks. The many thousands of miles of roads, built and maintained for timber cutting and fire control at public expense, also degrade U.S. national forests (see chapter 5). USFS accounting makes timber sales appear more profitable by showing roads as assets, when in reality the roads on hill slopes generally contribute even more to erosion and flooding than logging clearcuts.

To regenerate cut woodlands, the USFS plants and seeds cut-over acreage, *also* at taxpayer expense. Regeneration costs from 1977 to 1994 added up to nearly $2.5 billion,[97] which works out to $314 per acre planted—a total of about 1.2 billion taxpayer dollars.[98] Only about half of the planted acres could be certified as successful; that is, they regenerated timber. But USFS accounts misleadingly lumps both successful and failed plantations into the single category of new plantings.

Future Forests

In 2001 USFS reported that about half of U.S. timber is less than 50 years old, and two-thirds of the west's lumber volume resides in regrown trees less than 21 inches in diameter. Only a fifth is in older trees with diameters of 29 inches or more. Only 16% of western U.S. potential lumber volume is in 21- to 29-inch trees as much as 50 years old.[99] These statistics explain why regrown woodlands yield less timber per acre than do heritage forests, and why loggers so desire to cut the old forests. As a reminder, the remaining heritage forests grow on a mere 7% of all wooded U.S. lands.

In the 1990s, environmentally aware citizen groups endorsed the Sierra Nevada Forest Plan Amendment, called the "Framework," for managing Sierra Nevada national forests on the bases of science-based ecological principles and regionwide concerns about fire and clean water supplies. By 2004, environmental groups were facing logging proposals under HFI, also supposedly based on healthy forests concepts, but likely to give commercial loggers access to large, old trees in roadless reserves and perhaps even in national parks. Matching science and environmental law against politics and economic interests, these struggles illuminate the contradictions inherent in trying to both preserve natural functions and allow multiple human uses.

A growing demand for lumber in developing countries, especially China, sent timber prices skyward in 2004, a trend that could revive the west's lumber industry. Increasing paper consumption and a turn toward alternative biofuels also will add demand for regrown forest wood. New federal regulations to expedite tree cutting in national forests, while preventing public-interest lawsuits or appeals of logging permits under environmental laws, surely would fuel a logging revival targeted at the last of America's oldest trees. Agroforestry, cultivating and harvesting trees on plantations, might either add new lumber sources or supplant natural regrowth.[100]

Sierra Nevada heritage

The fiercest battles over managing and preserving heritage forests center on California's 400-mile-long Sierra Nevada range, the longest unbroken mountain belt in the lower 48 states. The Sierras and more northerly Modoc Plateau embrace 11.5 million acres of publicly owned forest, slightly more than 10% of the state's territory, including two national parks and 11 of the 155 national forests under USFS management. Forest types vary from foothill oak woodland to alpine trees.

During the 1970s and 1980s, many acres of Sierra Nevada forest outside the national parks and other preserves had lost their oldest and largest trees to either clearcutting or selective logging. Habitat losses to logging, fire, grazing, and spreading human developments threatened many species, including several spotted owl species.[101] In 1993, Congress acknowledged "growing recognition among scientists, land managers, conservationists, and other citizens that the Sierra Nevada was in deep ecological trouble." Both in 1995 and in 1996, the USFS tried to design a sound management plan based on a scientific review of the entire region, the congressionally commissioned and funded Sierra Nevada Ecosystem Project. These attempts proved futile and "[i]n the end, both plans proposed doubling the amount of logging in the range while offering little in specific protection measures for the owl" and other threatened species.[102]

By 1998, the USFS recognized that forest management throughout the Sierra Nevada needed redirection and had begun developing the Framework both to save habitat for endangered and threatened wildlife and to reduce wildfire potential. When publicly unveiled in May 2000, forest and biological scientists, lawmakers, business leaders, and citizens hailed the Framework for incorporating the best available science into forest planning. The Draft Environmental Impact Statement chose the most protective of eight proposed alternative regional management strategies for preserving wildlife and habitat. It combined strategies for defending habitat and research to improve protections for sensitive and endangered species with fire hazard reduction through selective tree cutting, log and underbrush removal, plus prescribed burning—all guided by likely impacts on wildlife. This scheme would have protected trees more than 20 inches in diameter from logging, thereby dropping timber production to as little as 200 million board feet from as much as 700 million board feet in the 1990s.[103]

Assuming that balancing extremely different and intense forest uses is even possible, the Framework certainly seemed like the right way to go about it. But everyone could see that it would severely limit logging in heritage forests. Barbara Boyle, Sierra Club's regional representative, proclaimed, "Basically the [Sierra Nevada] commercial logging program is not going to exist."

The 2000 presidential election replaced the Framework's political sponsors in Washington. In 2001, Dale Bosworth, the new USFS chief, received 276 appeals from land developers and commercial forest product interests. While praising the plan's sound foundation and "the hard work and dedication of the interested citizens, government agencies and many others who came together to help develop the Sierra Nevada Framework," Bosworth called for further review. The review and subsequent revision refocused plans for Sierran forests away from habitat and endangered species concerns and onto timber harvesting.[104]

Bosworth's review elicited much skeptical commentary from district rangers, the officials directly charged with implementing land management policies.[105] Nearly two-thirds of rangers' comments labeled the Framework's standard goals and guidelines as top-down management, likely to limit their responses to a continuously changing, complex, and dynamic forest environment. One wrote,

Place more emphasis on the desired condition over landscape and be very limited on prescribing how to achieve that desired condition…leave it up to us in the field to achieve that condition.…Nothing in nature is exact and uniform and when you try to apply standard prescriptions across…the Sierra Nevada, you are bound to run into problems.[106]

An even higher proportion of rangers worried that agency funding would be too low to support the Framework's complex management system, and another third noted that the restriction on clearing to no more than 10% of total area per decade would conflict with the emerging HFI. A few fretted about the region's timber industry and revenue and job losses for some mountain communities, but most agreed entirely with the wildlife protection and fire mitigation objectives of the Framework, affirming that "how to" rather than "why" is the issue.

Inevitably, the Framework did collide with HFI. In 2004, Regional U.S. Forester Jack Blackwell set the maximum diameter of Sierra Nevada trees to be logged at 30 inches—10 inches larger than the Framework limit. This policy would allow cutting of large trees on all 11.5 million acres of the Sierra Nevada's national forests, tripling the Framework's allowance.[107] Despite protests that heavily logged forest areas are fire prone, Blackwell also announced that the USFS would spend $50 million to promote logging in heritage forests on 700,000 national forest acres, ostensibly to protect trees, wildlife, and human settlements against large, intense wildfires.[108] When Blackwell opined that the Sierra Nevada Framework "was overly cautious," former Chief Forester Mike Dombeck retorted, "The original plan had input from our best scientists both inside and outside the Forest Service. Apparently now the efforts are due to commodity extraction."

The management policy for Sierra Nevada forests still is making feathers fly. Both HFI and the discarded Framework plan, plus similar state and local programs, have become lawsuit targets for massed environmental organizations. So far, political agitation has kept national and environmental laws in place, and courts have disallowed clearcutting in and near protected heritage groves. In 2006, a federal judge ruled that the USFS cannot allow commercial logging in California's Giant Sequoia National Monument, and a federal appeals court upheld the public right to review forest plans under NEPA. A 2005 suit against the USFS for omitting essential scientific information from the Framework revision still was pending in early 2007.

The new plantation

Modern agroforestry generally grows just one hybrid tree species per plantation (monoculture), hybridized by traditional protocols to enhance lumber yields and reduce rotation intervals. Ideally, plantations would allow previously logged areas to regrow throughout the west, but they tend to replace rather than augment natural forest stands. Monocultures starkly reduce the biodiversity necessary to preserve forest habitat for animal species. Monocultures might not threaten biodiversity overall if the total plantation area remained small, but replacing too many natural forests would considerably impoverish regional ecosystems. However, species monitoring, under Endangered Species Act constraints, could govern the pace of conversions. This would become yet another land management challenge.

In addition to traditional hybrid trees, agroforestry researchers also are bioengineering tree varieties. Potlatch Forest Industries is experimenting with cloning poplars from hybrids that can grow 60 feet tall in six years and can be harvested every six years.[109] Plantations of the fast-growing hybrids and clones take up water quickly, lowering water tables and potentially drying up springs and creeks. Compared to natural species' 15- to 20-year rotation, the anticipated six-year cutting intervals likely will cause greater erosion and greater sediment pollution, especially on slopes. Plantation road networks as dense as, or denser than, the clearcutting roads in natural forests inflict the same severe erosional damage of all other roads (see chapter 5).

Natural forests provide their own fertilizers, while fast-growing plantation trees rapidly deplete soil nutrients, requiring large fertilizer applications. Erosion carries fertilizer nitrate and phosphate into streams and ponds, promoting algal blooms that consume all the oxygen and kill fish and other animals (see chapter 9). Plantations of cloned monocultures eliminate the protection of tree species biodiversity. Since every tree contains the same genetic material, the artificial forests are highly susceptible to disease and insect attacks. They also require heavy pesticide and herbicide applications, which compound their pollution potential.

Plantations of genetically modified (GM) trees, created through directly altering a plant's cellular genetic material (DNA), pose many of the same threats as cloned hybrids. GM tree varieties under commercial development include Scotch pine, Norway spruce, silver birch, teak, apple, and cherry. GM Douglas fir plantations in western Washington State can grow to cutting sizes in half the time of wholly natural stands[110] and yield more useful wood and paper per tree than do their natural counterparts. The downside is that they probably use up soil nutrients even faster than the clones and so will need even more fertilizer.[111] Some GM trees are designed to make harvesting less energy consumptive, although more frequent harvests could offset that advantage.

Agroforestry proponents suggest the fast-growing cloned and GM trees might use even larger amounts of CO_2 than natural ones, helping to slow the rate of greenhouse warming. This is a spurious argument: Cutting forests releases large amounts of CO_2 from both soils and trees. And the CO_2 sequestered from regrowing them does not adequately compensate those losses.

Many environmentalists contend that agroforesters and government regulators have not adequately investigated potential bioengineering consequences, or taken them seriously, and that proposed safeguards are insufficient. GM poplars, pines, and fruit trees have a gene from the *Bacillus thuringiensis* (Bt) bacterium inserted into their cellular DNA, for example. They express insecticidal Bt toxin in all their tissues, including the edible parts,[112] which might be lethal to beneficial insects, such as bees and butterflies. Bt may kill the larvae of monarch butterflies and could threaten other beneficial insects. Herbicide-resistant GM trees have the potential to become noxious weeds. The environmentalists contend that these genetic modifications could spread to the natural gene pool and disrupt entire ecosystems in wholly unpredictable ways—akin to past "good-willed disasters" from diseased foreign plants.[113]

Supposedly, the Bt trees are engineered to be sterile, but "sterile" aspens growing in field trials in Germany began to flower after three years.[114] Pollen from Bt plantations could spread the genetic modification to natural trees, and their

Bt-bearing seeds could directly threaten the future of natural forest habitat. Another issue is an aspen variety with low lignin content in wood, developed to need fewer chemicals for making paper. Lignin is the main strengthening agent in tree trunks, and no one knows how lignin-poor trees will withstand winds. The effect the low-lignin characteristic might have on ecosystems if it spreads to natural aspen and related species is unknown.

Hot trees

North America's heritage forests endured the ice ages, but today's forests must adapt to rapid global warming to survive. Rapid climate change seemed like a science fiction scenario until 2000, but the reality is showing up in extreme weather patterns, unusually warm winters and broiling summers, melting ice caps, and rapidly eroding shorelines in far northern latitudes. Heat stress has visibly weakened the underpinnings of subarctic ecosystems, forcing both plants and animals to look for more comfortable conditions. Many climate scientists now believe that a new era of rapid climate change is already upon us and that future human generations will face worsening problems.

North American forests have responded to environmental shocks many times in just the past 60 million years. Pollen samples from the last glacial interval in what is now New York State indicates rapid temperature fluctuations about 12,000 years ago, jumping as much as 7°F over only 50 years. The sudden shift nearly eradicated such cold-adapted trees as birch, fir, and spruce. Between 12,000 and 7,000 years ago, mean 5–9°F global temperature increases forced white spruce and lodgepole pine species to shift their ranges by hundreds of miles. Over time, oak and white pine replaced tree species that could not adjust, and which simply disappeared.[115]

Global warming models now predict a similar temperature increase within the next *hundred* years, and this time the radical temperature change adds to all the human stresses on our woodlands. In the face of rapid climate change, intensive forest fragmentation, soil depletion, and groundwater decline, many or most forest ecosystems may not be able to reestablish themselves naturally (see chapters 9, 13). Under these kinds of pressures, the future may bring large-scale die-offs of many tree species and extinction of large stands of western temperate forest.

At advanced stages of global warming, severe, prolonged droughts are possible to likely. Droughts are fundamental causes of forest fires. A run of drought years is the immediate cause for the devastating 2000–2003 western fire seasons in southern California and the interior western United States, contributing to the nation's worst fire seasons on record (table 1.2). Dried-out trees, bark beetle infestations in weakened trees, and withering summer heat fed the disastrous, record-setting fires.

Globally, 95% of all wildfires are set by humans, and humans set about 50% of all fires even in lightly populated western states (about 75% in crowded California).[116] As the years grow increasingly warm, with human populations poised to rise another 50% over the next 50 years—perhaps far more in the tinder-dry, lightning-prone west—forest fires are likely to increase dramatically. Increased catastrophic crown fires will be the likely result of forest management practices that do not effectively thin forests because of funding issues, economic concerns, and political controversy.

Table 1.2   Recent U.S. Wildland Fire Acreage
Statistics

Year[a]	No. Fires	Acres Burned
2000	82,071	6,891,292
2001	63,737	3,265,574
2002	67,889	6,657,464
2003	50,022	3,161,924
2004	61,625	7,733,023
2005	54,051	8,175,432
2006	84,333	9,080,628
2007	74,031	8,245,535
5-Year Average: (2003–2007)	64,812	7,279,308
10-Year Average: (1997–2007)	67,557	5,736,924

National Interagency Fire Center Fire Information.
Available: ww.nifc.gov/fire_info/nfn.htm.

[a] Data for years 2000–2007 compares fire statistics between
1 January and date of inquiry (7 October 7 2007) for each
year, plus five-year and ten-year averages.

Both historically and currently, forest managers resist letting ground and understory fires burn and spread for fear of losing timber crop and the public's demand
that governments protect private residences. Nothing creates raging controversies
more than limiting where Americans can build homes or find playgrounds. But in
2003, newspaper editors questioned the wisdom of building subdivisions in forests,
in much the same way that frequent floods have raised questions over home and
town sites on river floodplains and coastal barrier islands. As global warming and
droughts increase fire dangers, insurance companies may impose limitations over
public and politicians' objections.

Forests in the Balance

A century and a half ago, the geographer and explorer Alexander von Humboldt
realized that human sustenance is inextricably tied to forests: "By felling the trees
which cover the tops and sides of mountains, men in every climate prepare at once
two calamities for future generations: want of fuel and scarcity of water."[117] We
have felled our mountain forests, opening them to soil erosion, adding habitat fragmentation from roads, urban developments, and recreational sites, and high fuel
loads due to counterproductive fire suppression. Our high fossil fuel consumption

has induced higher than natural global warming, which may be advancing too fast for either forest or human comfort (see chapter 12).[118]

Nobody knows how old forest remnants will respond to continual human fragmentation and human-caused extinctions, added to the inevitable natural disruptions from high winds, earthquakes, volcanoes, and fires. Many biologists and ecologists fear that old forests cannot survive rapid climate change without radical protection. Accumulating a few centuries of continued soil losses could, in addition, devastate even younger forests. The small, isolated forest remnants of the interior west will be especially challenged. Lending a sense of urgency, some ecologists warn that a variety of plant species could disappear across hundreds of thousands of square miles over the coming century.[119] While the oldest individual trees will die off naturally as time goes on, preserving their dynamic ecosystems is the best hope for preserving and extending North America's biodiversity. Each of the American west's heritage forests will need its own ecological management plan, on the lines of the Sierra Framework.

Given the predominant national outlook and the antiregulatory climate favoring commercial interests, there is no easy solution to the forest dilemma. Establishing forest management policies and practices that better preserve biodiversity will take a long time—longer than political cycles and climatic change allow. They would have to be sustained no matter what political party holds power. Any move toward better forest management must begin with widespread and fundamental changes in national and regional attitudes and awareness. Public pressure is needed to drive reforms from a broad-based consensus that ecosystem services are important natural capital, and that ignoring or compromising them too much will have disastrous consequences.

Achieving adequate USFS implementation would require strong political support for sound ecological management[120] and tough oversight of logging and replanting practices. Maintaining such a system requires public education and determined political leadership. State and federal forestry agencies would have to enforce laws conserving old trees, hill slope soils, and plant and animal species. Environmental organizations would have to finally reach agreement on goals. If asked, we would advocate a shift away from absolute preservation or restoration of an assumed "natural" state, likely an unsustainable goal in any case, toward reducing the human footprint and preventing impacts on natural processes that maintain well-functioning ecosystems.

To reach such a national consensus for saving our western forests, Americans would have to concede that, after decades of experimentation with negative results, the best way to manage forests may be to simply leave them alone, fires and all. Nature has been in the business far longer than we have, after all. We may need to accept a human retreat from forests on the basis that burgeoning residential and business developments are ultimately harmful.

We definitely will have to take conservation and consumption issues more seriously and more personally, and this is where individuals have the greatest power. Do we really need to consume so much wood? In addition to our very high lumber use, on average, each American uses almost 900 pounds of paper per year, more than double western Europe's per capita paper consumption (351 pounds per year). The U.S. paper consumption far exceeds the minimum amount essential for literacy and communication (80–120 pounds per year).[121] Much of the excess is in packaging and advertising.

We might need to accept some minor inconveniences to achieve big environmental payoffs. In many parts of Europe, shoppers routinely carry reusable cloth bags to reduce the wood demand for paper bags, and so could we. To address climate change as individuals, we might lobby for urban community and school gardens, plant or "adopt" small teaching forests to offset carbon emissions, and revive National Arbor Day with a new and greater sense of purpose. Getting children involved in all such programs would help to build conserving practices into the body of American traditions.

Most of these ideas clearly are incompatible with our consumer culture and deep political disagreements on values. But should rapid climate changes severely limit the forests' ability to adjust and regrow, they also could limit the free-oxygen–producing, carbon-dioxide–sequestering, water-purifying, and other critical life support services that forests provide us. Water scarcities already occur throughout the arid west, jeopardizing the future for thirsty, rapidly growing populations and industries (see chapter 9). Drought conditions linked to climate warming are likely to increase the potential for destructive crown fires.

We may have a limited window of opportunity to forestall a future of severe and rapid climate alterations made even worse by the destruction of our forest ecosystems, which are a key element of the west's natural wealth. Julia Butterfly Hill's tree-sitting, and political movements to preserve all remaining old trees in California, may seem radical, but if we Americans cannot reduce our resource demands, future generations will face an impoverished and dangerous future.

2   Harvesting the Future

We lose our health—and create profitable diseases and dependencies—by failing to see the direct connections between living and eating, eating and working, working and loving.
Wendell Berry, *The Unsettling of America*

For most of two centuries, the United States was a nation of small farms and many farmers, raising much of their own food along with one or more cash crops and livestock for local markets. Today, farms run by families of weatherbeaten farmers, pie-baking farm wives, and earnest 4-H offspring are disappearing. Americans live on supermarket or take-out food, mostly produced on extensive, highly mechanized and chemical-dependent industrial-scale "conventional" farms, raising single-crop monocultures or single-breed livestock. The larger farms cover tens of thousands of acres, too much for single families to manage. It is not agriculture, but agri*business*—an industry run by corporations.

Conventional industrial agriculture is highly productive, and supermarket food is cheap. So why should anyone worry about growing food with chemical fertilizers, expensive equipment, pesticides, and pharmaceuticals? The reasons, acknowledged even by the industry, are that agribusiness "saddles the farmer with debt, threatens his health, erodes his soil and destroys its fertility, pollutes the ground water and compromises the safety of the food we eat."[1]

Croplands presently encompass some 57 million acres in the 11 western states (table 2.1). Giant plantations consume huge amounts of natural resources—soil, fertilizers, fuels, and water.[2] Synthetic fertilizers keep overused soils in production, until they become too salty (salinated) and must be abandoned. Industrial farming has taken over large areas of wildlife habitat, including forest, scrub, desert, or prairie, to replace degraded croplands.[3] The clearings and massive pesticide applications threaten or endanger large and increasing numbers of plant and animal species in the western United States.[4] Pesticide exposures sicken family farmers and agribusiness workers in the fields, and add environmental poisons to our diet. Pesticides and other problematic agricultural chemicals accumulate in our bodies.

Agribusiness consumes especially huge amounts of increasingly costly, nonrenewable petroleum. "Every single calorie we eat is backed by at least a calorie of

Table 2.1   Cropland in the 11 Western States and the United States, 1997 (Thousand Acres)

State	Cultivated Cropland		Noncultivated Cropland	
	Irrigated	Nonirrigated	Irrigated	Nonirrigated
Arizona	905	77	229	1
California	5,090	1,130	3,191	224
Colorado	1,908	5,659	1,038	164
Idaho	2,822	1,719	624	352
Montana	929	11,598	1,234	1,410
Nevada	71	50	578	2
New Mexico	636	753	454	34
Oregon	829	1,848	851	234
Utah	329	376	922	51
Washington	1,022	4,556	757	322
Wyoming	448	530	886	310
Total Western United States	14,989	28,296	10,764	3,104
Percentage[a]	35%	65%	78%	22%
Total United States	48,878	277,906	13,253	36,962
Percentage[a]	15%	85%	26%	74%

U.S. Department of Agriculture. *Summary Report, 1997 Natural Resources Inventory* (revised December 2000), table 3.

[a] Percentage of total cultivated or noncultivated land.

oil, more like ten"[5] to run fleets of immense plowing, planting, cultivating, harvesting, and processing machines, plus countless irrigation pumps. Growing a pound of American beef consumes half a gallon of petroleum. A top executive of the giant agriculture-chemical corporation Monsanto has admitted that "current agricultural technology is not sustainable."[6] High-tech agriculture, such as cloning and genetically modifying crops, does not help conventional agriculture become more sustainable.

We taxpayers support these unsustainable farming practices, both at the grocery store and through our taxes. Especially in California's Central Valley, many large-scale farms irrigate crops with federally subsidized water, originally intended for small- and medium-sized farms. Huge dairies, cattle feedlots, and mass hog and poultry operations replicate this water-depleting pattern throughout the west. Tax money also goes directly to corporate farmers in the form of heavy federal payouts and price supports that destroy family farms.

A return to more traditional farming practices, especially small, ecologically balanced, local organic farms, supported by local communities, offers a way to rebuild healthy agricultural soils—the keystone component of America's natural capital—and a more sustainable future. If government programs shifted to encouraging smaller organic farms, we could begin repairing nutrient-depleted soils, restoring economic viability to small- and medium-scale farming, and improving American diets.

We Are What We Eat

We in the United States are predominantly urbanites or suburbanites who grow lawns, not vegetable gardens. In contrast to even citified Europeans, who tend to know how their food is grown, relatively few Americans understand the importance of nutrition or care to inquire where supermarket food comes from or how it gets there. Most of us get all our food information from television commercials or diet books. Some studies suggest that more than half of Americans eat mostly processed food—principally pasta, pizza, TV dinners, breads (and cakes), plus red meat and dairy products. Their main vegetables are onions and potatoes.[7]

Even those of us who vary their diets may lack access to fresh garden produce, unaware that the mass produced fruits and vegetables in standard salad bars or shrink-wrapped packages lack the succulence, texture, and flavors of garden-ripened produce. They may never have tasted the juicy sweetness of vine-ripe tomatoes or crunch of fresh garden lettuce, packed with vitamins and iron. Routine U.S. Department of Agriculture (USDA) reports show that the nutritional contents of fresh, conventionally grown, supermarket vegetables have declined from 1975 to 1997. But the USDA never publicized this information, so critical to public health, nor addressed the causes.

There's a simple reason we all need to know more about food than its taste and calorie content. What we eat, breathe, and drink—the substances that we take into our stomachs and lungs—link us directly to our environment. Each of us consists of more than a trillion cells, which are constantly dying and being replaced. "Living" means that bodily functions literally re-create all our organs, inside and out: skin, bone marrow, kidneys, livers, stomach linings, brains, reproductive organs—all the time. The air, food, and liquids that we consume are the raw materials, the building blocks, for this constant process. What is in them becomes part of us. Good food sustains health, and poor or contaminated food can undermine our minds and bodies, and even kill us.[8]

Both junk food junkies and people trying to "eat healthy" on the grains, fruits, and vegetables from large supermarket chains unknowingly swallow low levels of pesticide residues with every bite. The residues accumulate in fat cells, reproductive organs, and other body tissues—a process called bioaccumulation. A constant diet of chemical residues from conventionally grown foods can bioaccumulate in body tissues to concentrations many times greater than the amounts on foods or floating around in the environment.

Dairy and meat products contain detectable levels of hormones and antibiotics that industrial factory farms feed to livestock. Hormones can disrupt the body's critical endocrine systems, which run reproductive and other functions. Many pesticide formulations contain chemicals that imitate natural hormones. Bioaccumulation processes can build up excess hormones and other endocrine-disrupting chemicals, stored mostly in fat cells. The endocrine disrupters can impair adult reproduction, and new research suggests that they can disturb children's growth and sexual development.[9] Overusing agricultural antibiotics threatens the effectiveness of many antibiotic drugs that we rely on to control human diseases.

Corporate Farms at the Public Trough

Almost all U.S. farmlands are privately owned, yet large corporate farmers receive massive federal support. State and local government programs also pour money into private farms. In 2005, federal farm support payments totaled more than $25 billion—nearly 50% more than federal welfare payments. Farm support programs began in the New Deal, as loans to farmers facing low crop prices. The system kept grain prices from collapsing in the face of huge surpluses and "helped both to pay for the farm programs and smooth out the…swings in price."[10] Most farmers eventually repaid the loans and kept on farming.

Successive Republican administrations attacked the loans system. Ultimately, the 1970s corn sales to the then-Soviet Union drove food prices up, to a chorus of national discontent. Earl Butz, Secretary of Agriculture under Richard Nixon, hurried to lower prices and enacted programs that morphed government farm programs into flat-out handouts for large farmers growing defined commodities. The conservative Heritage Foundation contends that these farm subsidies overwhelm market forces, causing overproduction, low prices, and continual demand for more payouts. The low prices drive small- and medium-scale farmers out of business.[11] The subsidies also have increased total pesticide and fertilizer use, ruining natural lands downstream.

In the 1996 "Freedom to Farm" bill, and again in 2002, Republican-dominated conservative Congresses tried to wean farmers from government handouts and accustom them to free market risks and rewards. Both attempts failed. The 1996 bill awarded more than 70% of total tax-based farm subsidies to only about two-fifths of the nation's farmers, mostly running large operations in 12 states and growing wheat, corn, cotton, rice, and soybeans, with smaller amounts for sugar beets, peanuts, and livestock and dairy farm operations. Reported the USDA, "In real numbers, the top ten percent took home $48 billion out of $60 billion over a five year period."[12]

The 1996 law also changed the system from subsidies based on acres planted in qualifying crops to direct payments without restrictions. To give farmers flexibility in the face of shifting market trends, they now can decide what to plant, and even to not plant a crop in bad years. But somehow the law attached previously subsidized crop types to farm*lands*, independent of current use. This provision has put farms under houses—in particular, Texas rice farmers have made millions from selling their lands for housing developments.[13]

At the end of 2001, President George W. Bush instructed Congress to pass a farm bill based on free market principles as a way to expand international trade. An odd alliance of farm groups, free-market political groups, and environmentalists supported a draft Senate bill that capped farm payouts at $275,000 for any one recipient and increased funds for saving small farms as well as for soil conservation and pollution cleanups. The final bill earmarked less than a quarter of total farm spending—$12.9 billion over six years—for conservation programs that especially benefit small to medium farmers and farm owners.[14]

Agribusiness lobbyists carved up the draft, however, deleting the low subsidy cap that directed support to small farms, and dedicating most of the conservation money to cleaning up factory farm messes. The Environmental Quality Incentives Program (EQIP), originally intended to help farmers protect drinking water, became "a multi-billion dollar giveaway to a few industrial-type livestock companies" that "accelerates the consolidation of the livestock industry, harming family farmers and consumers." Most subsidy payouts went overwhelmingly to the same large operations growing the same set of overproduced crops. And payout amounts tripled.[15]

Agricultural census data show that fewer than 40% of U.S. farms, and only a fifth of farms in the 11 western states, grow any of these crops. Only 9% of farms in California, the largest farming state, get any support. Rather than address the bias of a system that has awarded subsidies to Fortune 500 companies, celebrities Ted Turner and David Rockefeller, basketball idol Scottie Pippen, and 14 members of Congress, the bill amended the Freedom of Information Act to restrict public knowledge of who gets these subsidies. Turning farm payments into state secrets sets a dangerous precedent of exempting federal programs from public disclosures to avoid embarrassing public officials and influential citizens.

Urging a veto, the Heritage Foundation calculated that the 2002 bill would pay out $30 billion annually in "guaranteed incomes and constant bailouts to a few of the wealthiest farm operators, unparalleled by those of any other industry...." Guaranteeing a minimum income of $32,652 for a family of four to every full-time farmer in America would be much cheaper—only $4 billion per year.[16] Farm payouts may keep farmer profits low, but they ratchet up grocery and restaurant prices. The Heritage Foundation estimated that the 2002 legislation will cost an average American household both $1,805 in higher taxes and $2,572 in inflated food prices over the 10 years the law is in effect, extracting at least $200 from every American household per year.

The 2002 bill also has had international repercussions. Brazil cotton growers' complaint to the World Trade Organization (WTO) resulted in a 2005 judgment against U.S. cotton subsidies—the first against a wealthy nation for domestic agricultural subsidies. In 2006, poor and developing nations' resistance to cheap subsidized U.S. crops flooding their markets threatened to undercut previous WTO agreements. In early 2007, Canada launched a WTO complaint about "trade-distorting" U.S. corn subsidies, which undercut corn prices for farmers worldwide.

Small farm squeeze-out

Writes *Metrofarm* author Michael Olson, "[C]ompetition for the consumer dollar is between the very big and the very small [farms]. The middle ground...is being squeezed into oblivion."[17] The main pressures are conventional farming's heavy costs for nonrenewable fuels, fertilizers, pesticides, and irrigation supplies—not to mention hugely expensive machinery. Crop prices have fallen under the pressure of subsidies, giving small-scale conventional farmers a crushing debt burden, which increasingly renders their operations uneconomic. One or two bad years in a row can force severely mortgaged farmers into bankruptcy.

Federal support goes to immense farms that can balance seasonal losses with income from other corporate divisions. Then the corporate farms buy out smaller neighbors with their federal payouts. Closing the circle of ruin, they then apply for additional support to compensate for low crop prices. The Heritage Foundation called it the "plantation effect," in which "family farms with less than 100 acres are...bought out by larger agribusinesses, which then convert them into tenant farms." "Freedom to Farm" flexibility destroys tenant farmers, however. The tenants collect support payments for their crops, but landowners take the money when land goes out of production. The owners get more from the government than from renters, so they kick tenants off the land. In 2006, the *Washington Post* reported that Texas rice-growing acreage has shrunk nearly 66% since 1981, with most rapid declines starting in the late 1990s.

Many small- and medium-scale farmers have to contend with low land prices and lower standards of living, pointing to the need for directing federal subsidies to those sectors instead of agricultural giants. Other pressures come from imported foreign produce, raised by low-paid labor under conditions unacceptable in the United States, which undercut both U.S. food safety standards and farm prices. California grape farmer John Baranek protested, "America demands that we farm clean and produce the highest quality food under the most strict standards in the world, and now we have our corporations buying crops from foreign countries using pesticides we banned 20 years ago."[18]

The small-farm squeeze also reduces farming efficiencies and contributes to degrading agricultural lands. Contrary to the widely held idea that bigger is more efficient, Institute for Food and Development Policy studies show that small farms actually produce far more per acre than do large ones. Compared to a large monoculture farm, smaller "integrated" farms generally raise more than a dozen crops and various animal products and so may have a lower yield per acre for one or several crops. But "the total production per unit area [on a small integrated farm]...can be far higher...[and] the commitment of family members to maintaining soil fertility on the family farm means an active interest in long-term sustainability not found on large farms owned by absentee investors."[19]

Small U.S. farmers also have a better conservation record than do the larger-scale farmers—they "devote 17% of their area to woodlands [compared to only 5% average for large farms]...and keep nearly twice as much of their land in 'soil improving uses,' including cover crops [to prevent soil erosion] and green manures [to add nutrients]." Factoring in the environmental costs of conventional farming enhances the significant economic benefits of medium- and small-scale organic farming alternatives.

Conventional Degradation

Throughout human history, farming inroads probably are most to blame for reducing forests and wildlife habitat to isolated remnants. Plowing fields always leads to soil erosion and degraded farmlands. In ancient Greece, the philosopher Plato lamented, "Once the land was enriched by yearly rains, which were not lost, as they are now, by flowing from the bare land into the sea. The soil was deep, it absorbed and kept the water in loamy soil, and the water that soaked into the hills fed springs and running streams everywhere." Agrees modern farm philosopher

Wes Jackson: "[T]he plowshare may well have destroyed more options for future generations than the sword."

America's late 1940s transition to modern "conventional" farming methods takes agricultural impacts far beyond erosion. Chemical agriculture and animal factory farms are the top dispersed sources of air and water pollution and major contributors to the 75,000 polluted miles of rivers and 4.5 million acres of polluted lakes, estuaries, and wetlands in the western United States (table 2.2).[20] In the words of Douglas Tompkins, founder of North Face sporting goods:

> Our conversion from agrarian, local, fully integrated food [raising] systems to industrialized, monoculture agricultural production has brought a staggering number of negative side effects, many of them unanticipated [including] soil erosion, poisoned ground waters, food-borne illnesses, loss of biodiversity, inequitable social consequences, toxic chemicals in food and fiber, loss of beauty, loss of species and wildlife habitat....[21]

The bad effects extend to all western lands—and to everybody who eats conventionally grown food or who breathes the air or drinks water from agricultural areas. Adds Tompkins, "To make the crisis even worse, we continue to export this destructive industrial system of food production around the earth."

Soil ruin

Practices that degrade soil, our least appreciated and most undervalued resource, destroy farmlands' fertility. Decades of abusive farm management have led to widespread

Table 2.2 Polluted Waters in the Western United States, 1998

State	Rivers and Streams (Miles)	Lakes, Estuaries, Wetlands (Acres)
Arizona	1,780	29,840
California	13,720	3,367,170
Colorado	1,750	11,080
Idaho	11,160	233,980
Montana	29,550	1,470,300
Nevada	2,600	109,340
New Mexico	2,330	66,340
Oregon	6,040	116,830
Utah	3,560	132,360
Washington	550	14,910
Wyoming	1,410	9,690
Total	74,450	4,461,900

U.S. Environmental Protection Agency. *Atlas of America's Polluted Waters* (EPA 840-B-00-0002, 2000).

soil erosion, stripping of western U.S. croplands, and growing numbers of abandoned farms. By the end of the twentieth century, eight million acres of degraded U.S. agricultural land either had been abandoned or taken out of production under the federal Conservation Reserve Program.[22]

Soil is the main medium in which most plants grow. Before plows broke the rich but fragile arid western soils, the land maintained its health by growing a vegetative mix, some plants adding nutrients as others extracted them. Farmers today plow constantly and grow the same set of nutrient-extracting crops every year. This is the best way to exhaust soils and open them to erosion.

Natural geological and biological processes developed soils over many hundreds to thousands of years. Rainwater interacts with carbon dioxide (CO_2) to release minerals from rock debris, and soil biota—abundant bacteria, fungi, algae, and other microscopic organisms from decaying living things—process the minerals into the nutrients that plants need to grow and thrive (see chapter 13). In undisturbed soils, the actions of earthworms, insects, bacteria, and fungi constantly process and exchange nutrients. Eventually, most natural soils develop a layered structure (profile), capped with protective and stabilizing biotic surface crusts. Upper soil layers develop openings that hold water and air, absorb raindrops, retard erosion-causing runoff, and enhance water seepage to natural groundwater-storage aquifers (recharge). Soil organisms, including plant and tree roots, remove natural waste substances, purifying the water (see chapter 1).

Rainstorms and windstorms shape the land in complex and continuous processes of soil erosion, transportation, and redeposition (see chapter 13). Soil crusts, plant covers, and other natural soil stabilizers minimize erosional effects, while human disturbances open soils to erosion. Topsoil, the upper layer that most effectively feeds plants, is the first to go and carries the nutrients with it, along with decayed litter and biota (figure 2.1). Removing topsoil also undercuts plant roots and kills the plants, exposing lower soil layers to faster oxidation and leaching by water. As erosion proceeds, the soil eventually loses all of its nutrients and larger biota.

After plowing, even gently sloping and flat lands undergo accelerated erosion—even where farmers employ erosion control techniques. Erosional effects and rates are much greater on slopes, because sheets of running water quickly carry soil away and cut shallow rills into the hillside. If not stopped, the rills develop into deep gullies (arroyos). Crawler tractors, developed in the early twentieth century, eased plowing on steeper slopes but greatly increased erosion rates—especially in fields of corn, cotton, and leafy vegetables.[23] A common modern tendency is to plow straight up a slope, increasing runoff and erosion even more.

Erosion turns in-place soil to unstable sediment and a costly problem. Conventional-farm soils store fertilizer and persistent toxic pesticides, but eroded sediments carry off significant amounts of the chemicals and deposit them downstream (see chapter 13). Sediments can bury undisturbed fertile soils and crops and damage equipment and trees. Overloading streams with sediment inevitably generates more and larger flooding episodes downstream. Sediment deposits are more easily eroded than the soils themselves and can continue polluting streams, rivers, lakes, and ponds for many years in succession.

Contour plowing and terracing reduce the worst erosion effects from rain runoff. But contouring slopes with huge machines can be dangerous, and terracing

Figure 2.1 A huge dust plume rises 5,000 feet above the southern San Joaquin Valley, California, at the height of the 24-hour December 1977 windstorm. Dust streams from canyons in the Tehachapi Mountains foothills (lower left). Before local wind-speed gauges failed, some had recorded winds up to 194 miles per hour, which uprooted or toppled powerline towers, destroying orchards, vineyards, and other crops. Photograph by Sam Chase.

restricts the planted acreage. Terracing also is costly and adds to a conventional farmer's huge debt burden. Neither technique can lower soil erosion on plowed lands to natural levels. No-till farming, which presses seed into the soil with mechanical drills, has less erosive potential and is gaining popularity.

Dust to dust

Wind storms carry dust particles thousands of miles, spreading pollution around the globe. To the 1930s economic depression, drought and strong winds added the Dust Bowl, which stripped soil from intensively farmed wheat fields on the plains of Kansas, Oklahoma, and Texas.[24] Some of the blinding and choking dust and sand storms buried crops still standing in nearby fields, severely abrading trees, houses, cars, and animals. Other windstorms rained Texas soils on Chicago or blew midwestern dust across New York City and Washington, DC, spreading pathogenic bacteria and fungi that cause severe respiratory problems and other illnesses.[25]

Acknowledging the threat of soils massively on the move, the 1930s federal government prescribed and funded soil-conserving practices, such as contour plowing and planting trees and bushes for windbreaks. But the Dust Bowl's hard lessons now seem forgotten on America's arid plains. Farms across the west yield dust abundantly, and windstorms again destroy soil fertility and threaten health and safety.[26]

Windbreaks are out of favor because they impede today's supersized farm equipment and popular center-pivot irrigation systems in rectangular fields. Temporarily increased wheat prices in the 1970s prompted farmers to tear out tree lines and hedgerows and plant every square inch of land. Taking out the windbreaks also eliminated valuable habitat for wild animals. By 1977, huge stretches of California's San Joaquin Valley lacked barriers to wind erosion altogether. When a spectacular wind storm swept down, it completely removed the soil, subsoil, and even underlying weathered bedrock across extensive tracts (figure 2.1). The soil fungus *Coccidioides immitis*, common in southern California, Arizona, and New Mexico, spread valley fever, a severe respiratory disease, in the dust. One fatality was a gorilla in the Sacramento zoo, some 300 miles away. The dust storm severely denuded areas on the ground, which still are visible barren patches. They are unlikely to return to full biological productivity unless left alone for thousands of years.[27]

Few data reliably estimate the extent of current erosional problems. The theoretical universal soil loss equation (USLE) and wind erosion prediction system (WEPS; see chapter 13, box 13.1) could be useful, but the data for making those calculations are inadequate. For 1992, the USDA conservatively guessed that a total of 2.1 billion tons of soil was lost to wind and water erosion, excluding gully erosion and mass wasting (see chapter 13). If piled on a single football field, that much soil would form a heap 200 miles high. The USDA figure was the most conservative of several soil loss estimates for that year—an unbelievable 1.1 billion tons less than the 1982 estimate.

Fertilizing and polluting

Such critical soil properties as thickness, layering, water storage capacity, and surface crusts cannot be restored in a human time frame (see chapter 13). To keep on growing crops in degraded fields, farmers add expensive fertilizers and other amendments. Assuming available resources and affordable fuel costs, fertilizers can restore plant nutrients and acid balances to upper soil layers, but nothing can restore subsoil acidity, soil textures, water storage capacity, or biota. In fact, synthetic fertilizers tend to kill the biota. Rising fuel prices also increase the costs of making fertilizer and mining already-depleted fertilizer raw materials, inevitably raising farming costs (see chapter 12).

Fertilizers are principally nitrogen (N), phosphorus (P), and potassium (K) chemicals.[28] Every year, U.S. industrial agriculture applies close to 30 million tons of NPK fertilizers to farmlands, about 13 million tons of nitrogen fertilizers—nine million tons are synthesized domestically, and the rest is imported. Agriculture in the United States largely depends on nitrate compounds made from ammonia, a nitrogen–hydrogen chemical (NH_3). The ammonia is synthesized in two laboratory processes from atmospheric nitrogen. One highly energy-consuming process generates hydrogen at temperatures as high as 660°F and pressures 3,000 times higher than at sea level.[29] About 38% (three million to four million tons) of annual U.S. hydrogen production goes to making fertilizer. Most of the hydrogen comes from natural gas, so fertilizer supplies are increasingly limited by natural gas costs and availability (see chapter 12).

Nitrogen-rich fertilizers also include animal manures. The United States produces approximately 1.6 billion tons of domestic animal manure annually; about half is used for fertilizing pasture and grazing land. In general, plants are able to absorb only about one-third to one-half the nitrogen in applied fertilizers, a wasteful and environmentally harmful excess.[30] The nitrates that plants cannot absorb go back to the atmosphere as ammonia gas or nitrous oxide, a potent greenhouse gas,[31] or they wash into streams or leach into groundwater. If manure is not buried but simply laid on the ground, it loses about half its nitrogen content in 24 hours. Overfertilizing has turned vast tracts of organically rich and fertile soils into a sterile compound that does little but hold plants upright.

The phosphorus and potassium in fertilizers come from nonrenewable phosphate rock and potassium-rich salt deposits, resources of uncertain availability. To make the fertilizers, phosphate rock and potassium salts are mined and treated—although potassium salts also can be manufactured in the lab from other potassium chemicals. Phosphate fertilizer consumption in the United States is about five million tons per year. Paradoxically, the United States supplies most of the world with phosphate fertilizers but also annually imports about two million tons of raw phosphate rock from North Africa and the Middle East for making the fertilizer. Economic geologists estimate that the United States will become a net importer of phosphate rock in 20–40 years. Since the world's remaining phosphate deposits come from politically and economically unstable regions, they probably are the fertilizer resources with highest potential for shortage shocks.

Potassium-rich salts are mined from "evaporite" salt beds, deposited as ancient lake or subsurface brines progressively dried up. The United States consumes about 11.5 million tons of potassium fertilizer annually and imports about nine million tons from Canada.[32]

Phosphate rock and fertilizer contain potentially toxic natural trace elements such as cadmium, selenium, and others that can harmfully concentrate in vegetable crops and in the animals that eat them, including us.[33] Other commercial fertilizers can be worse. Unbeknownst to farmers, urban back yard gardeners, and golf course and cemetery groundskeepers, until about 1997 many packaged fertilizers contained unidentified recycled hazardous wastes or mixed fertilizer and hazardous waste. Mining wastes can be packaged as fertilizer because federal laws uncritically exempt them from hazard labeling. "Ironite" was one such lawn and garden fertilizer product, consisting of mine tailings laced with high arsenic and lead concentrations from a Humboldt, Arizona, proposed Superfund site (see chapter 10). Revelations about the fertilizer plus wastes concoction inspired both Washington State and California to pass laws limiting heavy metal concentrations in fertilizers.

About 1.6 million tons of waste chicken manure, annually produced on megascale chicken farms, also wind up fertilizing croplands. The manure contains arsenic from chicken feed additives for controlling infections and increasing weight.

Most of the compounds remain unchanged through processing in the chickens' guts. Residual arsenic concentrations in chicken flesh increase the arsenic doses to people who drink water with elevated arsenic levels. Young chickens have three to four times the arsenic concentrations of other poultry.

Silent Spring *makers*

Pest and disease outbreaks are common in conventional agriculture. Huge doses of agricultural fertilizers poured on millions of acres have upset the natural balances that nature's diversity of life forms once maintained, opening new pathways to disease-causing bacteria and viruses and to nuisance and destructive pests. Farms growing single plant species and single animal breeds are highly vulnerable to disease organisms. A disease can wipe out vast crop acreages or whole livestock sectors, as has happened repeatedly across Britain and continental Europe in the past two decades. The crowded conditions inside factory meat-producing farms invite contamination threatening to human health, as *Salmonella* bacteria in undercooked chicken and eggs and *E. coli* in hamburger have demonstrated.

Conventional U.S. agriculture fights diseases and contamination hazards with agricultural pharmaceuticals and chemical pesticides—that is, drugs. They use about 500,000 tons of at least 600 different types of pesticides annually.[34] Despite this, plant pathogens and weeds still destroy over a third of all potential U.S. food and fiber crops every year.[35] Of the billions of pesticide tonnage applied to agricultural lands, 99.9% never reach a target pest, but disperse through the environment with unknown effects.[36] The U.S. Environmental Protection Agency (EPA) has the principal responsibility for regulating agricultural pesticides in the United States.

Pest control chemicals, principally insecticides, have been around for more than a century. Early ones were long-lasting inorganic copper sulfate, mercuric chloride, and lead and sodium arsenates. Organometal pesticides—organic chemicals combined with mercury, arsenic, tin and other substances—replaced the inorganic compounds, and some remain in use today. Organic compounds are based on carbon, oxygen, and hydrogen, the bases of life.

The history of dedicated pesticide use in American agriculture is an endless stream of glowing promises (figure 2.2), discoveries of public health and environmental problems, and introductions of new pesticides with equally unknown side effects. The insecticidal properties of the chlorine-containing organic compound (organochlorine) DDT went unrecognized from 1874 until the 1940s, when the U.S. military developed it to control mosquitoes and stop malaria and other insectborne diseases. It appeared in civilian markets by 1945, applied heavily to agricultural lands and forests. Usage peaked in the late 1950s at about 100,000 tons per year. Additional organochlorine pesticides—favored for their low solubility in water and their tendency to adhere strongly to soil particles and to resist physical, chemical, and biological degradation—appeared in the 1950s and 1960s. But these same special properties exacted unforeseen penalties, principally, high bioaccumulations in plant and animal organs, including in human blood, gonads, and mothers' milk.[37]

Organochlorines bioaccumulated in the organs of animals and plants low in the food chain, such as the soil bacteria, lichens, and the bees, ladybug beetles,

Figure 2.2 Advertisement in June 30, 1947, issue of *Time*, exalting the benefits of DDT.

and butterflies essential for pollinating crops. Animals at higher food chain levels consume very large amounts of plants and lower food chain creatures, increasing the concentrations in higher animals on the food chain. This is the natural process of biomagnification. Biomagnification was responsible for 1950s DDT poisonings of U.S. birds and other nontarget animals.[38] Laboratory research has identified many other bioaccumulative and biomagnifying organochlorine compounds as toxic to aquatic life, probable human carcinogens, and possible endocrine-system disrupters.[39]

Rachel Carson's seminal *Silent Spring* described the organochlorines' reproductive disruption, genetic damage, and cancers in wildlife and explained the threat to human lives, destroying their popularity. The late 1960s welcomed organo*phosphates*, derived from potent nerve gases used in chemical weapons.[40] More recent replacements include complex organic carbamate, pyrethroid, and nicotine-related compounds (figure 2.3). Many of the newer so-called "soft" pesticides are highly toxic, act quickly, and prove deadly to birds and fish. They kill an estimated 67 million of the nearly 700 million birds directly exposed to pesticides yearly. Agricultural use of all pesticides peaked around 1978 at nearly 1.2 billion pounds of so-called "active ingredients." Table 2.3 shows the most common pesticides used in 1987 and 1997 crop years.

Chemical companies advertise that the newer organophosphate and "soft" pesticides degrade more rapidly than organochlorines and do not bioaccumulate,[41] but their low solubility and environmental persistence signal a harmful potential (table 2.4).[42] Near a Geyserville, California, vineyard, avid birdwatcher Andrew Scavullo came upon 400 "dead or dying robins, twitching in a really ugly way." The birds had bathed in puddles of irrigation water laced with fenamiphos, an organophosphate targeted at nematode worms in grapevine roots. "Soft" pesticides also are largely responsible for 6 million to 14 million dead fish each year[43] and are highly toxic to frogs, crayfish, and agriculturally beneficial predatory and pollinating insects. Losing those species deprives birds and other wildlife of important food sources.

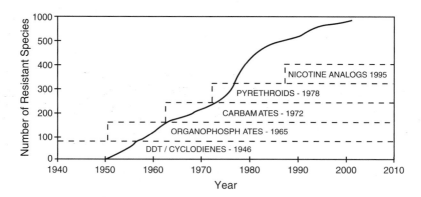

Figure 2.3 Graph showing the length of time that various insecticide types have been applied on farms (bars), and the year of first documented resistance in insect pests (dates). Solid line represents the number of species that have developed resistance. From I. Denholm et al. Insecticide Resistance on the Move. *Science* 2002;297:2222–2223. Copyright 2002, American Association for the Advancement of Science. Reproduced with permission.

Weed solutions

Controlling weeds on cropland once required labor-intensive plowing and disking, but rising 1970s labor and fuel prices drove farmers to chemical herbicides, which rapidly expanded total U.S. pesticide consumption. Herbicide use in America now hovers at around a billion pounds a year (figure 2.4), predominantly for treating seven crops: corn (45%), soybeans (18%), pasture (6%), cotton (6%), sorghum (5%), wheat (4%), and rice (3%).[44] Unfortunately, conventional farmers who reduce soil erosion with no-till planting methods have greatly increased their herbicide applications.

Currently used herbicides supposedly do not bioaccumulate, but some persist in the environment for long times (table 2.4), and very heavy applications can create serious problems. All are highly toxic to land and aquatic biota and pose threats to human health. The globally common insecticide carbaryl (Sevin) and herbicide glyphosate (Roundup) reduce species diversity in aquatic environments by 15% and 22%, respectively. In one area, Roundup completely eliminated two species of tadpoles and nearly exterminated a third, resulting in 70% reduction in the number of species. Roundup also kills land-dwelling amphibians.[45]

At the end of the last century, substantial and ongoing controversy centered on the environmental effects of methyl bromide fumigant and atrazine—the most heavily used herbicide in the United States. The evidence that atrazine can cause cancers is disputed, but it is irrefutably implicated in degrading human sperm quality and turning male frogs and other amphibians into females. A number of European nations have banned atrazine, but the EPA rejected a U.S. ban.

Gaseous methyl bromide, the preplanting fumigant for strawberries, tomatoes, grapes, almonds, carrots, various commercial flowers, and golf course turf grass, is best described as a biocide. Methyl bromide is indiscriminately lethal to soil

Table 2.3 Most Commonly Used Conventional Pesticides on U.S. Agricultural Crops, 1987 and 1997 (Million Pounds)

Pesticide	Type[a]	1987 Rank	1987 Applications	1997 Rank	1997 Applications	Toxicity Class[b]
Atrazine	H, T	1	71–76	1	75–82	III R
Metolachlor	H, A	3	45–50	2	63–69	III
Metam sodium	H	15	5–8	3	53–58	I
Methyl bromide	Fu		NA	4	38–45	I R
Glyphosate	H	17	6–8	5	34–38	II
Dichloropropene	F, Ch	4	30–35	6	32–37	II
Acetochlor	H		0	7	31–36	I R
2,4-D	H, Ch	5	29–33	8	29–33	I–III
Pendimethalin	H, D	10	10–13	9	24–28	III
Trifluralin	H, D	6	25–30	10	21–25	III
Cyanazine	H, T	7	21–25	11	18–22	II R
Alachlor	H, A	2	55–60	12	13–16	III R
Copper hydroxide	F	40	1–2	13	10–13	I
Chlorpyrifos	I, OP	14	6–9	14	9–13	II
Chlorothalonil	F, OC	19	5–7	15	7–10	II
Dicamba	H, M	23	4–6	16	7–10	III
Mancozeb	F	21	4–6	17	7–10	IV
EPTC	H, C	8	17–21	18	7–10	III
Terbufos-I	OP	11	8–10	19	6–9	I R
Dimethenamid	H		NA	20	6–9	II
Bentazone	H, M	15	6–9	21	6–8	III
Propanil	H	13	7–10	22	6–8	II
Simazine	H, T	28	3–4	23	5–7	IV
MCPA	H, Ch	25	4–5	24	5–6	III
Chloropicrin	I		NA	25	5–6	I R

A. L. Aspelin and A. H. Grube. *Pesticides Industry Sales and Usage—1996 and 1997 Market Estimates* (U.S. Environmental Protection Agency, 1999), table 8.

"Conventional pesticides" are exclusively or primarily used as pesticides. Table excludes multiple-use substances such as sulfur (50–75 million pounds in 1997) and petroleum oil and distillates (65–75 million pounds in 1997).

[a] Pesticide uses: H, herbicide; I, insecticide; F, fungicide; Fu, fumigant. Chemical groups: OP, organophosphorus; OC, organochlorine; A, amide; C, carbamate; Ch, chlorophenoxy; D, dinitroaniline; T, triazine; M, miscellaneous.
[b] EPA toxicity classification: Class I, highly toxic; Class II, moderately toxic; Class III, slightly toxic; Class IV, essentially nontoxic; R, restricted use.

Table 2.4 Pesticides Currently in Use (1990s) Predicted to Have Accumulative Potential in Sediment and Aquatic Biota, and Ecotoxicity

| Pesticide/Type[a] | Toxicity Class[b] | Human Toxicity[c] | Human Health Effects[d] | | | | | Ecotoxicity Effects[e] | | | |
			Re	Mu	Te	C	Or	Birds	Aquatic	Bees	Other[f]
Insecticides											
Chlorpyrifos/OP	II R	M					X	M-VH	VH	T	T; AP, wildlife
Dicofol/OC	II-III	S-M				C	X		S		
Endosulfan/OC	I R	H			X	E	X				
Esfenvalerate/P	II	M						S-M	VH	H	
Fenthion/OP	II R	M				I	X	VH	M	T	
Fenvalerate/P											
Lindane/OC	II	M				B/C	X	M	H-VH	H	
Methoxychlor/OC	IV					D	X	S	VH		
Permethrin/P	II-III R	S				C	X		VH	VH	T; SM, wildlife, some plants
Phorate/OP	I R	H					X	VH	VH	T	
Propargite/M						B					
Herbicides											
Benfluralin/D	n.d.										
Bensulide/OP	n.d.										
Dacthal (DCPA)/Ch						C			T		
Ethalfluralin/D						C					

Chemical/Group[a]	EPA class[b]	Ecotoxicity[e]	Human toxicity[c]	Health effects[d]	Carcinogen	Other wildlife concerns[f]
Oxadiazon					C	
Pendimethalin/D	III	S	H		C	
Triallate/C	III	S	H	X	C	
Trifluralin/D	III		VH	X	C	T; EW, Am
Fungicides/Wood Preservatives						
Dichlone	n.d.					
PCNB	III	S	H	X	C	
Pentachlorophenol/OC	II R	M	VH	X	B	T; cattle

Modified from U.S. Geological Survey. Pesticides in Stream Sediment and Aquatic Biota (U.S. Geological Survey Fact Sheet 092–00, 2000), table 1. Predictions based on persistence and solubility in water. Data on the identity and toxicity of "inert" substances are unavailable. Toxicity information is compiled from the Extension Toxicology Network. Available: ace.ace.orst.edu/info/extoxnet; the Agency for Toxic Substances and Disease Registry. Available: www.atsdr.cdc.gov; Irene Franck and David Brownstone. *The Green Encyclopedia* (New York: Macmillan USA, 1992), 24l; L. H. Nowell et al. *Pesticides in Stream Sediment and Aquatic Biota: Distribution, Trends, and Governing Factors* (Washington, DC: Lewis Publishers, 1999), table 6.15, 431–433; Allan Felsot. How Does EPA Decide Whether a Substance is a Carcinogen? *Agrichemical and Environmental News* February, 1997; 132; National Pesticide Information Center. Available: www.npic.orst.edu; U.S. Department of Agriculture Forest Service, Material Safety Data Sheets. Available: www.fs.fed.us/foresthealth/pesticide/material.shtml; and Northwest Coalition for Pesticide Alternatives, Pesticide Fact Sheets. Available: www.pesticide.org.

[a] Chemical pesticide groups: OC, organochlorine; OP, organophosphorus; P, pyrethroid; D, dinitroaniline; Ch, chlorophenoxy; C, carbamate; M, miscellaneous chemical groups.

[b] EPA toxicity classification: Class I, highly toxic; Class II, moderately toxic; Class III, slightly toxic; Class IV, essentially nontoxic; R, restricted use; n.d., no data.

[c] Human toxicity ratings: H, highly toxic; M, moderately toxic; S, slightly toxic. Rating depends on organs affected, pathways into the body, and formulation (e.g., solid pellets or liquid).

[d] Human health effects: Re, reproductive effects (lesions, atrophied testes, etc.); Te, teratogenic effects (prenatal deformities, fetal loss); Mu, mutagenic effects; Or, effects on organs, including kidney, liver, intestines, central and peripheral nervous systems, thyroid and adrenal glands, eyes, bones, lungs, brain, heart, testes, ovaries, endocrine organs. C, Possible carcinogenic effects; classes: A, human carcinogen; B, probable human carcinogen; C, possible human carcinogen; D, unclassified; E, no noncarcinogenicity evidence in humans; I, tests so far inconclusive.

[e] Ecotoxicity effects: S, slight toxicity; M, moderate; H, high; VH, very high; T, toxic but not rated.

[f] Other wildlife concerns: T, toxic to Am, amphibians; AP, aquatic plants; EW, earthworms; SM, soil microorganisms.

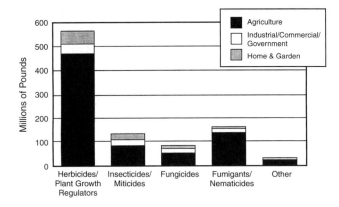

Figure 2.4 The three main sectors of conventional pesticide use, and the quantities of each pesticide type applied in 1997. From A. L. Aspelin and A. H. Grube. *Pesticides Industry Sales and Usage: 1996 and 1997 Market Estimates* (U.S. Environmental Protection Agency, 1999), 15.

biota—the good and useful, along with the bad and harmful. It is the most commonly used pesticide along the eastern rim of California's San Joaquin Valley. In 1999, farmers applied 15 million pounds under ground-covering plastic sheets. Inevitably, methyl bromide leaks into the air and attacks Earth's stratospheric ozone layer—it is considered the most powerful ozone destroyer in large-scale use.

In signing the 1987 Montreal Protocol, the U.S. agreed to phase out methyl bromide use by 2005, but has obtained annual exemptions that now extend through 2008, despite the availability of substitutes and European objections. The United States is maintaining a stockpile of methyl bromide as much as twice that in use, violating the protocol. The 2006 extension, in a year when the ozone hole reached record size, seems especially irresponsible.

The pesticide treadmill

Pesticide chemicals do kill many or most of the targets—but in one of nature's ironies, the genetic diversity within pest species eventually nullifies every pesticide's effects. Pest populations are so very large that they always include a few individuals able to survive. They generally breed prolifically, so survivors pass their poison-resistant genes to abundant offspring. Over time, this process builds pesticide resistance into the species. Then begins the treadmill of applying ever-increasing amounts of ineffective chemicals and developing ever more-potent substitutes.[46]

Plants also develop pesticide resistance. In August 2005, University of California researchers revealed that robust stands of herbicide-resistant mare's tail (horseweed) are invading irrigation canal banks, roadsides, orchards, and vineyards, from Napa County in the north to and throughout southern California. Horseweed grows 10 feet tall, competing with crops for sunlight, nutrients, and water. It is difficult to pull up and is a prolific seed producer, so mowing only helps it spread. Horseweed is resistant to even four times the recommended dose of glyphosate, the active ingredient in dozens of widely used herbicides. The same research suggests that hairy fleabane weeds also may be developing glyphosate resistance. So here we are on the treadmill, literally inventing weeds that can be only temporarily controlled.

Drugs for growth and profit

Factory farms in the United States pack thousands of single-breed turkeys, chickens, cattle, sheep, or hogs tightly together, feeding them largely on corn. Cattle cannot digest corn, so they weaken, and the crowded conditions invite chronic livestock diseases and catastrophic epidemics. Farmers add antibiotics to the feed to keep the animals going until slaughter. For poorly understood reasons, antibiotics help fatten some livestock on less food, adding an economic incentive for feeding excess nontherapeutic doses. Many of the drugs used on farm animals closely resemble the types used for treating human bacterial infections, the third leading cause of death in the United States after heart disease and cancer.

Agricultural antibiotics can be detected in mass-produced meats from supermarkets, which means that most Americans get a constant diet of them at low doses. Like agricultural pests, disease-causing bacteria are both exceedingly numerous and genetically variable, so at least a few always survive antibiotic doses. Reproducing swiftly and constantly, the survivors pass their resistance on to successive generations, creating new antibiotic-resistant bacterial strains along the way. Overusing antibiotics can only speed the process. Clearly, antibiotic overexposure undermines the drugs' effectiveness for treating infectious diseases at least, and at worst can render them useless. In other words, continuing to eat these drugs in our food eventually could make us very vulnerable to bacterial diseases. Public health authorities already have found pathogens in antibiotic-fed animals that resist many of our wonder drugs.

Soil microbes supply most of our antibiotics, and soils are intensively screened to find new antibiotic producers. But disease germs (bacteria) also develop antibiotic-resisting strategies. A recent study found that all 480 spore-forming microbes in three different soils were resistant to at least six different antibiotics. Two of the microbes are resistant to 15 and 21 antibiotics, respectively—including former gold-standard remedies for human bacterial diseases and ones recently approved for human use. This discovery raises major concerns that some poorly known group of natural processes could be transmitting resistance mechanisms from soil biota to human pathogens, a previously unimagined risk.[47]

Feeding antibiotics to livestock threatens to disarm critical defenses against human disease. For many years, antibiotics have been overprescribed for treating human illnesses, and more than 25 years ago, the U.S. General Accounting Office warned that feeding antibacterial drugs to animals could speed the development of dangerous antibiotic-resistant "superbugs." But factory-bred farm animals still eat antibiotics at eight times human treatment doses.[48] Bacterial resistances have appeared already, reducing the number of drugs that doctors can rely on. One now widespread is MRSA (methicillin-resistant *Staphylococcus aureus*), which does not respond to most penicillin-type antibiotics, particularly methicillin. MRSA appeared in Britain as early as 1961 and now poses a severe health threat, particularly in hospitals.

Children who drink milk from conventional dairies are ingesting bovine growth hormone (BGH) or the synthetic recombinant-gene versions (rGBH and rBST) that industrial dairies feed to cows. Growth hormones enhance milk production in dairy cattle, and BGH alters growth patterns in rats. Nobody knows what effect high growth hormone levels might have on either children or adults. Genetically modified hormones also raise levels of insulin-like growth factor-I (IGF-I) in milk by 25–70%, enhancing the milk drinkers' potential for developing breast and prostate cancers. This makes us all guinea pigs in an experiment on a largely unaware U.S. population (see chapter 7).

Bluntly put, abusing these drugs for commercial purposes could kill a lot of people. For that reason, European countries have stopped feeding antibiotics to livestock.[49] After the European Union banned one problematic antibiotic from animal feed, follow-up studies showed many fewer cases of bacterial resistance to drugs used against human diseases. But even though the federal Centers for Disease Control and Prevention identified a strain of *Salmonella* bacteria resistant to five different antibiotics in 1998, the United States still lacks effective regulations for these practices.

McDonald's Corporation, consumer of 2.5 million pounds of beef, pork, and poultry worldwide, received so many letters from customers about antibiotics in meat that it instructed their suppliers to stop feeding growth-inducing antibiotics by 2004. But if the suppliers still feed antibiotics to prevent diseases, and if the disease-preventing drugs are the same as the growth-inducing drugs, the McDonald policy will have little or no effect on Americans' antibiotic overexposure.[50]

Water Warnings

Roughly 220 million Americans live within 10 miles of a river, lake, or estuary too polluted to support normal fish communities or other water-dwelling creatures and plants, or to conform to the Clean Water Act's fishable and swimmable water quality standards (see chapter 10, appendix 2).[51] In the United States as a whole, approximately a third of the fresh water that people take from streams or groundwater supplies never returns to the source.[52] Much of the remaining two-thirds is contaminated and pollutes land or streams (see chapters 8, 10). Agriculture contributes a major proportion of those pollutants.

Pesticide-laden waters drain into streams, rivers, ponds, or lakes from agricultural fields and from forests sprayed extensively for bark beetles. In soils, persistent DDT and other organochlorine compounds tend to accumulate in middle-food-chain soil organisms. Eventually the insecticides reach streams and lakes and stay there for long periods, bioaccumulating in plants and animals[53] that metabolize DDT to DDE and other toxic breakdown products.[54] The pesticides have severely reduced the numbers and kinds of fish and bottom-dwelling creatures in fresh waters and in the ocean, where most stream waters end up.

Feeding dead zones

Excess fertilizer nutrients also can be toxic to water plants and animals.[55] The excess nitrate and phosphate promote an unbridled expansion of plants that foul waters with bad odors and tastes and clog water supply intakes and outflows. Since some soil bacteria convert nitrogen to highly toxic ammonia, even very low nitrate contributions from fertilizer can directly kill the larvae of several declining western frog species. Eutrophication results when nitrogen, phosphorus, and other nutrients become highly concentrated in a body of water. The nutrients feed excessive plant growth, which uses up most of the breathable oxygen in the water, suffocating fish and other aquatic animals. On one path to eutrophication, excess nitrates from fertilizers promote algal blooms in coastal bays and estuaries and freshwater lakes, ponds, and streams. The algae deplete oxygen and often spread harmful natural or synthetic toxins to shore birds, marine mammals, fish, shellfish, invertebrates, coral reefs, seagrass beds—and people.[56]

Each winter and spring, a *highly* eutrophic (hypoxic) "dead zone" develops in coastal Gulf of Mexico water, after winter rains flush excess fertilizers down the Mississippi River. The Mississippi River watershed covers about 40% of the continental United States, and its tributaries drain Montana, Wyoming, and other western states. It encompasses enormous crop acreages, most of them excessively fertilized, and is the obvious source of the dead zone's oxygen-eating chemicals. Fertilizer nutrients feed spring and summer algal blooms in the surface waters. When the algae die, they settle into colder and saltier bottom waters, where bacteria decompose them, using up dissolved oxygen until hypoxic conditions stress or kill any bottom-dwelling animals that cannot swim or crawl off.[57] The 1999 dead zone started at the Mississippi's mouth and stretched along the Louisiana and Texas coasts for 250 miles, covering 7,700 square miles of Gulf waters—about the size of New Jersey.

Lowering agricultural nutrients in the environment is the cheapest and fastest route to preserving clean drinking water supplies and aquatic food chains, both important to human health. Nitrogen that leaches into groundwater from agricultural soils and feedlots poses health problems for people who drink water from wells (see below).[58] A study of 1,255 domestic drinking water wells and 242 public supply wells drawing from major U.S. aquifers turned up 12% with one or more contaminants above EPA standards or other health safety limits. Nitrate was the most common contaminant.

Overwatering and flushing

Agriculture could not be successful in arid western states without irrigation. Ninety percent of western surface water and groundwater supplies irrigate about 15 million acres of cultivated cropland (35%), and nearly 11 million acres of noncultivated cropland (78%). Three-quarters of all western water comes from surface streams, stored in man-made reservoirs (see chapter 9). About a quarter of all water that humans use in the west, and a high proportion of domestic water supplies, comes from groundwater. But 86% of all western groundwater withdrawals are for irrigating farms.[59]

The story of how dam building came to support western U.S. agriculture is one of grand engineering achievements yielding enormously valuable farms in a land where normal annual rainfall levels are inadequate or unreliable. Taxpayer dollars originally built dams and reservoirs to support small farmers, but today the water largely goes to corporate farms. Along with federal farm commodity support, government subsidies also artificially lower the costs of agricultural irrigation water (see chapter 9). Another critically important engineering achievement, the immersion pump, allowed farmers to withdraw fossil groundwater from aquifers thousands of feet below the surface. The fossil water had accumulated over millions of years and cannot be replaced in human lifetimes. So irrigation can effectively mine deep groundwater (see chapter 9).

Together, damming and diverting surface water and pumping groundwater to irrigate agriculture deplete the arid west's few riverside riparian areas and dry up springs. It all contributes to widespread impairment and outright eradication of rare wetlands and spring habitats. Draining wetlands and diverting water from rivers to agricultural lands are responsible for most of the declines or extinctions of western native fish and other aquatic species (see chapter 9).[60]

The abundance of cheap irrigation water in the dry western environment has poisoned both land and water. A crop actually uses much less than half of applied irrigation water, and in places even less than a third,[61] so half to more than two-thirds is waste. Irrigation water leaches nutrients from the soil, and groundwater often is hard water already, loaded with dissolved mineral salts leached from soil and rock it has passed through. As irrigation water evaporates, it redeposits the salts near the ground surface (figure 2.5).

Overwatering can generate critical water shortages, while waterlogging farm soils speeds up the salination process. Overirrigation also increases seepage, thereby raising the water table (top of the groundwater zone) in the irrigated soil, which increases evaporation (see chapter 13). Some subsoils do not drain water efficiently, and both waterlogging and salination become acute problems if the water table rises to within about 10 feet of the surface. Both reduce farm productivity and ultimately can force farmers to abandon their land. For example, the ancient Babylonian story of Atrahasis told of farm fields turning white, undoubtedly from built-up irrigation salts. Overirrigating and salination destroyed the Babylonian empire, and today they are destroying farms in the western United States.[62]

Farmers try flushing cropland with excess water to remove salt concentrations, but flushing removes other chemicals, as well. Waste irrigation "return waters" are a toxic brew of fertilizers, pesticides, and natural soil nutrients. Return waters are a waste disposal problem, and the easier disposal options magnify the problems—California's "Kesterson Wildlife Refuge," in the San Joaquin Valley, is an ominous example. Although called a "wildlife refuge," Kesterson was created at public expense to solve the Westlands Irrigation District's waterlogging and salination problems. It actually was the disposal site for agricultural return waters laden with farm chemicals and excess natural selenium leached from the soils.

Very small and limited amounts of selenium are essential in most creatures' diets, but slightly larger concentrations are poisonous. The effects of excess doses vary from debilitating to lethal. At Kesterson, algae concentrated selenium out of the return water, poisoning lower animals and water birds that fed on it, plus animals higher in the food chain that fed on them. Selenium poisoning killed many animals outright, but also produced grossly deformed bird chicks (figure 2.6) and prevented many egg clutches from hatching.[63]

Figure 2.5 Cropland in southern San Joaquin Valley, California, degraded by salination. Photographed February 1985.

In 1985, Kesterson reservoir was drained and filled with dirt. But farms in the Westlands Irrigation District still have drainage problems. The Bureau of Reclamation launched a five-year "Grasslands Project" study, which irrigated 3,500 well-drained acres with the selenium-laced runoff. Selenium-contaminated bird eggs turned up by 2005. Promising solutions under review in 2007 included upgrading the district's irrigation technology to prevent over-watering; increasing use of high-quality groundwater, supported by coordinated regional groundwater management; reverse osmosis water filtration (see chapter 9); and taking land out of production. None is nature-proof or cheap,[64] and all are locally unpopular.

The powerful Westlands Irrigation District is trying to force another plan on the Bureau of Reclamation, which would stop irrigating most of the poorly drained farms (379,000 acres) and install high-tech filtration facilities to remove selenium from the irrigation water. It also would irrigate 19,000 acres with supposedly selenium-free wastewater and create more than 3,200 acres of new evaporation ponds to hold the return waters. While spending $700 million of taxpayer money, this project could create 3,200 acres of new Kestersons.

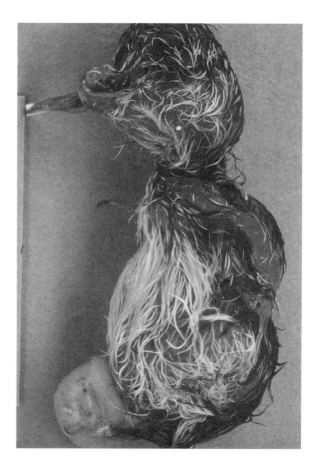

Figure 2.6 Deformed stilt chick with curled beak and no legs, born in the Kesterson Wildlife Refuge, an artificial wetlands fed by heavily selenium-contaminated agricultural drain water. Photographed 1984, courtesy U.S. Fish and Wildlife Service.

Manure on tap

The numbers and sizes of factory farms have increased dramatically since the 1990s.[65] By recent estimates, the United States has more than 450,000 large-scale industrial animal-raising operations, many in western states. A substantial proportion of them process more than 1,000 animals per year. To the water pollution from irrigated and conventional dry farms, these gigantic and poorly regulated factory farms and animal feedlots add huge volumes of animal wastes, replete with drugs and pathogens. The feedlot operations commonly spread millions of tons of wastes on fields or dump them into feedlot "lagoons"—essentially open cesspools. On the western Utah desert about 40 miles north of Zion National Park, Circle Four Farms processes 600,000 hogs a year and stores solid and liquid wastes in 92 noxious lagoons. The Circle Four Farms corporation plans to expand production to 2.5 million hogs annually, eventually generating as much raw sewage as the city of Los Angeles.

Where wastes are laid on fields, rain runoff combines with overflow from waste lagoons to pollute creeks and streams. Millions of gallons of animal feces and urine seep into groundwater storage aquifers from big feedlot manure pits, and eventually reach 35,000 miles of rivers. A 2003 report also found a "cocktail of antibiotics" in the dust blowing from hog-factory farms, plus high levels of micro-organisms, endotoxins, allergens, and antibiotic-resistant bacteria.[66] All the pathways—water, winds, and soil—take the toxins, pathogens, parasites, and endocrine disrupters into our food chain.

Like the cyanide ponds at open-pit gold mines (see chapter 4), toxic factory-farm waste ponds attract and can kill migratory birds. Unwary birds settle on the lagoons and eat poisonous algal biotoxins, then fly off to other locales, potentially spreading *Salmonella*, avian cholera, and avian botulism infections in their droppings. This can become a significant public health problem.[67]

Fish farms throughout the U.S. use feeds laced with harmful pollutants, including carcinogenic substances. The farms deliberately fatten salmon "from a global supply of fish meal and fish oil manufactured from small open-sea fish, which studies show are the source of polychlorinated biphenyls, or PCBs, in most farmed salmon."[68] The farmed salmon contain as much as 5–10 times more PCBs and other chemicals than do wild Pacific salmon, which feed in less polluted ocean areas. PCBs concentrate in fatty tissue and "even low concentrations of PCBs in fish meal can become a concern for human health" because farmed salmon bioaccumulate PCBs at "20 to 30 times the levels in their environment and their feed." As a result, a salmon dinner can deliver a large PCB dose. Exposures to PCBs and PCDFs (from burning PCBs) can cause human sperm malformations and reduce fertility in lab animals.

Protecting the National Drink

The 1974 Safe Drinking Water Act charged the EPA to set standards that maintain drinking water quality. To provide "an adequate margin of safety," the EPA first sets a maximum contaminant level *goal* (MCLG) for each identified contaminant, representing a level low enough to have no known or anticipated adverse human health effects. The legally enforceable maximum contaminant *level* (MCL) should be as close to the MCLG as is technically and economically feasible. Standards for some pesticides are set higher than MCLGs, however—probably in violation of the Safe Drinking Water Act.

Setting single-substance MCLs has caused much scientific and political ferment. MCLs depend on the exposure: higher for swimming than for drinking or fishing, for example. Also, MCL levels for one organism may be much higher or lower than for others. But setting a single MCL—low enough to protect the most sensitive user—could waste resources. It certainly has led to endless litigation. In 1972, following years when "the Cuyahoga River and the Houston Ship Channel caught fire, lakes the size of Erie were declared dead; fish kills choked the Chesapeake Bay; and Louisiana's Secretary of Agriculture declared Lake Providence...safe for humans so long as nobody went near it or ate the fish," Congress adopted the "best available technology" standard. The chosen technology limits toxic discharges or releases no matter what the harm or where the contaminants came from.[69]

The nitrate standards for drinking water quality are set to prevent excess nitrate in blood, which reduces its oxygen-carrying capacity and results in "blue baby syndrome," a condition potentially fatal to an infant or fetus. Among less well-defined risks to human health, epidemiological studies also suggest that nitrates in drinking water increase adults' risk for non-Hodgkin's lymphoma, a cancer of the lymphatic system. As yet, however, the EPA has not set a standard or desired goal for limiting nitrate in marine waters.

Under 1986 amendments to the Safe Drinking Water Act, the EPA was supposed to issue final MCLs and MCLGs for 83 chemical compounds by June 1989. But it has set only 28 MCLs to date, subject to change as knowledge and laboratory analyses improve. There are no phosphorus or potassium MCLs to address human health concerns. The EPA did, however, set a water quality MCLG for phosphorus of 0.1 part per million (ppm)—equivalent to two ounces in a water-filled railroad tank car—to prevent or slow aquatic plant growth in fresh drinking water supplies.

Reducing phosphorus in detergents lowered the level in surface waters, but agriculture and urban sewage treatment plants still disgorge substantial amounts. For instance, Denver's sewage treatment plants send about 1,200 tons of phosphorus per year into Colorado's South Platte River and its tributaries. For at least 150 miles downstream, phosphorus concentrations stay above the MCLG, supporting excessive algal growth in reservoirs. Dr. Fred Corson, retired Dow Chemical researcher, notes, "When public officials say that wastewater meets drinking water standards and all regulations for discharge into surface waters, they mean it meets criteria for about 75 of the about 87,000 chemicals currently being used."[70]

Hidden synergy

Pesticide pollution research generally addresses one chemical at a time, making analytical problems easier to solve but having little application to the real world. Out of more than 500 pesticides and additional breakdown products already in the environment, at least 84 have been detected in groundwater. Since MCLs and other water quality criteria are based on the effects of single pesticides, they do not account for the cumulative synergistic effects of combining multiple chemical compounds in water.[71]

The potential harm from combinations of most manufactured chemicals is largely unknown and unstudied (table 2.5). Mixtures of chemicals that all attack the same organ can be less toxic than the sum of the known effects of each (antagonistic), or can be more toxic (synergistic). If different chemicals in a mixture attack multiple organs, the toxic effects are even more complex. Many thousands of new man-made substances constantly enter the environment, generating extreme complexities.[72]

Human health standards are woefully lacking, but aquatic organisms fare far worse. For example, each of the common agricultural herbicides atrazine, metolachlor, and alachlor attacks tadpole development in a different way. Mixtures of all three delay maturation in frogs, far exceeding effects of any one by itself.[73] When exposed to 10 ppb atrazine, equivalent to two ounces in enough water to

Table 2.5 Interactive Effects of Pesticides Found in Aquatic Life Forms

Pesticide	Reported Synergists	Reported Antagonists
Chlordane	Aldrin methoxychlor Endrin, lindane Malathion, toxaphene	Lindane
Diazinon	Dieldrin	Toxaphene
Dieldrin	Hexachlorobenzene Diazinon	Hexachlorobenzene
Endrin	Chlordane Lindane	Lindane
Hexachlorobenzene	Dieldrin Lindane	Dieldrin Lindane Pentachlorophenol 2,3,7,8-TCDD
Lindane	Chlordane Hexachlorobenzene Mirex, toxaphene	Chlordane Hexachlorobenzene α-, β-, and δ-HCH[a] Mirex, toxaphene
Mirex	Lindane Toxaphene	Lindane Toxaphene
Toxaphene	Chlordane, malathion Lindane, mirex	Diazinon, lindane Mirex

L. H. Nowell et al. *Pesticides in Stream Sediment and Aquatic Biota: Distribution, Trends, and Governing Factors* (New York: Lewis Publishers, 1999), table 6.20, 460.
Each line in columns two and three represents a single experiment. Synergists are interactive mixtures more toxic than the sum of individual compounds. Antagonists are mixtures less toxic than the sum of individual compounds. Mixtures reported as both synergists and antagonists in this table may reflect different concentrations or the limited state of knowledge. Another class of mixtures has additive toxic effects (interactive mixtures as toxic as the sum of individual compounds).

[a] HCH, the insecticide hexachlorocyclohexane, also known as benzene hexachloride; α-, β-, and γ-HCH are different isomers, or chemical structures, of HCH that have different effects on living tissues. γ-HCH is the chemical compound in lindane.

fill three *Exxon Valdez* tanker ships, tadpoles metamorphosis to adults lagged 10 days behind normal. Adding 10 ppb of metolachlor doubled the maturation delay, and the full chemical cocktail more than tripled it. Increasing doses also increased such developmental abnormalities as contorted legs. Developmental endocrinologist Tyrone Hayes commented, "That's reality. No environment has just one chemical. This reality is a nightmare for regulatory agencies that regulate single chemicals, not mixtures."[74]

Known pesticide components are problem enough, but only the "active" ingredients in a formulation have to undergo toxicity tests. The labels, documentation that supposedly ensures safe use, do not have to disclose so-called "inert" ingredients (box 2.1),[75] which enhance the effectiveness of a formulation. Inerts can make up 90% or more of some brand-name pesticides. In addition, literally thousands of pesticide breakdown products (metabolites) are found in soils, waterways, and groundwater.

Analytical problems prevent laboratories from testing for many pollutants. So far scientists have looked for only a handful of breakdown substances in the nation's waters, mostly those the EPA labels "priority" pollutants (see chapter 12)—the ones having obvious effects. But many are as or more toxic than the parent chemicals.

Box 2.1 Actively Inert

Some pesticides' inert ingredients may be more toxic than the active ingredients. For example, research shows that synergistic (interactive) effects of "inert" components make Roundup herbicide twice as effective as glyphosate, its active ingredient, at damaging human placenta cells and inhibiting the sex-hormone-synthesizing enzyme aromatase.[a] Knight and Cox[b] reviewed "inert" ingredients in pesticide formulations, finding that 394 are or have been registered as active ingredients in other products. Some 209 of those active "inerts" are hazardous air and water pollutants—14 are rated extremely hazardous, 84 must be reported to the EPA's Toxics Release Inventory, 21 are known or suspected human carcinogens, and 127 are identified occupational hazards.

In 1987, the EPA devised a five-part scheme for classifying the full list of registered inerts, which then totaled about 1,200. It grew to more than twice that number by 1995. List 1 was to include substances of top priority concern for carcinogenicity, neurotoxicity, and adverse reproductive or ecological effects named on the product label. List 2 is for potentially toxic substances with high priority for testing, and List 3 for substances of unknown toxicity. List 4 substances are judged either minimal risks (4A) or have been studied and showed no adverse public health or environmental effects from current uses (4B).

This listing system incorrectly categorizes many substances known to be hazardous, polluting, or toxic, however.[c] The Unknown Toxicity list (List 3) includes chlorothalonid, a probable human carcinogen; coal tar, a known human carcinogen; naphthalene, a hazardous pollutant under the Clean Air and Clean Water Acts; chloropicrin, a severe respiratory tract irritant; and 1,1,2,2-tetrachloroethane—which must be reported to the EPA's Toxics Release Inventory. Improper listing means no pressure for further testing of the ingredients.

continued

Box 2.1 Continued

Until the list is corrected, no pesticides or any other product containing unidentified chemical components, including fragrances, can be declared safe.[d] Linda McElver, spokesperson for people with multiple chemical sensitivities, wrote,

> There can be no sound science or reasonable guarantee of safety when there are pesticides with secret untested and potentially more toxic inert ingredients. There can be no sound science when the full...product isn't tested...unless testing is designed specifically for an illness. These concerns are just the tip of the iceberg when considering the use of pesticides and other chemicals, for science is just discovering other ways that pesticides harm at levels considered safe today.[e]

To protect public health, the EPA should require full disclosure of all ingredients in chemicals for home and garden and for broad agricultural and industrial applications. Labels should disclose all "inerts" that were or are registered as active ingredients and other agencies regulate as pollutants. The EPA also should suspend registering pesticide formulations containing inerts of unknown toxicity until they—and all breakdown products—can be tested rigorously.

Notes
[a] Sophie Richard et al. Differential Effects of Glyphosate and Roundup on Human Placental Cells and Aromatase. *Environmental Health Perspectives* 2005. Available: dx.doi.org (DOI No. 10.1289/ehp.7728).
[b] Holly Knight and Caroline Cox. *Worst Kept Secrets, Toxic Inert Ingredients in Pesticides* (Eugene, Oregon: Northwest Coalition for Alternatives to Pesticides, 1998). Examples of active "inert" ingredients include nonyl phenol compounds, widely used industrially as dispersing and emulsifying agents in pesticide formulations. These toxic endocrine-disrupting compounds are persistent in the environment and abundant in food (Caroline Cox. Nonyl Phenol and Related Chemicals. *Journal of Pesticide Reform* 1996;16:15–20; Klaus Guenther et al. Endocrine Disrupting Nonylphenols Are Ubiquitous in Food. *Environmental Science and Technology* 2002;36:1676–1680). See also U.S. Environmental Protection Agency, *Inert (Other) Pesticide Ingredients in Pesticide Products—Categorized List of Inert (Other) Pesticide Ingredients.* Available: www.epa.gov/opprd001/inerts/lists.html (accessed September 2005).
[c] List 1 had 57 entries in 1987, but dropped to 2 by 1995; List 2 rose from about 60 entries in 1987 to 101 in 1995; and List 3, about 800 entries in 1987, went up to 1,981 in 1995. Of the 394 substances Knight and Cox (*Worst Kept Secrets*) identified as "active inerts," two were on List 1 in 1995, 30 were on List 2, 276 were on List 3, and the balance had gone to Lists 4A and 4B. Approximately 485 (19%) of the 2,500 inerts are regulated as hazardous substances under the Clean Air Act, Clean Water Act, Safe Drinking Water Act, and the Superfund Amendments and Reauthorization Act.
[d] For example, the broad-spectrum herbicide sulfometuron methyl breaks down to saccharin, which causes human genetic damage (Caroline Cox. Sulfometuron Methyl (Oust). *Journal of Pesticide Reform* 2002;22:15–20; see also Caroline Cox. Glyphosate (Roundup). *Journal of Pesticide Reform* 2002;18:3–17).
[e] Personal communication, 2006.

The government does not require abundance or distribution inventories, or studies of potential harm to humans or biota. Despite findings that soils and waterways contain higher concentrations of breakdown products than of the parent compounds, no standards yet protect us from them, whether singly or in combination.

Back to the Future

Long-term sustainable farming practices reduce or eliminate synthetic fertilizers and pesticides and also help rebuild soils and reduce erosion. They can provide wholesome human food with low impacts on natural habitats. The Amish in America's midwestern states never gave up the traditional farming of older times and are a model for today's popular "organic farming" movement, including "permaculture" and "biodynamic" variants. Integrated pest management (IPM), including integrated weed management, combines nonchemical pest and weed control methods with lower-than-conventional chemical applications. No-till farming methods reduce soil erosion, but monoculture row crops raised without chemical herbicides may face severe weed problems.

Ideally, organic farms are small- or medium-sized operations, growing diverse crops and also raising animals in farmyards and on pastureland. Small organic farmers practice crop rotation—planting major crops in a different area in successive years—to maintain healthy soils by balancing the nutrients one crop uses with nutrients another one adds. And organic farms generally avoid synthetic chemical fertilizers, mostly applying composts and composted manure and growing "green manure" legumes that fix atmospheric nitrogen in soil. Soil scientist John Reganold has visited 150–200 organic farms over 15 years and about the same number of conventional farms. "At every organic farm, the farmer has always shown me the soil," he said in an interview. "They always say, 'Look at the earthworms and look at the structure.' They are thinking the soil is part of the system." Reganold could not recall any conventional farmer that had showed off a farm's soil, and conjectured that soils on most Washington State apple orchards are malnourished from a diet of synthetic nitrogen fertilizers.[76]

Organic milk, eggs, meats, and vegetables now command premium prices in specialized supermarkets and direct-to-customer farmers' markets, enough to attract the interest of industrial farming corporations. Fully organic and rGBH/rBST-free milk producers cannot meet affluent-consumer demand, while free-range poultry and grass-fed and antibiotic-free beef are growing in popularity. The 2003 appearance of bovine spongiform encephalopathy (BSE), or "mad cow disease," in the United States enhanced the market for organic and grass-fed meat products over beef fed on slaughterhouse trimmings.

Our fathers' and grandfathers' farming methods were not federally regulated until 2000, when the USDA finally recognized organic farming as the fastest growing American food industry segment and began trying to define it. As small organic farms became suppliers to large organic food chains, they had to expand to fulfill orders. Many or most have had to largely abandon the polyculture ideal, while largely keeping the antichemical, anti-antibiotic faith. But big-organic growers and food processors have had the final word on how to qualify for a USDA "organic" label. As a result, organic cattle and chickens may be raised on polluting factory farms with pastoral views, and processed organic food can contain synthetic chemicals and food additives like conventional supermarket fare.[77] Many small organic farms, which adhere more closely to the ideal, spurn the USDA label.

Organic pest control ideally employs mechanical cultivation to combat weeds when they are small and most vulnerable, and biological controls—ladybug beetles, lacewings, spiders, wasps, and other beneficial insects—to suppress disease-carrying and plant-destroying pests. Crop rotation disrupts pest habitats, reducing diseases and pest damage. A minority of organic farmers use no chemicals other than dilute detergents, pepper sprays, and vinegar to control pests. The others, and particularly big-organic farmers, apply plant-generated (botanical) pesticides and derivatives, such as rotenone and pyrethrum, and the *Bacillus thuringiensis* (*Bt*) bacterium (see chapter 1). The USDA also approved antifungal sulfur and copper compounds for organic farming.

IPM methods, combining synthetic pesticides with organic farming's crop rotations, weed cultivation, and biological pest controls, are deemed more flexible for large farms that grow monocultures or only a few different crops. Although some critics claim that IPM means "include pesticides monthly," IPM practices both reduce the volume of pesticides on farmlands and opportunities for pests to develop resistance.[78]

Other farming alternatives include experimental "natural systems agriculture," favorable to grain cultivation, and management-intensive grazing. Propounded by the Land Institute's Wes Jackson, natural systems agriculture attempts to grow as many as four perennial grains together, which he calls "instant granola in the field." Jackson aims to mimic the biodiversity and sustainability of the original American prairies and eliminate annual grain monocultures. From 20 years of experiments, he believes that perennial grains can be bred to produce yields great enough for agricultural viability, perennial species can yield more when grown together than when planted alone, polyculture provides its own nitrogen and eliminates the need for chemical fertilizers, and "natural systems" fields produce their own defenses against weeds, pests, and diseases.

"Management-intensive grazing" is an intense organic system of rearing livestock under constant supervision, together with growing polyculture crops. Cattle graze real pastures but never bite grass in the same area on succeeding days. This system avoids both overgrazing and stranding cows in their own manure. The intense rotational schedule preserves herbal diversity to keep soil ecosystems healthy, and plant roots that bind soil and prevent erosion. Following one scheme, poultry come after the cattle and clean up insect larvae and parasites from the cow manure, reducing the potential for diseases. The complex management schemes emulate interlocking dependencies of plants and animals in a natural ecosystem and have the power to both heal and improve farmlands. This is farming at its most sustainable—and most demanding.

Organic comparisons

Only a few controlled studies have compared the quality and economic viability of IPM, organic, and conventional produce. A 22-year study of organic and conventional farming in Europe[79] showed that actual yields depend on the crop and its effects on the soil. For example, organic corn yields were about one-third lower than conventional corn crops in the first four years, while soybean yields were about the same for both organic and conventional methods. But over time, the organic crops produced higher yields, especially in times of drought when water and wind erosion severely degraded conventional farmlands. The soil on organic farms steadily improved in quality, yielding more soil microbes; better nutrient cycling, moisture retention, and soil structure; and three times the earthworm abundance of soils on conventional farms. Organic farming benefits also included reduced water demand, 30% less fossil energy use, and no groundwater contamination.

Other long-term studies have shown that organic crops take less than half the nutrients from soil than conventional crops extract from applied fertilizers, and organic farms use less than half the energy per bushel of produce than conventional farms.[80] Virginia Worthington analyzed the results of 41 previous studies on nutrients in conventional and organic vegetables grown from 1936 to 1987 in Britain, and from 1963 to 1992 in the United States. She made 1,297 comparisons of nutrient levels reported for equivalent weights of vegetables of the same type, grown both conventionally and organically in the same growing season. Statistical analysis of 12 nutrients in the five most frequently studied vegetables showed that the organic vegetables contain "significantly more vitamin C, iron, magnesium, and phosphorus and significantly less nitrates than conventional crops."[81]

Professor Reganold compared the quality of Washington State apples from separate organic, IPM, and conventionally farmed plots to assess each method's efficiency, sustainability, and economic consequences. All showed about the same rates of tree growth, apple production, and apple size, but the organic plot had the smallest negative environmental impact and ranked highest for sustainability and energy efficiency. The organic fruit commanded higher prices and so had the best economic return. Organically grown apples also ranked first for texture and flavor in subjective taste tests, ahead of either the IPM or conventionally grown produce, and had the highest sugar/acid balances in chemical tests.[82] Reganold concluded the "results show that organic and integrated apple production systems in Washington State are not only better for soil and the environment than their conventional counterpart, but have comparable yields and, for the organic system, higher profits and greater energy efficiency."

Happy palates are influencing California's premium wine makers to use organically grown grapes. Both the Benzinger and Davis Bynum family wineries in northern California have documented increased profits from reducing fertilizers and pesticides, employing farming methods that protect soils, creeks, and wildlife. But the profits owe mostly to the superior taste of the wines fermented in their vats.

The levels of pesticide residues on conventional, IPM, and organic produce indicate their relative health risks. In 2002, B. P. Baker and colleagues found that about 8 of 12 conventional vegetable crops sampled in markets had low but measurable pesticide residues. Ninety percent of conventionally grown apples, peaches, pears, strawberries, and celery samples also showed measurable pesticide residues. Chemicals and pesticides are abundant and carried around the world on dust particles, so all outdoor plants have some chemical residues—yet only about half of the IPM crops that Baker's group examined, and only a quarter of the same types raised by organic methods, showed any pesticide residues.[83]

In 2003, the USDA Pesticide Data Program reported that more than 80% of conventionally grown pears and bell peppers contain pesticide residues, identifying 42 different pesticides in 741 samples of bell peppers. Nearly all butter samples showed pesticide contamination. Overall, 46% of tested foods were contaminated with at least one pesticide.[84] Sources representing industrial agriculture have suggested that organic crops may carry similar amounts of botanical pesticide residues but so far lack supportive data. Botanical pesticides approved for use on organic farms could be just as hazardous to humans as synthetic pesticides, but that possibility has not been thoroughly investigated, either. Rotenone is a possible cause of parkinsonism in rats, for example, and botanical pyrethroids also may have negative health effects.

False Promises

American-style mechanized farming focuses on the bottom line, not on feeding people. But it allows the world to grow more food per capita than 30 years ago, despite the human populations' doubling over the same time. To paraphrase Wes Jackson, the United States grows increasingly more food on fewer acres than ever in history and exports billions of dollars of farm products a year, on a planet where people are hungry and starve by the millions. Organic farming cannot sustain huge and growing populations, but it provides a way to sustain farming. The gap between these two options defines a looming worldwide disaster, which entomologist Edward O. Wilson calls "The Bottleneck."[85]

The 1970s "green revolution" increased agricultural production worldwide and now is the model for technological solutions. It depended on crop varieties with higher yields and shorter growth cycles, with multiple cropping each growing season. But the efficiencies depended also on growing monocultures with irrigation and on huge and increasing petrochemical fertilizer and pesticide applications. It has bankrupted many, if not most, traditional farmers in Latin America, Africa, and Asia.[86] Along with America's small farmers, traditional farmers on other continents became tenants on corporate farms.

Genetically modified treadmill

Genetically modified (GM) food and fiber crops are supposed to be high-tech solutions for agricultural dilemmas in the United States. But GM agriculture uses as much or more energy and chemicals as conventional farming and so, if anything, may be even less sustainable than today's conventional agriculture. All chemical farming is a trap, spreading poisonous pollution to all animal and human populations—with growing evidence of severe effects, particularly on children. Climate warming and declining energy resources pose a tremendous challenge to U.S. agriculture, but industry and government generally tend to deny the problems, which go begging for attention.

At this writing, U.S. farmers grow transgenic crops that survive herbicide applications or secrete insecticidal *Bt* bacteria. Since the first U.S. commercial GM planting in 1996, GM crops accounted for about a quarter of the nearly three million acres harvested, dominantly the herbicide-tolerant varieties. Future GM varieties will resist frost to limit crop damage for some fruit growers, or will tolerate saline soils and reduce soil salination losses. Other transgenic plants synthesize a range of commercial compounds in their tissues, including human and veterinary vaccines, drugs, and other pharmaceuticals, or industrial chemicals for manufacturing paper, plastics, personal care items, and laundry detergents.

After three years of planting herbicide-resistant crops, weeds became hardier, and particular species developed pesticide resistance. So continuing to grow them requires more and more pesticide. By 2005, 122 million pounds of pesticides had been applied to the three dominant GM crops of corn, soybeans, and cotton— about a 4.1% increase.[87]

Crops engineered to grow faster and more efficiently also take more from the soil, degrading it faster than traditional varieties. Intensively cultivated GM crops will need irrigation in western states, further depleting surface and groundwater; will pollute waters and soil; and will add to erosion on tens of millions of top cropland acres. GM plants are more than likely to incite irreversible changes in natural systems, many still unimagined, with serious effects on people and wildlife. Inevitably insects and weeds will develop resistances to widely planted crop strains, forcing chemical companies to continually crank out new ones and new pesticides to go with them. Widespread *Bt* crop planting will generate insects with *Bt* resistance, eradicating organic farmers' most useful natural pesticide.[88]

Chemical traps

In *Silent Spring*, Rachel Carson warned, "the chemical war cannot be won, and all life is caught in its violent crossfire." The EPA's approvals for myriad persistent and potentially carcinogenic pesticides have caught us in their crossfire (table 2.4). Herbicides applied at unprecedented levels have unknown long-term effects on non-target plants and secondary impacts on animals. Supermarkets in the United States also sell foods bearing or containing antibiotics, excess hormones, and multiple pesticide residues. A single spinach sample can contain 14 different pesticide residues.[89]

Cancer researchers report increased risks of non-Hodgkin's lymphoma from pesticide exposures. Men exposed to atrazine, alachlor, and the nerve-attacking diazinon insecticide are many times more likely to have defective sperm and low sperm counts than unexposed men.[90] A Cancer Registry of California study found that Hispanic farmworkers have higher rates of brain, leukemia, skin, and stomach cancer than do other California Hispanics, and female Hispanic farmworkers also had more cases of uterine cancer compared to their larger, nonexposed peer group. As an indication of their exposure routes, in 2004 Kern County and Coachella Valley table grape pickers experienced vomiting and diarrhea after supervisors forced them to taste-test the ripeness of unwashed pesticide-treated fruits.

The effects of chemical mixes accumulating in human bodies at unknown rates and in a variety of combinations is largely unstudied and barely known. Even some conventional farmers avoid eating organophosphate-drenched foods. "I always plant a small area of potatoes without any chemicals," an Idaho potato farmer told author Michael Pollan. "By the end of the season, my field potatoes are fine to eat, but any potatoes I pulled today are probably still full of systematics [pesticides]. I don't eat them…I won't go into a field [until] four or five days after it's been sprayed."[91]

Pesticide exposures pose a particular threat to children. American children increasingly exhibit an alphabet stew of learning and behavioral disorders, which a study of indigenous Yaqui farm children suggests may be linked to toxic exposures. All Yaquis have a relatively low level of exposure to synthetic fertilizers and pesticides, but Yaqui Valley children live on conventional farms and eat conventional produce, so have greater exposure levels than do traditional Yaqui children living on nonchemical farms in nearby foothills. Yaqui Valley children have higher levels of problem classroom behaviors, and tests show higher learning impairments, compared with traditional Yaqui children. Other than pesticide exposures, no other differences exist between the study populations to explain these contrasts in the children's behaviors and intellectual development.[92]

The 1996 Food Quality Protection Act (FQPA) required an EPA review of pesticides initially registered before November 1984 for compliance with current scientific and regulatory standards—including human health and ecological effects. The EPA also is reassessing pesticide residue limits on food to meet FQPA safety standards. In 2000, the EPA's review of the organophosphate chlorpyrifos concluded that the "Dursban" formulation for home gardens should be taken off the shelves. The "Lorsban" formulation still is used on farms, however. EPA also ordered diazinon to be phased out by 2004, and FQPA review could force organophosphate insecticides—and possibly carbamate pesticides—off the market for food crops.

Manufacturers seem to feel that human test data might help them circumvent the FQPA's stringent standards. The EPA issued a directive against such tests in 2001, but a trade group for chemical companies filed suit, and in June 2003, a federal appeals court ordered the EPA to accept human test data. In 2005, Stephen Johnson, George W. Bush's appointee as EPA administrator, won Senate confirmation only by promising to scrap the Children's Environmental Exposure Research Study (CHEERS). Resembling chemical and biological warfare agent tests on unsuspecting civilians and military personnel (see chapters 6, 7), CHEERS would have paid Florida parents to expose their infant children to pesticides, including chlorpyrifos. Some 250 other human-testing programs are still in the works—one would pay young males to inhale methanol vapors and asthma sufferers to inhale potentially harmful ultrafine carbon powders.

Farm climates

At least in parts of the United States, increased "green revolution" crop harvests owed a lot to decades of unusually favorable climatic conditions. But climates are warming rapidly, and over the longer term, parts of the United States could become less favorable for agriculture. In particular, global warming could bring long dry spells. Droughts have the unparalleled power to focus attention on the equation of water supply versus water use and the human need for clean water (see chapter 9).

Recent droughts were midgets compared to many seen in tree-ring records that go back many thousands of years. But even without droughts, rising temperatures will reduce the abundance of surface waters essential for irrigating arid-land farms.

Snowpacks are the cheapest and best method for storing water with low evaporative losses. In the Pacific Northwest, where winter temperatures are close to snow melting, warmer spring temperatures already are reducing snowpacks during the spring or early summer and eventually may eliminate them altogether. Earlier melting lengthens the reservoir storage time of water that farms will need later in the summer. Longer reservoir storage means more evaporative losses and lower overall water supplies. Most western U.S. reservoirs lack the storage capacities for faster snowmelt runoff than "normal," and some western politicians are agitating for taxpayer money to build more dams. If even the lowest prediction of peak warming proves true, current major conflicts over water supplies among farmers, cities, reservations, and endangered species could literally explode (see chapter 9).[93]

Since plants live on CO_2, optimists suppose that high CO_2 levels will encourage crops and benefit agriculture. Crops also release oxygen, which should help moderate greenhouse gas buildup and slow global warming.[94] But atmosphere, soil, and plant interactions are not that simple.[95] CO_2 increases do not make all plant species simply grow faster—the actual effects depend on how plants, including crops, accomplish the photosynthesis processes that convert water and carbon to sugars. Most plants make simple ("C3") intermediate metabolite compounds, and others make more complex ("C4") metabolites.[96] Both groups include weeds as well as crop plants—so increased CO_2 concentrations will encourage *both* weeds *and* crops. Increasing CO_2 likely will favor C3 weeds (sicklepod) and crop plants (wheat, rice, soybeans) over C4 weeds (smooth pigweed) and crops (maize, or corn in U.S. parlance).[97]

Field experiments on increased CO_2 effects, and the drying effect on soils of higher day and nighttime temperatures, indicate that corn does not benefit from increased CO_2 and wheat and soybean yielded less than laboratory studies had predicted.[98] Increased CO_2 had no effect on sorghum yield in a weed-free environment but caused significant losses in the presence of both C3 (velvetleaf) and C4 (redroot pigweed) weeds. Increasing CO_2 appears to stimulate invasive weed growth and slow bacterial decomposition.[99] Worse, experiments suggest that increased CO_2 makes the particularly noxious Canada thistle resistant to commercial glyphosate (Roundup) and glufosinate (Liberty) applications.

Many other natural processes that affect farming success will change as the climate warms. Pest infestations likely will increase, ramping up pesticide applications with attendant health and environmental effects. Higher CO_2 may increase insect activity and also raise carbon-to-nitrogen ratios in crop leaves, stimulating some insects' feeding drives. Insects, nematodes, and wind will combine to spread plant diseases more broadly and endanger crops over wider regions.

The United States is turning to biodiesel and ethanol fuels to lower petroleum imports, and farmers growing biofuel crops are getting higher crop prices. The crops currently favored for fuels include corn and soybeans, both highly subsidized by federal taxpayers. Fuel crops suit the style of industrial farming that consumes land and energy, but burning these fuels in cars will contribute more greenhouse gases to the warming climate, while converting food crop acres to fuel-growing could compromise our ability to feed ourselves (see chapter 12).

Wild lands already are threatened with clearing for growing biofuels, and corn fields are especially prone to erosion. Food prices already are rising as a result of the conversions. If Americans understood that growing fuels means adding to global warming, raising food prices, and subsidizing erosion, they might not support biofuels. With author Michael Pollan, we would support government policies to turn 20% of cornfields back into grasslands, which sequester CO_2 even better than forests.

Crossing the Threshold

The surface of Earth is teeming with life—some of it is inimical to human welfare, but much of it is beneficial. American farming practices are careless with both. In a land enormously blessed for agriculture, American farming institutions have resisted learning the historic lessons of land abuse. Greek soils degraded over thousands of years, even under practices that the ancients considered good stewardship. Far from allowing us to escape a similar history, our advanced technology renders us capable of wrecking our soil's productivity much faster than the ancients could. Instead of respecting the natural forces that denuded Greek landscapes and converted the Fertile Crescent region to salt farms, U.S. agricultural schools teach a misplaced confidence that technology will fix any problem.

America's conventional farming practices have brought us to a crucial threshold. The direction that farming takes in the very near future will determine whether the nation, and fragile western lands in particular, have ahead of them an impoverished or enriched agricultural future. And it's not just America—worldwide soil degradation has lowered food and fiber crop productivity on at least 16% of agricultural lands.[100] Opening less-fertile lands to agriculture in compensation for those losses imposes an additional toll on wildlife habitat, ecosystem diversity, and clean water—all with major financial costs.[101]

The saddest story of the 2002 U.S. farm bill was losing the provisions that would have expanded and strengthened soil conservation and farmland preservation programs, largely to benefit medium- and small-scale farmers. But hope is alive in public-interest campaigns that preserve farmland from industrial farming and urban development,[102] and in the growing popularity of organic farming, which is spreading even to urban settings. These movements support each other because authentic organic and ecological farming methods, capable of restoring eroded farmlands and conserving water, are not suited to large plantations.[103]

Sadly, every taxpayer's money supports unsustainable conventional agriculture, mostly benefiting a very few large corporate farmers and chemical companies. Growing GM crops continues the same system, with additional unsustainable features. Powerful industrial farm interests are pressing for continuing large subsidies, even as biofuel crops push up food prices. And large-scale "organic" farmers team with rich conventional farming interests to subvert organic labeling. These movements could undercut the commercial viability of small and moderately sized organic farms.[104]

But conventional and GM agriculture face an uncertain future as the days of cheap petroleum come to an end (see chapter 12). As fertilizer and fossil energy resources decline, all energy costs will inevitably rise, increasing the costs of everything else—particularly conventional farming.

Will Americans fund this system to its collapse? Even if we do, the situation is not hopeless. If enough Americans really care and register their views, we can start solving these problems. In this case, the abused commons of Garrett Hardin includes our tax money, our food, and our health—so it's our business. A new federal farm bill is in draft stages (2007), giving land-conservation interests the chance to ally with fiscal political conservatives and again try to take American farming back to a more economically sound and resource protective system. We should all join in.

Scientific studies clearly point to the farming methods that can rebuild soils, while reducing (or even eliminating) erosion and pollution, and conserving water supplies. These methods do not work on large-scale industrial farms, so federal and state laws and policies ought to focus on bringing small and medium farmers back to the land. Initially, government programs might support smaller-farm conservation programs, helping them improve degraded lands and compete in the marketplace that 1970s and more recent subsidy programs have wildly distorted. We need to give most support to smaller farms that serve local markets, because they will be more sustainable as energy and fertilizer resources deplete.

Changing our farming system back to more traditional methods will not produce enough food for the world's growing population, but it could free up funds for helping other nations toward their own agricultural sustainability. Sustainability, and healthy foods from a healthy environment are the paramount issues. Farming methods that use more energy per calorie of food produced, and expand soil depletion and degradation, cannot be sustained much longer. We are more likely to save the future of agriculture by making the health and well-being of all life the bottom line.

3   Raiding the Range

Livestock grazing has proven to be the most insidious and pervasive threat to biodiversity on rangelands.
Reed Noss and Allen Cooperrider, *Saving Nature's Legacy*

"Home on the Range" evokes a western landscape "where the deer and the antelope play." But even at the song's debut in the 1870s, deer and antelope were declining in numbers and cattle grazing was degrading rangelands across the American west. In their natural state, arid North American lands are robust and productive, but they recover exceedingly slowly from heavy grazing.[1] By 1860, more than 3.5 million domesticated grazing animals were trampling arid western soils, causing severe erosion and lowering both water quality and water supplies in a water-poor region. The early start and persistence of grazing over such a long period of time invaded every nook and cranny of the public lands, making livestock grazing the most pervasively damaging human land use across all western ecosystems.[2]

Today, grazing affects approximately 260 million acres of publicly owned forest and rangelands, mostly in the 11 western states—about equivalent to the combined area of California, Arizona, and Colorado. Those acres include Pacific Northwest fir and ponderosa forests; Great Basin big sagebrush lands; the richly floral Sonoran Desert; magnificent high-desert Joshua tree forests; varied shrub associations in the low-elevation Mojave, Great Basin, Chihuahuan, and other southwestern deserts; and extensive Colorado Plateau pinyon–juniper forests stretching from northern Arizona and New Mexico to southern Colorado and Utah and decorating the arid inland plateaus of Washington, Oregon, and northeastern California.

Proponents of public lands grazing argue that cattle have not changed anything. They just replace the immense herds of hooved native herbivores—bison, deer, antelope, and elk—that once dominated western ranges. But in pre-European settlement times, natural forces, including unlimited predators and limited fodder, effectively controlled the native animal populations. Unlike cattle, the herds of deer, antelope, and elk wintered in generally snow-free lowland areas and used much less than their full range each year. And those animals were easier on the land, especially the rivers. Immense bison herds ranged over vast areas, never staying very long on any range.[3] Bison rarely visited the sites of today's major live-stock grazing problems in Great Basin and southwestern deserts, however. On northern ranges, bison obtained winter moisture from eating snow and did not cling to creeks and streams the way cattle do.

Grazing remains pervasive throughout the west, although very small numbers of cattle and sheep forage on western public lands any more. But even a relatively low level of grazing, under permits from U.S. Bureau of Land Management (BLM) and U.S. Department of Agriculture Forest Service (USFS), has helped to jeopardize at least 340 of the 1,207 plant and animal species listed as endangered, threatened, or proposed for listing under the Endangered Species Act. In fact, grazing has played a major role in taking nearly a quarter of those species to the brink of extinction.[4]

Today, the public purse protects private cattle and sheep herds from poor forage and scant water supplies on public lands in the arid western states. A powerful graz-ing lobby ensures that expensive government programs kill off livestock predators and even construct fences and water troughs for their herds. In this unkindest version of Garrett Hardin's tragedy of the commons, taxpayer subsidies pay for the privately owned cattle and sheep to overgraze the public's rangelands. Following Hardin's phi-losophy, many grazing opponents feel that more stringent regulation is the best rem-edy. Others want to get private cattle off sensitive allotments for prolonged periods.

Rawhide in the American Desert

The year 2000 opened a new era for America's public lands graziers, confronting them for the first time with serious challenges to their traditional subsidies and perks. On 15 May 2000, the U.S. Supreme Court ruled unanimously that domes-tic livestock grazing on public lands is not a right, but rather a privilege that can be withdrawn at the discretion of a federal land-management agency. That deci-sion affirmed the federal government's authority to plan the uses of public lands, including limiting or prohibiting domestic livestock grazing.

For some time before that high court ruling, environmental groups had attacked all aspects of cattle grazing on federally managed lands. In a precedent-setting 1990 case, a federal judge canceled seven grazing permits in four Arizona and New Mexico national forests because the USFS had not assessed grazing impacts on endangered species. In a departure from earlier precedent, the judge also required federal grazing-permit holders to remove their cattle from national forest lands. Following the hot and dry 2000 summer, environmental groups pushed the BLM to order Utah rancher Quinn Griffin and other permit-holding cattlemen to move their cows off the Grand Staircase-Escalante National Monument and let the land recover from drought and wildfires. When Griffin and supporters refused, the BLM impounded their cattle. Then some ranchers rustled the cattle back from a Salina, Utah, auction yard, raising the specters of nineteenth-century range wars.

In the spring of 2001, a coalition of environmental groups sued the BLM to better manage California's Mojave Desert lands, supporting 24 endangered or threatened desert plant and animal species. The suit highlighted threatened desert tortoises, charging that cattle eat the same plants as tortoises and deprive them of food—also trampling and generally degrading life-supporting tortoise habitat.

The BLM settled the suit by agreeing to enforce land-use restrictions to protect wildlife. In particular, BLM agreed to force seven cattle ranchers to take their cows off federal land in spring and fall, when tortoises eat and mate. State and national officials stampeded to the cattlemen's aid, while local sheriffs and county supervisors circled the wagons in opposition to the BLM. The ranchers appealed the agreement, slowing compliance while their cows kept on munching tortoise fodder.

At the same time, the Nevada BLM impounded 200 head of cattle for auctioning after ranchers Ben Colvin and Jack Vogt failed to pay the $1.35 per animal unit month (AUM) fee for the privilege of grazing cattle on public lands. That amazingly low fee makes the cost of feeding a cow and a calf for a month equivalent to the price of a can of dog food. Colvin and Vogt supporters included sheriffs of at least two Nevada counties and the Nevada Committee for Full Statehood, which does not recognize the BLM's authority over public lands. These and other grazing conflicts promise to spread.

Cowboys' free lunch

The events that took federal grazing policy to the Supreme Court reveal the political power of America's mythic frontier and its defining icon, the cowboy. Real cowboys, authentically known as *vaqueros*, occupied a historical interval of barely 20 years when cattle drives dominated the livestock industry.[5] The cowboy myth was born from sensational nineteenth-century journalism, nurtured in Zane Grey novels and distorted by Hollywood movies. It epitomizes Americans' enduring identification with frontier independence and courage in the face of hostile native people and natural forces.

A strong alliance of graziers and western congressional representatives keeps the myth from dying. That fleeting moment in U.S. history, and the images from literary inventions, covers up both the 200 years of gross abuses of the western ranges and the power that western grazing-permit holders wield over millions of acres of U.S. public lands. Former BLM Director Jim Baca remembers, "[W]hen I was BLM Director, if I wanted to move 10 cows in Wyoming, senators would get involved. I still marvel at it."[6]

Few or no local communities can base their economies on public lands grazing, which employs much less than 1% of western workers.[7] The grazing lobby benefits only a tiny minority of cattlemen—the few public lands grazing permit holders, who contribute essentially zero to the nation's economy but much to the declining quality of its lands. Some of the cattle ranchers rely entirely on public lands for livestock fodder.[8] Various estimates suggest that no more than 3% of all the nation's livestock graze those 260 million public acres.[9] The BLM oversees about 160 million acres, and the USFS about 100 million. In perspective, the total number of beef cattle in the 11 western states is smaller than the herds grazing just Texas and Oklahoma, where all cattle feed on private ranches.

Akin to the system for harvesting timber from public forests (see chapter 1), the cost of administering grazing permits is much greater than the total grazing fees collected. As a result, U.S. taxpayers subsidize domestic livestock grazing on public lands. In 2004 the public paid administrative costs of $115 million in excess of permit fees, due in part to the low grazing fees and the low percentage of fees returned to the U.S. Treasury that year—only 18%. Between 1980 and 2004, public lands grazing fees dropped 40% while private land fees rose 78%. To recover agency expenditures, the BLM would have to charge $7.64 per AUM and the USFS $12.26 per AUM.[10]

Munching and Trampling

Cows selectively graze the plant species that they enjoy most, giving less tasty ones more room to grow. Unfortunately, the cattle also trample most of the plants that they don't eat. Munching and trampling increase soil erosion, decrease water infiltration, and increase runoff on grazing lands. Livestock munching reduces plant foliage and also changes the plant species assortment, altering the vegetation's ability to hold soil and let water infiltrate (figure 3.1). Livestock trampling

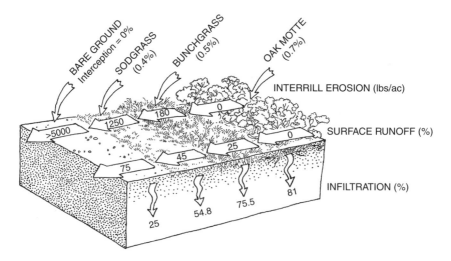

Figure 3.1 Effects of different plant species on rainfall interception, infiltration, and runoff. Each type of ground cover intercepts a different percentage of incident rainfall (parentheses). The near row of open arrows on the ground surface indicates the percentage of incident rainfall that runs off the surface; the far row, the level of soil erosion (in pounds per acre) affecting the different plant communities at the indicated level of incident rainfall. Numbers below the squiggly arrows represent the percentage of rainfall that infiltrates the soil. Amounts differ for sloping rangeland but effects are the same. Modified from W. H. Blackburn. Impacts of Grazing Intensity and Specialized Grazing Systems on Watershed Characteristics and Responses. In Developing Strategies for Rangeland Management, *National Research Council* (Boulder, Colo.: Westview Press, 1984), 927–983.

Table 3.1 Rates of Infiltration on Grazed and Ungrazed Lands

| State | Rate of Infiltration (Millimeters per Hour)[a] | |
	Ungrazed	Heavily Grazed
Arizona	40.6	30.5
Colorado	40.6–83.8	20.3–30.5
Montana	109.2–185.4	20.3–96.5
Wyoming	30.5–38.1	17.8–30.5

From Andrew Goudie. *The Human Impact on the Natural Environment*, 5th ed. (Cambridge, Massachusetts: MIT Press, 2000), 178 (table 4.7).

[a] One millimeter is 0.04 inches.

compacts soil, reducing both its capacity to capture and store water and its ability to exchange nutrients with plants. Trampling also denudes the land, promoting destructive rill and gully erosion, plus other environmental changes that cascade across all western ecosystems (see chapter 13).

In general, light to moderate grazing reduces a soil's water infiltration capacity to about three-quarters that of ungrazed soil, and heavy grazing reduces it to one-half or less (table 3.1).[11] Severely grazed soils may have only one-eighth the water infiltration of ungrazed land and lose twice as much sediment to erosion (figure 3.2).[12] Rainwater infiltration normally helps flush salts out of arid soils, but when grazing reduces infiltration, the salts and other toxic substances accumulate. Erosion greatly reduces essential soil nutrients and organic material in the uppermost layers of soil.[13]

Deforestation, agriculture, urbanization, and grazing—all add local climatic warming effects to global and regional climate changes. Plant cover and moisture contents regulate a soil's temperature, and normal plant evapotranspiration processes constantly release moisture, promoting cooler temperatures. Local temperatures are cooler where more plants grow than in denuded areas. Satellite images of the U.S.–Mexico border dramatically exhibit differences in vegetation cover between overgrazed Mexican lands and vegetated parts of Arizona north of the border. The overgrazed lands in Mexico reflect greater amounts of sunlight and so appear brighter than the Arizona side. Climate and geology are the same, and at one time the same vegetation mix grew on both sides, but overgrazing has severely reduced the plant cover south of the border.[14] Long-term climate data show the effects: Soil and air temperatures on the Mexican side are as much as 7°F higher than on the ungrazed Arizona lands.

Watershed woes

Rangelands are best characterized in terms of watersheds—the total land area that drains water to the same lake, or to an integrated system of streams and rivers. Watersheds include uplands and the crucial narrow strips of vegetation bordering

Figure 3.2 Advanced erosion on heavily grazed hill slopes, Temblor Range, California. Photographed November 1977.

streams and creeks (riparian zones). Both are inseparable parts of a watershed, and whatever affects one part affects all the rest. Riparian zones make up less than 1% of any watershed, but the sinuous water conduits are cradles of life in arid regions,[15] containing 25 times more vegetation than the much more extensive uplands.[16]

The health of riparian zones is crucial to the health of all surrounding lands, therefore, and all the land's inhabitants. More than three-quarters of vertebrate wildlife species in the southwestern United States—birds and rabbits, deer and antelopes, wolves and mountain lions, buffaloes and bears—depend on riparian zones for some or all of their life support. Plants, humans, and livestock also depend on the riparian zones.

Cattle and sheep cover a lot of ground, both in seeking edible plants and in finding their way back to water. The native riparian vegetation makes up as much as 80% of their diet. Unless restrained, livestock will beeline to the cool, green stream edges and do not voluntarily return to hotter, drier uplands, where food is harder to come by. The cattle strip stream banks of willows, sedges, and other bank-stabilizing vegetation that also provide shade. Once exposed to the sun, stream waters may become too warm for hatching fish eggs. Warm waters also kill fish and other water-dwelling species with heat stress and incubated diseases (see chapter 9). Bare stream banks are open to accelerated soil erosion and stream pollution, which also destroy the habitat's ability to support fish and other aquatic animals (figure 3.3).[17]

When livestock are rounded up and forced out of riparian zones, the damage from trampling expands to uplands where forage is more limited. Intensive grazing leaves hill slopes stripped of anything a cow could eat, eroding from the loss of

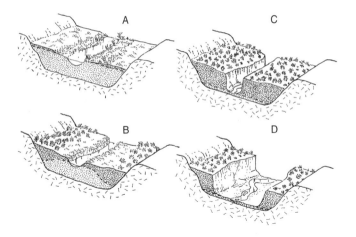

Figure 3.3 Progressive stream degradation from grazing. (A) Stream cuts into (incises) original vegetated floodplain (stippled) and develops a new floodplain (unpatterned). (B and C) Progressive stream incision deepens the channel and narrows the new floodplain. The unsaturated zone (dark stippling) thickens as the groundwater table (light stippling) declines with the incised steam channel. (D) Eventually, the incised channel widens at expense of the old floodplain. Modified from U.S. Department of the Interior. *Rangeland Reform '94* (Final Environmental Impact Statement, 1994), figure 3.2.

protective plants, and ringed with crisscrossing paths ("grazing-step terracettes")— the distinctive marks of hill-climbing livestock (figure 3.4).[18] The cattle's need to drink has not gone away, so ranchers pipe well and natural spring water to drinking troughs, depriving both native wildlife and the plants growing around springs and in spring-fed wetlands. Possibly tens of thousands of man-made water sources are scattered across western rangelands.

Contrary to popular management prescriptions, cattle severely degrade artificial water holes, and they provide little or no benefit to wildlife. Water troughs and surrounding areas quickly turn into biological deserts, with dying plants and trampled ground extending in all directions.

Critical crusts

Cascading rangeland damages start when livestock trampling destroys widespread biological surface crusts—complex mosaics of microscopic animal and plant life, including bacteria (especially cyanobacteria), green algae, lichens, mosses, and microfungi. The crusts enrich, protect, and stabilize the west's arid soils, insulate desert soils from extreme temperatures, and also preserve much of their fertility by fixing nitrogen and retaining moisture (see chapters 6, 11, 13).[19]

Biological crusts are vital to preventing soil erosion. Estimates suggest that nine million acres of western rangelands are increasingly vulnerable to wind erosion

Figure 3.4 Cattle and overgrazed hill slopes with conspicuous grazing-step terracettes, near Cholame, California. Photographed May 1985.

because of damages to natural protective soil coverings. Livestock trampling damages soil crusts as badly as the impacts of off-road vehicles (see figure 11.4). Full recovery for destroyed soil crusts requires centuries.[20]

Proponents of the currently popular "holistic" school of range management assert that the crusts swell when moistened, clogging soil pores, increasing rain runoff, and blocking water infiltration.[21] Studies show the opposite to be the case: Crusts tend to reduce runoff and enhance infiltration, and they prevent soil moisture from evaporating. In typical settings, biological crusts cover the ground surface between shrubs, and any crust-enhanced runoff is captured by the shrubs (also see discussion of vegetation enhancement in chapter 5, under "Upset Balances").

Rough surfaces enhance infiltration, more than offsetting the effects of swelling cyanobacteria. In the Colorado Plateau's cool deserts, winter frost-heaving causes surface roughening that impedes surface water flows and promotes infiltration all year round. Coarse lichen and moss components also contribute to roughening surfaces in cool environments, helping retain rain or meltwater. In some hot deserts, biological crusts reduce infiltration but also minimize soil water losses on soils of more than 80% sand with minimal lichen and moss cover.

Diseases and weeds

Grazing can spread diseases to both wild and other domestic animals. Domestic sheep, goat, and cattle diseases have been transmitted to bighorn sheep. In extreme cases, domestic sheep diseases apparently wiped out local native bighorn popula-tions. *Pasteurella* pneumonia is still a documented threat to the bighorn.[22] Many livestock-transmitted ailments in wild bighorn herds have become so wide-spread that the infection source no longer is clearly domestic—now the bighorn may be largely transmitting the diseases between and within their own herds. Unfortunately, the few studies of domestic–native animal diseases focus mostly on transmissions *from* wild *to* domestic herds.

Undesirable, invasive native and "exotic" (nonnative) plants—the ones we call weeds—have invaded 100 million acres of western rangelands, an area nearly the size of California. Wrote geographer Andrew Goudie, "The history of weeds is the history of human society…such plants follow people like flies follow a ripe banana."[23] A U.S. Geological Survey report concluded: "Ecosystem-level consequences of invasive nonindigenous [exotic] species have major ecological and economic implications and directly affect human health. However, little attention has been paid—and almost no progress has been made—in addressing the problem… ."[24]

Livestock grazing contributes enormously to spreading exotic plants across west-ern rangelands. Through accident or carelessness, exotic seeds may arrive in con-taminated agricultural seed and feed. Livestock also "plant" undigested seeds in their excrement, trampling seeds into soil along with the fertilizer. Many exotic plants were introduced intentionally; for example, extensive native shrub lands and juniper–pinyon woodlands across Utah and parts of Arizona and New Mexico still are being uprooted and replanted with exotic crested wheatgrass for cattle fodder.

Where exotic plant seeds can germinate, they tend to emulate fast-spreading and fast-growing pioneer plants (see chapter 1), which makes them invasive and able to outcompete slower growing native species, established over a long plant-succession history. In new environments, the weeds have no natural controls, and some release toxic compounds into the soil to suppress neighboring plants. Once exotics are established on grazed lands, they invade undisturbed areas nearby.[25] Not surprisingly, exotic weeds are expanding their territory at an estimated rate of more than 5,000 acres per day and cause billions of dollars of damage annually.[26]

The exotic weeds degrade western rangeland ecosystems because they reduce the variety of plant types and species that many different native animals rely upon for shelter and food, disrupting critical soil-enrichment cycles that restore plant nutrients (see chapters 2, 13). Even highly diverse native plant associations are vul-nerable to invading exotics. In the wake of grazing, broad areas that once grew a mix of native bunch grasses now support extensive stands of single weed species. For example, Colorado Plateau grasslands in Arizona, New Mexico, Utah, and western Colorado have yielded to rapidly expanding sagebrush scrub.[27] The wide-spread expansion of juniper woodlands into the Pacific Northwest and of mesquite in the Chihuahuan and southern Sonoran deserts are other dramatic examples of species replacements on overgrazed native grasslands. Today, mesquite dominates more than 28 million desert acres in southern Arizona and adjacent areas, and juniper woodlands occupy more than 2.5 million acres in the Pacific Northwest.[28]

Overgrazing has promoted invasions of plant species that cattle will not eat. The native perennial western bunch grasses could not survive the late nineteenth- and early twentieth-century livestock onslaught and were displaced by shrubs and cactus or replaced by annual cheatgrass, medusahead wild rye, and other unpalatable or poisonous plants.[29] The low-nutrition cheatgrass is palatable to cows for only a short period in the winter and early spring. Once the cheatgrass flowers, cattle avoid it because the seeds are surrounded by a long "awn" of bristly fibers that stick in their mouths. Other exotics with nasty side effects include saltcedar, or tamarisk, and Russian thistle, also called tumbleweed. Saltcedar has disrupted or destroyed more than a million acres of riparian vegetation.[30] Across the thirsty west, it sucks up an estimated 1.6 trillion gallons of water annually, more than enough to fill the Great Salt Lake to its present level twice over, *every year.*

Long-term monitoring suggests that weed-altered arid plant communities may require decades to recover if grazing completely stopped.[31] Some former shrub and grass communities may never recover.[32] Recent research indicates that the documented sharp rise in atmospheric carbon dioxide (CO_2) since 1950 has significantly aided weed invasions (see chapter 2). Experiments show that invasive cheatgrass productivity has increased under rising CO_2 concentrations in the past half-century, and future increases probably will continue fostering its success. The proportion of undigestable cellulose and lignin in cheatgrass and its resistance to bacterial decomposition also increase, providing another competitive edge over native plants on grazed land.[33] A number of other aggressive weeds also may benefit from increased atmospheric CO_2 and other global warming effects.

Grazing modifies the layered soil structure underfoot and substantially changes the above-ground interaction of plants and animals, as well as other ecosystem characteristics. How plants respond is complex and largely unknown—the outcomes depend on the pre- and postgrazing state of plants and their local environment.[34] Yet some believe that livestock grazing is such an important process in many arid environments that they predict ecosystem "destabilization" if cattle are removed, implying that exotic plant invasions actually have the desired effect of increasing diversity. To the contrary, studies show that the plant diversity reductions related to grazing-promoted exotic invasions has severely detrimental long-term effects on rangeland ecosystems.

Inconvenient Forests

Many of the western states' weed problems have resulted from federal programs that employ a number of immediately and widely destructive methods to "improve" rangeland, at taxpayer expense, by eliminating plant species that are not livestock fodder and that ranchers don't like. One species targeted for removal is the western juniper, which many western ranchers, range managers, and even scientists consider weeds. Opposition to pinyon–juniper forests is especially strong in the Modoc Plateau of northeastern California and adjacent parts of southeastern Oregon and southwestern Idaho—all sites of considerable juniper expansion over the past 120 years.

The juniper-forest detractors assert that junipers starve cattle and increase erosion; reduce rainwater infiltration into the ground, eventually drying up springs and streams; prevent rain from even reaching the ground and so promote evaporation; and withdraw shallow groundwater through their roots, reducing soil moisture and groundwater recharge even further. But U.S. Department of Agriculture research provides countervailing evidence that juniper shade reduces evaporation and that *removing* junipers increases erosion by exposing soils to sunlight and abrasive winds.[35] In addition, streams rise in arid and semiarid regions only during rare storm events, when junipers and pines are not likely to impede their flow.[36]

From the 1940s through the 1960s, through both publicly and privately funded programs, graziers in the Great Basin and southwestern states nonetheless converted hundreds of thousands of pinyon–juniper forest acres to nonnative grassland for livestock. These programs employed mostly mechanical means for removing trees and bushes—including bulldozers, cables, and chains (box 3.1). All were, and

---

Box 3.1 Cutting Arid Forests

Older methods:

- Cabling—uprooting 50–60% of pinyon and juniper trees by dragging 300-foot lengths of heavy cable between two large crawler tractors (1950s to 1960s)
- Chaining—dragging a heavy ship's anchor chain with welded hardened steel cutting edges through the forest (figure B3.1.1); more effective than cabling (1960s and 1970s to present)

Figure B3.1.1 Chained pinyon–juniper forest, east of Zion Canyon National Park, Utah. Photographed April 1989.

continued

Box 3.1 Continued

- Pushing—wheeled or crawler tractors have a bulldozer blade fitted with a pusher bar; the bar pushes trees over and the blade catches the roots, lifting trees out of the ground (1950s to present)
- Burning—includes selective incineration by helitorch, which has the same adverse effects as burning logging slash (chapter 1)
- Crushing—the LeTourneau Tree Crusher weighed 40 tons; three six-foot drums, with cutting blades 18 inches apart, crushed all trees taller than about 4 feet (first used in 1965 in New Mexico; persisted to the late 1970s using a Fieco brush roller, a heavy 16-foot-wide drum pulled by a large tractor)

More recent inventions:

- Roller chopper—modern modification of LeTourneau and Fieco devices, with stout blades fixed to a giant drum, dragged behind a bulldozer; crushes virtually any size pinyon or juniper
- Pettibone Hydromulcher—a giant lawn-mowing attachment on a large front-end loader "mulches" trees up to four inches in diameter; mainly for removing reinvading pinyon–juniper from chained and cabled areas
- Brush hog—a rotary mower as much as eight feet wide pulled behind a tractor, mainly for eliminating shrubs; cuts sagebrush 6–12 inches tall, kills 50–60% of the shrubs

are, intensely disruptive to soil and wildlife habitat. The BLM also aims eventually to eradicate the sagebrush and plant exotic grasses across arid ranges.

Chemicals on the range

Expensive federal programs continue killing pinyon–juniper and sagebrush ecosystems, mostly using herbicides and newer mechanical methods (box 3.1).[37] In 1991, the BLM approved a 10-year project in 13 western states, using 14 herbicides, two identified insecticides, and an unknown number of unidentified insecticides (table 3.2).[38] Since herbicidal preparations tend to kill many insects and other animals in addition to the target weeds (table 3.2), the reasons for adding insecticides are not clear. This eradication campaign used suspected carcinogens picloram and simazine and potential carcinogens 2,4-D and malathion. Four of the five combined chemical treatments contain potentially carcinogenic 2,4-D. The combination of 2,4-D and picloram can kill or sicken fish.[39]

In November 2005, the BLM issued a draft programmatic environmental impact statement on vegetation treatments in 17 western states, proposing annual applications of 18 herbicides on 920,000 acres of public lands. This proposal continues, without review, the use of 20 herbicides on grazing lands.

Assessing the potential effects of pesticides is problematic because generally only the identified active ingredient(s) are tested for toxicity even though many unidentified, so-called "inert" ingredients also may be toxic (see chapter 2, box 2.1). California restricted the use of diuron, atrazine, and simazine after they were detected in groundwater. Tebuthiuron also is suspected of having leached to groundwater and is on a monitoring list. Atrazine, the most heavily used herbicide in the United States, recently was fingered for feminizing male amphibians.[40] Evidence for atrazine's carcinogenicity is in dispute, but a number of European nations have banned its use. The U.S. Environmental Protection Agency (EPA) refused to do so in the United States, however.

Data are sparse on herbicides not known to be carcinogenic (table 3.2).[41] For example, nobody knows the potential human health effects of breakdown products from sulfometuron methyl in a formulation called Oust, an extremely potent broad-spectrum herbicide for controlling grasses, broad-leaf weeds, and woody tree species.[42] Under the wrong conditions, Oust applications can prove disastrous, but after sagebrush removal the BLM has few other options for preventing exotic cheatgrass invasions. The label specifies using Oust only if rain is expected. In 2000, the BLM picked normally rainy November for spraying Oust from airplanes across thousands of acres of burned public grazing lands in southern Idaho to kill cheatgrass and other noxious weeds. But rain did not follow the Oust application that year. Winds picked up fine herbicide-coated ash from the burned areas and dropped it on adjacent farmlands, killing or damaging crops over more than 100,000 acres.

Accountability

Pinyons and junipers have proved very hard to kill. The trees begin to regrow in only a few years, requiring more bulldozing, soil disruption, and poisoning to keep the land cleared. Even after a decade of such high-priced eradication efforts proved uneconomical and polluting,[43] a March 2000 BLM press release proposed a new assault against 18,000 acres of pinyon–juniper forest on Utah's Uncompahgre Plateau. The stated goal was to eliminate tree-shrub reinvasion on previously chained woodlands, paradoxically to "improve landscape health and function in general, and mule deer and elk winter range conditions." All 18 project applicants coincidentally hold grazing permits on those lands. Recent BLM proposals to remove pinyon–juniper in southeastern Oregon added such pretexts as improving the habitat for nesting loggerhead shrikes and bighorn sheep.

After contrasting the intentions and actual effects of juniper removal programs, one commentator concluded:

> It seems reasonable to suggest that before large sums of money are expended to modify a plant community, that baseline data be gathered to reflect existing conditions...and then, if change be initiated, that data be collected to substantiate whether or not any of the initial objectives were met. If baseline and post-treatment evaluation monies are not available, then the project should never be approved. This equates to professional accountability.[44]

Table 3.2 Toxicity of Rangeland Pesticides Used in Pinyon–Juniper and Sage Eradication

Pesticide	Acres	Toxicity class[a]	Human toxicity[b]	Human Health Class[c]					Ecotoxicity Effects[d]			
				Re	Te	Mu	C	Or	Birds	Aquatic	Bees	Other[e]
Herbicide												
Atrazine	5	III R	S-M				C	X	S	S-M		T; Am, m NTP
Clopyralid	13,535	III	S					X		H-VH		T; NTP
Clopyralid + 2,4-D	151											
Clopyralid + tebuthiuron	15,234											
2,4-D, all	26,019	I, III	S-M				I	X	S-M	S-H	H	T; SM, NTP, m animals
Dicamba	1,117	III	S					X		S		T; NTP, SM
Dicamba + 2,4-D	316									S		
Diuron	3	III	S		X	X		X		S	M-H	
Glyphosate	1,727	II	M	X	X	X	C, E	X	T	S-M		T; NTP, crops, BI, SM, ES
Glyphosate + 2,4-D	20											
Imazapyr	14		S-H	I			E	X	S	S		T; NTP, crops, ES
Picloram	46,256	III R	S-M			Poss.	C	X	S	S-M		
Picloram + 2,4-D	539											
Simazine	320	IV	S-M			Poss.	C	X		S-M		T; sheep, cattle; m NTP

Pesticide	Treated acreage	EPA class[a]	Human toxicity[b]	Health effects[c]	Carcinogen (possible)	Carcinogen class	Ecotox	Ecotoxicity[d]		Other wildlife[e]	Notes
Sulfometuron Methyl	23,895	III	M	X		(C)	X	S	S		T; AA, P, NTP, crops
2,4,5-T	2,200	Banned									
Triclopyr	77	III R	S				X	S-M			T; m NTP, mammals
Tebuthiuron	279,117	III	S-M	X			X	S	S		T; m NTP; trees, shrubs
Insecticide											
Malathion	11,235	III	S			I	X	M	M-H	H	T; NTBI
Sevin (carbaryl)	100	I-II	M-H	X	Poss.	B	X	T	M-VH	VH	T; NTBI; SM
Other	46,108										

Data on 10-year pinyon–juniper eradication program pesticide usage and treated acreage in 13 western States from U.S. Department of the Interior Bureau of Land Management, on Freedom of Information Act request. Remaining information is from the Extension Toxicology Network. (Available: ace.ace.orst.edu/info/extoxnet), Agency for Toxic Substances and Disease Registry. (Available: www.atsdr.cdc.gov), Irene Franck and David Brownstone (*The Green Encyclopedia* (New York: Macmillan USA, 1992), 24), L. H. Nowell et al. (*Pesticides in Stream Sediment and Aquatic Biota* (Washington, DC: Lewis Publishers, 1999), table 6.15, 431–433), and Allan Felsot (How Does EPA Decide Whether a Substance Is a Carcinogen? (*Pesticides and Environmental News* 1997;132). Information for sulfometuron methyl from Caroline Cox (Sulfometuron Methyl (Oust). *Journal of Pesticide Reform* 2002;22:15–20); for 2,4-D, from Caroline Cox (2,4-D: Ecological Effects. *Journal of Pesticide Reform* 1999;19:14–19); for imazapyr, from Caroline Cox (Imazapyr. *Journal of Pesticide Reform* 1996;16:16–20); for glyphosate, from Caroline Cox (Glyphosate (Roundup). *Journal of Pesticide Reform* 1998;19(updated September 2002):3–17); and for carbaryl, from Caroline Cox (Carbaryl. *Journal of Pesticide Reform* 2005;25:10–15). With the exception of sulfometuron and carbaryl (Sevin), the identity and toxicity of "inert" substances in the pesticides are unknown.

[a] Environmental Protection Agency toxicity classification: class I, highly toxic; class II, moderately toxic; class III, slightly toxic; class IV, essentially nontoxic; R, restricted use

[b] Human toxicity ratings: H, highly toxic; M, moderately toxic; S, slightly toxic. Toxic rating depends on organs affected, pathways into the body, and formulation (e.g., whether applied as solid pellets or liquid).

[c] Human health effects: Re, reproductive effects (lesions and atrophied testes, etc.); Te, teratogenic effects (deformities of prenatal origin, increased fetal loss); Mu, mutagenic effects; Or, effects on organs, including kidney, liver, intestines, central and peripheral nervous systems, thyroid and adrenal glands, eyes, bones, lungs, brain, heart, testes, ovaries, endocrine organs; C, possible carcinogenic effects: class A, human carcinogen; class B, probable human carcinogen; class C, possible human carcinogen; class D, unclassified; class E, no noncarcinogenicity evidence in humans; class I, tests so far inconclusive; class (C), carcinogenic effects from "inert" ingredients and breakdown products.

[d] Ecotoxicity effects: S, slight toxicity; M, moderate; H, high; VH, very high; T, toxic but not rated.

[e] Other wildlife effects: T, toxic to; P, potent to; m, many; NT, nontarget (NTP, nontarget plants); AA, aquatic animals (fish, algae, frogs); Am, amphibians; BI, beneficial insects (NTBI, nontarget beneficial insects); EW, earthworms; ES, endangered species; SM, soil microorganisms; sM, small mammals.

In spite of numerous Freedom of Information Act requests, we have been unable to obtain monitoring information on the effects of the 1991 BLM 10-year eradication program. It seems that there will be no evaluation of the earlier programs' successes—or lack of them—before launching yet another campaign. We can conclude only that taxpayer's money is being squandered without any professional accountability of grazing's externalized costs.

Half Healthy or Half Gone?

The present condition or "health" of public rangelands is controversial. The livestock industry reluctantly admits that public western ranges have sustained widespread damage but generally claims that the main injuries occurred long ago when cattle were greatly overstocked. Now, they say, graziers are good stewards and the condition of rangelands is improving. But neither the scientific literature nor various official and unofficial assessments support the contention that rangelands no longer suffer serious deterioration from grazing.

In the 1980s, the BLM judged that less than half of its lands were in fair to poor condition. Based on the same data, the National Wildlife Federation and Natural Resources Defense Council found *much more* than half of the BLM grazing lands in fair to poor condition.[45] When the U.S. General Accounting Office (GAO, now identified as the U.S. Government Accountability Office) investigators looked into the discrepancy, they discovered that BLM managers actually thought more than 40% of the BLM rangelands were in unsatisfactory condition and did not even know the condition of another 50 million acres under their jurisdiction. Even more staggering was the 55% of the BLM *riparian* lands in "unknown" condition. A 1988 GAO report concluded that the BLM's claim had no factual foundation and also reported more than half of USFS rangelands in unsatisfactory condition and nearly a quarter of USFS lands in "unknown" condition.[46]

In 1990, the BLM made the brash claim that western rangelands under its control were in better condition than ever before in the twentieth century. At the same time, academic researchers judged the critical riparian portions of those lands to be in the worst condition ever.[47] The GAO found BLM's upbeat assessment lacking in factual foundation. But this spurious claim is repeated frequently by grazing advocates and has even appeared in proposed federal legislation. In 1994, the BLM did an abrupt about-face under the Clinton administration and estimated that 43% of BLM-managed rangelands were either "functioning at risk" or nonfunctioning with regard to sustaining rangeland ecological processes and forage for livestock.

"Rangeland health"

The lack of agreement between government agencies' and nongovernmental organizations' 1980s rangeland-condition assessments suggests that the agencies use extremely narrow criteria. Rather than examining how the physical and biological components were functioning, they may have gauged only the condition and growth of one or two plant species that livestock prefer. Nor would comparisons between existing rangeland plant associations and supposed ideal "climax" (see chapter 1) or "potential natural" plant communities give accurate results.

In the mid-1990s, land condition assessments shifted to focus on the functioning ("health") of whole rangeland ecosystems—the condition of the soil, vegetation, water, and air.[48] Not surprisingly, the next decade of investigations revealed that complex natural systems are complicated. A fundamental problem for making assessments is the constant flux of natural plant communities (see discussion of gap-mosaic architecture in chapter 1 under "The Nature of Forests")—each plant species in an association may respond differently to climate change, resulting in species variations. In response to changes, new plants may be difficult or impossible to distinguish from the survivors of earlier, largely vanished, ecosystems. Those survivors may be barely hanging on and heading for extinction in areas where the climate has changed beyond their ability to adapt.

Another complication is that plant communities can take multiple pathways to achieving transient or long-term stability (see chapter 1). Such dynamic vegetation responses to climate change—whether natural or human influenced—undercut current concepts of range "condition" and "trend"[49] that relate only to human and livestock impacts. The poor quality of range "condition" data and lack of comparable conditions between areas turn such trend assessments into futile exercises.[50] The 1994 National Research Council's report on rangeland health concluded that an appropriate assessment requires more and different kinds of data than collected in the past, and new approaches to interpreting the information within an ecological framework. A realistic assessment of rangeland health requires recognition that indicators are location dependent and that different indicators could vary widely, even at the same sites.[51]

The new systems for analyzing rangeland health are elaborate and varied, and some may not be easily applied to practical range management.[52] One problematical BLM management policy has the worthy goals of maintaining "properly functioning" ecological processes in rangeland watersheds and habitats, or moving them toward proper functioning, and maintaining water quality to meet state standards.

The manner of assessing "proper functioning condition" draws on earlier methods for measuring forage production, primarily to benefit livestock, and has a number of problems.[53] A major issue is the need for baseline information on rangelands ecology, which is not generally available. Any ecosystem assessment baseline must include defined attributes of "normal" systems and habitats—yet, as suggested above, ecosystems may have multiple possible stable states. Formulating deterministic baseline definitions will surely create disagreements and inevitably interfere with assessing rangeland conditions. In addition, standard evaluation systems tend to become checklists, submerging observation and thought—the two most important elements of any assessment.[54]

Sludge on the range

Since the early 1980s, one approach to "restoring" degraded western public rangelands is applying sewage sludge, loaded with chemicals and heavy metal contaminants. In 1983, Pima County, Arizona, began spreading sludge on farm and rangelands surrounding the city of Tucson. Pima County's sludge is clean compared to New York City's—yet its sewers collect waste from 1,500 industries, containing more than 80 of EPA's "priority pollutants" with significant health concerns. These include dioxins, phenol, and toluene, plus high levels of cadmium, lead, copper, arsenic, and other toxic materials. When sludge spreading began, EPA regulations limited levels of only one of the heavy metal contaminants and only one of the potentially toxic chemicals found in sludge (see chapter 10).

Since the mid-1980s, the USFS has spread up to 40 tons per acre of municipal Albuquerque sludge on four experimental plots of highly degraded rangeland in the Rio Puerco Basin of north-central New Mexico. By the fifth growing season, plots receiving the highest applications showed that copper had increased 14 times and cadmium 20 times over the original levels in those soils. No data exist for other contaminants, nor does anybody know if plants will absorb these toxic substances and transfer them to animal and human food.[55]

Bohman and Taylor

In the 1950s, Utah private-land rancher Frank Bohman realized that the practices of earlier graziers, including his father, had severely overgrazed his family's ranch. "Pastures had been reduced to sagebrush, scrub oak, and dust by generations of free-ranging cattle and sheep. Erosion was severe, many of the springs [Bohman] recalled from his childhood had disappeared, and wildlife appeared to be in rapid decline. 'It broke my heart to see the land in such shape,'" Bohman said.[56] His heartbreak went beyond esthetics and wildlife habitat—the fodder was too poor to raise even 300 cattle a year, and he had to supplement their feed with expensive grain and hay.

Although Bohman could see that "his land, and western livestock ranching in general, were locked into a downward spiral," he believed that cattle grazing doesn't have to destroy arid lands. He proved his belief by intensively managing his cattle's grazing patterns, removing invader sagebrush and scrub timber, and reseeding native grasses. After 50 years of effort, he can now keep his cows on the range an extra two months and save a lot of the money that he once spent on hay. Even accounting for inflation, Bohman says he spent "only 20 to 30 cents per acre" on the reclamation.

The west's grazed public lands are just as ravaged by free-ranging cattle as Bohman's private ranch once was. Unfortunately, the public lands are not owned by visionary ranchers—they are owned by nearly 300 million Americans who don't agree on how to preserve them and managed by government bureaucracies that submit to aggressive permit-holding graziers in their local areas. The effects of overgrazing became undeniable in the 1930s and 1940s, when huge dust storms overwhelmed arid and semiarid lands in the western and southwestern United States. The ruin was particularly extensive and severe on public lands, and the federal government initiated range management and conservation programs to deal with the problem. The Taylor Grazing Act of 1934 was the most important of the programs, enacted "to stop injury to the public grazing lands by preventing overgrazing and soil deterioration" on 142 million acres.

Founded and projected on sound conservation principles, in its early stages the Taylor Grazing Act was administered by highly qualified and dedicated advocates. But after a promising start, California rancher Ian McMillan saw the new program run "squarely into the opposition of the grazing industry," which twisted and perverted the movement away from sound range management "into a creature working to foster and promote the very land abuse that it was originally designed to prevent." What McMillan saw was grazier desires translated into agricultural research at the university level, eventually easing apostles of "maximum range use" into academic positions once held by range-science leaders willing to confront overgrazing.[57]

The maximum range-use movement, today more benignly framed as "holistic range management" (distinct from "management-intensive grazing"; see chapter 2) promotes the idea that moderate, controlled cattle grazing encourages native plant diversity and productivity in a variety of ecosystems. Grazing may appear to enhance some plant species' growth and productivity—at least temporarily—but employing cattle grazing for regional land management is highly controversial.[58] Grazing has introduced exotic halogeton and knapweed into arid lands south of Montana and Wyoming, where buffalo never roamed.[59] Both plants are poisonous to both livestock and native animals. Even where native browsers did influence the types and diversity of native plants, we have seen that their grazing habits exerted far less pressure on ecosystems than does domestic livestock.

Conservation confrontation

In 1993, the Clinton administration formulated commonsense regulations for implementing the Taylor Grazing Act. The graziers' resistance to conserving rangelands takes us back to that 2000 Supreme Court decision. Proposed rule changes amended the permit process to make grazing uses conform to general land-use management plans and give equal footing to other land uses. The government proposed dropping a requirement that permit applicants have to be in the business of raising livestock, but made permanent range "improvements" on public land—fences, water developments, structures, and the like—the property of the government. News media reported the administration's proposal to double grazing fees as an outrageous increase, neglecting to mention that the proposed change would raise the fee by only $1.98, to $3.96 per AUM.

In challenging the new rules, pro-grazing groups argued that the Taylor Grazing Act gives permit holders the *right* to continue grazing their animals on public lands *indefinitely*, and a federal judge in Wyoming agreed. The 10th U.S. Circuit Court of Appeals reversed the decision, leaving grazing interests no other option than an appeal to the U.S. Supreme Court. They lost, but whether or not the public eventually will win remains in doubt. After 1994 congressional elections returned a Republican majority, the proposed raise in public lands' grazing fees dropped from view. Before long, Congress further reduced the fee to only $1.35 per AUM. When the George W. Bush administration took office, the BLM returned to protecting the interests of a few powerful ranchers.

Cows or Laws?

A comprehensive socioeconomic review of public lands grazing[60] found that removing all livestock from public rangelands would impose no hardship on local or regional economies. A 1992 GAO report supported that conclusion, finding little potential for economic disadvantage from discontinuing grazing on public lands in any area canvassed.[61] On the other hand, the deterioration of arid public rangelands from livestock grazing is a major public detriment. Another is the financial drain on American taxpayers.

In actuality, the public would *benefit* economically from discontinuing livestock grazing on public lands if the change also eliminated the large grazing subsidies paid out of public funds. For example, 1998 administration costs of the federal livestock grazing programs exceeded ranchers' permit fees by more than $100 million, due in part to $94 million in grazier subsidies. In 1998, another $14 million went for killing predators in 17 western states—thousands of coyotes, foxes, and bobcats, plus hundreds of badgers, mountain lions, and bears. Prairie dogs also are being poisoned for the benefit of cattle and are close to extinction. Prolonged over many decades as they have been, those activities in support of grazing are causing profound changes in the ecological balances of America's public rangelands, an important component of the west's natural capital.[62]

Grassroots protection

The BLM and USFS still do not or cannot exercise their Supreme Court–reaffirmed authority for preserving public rangeland. BLM's staff confided to the GAO that they commonly find themselves unable to protect the land as required by law:[63] in the words of one, "BLM is not managing the permittees; rather, permittees are managing BLM." Local BLM staffers also told the GAO of an area manager who "confronted a rancher he found cutting trees without authorization in a riparian area on BLM land and demanded that the cutting stop. Soon after, complaints about the incident through the rancher's political connections got back to the BLM district manager, who told the area manager to apologize to the permittee and deliver the wood to his ranch." Another area manager "documented numerous instances of riparian area trespass and fence-cutting by a permittee. The area manager said that when he asked the district manager to act on the matter, the district manager stated that he 'would not be a martyr for riparian [habitat].'"

Just how difficult it will be to beat the politician–grazier coterie became abundantly clear in May 2001, when the environmental coalition suit against the BLM, noted above, elicited a federal court order to seasonally remove cattle from 499,000 acres and permanently remove them from 11,000 acres of public desert rangeland in the Mojave Desert's California Desert Conservation Area (now the Mojave National Preserve). Eight affected ranchers protested the order with support from 28 California legislators and the San Bernardino County Board of Supervisors. Assemblyman Phil Wyman stated that the enforcement agreement between BLM and environmental groups "goes to a real core personal liberty issue," which he explained as violating ranchers' constitutional rights to use the land they lease for grazing cattle and sheep. Even though the court order does not infringe the ranchers' right to graze their own private land, their Wyoming-based attorney, Karen Budd-Falen, insisted that "forcing the cattle relocation...constitutes a legal 'taking' of the rancher's private property." The entity described as rancher's private property in this case also can be described as public money supporting the use, or abuse, of public lands.

President George W. Bush and his Interior Secretary Gale Norton also expressed solidarity with the cattlemen's attempt to convert a public lands grazing permit into a lifelong right of access, in spite of the U.S. Supreme Court's unanimous decision. The eight ranchers also had support from San Bernardino County Sheriff Gary Penrod, who terminated his department's agreement to enforce state laws on federal lands in the county, writing that the BLM's order to remove livestock "will directly and negatively impact the very livelihood of California cattlemen, and may result in physical resistance by cattlemen attempting to preserve their stock...Bureau of Land Management Law Enforcement personnel...may be precipitating possible violent range disputes through their official action."[64]

The cattle were to be removed from the disputed public lands on 1 March 2001. The BLM did not enforce the decision, however, and the cattlemen went back to court. All judicial decisions went against them—but when BLM Desert District manager Tim Salt tried to obey the court's orders, he was summarily assigned to a dead-end bureaucratic job in Washington. On 24 August 2001, after a hearing that let graziers present their case, an administrative judge upheld the legal and scientific conservation arguments and required BLM to implement the court order or face contempt charges. After further delays, in 2001 the BLM finally closed one large segment of range to grazing permanently and ordered other permit areas to close seasonally.

In June 2005, the BLM changed tactics, issuing regulations that give new rights to public lands graziers, while cutting out public input on grazing allotment boundaries and the processes of issuing, modifying, or renewing grazing permits and leases. The regulations were withdrawn after agency scientists protested distorting changes by BLM managers. Inevitably, some form of these regulations will be reintroduced.

Sunset on Overgrazing

Continuing current grazing practices likely will change the ecological face of western America in dramatic ways: expanding extreme erosion, drying up more and more streams and springs, extinguishing more species of fish and other riparian animals, replacing most native vegetation with exotic weeds, and threatening more and more bird, insect, reptile, and mammal species. Rancher Frank Bohman showed that cattle grazing does not have to destroy arid lands, however. Once cattle are removed, some areas—especially riparian vegetation—can recover in a matter of a few years. Monitoring riparian zones to prevent invasive species takeovers can return them to healthy, functioning stream courses.[65]

But taking proper care of the public land is more difficult because public lands grazing is unprofitable without taxpayer subsidies—even Bohman's 20–30 cents per acre would strain public grazing's economic viability. The subsidies hardly help small ranchers, the people represented by Marlboro brand's cigarette-smoking cowboy—some of whom actually may pursue the traditional lifestyle that Americans treasure.

In truth, many government handout recipients are billionaires: 10% of both the BLM and USFS grazing permit holders own roughly half of the livestock grazing public lands. To add salt to the wound, the BLM's own assessments show that nearly half of the public parcels that its 20 largest permit holders graze are in unsatisfactory condition, while only about a quarter of *all* BLM public grazing parcels are as badly used.[66] Unfortunately, just removing cattle from the other 99% of public grazing lands will not restore them to health. They are overrun by exotic grasses that replaced many food sources for native animals, creating starvation conditions throughout rangeland food chains.

Current successful efforts to take some upland ranges back from cheatgrass and other noxious invasive grasses may not survive the huge, now-common, state and federal budget deficits. As global climates become warmer, ecological systems could cross critical thresholds, changing plant propagation processes and survival rates abruptly. Those changes may be irreversible, rendering ecosystems unresponsive to any managerial protocols. Our ignorance seriously compromises attempts to judge how close our ecosystems may be to thresholds of rapid or irreversible change, in any case. Sadly, the previous "functioning at risk" and "nonfunctioning" ratings for severely degraded lands rarely changed the management policy on the ground.

Court decisions and appeals are too slow to save ecosystems at the brink. But a new "grassroots" attempt to save public lands from grazing began in Idaho, where state grazing permits are auctioned every 10 years. In January 1994, concerned citizen Jon Marvel offered the high bid of $30 per acre for grazing rights on a 640-acre allotment, for the purpose of keeping cattle off and letting the land rest. Under the usual interpretation of the Idaho law—that uses providing "the greatest long-term financial return" trump all others—Marvel's bid lost to a grazier who offered zero dollars. On appeal, the Idaho Supreme Court found in Marvel's favor, and conservationist bids for state grazing permits subsequently spread to Wyoming, Utah, New Mexico, and Arizona, supported early in the new millennium by the Arizona Supreme Court. The Clinton administration's proposed grazing regulations anticipated this movement, for they allowed nongrazers to apply for and hold federal grazing permits.

In 2005, "the myth-bound West [still insists] on running into the future like a streetcar on a gravel road."[67] The U.S. Supreme Court may have affirmed the managing agencies' authority over permits for grazing livestock on public lands, but any indication that the BLM has the courage to deny permits is yet to appear. If land managers on the ground are unable to exercise the authority granted by courts, the public has only one option: to press for eliminating livestock grazing from public lands. To forestall this level of protection, the BLM seeks instead to prevent public oversight of our incomparable lands.

Jon Marvel believes that Idaho public land "is worth more to people than $1.35 per animal unit month." This expresses public sentiment about federal grazing lands, also. Taking sacred cows and ghosts of cowboys that barely—if ever—existed off the public lands will require enormous pressure from committed citizens who represent the American people, the real owners of public rangelands. If the government succeeds in blocking public oversight, livestock grazing will continue to degrade the range. Only an informed and energized public, backed by its traditional right to protect lands belonging to the whole nation, can break the death grip of that powerful myth and the tragedy of America's grazing commons.

4   Digging to China

Leaving Nevada I thought about those dead lakes shining in the desert sun, the dead birds I had seen in Spokane, the hundreds and thousands of abandoned mines still leaking poisons into the west's water, the sprawling chemical filth of the flats below the Anaconda smelter stack, the blowouts that still corrupt rivers and water tables. At what ultimate cost...have we held so fiercely to this antique law, dreaming the long dream of treasure that I once saluted with such enthusiasm.
T. H. Watkins, "Hard Rock Legacy," *National Geographic Magazine*

Americans like to buy things and own them—barbecues and refrigerators, computers and iPods, cars and bikes, boats and even private planes. Some folks make their appliances last a long time, but manufacturers rely on most people to buy new ones every five years or so. The few critics of our system sometimes charge that items from appliances and vehicles are designed to break down relatively quickly, to prod consumption along. Walking through a showroom or past shop windows, how many people stop to wonder where all the stuff comes from or what happens there? Here is the short answer: Nearly everything you use every day is based on minerals mined somewhere, often leaving behind disfigured land and a toxic mess. Materials still mined in the western United States include metals, particularly gold, iron, copper, zinc, and molybdenum—plus gypsum, borates, and other salts, and most cement ingredients.

Mining is the prow of America's consumer-propelled ship. Its whole purpose is to dig up resources for transformation to consumer goods. But the resources are nonrenewable, so mining progressively eliminates and eventually exhausts them. The processes of exploring for and exploiting mineral deposits consume vast resources also, especially water and energy.

Natural processes spread mine pollution into water, soil, and air, at times killing all life in creeks, streams, and reservoirs. Geographer Lewis Mumford once estimated that "Mining's effects on the earth are now on the same scale as hugely destructive natural forces." He guessed the minimum amount of material moved by global mining operations at 28 billion tons in 1963—nearly twice the sediment all the world's rivers carry annually.[1] Determining just how much land may be affected by mine wastes, and how much farther the damage might spread, is more difficult. The massive scale of today's mining operations dwarfs Mumford's figure.

The dominant U.S. mining law offers wide swaths of U.S. public lands to any and all comers, whether foreign or domestic (box 4.1). The costs of claiming mineral rights and obtaining mining permits are very low, as are the sureties for later cleanup or environmental remediation. Mining companies pay no royalties for depleting the nation's natural mineral capital and polluting soils, water, and air, while today's mines actually drain groundwater from desert areas into the extremely deep pits. When faced with expensive cleanups, mining companies tend to go bankrupt, leaving taxpayers with the bills, the debilitating and degrading environmental effects, and zero potential for any other use of the lands.

Adding to what Garrett Hardin might have termed a *calamity* of the commons, western U.S. "hardrock" metal-ore mines long ago exhausted the richest lodes. To continue extracting metals from larger and larger masses with dwindling metal concentrations, miners depend on relatively low and stable energy prices. Now the era of stable energy prices is ending. Once cheap energy disappears, large-scale mining will go with it, leaving taxpayers with a load of toxic wastes and impossibly high remediation costs.

Everything Comes from Somewhere

Miners say, "If it isn't grown, it's mined," meaning that literally *everything* we use—and use up—is made out of naturally occurring resources (see appendix 3). Throughout human history, people have mined natural resources from the Earth for everything from shelters, tools and utensils, weaponry and vehicles, and art supplies and cosmetics to the fuels for cooking and powering machines. A three-bedroom, two-and-a-half bathroom house of about 2,000 square feet, with two-car garage, central air conditioning, and a fireplace, contains more than a quarter-million pounds of mined metals and other minerals.[2] An average car contains 2,117 pounds of steel (iron variously alloyed with manganese, nickel, chromium, and cobalt), 158 pounds of aluminum, 48 copper, and 18 zinc—all dug out of the earth. A car also contains 225 pounds of petroleum-based plastics, reinforced with 70 pounds of industrial minerals.

All the metals come from "hardrock" ores. Other materials for manufacturing, chemicals, and even foodstuffs come from softer sedimentary rocks—deposits of broken rock or mineral fragments from natural erosional processes, sorted by size and density in flowing water or wind. The commodity mined in the greatest amounts today is mixed sediment of gravel and coarse sand, called aggregate, essential for making concrete and asphalt. Glass, another major building material, is made from soft limestone (calcium carbonate) or trona (a sodium carbonate mineral), and silica derived either from fine quartz sand or feldspar minerals. Paint, paper, and plastics contain limestone, clay, and silica fillers. Cinder blocks are made of volcanic cinder, quarried out of extinct cinder cones, or of cinder and ash residues from human or natural coal burning. The rocks in building facings either are mined from quarries or collected on hill slopes.

We also mine and drill for coal and petroleum (oil and natural gas) hydrocarbons representing the preserved residues of ancient life turned to solid or liquid deep underground (see chapter 12). Petroleum also is the raw material for plastic, drugs, and synthetic fabrics such as spandex and polypropylene (see appendix 3).

Born in Thin Air

The mythic "Golden West" is synonymous with whiskered prospectors, boom towns, and gold strikes. Much of U.S. mineral wealth is gouged out of western mountains and deserts, and large western expanses remain open to mining on lands overseen by the U.S. Department of Agriculture Forest Service (USFS) and U.S. Department of the Interior Bureau of Land Management (BLM) (box 4.1). The public is drawn to deserted mine shafts, monumented mining sites, and ghost towns scattered across our western public lands—all that remains of the legendary age of gold striking, silver loding, rags-to-riches stories, claim jumpers, daring payroll robberies, saloon shootouts, Wild Bill Hickock, and the unsinkable Molly Brown.

Box 4.1 Free For the Taking

The free access principle, "Except as otherwise provided, all valuable mineral deposits in lands belonging to the United States, both surveyed and unsurveyed, shall be free and open to exploration and purchase, and the lands in which they are found to occupation and purchase..." is one of the 1872 Mining Law's most contentious provisions. Supporters argue that small miners and prospectors must have free access to all public lands to uphold mining's preeminence over other land uses. But the free access principle has been "in nearly full retreat almost from the moment it was adopted because of increased sensitivity to nonmineral uses."[a] Just two months after passage, lands were withdrawn to create Yellowstone National Park.

Other significant limits on free access include an 1881 U.S. attorney general ruling that the executive branch can withdraw lands from mining access without congressional authorization, the 1906 Antiquities Act designating national monuments, and President Taft's 1909 withdrawal of petroleum-bearing lands, later upheld by the U.S. Supreme Court. In addition, some 66 million acres are set aside for military and Native American reservations, protecting wildlife and water supplies, and other purposes. Among later actions:

- The 1920 Mineral Leasing Act removed coal, oil, natural gas, oil shale, and phosphates from free access under the Mining Law. In 1924, the Interior Secretary transferred large tracts of energy and fertilizer mineral lands to Mineral Leasing Act jurisdiction.
- In 1922 the Supreme Court ruled that the Mining Law allows disposal of title to federal lands under homestead and various other acts, excluding broad categories from free access. Congress previously had used this disposal element to deny mining access.[b] Some denials applied to whole states.

continued

Box 4.1 Continued

- A raft of laws removed free access for mining a variety of commodities between 1947 and 1955, creating the common rock and mineral deposits sales system under the Department of the Interior. In 1962, scientifically valuable petrified wood sites, mined for sandpaper grit, were withdrawn. The 1953 Outer Continental Shelf Lands Act set up a discretionary permit/leasing system for mining hardrock minerals under ocean water, thus denying free access.[c]
- The 1964 Wilderness Act denied free access to minerals in designated national forest sites but did not take effect for 20 years. In 1968 Congress withdrew designated wild river corridors.
- The 1976 Federal Land Policy and Management Act (FLPMA) reinforced executive authority to withdraw lands from mining access regardless of minerals types or values, and also incorporated restrictions for BLM lands, following the Wilderness Act's example.
- The 1980 Alaska National Interest Lands Conservation Act formally closed hundreds of millions of acres to free access, also regulating and restricting access to millions more.

Notes

[a] J. D. Leshy. *The Mining Law: A Study in Perpetual Motion* (Washington, DC: Resources for the Future, 1987), 45.

[b] For example, the 1911 Weeks Act to acquire logged lands for watershed preservation did not allow disposal and so denied free access. In 1888, Congress barred free access on all lands suitable for irrigation works, later reinforced by the Reclamation Act of 1902.

[c] In 1954, when the uranium boom brought the Mineral Leasing Act into conflict with the Mining Act's free access doctrine, the Multiple Mineral Development Act restored a limited form of free access to Mineral Leasing Act lands.

Few Americans realize that the romances woven around events of the ghost towns' glory days mostly were fictions, written and published by easterners for gullible eastern U.S. readers. The reality was grimmer. Few prospectors had the money for mining and smelting operations, even if they did hit a bonanza. Most prospectors never became rich from their own gold or silver strikes and had to sell their claims—very often to boom-town shopkeepers who had grown rich on prospectors' grubstakes. California's Leland Stanford, Sr., for example, made his fortune from a Gold Rush-era general store. He went on to become president of the Central Pacific Railroad, California's first Governor, and eventually a U.S. Senator.[3]

Most western lodes were mined out quickly and abandoned. Only a few of the west's mines remained active to the end of the nineteenth century, and most mining towns "went bust" within five years of their birth, leaving behind the now-romantic ghost towns. The miners and investors abandoned workings, wastes, tailings, and smelter slag piles to the forces of nature, along with the poorly constructed towns and trampled camp sites (see chapter 5).

American lifestyles constantly escalate consumption, driving increases in the scale of mining operations as resources diminish. The rich ores are long gone. The 1850-era miners produced an ounce of gold from two to three tons of rock—a yield of one-third to one-half ounce of gold per ton of waste—which translates to 64,000–94,000 times more waste than gold. When rich seams gave out, the miners dug and blasted into less mineralized wall rock zones. Likewise, copper ores contained about 8% metal four centuries ago. Even back in the 1880s, miners were forced to excavate large open pits to dig up lower grade copper ores with early steam shovels. By the start of the twentieth century, the lowest economically mineable grade of copper ore yielded only 3% metal, while today's ores typically contain less than 1%.[4]

By the mid-twentieth century, America's growing hunger for minerals had scooped out the less mineralized zones, too (table 4.1). Huge mining machines then gouged huge open pits into quite sparsely mineralized zones, greatly expanding the scale of mining. Today's gold miners excavate immense rock or sediment

Table 4.1 Projected Lifetimes of U.S. Mineral Reserves (Short Tons)

Material	Domestic Consumption	Domestic Production	Reserves[a]	Lifetime[b] (Years)
Bauxite	4,600,000	NA[c]	22,000,000	5
Chromium	550,000		0	0
Cobalt	12,000		0	0
Copper	3,400,000	1,600,000	50,000,000	14
Iron ore	83,000,000	67,000,000	3,000,000,000	40
Lead	2,000,000	529,000	7,000,000	3.7
Manganese	876,000		0	0
Nickel	174,000		47,400	0.3
Tin	68,000	9,800	22,000	0.3
Zinc	1,800,000	948,000	28,000,000	15
Gold	NA	360	6,200	17[d]
Silver	8,500	2,300	36,000	4
Gypsum	44,000,000	28,000,000	772,000,000	17.5
Phosphate	44,000,000	44,000,000	1,000,000,000	25
Potash	6,000,000	1,000,000	110,000,000	20
Sulfur	15,000,000	11,000,000	154,000,000	10.5

U.S. Geological Survey. *Mineral Commodity Summaries 2001* (2001). One short ton equals 2,000 pounds. Lifetime calculations assume complete reliance on domestic reserves at year 2000 consumption rates.

[a] Reserves are resources that can be extracted economically under current conditions.
[b] Projections based on 2000 consumption rate, except as noted. Because the trend of consumption was upward, it is conservative to assume that the 2000 rate will continue. More realistic lifetimes are shorter.
[c] NA, data not available.
[d] Consumption data not available—lifetime estimate based on domestic production.

masses containing meager amounts of gold in tiny-to-microscopic grains and flakes, which make yesterday's "low-grade" ores look rich in comparison. What now passes for gold ore produces one-hundredth (0.01) of an ounce or less per ton of waste rock. This means that 100 tons of barely mineralized rock produces an ounce of so-called "frog's hair" gold, too small to see without magnification. The mines now yield more than three million times more waste than gold.[5] Currently, the United States is digging up the rest of the world's mineral wealth (figure 4.1).

Partly driven by low ore grades, today's mining is colossal in scale. Machines larger than King Kong extract gold, copper, and other metal ores out of hard

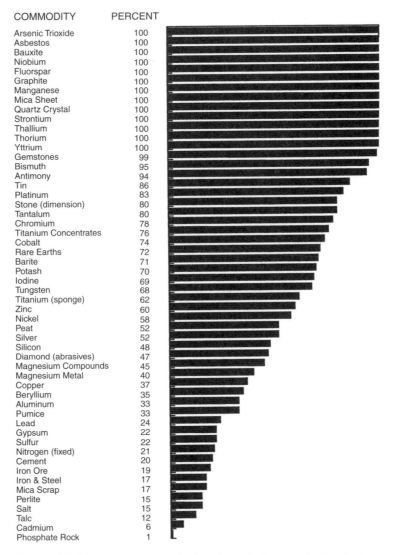

COMMODITY	PERCENT
Arsenic Trioxide	100
Asbestos	100
Bauxite	100
Niobium	100
Fluorspar	100
Graphite	100
Manganese	100
Mica Sheet	100
Quartz Crystal	100
Strontium	100
Thallium	100
Thorium	100
Yttrium	100
Gemstones	99
Bismuth	95
Antimony	94
Tin	86
Platinum	83
Stone (dimension)	80
Tantalum	80
Chromium	78
Titanium Concentrates	76
Cobalt	74
Rare Earths	72
Barite	71
Potash	70
Iodine	69
Tungsten	68
Titanium (sponge)	62
Zinc	60
Nickel	58
Peat	52
Silver	52
Silicon	48
Diamond (abrasives)	47
Magnesium Compounds	45
Magnesium Metal	40
Copper	37
Beryllium	35
Aluminum	33
Pumice	33
Lead	24
Gypsum	22
Sulfur	22
Nitrogen (fixed)	21
Cement	20
Iron Ore	19
Iron & Steel	17
Mica Scrap	17
Perlite	15
Salt	15
Talc	12
Cadmium	6
Phosphate Rock	1

Figure 4.1 Net U.S. imports of selected nonfuel materials, 2000. From U.S. Geological Survey. *Mineral Commodity Summaries* 2001 (2001), 5.

rocks, and phosphates, borates, clay, and potash from softer deposits. Giant machines quarry sand and gravel out of riverside terraces, dredge streams for gravel-sediment "placer" metal deposits, strip mountain tops for coal, and drill ever deeper oil exploration holes in even the most difficult and hostile surroundings. The pits go more than 2,000 feet deep, dismantling whole mountains and burying landscapes under millions of tons of waste rock and processed mine tailings.[6]

Mining pits have swallowed surrounding landmarks and Native American cultural sites, and even have eaten whole towns. Santa Rita, New Mexico, is one of the missing settlements. Trying to walk downtown today, you would find yourself flailing air over a miles-wide, man-made crater—much like Wile E. Coyote in a Roadrunner cartoon. Appropriately, Santa Rita natives call themselves "The People Born in Thin Air."[7]

Mining speculations

In theory, demand should determine the prices of mined materials and the pace of mining, but historically, metal price variations—especially for gold—have been driven more by politics and fear. Examples include a decline in western U.S. gold mining after 1971, when the U.S. government abandoned the post-World War II modified gold standard and the fixed price of $35 per ounce, to fight currency inflation. Middle Eastern political and economic unrest spurred rapid gold price increases, from $103 an ounce in 1976 to $850 in 1980. The price surge powered a huge expansion in open-pit gold mining and led to new techniques for extracting metals from extremely low-grade "ores." Similarly, wild copper price fluctuations in the 1970s sent a lot of people out to dig and explore for rock with only 0.2% metal or less.

Average gold prices generally have declined more recently—to $350 per ounce in 1990 and then to $280 or less by 2001. The 2001 low price of about $240 per ounce barely covered mining and processing costs. In 2003, depressed stock market conditions and the U.S. invasion of Iraq pushed the price of gold higher again, and in 2007, highly volatile gold prices have fluctuated between about $750 and $850 per ounce, reflecting world angst over the coming end to cheap oil supplies.

Mining is one of the most energy-consuming industries in the United States and most of the world.[8] Given no energy limitation, the bare traces of industrial minerals present in normal, unmineralized rocks eventually would become ores suitable for mining, everybody's back yard would become a mine pit, and we would all become "people born in thin air." But the behemoth mining machines run on fossil fuels, so going for ever-lowering ore grades is sustainable only as long as energy prices prop up mining operations. The United States now relies on imported oil and faces stiffer competition for it. World oil production no longer can expand readily to meet demand, making oil prices increasingly volatile. These energy pressures are fundamentally changing mining's economic equation (see chapter 12, appendix 9).

Mining the Environment

Excavating gigantic pits and stripping mountain tops have direct and potentially devastating impacts on land and water at mine sites. Natural processes spread indirect erosional and polluting impacts to surrounding lands and waters and degrade air quality over even greater distances. In many western states, digging volcanic cinders for construction aggregate and cinder blocks eventually obliterates small extinct volcanoes. In California's Mojave Desert, vehicles that collect blocks of decorative stone for ornamental building facings have imprinted hill slopes with extensive, eroding dirt trail networks. Historic dredge mining for "terrace placer" gold and other precious metals in dry river banks or abandoned river channels still disfigures large tracts.

Mining proponents commonly state, or imply, that mining damages only a tiny part of a vast public domain. "Approximately .06 percent of BLM lands are affected by active mining and mineral exploration,"[9] reads one National Research Council report. That figure omits an estimated 557,000 abandoned mines, however, including dozens of Superfund sites, many thousands of miles of abandoned mine roads— and thousands of abandoned and inactive underground mines (figure 4.2).[10] The National Research Council figure also excluded nonmetallic mineral exploitation and exploration, and extensive zones gouged by sand and gravel mining or polluted with smelting and acid mine drainage waters.

The massive scale of modern mining both consumes and pollutes immense amounts of water. Quarrying stream sediments for concrete aggregate pockmarks river and stream channels and floodplains, degrading water supplies and impairing natural stream processes (see chapter 13). Rains soak into immense piles of mine wastes, laced with toxic materials, and into the dams built to contain the

Figure 4.2 Abandoned coal mining lands in the 11 western states. From U.S. Department of Agriculture Soil Conservation Service. *America's Soil and Water— Condition and Trends* (1981), 18.

Figure 4.3 Subsidence pits and troughs over abandoned underground coal mines in Sheridan County, Wyoming. Photographed by C. R. Dunrud.

wastes—too often generating toxic mudflows that wipe out whole communities. Like the highly controversial "mountain-top removal" coal mining in eastern states, western strip mines also fill valleys with wastes and destroy creeks and rivers.

Yesterday's underground mines were notoriously dangerous to work in. By now the old ceiling props are decayed and the abandoned tunnels are prone to collapse (figure 4.3). Decades after abandonment, underground coal mines become sites of unquenchable underground fires when remaining coal seams come in contact with air and spontaneously catch fire. The open hardrock and coal mine pits also have unstable walls that constitute human and wildlife hazards, especially the ones abandoned without adequate safety precautions. Abandoned mines do provide alternative habitat for the west's beleaguered bats, however.

Exploring impacts

The gargantuan scale of modern-day prospecting for very low-grade "ores" matches the mines themselves. Environmental harm extends from large networks of seismic lines, exploration roads, drill pads, and utility corridors. No matter what the purpose, all roads strip surfaces bare of vegetation and have the multiple impacts of promoting accelerated erosion and flooding (see chapter 5). Roads also segment and destroy wildlife habitat and introduce weeds. Abandoned and uncontrolled mine roads open remote areas to recreational off-road vehicle trespass, expanding motorized vehicle damage (see chapter 11). The promise of riches still lures naive investors to scam mining operations, which leave scars on the land and investors alike. Even mines that never produce ore leave the enduring blemishes of roads, pits, trenches, wastes, and polluted water.

Mineral exploration areas must be claimed or marked in a prescribed fashion, and even this seemingly innocuous process can severely affect local environments. Claim markers (stakes) can be sticks, fence posts, metal rods, or piled rocks. Wooden posts formerly were the most common type of claim stake, but around 1970 some prospectors and mining companies began staking with cheap PVC pipe, including four-inch-diameter perforated sewer-line pipe. All stakes are driven into the ground and extend about four feet above ground level.

Biologist Lawrence LaPre discovered that the sewer pipe claim stakes were killing many species of wildlife. For example, he found clusters of as many as five dead birds in a single hollow claim marker—the victims all were small birds, probably seeking cavity-nesting sites. After entering the sewer pipe, a bird could not open its wings wide enough to fly out and became trapped inside. The exhausted creature eventually fell to the ground and died of starvation. LaPre also found dead lizards, various insects, and many beetles trapped in perforated pipes by the hundreds—also piles of dead bees (figure 4.4). He believes that tens of thousands of songbirds and small animals have died in plastic claim markers.[11]

Figure 4.4 Songbird that died in a PVC claim marker, California Desert Conservation Area. Photographed April 1991.

Gold Rush legacy

The wounds and wastes of old mining practices are still with us, especially from 1840s to 1850s California Gold Rush hydraulic "placer" gold mining. Powerful pumps and highly pressured water cannons blasted the streambeds and banks, reaming all potential gold-containing sediment and other precious metal concentrations (placers) out of streams and stream terraces and wreaking environmental devastation. All told, hydraulic gold miners dislodged 13 billion tons of rock, enough to fill 54 million of today's king-size 240-ton capacity ore trucks, each three stories tall. More than two billion tons of wastes flushed downstream into northern California's Sacramento River and into the upper parts of San Francisco Bay. Heaps of mine debris piled up at stream junctions, reducing life-sustaining habitat for wetland plants and animals. Countless stream dredging operations similarly disfigured smaller drainages over much of the west.

The masses of debris caused long-lasting changes to river shapes and flows,[12] raising the Sacramento River above its original bed an average of seven feet, and the tributary Yuba River 110 feet. These changes resulted in severe flooding. To protect themselves, Yuba and Feather river towns had to build levees to almost the height of their house rooftops. Eventually, hydraulic gold mining debris destroyed 400,000 acres of prime Sacramento Valley farmland in California, an area comparable to Hawaii's island of Maui, and took another 270,000 acres out of production. This collision with the greater economic and political force of farming halted hydraulic placer mining in 1884.[13] But after more than 100 years, the Yuba River bed still is 65 feet higher than before the age of hydraulic gold mining.

Pit Wounds

Each American consumes more than 1,300 pounds of metals and 21,000 pounds of nonmetallic minerals every year, on average.[14] Supplying enough low-grade mineralized rock to meet this demand requires miners to dig superlatively enormous pits. The British-owned Kennecott Copper Company's Bingham Canyon copper mine in Utah has removed more than 3.3 billion tons of soil, rocks, and dirt, for example. Digging the Panama Canal removed seven times *less* material. Kennecott now excavates waste rock and ore at a rate of about 500,000 tons daily, enough to fill 5,555 railroad cars on a train 50 miles long. To haul away the wastes, 30-foot tall ore trucks remove 2,100 loads every day. By 1980 the growing Bingham Canyon pit had swallowed two towns, and experts forecast that it eventually will raze the entire Oquirrh Mountain range.

Another huge pit, the Betze-Post Mine in Nevada, removed 159 million tons of rock in 1997 alone, yielding 1.6 million ounces of gold at 0.01 ounce per ton. When the mine runs out of ore in a few years, the Betze-Post pit will be nearly two miles long, three-quarters of a mile wide, and about a third of a mile deep— able to hide a 30-block patch of New York City inside it, including hundred-stories-tall skyscrapers.

Smaller open-pit mines provide nonmetallic silicate, phosphate, carbonate, clay minerals, and many other commodities. For example, the borate mine near Boron, California, is relatively modest at only a mile and a half long and 700 feet deep.[15] Pit mines for phosphates in Idaho and adjacent states may be 1,000 or more feet deep and several miles long.[16] These excavations wreak environmental havoc comparable to hardrock mine pits and also pose expensive cleanup problems.

Dewatering arid lands

Immense open-pit mines reach far below the water table, the top of the local zone of underground water—some Nevada gold mining pits go as much as 1,000 feet lower. This means that groundwater continually drains into the huge holes. To keep mining, water pumps must operate continuously to keep the water table below the pit bottom. At the larger Nevada gold mines, extensive well systems keep mine pits dry but use enormous amounts of energy. Dewatering a single mine pit can use enough electricity to power a city of 100,000—about the size of Provo, Utah.

Mine dewatering pollutes and depletes groundwater, decreasing the water flowing to spring-fed streams and wetlands—critical resources in arid lands (see chapter 9). Mine dewatering will dry up at least six streams in northern Nevada's gold-mining district, and more than 200 springs may soon disappear or suffer reduced flows. Diminished flows and dried river channels will destroy fish species and other aquatic life at or near the base of desert food chains. People, too, will see reduced water supplies. The five largest flood-prone Nevada mines pumped more than 770,000 acre-feet of groundwater between 1997 and 2000, enough to supply 1.5 million families for a year.

Pumping groundwater lowers subsurface water levels in a conical zone around every well—the so-called "cone of depression," with the lowest water level centered on the well (figure 4.5). Numerous wells in the same area, all pumping for long periods, tend to develop overlapping cones of depression, which can lower the water table across broad regions. For example, the overlapping cones of depression at a number of closely spaced northern Nevada mines have depressed the water table over hundreds of square miles. Groundwater monitoring at the Betze-Post Mine in 1993 showed that local water tables had fallen 800 feet below the regional water table. In 1998, Gold Quarry Mine dewatering had lowered the water table 600 feet in a three-mile-long elliptical zone beneath the mine.[17]

Polluted mine water can poison aquatic life and also threaten human water supplies and health. Mines take the cheapest available option for disposing of wastewater, including toxic water draining from mines and mine wastes (see below). Some of the water pumped out of flooding Nevada gold mines is reused for mine operations, and some goes to irrigate crop lands. Nevada also requires putting some toxic and acid mine water into infiltration ponds to let it soak back into the ground through salty soils. The acid, salt, and other toxic materials in this water can make the groundwater in shallow aquifers unfit for drinking or irrigation and destroy their future potential for potable water storage.[18] Any remaining water—generally the largest part—gets dumped into streams, which flow into the Humboldt River and then to the Humboldt Sink northeast of Reno. The water evaporates,

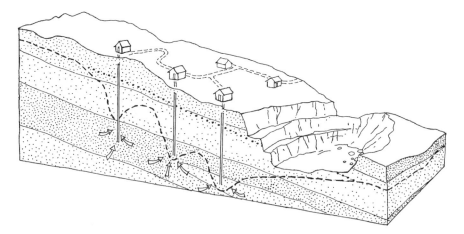

Figure 4.5 Sketch of groundwater drawdown cones (dashed line), which are depressions in the water table produced by pumping water out of wells faster than rainwater seeping from the surface can resupply (recharge) it to the water-storing aquifer. The dotted line is the original water table before mining and pumping. Arrows show groundwater flows toward active pumps.

leaving a surface coating of toxic salts for winds to blow around, spreading contamination farther.[19]

By 2020, these Nevada gold mines likely will run out of ore, and the pits will gradually fill with water to form lakes as much as 1,000 feet deep, some of them toxic. More than 150 years after dewatering ends, or maybe longer, the groundwater cones of depression will eventually disappear, but the dewatering effects will continue to deplete and pollute springs, streams, rivers, and wetlands for many hundreds of years. In Nevada's hot, dry climate, the huge amounts of water that evaporate from mine-pit lakes will keep water tables permanently depressed, preventing the groundwater from ever being replenished. Summarizing all these effects, University of Nevada Professor Glenn Miller concluded, "We are getting…20 years of tremendous economic benefits for Elko and enormous water problems for the next thousand years."[20]

Aggregate damage

Every person in the United States consumes an average of 1.7 million pounds (850 tons) of aggregate in a lifetime, dredged either out of active river and stream channels or floodplains, or from former floodplains (terraces). Aggregate also can come from quarrying and crushing massive rock materials; from talus, which is the broken rock mantle on steep, rocky hill slopes; from gently sloping alluvial fans linking mountain slopes with broad desert valleys; and from old volcanic cinder hills. Crushed stone sources once supplied about the same aggregate volumes as natural sand and gravel deposits, but over the last two decades the negative impacts of river channel mining on water supplies and the Pacific Ocean salmon fishery have lessened the amounts produced from rivers.[21]

A flowing stream possesses energy due to the water's motion. Normally functioning streams erode sand and gravel upstream and then deposit the sediments downstream where the water slows at bends or barriers. Coarser gravelly materials pile up in gravel bars or fill holes in the channel. When the streams rise out of their banks in floods, they deposit finer sediment across their floodplains (see chapter 13). Many western arid-land streams flow only during winter rains and carry low (or no) water during the summer. In-stream aggregate mining generally takes sediments directly out of active channels in the dry summer season, when stream water is most easily diverted around a quarry area.

Not surprisingly, removing gravels from active rivers and floodplains disrupts a stream's dynamic process of erosion, transportation, and deposition. When winter rains fill an active channel, the water flows into and through in-stream quarries, eroding both the upstream and downstream pit walls. Sediment from the upstream wall falls directly into the pit, giving the river extra energy for eroding the downstream pit wall. Over time, this process gradually smoothes out the holes in the river bed. But in the meantime, the river can pick up and carry more downstream sediment than usual from gravel bars, river banks, and even from its own bed—a process called lateral scour.

The accumulated impacts of in-stream quarrying operations degrade river channels regionwide long after mining ends. For example, by the mid-1990s, the erosion from large scale in-stream mining operations had eliminated all of the large migrating gravel bars from lower reaches of northern California's Russian River. Along many stretches, the river flowed directly on bedrock, while lateral scour carved away farmland and riparian (riverside) habitat and attacked major bridge abutments. In-stream sieving and crushing operations also increase water turbidity, which adds problems for fish and other aquatic life. During floods, aggregate quarries in floodplain terraces can have the same effects as in-stream mining. Flood waters surge into the quarry pits, eroding the upstream and downstream quarry walls and creating lateral scour on floodplains until the waters recede.

These days, aggregate miners must get permits for extracting sand and gravel on public lands. Depending on the state and particular site, in-stream gravel mining may require permits from local city or county agencies or from state fish and game regulators. In the case of streams where salmon spawn, the federal National Marine Fisheries Service must approve gravel mining permits. But numerous illegal aggregate-mining operations still devastate streams, floodplains, and older river terraces.[22]

Because active streams and rivers add sand and gravel to their channels, gravel miners commonly claim that in-stream aggregate is a renewable resource. This claim is misleading for most rivers, which have dams that catch most of the sediment. Monitoring data for dammed rivers show that the annual amount of naturally resupplied aggregate generally is much lower, sometimes by 10-fold or more, than mined amounts. Also, the stream channel and floodplain gravels are important water-storing aquifers. In the drier western U.S. climates, the losses of aquifer-storage capacity due to in-stream and floodplain gravel mining probably are enormous. For example, both in-stream and floodplain gravel mining in the Russian River has diminished its water storage capacity by an estimated 350,000 acre-feet since World War II.[23]

After prohibiting Russian River in-stream and terrace quarries in 2001, local officials temporarily allowed miners to skim sediment off the tops of gravel bars, an approach with lesser potential effects than digging pits. But taking sediment from gravel bars still reduces water storage capacity and contributes to lateral scour. Land holders in the Russian River floodplain are seeking permits to armor river banks with large "rip rap" boulders, hoping to stop erosion at their riverfronts. Obstructing the river from eroding one segment of its bank will only move the scour points, however, caving in downstream river banks and bringing more landowner pleas for intervention.

Mining talus deposits, and digging relatively deep pits in ancient alluvial deposits for aggregate, also can destabilize drier landscapes. In desert areas, the aggregate for road base and ballast has come from innumerable so-called "borrow pits," mostly on public lands along western roads, highways, and railroads.[24] The loan is never repaid, and many borrow pits are now abandoned to erosional forces. Some pits are small, ragged scars, while others cover many acres and are visible from roads and highways. Pits along some major highways have been reshaped to smooth out deep cuts but are largely barren of vegetation. In arid lands, the bare scars will remain centers of accelerated erosion indefinitely.

Mining Plagues

The nature and scope of mining and processing pollution depend on the type of ore and the methods for extracting target minerals. As early as 1550, Georgius Agricola described the indirect impacts and toxic consequences of mining metals:

> The fields are devastated...the woods and groves are cut down, for there is need of an endless amount of wood for timbers, machines, and the smelting of metals. And when the woods and groves are felled, then are exterminated the beasts and birds....Further, when the ores are washed, the water which has been used poisons the brooks and streams, and either destroys the fish or drives them away.

Agricola's words can be applied fairly to the west's major hardrock mining sites, where the wastes from nineteenth- and twentieth-century mining, and the expanding scales of modern mine operations, still poison the environment at large. Toxic wastes emanate from mines along Silver Bow Creek and other tributary streams in Montana's Clark Fork of the Missouri River headwaters, from Colorado's Summitville mining district among many others, and from New Mexico's Molycorp mine, Utah's Kennecott copper mine, California's Iron Mountain mine, the Coeur d'Alene mining district surrounding Kellogg, Idaho, and tremendous cyanide "heap-leach" gold mines in Nevada, Arizona, and California. The land, water, and air pollution from many of these mines have made them unpopular, even in tradition-oriented western communities.

Toxic headlines

Hardrock mining for metals contributes the highest levels and most severe types of toxic industrial pollution in the United States (figure 4.6). Metals mines account for 97% (464 million pounds) of the nation's total of arsenic compound releases, 92% of copper compound releases (1.3 billion pounds), 65% of zinc (706 million pounds), 87% of mercury (nine million pounds), 81% of lead (308 million pounds), and 64% of chromium (96 million pounds).[25] Mining for copper and other metals in Montana's Clark Fork region has spread more than 10 billion cubic feet of waste rock over nearly 2,500 acres—more than 2,000 football fields. This vast array of abandoned mine sites constitutes the nation's largest Superfund site.

According to the U.S. Environmental Protection Agency (EPA), 40% of western headwater streams are polluted with hardrock mine wastes containing the metals listed above. All can cause poisoning and chronic debilitation and a number of serious human diseases. In 1985, the EPA rated ecological risks from mining waste as second only to global warming and stratospheric ozone depletion, concluding "with high certainty" that mining waste releases "can result in profound, generally irreversible destruction of ecosystems." Cases in point include biologically dead rivers in Colorado and New Mexico. Of 10 states most affected by toxic chemicals from hardrock mining between 1998 and 2001, seven were in the western United States. In all three years, Nevada took first place, with Utah second and Arizona coming in third (table 4.2).

Metals are essential catalysts for the chemical transformations that support life but become toxic above the required levels for bodily functions. Even very small overdoses can disrupt an animal's or a plant's life processes.[26] Before industrialization, meager human diets provided very small amounts of the metals, so animal and plant organs

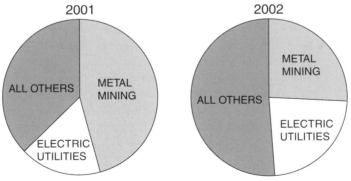

Figure 4.6 Proportions of toxic chemicals released by hardrock mining and electric utilities, 2001 and 2002 (in millions of pounds). The major 2002 decline in chemicals from metal mines is due entirely to a court decision that eliminated the category of untreated waste rock from federal inventories. From U.S. Environmental Protection Agency. *Toxic Release Inventory.* Available: www.epa.gov/tri.

Table 4.2   On- and Off-Site Toxic Releases from Coal and Hardrock Mining, 2001 and 2002 (Pounds)

State	Metal Mining		Coal Mining	
	2001[a]	2002[b]	2001[a]	2002[b]
Arizona	543,256,000	34,821,000		
California	1,447,000	2,510,000		
Colorado	22,019,000	5,916,000	2,427,000	2,447,000
Idaho	21,212,000	18,641,000		
Montana	24,682,000	18,536,000		
Nevada	773,436,000	486,842,000		
New Mexico	95,555,000	3,507,000	4,743,000	4,096,000
Oregon				
Utah	189,317,000	114,382,000	20	
Washington	199,000	93,000		
Wyoming			550	
Total, 11 states	1,671,000,000	685,248,000	7,171,000	6,543,000
Total, United States	2,270,000,000	1,299,000,000	16,166,000	15,599,000
% United States[c]	74	53	44	42

[a] U.S. Environmental Protection Agency. *2001 Toxics Release Inventory* (2003). Available: www.epa.gov/tri.
[b] Metals mining toxic releases omit toxics released in waste rock. They do not represent a reduction of actual toxic releases compared to 2001.
[c] Total mining production for the 11 states as a percentage of the U.S. total.

evolved efficient collecting and storage processes. Now that mining has vastly expanded metal sources, many animals and plants take in much more than their internal systems can handle—but we and other animals have no means of eliminating the excess. Not surprisingly, the Clark Fork region's metal-mining towns, such as Butte and Anaconda, Montana, have markedly high human death rates—due apparently to high rates of serious diseases, including cardiovascular and kidney diseases. In 1950–1951 and 1959–1960, the National Institutes of Health ranked Butte's mortality rates the highest of 480 U.S. cities. It ranked fifth highest in 1969–1971, as mining began winding down.[27]

Acid words and water

Even small-scale mining turns Earth's crust inside out, with disastrous environmental and health consequences. Ores of copper, gold, silver, lead, mercury, cadmium, and zinc formed underground, generally in combination with sulfur (sulfides).[28] Deep underground, the metals cannot easily combine with oxygen to contaminate surrounding rocks and groundwater. At the surface, complex sulfide interactions with oxygen or water determine the metals' fates and toxicity.[29] In an oxygen-rich environment, water most commonly leaches (dissolves and removes) sulfur from the ores, to form "acid mine drainage"—acidic water that drains from mines and mining wastes alike.[30] Acid mine drainage leaches many toxic metal compounds out of minerals even more effectively than do natural waters.

A solution's pH expresses its acidity or alkalinity—alkaline waters have pH values greater than 7, and acid waters have pHs less than 7. Acid drainage from residual silver, gold, copper, lead, and zinc sulfide minerals at California's abandoned Iron Mountain mine Superfund site generate some of the most highly acidified waters ever reported in a natural setting, with pH values as low as *minus* 3.6.

From Waste to Contamination

Metals mining's dominant *primary* contamination comes from waste rock—both untreated waste rock, too poor for processing as ore, and discarded tailings from processing anything from rich ores to slightly mineralized rock (figure 4.7).

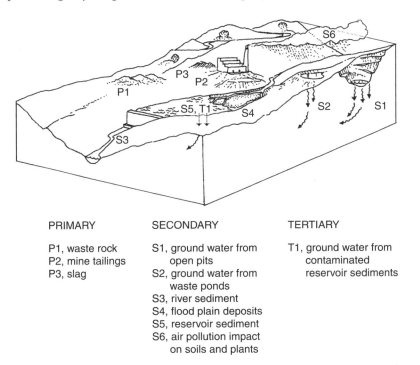

PRIMARY	SECONDARY	TERTIARY
P1, waste rock	S1, ground water from	T1, ground water from
P2, mine tailings	open pits	contaminated
P3, slag	S2, ground water from	reservoir sediments
	waste ponds	
	S3, river sediment	
	S4, flood plain deposits	
	S5, reservoir sediment	
	S6, air pollution impact	
	on soils and plants	

Figure 4.7 Primary, secondary, and tertiary contamination associated with mining and smelting of metallic ores. Primary contaminants come from mining and processing ores; secondary, from leaching and flood erosion of primary contaminants; and tertiary, from extended distribution of secondary contamination, here shown as seepage into groundwater from a waste impoundment. Arrows show seepage into groundwater at all stages. Modified from J. N. Moore and S. N. Luoma. Hazardous Wastes from Large-Scale Metal Extraction. *Environmental Science and Technology* 1990;24:1279–1285. Copyright 1990, American Chemical Society. Reproduced with permission.

Eroding exploration roads and small prospect pits are additional *primary* mining contamination sources (see chapter 5).

Natural chemical processes attach toxic metals to sediment particles, and earth surface processes spread them into soils, stream sediments, and surface and groundwater—yielding a legacy of *secondary* contamination. Rainwater flushes acidic water and polluted sediments into streams, especially during intense storms. The massive amounts of sediment eroded from mines and waste piles cause flooding along the rivers, and retreating floodwaters deposit contaminated sediments on river floodplains. Many rivers carry the contamination into downstream lakes and reservoirs. Acid waters also convey mine pollution into soils, and natural seepage adds them to groundwater. Winds lift dust particles off waste piles and blow them long distances suspended in air, along with acid smelter gases and soot particles. The result is extremely contaminated landscapes, soils, rivers, and groundwater in virtually every western mining district (figures 4.8, 4.9).

The wastes from soft-rock mining, such as phosphate-rich Phosphoria Formation in Idaho, also spread toxic metals. Phosphate mines contaminate surface and groundwaters, vegetation, fish, birds, and other animals with poisonous selenium and myriad other toxic trace elements.[31] The selenium is concentrated mostly in phosphate-poor layers, so it ends up in mine waste piles or escapes as vapor.

Figure 4.8 Erosion of waste rock dumps at a Globe, Arizona, copper mine. Photographed May 1984.

Figure 4.9 Wind erosion on abandoned uranium mill site north of Tuba City, Arizona. A light-colored plume of contaminated sand blowing from the right partly fills large and small impoundments, buries the fence line, and extends beyond site boundary (lower left). Photographed October 1984.

Great gray heaps

Digging immense rock volumes from hard-rock mining pits to get relatively tiny amounts of metal piles up literal mountains of waste rock, representing the many millions of waste ounces for every ounce of metal produced (a million ounces equal 31 tons). Throughout the United States, but mostly in the west, mines produce about 1.7 *billion* tons of solid waste each year (box 4.2), nine times the nation's already-titanic production of municipal solid wastes (see chapter 10). Mineralized ore or rock is crushed to the consistency of sand and then either roasted or soaked in hazardous chemical solvents, such as cyanide, to leach out the target minerals. The treated waste (tailings), together with untreated waste rock, account for 97–99.9% of all the rock and earth dug out of a mine pit. Both treated and untreated wastes are toxic.

The amounts of untreated mine waste rock vary but typically are about 50 to more than 200 million tons in a mine's lifetime. In many cases, these wastes represent as much as or more than three-quarters of all excavated material. The untreated rock generally gets dumped in huge piles that effectively bury landscapes, including wildlife and archaeological sites, beneath hundreds of feet of broken rock (figure 4.10). As early as 1912, Butte, Montana, mining geologist W. H. Weed noted that "great heaps of gray waste rock from the mines" had replaced vegetation as the "most conspicuous feature of the landscape."[32]

Box 4.2 Prize-Winning Polluter

The U.S. Environmental Protection Agency's (EPA) Toxics Release Inventory (TRI)[a] did not include mining contamination data until 1998, just after Kennecott Copper Company's inadvertent disclosure that its Utah mine had released more than 158 million pounds (79,000 tons) of toxic waste. That *single* mine ranked fourth among 19,000 polluters in the 1997 TRI.[b]

The 1998 TRI recorded more than three billion pounds (1.5 million tons) of toxic chemical releases from hardrock mining sites in the 11 western states—enough to fill more than 6,000 three-story-tall haul trucks. The western releases also represented nearly 90% of all toxic metal-mining releases in the United States, and more than 40% for all U.S. industries.[c] In 1998, the Kennecott mine ranked top for its reported release of 405 million pounds, then outdid itself in 1999 with more than a billion pounds of toxic releases. Kennecott worried, "Taken out of context, the TRI information will give a false impression of the mining industry" and inaccurately claimed that the releases "do not reflect any increased danger to public health or the environment."[d]

The mining industry has improperly claimed that the toxic releases in EPA reports represent the total of their untreated waste rock. In truth, the millions of pounds of chemicals in TRI reports reflect only toxic materials *contained* in waste rock, not the rock itself. Total waste rock volumes are greater by far. The industry and supportive politicians also have tried to minimize the releases by calling the toxic chemicals "naturally occurring" compounds, which cannot be recovered by extraction processes and constitute only a few pounds of material per ton of rock. These are half-truths, told to avoid the crucial fact that broken rock masses are highly vulnerable to natural leaching, which disperses toxic chemicals into the environment. In 2003, a federal court fell for the half-truths, ruling that mining companies do not have to report the "naturally occurring," "trace" contaminants arsenic, cadmium, copper, and beryllium in mining wastes to the TRI. Even worse was the EPA's decision that toxic releases from untreated mining waste rock need not be reported.[e]

In 2005, the EPA also proposed to eliminate reporting the releases of five persistent bioaccumulative toxins and to raise the reporting threshold for releases of 650 toxic substances in California from 500 to 5,000 pounds per year.[f]

Notes
[a] Congress created the Toxics Release Inventory in the Emergency Planning and Community Right-to-Know Act of 1986, following the pesticide factory disaster in Bhopal, India, December 3, 1984. The accidental release of toxic methyl isocyanate killed about 3,000

Box 4.2 Continued

people, injuring hundreds of thousands more (Francine Madden and Bettina Camcigil. *TRI Toolkit, Using the Toxics Release Inventory to Promote Environmentally Responsible Mining in Your Community* [Mineral Policy Center, 2000], 10).

[b] The EPA initially omitted mining from its Toxic Release Inventory even though the agency had reported in 1985 that asbestos and phosphates mining were generating one billion to two billion tons of waste annually, and that more than half of it could be hazardous to human health and the environment.

[c] All U.S. toxic releases in 1998 totaled 7.3 billion pounds. U.S. Environmental Protection Agency. 2000 Toxics Release Inventory (Public Data Release, EPA 260-R-02-003, Office of Environmental Information, 2002), tables 4-3, 4-5, 4-8, and 5-3. Available: www.epa.gov/tri.

[d] Bill Williams, Kennecott's vice president and general manager of environmental and engineering services, quoted in Jim Woolf. Kennecott Fumes over EPA's "Black List." *Salt Lake Tribune* 12 May 2000.

[e] Mineral Policy Center. New Ruling Exempts Mining Companies from Reporting Toxic Metals. *MineWire* 2003;6(10).

[f] Environmental Working Group. *Stolen Inventory* (Environmental Working Group, January 2005). Available: www.ewg.org; Kristan Markey and Bill Walker. *Stolen Inventory: Bush Proposal Would Allow Industry to Pollute California Communities Without Notifying the Public* (Environmental Working Group, 2006).

Figure 4.10 Liberty Copper Mine's waste-rock dump buries juniper woodland five miles west of Ely, Nevada. Photographed May 1977.

Contrary to mining industry claims, the untreated waste-rock dumps at mine sites are huge contamination sources. Broken-up rocks have large exposed surface areas, making the wastes vulnerable to leaching out acid mine drainage.[33] Even *after* the metal-extraction steps, mine tailings and smelter slag commonly have metal and toxic-chemical concentrations *thousands* of times greater than unmineralized rock or even untreated waste rock. In addition, tailings contain high amounts of toxic cyanide and arsenic from metal-extraction chemicals.

Mine tailings can be legally dumped into streams, but the tailings' contaminants must be reported to the EPA as toxic releases.[34] To keep wastes out of streams, mines throw up earthen dams to impound waste and tailings piles, many on steep slopes or in stream courses. All dams eventually become unstable and threaten to collapse. Sudden dam failures have resulted in landslides that extensively contaminated downstream floodplains and reservoirs[35]—but even without catastrophic failure, erosion of tailings piles extends the contamination in them far beyond the original mines and waste sites (see chapter 10, figure 10.4).

Coal miners' pollution

For centuries, coal mining took place underground in tunnels and pits. When the underground coal ran out, the companies simply abandoned mine excavations and waste heaps, making no or little effort to restore the land or reduce pollution. As a result, the 11 western states contain approximately 325 high-priority hazardous "coal-related abandoned mine problems," meaning open pits and collapsing underground mines (see figure 4.3). All the problem mines are unreclaimed despite a $1.8 billion surplus in the Abandoned Mine Land Fund established to address the problem. Thousands of other abandoned coal mines threaten "only" the environment, as opposed to human life and limb. Fewer than 10% of those are reclaimed, and no systematic inventory of the unreclaimed remainder has ever been attempted.

Coal mines contain methane and other volatile gases, which tend to catch fire when air seeps into abandoned mine tunnels (see chapter 12). Mine safety practices have greatly improved in recent decades, but methane flash fires in underground coal mines formerly claimed thousands of miners' lives—sometimes hundreds at a time. Spontaneous and uncontrollable fires in abandoned mine tunnels and cavities can burn through unmined coal seams for decades. Such fires are extremely difficult and dangerous to extinguish and can promote cave-ins of overlying lands, taking out roads and sections of farms and towns.

Coal-pit and strip mining both consume groundwater. From Black Mesa, Arizona, strip mines formerly piped slurries of coal plus water 273 miles to the Laughlin, Nevada, power plant. Until the plant closed at the end of 2005, the slurries used up 3,800 acre-feet of water—more than one billion gallons—every year.[36]

Coal deposits and coal-bearing rocks contain sulfide minerals, and so unmined coal remnants in underground coal mines, plus abandoned coal mine wastes, have emitted acid mine drainage for hundreds of years. Burning coal for heat and electrical generation releases large amounts of toxic contamination, principally deadly mercury from power plants, along with methane and carbon dioxide greenhouse gases (see chapter 12).[37]

Secondary legacies

Natural processes keep poisons spreading in the environment long after the miners have gone. Whether from land-based wastes or smoke stacks, the *secondary* spread of mining contamination damages cropland, soils, and farm animals in about 40% of all western U.S. watersheds. It also pollutes 12,000 miles of rivers, about the distance of four coast-to-coast trips across America. Secondary contamination processes spread both the waste mercury used in historic gold mining and the cyanide employed today far from mining waste piles and ponds. Arizona copper smelters rain acid-generating chemicals and toxic metals across hundreds of miles, including over the downwind populations of Phoenix, Mesa, and Scottsdale. Acid rain has been suspected of killing pines in southern California and promoting tree diseases elsewhere.

Montana's Clark Fork mining region still holds more than five billion cubic feet of contaminated tailings in ponds on more than 26,000 acres—about twice the size of Manhattan. Tailings ponds at Utah's Bingham Canyon copper mine alone cover about 9,000 acres. The tailings contain an estimated 10,000 tons of arsenic, more than 200 tons each of cadmium and silver, nearly 100,000 tons of copper, and 55,000 tons of zinc. Put together, the west's mining-contaminated lakes involve 180,000 acres, larger than Lake Tahoe by half.

Heavy pollution from mostly abandoned mines, dumps, and smelters covers more than 155,000 acres along Montana's Silver Bow Creek and more in the Clark Fork region. Metal mining at Kellogg, Idaho—now a Superfund site—has contaminated Idaho's Coeur d'Alêne River drainage, including beautiful Coeur d'Alêne Lake. The Coeur d'Alêne Indian Tribe's homeland now is loaded "from top to bottom" with poisonous lead, zinc, and cadmium compounds. Toxic metals in acid mine drainage and cyanide from Colorado's abandoned Summitville Mine has killed all the fish and other sensitive aquatic life over a 17-mile stretch of the Alamosa River. Near Questa, New Mexico, the Red River is now "biologically dead" for at least eight miles below the Molycorp Mine.[38]

Metals reach the rivers either as tiny particles in neutral and alkaline water or dissolved in acid mine drainage. Eventually, floods deposit the metals on river floodplains, killing vegetation. This process has made the Clark Fork and tributary floodplains into barren wastelands ("slickens"), encrusted with bright blue and green copper sulfates. Farther from mine and ore milling sites, the level of toxic contamination in river sediments drops off sharply, yet floodplain sediments 235 miles downstream from many mine sites still have 10 times the metal concentrations of less affected tributaries. Elevated concentrations can be detected as much as 340 miles downstream.[39] Even today, no plants can survive in many of these areas.

Research on Clark Fork sediments, benthic fish, and other biota along 135 miles of the tributary Blackfoot River near Missoula, Montana, demonstrate that the river's extensive marsh systems do not filter out metal contaminants any more, if they ever could. The benthic plants and animals living on river bottoms became poisonous from assimilating or ingesting metal-based contaminants in the sediments. Cadmium accumulated in food chain plants and animals most consistently, while nickel accumulated the least. The study's mathematical models predict that zinc contamination will be detectable in sediment 480 miles from a source and lead levels will be elevated 370 miles away, with traces of copper and cadmium still detectable at intermediate distances.[40] Benthic insects breeding in contaminated sediments as far as 235 miles downstream from the mines and smelters are trout food, and the likely reason for decreased trout populations in those streams.[41]

Gold Rush poisons

The 1840s California Gold Rush miners used some 26 million pounds of mercury (Hg) for extracting gold. Mercury can turn to a vapor at room temperature and, if inhaled deeply or repeatedly, can paralyze a person's central nervous system. Bacteria in water or wet soil convert mercury metal to highly toxic methyl mercury (CH_3Hg^+), which concentrates in water-based plants and animals' brains. Large bioaccumulations in fish sicken the people who eat them, and even very low methyl mercury levels are dangerous to children—especially the unborn.

The fate of all that Gold Rush mercury is mostly unknown, but mercury clearly has entered the food chain (figures 4.11, 4.12).[42] Much of it was mined out of California's coastal mountain ranges, at New Almaden and New Idria, where mercury-contaminated wastes poured into streams feeding lower San Francisco

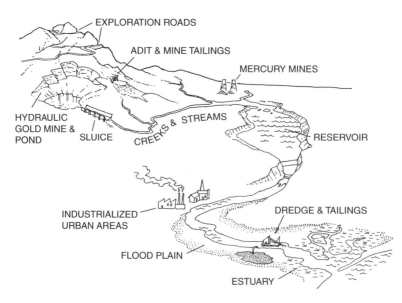

Figure 4.11 Transport of mercury-contaminated sediments by natural processes. Mercury comes from hardrock mercury mines and was used in hydraulic and hardrock gold mining to separate gold from unmineralized rock. Erosion of mine sites releases the mercury into streams, and dams accumulate it in reservoir sediments (upper left). At lower right, mercury leaches out of tailings into an estuary. Modified from C. N. Alpers and M. P. Hunerlach. *Mercury Contamination from Historic Gold Mining in California* (U.S. Geological Survey Fact Sheet FS-061-00, 2000), figure 9.

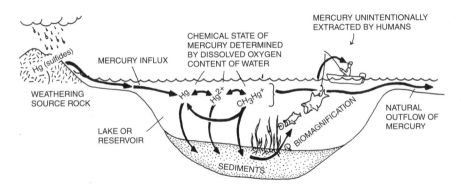

Figure 4.12 Natural pathways for mercury in an ordinary terrestrial pond or lake. Erosion of mercury-bearing rock (upper left) creates a flowpath to the pond, where mercury changes chemical form from metal (Hg) to mercury ion (Hg^{2+}) and then to methyl mercury (CH_3Hg^+). All can go into sediments and then into algae. It biomagnifies progressively through the food chain, from aquatic invertebrates to fish, and ultimately to humans. Mercury comes through inlet streams and leaves through outlet streams (at right). Modified from C. N. Alpers and M. P. Hunerlach. *Mercury Contamination from Historic Gold Mining in California* (U.S. Geological Survey Fact Sheet FS-061-00, 2000), figure 9.

Bay. The mercury concentrated in San Francisco Bay sediments has accumulated in the tissues of its fish. The mercury mined farther north, near Clear Lake, California, is contaminating local wells. Flooding on lands that previously were diked off for farming is likely to promote the conversion of mercury metal to methyl mercury, worsening the contamination hazard in fish for anyone or anything that eats them.

The widespread mercury contamination in San Francisco Bay sediments complicates plans to restore its fringing wetlands, a long-held dream of ecologists. Already, mercury contamination in old mine-waste gravels have frustrated attempts to use the wastes for restoring riverine fish habitat. Calls for removing dams also threaten to release mercury and other toxic elements trapped in sediment behind the dams.

Mercury still is pouring into the environment from such sites as eight Nevada gold mines, which produce wastes containing a total of nearly three million pounds of mercury and mercury compounds. Most of the contamination remains on site but is unlikely to stay there and represents a long-term threat. In 1998 the Jerritt Canyon, Nevada, gold mine released 9,400 pounds of mercury into the air—equivalent to mercury emissions from approximately 40 average-sized coal-fired power plants.

The cyanide contribution

Instead of mercury, today's mammoth "heap-leach" gold mines use cyanide to dissolve minute flecks of gold, silver, and some other metals from very low-grade "ore".

Masses of pulverized rock soak in cyanide solutions on "pads" the size of several football fields. The cyanide may be reused or drained into tailings ponds. Open ponds have generated deadly cyanide gas, killing thousands of animals—bats and birds, especially, but also larger mammals, reptiles, and amphibians—which inevitably seek out any watery expanse in a parched desert. Cyanide poisoned 900 birds at one Nevada gold mine in 1989 and 1990 alone.[43] Some heap-leach mines close to the U.S.–Mexico border have attracted desperately thirsty illegal immigrants, and mine operators eventually provided vats of drinking water to avoid poisoning them. Many have covered their tailings ponds to avoid attracting unfavorable attention to accidental wildlife poisonings, but the problem has not gone away.

Given lax state oversight, and many years of rain and snowmelt draining through abandoned heap-leach pads, both residual cyanide and other toxic chemicals find their way into streams. One example is Hecla Mining Company's abandoned Grouse Creek gold mine in Idaho, supposedly a "state-of-the-art" operation that played out after only three years. The dam for impounding cyanide wastewater is expected to collapse soon, and Idaho proposes to gradually release it into the Salmon River.

The gold mining industry maintains that cyanide in tailings-pond waters quickly changes to harmless compounds and that cyanide breakdown products are less toxic than the cyanide itself. But many of the cyanide products are toxic enough to kill fish, particularly in the trickles that constitute desert stream flow. If breakdown compounds persist for long periods of time and bioaccumulate to become highly concentrated in fish tissues, they threaten any higher animals that eat them, including people. The Zortman-Landusky gold mine, Montana's largest, adjacent to the Fort Belknap Indian Reservation, yields an instructive example of cyanide's toxic potential. Having won lawsuits against the mine, one tribe representative described its devastating impacts: "[T]he company has torn away our whole mountain, and there is acid coming out of every drainage. The damage that is done is incomprehensible to anyone who has never seen a cyanide heap leach operation… it is an environmental nightmare."[44]

Mining the Law

Over more than a century and a quarter, the 1872 General Mining Law, governing metals extraction on public lands, has promoted the hard-rock mining industry.[45] In the mid-1800s, nobody was barred from entering public lands, either to search for or extract minerals. To the public's lasting detriment, the 1872 law and predecessors legalized all "entry" on public lands for mineral prospecting and gave extraordinary occupancy rights to claims for mineral extraction (box 4.1). The law also gave the U.S. Department of the Interior responsibility for public resources protection. Its rules can obstruct companies trying to acquire sufficient property for economic mining in today's conditions, however.

Almost from the day it passed, the 1872 Mining Law has drawn repeated calls for reform[46]—yet, "It is difficult to imagine a law more frequently evaluated, with so little in the way of results."[47] The most obvious provisions crying out for reform include the "right to mine," which makes mineral extraction the highest and best use of mineral-bearing lands; open access to federal lands for exploration and claim staking without oversight; and the right of a claim owner to patent (i.e., purchase) federal lands for a pittance after proving the deposit's value.[48] Now much amended, the Mining Law is an open invitation to abundant abuses. As George Julian wryly commented in 1884, the "business of mining…sharpens the faculties and dulls the conscience." Unpatented claims often take on unauthorized uses as summer and permanent residences, resorts, junkyards, marijuana fields, and shopping centers. One even supported a house of prostitution.[49] Another significantly large use of mining claims is for tax write-offs.

Many large modern mines straddle both public and private lands. Private landowners receive royalties for letting miners dig up their land and take the minerals. In contrast, the U.S. Treasury collects no royalty payments to compensate the public—the actual owners of public lands and their minerals—for extracting nonrenewable resources. Among the many ironies of the mining law is that pinyon nut gatherers on national forest lands must buy a permit, while gathering nonrenewable gold and silver is free.[50]

American types

Like public lands grazing (see chapter 3), romantic myths and a powerful clique of western Congressmen have kept the 1872 Mining Law in place.[51] Every reform effort runs up against a lone prospector and his burro, the ghosts of mining past. One supporter declared, "When you talk about wiping out the prospector, you're talking about wiping out just about the most American type there is."[52]

But even mining industry sources agree that the little guy has scant relevance to modern prospecting or mining and no longer contributes significantly to the national or local economies, let alone to the public's interests. As early as 1922, a mining engineer told Congress, "There still remains a little room, and only a little, for the prospector." By 1941, Interior Secretary Harold Ickes told a Senate Committee, "[T]he individual prospector no longer exists as a significant factor in the mining industry." A small window of opportunity for small prospectors appeared in the brief 1950s uranium boom, and then closed again.[53]

Today, the antiquated mining law protects foreign interests more than American ones. Canadian and British companies, especially, own mining rights on more than a million western U.S. public acres (table 4.3).[54] In a 1999 debate on proposed Mining Law amendments, Congressman Nick Rahall remarked,

> It is our very own Mining Law of 1872 which continues, with reckless disregard to our economy and environment, to turn over federal assets to the control of foreign nationals. Our lands, our resources, owned by all Americans, are being claimed by foreign entities. The hard-rock minerals on these lands are being privatized by foreign entities for a mere pittance—$2.50 an acre.[55]

Many of mining's most grievous ills may be corrected by H.R. 2262 Hardrock Mining and Reclamation Act of 2007, a strong reform bill that the U.S. House of Representatives passed in late 2007. But the bill has yet to pass big hurdles in the Senate and the White House.

Table 4.3   Foreign Companies' Mining Rights on U.S.
Public Lands

State	Estimated Acreage
Arizona	101,618
California	48,109
Colorado	14,693
Idaho	10,686
Montana	60,063
Nevada	787,341
New Mexico	5,867
Oregon	1,818
Utah	25,184
Washington	11,804
Wyoming	83,215
Total	1,150,400

Environmental Working Group. *Who Owns the West? Section 3.
Foreign Control of Mining Rights on Public Land* (2004). Available:
www.ewg.org/mining/report/index.php.

Challenging claims

Unsurprisingly, the 1872 Mining Law did not mention environmental protection, and it still invites all and sundry to claim and dig on the still-open public lands, whether or not they can show substantial mineral values. But the law's establishment of mining as *the* preeminent use for public lands met with early and frequent challenges, resulting in additional laws, court decisions, and administrative rulings, which recognized other valued public land uses and limited free entry for mining (box 4.1).[56]

A high proportion of claims do lack any mineral value. Some still are commodity claims—clay for making kitty litter, for instance. Even areas proposed for withdrawal and protection for other purposes can be damaged in the early stage of mineral exploration, because the law allows mining claim filings before final approval for the withdrawal. Since 1970, the National Environmental Policy Act has required evaluations of mining projects affecting more than five acres on federal lands, to at least determine if claimed lands have a potential benefit.

On BLM lands, approximately 1.2 million individual unpatented lode claims (20 acres or less) and placer claims (160 acres or less) continue to tie up the mineral rights on about 25 million to 35 million acres of public land—roughly equivalent to the areas of Ohio and Michigan, respectively. Unpatented claims can be held indefinitely for a mere $100 annual assessment fee, a sizable sum in 1872 but insignificant today.[57] By continuing such claims, the 1872 Mining Law prevents other public uses while tolerating environmental damage that will be much more costly to fix than the accrued $100 per year fees.

In 2000, following years of debate and major public involvement, the federal government tried to improve environmental protection with modest changes to 1980s regulations governing hardrock mining on federal lands.[58] At nearly the same time, then-Secretary of Interior Bruce Babbitt formally withdrew federal lands in Imperial County, California, from mineral entry for a gold mine. Babbitt's action was the first-ever Department of the Interior denial of a mine permit.

Canadian Glamis Gold Ltd. had tried to claim lands in the California Desert Conservation Area for a huge open-pit gold mine. The pit would eradicate a multitude of archeological sites in many areas, including fragile geoglyphs (patterns that ancient humans etched in the ground), traditional trail crossings, and cremation sites with sacred and cultural importance to the Colorado River Indian Tribes.[59] The agency's annual report stated: "For the first time BLM can reject a mining proposal if the proposal will cause substantial irreparable harm to significant scientific, cultural, or environmental resource values of the public lands that cannot be effectively mitigated."[60]

Less than a year later, the George W. Bush administration's BLM appointees proposed rule 66 FR 16162 that suspended the new regulations. This proposed reversal attracted nearly 50,000 public comments, the great majority requesting BLM to retain the modified rules. In October 2001, the BLM issued the Final Rule, which eliminated protective bases for denying mining permit applications and reinstituted old standards for waste disposal and revegetating mined lands. That leaves us with weak and demonstratedly ineffective standards for protecting land surfaces, groundwater, wildlife habitat, wetland and riparian areas, and other resources.[61]

The message of the "new" rules is free rein for mining companies, bare or no protection for the environment, and the bill for cleanup and restoring mined lands sent to the public purse.[62] Significantly, in 2003 a U.S. district court held that the federal government does have the power to deny proposed mines if they will cause "undue degradation" to public lands.[63]

Regulatory Cesspool

Many miners have left behind poor past practices such as dumping tailings into streams or leaving them in unprotected piles, and today a mining company must post a reclamation bond to mine on public lands. In practice, however, the bond amounts are much too small to cover the actual costs of restoring the lands to a reasonably productive state.[64] The task and cost of mine cleanup both fall mostly on states—but in states that lack strong reclamation bonding laws, the messes don't get cleaned up at all. The public inherits either degraded land or large cleanup bills, or both.

The scope and severity of hardrock mining's extended environmental damages call for regulations to prevent or control spreading secondary contamination. But the mining industry long has claimed to be in jeopardy from environmental regulations and fights even commonsense cleanup rules, such as setting aside a portion of profits for remediating mining damage before it can spread. Since 2001, the Department of the Interior has even encouraged the mining industry to lobby for provisions that severely weaken even ineffective bonding programs. A bonding task force sponsored by the Department of the Interior—stacked with industry representatives—had several closed meetings with industry representatives, but only one with concerned environmental groups. Former BLM Director Jim Baca argued that it is not "too much to ask that if the mining companies want the public's resources, they should pay for what they get and clean up after themselves."[65]

Dirty cleanups

The improved tailings disposal methods do not address or control other common mine waste problems. A general lack of laws with teeth and underfunded enforcement allow mining companies to emulate the former owners of Colorado's Summitville Mine. The Canadian company first declared bankruptcy—transferring environmental cleanup costs to the public—and then sold the mine in the winter of 1992. Soon after, a cyanide-laced tailings pond overflowed, turning the downstream Alamosa River into a dead zone. Official estimates put the cleanup cost at more than $170 million, but Colorado accepted only $30 million to settle a 2001 lawsuit. Long-term problems remain, including acid mine drainage from a complex warren of surface and subsurface mine tunnels, which will poison the Alamosa River indefinitely.[66]

Off-site contamination from numerous failed fixes cry out for regulatory reform and enforcement. For example, New Mexico's Molycorp Mine allowed—or was unable to prevent—more than 100 documented slurry spills between 1986 and 1991, poisoning a stretch of the Red River with acid mine drainage. Federal funds built an earthen dam 150 feet high at California's Iron Mountain mine to keep acid mine drainage out of the Sacramento River, which mingles many northern California river and reservoir waters to feed the State Water Project and eventually southern California farms and Los Angeles homes (see chapter 9). Congress also appropriated $800 million to operate and maintain the contamination-mitigating Iron Mountain dam "in perpetuity." As already noted, earthen dams do not last forever. The appropriation is a paltry sum for maintaining a dam literally forever.

Another example is a 700-foot-high dam at Idaho's Thompson Creek molybdenum mine, one of the largest mine impoundments in the world. Holding more than 100 million tons of tailings, that dam's failure would cause extensive downstream damage and loss of life. Today acid mine drainage from the tailings is uncontrolled and already contaminates local surface and groundwater resources. It, too, will remain a public problem in perpetuity.

High in the annals of indiscriminate disposal are the more than 10 million tons of Atlas uranium mine tailings, still in heaps located 750 feet from the Colorado River's edge and only five feet above mean water level (figure 4.13). Uranium is both chemically toxic and radioactive. Even high-grade uranium ores contain less than 1% of the metal, so uranium mining always leaves masses of radioactive tailings that emit radon gas and many other breakdown products that need to be isolated from the environment (see chapter 13). Yet the Atlas mine wastes have leaked radioactive materials and other toxic pollutants for many years. Every day they still release an estimated 28,800 gallons of radioactive pollutants and toxic chemicals into the only major river draining the southwestern United States. Wave action in the river constantly stirs up these toxic sediment particles. Downstream from the mine dump, ammonia levels are several *hundred* times higher than state water quality standards allow and eight times the level considered lethal to fish. The Nuclear Regulatory Commission initially approved plans to cap the Atlas mine wastes rather than move them farther from the river, but subsequently reversed that ill-advised decision.[67]

Recycling metals is a leading alternative to mining poorer ores, but recycling mine wastes is questionable. Many attempts have created surprising, even threatening, problems. Examples include using arsenic-bearing mining slag for stabilizing unpaved roads at local sawmills and for sandblasting ship hulls. This recycling project contaminated sediments in the Hylebos Waterway, Washington, with arsenic.[68] Arsenic-laden mine waste formerly was recycled as "Ironite" fertilizer and sold at major gardening supply outlets for all kinds of gardening purposes (see chapter 2). Also, uranium mine wastes were recycled into house-site fill beneath Grand Junction, Colorado, housing developments, exposing men, women, and children to dangerous levels of radioactive radon in their own homes.[69]

Figure 4.13 Atlas uranium mine tailings dumped on the Colorado River floodplain within 750 feet of the water. River floodwaters and rainfall leach the dump, and leachates carry toxic and radioactive materials from the tailings pile to the river. Photographed July 2000.

Of coal and water

The history of coal mining regulation demonstrates the political power of mining companies and weakness of environmental protections against gigantic mining enterprises. Coal mining regulations first emerged as recently as 1972, after a West Virginia strip mine impoundment failed, killing 125 people. But President Gerald Ford vetoed the first two bills through Congress—one, the Surface Mining Control and Reclamation Act (SMCRA), called for simultaneous reclamation, meaning that miners lay waste rock from each strip of mined coal onto the previously mined strip. This method greatly reduces the work of reclamation.[70]

President Jimmy Carter signed SMCRA in 1977 and created the Office of Surface Mining (OSM) to implement the law.[71] Just a few years later, however, industry lobbyists influenced Congress to cut OSM's enforcement budget and headquarter enforcement staff far from mining country. Severe budget cuts led to a demoralizing 1995 reduction in force in the agency, primarily targeting enforcement agents. OSM remains a political orphan, and legitimate fears of firing or layoffs have sharply reduced the number, quality, and thoroughness of inspections.[72] Politically appointed managers set aside the few noncompliance citations.

Efforts to weaken coal mining regulations enforcement continued into the twenty-first century, along with relaxed environmental standards for coal burning. Although the Clean Water Act and mining regulations once supported successful lawsuits to stop mountain-top coal removal, alarming administrative rule changes now let the U.S. Army Corps of Engineers permit mines to dump a variety of wastes into the nation's rivers and streams.[73]

Colliding Realities

It is easy to see that mining is essential to the American way of life today and what passes for culture. The U.S. economy hinges on continuously increasing our consumption level and expanding our markets. For this latter purpose, the United States has encouraged other, more populous nations to develop as we have, and now they compete with us for resources. Ominously, Earth's population doubled twice in the twentieth century (from 1.5 billion to 6 billion) while per capita copper consumption doubled four times—from a quarter of a pound per person in 1900 to four pounds in 2000. While the U.S. population increased only 65% between 1950 and 1990, American industrial minerals, metals, and plastics consumption increased 130% (see appendix 3).[74] These trends are heading our economic wants toward a collision with geologic and economic realities.

Feeding high growth rates from already high per capita consumption levels requires continually finding, producing, and consuming all manner of natural resources worldwide at increasingly higher levels. Resources expert Brian Skinner has estimated that sustaining present population and consumption trends over the next 50 years will require miners to locate and produce larger amounts of previously unknown mineral resources than *in all of human history*. But new sources are increasingly hard and costly to find, even with advanced exploration technologies.

To restate our now-threadbare theme, *mineral exploration and mining companies are not finding any more rich ores.* That part of America's natural capital is gone. Mining companies now are seeking and extracting very slightly mineralized materials, too poor to really call "ores." This makes Skinner's projection a "Mission Impossible," with no final victory. The abysmally small amounts of gold mined from immense masses of rock today prophesy the future of all metals, many nonmetals, and also fuel minerals. Former mining wastes have even been scavenged for trace minerals in recent decades. This means that world consumption of scarce metals such as copper, zinc, and platinum at the rate we are using them is not sustainable, even with recycling.[75] Bottom line: Americans cannot hope to continue enjoying the world's highest-consumption lifestyle for very much longer.

Mining the dumps produces even more acid mine drainage and other pollution. To maintain America's prodigiously consumptive lifestyle, mining exploration teams will destroy ever larger amounts of land as they seek mineral values higher than in common rock. The mining will extract ever greater amounts of rock and soil, sand and gravel, and leave bigger holes. All this promises to accelerate mining's destructive drain on groundwater supplies and extensively poison soils and water resources, unmitigated by modest reclamation requirements. Already, many western wildernesses, wildlife refuges, and national parks and monuments are in the sights of oil and mineral exploration companies.

"Energy is the key which unlocks all other natural resources,"[76] and mining is one of the most energy-consuming industries, nationally and globally. Finding and extracting metals and other materials from very slightly mineralized rock today requires heavy energy inputs—not least for manufacturing the gargantuan mining machines. Mining ever-poorer ores, down to ordinary unmineralized rocks, will ratchet energy demand to astronomical levels. We deeply doubt that the United States or the world can find enough *affordable* energy for that level of mining.

Along with a declining ability to locate mineable lodes, the rate of finding fossil energy resources also declined in the latter twentieth century. In 1971, U.S. petroleum production went into irreversible decline, and a growing chorus of oil experts now predict that worldwide production of the most easily extracted oil resources will follow within a decade. Most known or proposed energy alternatives do not completely substitute for petroleum-based fuels (see chapter 12, appendix 9). In addition, rising petroleum costs will increase the price of everything. Eventually, rising energy costs are likely to prevent us from mining our back yards.

Hi-tech cleanup

Even if mining comes to a stop, we will still have polluting pits, tunnels, and wastes. But mine cleanups consume energy and other resources, and those future shortages could eventually halt even the easier cleanups. Bioremediation, using bacteria and other microorganisms to convert toxic compounds to nontoxic ones, and phytoremediation, using plants to remove contaminants, may offer less energy-intensive cleanup modes under limited conditions. Bioremediation transforms organic chemicals to simple water, carbon dioxide, and nitrogen and sulfur oxide compounds. Because the process requires an oxygen supply and is most effective in well-defined contamination zones, bioremediation has proved most useful for reducing sewage and petroleum pollution. So far, the applications to cleaning up mining wastes are more limited.

Phytoremediation processes mainly extract and concentrate toxic heavy metals, such as cobalt, nickel, and zinc, in plant tissues. But the basic problem remains—if the plants decay where they grow, the toxic metals simply return to the soil, and in higher concentrations. If the plants are removed—say, for reseeding a new crop of remedial plants—no obvious disposal sites exist that would keep the metals in plant materials out of the environment.

Bioremediation and phytoremediation are still very new approaches to the problem of mining wastes and may yet prove generally useful. But to develop them, scientists will have to find out much more than they now understand about complex physical and chemical interactions between toxic mine wastes and the natural environment.[77] More knowledge also is needed about the natural processes that take sediment-bound and dissolved metals into the food chain, the full range of adverse effects metals have on aquatic plants and animals, and especially their effects on us, the vulnerable mammals at the top of the food chain.

There is much left to learn about all these topics, so the only possible line of defense is to bring changes in problematic mining practices and clean up today's huge mining messes as much as possible for the benefit of future generations. Under 2002 revisions to federal mine waste disposal regulations, very little is proposed for impeding the spread of toxic contamination ever farther from mine sites. And where long-term contamination sources will release toxic pollution for centuries to come, mine waste cleanups may be futile. The costs of cleaning huge regions, including riverbanks and river sediments, probably cannot even be estimated reliably.

Saving the Future

The future will crowd Americans into a tight corner unless we can change our consumptive lifestyle, which drives mining and mining pollution while generating endless wastes and waste-dump pollution (see chapter 10). This may seem equivalent to saying that we have to stop being Americans—yet our grand overconsumptive spree began only about 60 years ago. America has a wealth of other traditions from much earlier times to fall back on—among them, fabled, but real, American ingenuity.

Recycling metals can partially alleviate the supply problem—and all forms of recycling will have to grow in the future as energy costs mount. The Mineral Information Institute reports that half of the 5.5 million metric tons of aluminum the United States uses each year already comes from recycled products. Even higher proportions of gold and other precious platinum group metals are recycled, as well as iron, steel, and lead. Although recycling gobbles energy, much of it uses electricity, which can be generated from a number of alternative sources (see chapter 12). Unfortunately for the growth economy, neither declining petroleum resources nor recycling allow for expanding per capita consumption unless population levels decline substantially, starting right away. So per capita consumption seems fated to decline.

Right now, the United States should concentrate on cleaning up mining damages before it becomes too expensive. Still strongly promoting the tragedy of our commons, the outmoded General Mining Law does not give mining companies incentives to invest in cleanup. And the current political climate is unsupportive of mining royalties, regulations, or effective regulatory enforcement. Yet imposing royalties for mining on public lands would give relief to us, the taxpayers, who have had to provide funds both for cleaning up huge mining messes and maintaining mining waste impoundments forever.

Americans first have to stop accepting both the financial burden and the health risks. But how can we make that change? Gold is an important constituent of computer and other hi-tech components, but most mined gold goes into jewelry. Raising the public's consciousness about gold mining's immense pollution might help. For example, author Rebecca Solnit wryly imagined a consumer's attitude if purchasing a gold watch also meant bearing responsibility for the waste: The consumer sees a large truck pull up to her modest suburban home; the driver knocks at her door and asks whether to dump her seventy-nine tons of waste rock on the front lawn or the back, warning, "keep the kids and the dog off 'cause of the acid and arsenic."[78]

Some states are limiting mining pollution. People living downstream from some western mines have decided that they don't like cyanide, arsenic, and acid in their back yards. Contrary to the nation's supposed antiregulation climate, land and water pollution from cyanide heap-leach gold mining has affected politically conservative Montana so heavily that the voters banned it in 1998. Colorado is considering a similar ban. A more reachable goal is building concerted and sustained public pressure to make the EPA restore the category of untreated mining wastes to its Toxics Release Inventory, just so we can know how much the mines dig out every year.[79]

A ray of hope may be the California State Board of Mining and Geology's 2003 adoption of requirements for backfilling all new open-pit metallic mines, which will not issue a permit to proposed mines unless they can be economically backfilled. Rejecting traditional arguments against filling pits,[80] the board requires miners to face the full costs of mining, including environmental costs. Operating under a aggressive, but fledgling, 1993 Mining Act, Montana and perhaps New Mexico are ramping up bonding requirements to make companies bear at least part of the costs of mine cleanups.

We will face a terrible dilemma if we cannot soon face up to Earth's limited ability to satisfy growing demands for metals and other mining products, and begin adjusting. Real and lasting solutions will be found only after we begin addressing the root problems of high consumption and population growth. Developing a serious national debate about the real need for gold mining might be a place to start. Open-pit gold mining, in particular, encompasses nearly all the ills foisted on public lands: antiquated laws with gaping regulatory and legal holes that give U.S. resources to foreign nationals and no return to public coffers, enormous cleanup costs laid on an unknowing public, permanent scars on public lands, and spreading toxic contamination.

A start at lowering our consumption expectations might begin with a national examination of what gold mining really contributes to our society compared to its land-devastating and pervasive water-polluting impacts. Of the inherent conflict between clean water and jewelry,[81] Rebecca Solnit mused, "To value gold over water is to value economy over ecology, that which can be locked up over that which connects all things."[82] A simple willingness to shift focus away from gold as a need or want and look at its water-depleting effects, could better connect America's hedonist present to a less stark future than we are preparing right now. And on the way we might mitigate some of the perils that our heirs must brave to get there safely, with most of our Founders' most cherished ideals intact.

5   Routes of Ruin

There are few more irreparable marks we can leave on the land than to build a road.
Mike Dombeck, U.S. Forest Service Chief, 1998[1]

The United States is more wedded to vehicles than is any other nation, and "freedom" to many Americans seemingly means driving their individual vehicles anywhere they choose. Opinion polls commonly show high proportions of U.S. citizens more concerned about gas prices, potholed highways, or restrictions on vehicle access to backcountry washes and dirt roads than about government scandals, stolen elections, or environmental damage.[2] Unfortunately, vehicles and roads exact a huge toll on lives and health and threaten our future well-being. Driving wheeled vehicles, and constructing roads to support them, comes close to topping the list of humankind's most environmentally damaging activities.

On most soils, even foot traffic creates tracks, trails, and roads. After ancient people invented wheeled vehicles to carry their burdens and themselves, they found that running water quickly rutted and potholed the cart tracks, and gully erosion chopped them up on slopes. Rainstorms eroded the tracks, flooding the dislodged sediment into streams and creeks and burying downslope croplands. Rutted tracks prevented Roman chariots from driving as fast as they were designed to go, so the talented Roman engineers quite naturally invented paved roads—some with better staying power than asphalt highways. But Roman paving did not solve the erosion problems that roads created, and in some ways made it worse.

Today, some parts of the United States contain more motorized vehicles than people. The varied vehicle uses, including military training, have vastly proliferated roads and roadlike corridors—especially numerous utility routes—across every type of American landscape. Erosional forces and their effects have not changed since Roman times, but modern engineers still fail to choose transportation routes or build roads to minimize environmental damages. The roads spread severe erosional effects everywhere, along with pervasive pollution. On top of it all, television images encourage Americans to take recreational cars, trucks, motorcycles, and all-terrain vehicles anywhere we wish. The naked ruts they create are an insidious form of road building.

Plots abound to impose roads on protected American lands, including in national parks and wilderness areas. The intent is to rid the nation of its public lands, its commons, which include its natural lands. The effect will be to eradicate what is left of our best natural lands, what is left of America's original natural capital.

Got Roads?

The contiguous United States contains seven million densely packed miles of paved and unpaved, public and private roadways. Especially across the west, these roads include major transportation routes, ranch roads, and roads for logging, recreation, and mineral and energy exploration and development. Extensive regional to transcontinental utility networks also consist of, or include, unpaved roads. Other routes are created by or for cross-country recreational driving (see chapter 11). Whether constructed or haphazard, roads and corridors segment the lower 48 states into ever smaller parcels, creating formidable barriers to free movement of both humans and wildlife. All conspicuously scar scenic vistas, and all have the same severe erosional and polluting impacts on land and water—particularly on the life supporting habitat of native wildlife.

Four million miles of America's roads are publicly owned and maintained. That's the same as 160 trips around Earth at the equator. More than 2.5 million miles of graded commercial corridors also crisscross the nation: two million miles of fuel-carrying pipelines, 271,000 miles of electricity transmission lines, and 174,000 miles of railroads. The western United States also has more than 100,000 miles of canals and river diversions that have the same effects as roads.[3] Road proliferation in the 11 western states includes nearly 786,000 miles of public roads—"only" 31 circuits of the equator.

Roadways in the United States vary from 10 feet to more than a hundred feet wide, covering close to 150,000 square miles. But a road's erosion and pollution impacts are much broader, extending ecological and physical damage to more than four times that much land, plus streams, ponds, and lakes—in total, affecting a region larger than Washington, Oregon, California, Arizona, and Nevada combined (figure 5.1).[4] The total western U.S. area directly and indirectly damaged by public roads alone may approach 100,000 square miles, comparable to the size of Arizona.[5] More than 80% of the west's public roads are unpaved rural routes, which open land to severe degradation. Paved surfaces protect the graded track, but extend erosion and pollution even farther.

National forest lands in the United States, mostly in the western states, contain nearly 450,000 miles of primarily dirt roads built for timber cutting and fire control at public expense.[6] The U.S. Department of Agriculture Forest Service's (USFS) annual road maintenance costs exceed $8 billion—a huge taxpayer subsidy for a private enterprise industry operating on public land (see chapter 1). The U.S. Bureau of Land Management (BLM) issues official permits for most of the 80,000-plus miles of unpaved road and corridors on its lands, including access for private mines and ranches; gas and oil pipelines; and electric powerlines and aqueducts. Authorized and unauthorized off-road recreational vehicle (ORV) roads and tracks add even more.

In Arizona, California, Nevada, and New Mexico deserts especially, the combination of road and utility corridor clearings, ORV playgrounds, and abandoned remote

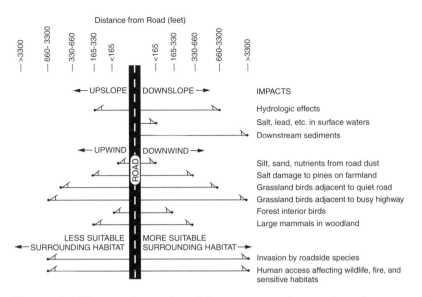

Figure 5.1 The envelope of road impacts, extending various distances beyond the roadbed, both upslope (left) and downslope (right). Modified from R. T. T. Foreman et al. Ecological Effects of Roads: Toward Three Summary Indices and an Overview for North America. In K. Canters, ed. *Habitat Fragmentation and Infrastructure* (Delft, The Netherlands: Ministry of Transport, Public Works, and Water Management, 1997), 40–54.

subdivisions (see chapter 8) enormously degrade western lands. And major interest groups opposing the existence of an American commons try to expand road systems and gain even greater access to America's remaining wild and natural public lands.

Reshaping the Earth

The wheels of vehicles we love to drive, together with running water, have a tremendous power to reshape Earth's surface. Ordinarily, much of the rain falling on an exposed soil can seep (infiltrate) into the ground. But our heavy cars, sport utility vehicles, and trucks compress the soil so much that dirt roadways become nearly impervious to rainwater. The water ponds in flatter areas and runs off of slopes, eroding them. Paved roads shed even larger volumes of water, with even greater power to erode downslope areas. Highways and roadway wear also contaminate land and streams with fuel and lubricants leaks, road salt and dust-abating chemicals, and weed-reducing herbicides. Accidental road spills and pipeline leaks contribute even larger chemical cargoes.

Roads threaten wildlife in myriad ways: They notch hillsides, fragment forest, savannah, or scrubland habitat into smaller and smaller pieces, and cut across creek banks and channels (see chapters 1, 6, 11). Roads also bring a large human presence into remote areas, along with exotic plants that may replace native vegetation (see chapter 3).[7] Both during road construction and in collisions with vehicles, road kills take an enormous toll on birds and ground-dwelling wildlife. All these changes degrade the habitat that indigenous wildlife depend on. To survive, many animals must move or find ways to get by in degraded habitat, where they may be increasingly exposed to predation and have fewer opportunities to find potential mates.

Base damage

A very large scientific literature documents the erosional and polluting effects of paved and unpaved roads.[8] After blades scrape away the porous upper soil layer, rain no longer sinks easily into the ground. Instead, the water collects on level surfaces and flows down slopes. Military and recreational vehicles traveling across untracked land compress the soil beneath their treads or wheels, with the same effect (see chapters 6, 11). Over time, uneven soil compression from vehicles of various weights and sizes creates jaw-chattering "washboards"—or "whoop-de-dos" if really large. The road corrugations localize deeper potholes and increase roadway erosion.

In rainstorms, road segments link up with natural creeks, streams, or rivers at crossing points.[9] Concentrated water flows in roads, roadside ditches, and culverts all erode new steam channels or deepen natural ones. Roadside gullies and constructed drains increase the density of hillside drainage channels by as much as 6%.[10] Roads and vehicle tracks in dirt roads or trails can capture the rainwater and become stream channels themselves. Rainwater flowing down a sloping dirt road tends to concentrate in channels or spread out as sheetflow, rapidly excavating small, closely spaced rills. Rills can quickly grow to become gullies—channels so wide and deep that they cannot be crossed by a wheeled vehicle or taken out by a plow (see chapter 13). On hill slopes, the captured water from a paved road can lead to landslides, mud flows, and debris flows (see below).

People in hilly regions know that they suffer greater erosion than do flatlanders, and effects increase with the pitch and length of sloping road segments. The type and texture of soil generally determine how badly the rainwater can erode the surface of an unpaved road, but increased rainfall intensity and duration also increase the erosional effects. Heavy rains tend to strip soil and rock sediments from hillsides and deliver it to the nearest creeks, increasing flooding in streams and rivers at lower elevations. Studies of logging roads reveal that large runoff amounts from road surfaces can increase annual flood levels by 2–10%. In the Pacific Northwest, where big storms tend to recur at 10-year intervals, the growing numbers of logging roads are likely to increase annual flood levels by 3–12%.[11] Eventually, people living in valleys find that all the sediment has wiped out fish in local streams.[12]

Dust from roadways has become a major source of air pollution in the arid west. Car wheels abrade the surfaces of dirt roads, kicking up characteristic dust clouds that can be seen for many miles across the high plains and desert flats. Researchers from the U.S. Geological Survey have catalogued severe illnesses from breathing both the inorganic materials and pathogens in western dust clouds into the lungs. Some pathogens can harm people and animals alike—one example is *Coccidioides immitis*, a fungus living in soils across central California, Arizona, and New Mexico and into Mexico, which causes valley fever.[13] Every new road expands the number of sites where whipping winds can pick up fine-grained sediments and incorporate them into dust or sand storms, threatening public health and safety. Applying dust-suppressing chemicals to unpaved roads only adds pollution to land and water.[14]

Paving roads cannot provide a solution to these problems. For a while, the paving protects scraped subsoil under the constructed road base, but it is expensive to install, requires constant maintenance, and uses large amounts of natural resources. Hot and cold weather conditions constantly expand and contract the pavement cap, and traffic wear dislodges fragments. Pavements capture even more water than unpaved roads and tracks. And wherever that water flows off the pavement, it can cause severe erosion on adjacent land.

Collapsing hills

Building a level roadbed on a steep hillside requires cutting notches into the slope. The sides of the cuts tend to remain bare, and many expose raw, fractured bedrock. In the most popular "side casting" construction method, rock and soil from the notch may be either dumped on lower slopes or used as "road fill," to widen and smooth the cut notches and fill stream channel crossings. Cutting the notches tends to dislodge large rocks, which damage trees and underbrush as they roll and bounce down steep rocky slopes (figure 5.2). The vegetation formerly held soil in place and helped control mass wasting, one of nature's most dramatic erosional processes (see chapter 13). Severely damaged vegetation provides much less in the way of control.

Figure 5.2 Side cast road and windmill pads in a wind-energy development, Tehachapi Mountains, California. The pad and road builders dug up coarse materials and dumped them far downslope. Photographed April 1986.

Landslides and debris flows, including mudflows, are the most obvious forms of mass wasting. In combination with the less conspicuous slow downslope movement of "soil creep," they dominate erosion processes in steep terrain. Rock slides and landslides are common in bare road cuts, especially if the climate is cold enough for water to freeze and expand in fractures, a process that dislodges soil and rock fragments. Over time, mass wasting processes may effectively remove unmaintained roads (see chapter 13).

Probably because they collect rainwater, roadbed fills are especially vulnerable to collapsing and creating landslides or debris flows.[15] Typically, hillside roads shed runoff into ditches and culverts. But any blockage will back up water beside or beneath the road, where it can seep slowly into the relatively porous road fill. During intense rainstorms, pressure builds up from water collecting in the fill, weakening the road mass and preparing it for collapse. Eventually, the road fill can turn into a mass of soupy mud with the consistency of runny concrete, which pours rapidly downslope in a debris flow. Sometimes preceded by a rush of air, debris flows strip vegetation and can scour soil down to bedrock as they careen downslope. Like snow avalanches, they carry rocks, boulders, and anything else in their irresistible paths, threatening forests, houses, and people's lives (see chapter 13).[16]

Roads in steep terrain actually increase the likelihood of mass wasting and yield 10–30 times as much sediment as all other road surface erosion processes combined.[17] Road-initiated mass wasting delivers heavy sediment loads to stream channels, modifying the water flows and severely disrupting streamside and in-stream (aquatic) habitat. The soil, rocks, and rubble that debris flows carry choke streams, killing fish and destroying gravelly pools that they need for spawning. But less dramatic forms of sheet and rill erosion on unpaved roads, and on pipeline and transmission line corridors, also can produce enough sediment to diminish or destroy aquatic animals and plants.

Frequent hill slope collapses throughout the Rocky Mountains and in the Pacific coast ranges of California, Oregon, and Washington mainly originate from roads on steep logged slopes with crumbly soils or landslide-prone rocks and from dwelling sites on unstable hillsides.[18] Road-initiated landslides and other slope failures yield anywhere from 25 to 340 times more sediment than roadless areas in natural forested areas with similar slopes, vegetation, and weather.[19] Studies in Idaho's Clearwater National Forest following the major storm years of 1974 to 1976 and 1995 to 1996, showed that more than half of debris flows started as hill slope collapses along logging roads (see chapter 1). Three-quarters of the 1995–1996 road failures were road fill collapses, and the rest were roadcut collapses.

New roads have higher failure rates than older ones, and most hillside collapses occur between 4 and 10 years after a road's construction. Some studies suggest that road failures can decline to the frequency of natural landsliding by the time the road is 20 years old, but others show little improvement after even 25 years. Because high-intensity rainfall promotes slope failures,[20] and intense storms tend to recur on roughly 20-years intervals (but twice that rate in the Pacific Northwest), most of the studies were too short for assessing the real recovery rates of debris flow-prone roads.

The USFS argues that road building methods have improved significantly since the 1970s and have reduced mass wasting.[21] But the Clearwater National Forest studies do not support that contention. Unfortunately, destabilizing side casting remains the construction method of choice for logging, pipeline, and other access roads for short-term use. It is also the preferred method of construction for wind energy, military, and mining uses.

Maintaining damage

There is no question that fresh dirt roads erode rapidly or that the rate of erosion decreases rapidly over time (figure 5.3). The falling rates are due partly to plants revegetating roadway ditches and roadbed notches. Ironically, wind and water erosion also help reduce erosion rates because they preferentially remove fine sediment particles and leave a thin protective layer of coarser particles. The coarse lag material armors the road surface and slows the rate of erosion.[22] Protective inorganic crusts and coatings also may form on steep cut slopes. But until natural processes can virtually obliterate them, erosion rates remain higher on dirt roads than on natural, undisturbed surfaces. The obliterating processes can take decades to centuries, or longer.

Speaking of "maintaining" dirt roads is an oxymoron, because clearing vegetation out of ditches can increase erosion and sediment yields as much as seven times.[23] Grading to remove gullies and rills destroys surface armoring and pushes erosion rates back to high values—up the curve in figure 5.3. Just letting vegetation recover on roadsides would help reduce blowing sand and dust, but road shoulders commonly are regraded far beyond any obvious need. For example, figure 5.4 shows the extremely wide graded shoulders of a road on sensitive soils in an area prone to wind erosion. Many times, road shoulders and cuts are even cleaned of vegetation across sand dunes, providing additional sources to the scourge of blowing sand.

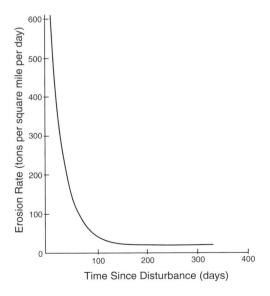

Figure 5.3 Erosion rates on roads plotted against the time since the most recent disturbance. Erosion falls steeply for the first 100 days, then maintains a lower, relatively constant rate that still exceeds the natural rate of an undisturbed surface. From W. F. Megahan. 1974. *Erosion over Time on Severely Disturbed Granitic Soils: A Model* (General Technical Report INT-156, U.S. Department of Agriculture Forest Service, Intermountain Forest and Range Experiment Station).

Figure 5.4 Excessive shoulder grading along a road in Eureka Dunes, Eureka Valley, Inyo County, California, in an area highly vulnerable to wind erosion. Photographed January 1976.

"Maintaining" culverts under stream crossings involves digging out stream channels both upslope and downslope from a road, which creates a step (knickpoint) in the stream channel. The stream adjusts to the knickpoint by eroding deeper into its channel below the road crossings. The stream also erodes the knickpoints, pushing them upstream and undermining the culverts.[24]

Spills along the Way

In addition to the physical disturbances from building roads, railroads, and utility corridors, every mile of a road or utility corridor spreads pollution into urban and rural soil and water supplies. Pollution sources include hazardous petroleum-based compounds in asphalt from road construction, hazardous toxic metals from cars, dust-controlling oils, synthetic fertilizers, and weed-killing herbicides on road shoulders, and other chemicals.[25] In winter, deicing salts wash off roadways and pollute nearby streams. Roads also help spread weed invasions, insect and rodent pests, pathogens, and noise far into the surrounding lands and habitats.[26] In addition, the vehicles, trains, and pipelines all suffer from spills and leaks.

Traffic exhaust emissions and normal road wear on vehicles deposit toxic metals along roads, such as lead from gasoline combustion, nickel and copper from worn clutch linings, and zinc from tire debris.[27] Lead deposits from automotive exhaust once broadly contaminated roadside soils. Taking the lead out of gasoline greatly reduced lead contamination, but recent studies show that lead can be remobilized from already-contaminated soils, and it still causes lead poisoning in urban children.[28] Deteriorating catalytic converters release large amounts of platinum-group metals along many major travel routes. The concentrations of those metals in some roadside soils are similar or higher than the ore deposits.[29]

The United States generates more than 13,000 major oil spills every year, dumping an average of 15 million gallons of crude oil and refined products along roadsides. Nearly a quarter are vehicle or pipeline spills, while less than a fifth happen at industrial sites or oil fields and refineries. The National Highway Safety Administration records an average of seven million traffic accidents annually, each releasing small amounts of many thousands of petroleum products and other hazardous substances. The nation's main highways also experience about 3,000 industrial chemical spills per year, involving explosives, agricultural and industrial chemicals, and hazardous wastes. In the near future, radioactive wastes will join this list (see chapter 12). Pipelines carry such extremely hazardous liquids as crude oil, diesel fuel, gasoline, jet fuel, anhydrous ammonia, and carbon dioxide. Both the spills and the leaks can add significant amounts of greenhouse gases to the atmosphere.

Railroads generate 1,900 spills annually. A major 1991 spill resulted from an accident in northern California, which dumped large volumes of herbicide into the Sacramento River, wiping out aquatic life for 38 miles downstream. The river has since recovered, but California's attempt to enhance state regulations and prevent repeats have not fared so well. In 2000, a federal judge threw out 7 of 10 safety rules the state tried to impose after the accident.

Since 1986, leaks from U.S. pipelines have caused nearly 400 deaths, more than 3,000 serious injuries, and more than $1.3 billion in property damage. From 1986 through 2002, more than 3,000 hazardous-liquid pipeline accidents caused 36 fatalities and $811 million in property damage and released over six million gallons of toxic substances into the environment annually.[30] A particularly terrible 1999 gasoline leak from the Olympic Pipeline, near Bellingham, Washington, killed two 10-year-old boys and an 18-year-old man. About 250,000 gallons of gasoline flowed down a creek to a city park and then exploded in flames, scorching a mile and a half of riverbank and setting a house on fire. Public reaction to the leak wrecked plans for another pipeline along a route crossing three state parks, at least temporarily. Thousands of minor spills release even more.

Nearly 3,500 accidents along natural gas pipelines through 2001 caused about 340 fatalities and more than $500 million in property damage. Natural gas leaks alone led to more than 3,000 major pipeline accidents between 1986 and 2000. In the summer of 2000, a buried El Paso Natural Gas Company pipeline exploded at a campground near Carlsbad, New Mexico, leaving a house-sized crater. Flames from the leak killed 12 campers, including five children.

Instead of proposing enforcement actions with proposed fines, the federal Office of Pipeline Safety issued only letters of concern or warning throughout the 1990s. Whether the letters can protect the environment or people's lives remains to be seen. The nation's pipelines are aging, and some have been in the ground since the 1920s. The United States has no national policy to repair or rebuild pipelines, and pipeline companies show little interest in keeping them repaired, so the spills and leaks are likely to increase.

Since 1989, when the *Exxon Valdez* ran aground in Prince William Sound, Alaska, the severe health threats from massive oil spills have become well known. The problems arise from toxic components in petroleum and their tendency to stay in the environment. Refined gasoline contains more than 200 hazardous elements and compounds, representing many different potentially adverse effects on plants and animals (see chapter 12, box 12.1). Potential health threats from small spills and continuing or recurring leaks are less clear and poorly studied.

Rains wash a large but unknown amount of petroleum-based pollution from highways into waterways, including semivolatile organic compounds from oil, gasoline, and grease; polycyclic aromatic hydrocarbons (PAHs) from crankcase oil and vehicle emissions; and volatile organic compounds (VOCs). There are few data on the national abundance of PAHs and semivolatile organic compounds in highway runoff, and VOCs are even less well documented. PAHs in waterways commonly exceed both U.S. Environmental Protection Agency drinking water standards and the standards and guidelines for safeguarding aquatic habitats. Three VOCs, including carcinogenic benzene, also commonly exceed drinking water standards, and benzene recently was found in some soft drinks.

Upset Balances

Roads and utility corridors divert rainwater, carve up habitat, and destroy plants and animals in a number of different ways. All these impacts change natural ecosystem balances in remote landscapes, affecting many species' prospects for survival. The "vegetation enhancement" effect of roads and utility corridors in western deserts is one example of water diversion effects. Anyone driving across western deserts will see larger and greener shrubs growing close to the road than farther away (figure 5.5). This is due to the roads' effect of capturing rainwater and bestowing more on roadside plants, which helps them grow larger than they could under natural conditions.[31] Since no plants grow in the road itself, the roadside plants benefit from limited competition for the excess diverted water. Plants in the enhanced zone have greater mass than the total formerly growing in the roadway and also support larger insect populations than the removed plants could have supported.[32]

The water capture of roads also blocks stormwaters from flowing farther downhill beyond a road, significantly depleting both the normal water supply of downslope plants and the amount of water stored in downslope soils between storms.[33] As a result, downslope plants distant from the road cannot grow as large as they normally would. Vegetation differences dramatically illustrate these effects above and below extensive zigzag levees that protect vulnerable roads, railroads, and aqueducts in the southwestern United States (figure 5.6). The roads cross sloping alluvial fan sediment deposits, which connect steep mountain fronts with adjacent desert basins. Stormwaters flowing from the mountains and across the fans have cut countless shallow channels, which constantly shift position from one season to the next. The levees dam the natural channels and direct captured stormwaters into the levees' sawtooth elbows, where man-made drains carry them under or over the road or aqueduct.

Figure 5.5 "Roadside enhancement" effect, showing larger and greener creosote shrubs beside Kelbaker Road, southeast of Baker, California. Photographed December 1980.

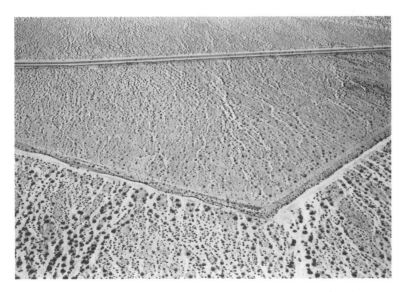

Figure 5.6 Aerial view of zigzag levee protecting the Colorado River Aqueduct (background), southwest of the Turtle Mountains, California. Plants between the levee and the aqueduct are denied surface runoff and are smaller and sparser than in upslope areas (foreground). Photographed December 1980.

Because the levees deny downslope vegetation their normal water supply, those plants get water only from rains falling directly on them. A person standing on the ground can see that plants downslope from levees protecting the Colorado River Aqueduct are smaller and fewer than the upslope plants, having suffered this water-starvation effect for more than 50 years. The contrast is even more vivid from the air (figure 5.6). Studies comparing the size and number of plants on each side of the levees have confirmed the apparent differences.[34]

Roads and utility corridors dissect (fragment) virtually every natural western land-based habitat into ever smaller pieces across shockingly large regions. The smaller the habitat fragments, the less able they are to support either wildlife or humans.[35] Habitat fragmentation enlarges sunnier, drier edge habitats at the expense of cooler, damper interior habitats and the animals that depend on them (see chapter 1). Roads reduce foraging areas for small- and medium-sized animals, but large animals can be even worse off—a density of one road per square mile of land cuts effective elk habitat nearly by half, for example. A density of six road miles per square mile *eliminates* elk habitat.[36] Nearby traffic noises interfere with the hearing of wildlife species, critical both for predators trying to find their prey and for potential prey trying to protect their lives. Sounds often are important for mating, and later for protecting young in nests or burrows—so road noises can reduce the animals' reproductive success.

Constructing corridors for pipelines and electric transmission towers has been highly disruptive to habitat, especially in southwestern deserts. Natural desert plants recover very slowly in clearings,[37] and attempts to revegetate disturbed zones with native plants have achieved only limited successes. The BLM, USFS, and Federal Energy Regulatory Commission (FERC) have seen enough documented problems to insist on vegetation-conserving construction practices. But the agencies do not provide focused or effective supervision, which effectively devalues the federal lands used for utility corridors as habitat and groundwater recharge. Most of the 176 general and specific mitigations for pipeline corridors, which the FERC recommended in 1990, are requests and not mandates. They are routinely ignored.

Energy pipelines corridors crossing the Mojave Desert include the All American crude oil pipeline built in 1987, and Kern River and Mojave natural gas pipelines built in 1991. All were bulldozed to excessive widths and stripped of vegetation, in violation of conservative construction practices.[38] For the Kern River and Mojave pipelines, the FERC had required that "Joshua trees, sensitive cacti and perennial species...shall be carefully removed from the rights-of-way, held in secure locations, and replanted...near their original locations" after construction. But the builders did not make an effort to meet those requirements.

The Mojave pipeline builders salvaged and transplanted a mere 140 barrel cacti. In violation of accepted (and required) erosion-reducing engineering practices, pipeline construction did not restore mountain-front drainage channels at right-of-way crossings. The corridors diverted runoff and caused unnecessary erosion, resulting in public health and safety hazards.

The Kern River pipeline plowed through hilly Joshua tree forest in an area of the Clark Mountains under BLM jurisdiction, causing massive damage (figure 5.7). In apparent violation of the National Environmental Policy Act, the contractors

Figure 5.7 Bulldozed Joshua tree forest along the Kern River natural gas pipeline, north side of the Clark Mountains, California. Joshua trees, other yucca, and cacti were not salvaged and transplanted back on the pipeline right-of-way, as required by the construction permit. Photographed May 1991.

proposed salvaging only 100 Joshua trees and 100 cacti in the pipeline corridor for the entire state of California. BLM set a higher transplanting goal, but the builders did not meet even that minimally acceptable level. In 2003, Kern River operators added a larger capacity pipe, widening the corridor by upward of 25 feet and clearing two entirely new miles of corridor. The cleared areas yielded many yucca and cactus plants for transplanting, but a large number were mature plants and less likely to survive transplanting than younger ones. Many of the more viable transplants were not placed in areas where similar species grow outside the corridor, which bodes ill for their survival.

Every day, road and railroad collisions kill more than one million vertebrate animals in the United States.[39] Many recreational drivers follow pipeline and other utility corridors, but ORVs can convey joyriding hordes into even the most remote areas (see chapter 11). The increased access wipes out native animals in their own habitat, exposes archaeological resources and cultural artifacts to theft and vandalism, subjects threatened or endangered plants and animals to illegal collecting, and opens backcountry areas to illegal dumping of nonbiodegradable wastes. The increased road and motorized recreational access to natural habitats also brings heavy hunting pressure, including "recreational shooting" and "varmint shooting" of native animals and plants. The cumulative effect of road kills, hunting, and poaching is a major obstacle to preserving the natural diversity of animal species, required for high-quality habitat.

The environmental consequences of recreational shooting include very high concentrations of toxic lead in soils—as much as 3 *pounds* per square foot have been measured at a national forest shooting range. Lead can dissolve in forested environments, contaminating surface and shallow groundwater. Trees at the edge of many shooting ranges have 45–140 bullet wounds per square foot of exposed trunk. Some intensive recreational shooting sites, such as California's Texas Canyon in Angeles National Forest, have been closed for at least a decade. Sadly, the soils will remain lead-contaminated indefinitely, threatening both wildlife and people who might drink the groundwater.

No Place Untouched

Road systems to serve more logging, grazing, and new urban/suburban develop-ments are constantly expanding across relatively untouched western lands that sup-port healthy native ecosystems. The 1963 Wilderness Act was intended to preserve native habitat and wildlife on defined roadless areas, and most of the nation's few remaining roadless areas are public lands in the western states. During the Reagan and first Bush administrations, ORV activists, advocates for extractive industries, some western counties—and even a few state government bureaus—began multipronged efforts to push roads through roadless areas and previously designated wilderness areas, to preempt official wilderness designations. Some of the same groups also advocate building more roads in national parks, preserves, and monuments.

Partly to stop roads from fragmenting national forests, the federal government tried to protect 58 million roadless acres in the 1990s. In 2004, however, federal policies began reversing those protections.

Gutting rules

The USFS began preparing inventories of roadless areas for potential wilderness designation on national forest lands in the 1970s. The agency now lists more than 2,800 roadless areas totaling about 54.3 million acres. But some of the identified areas "grew" roads during the inventory period, reducing truly roadless USFS lands to about 51.5 million acres—only 2% of U.S. land surface.

In 1998 USFS prepared an environmental impact statement that assessed the potential impacts of protecting roadless areas. The preferred alternative restricted road construction and reconstruction but allowed "stewardship" logging, includ-ing clearcutting, ignored soil damage and erosion problems, and allowed unabated access to ORVs. In short, it proved inadequate for preserving the natural charac-ter of roadless areas. The alternative also proposed a five-year delay in protecting Alaska's Tongass National Forest roadless areas, leaving 2.5 million acres of the nation's last large unspoiled temperate rainforest open to immediate exploitation. And it left management decisions for another six million roadless acres to the discretion of local officials, who commonly respond more favorably to consumptive interests than to ecosystem protection.

The environmental impact statement elicited a record-setting 1.6 million public comments, more than 90% strongly favoring maximum protection. This overwhelming support for ecosystem protection led to a final Roadless Area Conservation Rule, implemented in January 2001, which should end virtually all logging, road building, and coal, gas, oil and other mineral leasing on 58.5 million acres of unspoiled land. The final rule did not end the ravages of ORVs, however. Nine lawsuits quickly challenged the new rule, and a preliminary court injunction suspended its implementation. Then-Attorney General John Ashcroft declined to appeal the suspension, breaking a promise to defend the rule in court that he made in his confirmation hearing.

The Department of Agriculture delayed the rule's effective enforcement date by 60 days and started a formal process that allowed individual forests to opt out. This same approach has gutted mining reform (see chapter 4) and advanced industry plans to invade roadless areas for oil and gas exploration (see chapter 12). At the end of 2002, the federal Ninth Circuit Court of Appeals revoked the rule's suspension. Mark Rey, U.S. Department of Agriculture undersecretary in charge of forest policy, responded: "[I]t doesn't fundamentally change our course of action. [We will just be] obliged to pull the old rule back to do a new one."[40]

While the September 11 terrorist attack riveted the nation's attention, the USFS did issue new rules that gutted protections for roadless areas, particularly in Tongass National Forest. The new rules removed mandatory environmental impact analyses for any activities in roadless areas, including allowing regional forest officials to end environmental reviews and public comments.[41] For the next two years, interim directives largely pulled back more regulations, and on December 16, 2003, the USFS effectively opened Inventoried Roadless Areas and contiguous unspoiled areas to road building.[42] Soon after, a piecemeal process began slicing up these areas with roads, opening 300,000 acres in Tongass National Forest alone.

In keeping with Rey's promise, in 2004 the George W. Bush administration proposed a rule allowing state governors to petition for state management of roadless areas. The Department of Agriculture can permit the requested changes for reasons of *either* better protection or exploitation. Since one governor's authority is not transmitted to each succeeding administration, this new rule has made planning a shambles.[43]

In 2006, California exercised the option to protect its roadless areas from all developments.

Obscure inroads

While the USFS gutted roadless area protections, former 1970s "sagebrush rebellion" leaders aimed to undercut federal authority over all public lands and open roads across national parks, national monuments, and already-designated wilderness areas. The point of attack is Revised Statute 2477 (R.S. 2477), a deceptively simple provision of the obscure 1866 Lode Mining Act. R.S. 2477 grants "right of way for the construction of highways over public lands, not reserved for other uses...." The statute had encouraged western settlement and development by allowing highways across public lands in the latter nineteenth and early twentieth centuries. It went into disuse by the 1960s, when the nation's transportation infrastructure was largely built out. The 1976 Federal Land Management and Policy Act, which empowered BLM, largely repealed R.S. 2477.[44]

Using century-old legal precedents, groups that question the entire concept of federal public lands now claim virtually any route a prospector, rancher, hunter or vagabond may have used—even if long abandoned, obscure, and impassable—as a public highway. They also argue that R.S. 2477 allows claimants to upgrade any obscure track into an improved two-lane road with no regard for environmental, habitat, or other values. Many roads currently claimed under R.S. 2477 are long-abandoned cattle tracks and ways, and most lead to no identifiable or existing destination. Some are barely visible on the ground. These claims have enormously expanded the number of roads across the western United States.

In 1988 Donald Hodel, interior secretary under Ronald Reagan, initiated the practice of accepting upgraded obscure routes as legitimate claims under R.S. 2477. Hodel's action threatened to convert some of the most spectacular and ecologically important public lands in this country, including every hiking trail in Zion National Park,[45] into a highway. He also agreed that "constructing" routes could include removing high vegetation and large rocks and that routes could be deemed "constructed" if only a few vehicles ever had driven on them. His ruling opened a Pandora's Box of claims on anything from cow paths to intermittent stream channels and invited challenges to federal government jurisdiction over public lands. Since 2000, the process for filing R.S. 2477 claims has become even easier.

The scope of the R.S. 2477 claims from counties and ORV groups is breathtaking. As of 1993, the National Park Service estimated that 68 of its management units, comprising 17 million acres around the country, could be affected by the claims with devastating impact on national parks. Claimed areas include 21 designated California Desert Conservation Area wilderness areas in San Bernardino County alone, 17 of which contain more than 800 claimed miles (figure 5.8). More than 600 miles of claims are in 11 designated National Park Service wilderness areas in the Mojave National Preserve. Moffat County, Colorado, filed R.S. 2477 claims on 2,000 miles of rights of way across eastern Dinosaur National Monument, four designated wilderness study areas, and the Yampa River Recreation Area.[46] And several counties in Montana, Idaho, and Oregon have claimed essentially every road on national forest lands within their boundaries.

Almost every state harbors potential R.S. 2477 claims and no national treasure is immune. To date, Utah counties have filed perhaps 10,000 to 15,000 R.S. 2477 right-of-way claims through national parks, national monuments, and wilderness study areas, in addition to multiple-use public lands. In 1996, the BLM allowed two Utah counties to grade a road on faint tracks near Canyonlands National Park, prompting a Southern Utah Wilderness Alliance and Sierra Club lawsuit.[47] The 2001 court ruling on that suit denied the counties' claims, stating that an R.S. 2477 route must be a "highway" open for public use and must connect the public with identifiable destinations or places—and it must have been created for public purposes before 1976, when the land was reserved for other uses.[48]

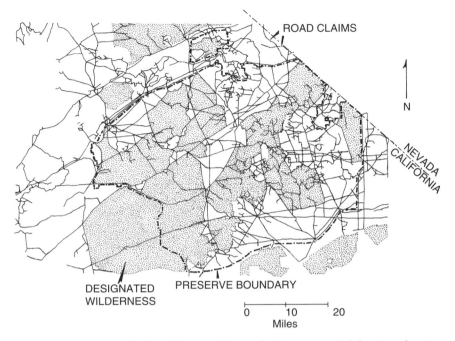

Figure 5.8 Map of the Mojave National Preserve, California, showing wilderness areas (stippled) and the network of R.S. 2477 route claims (thin black lines). Bold dot-dash line is the preserve boundary. Eleven wilderness areas in the preserve are threatened by more than 600 miles of R.S. 2477 claims. Modified from Amanda Dranginis, *Department of the Interior Facilitates Giveaway of Parks and Wilderness: R.S. 2477 Rule Takes Effect* (California Wilderness Coalition, Spring 2003), 7. Reproduced with permission.

In 2003, George W. Bush's first interior secretary, Gale Norton, gave Utah's bogus R.S. 2477 claims more kindly treatment in closed door negotiations. The Department of the Interior (DOI) signed an agreement with the counties as a basis for transferring land ownership from BLM to a state or county claimant.[49] The U.S. General Accounting Office (now called the U.S. Government Accountability Office) subsequently determined that the agreement explicitly violated a 1996 Congressional prohibition against final agency rule-making on R.S. 2477 without express Congressional approval. Nevertheless, in 2005 DOI directed federal agencies to assume that an R.S. 2477 right-of-way claim may be valid, even if unproved in court, and encouraged agencies to ignore unauthorized road-building on public lands. This policy further reduces the burden of proof on a claimant—but even worse, it allows claims in wilderness areas, national parks, and national wildlife refuges.

Government recognition for all claimed R.S. 2477 rights-of-way could effectively eliminate millions of acres of federal wilderness in some states. This attempted give-away is not likely to survive legal challenge, and administrative actions don't outlast an administration's tenure—so they can be undone by Congress or future presidents. But before the R.S. 2477 attack subsides, these claims will have inflicted huge damages on remote and reserved lands belonging to, and supporting the health of, the whole American public.

Better Routes

Roads and pipelines are firmly entrenched in American culture, and we certainly need many of them. But today's plethora of roads is not healthy for the western United States, or the country. We suggest four questions that can guide road-addiction recovery in the western states, and also provide answers:

1. Do we have enough, or more than enough roads and utility corridors?
2. Are the roads and corridors built to minimize damages to land and water?
3. Do we really need to be able to drive everywhere?
4. Should we allow dense road systems on public lands?

The clear answer to the first question is that the nation has many more than enough roads and utility routes. In spite of slow recovery rates (box 5.1), removing roads makes sense, and the road reclamation process is relatively cheap. The well-known adverse environmental impacts of roads already have generated proposals to remove dirt and unimproved roads, and also railroads, but a serious program for removing many more unneeded roads is past due. In the Pantheon of difficult environmental issues with difficult or improbable solutions, restoring roads is technically the easiest, cheapest, and may in time become the most popular.

Although building a diverse ecosystem of plant and animal species will take hundreds of years to millennia, when the land is left alone natural processes reduce or eliminate the marks of mild disturbances on dirt roads and town sites to a remarkable degree. But people have to stop disturbing the roads so that recovery can begin. Roads can be closed with barricades, ripped, partly revegetated to obscure them, and then left to complete the process at their own rates. In some desert areas, the BLM already obscures illegal roads and ORV tracks with artfully repositioned rocks and dead vegetation. A project of erasing roads as part of a mine reclamation project in the Mojave Desert shows much promise, even on steep slopes in an arid climate. Spreading local native plant seeds also has achieved good preliminary results.

The answer to the second question is overwhelmingly negative. Existing engineering standards can reduce the impacts of roads and utility corridors, but construction standards are generally ignored or not enforced. The nation, and the western states especially, would benefit from greatly expanded public and management agency oversight in the future. We also need stronger federal, state, and local laws and regulations requiring more rigorous pipeline and shipping safety measures, plus better enforcement of all safety laws. But a lot more Americans would have to demand programs for improving new roads and corridors and for minimizing their erosional and contaminating impacts.

Box 5.1 Slow Road to Recovery

Reclamation can easily erase the visible marks of roads, but not the deeper damage. Research on the condition of abandoned western mining boom towns, unoccupied remote subdivisions, and military camps and tank tracks show that desert surfaces recover very slowly, long decades after roads and corridors fade from view. Abandoned mining towns in Nevada and eastern California, unoccupied for 45–77 years before recovery studies (but grazed by cattle), still show significant differences in plant types, plant cover, and the variety of different plant species, compared to undisturbed control areas. Invasive cheat grass and other exotic grasses dominate in roads and trampled areas.[a]

Along abandoned southern Nevada roads, compacted soils and cleared vegetation have not fully recovered 88 years after the last significant use.[b] Studies of World War II tank maneuver areas and roads at briefly occupied 1940s military camps in the Mojave Desert provide further evidence that disturbed desert areas need long recovery periods (see chapter 6).[c]

One 1980 study compared human impacts in the mining boom town of Wahmonie, Nevada, with undisturbed lands nearby. Both erected and abandoned in 1928, Wahmonie's human impacts consist of a foot-trampled tent site, a bladed but never used street system, and a main street that carried a low level of Nevada Test Site traffic until 1961. Soils in the briefly used roads and living sites were still compacted as of 1980.[d] A 2002 study examined 31 briefly disturbed locations, including Wahmonie, Mojave Desert mine towns, and 1940s military campsites (see chapter 6), abandoned as much as 91 years ago. Of all the study sites, only two had fully recovered after 70 years. The original Wahmonie site vegetation still had not recovered after 75 years.[e]

Assuming that disturbed desert sites recover at a constant rate (linear model), the time needed for full recovery in the 2002 study sites calculates out to 92–100 years. But assuming instead that the processes promoting recovery are fastest immediately after occupation and then progressively slow down over time (logarithmic model), it all takes much longer. The logarithmic model predicts that even 85% recovery will take 105–124 years.[f] Applying the measured recovery rates to studies of five abandoned mining towns in Death Valley National Park, California, and two other Nevada sites suggests that they will require 80–140 years for full recovery.[g]

Notes
[a] P. A. Knapp. Secondary Plant Succession and Vegetation Recovery in Two Western Great Basin Desert Ghost Towns. *Biological Conservation* 1992;60:81–89; P. A. Knapp. The Response of Semi-arid Vegetation Assemblages Following the Abandonment of Mining Towns in South-Western Montana. *Journal of Arid Environments* 1991;20:205–222.

continued

Box 5.1 Continued

[b] Cheat grass is *Bromus tectorum*. Soils and vegetation recovery stages showed no significant correlation with the age of the roads but are markedly different from nearby control areas (J. D. Bolling and L. R. Walker. Plant and Soil Recovery Along a Series of Abandoned Desert Roads. *Journal of Arid Environments* 2000;46:1–24).
[c] D. V. Prose et al. Effects of Substrate Disturbance on Secondary Plant Succession, Mojave Desert, California. *Journal of Applied Ecology* 1987;24:305–313; J. E. Lovich and David Bainbridge. Anthropogenic Degradation of the Southern California Desert Ecosystem and Prospects for Natural Recovery and Restoration. *Environmental Management* 1999;24:309–326; Jayne Belnap and Steve Warren. Patton's Tracks in the Mojave Desert, USA, an Ecological Legacy. *Arid Land Research and Management* 2002;16:245–259; Anja Kade and S. D. Warren. Soils and Plant Recovery After Historic Military Disturbances in the Sonoran Desert, USA. *Arid Land Research and Management* 2002;16:231–243.
[d] Now enclosed by the Nevada Test Site and off limits to normal use or visitation, Wahmonie has remained largely free of disturbance since its brief 1928 occupancy. Until 1961, the test site used Wahmonie's main street as an access road. A modern dirt road was in use at the time of study (R. H. Webb and H. G. Wilshire. Recovery of Soils and Vegetation in a Mojave Desert Ghost Town, Nevada. *Journal of Arid Environments* 1980;3:291–303).
[e] Undisturbed (control) and disturbed test plots, established at the Wahmonie site in 1963, were remeasured in 1967, 1970, 1975, and 2003. Comparing the disturbed site (not including roads) with the control in 2003 shows dramatic residual disturbance effects, 86% lower numbers of creosote (*Larrea tridentata*) and 51% lower Anderson boxthorn (*Lycium andersonii*), among other differences. In 2003, the total of live plants was lower by 13% than in the control area, and the percentage of cover and biomass index also reflected those numbers (R. H. Webb et al. *Perennial Vegetation Data from Permanent Plots on the Nevada Test Site, Nye County, Nevada* [U.S. Geological Survey Open-File Report 03-336, 2003], 234–238). This immensely valuable report updates measurements on 68 permanent Nevada test site vegetation plots, involving various kinds of human and natural disturbances. The studies' results also indicate important controls of drought, higher than normal rainfall, and frosts, as well as the effects of human disturbances and natural disturbance such as fires, on the species composition, sizes, and abundance of plants. Repeat photographs show dramatic visual recovery, but "the species composition is not even close to the undisturbed plots, and extrapolation suggests that as long as a millennium will be required for recovery of species composition" (Webb et al. *Perennial Vegetation Data*, 12); see also U.S. Geological Survey. *Recoverability and Vulnerability of Desert Ecosystems* [Fact Sheet 058-03, 2003]).
[f] R. H. Webb. Recovery of Severely Compacted Soils in the Mojave Desert, California, USA. *Arid Lands Research and Management* 2002;16:291–305. Assuming that soils were fully compacted and abandoned with no subsequent disturbance, recovery times are calculated from the equation $IR = (P_d - P_a)/(P_u - P_a)$, where IR is the recovery interval, P can mean soil density or penetration depth: P_d represents historically disturbed soils, P_u is undisturbed soil, and P_a is soil in active roads (representing high current compaction). TF, the amount of time in years estimated for full recovery, is given by TF = TR/IR, where TR is the recovery time in years.
[g] R. H. Webb et al. Recovery of Compacted Soils in Mojave Desert Ghost Towns. *Soil Science Society of America Journal* 1986;50:1341–1344; P. A. Knapp. Soil Loosening Processes Following the Abandonment of Two Arid Western Nevada Townsites. *Great Basin Naturalist* 1992;52:149–154; see also D. V. Prose and S. K. Metzger. *Recovery of Soils and Vegetation in World War II Military Base Camps, Mojave Desert* (U.S. Geological Survey Open-File Report 85-234, 1985).

Selecting appropriate routes for new utility corridors can reduce negative impacts immediately. But instead of careful planning, in 2006 a crash program began designating hundreds of miles of new utility corridors across the west. Meanwhile, existing roads and corridors are poorly managed: While paved road systems require continuous maintenance, dirt road maintenance should be greatly reduced, especially in desert lands. Since pipeline accidents have raised citizen opposition to new pipeline corridors, repairing or restoring old pipelines should become a priority.

Eventually, no new roads should be permitted without thorough reviews of present routes. High penalties should be exacted for violating strict construction standards. Recreational vehicles could be restricted to stadiums or existing designated routes, which eventually will need to be paved or reclaimed. Other critical environmental protections should include requiring that pipelines with similar routes share construction corridors, as recommended for lands of special value in the Carlsbad Resource Area, New Mexico, and Kofa Game Reserve, Arizona. In addition, blading should be avoided during construction except where absolutely necessary. Just driving heavy equipment across the land does not kill the more resilient plants and gives all plants a better chance at recovery, compared to blading, which generally removes root crowns and the seed bed.

Our answer to the last two questions is a resounding "No!" But only an aroused public can protect public park and forest lands. Without broad agreement that we must not extend roads into wilderness, and strong public intervention, public lands opponents will claim even more R.S. 2477 roads. The federal government also may open inventoried national forest roadless areas and expand road building on habitat-rich lands to develop marginal or nonexistent energy resources (chapter 12).

Arousing a public in love with cars and road trips is a big challenge. Perhaps only the end of cheap petroleum will deter American car lovers from ruining public lands. A movement for road obliteration and restoration already is having an effect,[50] but many more must speak up for substantial route limitations and reclamation if we are to save wilderness—and ourselves.

6　　Legacies of War

After September the 11th, I made a commitment to the American people: This nation will not wait to be attacked again. We will defend our freedom. We will take the fight to the enemy.
George W. Bush

The problem with defense is how far can you go without destroying from within what you are trying to defend from without.
Dwight D. Eisenhower

Since 1900, United States troops have fought in more foreign conflicts than any other nation on Earth. Most Americans supported those actions, believing that they would keep the scourge of war far from our homes. But the strategy seems to have failed—it certainly did not prevent terror attacks against the U.S. mainland. The savage Oklahoma City bombing in 1995 and the 11 September 2001 (9/11) attacks on New York and Washington, D.C. were not the first to inflict war damage in America's 48 contiguous states, however—nor were they the first warlike actions to harm innocent citizens since the Civil War.

Paradoxically, making war abroad has always required practicing warfare in our own back yards. Today's large, mechanized military training exercises have degraded U.S. soils, water supplies, and wildlife habitats in the same ways that the real wars affected war-torn lands far away. The saddest fact of all is that the deadly components of some weapons in the U.S. arsenal never found use in foreign wars but have attacked U.S. citizens in their own homes and communities.

The relatively egalitarian universal service of World War II left a whole generation of Americans with nostalgia and reverence for military service. Many of us, perhaps the majority, might argue that human and environmental sacrifices are the price we must be willing to pay to protect our interests and future security. A current political philosophy proposes that the United States must even start foreign wars to protect Americans and their homes. But Americans are not fully aware of all the past sacrifices—and what we don't know *can* hurt us.

Even decades-old impacts from military training still degrade land and contaminate air and water, particularly in the arid western states, and will continue to do so far into the future. Exploded and unexploded bombs, mines, and shells ("ordnance," in military terms) and haphazard disposal sites still litter former training lands in western states. And large portions of the western United States remain playgrounds for war games, subject to large-scale, highly mechanized military operations for maintaining combat readiness and projecting American power abroad. These military operations continue to damage millions of acres of public land.

Unfortunately, the Department of Defense (DOD) has failed to clean up the old contamination or prevent additional environmental damage to western lands, while natural processes constantly spread them to populated areas. At the same time, residential developments are expanding into formerly remote areas, including decommissioned training sites (see chapter 8). The less remote military reservations are being converted to civilian uses, exposing families to unexploded ordnance (UXO). The DOD also has been extremely reluctant to reveal the variety of past weapons-related experiments using biological and other harmful substances, some of which purposely exposed hundreds to millions of Americans to potentially lethal toxins. Especially vulnerable populations are those living downwind from weapons development or test sites in Washington, Idaho, Nevada, and Utah ("downwinders").

Understanding the long-term toll of modern weapons and mechanized warfare is especially important in a time of declining energy resources and a destabilizing U.S. war in the world's main oil-exporting region. Studying past practices can help us change the destructiveness of military training, or determine that some weapons actually make us more vulnerable to the newly perceived threat of global terrorism. Instead of planning new rounds of weapons production and testing, the United States needs a public debate about the national security strategies that can destroy, and that have destroyed, American lands and civilian lives.

World War II at Home

World War II spread army and air force training and testing ranges across public lands in the 11 western states, encompassing dry, barren salt pans, sloping alluvial fans, broad, harsh-appearing grass- and shrub-covered intermountain valleys, and tree-clad mesas. Before the 1940s, these landscapes supported vibrant ecosystems of interdependent plants and wildlife, in spite of meager rainfall. But wartime and subsequent Cold War military operations have dramatically altered those lands, essentially forever in human terms. Of all military activities, nuclear weapons manufacturing and testing have had the longest lasting effects, leaving great, gaping holes in the earth (see chapter 7, figure 7.7) and widespread radioactive contamination (see chapter 7 for a discussion of the effects of radioactive weapons development and testing).

Many former training, testing, and weapons and waste disposal sites were never cleaned up and now are abandoned or turned to other uses. Large expanses remain tracked and barren of most vegetation, pocked with bomb craters, and littered with UXO—artillery shells, bombs, bullets, and land mines.

The most remote of the damaged and polluted WWII training and bombing sites are shut away inside military reservations. But natural processes that we cannot control still connect the trampled, bombed, and dumped-on sites with populated areas. Both winds and water have carried pulverized soils[1] and plants long distances, including across cities, towns, and suburbs. Even if acutely damaging military activities halted today, dangerous pollutants from past thoughtless or reckless decisions and actions in the name of national security will continue to spread widely.

Patton's playground

During WWII, training exercises and concentrated weapons testing took place on both permanent and temporary bases scattered across the western states. Starting in April 1942, General George Patton prepared more than a million U.S. Army troops to fight in North Africa at temporary bases on the "Desert Training Center" (DTC). About the same size as South Carolina, the 18-million-acre DTC extended 350 miles east from Pomona, California, to Phoenix, Arizona, and 250 miles north from Yuma, Arizona, into southern Nevada.[2]

Preparing the DTC to handle large troop assemblies, plus a variety of motorized military maneuvers and operations, meant building forts, air fields, and more than a dozen temporary base camps. Each camp was a good-sized town, sprawling across 2,500–6,000 acres. Bulldozers often cleared camp sites for tent rows, road networks, and vehicle parking lots, but at many places the troops hand-cleared rocks from natural desert surfaces and soils (figure 6.1).[3]

Patton's real job was at the front, and he soon departed. In 1943, his brigades joined with British forces to defeat the German Panzer divisions in North Africa. By May 1944, the DTC's roads, barren camp sites, and war litter—including ordnance—were all abandoned to desert scavengers and natural forces. The human

Figure 6.1 Aerial view of Camp Iron Mountain looking north, showing the WWII-era road system, large light patches of former vehicle parking areas (to the right of third road from right), and light tent areas (between third and fourth roads from right, also to right of fifth road). Many other disturbances also remain visible, and some roads are used by recreational vehicles. Photographed April 2003, 60 years after the camp was abandoned.

scavengers can attest that discarded WWII hardware represents a significant proportion of America's once-rich lead and copper ore deposits (see chapter 4). Many of the DTC training sites remained isolated and now provide opportunities for studying how (or if) a variety of desert ecosystems can recover naturally after such severe use.

Former WWII trainees can remember the immediate effects of dozing and trampling desert surfaces at Camp Hyder, Arizona: "[W]hen the Division was called up to maneuver as a whole on a broad front, clouds of rolling dust engulfed the landscape for days," one recalled 50 years later. Detailed studies of Mojave Desert Camps Iron Mountain, Clipper, and Ibis show the long-lasting impacts from military training, even after four to six decades of disuse.[4] At Camp Iron Mountain, rock borders still mark the edges of roads and pathways, and rock circles even surround shrubs. The trainees also used naturally colored rock chips to create elegant unit insignia mosaics, many still visible. Also at Camp Iron Mountain, General Patton ordered construction of a 75 × 150-foot DTC model, accurately depicting every mountain and valley (figure 6.2). Even ravaged by wind, water, and burrowing animals, this artificial terrain remains vivid. Visitors can imagine the booted general stomping around the elevated board walkway, snapping out orders.

Figure 6.2 Aerial view of General Patton's World War II Desert Training Center (DTC) base camp at Camp Iron Mountain, east of Twentynine Palms, California. Flag circle and relief model are near the wide cleared road segment nearest viewer. The dark rectangular area is the DTC scale model: Parts of the model's elevated board walkway are visible at left and in front. Roads have diverted some of the stream channels. Photographed May 1982.

The Mojave camps were built across sloping alluvial fans—overlapping conical sediment deposits at the base of steep mountain fronts. Myriad small runoff channels, which carry rainwater from mountain canyons to valley floors, give natural alluvial fan surfaces a wrinkled appearance. During camp occupancy, intense summer rainstorms nearly obliterated several camps located too close to these part-time drainage channels.

The army engineers created level roads by cutting through ridges and filling in the channels, but did not install culverts to convey water under the roads. Generally, the fill material formed dams, which trapped the rainwater and prevented it from reaching downslope ecosystems. After road construction, each runoff-generating storm deposited a sediment layer on the upstream side of the dams. In 1985, 42 years after the camps' abandonment, one dam had accumulated sediment from 22 runoff events. Some campsite roads and pathways captured rainstorm runoff,[5] diverting water from natural channels (figure 6.2) and promoting washouts or flooding farther down the fan. No studies have assessed the long-term effects of these dams and diversions on plants and animals in the camps or farther downslope.

Desert soils are generally fragile, and the plants few and far between, but they develop biological crusts—complex mosaics of microscopic bacteria ("cyanobacteria"), lichens, mosses, and microscopic fungi—that protect areas between plants from wind and rain erosion, enrich the soil in those areas, and enhance soil moisture (see chapters 3, 13).[6] Most training operations and camp activities, including hand clearing and trampling in tent areas, road grading, and driving and parking tanks and other military vehicles, wiped out the plants, crusts, and other natural stabilizing soil covers.

Soldiers' trampling feet, as well as their jeep and truck tires, tank treads, and dozers, also compress and compact the desert soils. Soils at Camps Iron Mountain, Clipper, and Ibis still were compacted 42 years later and had as much as 15% greater soil densities than in nearby undisturbed "control" areas. The compaction effects extend as much as 12 inches below the ground surface. Wind and water erode compacted soils faster than natural soils, and compaction slows seed germination for some plant species while starving established plants of needed soil nutrients—nature's plant food.

Scraping, and even hand-clearing plants and rocks, slows or obstructs the comeback prospects for most native plants. Slow plant recovery means that the original diversity of plant and animal species and ecosystem functions recover even more slowly. Many perennial plants have not regrown in the disturbed areas at Camps Iron Mountain, Clipper, and Ibis, signifying long-term habitat degradation.[7] Before disturbance, the important perennial species included disruption-tolerant burroweed and more sensitive creosote bush, which put out extensive soil-stabilizing root systems.[8] Even after more than half a century, the machine-cleared Camp Ibis airfield grows mostly burroweed, to the near exclusion of other long-lived native species (figure 6.3). Short-lived annual plants have not made much of a comeback, either.[9]

Mechanically graded tent areas on an early 1940s military camp site at the U.S. Army Yuma Proving Ground in Arizona showed virtually no soil crust or perennial plant recovery, even 56 years after abandonment. But half the perennial plant types

Figure 6.3 Aerial view of Camp Ibis airfield west of Arrowhead Junction, California. Dark plants in the surrounding undisturbed areas are creosote and yucca. *Ambrosia* plants predominate in the airfield, nearly excluding all other species. Photographed May 1999.

had regrown substantially at a nearby ungraded motor pool site, yielding plant coverage and densities more comparable to undisturbed areas. Grading in the tent site had removed soil, seed bed, and plant root crowns, which slowed recovery much more than localized vehicle compaction at the parking lot.[10]

All studies point to the conclusion that soils in military campsites will take about a century to regain normal soil densities and healthy plant coverage (see chapter 5, box 5.1).[11] The very delicate biological soil crusts probably recover even more slowly. Enough crusts may grow in a few decades to help stabilize the soils, but the complex aggregate of species making up the biological crusts probably takes much longer to recover, perhaps millennia.[12] Soil denuded of these stabilizers erodes faster than it did before and provides less support for plants.[13]

The tracks of our tanks

The DTC camps were concentrated areas of human impact, while the tanks ranged widely across the valleys and gentler hill slopes. Any valley floor that hosted tank exercises still carries their scars—the heavy machines destroyed biological soil crusts, compacted and deformed the soil, crushed small animals or caved in their burrows, and damaged any plants they rolled over. Even a single tank, passing only once across a desert surface, severely compacted and deformed the soil.[14] Like trampling—and even more like grading—the tanks' soil-compacting effects reduced rain infiltration and increased the daily temperature range within the soil. They also increased the albedo, the amount of light and heat reflected off the ground surface. All these changes create conditions that tend to exceed the comfort level for many desert plants and animals, damaging their prospects for thriving, or even surviving.

Tanks and other heavy machines also destroyed the desert pavements that protect many undisturbed desert soils. The pavements form over hundreds or thousands of years, as wind and water remove fine silt and sand particles so that pebbles and larger stones dominate the ground surface. The stones' exposed surfaces accumulate clay, along with iron oxide and manganese oxide minerals that bestow a dark coloration (desert varnish).[15] Vehicles cast the dark pebbles aside, exposing light-colored silt immediately below the surface (figure 6.4). This effect of displacing desert varnish is what has kept the WWII tank tracks visible for so many years.

The compacted tank tracks were and are surface depressions that collected windblown sand and have developed a generally contrasting set of plant species in comparison to undisturbed soil outside the tracks. Plants with spreading root systems grow easily in the loose sediment in the tracks, but plants with vertical tap roots grow mostly outside the tracks because they have difficulty penetrating the compacted soil.[16] Annual plants in the tracks tend to be stunted compared to the same species on undisturbed land nearby, however. These differences in plant sizes also help make the tracks visible.

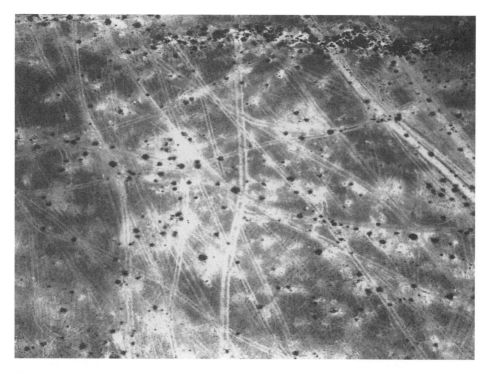

Figure 6.4 Aerial view of 1942–1943 tank exercise tracks (narrow tracks), with overprints of wider tracks from 1964 tank exercises. Photographed April 2003.

Shooting up the place

So many WWII aerial gunnery (strafing) exercises took place across the DTC in 1942 and 1943 that scavengers still find widely distributed 50-caliber shell casings and bullets even many miles from target areas. Some of the shooters aimed at bull-dozed DTC strafing targets in the Mohave Mountains and adjacent Dutch Flat of western Arizona. The targets are rectangular clearings on moderately sloping ter-rain, arranged in three subparallel rows, each about 20 miles long. Each clearing is roughly 1,000–2,000 feet long and 300 feet wide and spaced about 2,000 feet from its nearest neighbor. At upland sites, the bulldozer cuts still reveal every move the blade made in clearing the target. The cuts are so shallow that they often left patches of intact varnished desert pavement between dozer berms of gravel and plant debris (figure 6.5). These desert targets were largely abandoned after 1943, making them useful for investigating natural desert vegetation recovery modes and rates.

The degree of recovery depends on local soil conditions and the type of desert surface.[17] Creosote shrubs fully recovered on the targets carved into older Mohave Mountain alluvial surfaces, probably because the shallow bulldozer cuts left root crowns intact. But strafing targets in broad Dutch Flat valley show poor creosote recovery, despite substantial runoff coming from the forested Hualapai Mountains. After nearly 60 years, both air photos and ground observations show little or no native Joshua trees, nor any species of cacti, ocotillo, or paloverde on the clearings.[18] All of

Figure 6.5 Aerial view of WWII strafing target, Mohave Mountains, Arizona. Berms of soil and plant debris reveal the bulldozer's movements 40 years later. Photographed May 1982.

these plants grow throughout the region, including on natural uncleared lands close to the targets. Biological soil crusts are beginning to recover in the targets, reducing the color contrast with surrounding lands and restarting the natural ecosystem functions.

Permanent War

Intensive military training continued across the 11 western states throughout the Cold War and will go on for the foreseeable future. The DOD currently controls more than 12 million total acres in the west, twice the size of Vermont. Patton's DTC evolved into two large permanent California training bases: the U.S. Army National Training Center (NTC, formerly Fort Irwin), near Barstow, and the Twentynine Palms Marine Corps Base, near Palm Springs—both under continuous and intensive use. Other current western training sites include the Chocolate Mountains bombing range in Arizona; Yuma Proving Grounds in California and Arizona; coastal California's Fort Hunter Liggett and Camp Pendleton; central California's Camp Roberts and Camp San Luis Obispo—also Fort Carson, Colorado; Nellis and Fallon, Nevada, air bases; and Fort Lewis in Washington. The airspace over another 61 million acres or so of Arizona, California, New Mexico, Nevada, and Utah—in aggregate, the size of Oregon—is restricted for military uses (figure 6.6).[19]

There is no doubt that all the impacts from two years of DTC use apply also to current military training and weapons testing, although modern tanks, troop carriers, Humvees, and other vehicles are larger and heavier than WWII equipment. One example is the massive 1964 Desert Strike tank training exercise across half a million Mojave Desert acres, somewhat incongruent with a contemporaneous war in Southeast Asia. The Army's Desert Strike tanks were larger and heavier than Patton's

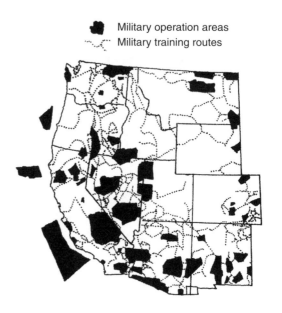

Military operation areas

Military training routes

Figure 6.6 Map showing restricted military airspace over the western United States and training flight routes. From National Air Coalition, Rural Alliance for Military Accountability.

and destroyed biological soil crusts over much wider areas. Their wider treads, however, spread out pressure effects and so compressed the soils less severely.

The 642,000-acre NTC, an area the size of Rhode Island, has been a mechanized troop training and heavy weapons gunnery site for many decades and obviously has sustained much more severe impacts than did the DTC. It is scheduled to be active for the foreseeable future and so will continue to sustain heavy damage. Both NTC and the Twentynine Palms base are devastated landscapes of rutted passes and valley flats plowed wall-to-wall with tank tracks. Since the Persian Gulf War, nearly every NTC hillock is a Humvee playground. The bases' churned up soils are sources of major dust clouds that choke the downwind towns and ranches. Dump piles randomly litter slopes and arroyos with everything from targets and boxed "Meals, Ready to Eat," the standard military rations, to razor wire and many forms of fake and real ordnance.

The military's larger vehicles and greater reliance on remotely controlled drones and missiles, and other technological innovations are expanding military training demands on the west. In 1978, the U.S. Army claimed it needed 21 times more land for training one mechanized battalion than it did in WWII (figure 6.7).[20] In 1993 the army proposed enlarging the mountainous NTC by 220,000 acres of land usable for maneuvers—meaning land with low slopes. To annex more than 200,000 relatively level acres to NTC would mean adding more than 300,000 total acres of contiguous lands.

Until the 9/11 terrorist attacks, both local residents and environmental groups protested the NTC expansion and tried to limit it. That vigorous opposition suggests that the military's perceived need for training space might have outstripped the supply of remote U.S. lands expansive enough to meet additional demand. As 9/11 dampened public opposition, late in 2001 the U.S. Department of the Interior's Bureau of Land Management, overseer of most public desert lands, offered the Department of the Army a compromise—150,510 acres for NTC expansion, and another 200,000 acres preserved as wilderness.

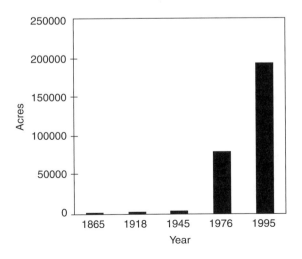

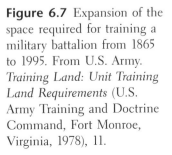

Figure 6.7 Expansion of the space required for training a military battalion from 1865 to 1995. From U.S. Army. *Training Land: Unit Training Land Requirements* (U.S. Army Training and Doctrine Command, Fort Monroe, Virginia, 1978), 11.

If the lands used only briefly for military training will not return to a more natural condition for periods of many decades to a thousand years or more,[21] continued intensive military training and testing can only add severe damage to sensitive desert ecosystems and forestall recovery essentially forever in human terms. Some efforts to rehabilitate old town sites and lands used for massive motorized recreational pursuits (see chapter 11)—showing damages similar to Patton's camps—have claimed success (see chapter 5, box 5.1).[22] But the real test is what happens over longer times, as global climate warming stresses desert water sources and plants. For now, neither the military nor the public are able to offer practical, reliable, or effective strategies for repairing the lands abused by perpetually practicing for war.

Old bombs' network

Modern wars and tactics rely more and more on aerial gunnery and bombing than ever before. Active bombing sites cover nearly eight million acres in the western United States, including California's Twentynine Palms (600,000 acres) and Chocolate Mountains Gunnery Range/El Centro Naval Air Facility (125,000 acres), Nevada's Nellis Bombing and Gunnery Range (1.6 million acres) and Fallon Naval Air Station (125,000 acres), Arizona's Barry M. Goldwater Range (2.5 million acres) and Yuma Proving Ground (one million acres), Utah's Test and Training Range (1.7 million acres), Idaho's Saylor Creek Bombing Range (100,000 acres), Oregon's Boardman Bombing Range (45,000 acres), and New Mexico's Holloman Air Force Base (50,000 acres). Artillery practice occupies additional large areas—such as the 40,000-acre Bravo-20 area of Fallon Naval Air Station, battered for many decades by bombing and strafing training. Gunnery and bombing practice, along with unplanned, unrecorded, and unmonitored waste disposal, have contaminated soils and surface and groundwater on every one of the practice ranges.

From the bombing ranges, windstorms carry dust clouds long distances, exposing fast-growing western populations—the new-century downwinders—to a wide variety of contaminants. Lands surrounding the bombing ranges are prone to accidental bombings, so both the Bravo-20 area and surrounding lands are intensely cratered, littered, and polluted by targets, weapons debris, and explosive residues.

Unknown amounts of UXO, about 10% of fired rounds, lurk in the western valleys and slopes on formerly utilized defense sites (FUDS), used for bombing and gunnery practice during and since WWII. Whether fired or abandoned, the UXO that penetrated beneath the surface presents grave dangers to wildlife and humans. In a sandy wash, its presence may be indicated by no more than a dimple in the sand, escaping the notice of a walker or driver. Complicating efforts to detect live bombs and shells are the hundreds of nonexplosive "practice" rounds used in the same exercises. Both duds and live ordnance add greatly to soil and water contamination. Another source of UXO is "excess" (unused) live ammunition, commonly abandoned, buried improperly, or dumped in a lake, wetland, or the ocean.

Toxic substances, both radioactive (see chapter 7) and nonradioactive, leak from corroded UXO (table 6.1; see also appendix 4),[23] and even removing UXO carries the risk of explosions or fires. When ordnance is detonated or burned in place, the chemical residues must be cleaned up if the land is to be used safely for other

Table 6.1 Hazardous Constituents of Munitions

Constituent	Potential Toxicity Effects
TNT[a]	Possible human carcinogen; affects liver; skin irritant; cataracts
RDX	Possible human carcinogen; causes prostate problems; affects nervous system; causes nausea and vomiting, and organ damage in lab animals
HMX	Lab animal tests indicate possible damage to liver and central nervous system
Perchlorate	Causes itching, tearing, pain; ingestion may cause gastroenteritis, abdominal pain, nausea, vomiting, and diarrhea; systemic effects may include ringing of ears, dizziness, elevated blood pressure, blurred vision, and tremors; chronic effects may include metabolic disorders of the thyroid
White phosphorus	Reproductive effects; liver, heart, or kidney damage; death; skin burns, throat and lung irritation, vomiting, stomach cramps, drowsiness
Lead[b]	Low levels can slow children's mental development and cause learning and behavioral problems; high levels may cause lasting damage to the developing nervous system and the reproductive system; kidney damage, anemia, deafness, blindness, coma, and even death can also occur; adults can suffer from many of the same effects as children and may also be affected by loss of hand/eye coordination, hypertension, high blood pressure, and stroke

U.S. General Accounting Office. *Military Munitions: DOD Needs to Develop a Comprehensive Approach for Cleaning Up Contaminated Sites* (GAO04147, 2003), table 3; Jim Giles. Collateral Damage. *Nature* 2004;427:580–581.

[a] Compounds: TNT, trinitrotoluene; RDX, hexahydro-1,3,5-trinitro-1,3,5-triazine; HMX, octahydro-1,3,5,7-tetranitro-1,3,5,7-tetrazocine; white phosphorus, N-2,4,6-tetranitro-N-methylaniline.
[b] Lead is used as a primer for detonating explosives. Most common types are lead azide and lead styphnate, responsible for dangerous levels of lead in soils at firing ranges and heavily hunted areas. It may also be released from deteriorating dumps of excess munitions.

purposes.[24] As military bases are decommissioned, people should not be invited on them until and unless cleanup programs can guarantee safety. Unfortunately, weapons cleanups are very expensive and rarely or never attract funding for such large areas (see chapter 10).[25]

Far from managing or cleaning up contaminated areas, DOD has not yet discovered the full extent of known or possible UXO contamination on U.S. military training ranges. Most FUDS remain unsurveyed,[26] but DOD estimates they cover more than 39 million acres, approximately the size of Washington State. Many sites are found only accidentally because records of their existence—let alone the types of ordnance—never were made or have been lost.[27] Many munitions burials were not reported, while the records that do exist tend to omit critical details—such as locations or burial depths. From time to time, live weaponry turns up on public lands with no known military history, such as the dozen live land mines found along the Johnson Valley, California-to-Parker, Arizona, cross-country motorcycle race route in the 1980s, and the four land mines uncovered in Ward Valley, California, during 1990s site studies for a proposed low-level radioactive waste dump (see chapter 10).

The U.S. Army Corps of Engineers is responsible for identifying, investigating, and cleaning up DOD ordnance-related hazards. The corps has estimated that more than 9,000 FUDS are potentially eligible for cleanup and that more than 4,000 of those sites require no cleanup. The U.S. General Accounting Office (GAO, now the Government Accountability Office) counterestimated, however, that more than a third of the 4,000 site evaluations were inadequate.[28] The corps has a better record when it comes to more recent training ranges—as of 2000, it listed 1,600 of those FUDS as potentially contaminated with UXO and other explosive wastes, and 750 as associated with former military training ranges and therefore potentially UXO contaminated. But the GAO has identified an additional 200 potential past training range sites.[29]

Under congressional pressure to assess cleanup costs, the DOD finally issued a UXO site inventory in June 2003. At least, DOD *said* it released the inventory. The public has yet to see it, so the inventory's accuracy and completeness remain unverified.

The legacy

In the twentieth century's waning stages, DOD finally recognized a duty to protect identified natural and cultural resources on its permanently restricted bases. It established the "Legacy" program (now the "Legacy Resources Management Program") to help military services comply with the National Historic Preservation Act, National Environmental Policy Act, Endangered Species Act, Clean Air and Clean Water Acts, and other environmental laws and standards (see appendix 1). At first, military base commanders were charged with implementing the Legacy Program, but by the mid-1990s the Army had transferred implementation to the Army Corps of Engineers, even for desert lands.

Under the Legacy Program, a few limited site-specific scientific studies looked at ecosystem impacts on military training lands. These included studies on both the long-used Fort Carson and newer Pinyon Canyon training sites, both in Colorado,[30] and a generalized study on plants and wildlife habitat impacts of current armored training at NTC.[31] But there are almost no data for evaluating the potential impacts on soils, land surfaces, air and water erosion, and microscopic ecosystem components from expanding NTC and other bases. All evidence, however, points to the conclusion that expanding military training into largely pristine natural areas cannot avoid inflicting severe damage on desert lands and wildlife.

The NTC study showed that constant intensive training has reduced both regional air quality and local water quality, sending ecosystems and wildlife habitat into long-term decline. Motorized training significantly lowers the proportion of coarser pebbles and cobbles in surface soils and increases the proportion of sand particles (less than 1/9 of an inch), greatly enhancing soil erosion. The frequent strong desert winds carry the sand and smaller silt particles far beyond the military sites' boundaries. The NTC study also showed that training activities reduce the density, bulk, and mean height of dominant woody creosote and burroweed shrubs. Several plant species simply disappeared from intensively used areas. Cheesebush and other highly disturbance-tolerant native species thrive in these areas and have increased in number.

All the big training bases have lost their taller plant populations and now mostly support shorter species. Small plants squashed under wheels and tank tracks tend to survive better than the tall plant species overturned by tanks. Similarly, four-wheel drive recreational vehicles disproportionately destroy taller desert shrubs compared to effects of motorcycles (see chapter 11). Many birds and small mammals species cannot tolerate military activities, even at moderate levels,[32] and have to abandon the intensive-exercise areas. The fate of those animals remains unknown, but they may have tried moving to relatively undisturbed areas outside the base perimeters. There is no way to assess whether the dispossessed creatures find homes in those areas, which likely were fully occupied already.

The NTC expansion extends military impacts into healthy desert tortoise and Lane Mountain milk-vetch populations—both officially threatened or endangered species.[33] The expansion also threatens Mojave ground squirrels, which the U.S. Fish and Wildlife Service has tried to list for protection against much political resistance. No adequate, objective scientific study has evaluated the potential effects of military training on these keystone desert species. In 2000, DOD faced opposition to the effects of NTC expansion and of Navy submarine-locating tests using high-decibel signals on endangered and threatened species. Ignoring the Legacy Resources Management Program, the Pentagon asked Congress for exemption from Endangered Species Act requirements on military lands.[34]

The 2004 DOD appropriations bill granted DOD exceptions to some environmental laws; for example, it exempted the military from aspects of the Marine Mammal Protection Act and other aspects of the Endangered Species Act. The bill also allowed military operations near endangered species' critical habitat with little oversight, and included a looser definition of what constitutes "harassment" of marine mammals.[35] To prevent U.S. Fish and Wildlife Service interference and the Department of the Interior from designating "critical habitat" for threatened and endangered species on military lands "in some cases," the bill lets the military define and manage critical habitats. But the bill does not necessarily allow DOD to reclassify previously designated critical habitat, such as the tortoise habitat that the NTC just engulfed. Military assaults on environmental regulations in the halls of Congress continued into 2005 and 2006.

In Our Back Yards

Widespread chemical contamination from weapons research, manufacture, and testing may affect as much as 870,000 acres at 59 western military installations.[36] The contamination emanated from repeated, poorly conceived and poorly documented processes and policies—all expressing reckless disregard for, or abysmal ignorance of, the intimate connections among soils, water, living plants and animals, and people. No one has yet invented a feasible method for cleaning up or disposing of many of the hazardous pollutants. Some lands will remain unsupportive of wildlife breeding and survival at best, and at worst will hurt people, and their economic interests and recreational pursuits, far into the future.

The huge list of contaminated former military facilities includes storage tanks, buildings, landfills, dam/reservoir impoundments, waste treatment plants, and hazardous waste reservations.[37] The U.S. Environmental Protection Agency (EPA) lists 45 Superfund (National Priority List) sites on western military bases, all needing cleanup. Hundreds more non-Superfund sites may or may not get cleaned up, ever.[38] Between 1985 and 2003, DOD environmental restoration costs totaled $21.7 billion, and base realignment and closure cleanup came to another $8.5 billion. Annual costs have come down about $440 million over the past decade, but remain close to $2 billion. And much work remains to be done.[39]

Groundwater contamination beneath the military sites is a widespread and growing problem, far greater than authorities care to admit. The contamination is everywhere—in surface water, soil, and groundwater and in the sediment and rock of relatively thick unsaturated zones, between the ground surface and the top of the groundwater zone. In time, the contaminants will seep into presently uncontaminated soil and water supplies. A great deal of money has gone into developing potential techniques for remediating soils and groundwater contamination, but no economically feasible methods are available for treating the worst and most intractable groundwater pollution (see chapter 10). For some contaminated weapons-manufacturing sites under the Department of Energy, the "solution" is to "maintain institutional control in perpetuity"—murky official lingo for "keep people out of those areas forever."[40]

UXO at home

Originally far from populations, many FUDS have sprouted neighborhoods or morphed into recreation areas, in spite of their potential to harm people where they live and play. Unfortunately, DOD's desire to unload financially burdensome bases raised a clamor to decommission military bases for accessing their prime developable lands. As a result, something like 16 million potentially contaminated FUDS acres have been transferred to other agencies' jurisdiction or private developers, without cleanup or other precautions.[41] More than seven million acres of public lands now under various U.S. Department of the Interior agencies have known or suspected UXO contamination. The Bureau of Land Management alone now oversees more than five million FUDS acres in 10 western states, with unrestricted public access for the most part.[42]

Turning over UXO-laden training lands for human uses without appropriate mitigations "results in the public being put at greater risk of sickness, injury, or even death from unexploded ordnance or its constituent contamination."[43] Demilitarization has exacted a toll already—one EPA survey of 61 UXO sites listed five accidental explosions, including three fatalities and two injuries, plus "incidents" at 24 other former DOD facilities.[44] Another EPA document reportedly listed 126 close civilian-UXO encounters over 83 years, including 65 fatal accidents and more than 100 injuries. Examples include a hiker who found ordnance on Camp Hale, Colorado, in 1999. The camp was used for WWII mountain training and now is under U.S. Forest Service management. Camp Hale closed to public access after a 2000 discovery of unexploded grenades. In 1983, an artillery shell explosion killed two children housed on the former Camp Elliot, California, gunnery range. These incidents have made headlines, but accurate figures on UXO accidents are not available.

Cleaning up military training hazards is a low government priority, and full cleanups would require millennia at proposed appropriations levels.[45] We can expect future disasters related to two new schools and a stadium beside Lowry Bombing Range, near Denver, Colorado; on or near homes, farms, parks, and a wildlife refuge at the former Baywood Park military training area, Morro Bay, California; and in a planned development of 5,000 homes and a golf course at former Camp Beale—a UXO-laden WWII combat range in Yuba County, California.

With cleanup funding in significant decline,[46] programs are focusing on the desired end instead of traditional "milepost" goals. The end-state focus could lead to large-scale containment, but treatment or removal is the better solution for protecting public health and the environment.[47] Given the lack of cleanup zeal or knowledge about where troops discarded or buried live munitions, the number of deaths and injuries to date and the high costs of locating buried UXO should block planned conversions of many military lands to civilian use. But there is no guarantee that the Pentagon will protect civilians better now than it has in the past.

Dumping razor blades

During WWII, the United States accelerated top secret programs to develop agents and weapons of "unconventional warfare."[48] Throughout the Cold War, remote saline basins of Nevada and Utah desert lands served as both testing and dumping sites for these chemical and biological weapons. The remote lands supported sparse Native American populations and ranchers of European descent, who could raise little but sheep on the thin vegetation. The military authorities dubbed them "a low-use segment of the population" and called central Utah "a damn good place to dump used razor blades."[49] The dumping went far beyond razor blades—it exposed downwind sheep herders and more distant city-dwellers alike, some of the original downwinders, to chemical and radioactive clouds (see chapter 7).

Records show that the military detonated more than 55,000 chemical rockets, artillery shells, bombs, and land mines at the former Dugway Proving Ground west of Salt lake City, now the Utah Test and Training Center. The detonations released deadly nerve agents into the air, as well as the biological agents that cause anthrax, Q fever, parrot fever, rabbit fever, undulant fever, and valley fever. The agents sometimes drifted over nearby ranches and reservations.[50] One accidental VX nerve agent release in 1969 killed sheep more than 35 miles to the east,[51] more than halfway to the densely populated Utah cities of Salt Lake City, Orem, and Provo. Sheep rancher Ray Peck and his family were the most highly affected people in the VX cloud's path. The Pecks had been a largely healthy family before VX exposure, but afterward suffered from headaches, numbness, and burning sensations in their limbs. With no previous history of reproductive problems, after VX exposure Mrs. Peck experienced two stillbirths and the couple's grown daughters have suffered several miscarriages.[52]

The military's insecure storage and disposal of pathogens, nerve gases, and other chemicals and weapons also have contributed long-lasting surface and underground contamination. The pollution's nature and scope are veiled in secrecy. Weapons containing mustard gas and nerve agents have contaminated storage and disposal sites on the Pueblo Logistics Base in Colorado and Nevada's Hawthorne Army Ammunition Depot.[53] Similar contamination is likely everywhere that weapons were both tested and stored, including Tooele Army Depot, Utah; China Lake Naval Base, California; Umatilla Army Depot, Oregon; and offshore of California, near the Farallon Islands.

The Tooele Depot stores more than 42% of the nation's chemical weapons, and the Umatilla Depot 12%—about seven million pounds in all. Tooele Depot personnel have the hazardous task of dismantling and disposing of it, in an ongoing project (see chapter 10). Anthrax-bearing weapons contaminated Dugway soil in the late 1960s, when the area still seemed relatively remote and undeveloped.[54] Former Dugway workers' tales suggest the Test and Training Center still harbors chemical and biological UXO that demolition teams never found.[55]

In the summer of 2005, a clam dredge off the coast of New Jersey came up with a World War I–era artillery shell filled with solidified mustard gas. Subsequent Newport News *Daily Press* investigations revealed widespread weapons dumping in U.S. coastal waters between 1944 and 1970. The Army admits to disgorging 64 million pounds of nerve and mustard gas, along with 400,000 chemical-filled bombs, land mines, and rockets—but can't identify all of the sites or what they contain. Until the international ban on ocean dumping took effect in 1972, military and other federal agencies dumped undocumented hazardous wastes off the Farallon Islands north of San Francisco.

The federal government has side-stepped the cleanup problem by ignoring it and violating treaties. In late 2003, the federal government reversed a 25-year ban on selling or disposing of properties contaminated with carcinogenic polychlorinated biphenyls (PCBs). PCBs spread easily and resist chemical breakdown, becoming widespread contaminants in air, water, soils, plants, and animals. In 2002, President George W. Bush reversed a moratorium on sending obsolete, PCB- and diesel oil-contaminated Navy ships to countries that have no protective environmental laws. The Navy ships were to be scrapped or sunk at sea to serve as artificial reefs, recklessly exposing ocean life and people who eat seafood to PCBs.

Rocket cocktail

In this age of aerial warfare, the chemical perchlorate has shown up as a widespread source of military pollution in western waters. Perchlorate is a component of fireworks, matches, and emergency flares, but 90% of perchlorate production goes into rocket and missile fuel. It is naturally present in some potash deposits and in nitrate compounds imported from Chile.[56] Perchlorate rarely showed up on contaminant lists of concern until the mid-1990s, but today it is the target of investigations and cleanup projects nationally, and extensive research projects on its health effects are under way.

Substantial perchlorate production, amounting to some 20 million pounds per year, began in the early 1950s near Las Vegas, Nevada. Over many years, perchlorate leaked into lower Colorado River reaches that provide drinking water to 21 million people, including many California farmers, and the cities of Los Angeles, San Diego, and Tucson, and parts of Mexico. Colorado River waters contain four to eight parts per billion (ppb) perchlorate—about two ounces in enough water to fill three *Exxon Valdez*-sized tanker ships.[57] Farmers irrigating crops with Colorado

River water spread the pollutant to food crops.[58] Not surprisingly, perchlorate is found in supermarket milk, cheese, and lettuce at levels exceeding some current standards. The perchlorate levels in foods pose a significant threat to the health of unborn babies and young children.[59]

Other perchlorate sources include deteriorating rocket fuel stockpiles at test and training bases, and a former flare factory in Morgan Hill, California, which has contaminated 389 wells, including city water supply wells, with 10–100 ppb perchlorate. Two contaminated city wells are located upslope from the source, showing that perchlorate from the flare factory has migrated opposite the expected natural groundwater flow direction for distances of more than half a mile to greater than one mile.[60] In an area of complex geology, the upgradient contamination probably reflects pumping at local wells, which pulls the groundwater in directions it does not flow naturally. Ironically, perchlorate also contaminates well water serving neighborhoods close to Pasadena's Jet Propulsion Laboratory—home to some of the rocket scientists responsible for America's space program and their families. Neighborhoods in Simi Valley, California, also have perchlorate in their drinking water, likely leaked from the nearby former Rocketdyne test site for the Jet Propulsion Lab's rocket engines.[61]

The EPA proposed a 22.5 ppb maximum safe contaminant level (MCL; see chapter 2) for perchlorate in drinking water but had not adopted it as of November 2006. Under political pressure, the initially recommended standard of 1 ppb quickly disappeared. Rather than setting a fixed drinking water standard, the National Academy of Sciences (NAS) recommended a reference dose (evaluating the risk in terms of dose for a given body weight). California has proposed, but not adopted, a 6-ppb drinking water standard, claiming consistency with NAS findings. Although widely misinterpreted, the NAS recommendation actually equates to a fixed water standard of less than 3 ppb. Massachusetts has adopted an MCL of 2 ppb.

Perchlorate contamination is so widespread that the cleanup costs will be enormous—a matter of concern to the DOD. Assistant Deputy Undersecretary of Defense for Environment, John Paul Woodley, Jr., issued a 2002 draft guidance document, stating "At this time, it is premature to take further action in absence of promulgated regulatory standards. I am not authorizing any environmental restoration study or cleanup beyond sampling and analysis without a regulatory driver."[62] The proposed 6 ppb California standard signals a political deal that avoids cleaning up the Colorado River.[63]

Secret Attacks

Under congressional pressure, in 2002 the DOD eventually released information about test programs that exposed soldiers and civilians to lethal nerve agents and other hazardous substances.[64] Just one of these programs, Project 112, consisted of 134 "tests" of chemical and biological warfare agents, or proxy "simulants," on military personnel between 1963 and 1974.[65] The official explanation that "simulants replaced actual chemical and biological warfare agents in most of these tests, but some plans involved the use of actual chemical and biological warfare agents"[66] minimizes the extent of human exposures. It doesn't reveal that some of the "simulants" also can harm people.

Project 112 actually tested at least 33 different chemical and biological substances, including 17 able to cause serious diseases or death in people, in 60 different trials (see appendix 4). Twenty-three trials used the toxic to extremely lethal
nerve gas agents VX, sarin, soman, and tabun. In four tests, the military used
simulants instead of the toxic warfare agents, but those substances can cause cancer in test animals. One of the carcinogenic simulants was a radioactive phosphorus isotope. And although government fact sheets consistently label the bacterium
Serratia marcescens used for 24 of the trials as harmless to humans, the Centers for
Disease Control and Prevention describe it as "not *known to consistently cause disease* in *healthy* adult humans" (emphasis added)—a rather different consideration.[67]

Since WWII, the DOD also conducted other classified tests with chemical and
biological agents. The GAO reports hundreds of radiological, chemical, and biological
tests on hundreds of thousands of citizens. More than 7,000 Army and Air Force personnel underwent these tests. The Army Chemical Corps conducted some of them
as part of a classified medical research program to develop incapacitating chemicals,
such as nerve agents, nerve agent antidotes, psychoactive chemicals, and irritants.

At some point or other, every U.S. community became "low-use" populations
for military purposes. In 1977 U.S. Senate hearings, officials admitted to completing 239 open air chemical and biological tests under Operation LAC (Large
Area Coverage) from 1949 to 1969. Military experimenters purposely dispersed
chemical and biological agents in and over huge regions, including dense civilian settlements. Few regions were left untouched. Operation LAC sprayed "simulants" and relatively nonhazardous biological and chemical warfare agents over
the San Francisco Bay region, National Airport terminal in Washington, D.C., and
Pennsylvania Turnpike.[68] In the New York subway system, proxy-terrorists tried a
more limited dispersal method, throwing bacteria-filled light bulbs on the tracks.[69]

Simulants tended to be bacteria rather than poisons—yet military authorities
knew that some sensitive individuals could suffer ill effects. When testers sprayed
the *Serratia marcescens* bacteria over San Francisco, six people were hospitalized
immediately, one died three days later, and many others continue to suffer from
chronic, untreatable infections. As recently as 1975, medical laboratories in San
Francisco were still finding *Serratia* in the population. The military never admitted a connection but stopped using certain of the simulants, in tacit acknowledg-
ment of the potential for harm.

These targeted populations were deliberately not warned, nor were they monitored for adverse effects. Belatedly, the DOD made an effort to contact people that
Project 112 had affected, but as of 2004 many test subjects had not been notified.[70]
The decisions to expose chemically or bacterially sensitive people were taken by
scientists and bureaucrats willing to sacrifice a small proportion of the population
considered a statistically "acceptable risk." Having made decisions that destroyed
lives, DOD covered them up for more than 30 years, denying their accidental victims any recompense.[71]

These inexcusable actions are not a thing of the past—open air testing of supposedly harmless simulants continues today at the Utah Test and Training Center.
The puzzling illness clusters of "Gulf War" syndrome have been studied for
more than a decade at undefined cost, to the usual chorus of Pentagon denials.
Although still under debate, many experts see in the Gulf War data a convergence
between the soldiers' exposures to largely untested and haphazardly administered

vaccines and other novel pharmaceuticals and hazardous chemicals on the battle-field. The hazards include radioactive smoke from depleted uranium (DU) in the artillery shells and bombs used in the Gulf War and now in Iraq (see chapter 7). The variety of symptoms muddies the search for causes, but may be simply an expression of widely variable chemical sensitivities in the exposed population. In addition, the effects of most chemical mixtures in all modern environments are largely unstudied and unknown (see chapter 2).

The U.S. government's lack of concern for American lives and health after the devastating World Trade Center attacks added to Al Qaeda's casualties. Following its own Inspector General's report that the EPA had too little information on whether or not the air at "ground zero" was safe to breathe, the agency still assured mournful and heroic rescue and salvage workers that the devastated ground zero site posed no health risk. We now know that the smoldering World Trade Center ruins released massive amounts of air pollution, including "very fine metals, which interfere with lung chemistry; sulfuric acid, which attacks lung cells; carcinogenic organic matter; and very fine insoluble particles such as glass, which travel through the lungs and into the bloodstream and heart." Exposure to ground zero air degraded many work-ers' health for several months and left others with long-term illness.[72]

Convention's Edges

By 1969, when President Richard Nixon halted biological weapons production, the U.S. arsenal held more than two *million* biological bombs, ready to go.[73] In spite of past U.S. agreement to abide by international weapons conventions that prohibit them, the United States goes on researching and developing chemical and biologi-cal weapons of mass destruction (WMDs), designed to kill thousands to millions of people. The 1972 international Biological Weapons Convention (BWC) set a limit on this unthinkable arsenal, at least theoretically,[74] by requiring that any U.S. bio-logical and chemical weapons research had to be purely for defensive measures.[75]

Even though WMDs are not highly effective against terrorists, the 9/11 terror-ist attacks spurred new and proliferating academic and secret military research programs, creating eight new study centers for weaponizing an array of biological and chemical agents—including genetically modified (GM) bacteria and viruses.[76] Some "defensive" programs have raised legitimacy questions under the treaty. One such program is "Clear Vision," which constructed small bomb clusters (bomblets) to test disseminating biological agents under varying atmospheric conditions. Clear Vision came so close to the limits that the government abandoned it in 2001, but the U.S. Army's patent on a "rifle-launched nonlethal cargo dispenser" (a grenade), which can be adapted easily to delivering biological agents, clouds America's com-mitment to the BWC.[77]

The United States and other nations are supposed to accept assurances that secret U.S. biological and chemical weapon experiments actually comply with the BWC's restrictions. But U.S. bioweapons defense research is secret, so the public cannot know whether or not the government still (or ever) honored the thin line separating defense from offense. Once the research reaches development, weap-onized substances will need testing, which is likely to contaminate all the old, and probably some new, test areas in the process—along with research labs and manufacturing sites. We should wonder whether rapidly increasing "biodefense" work might further erode our sense of safety and our actual security.

Recent bioweapons advances include GM forms of mousepox, related to the smallpox virus. Supposedly, mousepox virus kills only mice, and as yet the new strain is not apparently contagious. But GM rabbitpox and cowpox viruses do affect humans and a number of other animals, so the goal for developing them seems far from benign.[78] If, for instance, genetic engineering Project Jefferson, revealed in 2001, were to create a potent vaccine-resistant anthrax bacterium, the world would suddenly harbor a new threatening and uncontrollable disease.[79]

Current U.S. nonlethal chemical weapons research[80] pushes the nation's Chemical Weapons Convention (CWC) adherence to the edge of a gray zone, if not beyond. The CWC prohibits developing and using chemical agents that cause "temporary incapacitation or permanent harm to humans or animals." Yet DOD's Joint Nonlethal Weapons Program is set on weaponizing Valium, Prozac, Zoloft, and other sedatives, as well as "drugs of abuse" and common convulsants, such as the active ingredients of rat poison. DOD avoids prohibited research areas by having work done at other agencies through "memoranda of agreement," in "a pretty clear intent to violate the treaty."[81]

The CWC's implementation has been delayed, heightening international concerns about proliferating chemical weapons.[82] The lethal effects of an opiate the Russians used in September 2004 to control Chechen terrorists holding hundreds of Beslan children and teachers hostage suggests the validity of this concern. New threats are coming, in the form of credit-card–sized to notebook-sized "microreactors," which can synthesize such lethal chemicals as hydrogen cyanide, phosgene, and methyl isocyanate in the middle of a crowd. This "wonderful" technology opens powerful new avenues for proliferating chemical weapons with no effective antidote.[83]

Attackers using biological and chemical agents could pollute our air, water, and soil. The GAO has noted that bioterrorist attacks "could be directed at many different targets in the farm-to-table food continuum, including crops, livestock, food products in the processing and distribution chain, wholesale and retail facilities, storage facilities, transportation, and food and agriculture research laboratories." Terrorists aiming for severe economic dislocation would make livestock and crops the primary targets, but if the primary goal were to harm humans, processed food products would be the target.[84]

Is developing our own biological and chemical WMDs a way to defend against these kinds of attacks? Or are we spending time and resources to show terrorists the way? Richard H. Ebright, lab director at the Rutgers University Waksman Institute of Microbiology, has likened creating GM germs to see if we can kill them to "the National Institutes of…Health funding a research and development arm of Al Qaeda."[85] History suggests that the most likely people to be harmed by U.S. biological and chemical weapons programs will be American citizens.

Full Costs

In a nation that holds property ownership nearly sacred, many citizens feel that national security may require land seizures and sacrifices, no less than sacrifices of human and animal lives. But very few Americans understand the negative effects of our foreign policy on American lands and on themselves. Most of those who trained at Patton's DTC in 1942–1943 saw the desert as a hell, but some remembered its beauty. Few noticed the impacts of training exercises on the ecosystems—impacts that still can be observed decades later. Scientists are only beginning to understand how permanent and pervasive so many of the military damages are, and the potential of natural processes to spread them to our ranches, farms, and suburban neighborhoods.

Few of us stop to question what "national security" really means. We may know what it feels like but find it hard to define. When the words and the feeling are used wantonly, disregarding health and safety precautions, outrages can be perpetrated in the name of national security. This disregard let experimentation on uninformed American citizens become a hallmark of military chemical and biological weapons tests. Bomb plant workers and downwinders fought decades-long battles to win even minor reparations from the U.S. government, while suffering debilitating illnesses unto death. There is no national recognition of the total costs in innocent human lives from the unregulated and secret experiments that used average Americans, in and out of uniform, as guinea pigs—and no estimates of whether the nation really benefited from those sacrifices. There certainly has been little effort to make recompense.

It seems time to examine the basic components of "national security" more closely, as the shrinking supply of remote American lands takes military uses closer to neighborhoods than ever before, and into conflict with conservational and recreational land uses and safety considerations. We need to extend President Eisenhower's warning, quoted at the opening of this chapter, to our native soil and habitat and consider how dear our land is to us. Can we defend the nation as we destroy vast stretches of land and pollute the waters that flow from them? How much of our land and how many species are we willing to destroy? Can we justify exposing ordinary Americans in their own homes to unannounced weapons tests? What of value will be left, and how will the destruction affect our lives and our children's futures?

Clearly, military force is sometimes required, so maintaining and training troops are essential parts of national security. But military land use policy must give the long-term devastation from vehicle and bombing impacts high priority and close scrutiny. We need to question whether we must continue huge-scale training maneuvers and whether we can continually expand active training at the NTC and other sites. The answer to such questions will be especially critical if the United States is forced to depart its many bases on foreign soil.

Thinking "innovatively," former Defense Secretary Donald Rumsfeld attempted to transform the Army into a more strategically responsive force. His "Future Combat Systems" program coupled nontraditional fighting tactics with an extensive "secure" information network, "to see, engage, and destroy the enemy before the enemy detects the future forces," so compensating for "loss of size and armor mass."[86] The land-use implications of this "strategically responsive forces" concept remain unclear—but Rumsfeld's smaller, highly trained armies have failed to overcome Afghan and Iraqi insurgencies. Desert training undoubtedly helped U.S. troops race to Baghdad but provided no preparation for years of urban warfare.

Even before the problematic Iraq invasion, many critics within and outside the U.S. military had argued that future wars will be fought in urban environments, reducing or eliminating the need for huge-scale training. If the logic behind the Army's NTC-expansion rationale were extended to urban warfare training, it would mean commandeering large sections of American cities for search-and-destroy training. This is unlikely to happen, for the same reason that Marines were not allowed to practice landings on west coast beaches.

No one has explained how maintaining an immense stockpile of nuclear, chemical, and biological WMDs will protect U.S. lands or populations from future terrorist attacks. In 2001 the ineffectual response to anthrax-laced letters suggests that nearly three decades of "defensive" biological and chemical weapons research has not delivered much protection, either for Congress or for the general public.[87] Instead of lowering WMD inventories and threats, the nations named as U.S. enemies have scrambled to develop their own WMD systems, raising the threat levels worldwide.

Americans are intensely patriotic and tend to give the military the benefit of every doubt—but even military thinkers understand that the armed forces cannot overtly practice on unsuspecting citizens going about their business without undermining lives and national well-being. The same recognition must extend to the health of our lands and water, which fundamentally underpins our economic well-being and may be more fundamental to the meaning of "national security" than picking targets for strategic wars. To avoid repeating old mistakes in the face of national security alarms, patriotic American citizens will have to make the government and military authorities start prioritizing land, water, and public health protection.

We must recognize that U.S. citizens already are overexposed to myriad threatening mixes of industrial, agricultural, and domestic chemicals in the natural environment. We must also understand that people's response to chemicals and pollution vary widely, and some may be especially sensitive to one or more chemical mixes. Adopting a precautionary, rather than an "acceptable risk," attitude toward chemical exposures should be a first step. Prohibiting the military from adding exposures should be a close second. A precautionary policy would make the government more able to protect Americans against future hostile attacks. For example, instead of letting the valiant World Trade Center rescue workers feel safe while breathing polluted air, day after day, a precautionary policy would equip them with air filtration equipment and insist that workers limit their exposure times.

It is too soon to conclude that any on-land military training is a modern anachronism—but the U.S. quandary in Iraq points to war as an untenable democratizing force in formerly autocratic countries. The Afghan and Iraq wars apparently have not discouraged terrorists, as President Bush had hoped, but have instead provided terrorist training opportunities. Even U.S. military leaders have opined that military force, however necessary in some instances, is not the best response to every threat. Choosing the wrong battles can weaken us more than choosing to refrain from conflicts.

As our conflicts of choice go out of control, Americans might conclude that making and keeping international alliances and fostering cooperative partnerships wherever possible give better protection. Perhaps, as in the past, our best defense against terrorism may be to live up to our ideals and set an example of governance that all want to imitate, in spite of cultural differences. Policies based on these outlooks could also lead to better stewardship of U.S. military lands, and of all American populations.

7   Creating the Nuclear Wasteland

No one disputes whether atomic radiation is harmful or harmless as they did in the fifties because atomic downwinders wrote the answer on their deathbeds.
Chip Ward, *Canaries on the Rim*

In a very real sense, whenever weapons of mass death are unleashed, all humanity is downwind.
Lewis M. Simons, "Weapons of Mass Destruction"

"At the heart of the matter nuclear weapons are simply the enemy of humanity"— retired U.S. Air Force General Lee Butler, former Commander of Strategic Nuclear Forces, spoke these words in his testimony to a 1999 Joint Senate–House Committee on Foreign Affairs. They probably express the deep feelings of most of the world's people, including most Americans. Towering mushroom blast clouds and the shapes of atomic weapons are common symbols of doom. The specter of nuclear weapons in the hands of terrorists haunts us, and the possibility of attacks on U.S. citizens with "dirty bombs"—a bomb made of conventional explosives that scatters radioactive materials—raises major concerns. As it should.

Nuclear weapons and the nuclear waste that they generate truly are destructive to all life and must be controlled. If we fail to prevent their proliferation in the world and stop generating them ourselves, they could destroy us without respect for national boundaries—even without a real nuclear war or dirty bomb terrorist attacks. They already have poisoned great expanses of American lands from coast to coast.

American soil, water, and air started accumulating radioactive pollution during the World War II race to build an atom bomb. Radioactive contaminants spread into the environment at every step in the process, from mining the uranium for bomb fuel and purifying and enriching the uranium to make plutonium, to detonating bombs to test them and disposing of the wastes. Radioactive materials currently contaminate buildings, soil, sediment, rock, and underground or surface water within more than two million acres administered by the U.S. Department of Energy in the 11 western states.[1]

All sorts of Americans were carelessly exposed to radioactive bomb fuels during WWII and the Cold War, but especially the atomic scientists, uranium miners, and bomb plant workers who were exposed to them every day. For nearly two decades, U.S. atomic bombs blew up and contaminated American lands. Both

American soldiers at the test grounds and civilians on ranches or farms and in homes were exposed to the dangerous radioactive fallout (see appendix 5). Perhaps unknown to most Americans is the fact that radioactive contamination from U.S. atomic weapons tests also spread across the whole country and far beyond U.S. borders.

Americans desire to protect our families and communities from radioactive wastes—this is the atomic age's primary unsolved problem. Their concerns have slowed or obstructed some 1990s industry and government disposal proposals, such as mixing radioactive waste with sewage sludges for agricultural fertilizers (see chapters 2, 10) and "reindustrializing" waste radioactive metals into steel for making personal and household implements (see chapter 10). But nuclear industry advocates attribute our fears to ignorance. If Americans were better trained in science, they suggest, we would be less worried about nuclear materials or radiation effects.[2]

Even the most rudimentary information about radioactivity's health effects gives plenty to worry about. More alarming, however, is the U.S. government's historic carelessness with atomic energy, which has destroyed hundreds of thousands of our own citizen's lives. As proposals for new U.S. bomb building and testing programs surfaced, the 2006 polonium-210 poisoning of a former Soviet intelligence officer in London, England, reminded the world that very tiny doses of some radioactive isotopes can kill quickly and horribly. Fears for the U.S. soldiers currently exposed to waste "depleted" uranium in bomb and artillery shell warheads give General Butler's words even more force.

Hazard Life

Atoms make up the cells of all living things. Once radiation enters the body, it can knock electrons out of the atoms in cells, making the atoms into electrically charged ions (box 7.1). Ions in living tissue set off "a chain of physical, chemical, and biological changes that can result in serious illness, genetic defects, or death."[3] The immediate health threat from nuclear bomb testing is "hot" beta-particle and gamma-ray radiation, which can go through anything except heavy steel or lead shielding.[4] Before decaying away in minutes to weeks, hot radiation can produce severe external skin burns and lesions, followed by hair loss, nausea, and dizziness. Miscarriages are common among radiation-exposed pregnant women, and their children often suffer from birth defects, chronic health problems, reproductive system diseases, and cancers.[5]

Larger and slower moving alpha particles cannot penetrate the skin but are dangerous if ingested. When inhaled, even tiny amounts of alpha-emitting substances can be fatal—less than a millionth of an ounce of inhaled radium killed early radiological chemist Edwin Lehman.[6] Internal alpha-emitting substances concentrate in organs, including sex organs and bones (see appendix 6), and they continue to decay, emitting highly ionizing beta particles.

Internal alpha radiation promotes internal cancers and can cause life-long health problems. A sad example is the case of women who painted watch dials

Box 7.1 Atomic Age Primer

Radioactivity arises naturally from atoms' complex internal structure. All atoms are made of smaller particles, including electrically charged electrons and protons and uncharged (neutral) neutrons. Protons might be thought of as atomic DNA, because their number defines the characteristic atomic configuration of each *element*, such as hydrogen, oxygen, and iron. Each and every hydrogen atom has only one proton, each oxygen atom has 16, and so forth. Every neutral atom contains the same number of electrons as protons, but most can lose or gain electrons to become positively or negatively charged *ions*.

The electrons buzz eccentrically around the central nucleus of protons plus neutrons.[a] Electrons are virtually weightless, so the total count of protons and neutrons gives atoms their weight.[b] The numbers of neutrons can vary, yielding differently weighted atoms of the same element, called *isotopes*. Different isotopes might be thought of as identical twins, triplets, or quadruplets: All bear the same DNA but have different weights. Isotopes are named for their total weight, obtained by adding the number of neutrons and protons—for example, uranium-235 has 92 protons and 143 neutrons, while heavier uranium-238 has 92 protons and 146 neutrons.

Very heavy atoms, such as uranium, have such large numbers of neutrons that they fly apart spontaneously, releasing heat and radioactivity, mostly of charged ionic particles—*alpha particles* (two protons and two neutrons, with +2 charge)[c] and *beta particles* (electrons by another name, with –1 charge)—plus X-ray-like bursts of energy, called *gamma rays*. Since radioactivity comes from the breakup of nuclei, radioactive isotopes are also called *radionuclides*. Spontaneous nuclear disintegrations create the natural "background" radioactivity that we unknowingly live with. It cannot be detected by human senses.

In the 1930s, physicists fired neutrons at uranium-235 atoms, splitting (*fissioning*) them into atoms of several lighter elements and releasing a large amount of energy as heat. In atomic fission bombs, two uranium or plutonium masses suddenly merge to create a "critical mass" of atomic fuel,[d] which suddenly releases swarms of neutrons. The neutron swarms increase neutron bombardments suddenly, setting off runaway series of atomic disintegrations, and explosively releasing the huge reserve of atomic energy in the merged mass. The explosion also yields intensely radioactive "hot" gamma rays and beta particles, plus longer lived radionuclides—including radioactive "transuranic" elements, such as plutonium and polonium.

continued

Box 7.1 Continued

All radioactive emissions are called radioactive decay because they change the nuclei of radioactive atoms to the nuclei of different—often nonradio-active—atoms. The decay progressively reduces, and eventually uses up, the radioactive material. Physicists talk about decay rates in terms of *half-life*—the time for half of the radioactive material to disappear. Each isotope has a characteristic decay rate—for example, "hot" radioactive substances, such as iodine-131, can decay away completely in a few months or years at most. Very long-lived radioactive substances, such as uranium and plutonium, have extremely slow decay rates of tens of thousands of years or more and create other radioactive substances as products of their decay, such as radium and radon.

Notes
[a] Werner Heisenberg substantially modified Niels Bohr's simple model of atoms as minia-ture solar systems, with planetlike electrons regularly orbiting the nucleus, to one of vari-ously shaped but symmetrical electron clouds representing all possible locations of electrons around the nucleus at any moment.
[b] A neutron is a proton with an embedded electron.
[c] Heavy and positively charged alpha particles are equivalent to the nucleus of a helium atom.
[d] "Atom bombs," such as the ones developed and used in World War II, are fission bombs, having a core of fissionable material such as uranium-235 or plutonium-239. "Fissionable" means that the atomic nuclei can be split when bombarded with neutrons.

with luminous radium in the 1920s. They ingested tiny radium amounts over ten years and more from licking their paintbrushes to maintain a fine line, as supervisors had instructed them to do. All the women suffered anemia, tooth and jawbone deterioration, stomach cancer, fatigue, and heart problems before dying of radiation poisoning. Their highly radioactive bodies had to be buried in lead coffins. Other health effects include blood infections, poor circulation, arthritis, neuromuscular disorders, and kidney and thyroid disease.

The hazard life of a radioactive material, arbitrarily set at 10 times the half-life, is the length of time it remains hazardous. Radionuclides used for medical treatments generally decay to nonradioactive forms in hours, days, or weeks, while the hazard life of plutonium-239—mainly used in weapons and power plants—is 240,000 years. This means that much of the waste plutonium we create today will be around for 240,000 years, essentially forever in human terms. A pound of plutonium-239 will decay to half a pound over 24,000 years. After another 24,000 years of decay, a quarter pound of that plutonium will still be with us, retaining "every bit of its original potency per unit of weight.... Hundreds of thousands of years later there will still be a dust-like speck of plutonium...."[7]

Exposures

With all of our experience, the federal agencies cannot agree among themselves what level of human-generated radiation exposure is safe enough.[8] Debates still rage about whether low radioactivity levels can harm people—including exposure to the widespread traces of natural radioactive materials in rocks and soils.[9] Air and space travelers, and even high-altitude communities, have higher cosmic ray exposures than do people living at sea level under a thicker layer of protective atmosphere. Government and industrial health researchers estimate that natural radiation probably causes somewhere between 5% and 50% of all cancers.[10] Radiation exposures are cumulative, so every new source adds to the natural burden. For example, the miners working in unventilated uranium mines developed lung cancers at a rate five times higher than that of the general population.[11] Heightened airline security recently added whole-body X-ray scans to airline passenger exposures.

Even one ionizing alpha particle has the potential to destroy cells, so without any unequivocal evidence to the contrary, authorities have assumed that each and every particle of radioactive material can harm living tissue. At the end of its final half-life, 240,000 years from now, even the last particle of man-made plutonium-239 will be capable of causing cancer.

Starting with the 1945 Trinity test, the United States began increasing public exposures to radioactive materials. The government exploded atomic bombs in the atmosphere until 1963, then switched to underground testing (see appendix 5). Atmospheric tests dropped bombs from airplanes or detonated them on towers and balloons, releasing dangerous radioactive "fallout" clouds of hot beta particles and gamma rays and raising radiation exposures everywhere they fell. Some underground explosions also vented radioactive gases to the surface. Winds carried all the bomb radiation clouds far and wide.

Longer lived radioactive waste comes from bomb fuels and fission products, including radium, uranium, plutonium, and their radioactive byproducts. All mostly emit alpha radiation. Both uranium and plutonium also are highly flammable and extremely toxic—plutonium is one of the most toxic substances known. Natural processes yield very little plutonium, but bomb building and power plants create large quantities. The global plutonium inventory has risen from a few milligrams in 1941 to 1,500 tons in 1995 and is increasing by more than 75 tons every year.[12]

Wherever the nuclear bombs went off, they contaminated vast quantities of soils, rock, and water with longer lived radioactivity. These dangerous materials are not identified as waste and do not appear in contamination inventories. Plutonium still is common in the dust at the test sites. High winds still pick up the radioactive dust and carry it all over the United States and beyond.

The Full Sacrifice

For four decades of Cold War, the heart of America's foreign policy lay in nuclear bomb factories and threats of retaliatory nuclear holocaust upon the Soviet Union.

Many Americans still argue that the wartime nuclear program saved American soldiers' lives. Most also believe that the massive U.S. arsenal of weapons of mass destruction was the major force for maintaining the Cold War standoff and eventually disintegrating Soviet power. No nuclear bomb ever fell on the Soviet Union, but wars always exact a terrible price, and the Cold War was no exception. That price was largely exacted at home. War memorials do not mention the home-front sacrifices, and only rarely do movies or television programs focus on the Cold War's full price in tragic human lives and untimely deaths on American soil. Too few of us are even aware of them.

Managers for the nuclear industry, and their government overseers, knew that radiation exposures could kill people. But DuPont, General Electric, Kerr-McGee, and other companies, as well as the Atomic Energy Commission (AEC)—and its successor agencies, the Department of Energy and the Nuclear Regulatory Commission—put production above safety. The AEC never required companies to warn the inexperienced and poorly trained atomic workers that making atom bombs could take their lives.

No authority ever warned Navajo uranium miners against highly toxic radioactive particles from radon gas decay in underground uranium mines, for instance.[13] Even before uranium miners developed lung cancers and lodged worker's compensation claims, AEC's scientists and managers knew that installing ventilation systems could substantially protect them but did not insist on proper mine ventilation— nor did they make the miners wear respirators to protect themselves. Instead, the officials assured miners, state officials, and even mine owners that radon would leave miners' lungs an hour after they had returned to the surface. This was a misleading truth, because it is not the radon that kills, but toxic radioactive polonium-210 dust from radon decay. Polonium-210 particles could not leave the miners' lungs, and destroyed their organs from within.

In time, the inevitable death toll began, taking one miner named Raymond Joe along with many others. Like former KGB spy Alexander Litvinenko, Ray Joe died horribly—not from malevolent exposure, but from the careless negligence of his own government. Six years after his death, the Joe family received $100,000, "to equal the life of Raymond Joe, who scraped radioactive rock…to fuel the Cold War. It was never fought, but it killed Ray Joe just the same."[14]

Bomb plant officials and managers actually *assured* workers that the radioactive materials they handled every day *could not hurt them*. "They told us it was low level and wouldn't harm us unless we held it in our hands or took a bite out of it," remembered Jonathan Garcia, who had buried waste uranium and plutonium from bomb fuel enrichment in Los Alamos National Laboratory's "hot dump."[15] The AEC managers had urged him and co-workers not to don protective clothing: "They expected you to just hold your breath and run past areas where there was radiation…. [It] saved the time you would use up putting on the protective suits."[16] Those saved minutes eventually subtracted years from many bomb plant workers' lives.

Unknown numbers of excessively exposed workers, from blue-collar workers to scientists, died from cancers long after their daily radioactive exposures had ended. Jonathan Garcia, Los Alamos National Laboratory's hot dump operator, was one of them. People who suffered inordinately from radiation-related diseases had to retire on disability before reaching retirement age and lived out their last years in poverty. Their children were plagued with genetic damage and deformities.[17] About 98,000 eventually filed official claims, but probably thousands had died before the government acknowledged any responsibility for their illnesses. The toll also includes pioneer radiation chemist Marie Curie, and probably atomic physicists Enrico Fermi and Richard Feynman.[18]

Testing, testing

During the Cold War, the U.S. government was avid to more precisely gauge the deadly nature of radioactivity. After 1990, the government opened top-secret files, revealing the details of scientific experiments that purposely released radioactivity into the air over populated regions or served up deadly radioactive plutonium in food to prisoners and impoverished patients—all to observe the health effects on civilians.[19] United States soldiers and sailors were purposely marched or sailed close to tests of both nuclear fission "atomic" and nuclear fusion "hydrogen" bombs, exposing them to hot radioactive fallout clouds at remote Pacific locales and in Nevada (see appendix 5).[20]

Radioactive fallout from bomb plants and test blasts also rained on towns full of civilians—men, women, and children, the original "downwinders." They were not formal test subjects, and government officials assured them that they were contributing to or witnessing a great enterprise. Ranchers were irradiated along with their crops and livestock,[21] and the food they grew spread radioactive contamination to the public at large. The immediate injuries and subsequent life-long illnesses and disabilities of these people and their descendants are not as easy to justify as an expense of war. The largest civilian exposures emanated from bomb-fuel and bomb-test sites close to population centers—principally the Hanford Nuclear Reservation next to the city of Richland, Washington, surrounded by farmland on both sides of the Columbia River, and the Nevada Test Site (NTS) near populous Las Vegas, Nevada, playground for Americans of all walks of life.

From the mid-1940s, through the 1950s, and into the 1960s, Hanford continually released strontium-90 and iodine-131 into the air. The hazard life of iodine-131 is about three months, and that of strontium-90 is around 300 years. These dangerous atomic isotopes swept across wide Pacific Northwest regions, falling on and contaminating grass and hay in fields where cows and other domestic animals could eat it. The Hanford downwinders—farmers and townspeople living east (downwind) of Hanford Nuclear Reservation—unknowingly absorbed dangerous levels of hot radiation, equivalent to thousands of chest X-rays per person.

From soil water, vegetable crops take up iodine-131,[22] along with much longer lived iodine isotopes. Anyone who ate vegetables grown in the Hanford area in the late 1940s unknowingly risked thyroid disease from concentrated iodine-131 exposures. Hanford health physicists secretly measured radioactive iodine fallout in the region, but neither they nor Hanford's managers told the thousands of local residents, including their own employees, that they were being exposed to radioactivity far beyond the maximum recommended human dose. When insiders urged them to distribute potassium iodide pills, which would protect the public from excess radioactive iodine, Hanford managers refused.[23]

Bones are made of calcium, and we encourage children to drink calcium-rich milk to support rapid bone growth. Strontium and calcium are chemically similar, so animals absorb them both from foods, especially milk. For many years, the unwarned Hanford-area parents gave their children locally produced milk laced with radioactive strontium-90, which can cause childhood leukemia. Eventually, a long-buried study of the Hanford Nuclear Reservation region's population revealed a "startling" increase in infant mortality in 1945, compared to 1943 just before bomb fuel manufacturing started.[24]

In December 1949, Hanford purposely released radiation from the still-mysterious "Green Run" experiment. Hot radiation carpeted farms, towns, and people—not to mention the soil and water—across parts of Washington, Oregon, and Idaho. Green Run's total radioactive release is not public knowledge, but reconstructions suggest as much as 11,000 curies of iodine-131 and 20,000 curies of xenon-133[25]— more than 700 times the Three Mile Island accident's radiation release (see chapter 12).[26]

A very large number of sheep died on nearby farms in Franklin County, Washington, that winter. In the spring of 1950, a large number of ewes delivered deformed or stillborn lambs. Both before and after Green Run, families on those farms gave birth to babies with deformities and other birth defects. When grown, many of those children suffered from multiple health problems, including sterility. More than 60 men and women eventually died from heart attacks or cancers before the age of 60 in a thinly populated area closest to Hanford, later known as "Death Mile."[27] Hanford downwinders did not learn the probable cause of their hair losses, anemia, unexplained fatigue, and reproductive disasters—all common symptoms of radiation poisoning—until 1986.

Some Hanford victims, such as June Stark Casey, lived far from the reservation. She was June Stark then, a student at Whitman College in Walla Walla, Washington, 50 miles distant. Fatigue and chills struck her suddenly during the 1949 Christmas break, followed by permanent loss of her long, naturally curly hair. She has suffered ever since from a variety of illnesses: severe hypothyroidism, miscarriage and stillbirth, multiple tumors in various organs, skin and breast cancers, and a chronic degenerative spine disorder. Only in 1986 did June Stark Casey learn about Green Run and that its radiation had blanketed Walla Walla on December 2, 1949.[28]

Low-use populations

Largely to save money, the government relocated Cold War atomic bomb tests from Pacific islands and atolls to the U.S. mainland.[29] In 1951, the Pentagon located the NTS on 865,000 acres of ancestral Western Shoshone lands (figure 7.1). Sparsely populated southwestern Nevada must have seemed remote enough in the late 1940s to absorb giant explosions and dissipate threatening radiation clouds. The U.S. government had formally ceded those lands to the Shoshone in the 1863 Treaty of Ruby Valley—but "[t]hey were looking for open space on public lands, and as far as they were concerned Indian lands were free for the taking," opined Janet Gordon of the Western Shoshone.[30]

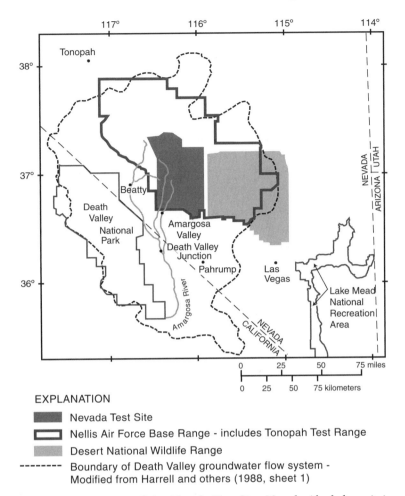

Figure 7.1 Location of the Nevada Test Site, Nevada (shaded area), in relation to Death Valley National Park, California. Dark outline is the boundary of Nellis Air Force Base, which surrounds the Nevada Test Site. Dashed line surrounds the Death Valley groundwater basin. From R. J. Laczniak et al. *Summary of Hydrogeologic Controls on Ground-Water Flow at the Nevada Test Site, Nye County, Nevada* (U.S. Geological Survey Water-Resources Investigations Report 96-4109, 1996), figure 3.

Uninformed by the government at first, and then assured that there was no danger, for more than a decade the holdout Western Shoshone and newer southern Nevada settlers, plus residents of southwestern Utah—and their cars, homes, ranches, and businesses—became the selected atom-bomb downwinders. An AEC memo had characterized these people living downwind from the nuclear bombing ranges as "a low-use segment of the population."[31] Tests were fired in times when weather patterns would carry the fallout eastward, to spare the urban populations of southern California.

Relatives of Janet Gordon were herding sheep on horseback close to the NTS in Nevada in 1953, when Gordon's brother rode his horse through an apparent ground fog. Actually it was fallout from the Upshot-Knothole series of bombings at the NTS (see appendix 5). Gordon remembered, "[M]y brother was sick all night…his hair came out in patches, the wool came out on the sheep, the horse died. Of the eight men that were at the camp with my brother, six of them have died from some cancer."[32]

After the 1953 Upshot-Knothole bombings, the people living in Cedar City and St. George, as well as Utah sheepherders and their sheep, suffered from burned skin and hair loss. The usual plagues—anemia, gum and jaw deterioration, miscarriages, and stillbirths—followed. Live-born human children as well as farm animals exhibited high rates of physical defects. Before 1953, leukemias had been virtually unknown in southwestern Utah. After 1956, the leukemia incidence soared, devastating St. George and Cedar City populations.[33]

Later, the authorities did issue warnings, of a sort. Utah downwinder Wayne Simister wrote us, "I well remember as a child the announcement coming over the Salt Lake [City] radio one morning: 'DON'T have your kids eat the snow today! It has radioactive fallout in it!' Horrid memories like this still make me very mad."

Although Nevada bomb tests were scheduled to avoid sending radioactive clouds toward Los Angeles, wayward air currents still spread contaminants nationally and globally.[34] Wind shifts even carried radioactive clouds over Los Angeles, many times. Both western and eastern United States seaboards—and many areas between—have been irradiated, making downwinders of all Americans (figure 7.2).

Figure 7.2 Paths taken by more than one nuclear bomb test radiation cloud across the United States, carrying radioactive fallout from both above-ground and below-ground detonations. From Richard L. Miller. *Under the Cloud: The Decades of Nuclear Testing* (Legis Books, 1999). Copyright 1991. Reproduced with permission.

Waste Lands

The government has released few documents showing the enormous level of pollution on and in American lands from producing and testing nuclear weapons. The tests were not only for deterring attacks. The ill-founded "Plowshare Program" tested nuclear bombs for a variety of "peaceful" projects such as digging canals and harbors (see appendix 7). The hot bomb test radiation mostly has decayed away, but longer lived radionuclides remain in soils and waste dumps, still able to harm Americans on their own soil. For example, people in southern Nevada and Utah may still have radiation fall on them every time contaminated soils blow off the NTS (see appendix 6).

The nation as a whole harbors at least 1.3 billion cubic feet of radioactive wastes, emitting 1.01 billion curies of radiation[35]—about 10 times the official estimate of radioactivity released in the Chernobyl accident.[36] That level is enough to kill 99 trillion people—more than ever have lived on the Earth.[37] Hazardous contamination from either bomb making or testing—or both—directly affects 44 areas in the western United States[38] (figure 7.3). Some three billion cubic feet

Figure 7.3 Map of radioactive contamination sources arising from nuclear weapons manufacture and testing in the 11 western states. H, Hanford Nuclear Reservation; INEEL, Idaho National Engineering and Environmental Laboratory; LLNL, Lawrence Livermore National Laboratory; NTS, Nevada Test Site; LA, Los Alamos National Laboratory; RF, Rocky Flats weapons facility. From U.S. Department of Energy. *Closing the Circle on the Splitting of the Atom* (DOE/EM-0266, 1996), 74.

of radioactive mill tailings left over from uranium ore processing still lie about, mostly in the 11 western states.[39] For decades, piles of radioactive uranium mine tailings, also loaded with nonradioactive toxic heavy metals, have laid on the banks of the Colorado River in Utah, exposed to wind and rain erosion (see chapter 4, figure 4.13; and chapter 10).

From nuclear bomb production alone, Department of Energy sites and facilities store nearly one million tons of hazardous nuclear and nonnuclear materials[40]— enough to fill more than 11,000 railroad cars, on a train 100 miles long (see chapter 10). In addition, the department manages lands holding more than 67 billion cubic feet of contaminated water, sediment, soil, and rock, and approximately 5,100 contaminated facilities.[41] The contaminated water amounts to more than 63 billion cubic feet—enough to fill the Great Salt Lake six times.

The largest amount of bomb-waste contamination lies on and beneath the large nuclear reservations, particularly the 375,000-acre Hanford Nuclear Reservation, upstream from the populous city of Portland, Oregon, and smaller towns of Kelso and Longview, Washington; the NTS, adjoining World War II Nellis Air Force Base, near Las Vegas; and Rocky Flats Environmental Technology Site, upslope and upwind from growing Denver, Colorado, and suburbs. Other areas of concentrated radioactive contamination include Idaho National Engineering and Environmental Laboratory, Idaho; Los Alamos National Laboratory, near Albuquerque, New Mexico; and Lawrence Livermore Laboratory, plus numerous smaller sites near San Francisco, California.

Wasting Hanford

During WWII and Cold War bomb-fuel production, the Hanford Nuclear Reservation released 25 billion cubic feet of dangerously radioactive wastes to the environment,[42] equivalent to the worst contamination at notorious Ural Mountains and Siberian Soviet bomb factories. In addition to the 1940s airborne radioactive releases, discussed above, Hanford site experimenters also knowingly dumped uranium and plutonium radionuclides into the Columbia River. The amount of radioactive liquids released to the soil or surface waters would be more than enough to fill two Great Salt Lakes.

After taking over from DuPont, the General Electric Corporation ran Hanford Nuclear Reservation during the 1946–1965 period of its worst intentional and accidental radiation releases. Hanford recorded approximately 270 other unplanned releases and spills, but information about them is scanty—one spill involved dumping an estimated 25,000 curies of radioactive cesium-137 into the ground.[43] As recently as May 1997, a chemical explosion in a Hanford waste tank exposed workers to plutonium and other hazardous chemicals. The result of all these disposal and storage failures is an estimated 370 billion gallons of groundwater, with various radioactive and other hazardous contaminants, beneath the Hanford waste site—enough to fill the Great Salt Lake nearly five times.[44]

The planned releases relied on natural processes ("natural attenuation"; see chapter 10) to eliminate or diminish the toxic and radioactive threats that they posed. But the scientists and managers acted without pretesting that reliance to be sure that it was well founded.

Dilution solution

Fissioning uranium in nuclear reactors to make plutonium also releases great heat (see box 7.1), so reactors require a great deal of cooling. Hanford's nine weapons reactors were built on the banks of the Columbia River for access to cooling water. Huge volumes of water passed through the reactor vessels, carrying off traces of uranium, plutonium, and various other radioactive waste products along with the heat. The water was pumped directly back to the Columbia River without any treatment,[45] taking along unknown quantities of radioactive liquid and other toxic reactor chemicals. The single-direction-flow cooling design assumed that large volumes of Columbia River water would dilute the smaller amounts of reactor contaminants. Eventually, the chemicals were supposed to slip downstream and disappear into the even larger volume of Pacific Ocean water.

Unfortunately, natural processes do not offer a "dilution solution," either for radioactive materials or most other wastes in surface waters (see chapter 10). By about 1960, environmental scientists had begun to realize that plants and animals are pollution collectors and that natural functions actually *prevent* dilution. The tiny river-bottom plants and animals absorbed heavy toxic and radioactive chemicals, which became concentrated in their organs and tissues.[46] Larger crayfish, and eventually bony fish higher up the food chain, eat quite large amounts of the contaminated smaller plants and fish. So the flesh of Columbia River fish accumulated astonishing levels of radioactive and toxic concentration—as much as 170,000 times greater than in the river water.[47]

Hanford's wastes are a risk to fishermen—especially to traditional Native Americans who live by fishing the Columbia River and might eat the highly radioactive fish.[48] In the late 1990s, the public interest Government Accountability Project (GAP) also found high levels of radioactive strontium and thorium in riverbank mulberry plants along Columbia River tributaries upstream from the pipes that take Richland, Washington's drinking water out of the river.[49] This means that Richland's population could be concentrating radioactive isotopes from the river water in their bodily organs.

Soil binding

In the early 1950s, bomb scientists blindly believed that soils would absorb and hold plutonium and the other highly radioactive reactor wastes. On this premise, Hanford managers approved pouring more than 340 billion gallons of so-called "low-level" radioactive liquid wastes, equivalent to 7,000 *Exxon Valdez* tanker ship loads, directly into the ground at more than 1,200 individual sites. Hanford ditches and ponds apparently received long-lived plutonium and other highly toxic wastes. Some wastes went into "disposal" ponds and ditches near the riverside reactors, where the groundwater table lies close to the surface.[50]

"Scientific" calculations predicted that the reactor wastes would stay in soil near the surface and not seep through the region's very thick and relatively dry "unsaturated zone" (see chapter 13) to groundwater.[51] These calculations were based on untested beliefs and again were wrong. Leakage from the riverside disposal sites has contaminated both soil and groundwater with radioactive tritium (hydrogen-3), as well as with chromium and other toxic materials. Groundwater does not sit still but moves through soil and rock under the influence of gravity, eventually surfacing at a spring or flowing into the closest stream or river. Contaminated groundwater rapidly flowed the very short distance from Hanford's plutonium production reactors to the Columbia River.[52]

Through so-called "reverse wells," operators also injected waste plutonium *directly* into groundwater. The ways that plutonium might spread, and the risks it could pose in future, are largely unknown[53]—but it could go well beyond reservation borders in water, animals, and plants. All of these carriers could expose human populations to the toxic hazard (see chapter 13).

Fast river route

In the 1950s, Hanford operators sealed really hot high-level wastes in buried tanks to isolate them from the environment. One hundred seventy-seven of the radioactive waste-filled tanks lie beneath the "200 East and West sites" near the Hanford Nuclear Reservation's geographic center (figure 7.4). The tanks contain more than 55 million gallons of waste, more than enough to fill 400 railroad tank cars. Sixty-seven of the 200-site tanks have leaked, draining between 600,000 and 900,000 gallons of radioactive pollution into unsaturated-zone soils and fractured rock.[54] Toxic plumes, carrying radioactive technetium-99, cesium-137, cobalt-60, strontium-90, europium-154, as well as uranium-235, uranium-238, and many noxious nonradioactive substances, still are leaking into the 300-foot-thick unsaturated zone. From there they will ultimately migrate into groundwater and then to the river (figure 7.5).

Gathering accurate information about the leaks requires an adequate number of well-located monitoring wells, deep enough to intersect the lowest level of pollution.[55] But the waste site's monitoring wells are only 100–150 feet deep, just halfway through the unsaturated zone, so the plumes' full reach is largely unknown.[56] Groundwater samples from local wells show that plumes of technetium-99, cesium-137, and cobalt-60 have made it all the way to groundwater and spread at least 500 feet away from the leaking tanks. The contamination plumes are tracked by detecting gamma-emitting radionuclides, which move more slowly through soil than such alpha-emitting materials as technetium-99, so the position of faster moving pollutants is unknown.

Adding to the accidental leaks, between 1946 and 1966 Hanford bomb makers ran out of storage room in tanks and intentionally poured more than 120 million gallons of liquid wastes directly into ditches called "specific retention trenches." The wastes contained more than 150 million pounds of corrosive chemicals, enough to fill 75 railroad cars, along with materials having more than 65,000 curies of total radioactivity.[57]

All the failed assumptions about retaining soils, sealed tanks, and unsaturated zone barriers had contributed to the untested conclusion that Hanford's hazardous wastes could not reach groundwater in less than 10,000 years. Scientific assumptions are meant to be tested—and that one should have been exposed as false very quickly, since tritium showed up in groundwater beneath Hanford's 200 East and West sites as early as the 1940s.[58]

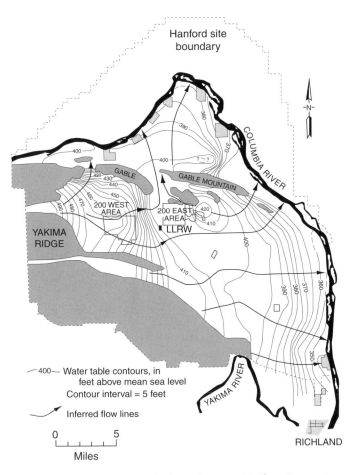

Figure 7.4 Map of the Hanford, Washington, Nuclear Reservation, showing the water table in 1992 (contours give elevations). Arrows at right angles to water table contours represent groundwater flow paths, showing that any contaminants reaching the groundwater will migrate to the Columbia River. Modified from K. A. Lindsey et al. Geohydrologic Setting of the Hanford Site, South-Central Washington. In D. A. Swanson and R. A. Haugerud, eds. *Geologic Field Trips in the Pacific Northwest* (Geological Society of America, 1994), chapter 1C. Copyright 1994, The Geological Society of America. Reproduced with permission.

Radioactive tritium combines with oxygen in water molecules, just like any other hydrogen atom, making tritium the most mobile radioactive contaminant in groundwater.[59] Tritium plumes entered the Columbia River by 1983 (figure 7.6), but as early as 1963 monitored wells close to the Columbia River had detected plumes of tritium and ruthenium-106, with tritium concentrations twice the safe drinking water standard. Tritium concentrations in those wells increased to 10

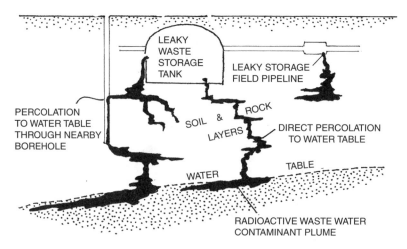

Figure 7.5 Diagram of leaking single-wall storage tanks and pipelines, which either hold or convey high-level radioactive waste at the Hanford, Washington, site. Not to scale.

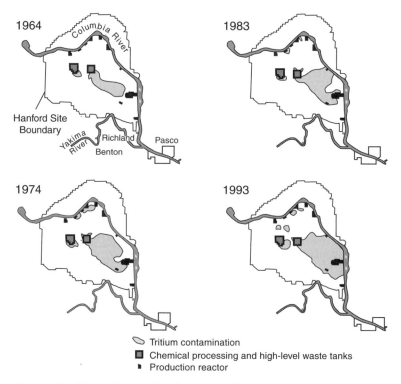

Figure 7.6 Maps showing development of the tritium groundwater plume (light shading) from the 200 East and 200 West facilities (dark outlines) at Hanford Nuclear Reservation, Washington, from 1964 to 1993. The smaller plumes adjacent to the Columbia River come from reactor sites that feed directly into the river. From *U.S. Department of Energy. Closing the Circle on the Splitting of the Atom* (DOE/EM-0266, 1996), 73.

times the standard by 1982.[60] Closer to their origin at the 200-sites, the tritium concentrations are a thousand times greater than safe drinking water standards.

The tritium apparently moved the 14 miles from contamination source to river in no more than nine years. The time for ruthenium to migrate 14 miles now is estimated at seven to eight years, and the faster traveling tritium at only six to seven years (see chapter 13).[61] The pollutant "travel times" require speeds of about 0.5–2.5 miles per year—much faster than Department of Energy scientists ever had believed contaminants could reach groundwater and flow long distances through rock and soil. Ironically, the department has attempted to cover its poor record by discrediting reports of dangerous radioactive contamination in the Columbia River.

Sadly, Hanford's tritium plumes are only the first arrivals in a gruesome groundwater race. Slower moving plumes are coming behind the tritium and ruthenium, carrying undetermined mixtures of highly hazardous alpha- and beta-particle emitters—cesium-137, iodine-129, strontium-90, technetium-99, and uranium isotopes, along with such toxic nonradioactive substances as arsenic, carbon tetrachloride (CCl_4), chloroform, chromium, cyanide, nitrates, and trichloroethylene.[62] This means that the Columbia River's contamination level can only increase.

Nevada Wasteland

On the riverless NTS, radioactivity contaminates approximately 565 million cubic feet of soil, sediment, and rock, plus 280,000 cubic feet of groundwater overall.[63] One hundred nuclear bombs, exploded either at the ground surface or in the atmosphere, have irradiated vast areas within and beyond NTS boundaries. In addition, at least 33 early NTS nuclear explosions in the atmosphere, called "safety" or "equation of state" experiments, blew up packages of plutonium and uranium with high explosives (see appendix 6). And more than 800 underground tests have turned vast tracts into a radioactive form of whole-earth Swiss cheese (figure 7.7).

The NTS also harbors buried radioactive wastes deliberately dumped in landfills and boreholes, amounting to nearly 10 million curies by January 1996. The total is four times the minimum estimate of radioactivity released from the Three Mile Island meltdowns. An unknown amount of radioactive and chemical wastes also is buried at 1,800 so-called "industrial sites," including leach fields, sumps, disposal wells, and leaking tanks. The Department of Energy expects to find another 1,500 industrial site disposal areas.[64] Failed radioactive waste "disposal" is a source of extensive radioactive contamination, both above and below ground.

The years before 1963 were heady times for atomic scientists, and no one was watching (see appendix 5). The "safety" test experimenters chose to dust more than 3,000 acres with plutonium and other dangerous contaminants, distributing more than 40 picocuries per gram (pCi/g) of radioactivity across the landscape.[65] Today, 10 pCi/g is considered a fatal dose. The scientists wanted only "to determine the size and distribution of plutonium particles which might result from fires and conventional explosive accidents involving nuclear weapons" (see appendix 6). A more responsible choice for tracing plutonium dispersal patterns might have been to use nonradioactive material with similar characteristics.

Figure 7.7 Aerial photograph of Yucca Flat bomb craters, Nevada Test Site, Nevada. A Plowshare program test (see appendix 7) excavated Sedan crater (foreground) with a 104-kiloton hydrogen bomb blast at 630 feet below the surface, which ejected eight million tons of rock debris out of the 1,280-foot-wide × 320-foot-deep crater. Four million tons fell back into the crater. Other craters in the middle and background resulted from deeper or smaller nuclear bomb detonations that did not eject material but let fractured rocks collapse into deep blast cavities (see appendix 8, figure A8.1). U.S. Department of Energy photograph.

For some "safety" tests, animals were held in cages or chained near or downwind from the blast sites "in areas certain to be contaminated by radioactive fallout...to determine biological uptake of dispersed plutonium particles by various species...."[66] Other experiments put caged test animals into contaminated areas (see appendix 6). Their pleasant dispositions made beagles favorite experimental subjects, only to be killed later and dissected. Radioactive beagle carcasses posed a decades-long waste disposal problem.[67]

An unspecified number of NTS blasts, called "hydronuclear" tests for undisclosed reasons, were detonated in boreholes less than 100 feet deep. Like the above-ground "safety" tests, the hydronuclear explosions contaminated the sur-roundings with plutonium. The test results are still classified, so nobody ever has investigated the extent of pollution, contaminant migration, or possible remedies. In addition, 828 below-ground nuclear tests blasted out large subterranean cavities, which remain thoroughly permeated with radionuclides (figure 7.8).[68] Either inad-vertently or by design, some of the later underground tests blew away capping rock

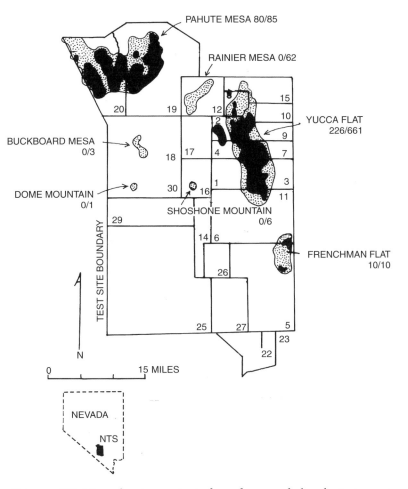

Figure 7.8 Map showing principal underground bomb test areas within the Nevada Test Site. Stippled pattern shows areas where bombs detonated above the water table, and black areas are where bombs detonated below the water table. Figures separated by a slash (e.g., 80/85) give the number of tests above/below the water table (other numbers identify test areas). From R. J. Laczniak et al. *Summary of Hydrogeologic Controls on Ground-Water Flow at the Nevada Test Site, Nye County, Nevada* (U.S. Geological Survey Water Resources Investi-gations Report 96-4109, 1996), figure 2.

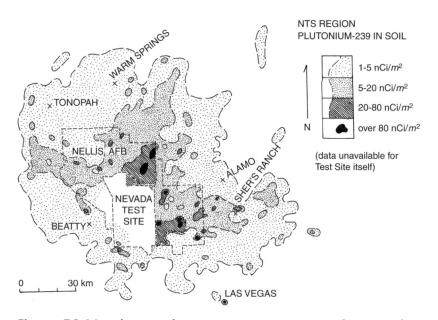

Figure 7.9 Map showing plutonium concentrations in soils surrounding the Nevada Test Site, deposited in fallout from atmospheric Nevada Test Site bomb tests. The highest concentrations spread to the east and northeast, generally opposite the direction of Los Angeles relative to Nevada Test Site, showing the wind directions during most tests. From W. A. Bliss and F. M. Jakubowski. *Environmental Plutonium Levels near the Nevada Test Site* (Nevada Operations Office NVO-181, U.S. Energy Research and Development Administration, 1977), 187–205.

and soil, contaminating air and ground surfaces. And underground "subcritical"[69] tests continue to contaminate rock and soil at the NTS.

Inventories of NTS surface soils' radioactivity, as of 1 January 1990, suggest a total greater than 2,000 curies.[70] More than half is plutonium from nuclear bomb tests and the safety experiments.[71] Surrounding the NTS, soils contain plutonium dust from atmospheric fallout over an area at least 78 miles wide by 96 miles long (figure 7.9; see also appendix 6). In 2006, the military planned "Divine Strake," a test explosion of 700-pounds of TNT over an old tunnel on the NTS. The bomb test was canceled early in 2007 after widespread protests. The state of Utah and many descendents of original test site downwinders, particularly, argued that the large-scale explosion would propel huge masses of plutonium-bearing soil into the atmosphere, once again raining radioactive materials on farms and people downwind.

Radioactive contamination in the "safety" test areas precludes most human uses of Shoshone lands on the NTS for hundreds of generations.[72] The U.S. Energy Research and Development Administration (predecessor of the Department of Energy) acknowledged this sacrifice of sacred Native land with a statement exemplifying the bomb scientists' hypocrisy: "The above ground areas where safety experiments have been conducted in the past offer unique sites for studies of the behavior of plutonium in the natural desert environment. Recognizing this, the Nevada Operations Office intentionally has preserved these sites."[73]

Wasted groundwater

A direct result of the 828 underground bomb explosions is radioactively contaminated groundwater at as many as 316 of the detonation sites. Bombs exploded close to or below the water table at depths of 500 to more than 2,000 feet below the surface. An estimated 110 million curies of radioactive contaminants, about equivalent to the minimum Chernobyl release estimate, were welded into melted rocks around explosion cavities—both in water-saturated rock below and unsaturated rock immediately

---

Box 7.2 Colloids Bearing Gifts

Underground weapons tests left unfissioned plutonium bomb fuel fused into rocks beneath the Nevada Test Site (NTS). Plutonium is essentially insoluble, so groundwater migrating through the contaminated rocks should be unable to carry it away. More soluble bomb-produced radioisotopes decay to plutonium isotopes, however, and can disperse tiny amounts of plutonium in water—soluble curium-244 decays to plutonium-240, for example.[a] The abundant zeolite minerals in altered volcanic NTS rocks seemed to offer another barrier to any plutonium releases. Zeolite minerals are used for industrial filters because they readily catch and hold (absorb) a wide variety of ions (box 7.1), including plutonium. But the untested, perhaps even subliminal and unexamined, expectation that zeolites would *stay within* the rocks turned out to be wrong.

Wherever the NTS's common volcanic rocks are rich in zeolite minerals, zeolite "colloids" also are abundant in the soils and groundwater.[b] Colloids are dust-sized particles that resist settling out of water and can be carried long distances. They are one of nature's foremost inorganic scavengers (see chapter 13). Repository scientists failed to anticipate the possibility that NTS's zeolite minerals might shed colloids abundantly into water seeping through soils and rocks. More recent studies at NTS, other nuclear reservations, and radioactive waste sites, have discovered that zeolite colloids are carrying plutonium and other radionuclides long distances through the unsaturated zone and into very deep groundwater.[c] The colloids in groundwater can come to the surface at springs or in wells. Once dried out, strong winds can blow the tiny particles and their hitch-hiking radionuclides for hundreds of miles.

Ironically, in 2002 Congress approved a tract of land at neighboring Yucca Mountain to become a national high-level radioactive repository for nuclear power plant waste.[d] In part, the Yucca Mountain site was selected because it, too, is made mostly from zeolite-rich volcanic rock (see chapter 10).

continued

Box 7.2 Continued

Notes
[a] Minhan Dai et al. Sources and Migration of Plutonium in Groundwater at the Savannah River Site. *Environmental Science and Technology* 2002;36:3690–3699; D. I. Kaplan et al. Actinide Association with Groundwater Colloids in a Coastal Plain Aquifer. *Radiochimica Acta* 1994;66/67:181–187.
[b] R. W. Buddemeier and J. R. Hunt. Transport of Colloidal Contaminants in Ground Water, Radionuclide Migration at the Nevada Test Site. *Applied Geochemistry* 1988;3:535–548. A. B. Kersting et al. (Migration of Plutonium in Groundwater at the Nevada Test Site. *Nature* 1999;396:56–59) demonstrated plutonium migration in groundwater over a distance of 0.8 miles in 30 years at the Nevada Test Site.
[c] D. I. Kaplan et al. *Enhanced Plutonium Mobility During Long-Term Transport Through an Unsaturated Subsurface Environment* (Westinghouse Savannah River Company WSRC-MS-2003-00889, 2003). Colloidal migration of plutonium and americium in unsaturated and saturated zones at the now-closed Rocky Flats, Colorado, facility also pose a problem (A. B. Kersting. *The Role of Colloids in the Transport of Plutonium and Americium: Implications for Rocky Flats Environmental Technology Site* [UCRL-TR-200099, University of California Lawrence Livermore National Laboratory, 2003]). W. R. Penrose et al. (Mobility of Plutonium and Americium Through a Shallow Aquifer in a Semi-arid Region. *Environmental Science and Technology* 1990;24:228–234) demonstrated plutonium transport as much as two miles from its source at Los Alamos National Laboratory, and A. P. Novikov et al. (Colloid Transport of Plutonium in the Far-Field of the Mayak Production Association, Russia. *Science* 2006;314:638–641) demonstrated transport of plutonium and other actinides (radioactive elements heavier than actinium) in groundwater over nearly two miles from its source in a Russian nuclear production site.
[d] More than 55,000 tons of high-level nuclear waste, in the form of spent fuel rods from commercial reactors, currently were being stored at 72 sites around the country as of 2006 (U.S. Government Accountability Office. *Yucca Mountain: Quality Assurance at DOE's Planned Nuclear Waste Repository Needs Increased Management Attention* [GAO-06-313, 2006].

above the water table.[74] The radioactive materials include tritium, plutonium, and myriad plutonium fission byproducts. The other 512 underground blasts left nearly 200 million curies of tritium and plutonium, and approximately 4,400 *pounds* of unfissioned plutonium, in unsaturated rock and soil above the water table.[75]

About two-thirds of the radioactivity from underground tests lies in rocks beneath Pahute and Rainier mesas, in the northwestern NTS (figure 7.8). Scientists at the NTS expected that any plutonium molecules getting into groundwater would be held (sorbed) by "ion-exchanging" zeolite minerals in the deep volcanic rocks where the bombs exploded.[76] These scientists never anticipated that groundwater would or could carry it away from test sites. Quite unexpectedly, the zeolites have provided an escape route, which has put radioactive contamination into deep groundwater (box 7.2). The mesas provide the NTS's shortest groundwater pathways to populated areas, but flow paths are not well defined.

So far, the easily tracked radioactive pollutants are headed southwest, toward Oasis Valley, where community wells provide water for drinking and crop irrigation. Continuing on the same heading will take the plumes to Death Valley, a tourist mecca where all drinking water comes from wells.[77] Plutonium also is reported in groundwater close to the northwestern edge of the NTS, along with huge tritium concentrations.[78]

The underground tests also added massive amounts of hazardous nonradioactive materials to NTS rock and soil. No comprehensive inventory is available, but a single underground test can yield more than 125,000 pounds of lead, along with a wide variety of other toxic metals, plus organic chemicals. A subterranean material is hazardous only if humans can come into contact with it—and some argue that even in large amounts the lead is relatively harmless until it enters groundwater or is mined and exposed to rain and wind.[79] But many, including the authors, would argue that toxic metals spread intentionally in the environment cannot be judged harmless. And the experiences summarized in this chapter indicate that even the toxic materials not actually *blasted* into a groundwater zone eventually will get into groundwater through natural processes. These hazardous materials remain toxic forever.

So Many More

More than 12 million cubic feet of radioactive waste was pumped into the ground or buried in trenches at the other major western atomic sites, principally Idaho National Engineering and Environmental Laboratory (INEEL); Rocky Flats Environmental Technology Site, Colorado; Los Alamos and Sandia National Laboratories, New Mexico; and Lawrence Livermore National Laboratory, California (figure 7.3).[80] Nuclear weapons massively polluted land and water at these sites, but little information is available about the actual locations of many highly toxic materials. For example, the amounts and even the location of buried transuranic wastes—radioactive materials heavier than uranium (box 7.1; see also chapter 10)—are unknown at many nuclear weapons facilities.[81]

Unclassified estimates of transuranic waste volumes and radioactivity at three of the nuclear weapons facilities in the western United States vary wildly (figure 7.10). Nobody knows how much plutonium is in the ground at INEEL—or where it is, where it might be headed, how fast it might get there, or where it is likely to turn up hundreds of years from now. The only comprehensive attempt to track historical records on buried transuranic waste at INEEL[82] revised a prevailing estimate of 73,300 curies upward to a figure between 640,000 and 900,000 curies. In comparison, the Hiroshima and Nagasaki bombs together produced one million curies of radiation. Like Hanford and NTS, an unknown but almost certainly large amount of radioactive and hazardous wastes is in the unsaturated zone above groundwater at INEEL and all other weapons sites, episodically migrating to aquifers or fracture zones and spreading long distances.[83] Plutonium and other contaminants probably have left the INEEL reservation in rainstorm runoff, on dust particles carried in windstorms, and attached to ash from wildfires.[84]

Plutonium and other contaminants continue to move beyond reservation boundaries at Los Alamos National Laboratory, New Mexico, and Rocky Flats, Colorado, about 15 miles northwest of Denver. Tests showed that soil over 19 or more square miles around the Rocky Flats reservation can have plutonium concentrations as much as 380 times above natural background levels. Fires at Rocky Flats wafted plutonium to the atmosphere in 1957 and 1969. Accidental spills and leaks from corroding waste barrels, stored outdoors until 1968, also have released plutonium to the environment.[85]

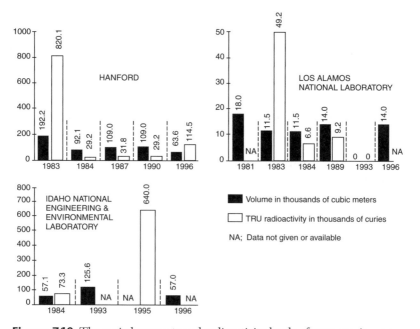

Figure 7.10 The varied amounts and radioactivity levels of transuranic wastes (TRU) buried at the Hanford Nuclear Reservation, Los Alamos National Laboratory, and Idaho National Engineering and Environmental Laboratory, according to different estimates. The apparently irrational variations may be related to changes in the definition of "transuranic waste" from more than 10 to more than 100 nanocuries per gram, the recognition that some "retrievable stored waste" is not readily retrievable, reexamination of old records, and mistakes. From Marc Fioravanti and Arjun Makhijani. *Containing the Cold War Mess: Restructuring the Environmental Management of the U.S. Nuclear Weapons Complex* (Takoma Park, Maryland: Institute for Energy and Environmental Research, 1997), figures 2, 4, and 5. Available: www.ieer.org/reports/cleanup/index.html. Reproduced with permission.

Hallmarks

Hallmarks of the nation's nuclear weapons program have been and still are extraordinarily sloppy record-keeping, careless handling of deadly materials, and blind faith in nature's ability to absorb anything. The AEC originally served as a watchdog and industry booster. The dual roles conflicted, so the Department of Energy and the Nuclear Regulatory Commission were created to separate them. Both maintain cozy, mutually supportive relations with the industry, however, and neither is known for trying to protect public health. Past plans for monitoring buried wastes, as well as responses to public concerns, reflect unscientific attitudes of minimization ("It won't hurt you," "It won't go far"), denial ("We have no evidence of contamination," "We know what we're doing"), and rationalization ("War is hell, it requires sacrifices").

The government also found tricky ways to cover up some bad mistakes. Chemical engineer Billings Brown discovered one telling example: During the years of atmospheric testing, Dr. Brown had recorded high radiation levels in northern Utah that exceeded children's tolerance levels. Working for the Pentagon's Institute for Events Analysis in 1968, Dr. Brown discovered that government-sponsored cancer studies defined northern Utah as a low-fallout area and used it as the normal incidence "control" for studies on southern Utah cancer rates. Choosing northern Utah for the cancer study control cleverly hid the impact of bomb fallout on northern Utah and allowed the government to claim a lower cancer rate for southern Utah than if a truly normal rate had been used. This falsification of southern Utah exposures left the threefold jump in southern Utah leukemias after 1955 unexplained.[86]

When cover-ups proved inadequate, government authorities denied that the radioactive fallout could have harmed downwind populations, forcing disabled and dying radiation victims into decades of costly lawsuits. Only in January 2000 did the United States admit that radiation had killed nuclear weapons workers. Similarly, until confronted with high cancer rates among the military's "atomic veterans," the U.S. government would not acknowledge the link between their radiation exposures and their illnesses until the late 1990s and even later. Even now, many irradiated soldiers with lifelong disabilities remain unacknowledged victims of atomic bomb radiation. Troops currently exposed to depleted uranium weapons may be adding more atomic veterans (see below).

Until recently, the Department of Energy continuously denied the widespread dumping of radioactive materials that went on during the Cold War. Only under Hazel O'Leary, Secretary of Energy from 1993 to 1996, did the department release factual documents on unanticipated radioactive waste dispersal in the environment, to the public's great benefit.[87] We authors obtained most of the information in this chapter because of Secretary O'Leary's intent to fully inform Americans about all aspects of the atomic century. The pattern of obscuring nuclear projects and their consequences resumed when she left office during the Clinton administration's first term.

Dirty science

The nation's most eminent scientific organizations did not succeed in applying good science to waste disposal problems. In 1954 the National Academy of Sciences (NAS) created a Committee on Waste Disposal to advise the AEC, with AEC funding. The NAS initially selected a mostly independent panel, dominated by a stellar group of geologists.[88] Barely a decade later, the panel had become subservient to the nuclear industry,[89] and even after the AEC split and merged with broader energy agencies, the panel's recommendations supported industry positions. Today's equivalent NAS Board on Radioactive Waste Management is overloaded with representatives loyal to the Department of Energy, Nuclear Regulatory Commission, and nuclear industry.

The original Committee on Waste Disposal asked the AEC tough questions about ongoing disposal methods and demanded to know the effects of dumping radioactive waste in unlined trenches.[90] An independent 1957 report on waste leaking from storage tanks completely discredited the old presumption that Hanford soils would adsorb and contain contaminants, such as cesium-137, near the tanks.[91] Hearings transcripts show committee members' concerns about potential waste disposal problems for unsaturated zones, groundwater, rivers, and streams.[92] When the AEC could not substantiate its contention that more than a minor proportion of contaminants had been trapped in "soils," some panel members recommended discontinuing those waste disposal methods. The 1957 panel also protested the "popular idea...that dilution is easy to obtain if you have large masses of water."[93]

Subsequent committee discussions addressed whether plants and animals could take up contaminants, contaminant concentrations in water-rich environments, and the possibility that potentially toxic contaminants could enter the human food chain. In a 1960 report, the same NAS committee concluded, "The movement of fluids through the [unsaturated] zone and the consequent movement of the radio-isotopes are not sufficiently understood to insure safety,"[94] and labeled the belief that radioactive contamination would not get into groundwater at arid sites "unproved." The committee repeatedly criticized AEC's disposal practices, finding none of the principal weapons production and reactor test sites suited for safe radioactive waste disposal. Their first (1957) and last (1966) reports stated, "[T]he hazard related to radioactive waste is so great that no element of doubt should be allowed to exist regarding safety,"[95] and "Safety is a primary concern, taking precedence over cost."[96]

If the NAS panel had influenced AEC oversight of nuclear waste disposal, the west's nuclear reservations might not be so highly contaminated. But the AEC suppressed the committee's critical reports until Congress forced public release in 1970. AEC also threatened to withdraw the committee's funding until NAS "compromised" in 1967, dismissing the independent committee members and letting AEC both help select future members and review and control the distribution of committee reports.[97] Under Department of Energy control, subsequent NAS committees on radioactive waste management failed to seriously examine high-level radioactive waste leaks from tanks and other problems.[98]

The original NAS committee's reports have been proved correct, unfortunately. The U.S. policy of putting long-lived, high-level nuclear waste in dry western soils has not kept radioactive pollution out of groundwater. Nature has showed that it can move water from the surface, and through the supposed barrier of a thick unsaturated zone, to depths below the level of the proposed Yucca Mountain high-level waste disposal site in Nevada (see chapter 10). In a case of déjà vu, nearly 40 years later a 1998 U.S. General Accounting Office report[99] concluded that unsaturated zone processes at Hanford still are too poorly understood to ensure safety (the General Accounting Office has now been renamed the Government Accountability Office). Recently, hydrologists and soil scientists finally have begun to admit that water can move from the surface along fast pathways—cracks and fractures—through thick unsaturated zones, taking contaminants quickly to groundwater (see chapters 10, 13).

Costs and cleanup

Nuclear science never has found a safe way to deal with long-lived high-level radioactive wastes. The costs of site cleanups and hitches in performing the cleanups reflect their difficulty. High-level and transuranic wastes must be permanently isolated in "geologic" repositories, but science has not been able to impose a geologic criterion for repository selection. Low-level radioactive waste burial sites also are problematic, whether in arid or in nonarid regions. All have leaked and many contaminate groundwater (see chapter 10).

The Department of Energy has estimated the cost of cleaning up the huge amounts (table 7.1) of radioactive and hazardous nonradioactive materials from more than half a century of nuclear weapons production at $142 billion, with a completion date of at least 2035. These figures do not include the $60 billion spent since 1989 and address only wastes that the government actually acknowledges. The estimates

Table 7.1 Types, Estimated Quantities, and Planned Treatment and Disposal of Nuclear Wastes

Description	Estimated Quantities[a]	Planned Treatment, Disposal
Enriched uranium	8,400 containers	Blend high-quality material for reactor use
Plutonium	5,800 containers	Stabilize, package, dispose in A or B site[b]
Plutonium/uranium residues	119 tons	Dispose in A or B site
Depleted uranium and uranium	755,000 tons	Convert to stable form, package, dispose in C site[c] or reuse
Storage tank waste	88 million gallons	Separate high- from low-activity parts, stabilize and vitrify high-activity waste package and dispose in B site; treat low- activity part, bury in shallow pits at current location
Sodium-bearing tank waste	900,000 gallons	Treat, package, dispose in geologic repository
Vitrified waste	18,700 canisters	Dispose in B site
Solidified low-activity waste	35 million cubic feet	Bury in shallow pits at current location
Dried high-level waste	55 million cubic feet	Treat, repackage, and dispose in B site
Transuranic materials[d]	5 million cubic feet	Treat, repackage, and dispose in A site
Low-level and low-level mixed waste[e]	42 million cubic feet	Treat, repackage and bury in C site
Hazardous chemicals, heavy metals, others	No comprehensive estimate	

continued

Table 7.1 *Continued*

Description	Estimated Quantities[a]	Planned Treatment, Disposal
Spent nuclear fuel	2,700 tons	Treat, package, and dispose in B site
Reactors and other contaminated facilities	4,400 facilities	Decontaminate, decommission, and then destroy[f] or reuse
Underground radioactive waste tanks	241 tanks	Empty tanks, fill with grout or other material
Waste burial grounds, dried ponds, spills, other types of sites	10,400 sites	Treatment varies from removal and disposal to release of site for other use

U.S. Government Accountability Office. *Nuclear Waste: Better Performance Reporting Needed to Assess DOE's Ability to Achieve the Goals of the Accelerated Cleanup Program* (GAO-05-764, 2005), 8–9.

[a] Quantity estimate rounded and converted to English units. Container, canister sizes not provided.
[b] A, Waste Isolation Pilot Project transuranic waste disposal site in New Mexico; B, Yucca Mountain high-level waste site in Nevada.
[c] C, Any existing DOE or commercial low-level waste site.
[d] Various materials contaminated with plutonium and other transuranic elements, with or without hazardous substances; includes soil. This figure is a gross underestimate.
[e] Various waste materials, including building materials and soils. Mixed waste contains both radioactive and hazardous substances.
[f] Disposal method not specified.

are likely to be on the low side.[100] Cleaning up Hanford's enormous mess will cost dearly—and the costs are rising. Just stabilizing the radioactive and hazardous wastes stored in Hanford's underground tanks and *preparing* for disposal was estimated at $11 billion in 2006, an increase of 150% in five years. And this estimate is only preliminary, since the project still is stumbling through its startup phase.[101]

Recent NAS Boards on Radioactive Waste Management worry more about costs than public safety, and mostly ignore U.S. Government Accountability Office and public concerns.[102] Examples include 1994 approval of a poorly characterized low-level radioactive waste site (see chapter 10)[103] and a 1998 hearing where panel members emphasized budgetary limitations and discouraged "open-ended" research projects on fluid movements in the unsaturated and groundwater zones at Hanford.

Of the 144 sites in the United States that played roles in developing and producing nuclear weapons, 109 will never be cleaned up to a condition that allows unrestricted uses. The large reservations covered in this chapter, as well as Oak Ridge, Tennessee, and the site at the Savannah River in South Carolina, among others, are sacrifice zones. Isolating the reservations will create more problems, because of immensely long radionuclide hazard lives. The Department of Energy's initial plans for closing Colorado's Rocky Flats site called for covering much of the industrial area with clean soil caps. But ultimately the contaminated soil was removed for disposal at the relatively well-designed Waste Isolation Pilot Project transuranic wastes repository in New Mexico.

Contaminated buildings at Rocky Flats were demolished and sent to low-level radioactive waste sites. Site cleanup was certified "complete" in December 2005 at a total cost of more than $7 billion, and in July, 2007 4,000 acres were transferred to the U.S. Fish and Wildlife Service for a wildlife refuge. Residual plutonium contamination will remain high (a maximum of 50 pCi/g), limiting public access, and there is no guarantee that it will not migrate offsite.[104]

The United States still has not consolidated highly toxic and long-lived waste plutonium from myriad disposal sites, including a substantial amount in 12-foot-long fuel rods that cannot remain stored at the Hanford waste site. As yet, there is no plan for processing plutonium into a permanent form for disposal, preventing shipment to Savannah River. Worse yet, the Department of Energy has no plan for monitoring any plutonium disposal site to ensure its integrity.[105] The Waste Isolation Pilot Project site is not designed for isolating high-level waste and is still untested by nature. It may contain transuranic wastes safely but has too small a capacity for all the wastes needing disposal (see chapter 10).

Cleaning up the regionally extensive volumes of radioactively contaminated groundwater probably is impossible. The Department of Energy's Environmental Restoration Division evaluated some of the most bizarre "cleanup" schemes ever imagined for controlling the spread of underground bomb-test radioactivity at the NTS, however. One proposed drilling hundreds of wells and moving ground-water around the contaminated area on a regional scale—pumping out clean water upslope, then pumping it back into the ground downslope from the contamination. Each rainstorm that recharges groundwater in the test areas moves more contamination toward deep aquifers, so the pumping would have to continue for thousands of years. Another proposal would excavate approximately 3.5 trillion cubic feet of contaminated materials from the bomb sites, leaving a multitude of open pits, and then "treating" the materials and eventually reburying them in the excavations. This idea penciled out to $7.3 trillion *in 1990 dollars*.[106]

The Department of Energy's futile "solution" is to "maintain institutional control in perpetuity" over contaminated groundwater,[107] to prevent anyone in the future from tapping a contaminated water source. It may seem better than making Americans eat radioactive wastes, but groundwater does not sit around waiting for solutions—it flows through rocks, carrying contaminants along. As the waters flow, the boundaries of "institutional control" must continually expand.

In your own back yard

The recent history of radioactive wastes at Hanford Nuclear Reservation alone shows that the United States has no adequate method for safe disposal or retention of moderately to highly radioactive wastes and does not want to pay for onerous cleanups. The Department of Energy's solution is to simply reclassify stored high-level waste as low-level waste so that nothing needs to be done with it, or to "reindustrialize" and "remanufacture" it (see chapter 10).

Colorado's Rocky Flats site illegally sent radioactive wastewater, probably containing plutonium, to the nearby Lowry landfill. In 2000, the U.S. Environmental Protection Agency addressed the illegality by proposing to downgrade the hazard level *classification* for that waste. The reclassification would not change the waste's radioactivity level one whit, but it would allow radioactive liquids to be dumped into public sewers and flow into nearby rivers—or would mix them directly into irrigation water, putting them into food and groundwater. Solid radioactive wastes are to be mixed with sewage sludge and sold as fertilizers, which could be spread on thousands of farms, efficiently delivering radioactive contamination to our door-steps and dinner tables (see chapter 10).

The Hanford and Idaho (INEEL) sites still store approximately 65 million gal-lons of high-level liquid, part-liquid, solid, and part-solid sludges in tanks, with a total radioactivity of 399 million curies—nearly four times the officially reported Chernobyl release.[108] Some of what remains soon could be redesignated as "low-level" waste at the scratch of a bureaucratic pen. In 2006, the Department of Energy issued a proposal to downgrade the classification of high-level wastes from reprocessing spent nuclear fuel, which now are stored in INEEL tanks.[109] What might come of that reclassification is anybody's guess.

In 2000, the U.S. Department of the Interior floated a reindustrialization plan to sell 6,000 tons of waste radioactive nickel, with no cheap disposal option, for remanufacture into steel on the dubious premise that other steel ingredients would sufficiently "dilute" the nickel to make it safe for human use. In the absence of safety standards to protect human health, the steel could be made into I-beams, automobiles, jewelry, silverware, and even leg braces and hip replacements.[110] Popular outcry greeted the plan, and supposedly it was abandoned. But it may rise again. In 2001, a lawsuit prevented the U.S. Department of Transportation from adopting a rule that exempted radioactive materials and wastes from labeling and regulation in interstate commerce. But the Department of Transportation lets the radioactive wastes enter into interstate commerce, and come closer to our everyday lives, through a loophole, nonetheless. Undoubtedly, the government will again try to grandfather the practice into its regulatory codes.

Some nuclear bomb waste has been recycled into new weapons. "Depleted ura-nium" (DU) is uranium hexafluoride gas waste left over from enriching nuclear bomb fuel.[111] Like all uranium, it emits alpha particles and is highly flammable. Uranium is a very dense metallic substance, 70% denser than lead, which can pierce armored vehicles. DU shells and bombs start fiercely hot fires upon impact, spreading thick black smoke with breathable alpha-emitting uranium, far and wide. The United States has a huge DU supply, with few disposal options, and manufac-turing it into "dirty" conventional bombs and shells saves storage costs.

DU ordnance was fired in Kosovo in 1999, in the 1991 Gulf War, in the post-9/11 assault on the Afghanistan Taliban government, and in the 2003 invasion and current (2007) occupation of Iraq.[112] Tests of strategic DU weapons are adding radio-active dust, with a half-life equivalent to the age of the Earth, to Nevada's Nellis Air Force Range and wherever the winds can blow it. News stories began reporting the possible consequences in 2005, as some returned soldiers discovered themselves to be highly contaminated with DU. The wife of at least one gave birth to a child with "atypical syndactyly"—a common deformity related to radioactive exposure.

Diabolical Genie

Once released, atomic radiation is a diabolical gift that keeps on giving. The American public does not fully comprehend the immensely long times that radioac-tive contamination from military activities will persist on western U.S. lands, and in its water and blowing dust. What we don't know can hurt us. Even scientists are only beginning to realize that humans, like other animals, continually interact with the physical forces that shape the Earth's surface—with water and soils, and with the plant and bacterial life that water and soils support. For this reason alone, radioactive pollution and other atomic bomb damages are not contained and never can be.

The financial cost of the U.S. nuclear weapons program from 1940 through 1996 is nearly $5.5 trillion (in constant 1996 dollars).[113] The future costs of storage and disposal of accumulated wastes, plus dismantling nuclear weapons and disposing of dangerous surplus materials, will bring the total to at least $5.8 trillion. But even this astronomical figure still omits the cost of solving the worst environmental problems arising from weapons manufacture—to keep the radioactive contamination from spreading and to clean up land and surface and underground water.[114]

We cannot predict future risks from the toxic and radioactive wastes in the environment with any confidence because we do not know all of nature's pathways. For example, researchers have uncovered unexpected radioactive concentrations in land and marine sediments worldwide, as wind and water redistribute bomb fallout and other radioactive substances.[115] Seabirds become contaminated with radioactive isotopes from eating contaminated fish, shrimp, and small crabs that live in oceanic sediments. The radioactive material concentrates in the bird's organs and comes back to shore in the droppings. The bulk of that radioactivity must come from nuclear power plant accidents and radioactive materials dumped at sea, especially in the Arctic.[116]

The techniques and knowledge needed to protect nuclear-waste repositories from public access for centuries to millennia simply do not exist. Who knows where future Americans will live and congregate, or what resources they will seek out in a resource-depleted future—especially if we continue to destroy our natural capital? Institutional land uses, or physical barriers such as fences, signs, monuments, and the like, are notoriously prone to failure.

We can try to limit access to contaminated lands as long as our culture is continuously stable, but limiting access to radioactive water is more difficult. Dumps that received uranium and plutonium will be hazardous long enough for contaminated groundwater to travel great distances, and even into our drinking water supplies. Nobody can predict the potential effects of today's rapid climate change on the deep repositories for isolating plutonium and uranium wastes. Even the Yucca Mountain high-level waste repository in Nevada is expected to release radioactivity into groundwater (see chapter 10). Once we forget the locations of former atomic test and bomb factory sites, future populations could be subjected to food grown with contaminated irrigation water on or near those contaminated lands.[117]

Within 100 days of taking office in 2001, the George W. Bush administration proposed expanding nuclear testing and nuclear power plant construction, and also hurrying the schedule for destroying older nuclear weapons.[118] The government soon began circulating classified plans for new nuclear weapons and shifted the official U.S. stance on nuclear weapons, from deterrents to tactical weapons for preemptive strikes against perceived enemies (see appendix 8). Plans for atomic warhead reductions soon vanished (see chapter 12), and in 2003 the government proposed renewed nuclear weapons testing, in violation of the Comprehensive Nuclear Test Ban Treaty,[119] to maintain arsenal "flexibility" and develop tactical nuclear weapons (so-called Bunker Busters), for deeply buried targets (see appendix 8).

The twin atomic devastations of Hiroshima's and Nagasaki's civilian populations still tarnish the World War II record of the United States. What, if anything, could be said retrospectively about unprovoked nuclear bombing of civilian populations? The example of the United States using nuclear weapons for preemptive strikes could unleash the Cold War's feared nuclear holocaust. And what would new nuclear weapons mean for our own troops and civilians? So far, the only victims of atomic war have been the Japanese and our own people, destroying the lives of so many.

Nuclear weapons will not obstruct terrorists and thus do not reflect clear national needs. If we fail to solve the twentieth century's nuclear problems, additional lives and lands will be sacrificed. Developing new nuclear weapons will mean more radioactive fuel mining, more fuel processing, new testing programs, and a lot more wastes. Testing nuclear bunker busters and bombs for the "Star Wars" antimissile defensive system is likely to swell nuclear exposures and prolong them. The proposed "Divine Strake" bomb test shows how even nonnuclear bomb tests will increase radioactive exposures.

The devastating illness dubbed "Gulf War syndrome" and cancers that NATO troops exhibit may be traceable, in part, to DU weapons.[120] A decade of official refusals to issue either warnings or admissions about Gulf War syndrome indicates that the U.S. defense establishment still cares little about individual Americans. "A central principle of environmental health protection—protecting those most at risk—is missing from much of the U.S. regulatory framework for radiation," wrote Arjun Makhijani and colleagues of the Institute for Energy and Environmental Research.[121] Gulf War syndrome II, from the effects of exposure of U.S. troops to radioactive and other chemicals in Afghanistan and Iraq, may be developing as we write. Meanwhile, the U.S. government has been studying ways to cut the federal programs that compensate Cold War radiation victims.[122]

We cannot put the nuclear genies back into their bottles, but we can stop mining uranium and fissioning it and stop making nuclear bombs and radioactive ordnance. We can end nuclear and "safety" tests. And we can start raising public consciousness, isolating dangerous wastes more conscientiously (see chapter 10), and protecting the public from contaminated lands. In mid-2005, the Navajo Nation, which suffered so many Cold War losses, showed the way by placing 17 million acres off-limits to uranium mining in spite of its potential to profit from revived nuclear industries.

American citizens must understand that all of us are downwinders. We need to join with the most ravaged atomic victims and make national leaders focus on the critical long-term human health consequences of bomb making and of bombing our lands. Before the nation engages in another nuclear buildup that could again compromise the lives of American citizens and soldiers, the past sacrifices should be thoroughly and publicly acknowledged, and the continuing threats revealed.

Rejecting radiation exposures by exposing these facts is the only way to prevent another grim toll of ordinary Americans, who just happen to be in the way, terrifyingly sacrificed to a so-called "war on terror." As the late Western Shoshone Spiritual Leader Corbin Harney said: "It's in our backyard...it's in our front yard. This nuclear contamination is shortening all life. We're going to have to unite as a people and say no more! We, the people, are going to have to put our thoughts together to save our planet here. We only have One Water...One Air...One Mother Earth."[123]

8   No Habitat but Our Own

"Smart growth" [is] a euphemism for predictable and voluntary disaster.
A. R. Palmer, *GSA Today* March, 2000

Americans tend to think of the western United States as open spaces and the east coast as urban and crowded. After all, the northeast corridor from Washington, DC to Boston, Massachusetts, exemplifies the modern "mega-conurbations" of cultural historian Lewis Mumford—"nearly unbroken belt[s] of residential and commercial development, dotted with isolated parklands but little actual countryside."[1] Ironically, the eastern urban centers melded together in imitation of Los Angeles, California, that haphazard collection of zoning-defiant industrial-residential-commercial mélanges.[2]

By now, Los Angeles's cement-and-asphalt environment has become the very model of a modern human habitat and the nation's poster child for suburban sprawl. In an attempt to emulate its glittery lifestyle, every prosperous American town has snaked strip developments out along major highways, spraying cheap commercial-residential urban–suburban developments in all directions. Supported and encouraged by enormous public investment in roads, highways, and other infrastructure, the sprawl constantly expands until it displaces all other land uses and human habitat becomes the dominant or only habitat.

We seem to have little concept that clean environments, and clean air and water in particular, support the physical, mental, and economic health of human societies (see chapter 1). This is why environmental guru Paul Hawken and co-authors termed them "natural capital." Sprawling urban–suburban habitats are not very healthy because they foul the air and make numerous contributions to water pollution. Developments are dominated by gas-belching automobiles, gas stations with leaky underground storage tanks, and asphalt roads and parking lots. Residential suburbs shed megatons of lawn fertilizers and pesticides into local streams and lakes. All these relatively uncontrolled chemical releases make cities and suburbs into sources of land, water, and air pollution, which damage both human health and livelihoods. Urban wastes come back to haunt us through our air and water and also come floating onto our beaches.

Urban and suburban areas depend on nonurban areas for food, clean water and air, and raw and manufactured materials. Our Earth simply cannot support human life if urban growth continues wiping out all its agricultural land, isolating wildlife in limited preserves, taking clean water from rural areas, and spreading pollution from the mountains to the shore. All of these habitats people need for survival.

Expanding Footprints

Current estimates of total urban areas nationally are somewhere between 51 million and 98 million acres. The lower estimate could almost cover Utah with population, while the higher is nearly the size of California, the third largest state.[3] Especially in the west, sprawl increased by 33% from 1982 to 1997, and six of the nation's 10 fastest-growing urban centers on the 2000 U.S. Census list are in western states. In the 11 western states, more than 16.5 million nonfederal acres (4.2%) are urbanized. Besides Los Angeles, major urbanizing areas include San Francisco–San Jose–Oakland–Richmond, California, circling San Francisco Bay; Denver–Evergreen–Littleton, Colorado; Salt Lake City and Provo–Orem, Utah; Phoenix–Mesa–Tempe–Scottsdale, Arizona; Seattle–Tacoma, Washington; and Portland, Oregon's advance into the fertile Willamette Valley. Figures 8.1 and 8.2 depict the progress of western sprawl from the 1800s to 1900s in northern California and Oregon.

Population growth and increasing per capita land consumption are the drivers of urban sprawl. Western populations are increasing at very high rates—Nevada's 66% population growth led the nation from 1990 to 2000.[4] Between 1970 and

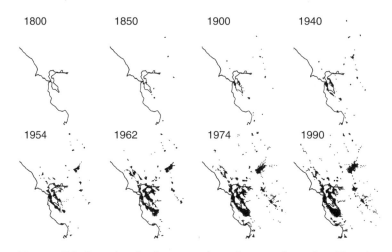

Figure 8.1 Growth of urban sprawl in the area from San Francisco Bay to Sacramento, California, between 1800 and 1990. From C. A. Bell et al. Dynamic Mapping of Urban Regions: Growth of the San Francisco/Sacramento Region. In *Proceedings of the Urban and Regional Information Systems Association* (San Antonio, Texas, 1995), 723–734.

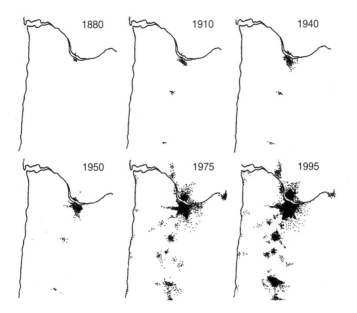

Figure 8.2 Growth of urban sprawl from Portland to the Willamette Valley, Oregon, from 1880 to 1995. From U.S. Geological Survey. *Analyzing Land Use Change in Urban Environments* (Fact Sheet FS 188-99, 1999).

1990, the growing populations of 12 western cities devoured at least 65,000 more acres each, as tract sizes for housing and other urban development also expanded.[5] Los Angeles grew an average of nearly 35 acres per day between 1970 and 1990, while San Francisco added 125,000 urbanized acres, equivalent to 111 Golden Gate Parks.

Population growth itself does not lead to urban sprawl if the amount of land each person occupies stays relatively low—but since 1980, U.S. development has mushroomed into outlet malls and megamalls, "big box" stores that rival football fields, and ballooning house sizes. The expanded scale of development covered so much land from 1990 to 2000 that even cities with no population growth paved over another 26% of surrounding lands. Where populations also increased, urban sprawl grew even faster. Cities with more than 50% population growth expanded an average of 112% across previously undeveloped lands.[6]

The nation's typical home has grown from 1,560 square feet in 1974 to 2,149 square feet in 2004—more than 35% in 30 years. In the western United States, the average size of single-family residences grew nearly 12% between 1995 and 1999, to 2,235 square feet.[7] Massive 5,000- to 6,000-square-foot single-family McMansions ("starter-castles," "trophy houses") sprout across the west by the tens of thousands (figure 8.3). In less than a decade, the average size of new houses along California's San Mateo County coast surged 400%, from 2,500 square feet in 1993 to 10,000 square feet in 2002.

Unbounded extravagance inspires *monster*-house communities of 8,000-square-foot abodes—six bedrooms, nine bathrooms, two home offices, plus wine cellar, media

Figure 8.3 A modest home near Santa Rosa, California (foreground), and two of many "starter castles" sprouting on area farms—in this case, vineyards. The new homes commonly exceed 5,000 square feet of living space. Photographed October 2001.

room, and more—for families of four.[8] The monster house craze, some reaching 21,000 square feet, prompted the San Mateo County Board of Supervisors to consider a cap at 5,000 square feet. So far they have not acted on it. The west boasts monster home developments in Denver and Salt Lake City.

Building and maintaining gargantuan single-family dwellings devours materials and energy. McMansions and monster houses drive American's per capita consumption—already outstripping the rest of the world—to even greater heights. More than a quarter of a million pounds of mined minerals and metals goes into a 2,000-square-foot house with a two-car garage, three bedrooms, two and a half bathrooms, central air conditioning, and a fireplace (see chapter 4, and appendix 3).

Critical Habitats

Cities and suburbs do not feed themselves or provide their own water. All their support must come from farms, forests, mines, rivers, and reservoirs beyond their boundaries. Maintaining dense human habitats takes water out of streams and underground water storage aquifers, which formerly supported natural ecosystems and wildlife habitat or farms. Impermeable concrete and asphalt cover up soils, block seepage pathways to groundwater-storage aquifers, and make rainwater run off faster to streams, limiting the water supply, accelerating erosion, and increasing flood rates and heights.

Development devastates wildlife habitats, endangering so many native animal and plant species that many are at the edge of extinction. Once houses and malls cover as little as 5% of a natural area, sensitive water-based (aquatic) ecosystems decline quickly. The scope of change is greater for forested or range lands than for farmland (see chapter 1).[9] Ironically, urban development competes with agriculture for the prime farmlands that feed urban populations—especially the flat, sparsely hilly, or gently rolling lands most easily converted to tract housing (figures 8.4, 8.5). By 1982, metropolitan areas had engulfed 69 million acres of the nation's prime farmlands, and only ten years later reached 82 million acres.[10] This competition comes as the demand for locally grown foods has increased.

While supporting only about 5% of the world's population, the United States has managed to imperil 69% of the world's nearly 2,000 endangered and threatened plant and animal species. Our largest habitat-destroying land uses are industrial

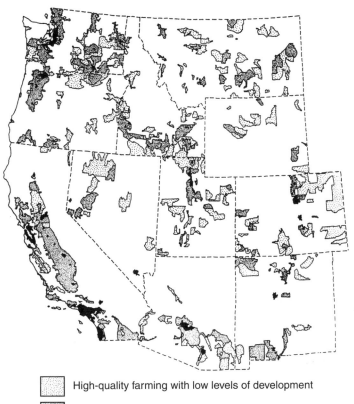

■ High-quality farming with low levels of development

■ High-quality farmland threatened by high development

◆ Urban areas

Figure 8.4 Map showing areas where urban–suburban sprawl threatens prime farmlands. Modified from Sprawling Development Threatens America's Best Farmland. *American Farmland Trust* Winter 2003;10.

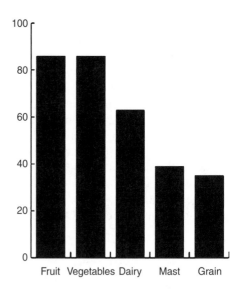

Figure 8.5 Percentage of U.S. food production in areas of urban sprawl. Modified from Sprawling Development Threatens America's Best Farmland. *American Farmland Trust Winter* 2003;10.

agriculture and urbanization.[11] Between 1970 and 1990, the 100 largest urbanized areas in the United States wiped out more than nine million *additional* acres of farmland, natural habitat, and rural open space, replacing them with housing, roads, shopping and entertainment malls, and industrial parks.[12]

Both public and private lands are under constant assault from the growing wants, which Americans have been taught to call the "needs," of urbanizing sprawl. Urban–suburban water demands are a major cause of species decline in the arid west. Developments drain or fill wetlands habitats and overuse the west's few perennial rivers, both great and small. The Colorado River's waters were overallocated many decades ago (see chapter 9), largely to irrigate farms, flush urban toilets, and sustain lawns. Even in high-rain years, the Colorado river delta, once a chain of fabulously rich habitats, has received little more than a trickle for many decades.[13] Urban thirsts also are drying extensive reaches of New Mexico's Rio Grande (see chapters 9, 11). In 2001, the National Wildlife Federation named urbanization the primary menace to 66% of threatened and endangered native plant and animal species in California alone.

The 1973 Endangered Species Act (ESA) was enacted "to provide a means whereby the ecosystems upon which endangered species and threatened species may be conserved" (see appendix 1). Sections 3 and 7 of the ESA require protection for ecosystems that provide suitable habitat for threatened or endangered species. Section 3 calls for special management or protection of "critical habitat," including the sites a species occupied when it was listed as threatened or endangered, and other areas essential to their conservation. Section 7 provides additional protection by disallowing any federal action that could "jeopardize the continued existence of any endangered species or threatened species or result in the destruction or adverse modification of critical habitat."

The U.S. Fish and Wildlife Service (USFWS) and National Oceanic and Atmospheric Administration's National Marine Fisheries Service, responsible for implementing the ESA, have been extremely slow to designate critical habitat for listed species. For most species, this lack of designated habitat limits ESA protection to avoiding "jeopardy" (extinction) of a species in its occupied habitat, without considering the potential for destroying or harming current and other habitats that could be critical for a species' recovery. The jeopardy standard is needed for species on the edge of extinction, while saving unmodified habitat is most important for recovering or declining species.[14]

Contrary to the ESA's intent, the USFWS and National Marine Fisheries Service have issued a joint regulatory definition that makes the adverse modification standard virtually identical to the jeopardy standard. The USFWS continually tells the public that critical habitat is of little or no benefit to endangered species, failing to mention that federal courts, including the U.S. Supreme Court, have repeatedly rejected this argument. Appeals of some court decisions in 2003 claimed that critical habitat does not benefit endangered species and may even thwart their recovery. This is like saying that people don't benefit from having homes and that shelters and soup kitchens may even endanger homeless people.

The USFWS's data show the opposite: Species with access to critical habitat are more than twice as likely to have improving population trends than species cut off from habitats that best support them.[15] Maintaining critical habitat requires constant vigilance, because those areas suffer both large and small inroads on a daily basis (see chapters 1, 11).

Amendments to the ESA promote "creative partnerships between the public and private sectors" for conserving species and habitat on nonfederal lands. To avoid full involvement of the USFWS, ESA section 10 allows private landowners to apply for an Incidental Take Permit, which removes penalties for otherwise lawful activities that harm (displace or kill) listed animal species. No similar formula protects endangered plant species on nonfederal lands unless they are state listed. Applicants must create a habitat conservation plan (HCP) to demonstrate how they will "minimize and mitigate" the planned activity's impacts on the endangered animals.

HCPs may apply to more than half a million acres, and some cover more than one threatened species. At the end of 2000, outgoing Interior Secretary Bruce Babbitt approved one HCP addressing some 200 threatened and endangered species in five million acres of Clark County, Nevada. At the same time, Babbitt let megasprawl continue on 135,000 acres of desert tortoise habitat around Las Vegas—likely allowing huge "takings" of the threatened tortoises. Where critical habitat designations appeared to impede development, Babbitt's successor, Gale Norton, began reversing them.

Clearly, HCPs work well for developers, but the harried wildlife make all the compromises and take all the hits. In theory, the "takings" are compensated or mitigated, but dead animals cannot be brought back to life, new ones may be difficult or impossible to breed, and artificial habitats are never as rich or functional as nature's own.[16] HCPs can be amended, but amendments tend to become a chain of compromises that lead to ever greater habitat losses. A recent court ruling on an early HCP, which served as a national model for species protection, is likely to be substantially compromised in a required reevaluation of critical habitat designations for a threatened checkerspot butterfly.[17]

Oh, Pioneers

Even in the spacious west, land prices grow along with the costs of supporting sub-urbs, driving lower income people into lands that once seemed uninhabitable. The 1938 Small Tract Act designated 457,000 federal acres for settlements during that time of drought and depression, seeding the first of many isolated, speculative sub-divisions across southwestern states. The southern California desert is dotted with such failed developments—originally just bulldozed grids of "streets" framing small land parcels. Getting title to a five-acre parcel required little more than building a small shack, but staying in it defeated most of the owners. The 1976 Federal Land Policy and Management Act (see appendix 1) repealed the Small Tract law, but too late to prevent extensive street systems from fragmenting desert habitat and accelerating erosion on abandoned plots with their decayed shacks. The bulldozer scars remain visible more than half a century later and will continue eroding for a very long time (see chapters 5, 6).

Grander development schemes proliferated during the 1950s. California City, billed as California's "third largest" (in acreage), sits on formerly prime desert tortoise habitat, now denuded and open to increased water and wind erosion.[18] The "city" consists of a small core development surrounded by 130,000 acres of unused streets and lots, sustaining a 2002 population of only 11,450. Hoping to force land owners to sell out, in 2004 the city's redevelopment agency invoked its power of eminent domain and declared 15,000 acres of open desert "urban and blighted." To improve the city's tax base and increase population, a 4,340-acre Hyundai automobile proving ground, and its workers, may replace the earlier settlers.

Salton City, another 1950s scheme, is an intricate web of "streets" scraped into desert vegetation beside the Salton Sea and advertised as the working man's Palm Springs. With the accidental "sea" evaporating to dryness, decaying relics of the failed resort scatter along a largely unoccupied, and often smelly, shorefront. With a population of about 100, the "city" provides 128 acres per resident—over a thou-sand times more land per capita than Los Angeles. Intense winds rake the sur-rounding barren intentional construction sites, whipping up dust and sand storms as they erode the surface.[19]

Great schemes

Some grand development schemes did not fail. Originally a dingy cluster of casinos and diners, Las Vegas, Nevada, attracted out-of-towners but relied on the Hoover Dam company town of Boulder City for workers and energy supplies. Today, Las Vegas is expanding wildly, defying water limitations in such a dry region. Water has proved a major headache as the city ballooned through the 1980s and to the present. A 2004 study for the Southern Nevada Water Authority was expected to focus on limiting growth to manage drought. But the study report warned instead that interrupting growth could lead to economic catastrophe and encouraged the belief that growth and drought are unconnected. More likely, unrestrained growth eventually will demonstrate the reverse.

Neither Laughlin, Nevada, nor Lake Havasu City, Arizona, existed before 1963, but today they cover many acres of once-rich desert habitat along the lower Colorado River. Emulating the Las Vegas model, gambling promoter Don Laughlin bought land on the Nevada side of Lake Mojave reservoir in 1964 for a gambling mecca, with the uncannily prescient slogan "Destined to become a truly great American Dream." Naming the town after himself, Laughlin relied on the services of Bullhead City, Arizona, across the river—an unassuming winter vacation spot for elderly midwestern "snow birds." By 2000, Laughlin had grown to about 7,000, with a new $3.5 million bridge connecting it to the more viable Bullhead City of more than 27,000 souls. Today Bullhead City and the Laughlin casinos dominate the night sky across the lower Colorado River Valley between Las Vegas and Needles, California.

In 1964, promoter Robert McCullough also bought 26 square miles of land beside Lake Havasu reservoir and proceeded to denude the whole acreage of unusually rich desert habitat, grade the usual road grids, and survey thousands of potential house and condo-complex lots. The Lake Havasu City speculation attracted only a few retirees and vacationers until 1968, when McCullough purchased the Jane Austen-era London Bridge and built it into a major western tourist attraction. In spite of gullible tourists expecting to see London's iconic Tower Bridge,[20] and the flocks of "snow birds," Lake Havasu City remained too small to service major tourism until McCullough added chain saw and boat motor assembly plants. When Arizona legalized dog and racetrack gambling in 1978, Lake Havasu City attracted Las Vegas gaming interests, which finally brought enough population to fill in empty plots and fuel expansion.

The grandiose schemes for adding sprawl never stop. The 1997 "Stagecoach Trails" development of 40-acre "ranches" occupies part of Dutch Flat, a few miles east of Lake Havasu City. It consists of 500 miles of named "streets" that fragment more than 130,000 richly vegetated desert acres, once abundant with wildlife. Less advanced projects include "Coyote Springs," a modern city of 50,000 to be carved out of desert habitat north of Las Vegas. Other proposed desert sprawls include:

- "Joshua Hills," on 9,000 fragile desert acres in threatened desert tortoise habitat east of Palm Springs, California. Strong public opposition has forced the developer to sell out to the Nature Conservancy. Envisioned as a city of 7,000 housing units, three hotels, two country clubs, a school, shops, restaurants, a "World Trade Center University," and Technology Center, Joshua Hills would have severed the biological corridors between Joshua Tree National Park and the Coachella Valley Preserve, home to many endangered species. Just to keep 12 planned golf courses green would have used six million gallons of water per day (see chapter 11). Since Joshua Hills died, another area bordering Joshua Tree National Park is proposed for development.

- The very slowly growing Community of Civano, Arizona, originally planned as a solar city on 818 acres southeast of Tucson, now is only optionally solar.

- Silver Bear Ranch, on a scenic 250-acre ranch outside Silver City, New Mexico, lets residents have their cow and eat it too. Each landowner shares in a luxury bunk house, full-time horse wrangler, and permanent herd of cattle. The cattle likely will graze federal lands, under taxpayer-subsidized permits (see chapter 3).

- The planned Mesa del Sol development of 100,000, south of Albuquerque, New Mexico, will occupy 12,000 acres on a former Sandia Laboratory "mixed waste" dump, containing both chemical and radioactive wastes (see chapter 10).

- Charleston View, the most remote proposed development, plans to house 150,000 people on an overdrafted groundwater aquifer, with water levels dropping a foot a year. Las Vegas, the closest transportation hub, is a long trip away, through Death Valley. Hopes for Charleston View's water supply depend on deeper carbonate aquifers in a discombobulated geologic terrain. A 1,500-foot dry test well dimmed water hopes in 2005.

Inhuman Habitats

Developers and many citizens sometimes declare that humans are the threatened species. This may more true than they know, because the pollution, diseases, and human-introduced invasive species in and from urban and suburban developments can be dangerous to human health.[21] City and suburban pollution also foul the air, change local weather patterns and stream flows, and contribute to flooding and erosion in countryside far beyond city limits.

Like central cities, suburbs and commercial zones form ozone—the driver of urban air pollution (see chapter 12), which damages human hearts and lungs. The pollution can extend for hundreds of miles downwind, eliminating rural fresh air havens. Suburban ozone actually degrades rural air quality more than in cities, where the abundance of nitrogen chemicals in smog can reduce ozone levels; for example, studies show that cloned cottonwood trees grow better in smoggy central cities than in ozone-impacted countryside. Both urban and suburban air pollution also influence local weather patterns by increasing rainfall.

For nearly 200 years, cities have been known to be heat islands, far warmer than surrounding rural lands, where tree shading and moisture transpiration (see chapter 1) help cool the surroundings. Developments replace soils and plants with broad expanses of artificial surfaces on streets, parking lots, and buildings—all heat up faster and hold more heat than do natural surfaces. The temperatures on black-top–coated city roofs can reach 170°F in summer, and both summer and winter urban temperatures range from 2°F to 10°F higher overall than surrounding rural areas.

The combination of concentrated urban heat and air pollution affect regional precipitation patterns, suppressing rain or snow in some places and increasing it elsewhere.[22] Hot air rises, drawing cooler air from surrounding areas into the urban heat islands at street level. Warmer air can hold more moisture than cooler air, and when moist, warm air rises, it cools and releases the moisture in rain showers.[23] The heat-island effects of both Atlanta, Georgia, and Houston, Texas, result in higher levels of warm weather rainfall for areas downwind from the urban core.

In contrast, air pollution in California's Central Valley suppresses rainfall and snowfall on the neighboring Sierra Nevada's west-facing slopes, depriving the mountains of a trillion gallons of water per year. Clouds are water droplets that condense onto dust and other particles in the air. Pollution introduces many more particles than nature creates, so polluted clouds contain more—but smaller—droplets. The smaller droplets fall much more slowly in the cloud and evaporate more rapidly in dry air than do larger droplets, so they never grow large enough to produce rain. Downwind from pollution sources, these processes can cause annual precipitation losses of 15–25% over topographic barriers such as the craggy Sierra Nevada.[24] Another example is the 24% drop in annual rainfall at Cuyamaca, California, which appears linked to aerosol pollution from San Diego, 40 miles away.

The 2003 California and Colorado fire season once again demonstrated the folly of sprawl on remote lands, in rough terrain where firefighting is difficult (see chapter 1). In drought-stricken woodlands, gates do nothing to stop large fires. Pouring billions of dollars into rebuilding burned-out McMansions simply sets them up for future wildfires. A report released in 2001 predicted possible losses of 250–300 homes and potential destruction of "biblical" proportions from inevitable wildfires in southern California. Less than two years later, the October 2003 Scripps Ranch fire turned this nightmare into reality, destroying 312 homes. More than half were roofed with highly flammable shake shingles.[25]

Bad water and air

Impervious urban surfaces increase rainwater runoff that soils might otherwise absorb. Rainwater and snowmelt water cannot easily seep through concrete or asphalt to replenish (recharge) the groundwater-storing aquifers that many urban and rural areas rely on (see chapter 13). Creeks, streams, floodplains, ponds, lakes, and other wetlands are especially important recharge areas. Reducing recharge lowers the groundwater table, even if no wells are pumping water out of the aquifer. Falling water tables in turn reduce springs and diminish stream flows, even far from the urban center. In some desert areas, freshwater aquifers lie beneath rock layers filled with salty brine. Excessive fresh water pumping can draw the brine downward and into a community's drinking water supply.

In coastal zones, pumping large amounts of water to supply urban developments can deplete fresh water in the ground, letting salty ocean water invade the aquifers (figure 8.6). Saltwater intrusion increasingly threatens fresh drinking water supplies for some of the nearly 10 million Los Angeles County residents dependent on groundwater. The complex water pathways beneath southern California have thwarted expensive attempts to block the seawater.[26]

Southern California development continually outpaces new water supplies, while new growth stimulates water hunger. The water supply is never adequate for very long. Starting in 1913, Los Angeles diverted water for its own use from streams that once replenished California's Owens Lake. In addition to destroying habitat for Owens Valley wildlife and cutting off the water supply to local farmers, the lake became the world's second worst source of air pollution. From the dry lake bed, frequent windstorms loft far-traveling clouds of tiny, breathable particulates—including high arsenic and sulfate salt concentrations. The small particle sizes of Owens Lake particulates, called "Keeler fog," and the large volumes of poisonous salts are health hazards for humans and wildlife alike.[27]

Federal pressure in 2003 forced some southern California farmers to sell their federally subsidized Colorado River irrigation water to San Diego and Los Angeles, largely to support urban sprawl (see chapter 9). In semidesert Temecula, California,

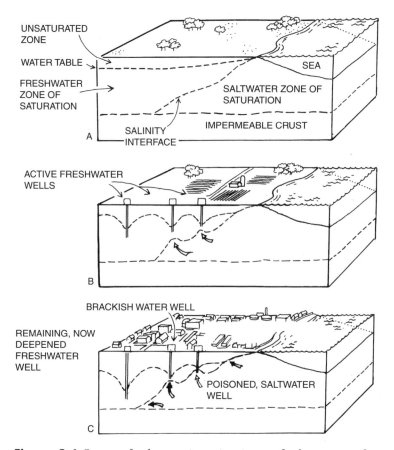

Figure 8.6 Stages of saltwater intrusion into a freshwater aquifer. (A) Presettlement, a freshwater wedge of coastal groundwater floats on salt water. (B) Pumping from initial settlement and groundwater resource development produces drawdown cones in the freshwater zone. Rising intrusions of saltwater replace the fresh water. (C) Economic growth takes too much of the groundwater resource, consuming the freshwater wedge and drawing saltwater into wells.

between the two urban behemoths, this new water supply promotes habitat-destroying McMansion developments. The monster houses are landscaped with water-thirsty iceplant, stuck in a gravelly soil that passes water like a sieve.

During rainstorms impervious roofs and paving inflate the level of maximum, or peak, runoff amounts (figure 8.7) and also increase the number of peak runoff episodes. Higher and more frequent maximum runoff levels increase the likelihood of flooding. Larger stormwater volumes flow rapidly into what remains of the natural creeks and rivers—the surface drainage system—giving the water greater erosive power to remove soil and carve the channel deeper or wider, or both. Floods happen when the stormwater runoff exceeds the drainages' ability to carry the flowing water.

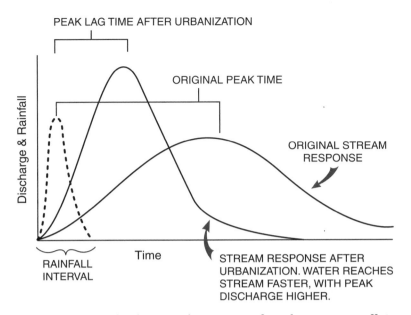

PEAK LAG TIME AFTER URBANIZATION

ORIGINAL PEAK TIME

ORIGINAL STREAM RESPONSE

Discharge & Rainfall

Time

RAINFALL INTERVAL

STREAM RESPONSE AFTER URBANIZATION. WATER REACHES STREAM FASTER, WITH PEAK DISCHARGE HIGHER.

Figure 8.7 Graph showing the timing of peak storm runoff in urbanized areas compared to predevelopment conditions, for the same size of storm (dashed line). More water runs off urbanized areas than natural areas, which also release the water more slowly. Thus, following storm events the maximum stream discharge from urban areas comes earlier and flood levels are higher than from undeveloped land. From C. W. Montgomery, *Environmental Geology*, 5th ed. (Boston: WCB McGraw-Hill, 1997), 135 (figure 6.14). Reproduced with permission of McGraw-Hill Companies.

The sediment eroded from stream channels initially pollutes the water but eventually it comes to rest, either in the stream channels or beside the channels on stream and river floodplains. Over time, the sediment deposits coming from developed lands reduce a stream's ability to carry water and restrict the floodplains' capacity to transmit or store flood waters. With each new storm, the areas that can hold or absorb water decrease in size, expanding the size of floods and increasing future flood risks (see chapter 13).

The 1993 Mississippi River floods and hurricane Katrina in 2005 both demonstrated that building in river floodplains invites flooding disasters. Although generally classified as acts of God, reckless human choices are the real causes of such calamities. To prevent successive disasters, communities pay for various engineering controls, such as levees—but levee failures tend to cause a high proportion of flood disasters. Levees actually increase flood levels and damages by preventing flood waters from spreading onto natural flood plains where water velocity diminishes, and by releasing high-velocity flood waters through levee breaches.

The great 1993 Mississippi Valley floods temporarily influenced the federal government and local residents to move development away from highly flood-prone areas, as a far better option than trying to control floods. But management largely bows to short-term economic gain. Like water flowing through cracks in a levee, human developments still are spreading rapidly into floodplains and developments are built below levees, preparing the next costly New Orleans–type disaster.[28]

Urban wastes

Urbanization contaminates water supplies for long distances beyond city and suburb boundaries. The pollution includes lawn pesticides and fertilizers, pet wastes, and a growing variety and abundance of garbage. Sewage, whether treated or not, carries everything from pathogens, pharmaceuticals, and hormones to an astonishing variety of toxic industrial and household chemicals (figure 8.8).[29] These dangerous substances, with demonstrated toxic effects on all life forms, pollute streams, rivers, coastal ocean waters, and groundwater.

The combination of chemicals and pathogens in our waters have virtually unknown potential negative effects, making today's standards for limiting soil, air, and water contamination simplistic and misleading. Authorities still have little

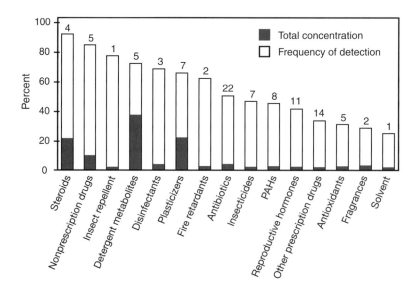

Figure 8.8 Graphs showing organic contaminants in wastewater by category, their concentrations (dark bars), and frequency of detection (open bars) for each type. Numbers indicate the number of compounds in each category. PAHs, polycyclic aromatic hydrocarbons, often produced as byproducts of combustion. From D. W. Kolpin et al. Pharmaceuticals, Hormones, and Other Organic Wastewater Contaminants in U.S. Streams, 1999–2000: A National Reconnaissance, *Environmental Science and Technology* 2002;36:1202–1211. Copyright 2002, American Chemical Society. Reproduced with permission.

sense of the extent of urban contamination because current laboratory techniques cannot detect many pollutants.

Stormwaters propel every conceivable type of flotsam and jetsam into city and suburb drains and sewers, rafting it on streams and into bays, estuaries, and the ocean. Satellite images vividly show urban effluent entering oceans.[30] From 60% to 80% of total marine litter comprises plastics of all kinds. Plastic garbage, especially—much of it from coastal cities and seaside developments—is loading the ocean's environments with indigestible garbage unknown on Earth before 1950. The plastics can choke or asphyxiate animals that mistake them for food; poison them with polychlorinated biphenyls (PCBs) and plasticizing phthalates, which ooze out of plastic; and entangle them in packaging straps. The innards of autopsied sea animals commonly are crammed with cigarette lighters, plastic bags, tampon applicators, toy soldiers, and a host of other plastic objects—particularly packaging.[31] Floating plastics acquire living coatings and crusts of bacteria, diatoms, algae, barnacles, hydroids, and tunicates and carry them into new and unfamiliar habitats.

Other urban contaminants include perfluorinated chemicals (PFCs). Such common household items as nonstick cookware, cosmetics, furniture, cleaning compounds, clothing, rugs, and food containers release PFCs into air, water, and landfills. When introduced more than 50 years ago, PFCs were considered biologically inert, but studies show them to be toxic to humans and wildlife.[32] PFCs are completely unregulated, so industrial releases go unreported. Like DDT, PCBs, and dioxins, PFCs are extremely persistent and so pervasive in the environment that they contaminate the blood of humans and wild animals the world over. At last, the U.S. Environmental Protection Agency (EPA) is taking steps toward a semblance of control over these chemicals, and U.S. manufacturers and users have agreed to reduce emissions by 2010. But hundreds more enter the environment every year.[33]

Cleaning Water

The EPA identifies stormwater runoff from urbanized areas as a leading reason for poor U.S. water quality, causing pollution of all important drinking water sources, whether streams, lakes, or groundwater aquifers. The fire that raged on Cleveland, Ohio's Cuyahoga River inspired the 1972 Clean Water Act (see chapter 10 and appendix 2), with the intention of making the nation's waters "swimmable and fishable" by 1983 and eliminating pollutant discharges into our waterways by 1985. The EPA-enforced law has reduced pollutant discharges substantially—yet about 40% of U.S. waters remain unsafe for swimming or fishing 30 years after enactment, and 19 years after the initial goal went unfulfilled.

In 2005, the EPA seemed set to turn its back on the law and allow urban sewage discharges during storm runoff, after solids (but not pathogens) are removed. This rule change legalizes sewage dumping instead of trying to eliminate accidental failures at aging and deteriorating sewage treatment systems. With an EPA budget that has shrunk from $9 billion a decade ago to $700 million in 2005, U.S. water quality levels soon will revert to pre-Clean Water Act levels.

For 30 years, most states ignored the Clean Water Act requirement to identify and list polluted waters and establish criteria for allowable levels of solid and dissolved materials, called total maximum daily loads (TMDLs). The provisions called for assessing urban stormwater, as well as agricultural and logging runoff. Each state also was required to set pollution and sediment reduction goals to meet their own water quality standards. In the 1980s, a spate of environmental lawsuits forced the states to begin TMDL assessments. Thirty-four have taken some actions, but so far with few tangible results.[34]

The Clinton administration waited until July 2000 before charging states to determine TMDL levels and clean water criteria, giving them 10–15 years to develop regulatory plans for cleaning up water to meet the targets, subject to EPA approval. Calling this too inflexible, in 2002 the EPA under George W. Bush appeared intent on letting states set their own timetables and standards. A Clinton administration official mused: "We tried to put teeth into the program so it couldn't just drift anymore. I guess now the program is going to the dentist to get its teeth removed."

The United States will not meet Clean Water Act goals as long as the EPA continues to issue permits to polluters, allows widespread illegal pollution discharges, and fails to enforce its own permit conditions on industrial, municipal, and—worst of all—federal facilities.[35] For the most part, the long-term effects of lower-than-toxic chemical doses on either humans or wildlife remain unknown[36]—without exception, authorities emphasize these knowledge gaps. The EPA has no human health standards for many contaminants and even fewer standards for protecting wildlife. Even when health standards are established for individual pollutants, virtually no information exists on possible *combined* effects from myriad contaminants mixing together in the environment (see chapter 2, table 2.5). Yet the government continues to permit new and exotic chemicals and to let them be dumped into soils, rivers, lakes, and oceans—as though the environment has an infinite capacity to absorb them with no harm to people or anything else. Researchers know this is not so, and the warning signs are everywhere.

Urban Roulette

Human settlements have proved convenient for many human purposes, concentrating industry, markets, and the labor force at transportation hubs and focusing cultural resources and entertainment. But today's smoggy and chemically burdened urban areas do not constitute healthy human habitat. Growing western populations and increasingly consumptive land uses have grown like cancers, wiping out farms and natural areas, eating up huge areas of fertile agricultural land, and destroying the land's capacity to support wildlife—and ourselves. Natural habitats are not just for recreation; they support human life, as well, and also support economic life, in addition to the quality of life. China—where urban and industrial dumping have poisoned river and ocean fisheries and bad air and bad water sicken workers, causing hundreds of thousands of premature deaths each year—exemplifies the dire consequences of nearly unregulated industrial pollution and demonstrates how development can undercut human life and health.

We can start choosing to not live at such risk. The first, easiest step is to allow only developments that decrease house sizes and increase the population density of existing urban areas, following a European model. A second step is to truly enforce clean air and water laws, to make them fulfill their purposes. An urgently important change would require in-depth research on the potential environmental impacts of new chemicals—singly and in combination with existing ones—before approving them for any use and approving no more toxic chemicals for sale or distribution until the effects of previously approved formulations are well known.

A critical third step is curbing population growth and reducing per capita consumption (see chapter 4). In the face of ever-increasing population, Americans must face the unpleasant fact that, unless one area's growth is balanced by another's decline, there is no such thing as "smart growth," and that "sustainable growth" is a contradiction.

The enormous twentieth-century investment in urban infrastructure created unsustainable population distributions, particularly in the western states. As cheap oil declines and fuel alternatives prove inadequate to support our current lifestyles (see chapter 12), the systems for transporting food, water, and goods long distances to far-flung populations, and long-distance commuting to work, will become insupportable. Local food growing and local water supplies will become a major need. Communities will have to become much more self-sufficient to survive, greatly reducing per capita land use for housing and industry.[37] The time for making these changes is now.

Land and habitat destruction, combined with toxic air, water, and soil contamination, is urban roulette of the worst kind. Spiraling development, and the population growth and consumption that promote it, ultimately are unsustainable. In terms of living things, runaway growth is cancerous, killing the organism that supports it. The Earth is the organism that supports us with its varied habitats and free services. If we succeeded in overwhelming and compromising them all, there would be no support for human life.

9   The Last Drops

Water dries up in arid country but controversy over it, never.
Sam Bingham, *The Last Ranch*

The western United States has low overall rainfall and snowfall levels, few rivers, and many deep groundwater basins. Small Native American populations once lived within the restraints of aridity by seeking harmony with nature. But owning land in such an arid region means little or nothing without a supply of fresh water. Instead of limiting population growth in the face of scarce and unpredictable rainfall, however, the west's aridity challenged the newcomers to redirect water supplies and make the rich desert soils bloom. The region's localized precipitation, generally doled out on boom-and-bust schedules, has made water "the most essential and fought over resource in the western United States."[1]

Raising a lone voice of warning in 1893, western explorer John Wesley Powell foresaw that irrigating western lands would pile up "a heritage of conflict and litigation over water rights for there is not sufficient water to supply the land."[2] That Powell was right about conflicts goes without saying, for the west's bitter heritage of water wars speaks for itself.[3] Invading Americans used legal doctrines of first appropriation and "beneficial use" to take water from Indians' lands and then turned to taking it from each other, oblivious to the effects on wildlife and natural habitats. Today's depleted river flows and overpumped groundwater basins indicate that Powell probably was right about water supply limits, too.

Expanding populations and increasing water contamination have strained supplies of fresh, clean water, even as per capita water demands decrease. By the 1970s, degraded natural settings, rising water pollution, and disappearing native fauna had lowered the quality of western life and built a constituency for environmental protection. But the 1970 National Environmental Policy Act and 1973 Endangered Species Act simply pitted environmental groups and courts against irrigators, cities, and states. In an ironic reversal, recently enriched Native Americans are poised to exercise their primary legal claims to many western rivers.

Since 1999, drought has stalked western states, more sharply focusing past water overallocations, current problems, and looming peril. The future taking shape is one of endless struggles for control of dwindling water supplies at increasingly higher prices. In spite of the high energy and dollar costs of diverting streams, treating large wastewater volumes, and desalinizing seawater, finding "new" water sources remains a western obsession. The only realistic solution is to reduce water consumption to sustainable levels. Otherwise, polluting effects, evaporative losses, and competitive pressures ultimately will force this decision upon us.

Courts still do not recognize such concepts as natural capital or the need to protect the common resource of water as a human benefit, but they increasingly recognize nature's claims on water. Leaving it in a stream no longer constitutes abandoning a water right,[4] and western states now can use water rights for fish and habitat preservation. If normal climate cycles plus global warming bring long western droughts, our descendants will need to adopt more conserving doctrines and respect the water budget nature provides—also recognizing that surface and groundwater are interconnected and that protecting natural ecosystems provides free to low-cost air, clean water, stable soils, and healthy river and ocean fisheries, while also lowering floods in wetter years.

Correcting Nature's "Mistakes"

European-origin settlers discovered that nature's water spreader had provided unevenly for the western United States. Too much precipitation fell in the cold north and on mountains relatively unsuited to farming. Too little rain and snow graced the hot southern basins and high plateaus, where farmers could grow crops more easily. But mountain snowfields fed Columbia, Colorado, and Rio Grande river systems (figure 9.1) all year, in spite of seasonal and annual variations (box 9.1). As they do today, most of the smaller western rivers—California's Sacramento, Klamath, Owens, Eel, Russian, and Trinity; Oregon's Deschutes; Arizona's Verde, Gila, and Salt; Utah's Sevier; and New Mexico's San Pedro and Chama—ran heavily only during spring snowmelt and mid- to late-summer monsoon rains. Shallow hand-dug groundwater wells close to streams and marshes tended to dry up in the summer's heat.

The west's water limitations did not stop settlers from "winning" the west; instead, they prompted ingenious plans for correcting nature's "mistakes." For the last three-quarters of a century, the emphasis has been on accelerated groundwater pumping, damming the west's major rivers, and diverting the captured underground and surface waters to farms, ranches, towns, and mines. Water developments have been the key to economic development in the 11 western states. The mighty dams that block the west's major rivers in their profound canyons are America's version of the Egyptian pyramids.

Knowing that the costs of diverting the Colorado River for irrigation would be far more than individual farmers or ranchers, agricultural associations, cities, or even states could afford, John Wesley Powell had urged Congress to freeze land handouts under the Homestead Act until surveys could determine which areas were best suited for farming.[5] Congress ignored him, eventually providing many

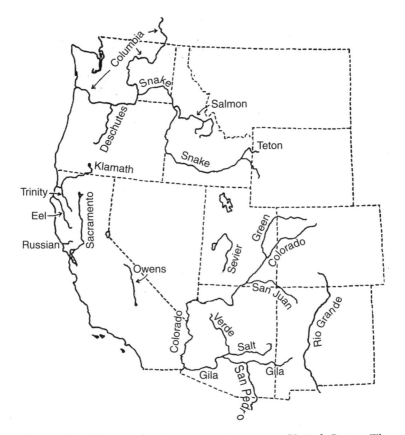

Figure 9.1 Major and minor rivers in western United States. The Missouri River, which rises in Montana, gaining tributaries from many western states as it becomes the Mississippi River's greatest tributary, is not shown.

billions of taxpayer dollars to dam the west's great rivers. Agricultural irrigation now uses the largest amounts of fresh water in the western states (table 9.1).

With all that federal money for water supplies, wrote author Marc Reisner, "the American West quietly became the first and most durable example of the modern welfare state." The U.S. Bureau of Reclamation (BOR), part of the U.S. Department of the Interior (DOI), contracts with conservation districts to provide the federally subsidized irrigation water to farmers. Originally intended for small family farms, the allotments now go mostly to industrial agriculture and marketing interests. Western power brokers control the water flows, garnering substantial fortunes for themselves, other land speculators, and the companies that build dams and water distribution systems.[6]

The west's urban–suburban populations mostly get their residential, commercial, and industrial water from public water systems, which tap both surface and groundwater supplies. With by far the largest population, California uses the most water and moves it the farthest of any state. California's average per capita use of

Box 9.1 Mountain Snows in Desert Rivers

The west has only three big rivers,[a] all originating from tributaries rising in the Rocky Mountains, on North America's Continental Divide. At the time of European settlement, mountain snowmelt fed the rivers' flows all year round, even across broad stretches of downright desert.

- The Columbia River starts from confluence of the Snake and Salmon Rivers along the Idaho–Oregon boundary. Snowmelt waters sustain the Columbia across several hundred miles of arid plains and through deep and broad canyons, now stair-stepped by dams, to the rainy Pacific coast. It is the only major western river that still adds large water volumes to the sea.
- The Colorado River starts as a small stream in the Colorado Rockies and then gathers waters from Wyoming's larger Green River and Utah's San Juan.
- At 1,865 miles long, the Rio Grande is the world's 24th longest river. It rises in the San Juan Mountains of southern Colorado, on the Continental Divide's relatively dry eastern side. Exiting New Mexico's gorges, it flows east, dividing Texas from Mexico. Today, Rio Grande water flows into the Gulf of Mexico only in exceptionally wet years.

The west's big rivers contain various forms of pollution. The lower Columbia and Colorado rivers carry radioactive wastes dating back to World War II and the Cold War arms race (see chapter 7). The Colorado River also contains perchlorate contamination from rocket fuel factories (see chapter 6), and the lower Rio Grande is polluted by chemicals from multiple sources, dominantly agriculture.

Note
[a] The Missouri River rises in the Rocky Mountains but flows eastward through Montana, Wyoming, and eastern Colorado, forming the Mississippi River's largest tributary. So far, its waters have not been diverted westward to irrigate America's driest lands.

publicly supplied fresh water has fallen from around 250 gallons per person per day in 1970 to 181 gallons in 2000, while faster growing Arizona, Nevada, and Utah use much higher levels per capita (table 9.1). In 2000, Nevada had the highest per capita consumption of 314 gallons per day, followed closely by Utah at 286, while Colorado, Arizona, and Wyoming averaged 209 to 217.[7]

For more than 50 years the wells, dams, reservoirs, aqueducts, and pipelines have provided seemingly endless water, allowing western populations to swell far beyond levels that arid regions can support through protracted droughts. In only about a century, overconsumption, water contamination, and increasing population

Table 9.1   Total and Per Capita Fresh Water Use in the 11 Western States, 2000

State	Population	Public Water Systems[a]		Irrigation Fresh Water	
		Gallons/Day	Gallons/Person[b]	Gallons/Day	Gallons/Person
Arizona	5,130,632	1,080	211	5,400	1,053
California	33,871,648	6,120	181	30,500	900
Colorado	4,301,261	899	209	11,400	2,650
Idaho	1,293,953	244	189	17,100	13,215
Montana	902,195	149	165	7,950	8,812
Nevada	1,998,257	629	315	2,110	1,056
New Mexico	1,819,046	296	163	2,860	1,572
Oregon	3,421,399	566	165	6,080	1,777
Utah	2,233,169	638	286	3,860	1,728
Washington	5,894,121	1,020	173	3,040	516
Wyoming	493,782	107	217	4,500	9,113

U.S. Census Bureau (1 April 2000), and U.S. Geological Survey Circular 1268 (March 2004).

[a] Public water system and irrigation fresh water figures in millions of gallons per day. Public water systems can include both surface and ground water supplies. The data do not include private well-water withdrawals.
[b] Per capita figures in gallons per person per day.

pressures have depleted the west's major rivers and groundwater basins, turning fresh water into a declining resource.[8] Along with residential uses, groundwater withdrawals for agriculture, mining, and other industries have severely depleted many aquifers (chapters 2, 4). The ambitious water projects overestimated their supplies and promised too little to too many, fomenting interminable and passionate confrontations.

Supplying endless water to ever growing populations carries a toll of water pollution, salinated soils, endangered wildlife, abandoned farms, and displaced people. These problems are gargantuan already, and the series of dry years from 1999 through 2006 exacerbated them. If the current drought lasts much longer, the problems will become acute.

Dam it

To date, the United States has built more than 75,000 dams higher than six feet. Many rivers' usable water supplies are limited by diversions, including water transfers from one river basin to another in short-sighted efforts to fulfill regional agreements and satisfy local demands. The reservoirs that trap and store stream flows cover about 3% of the land surface[9] and store something like 60% of the total water supply flowing through western American rivers and streams in a typical year.

Dams blocking the flow of rivers significantly disturb whole river systems.[10] They create artificial base levels (knickpoints) that alter river processes and change the shapes of river channels. The water held in reservoirs behind the dams drown biologically rich "riparian" river floodplains[11] and impede the life cycle commute of anadromous salmon and steelhead species, which live in the ocean but breed in coastal streams. Overall, dams reduce downstream water flows, so the sediments coming from lower tributaries build up in downstream river reaches, choking gravely pools that fish need for spawning. The combination of choked pools and sediment starvation has damaged both salmon and shrimp fisheries.

Dams trap virtually all of the river sediment in reservoirs, along with the fertilizers, toxic pesticide residues, and nutrients that cling to sediment particles. A great amount of wind-borne sediment also ends up in reservoirs. Decaying organic material gets trapped behind dams and releases methane, a powerful greenhouse gas. The trapped river sediment once fed nutrients to coastline marshes, mudflats, and beaches, and to ocean plankton—so the dams starve coastal and ocean ecosystems. Because of dams, many coastal areas must import sand to save their beaches.

Dams release reservoir water on a regulated schedule for generating power or preventing overflows. The sediment-free reservoir water has a greater ability to erode downstream riverbanks and beaches than does sediment-laden river water (see chapter 13). The reservoir water releases can drastically erode the downstream channel and banks for many miles unless tributaries deposit sand and other fine sediments close to the dam. Aerial photographs show that river beaches along a 60-mile stretch of the Snake River below Hells Canyon Dam were 75% smaller in 1982 than in 1955, before dam construction, for instance.[12] As water releases erode away most or all fine sediment downstream from a dam, the river channel develops a cover of mostly boulders and pebbles—a process called armoring. Armoring effects first appear a short distance below the dam and slowly progress downstream.[13] Heavy erosion generally continues until the channel is covered with boulders too large for the river to sweep away, or until the underlying bedrock is exposed.

Although many dams are built to control flood levels, flood mitigation effects extend only a few tens of miles downstream. Below that, sediment from numerous tributary streams accumulates in the main channel and makes the lower river reaches shallower. The dam prevents natural floods that ordinarily clean excess debris out of the channel, so the sediment accumulations limit the amount of water that it can hold, raising floods above natural levels. After the 1916 construction of New Mexico's Elephant Butte Reservoir on the Rio Grande, sediment began clogging the channel at El Paso, Texas, 125 miles downstream. The sediment accumulations caused far higher flood levels in 1942 and 1987 than were recorded before the dam was built.[14]

High evaporation losses from reservoirs diminish the water supplies in arid regions, particularly during the hot summers. And pollution flows into them from every human activity—road building, logging, irrigation, grazing, hard rock mining and energy development, urban runoff, and waste disposal of all sorts.[15] Eventually, sediment deposits fill up the reservoirs, eliminating water storage capacity. Even such very large reservoirs as Lake Mead, behind Hoover Dam on the Colorado River, already require continual dredging. Sediment-filled reservoirs have to be abandoned, and many thousands of small reservoirs have been abandoned already.[16]

Hidden waters

Many western communities depend on groundwater from wells for their water supply, and often it is a cheaper and more reliable water source than surface water. Underground water is surface water that sinks into the ground, however. It can come directly from rainfall, or from water in streams and rivers, lakes, and marshes. In the most highly arid states, the water may have to seep long distances through relatively dry, unsaturated (vadose) zones to reach natural underground storage reservoirs, called aquifers. Over time, and depending on local pressures, water can penetrate through rock layers, even ones with poor capacity for transmitting water, to deep underground levels- (see chapter 13).

Nineteenth-century pioneers often settled near springs and seeps, where groundwater flows back to the surface, or they hand-dug shallow wells. By the time their shallow wells dried up year round, powerful machinery could drill into much deeper aquifers, and submersible pumps could tap very large groundwater volumes that had accumulated over millions of years. Today's well pumps can withdraw groundwater from thousand-foot depths at whatever rate the operator can afford, but we still depend on rainfall and stream water to refill (recharge) the aquifers. Groundwater recharge is slow, and the connections between surface water and underground water are not easily seen or measured (figure 9.2). In arid lands particularly, human practices have eliminated or greatly reduced connections that existed in the past. The laws of some states don't even recognize their existence, provoking ceaseless litigation.

What happens at the land surface does affect groundwater, however. Lowering stream flow levels can drain water from aquifers, for example, especially in arid regions.[17] Streams cutting into their channels as a result of dams and river diversions, described above, also lower groundwater levels and drain supplies. Conversely, overpumping groundwater can draw river water into drained aquifers and reduce surface flows.

Only relatively shallow aquifers get recharged reliably in wet years, especially if impermeable paving and other structures have blocked natural recharge routes. The deeper aquifers take hundreds to thousands of years to fully recharge through dry soils and rocks. Withdrawing water from many thousands of feet below the surface removes the fluid pressure that preserves pore spaces in deeply buried rocks.[18] As the water is withdrawn the pore spaces collapse, and the rocks can no longer store water—so even if pumping stopped, such fossil water cannot be replaced.

Withdrawing excessive groundwater amounts (overdrafting) from deep aquifers is the same as mining a nonrenewable resource, like petroleum. The faster we pump it out of the ground, the faster we run out. Overdrafting deep groundwater also can lead to permanent ground collapses at the surface. Such depleted aquifers have collapsed under Tucson, Arizona, and the Santa Clara and San Joaquin Valleys of California.[19] In some places the ground surface has dropped dramatically— more than 25 feet over 50 years, damaging roads, and destroying buildings and

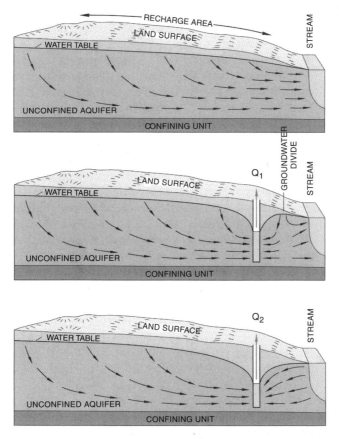

Figure 9.2 Surface and groundwater interconnections. (Top) Groundwater flows into gaining streams. (Middle) Well pumping depresses the water table, creating a divide that directs groundwater flow both to the well and the stream. (Bottom) Well pumping lowers the water table deeper than the stream or other surface water, and the losing stream water replenishes groundwater. From W. M. Alley et al. *Sustainability of Ground-Water Resources* (U.S. Geological Survey Circular 1186, 1999), 31.

water and sewage lines.[20] On the gambler's assumption that we can always find more water supplies, falling groundwater levels, groundwater depletion, and over-draft have attracted little attention in many parts of the west, however.

In coastal zones, fresh groundwater meets saline underground water seeping inland from the sea. The fresh groundwater cannot move into the dense, salty ocean water, but floats on top of it. Overdrafting fresh groundwater in coastal zones causes the salty groundwater to move farther inland, replacing freshwater supplies with ocean brine (see chapter 8). But not all inland ground water is fresh. Saline water leaks from underground salt deposits and oil-bearing zones in some areas, and in those places poorly constructed wells, and poorly or unsealed abandoned wells, can allow saline waters to contaminate freshwater aquifers and functioning wells. Underground chemical storage tanks, especially gasoline-storage tanks, have contaminated groundwater nationwide, including municipal and private drinking water supplies.

Irrigation games

Federal water projects supported the mid-twentieth-century transition to huge irrigation-dependent industrial farms (see chapter 2)—many tilling the marginal or unsuitable lands that John Wesley Powell warned about. All irrigated farming carries significant negative environmental impacts, principally salt buildup in soils (salination). Plants take up irrigation water and transpire it through their leaves, leaving the salts behind, so constant irrigation continuously adds salt to the soil, whether the water comes from rivers or underground aquifers. Eventually, the salt content builds up to a level that kills plants and will degrade even the best agricultural land (see chapter 2). Irrigated farming on unsuitable lands damages the soils for growing any plants over even shorter time frames.

Salination from arid land irrigation is far from a minor problem. It has spelled doom to whole civilizations, including ancient Mesopotamian Babylon and Sumer, and south Asia's Indus Valley settlements of about 2,000 years b.c. It still has the power to damage regional economies and is difficult to repair. We should take it very seriously.

Adding more salt to the wound of general irrigation practices, illegal "water spreading," the practice of diverting federally subsidized water to unsuitable lands for irrigation, has grown common.[21] Lands that develop excessive soil salt accumulations require increasing amounts of water to flush the salt out. When salt builds to a point where the available water supply cannot flush it out, the lands must be abandoned.

Irrigation wastewater flows from irrigated fields back to streams, rivers, and reservoirs. This water is laden with polluting salts, fertilizers, and pesticides, which can poison both wildlife and people. A prominent example is California's Westlands Irrigation District, source of the selenium contamination that poisoned Kesterson Reservoir waterfowl. Kesterson was a sump for agricultural irrigation wastewater, doubling as an artificial wetland until the discovery that selenium salts in the water, leached from Westlands' soils, were poisoning wildlife (see chapter 2). The reservoir had to be filled in and closed, and solving the basic problem will be costly for taxpayers (see chapter 10).[22]

Compacts for Conflict

The vast southwestern water projects were built under the negotiated framework 1922 Colorado River and 1938 Rio Grande compacts, which codified interstate and international water agreements in federal law. The compacts underlie many of the west's ongoing water conflicts and certainly will generate many future ones.[23] Neither compact took account of the natural weather variations that control water levels in western rivers, and both focused solely on economic benefits for new settlers, largely ignoring Native American claims or the needs of native plant and animal species. These agreements have promised far more water to competing users than the great rivers can supply.

The Colorado River Compact parceled out 15 million acre-feet per year (Mafy) of water to seven states that border or straddle the river and its major tributaries (one acre-foot is nearly 326,000 gallons, about twice the average yearly water use for a family of four). The water must be used principally for irrigation, municipal uses, and less prominently for hydroelectric power generation. The compact allocated 7.5 Mafy to the Upper Basin states Wyoming, Utah, Colorado, and northern New Mexico, and a similar amount to the Lower Basin states California, Nevada, and Arizona, to be shared out by mutual agreement.[24]

The Compact gave Lower Basin states the right to increase their water deliveries by 1 Mafy and directed both basins to share equally any burden from future water treaties with Mexico. In the 1945 treaty, the United States promised to deliver 1.5 Mafy of Colorado River water to Mexico in "normal" years, less in drought years, and 1.7 million Mafy in surplus years. Because the Colorado Compact allotments were based unknowingly on 18 years of unusually high river flows, the surplus years never materialized. Long after the compact's ratification, it was clear that the negotiated allocations were not deliverable, even in wet years.[25]

California's share of the Lower Basin allocation was set at 4.4 Mafy. But California has taken more than that amount every year since Hoover Dam's 1936 completion, and California's water hunger has fomented many of the conflicts over Colorado River water. The conflicts generate more promises, and more federal water projects. Irrigation return flows render Colorado River water too saline and polluted for drinking or irrigation by the time it reaches the California–Mexico border. Mexican protests over the poor water quality led to a 1973 U.S. commitment to control the salinity, and the United States has built desalination plants to fulfill that promise.[26] Those plants produce extremely expensive water.

Before Colorado River Compact dams went up, federal treaties had promised water rights to the west's Native American tribes, but prayers and ceremonies supplied water more reliably. In 1908, the U.S. Supreme Court *Winters* decision[27] recognized the prior water rights of Indians on reservations along western rivers and required that pre-Colorado River Compact water rights be served before other claims—no matter if Los Angeles, Phoenix, and Albuquerque become giant ghost towns. After government forces had destroyed their way of life as hunter-gatherers, driven them onto reservations, and forced them to compete as farmers or become wards of the state, author Marc Reisner found the bestowal of senior water rights on Native Americans an "exquisite irony."

Crossing divides

New Mexico also faces complexly interwoven water conflicts arising from the Rio Grande Compact. These conflicts involve the disparate claims of established agricultural irrigators, expanding urban areas and industries, Native Americans, international treaties, and endangered species, on approximately the same water. The Rio Grande Compact divides the river into an upper section, from its Colorado origin to a point near Santa Fe in northern New Mexico; a middle section between Santa Fe and Socorro, New Mexico; and a lower section from Socorro to the Gulf of Mexico. It requires minimum water amounts in the Rio Grande at the Colorado–New Mexico and New Mexico–Texas borders, promises 60,000 acre-feet per year to Mexico and protects "prior and paramount" water rights for six small independent Indian Pueblos.[28] To meet these commitments plus provide water for towns and farms, both Colorado and New Mexico must transfer water into the Rio Grande from other river systems. Under the Colorado River Compact, New Mexico imports its supplemental water from a Colorado River tributary.

In the 1948 Colorado Compact settlement, New Mexico's allotment was 11.25% of the Upper Basin's portion—about 650,000 acre-feet annually. The Colorado River tributaries drain westward from the Continental Divide to the Pacific Ocean, while New Mexico's major population centers lie east of the divide, in the Rio Grande basin. So after ratification, New Mexico quickly protected its Colorado water claims with a major plan to divert water from the Colorado watershed's upper San Juan River, through tunnels drilled under the formidably mountainous divide to the upper Rio Chama, a Rio Grande tributary.[29]

Water from the San Juan–Chama project goes entirely to Middle Rio Grande (MRG) water users. The MRG includes all of New Mexico's significant irrigation districts and population centers—particularly Albuquerque, the capital and most populous city, plus a number of high-tech industries, the Indian Pueblos, and critical habitat for endangered and threatened native species.[30] The original federal allocation of water rights entirely to human uses eventually created head-to-head combat between state and federal laws.

Before 1990, Albuquerque had no interest in the more expensive San Juan–Chama water, believing that its regional aquifer held a virtually infinite supply of lower cost groundwater. When San Juan diversions began in 1976 the city "leased" (i.e., sold) its share to the BOR for irrigating farms.[31] Between 1990 and 2000, Albuquerque's population approached three-quarters of a million, and a study proved the infinite groundwater assumption incorrect.[32] Having blithely overused the groundwater, the city suddenly needed its San Juan–Chama allocation. Projects for diverting Albuquerque's share of imported river water for city use include diversion and pumping facilities and pipelines, which were to be completed 2007. A water treatment plant is scheduled to begin operating by 2009.

Despite receiving only 8–12 inches of rainfall per year, Albuquerque did not plan to lower its per capita water consumption—193 gallons per day in 2004, much higher than the average 140–165 gallons per day of other southwestern cities. Forty percent of Albuquerque's residential water demand goes to landscaping, including Kentucky bluegrass lawns that drink 45 inches per year. If per capita water use did not shrink and population growth continued at the 1990s rate of 21% per year, Albuquerque would have 1.5 million people in 2040, and a residential water demand four times larger than its San Juan–Chama allocation. In 2004, the state engineer ordered the city to cut per capita water use to 175 gallons per day. Albuquerque met the goal in 2005 and continues improving through a variety of conservation incentives.

The rub comes when Albuquerque takes its San Juan–Chama allotment. Downstream MRG irrigators no longer will get that water, and the Rio Grande Compact could go into real crisis. Albuquerque plans to take 94,000 acre-feet per year from the river, withdraw less of its dwindling groundwater, and discharge 47,000 acre-feet per year of treated wastewater to the river.[33] Evaporative losses from the upstream reservoirs cut into the withdrawal amounts.[34] No one can accurately predict the effects on Rio Grande flows, but city officials expect this scheme to reduce the river water less than the groundwater extractions.[35]

There is little room for error. The water available for lands, humans, and habitat south of Albuquerque may decline by 50,000 acre-feet per year—enough to submerge 50 football fields under a thousand feet of water. Average MRG flows now are barely adequate to supply current demands, let alone meet treaty commitments to Texas and Mexico.

One-way streets

New Mexico has long felt a special need to defend its San Juan River water claims against the powerful Navajo (Diné) Nation, and its Chama River water claims against the Jicarilla Apache Nation. The Diné reservation, largest in the United States at 17 million acres, sprawls across four states.[36] In 1930, the Bureau of Indian Affairs began planning the Navajo Indian Irrigation Project (NIIP) for Diné lands straddling the San Juan River in northwestern New Mexico. Much of the land is very poor for farming and requires large amounts of irrigation water, especially in times of drought, so the federal government implicitly has attached large water rights to the reservation.

The state of New Mexico opposed NIIP, forcing it into partnership with the San Juan–Chama Project. Both were authorized in 1962, but House Interior Committee Chairman Wayne Aspinall saw to it that the BOR obtained San Juan–Chama Project funding first. When San Juan–Chama went into full operation in 1976, NIIP was only about 10% complete, and since then its authorizations progressively dwindled—from 508,000 acre-feet of water to irrigate 110,000 acres of land to 370,000 acre-feet on 60,000 acres. By 1992, only 50% of irrigable Diné lands were receiving water,[37] and even today 40% of Navajo Nation residents still must truck water to their homes. Eventually, the Diné sued in federal court to secure firm water rights for its farmers.

Ironically, Albuquerque did not want the San Juan–Chama water at first, and some other contractors did not take their allotments, either.[38] After 1976, San Juan–Chama water diversions took place every year, but water not sold for irrigation simply sat in reservoirs, evaporating at a high rate and collecting pollutants from farmlands, cities, boats, and campsites. In opposing the Diné suit, Albuquerque argued that it had to keep the interbasin water in reservoirs for recreational purposes. A 1981 court ruling declared letting San Juan–Chama water evaporate from reservoirs an illegal waste, however, and did not recognize storing water for recreation as an authorized beneficial use. In response, the U.S. Congress immediately passed a law approving both practices—a pattern of duplicity that Native Americans find all too familiar.

In spite of the Jicarilla's superior right to Rio Grande water, and their beneficial uses for irrigating crops, domestic uses, and a prosperous recreational fishery, New Mexico negotiators gave them barely a nod—and none of the water. It could be said fairly that the average 110,000 acre-feet per year San Juan–Chama diversions are stolen from the Jicarilla Apaches.[39] Yet New Mexico settled the Jicarilla's claims in 1999. The Navajos didn't even get to court until 2003, and their claims' adjudication may well drag on.

CAPped

Arizona did not obtain congressional authorization to take its Colorado River allotment through the Central Arizona Project (CAP) until 1963.[40] Building CAP ultimately cost $4.7 billion, much more than anticipated, and still is very expensive to operate. Using a lot of energy, the water must be raised nearly a quarter of a mile, from the bottom of a moderately deep gorge up to the Granite Reef Aqueduct. Then the water must be pumped east more than 300 miles.

The CAP was supposed to support Tucson's growth and remedy 50 years of groundwater mining to irrigate southern Arizona farms. But the 1980 delivered price turned out to be more than three times the price of pumped groundwater. Even given federal subsidies for building the distribution systems, Arizona farmers faced bankruptcy if they had to use CAP irrigation water. Marc Reisner commented, "[A]bout the only crop you could raise [profitably] with [such expensive] water…was marijuana."[41] Arizona cities had to build their own water distribution systems, so CAP water was even more expensive for them. The high costs turned Tucson toward greater water efficiency.

CAP water also required disinfecting, although treating it with a combination of ozone and chloramines makes the water toxic to kidney dialysis patients and fish. The treated water was highly acid, also, and corroded the city's pipes. Foul CAP water was delivered to Tucson households in 1992, forcing the city to sell its allotment to farmers. In 2001, the city began to blend CAP water with groundwater for household use. The CAP water allocations of 1.4 Mafy have the lowest priority for Colorado River diversions, so are very vulnerable to drought.[42]

Drying Trends

Along with fundamental flaws in the distribution systems, drought is bound to intensify western rivers' overallocation problems. Dry years mean lower snowfall levels in the Rocky Mountains, which also lowers the flows of western rivers and reservoir levels. From 2001 to 2003, average Colorado River flows became the lowest on record—only 5.4 million acre-feet, close to half of Dust Bowl era levels.[43] In 2004, the five-year drought became the seventh worst in a 500-year-long tree-ring record for the Colorado River basin. Shifting in position and severity, the situation ameliorated somewhat in 2005, but severe drought conditions persisted in many western states through August 2007.

The enormous Lake Mead reservoir behind Hoover Dam, 85% of Nevada's water supply, had dropped 80 feet by the spring of 2004. Upstream, leaky Lake Powell behind Glen Canyon Dam was down 117 feet and still dropping, reaching its lowest level in early 2005 (figures 9.3, 9.4).[44] By March 2005, water storage in both Lakes Mead and Powell had shrunk by 37% and 70%, respectively—the lowest in 22 years, and not for want of storage space. For the years 2000 through 2006, Lake Powell received between 25% and 105% of prior average water volumes.

In the sixth year of severe drought, Montana's 2006 mountain snowpacks were nearing record lows. In the usually overwatered Pacific Northwest states, Washington's Cascade Mountains snowpack was only 20–30% of average in the balmy January of 2005. Pitched battles erupted between irrigation farmers and communities dependent on water from the 10,000-square-mile Eastern Snake River Aquifer. Irrigation companies long ago had established rights to the spectacular springs issuing from the aquifer's exposed western end, and during previous droughts had curtailed water distributions to surface users with lesser rights. In 2006 the irrigators made the unprecedented demand for curtailed groundwater pumping to prevent declining surface water flows. Resolving this problem will have far-reaching social and political implications.

Figure 9.3 Far northeastern end of Lake Powell, in Utah, 29 June 2002. Arrow 1 points to the Utah Highway 95 bridge across the Colorado River, and arrow 2 points to a high-water mark near the confluence with the Dirty Devil River, entering from left.

Figure 9.4 Same area as figure 9.3, 22 May 2003, showing Lake Powell, shrunken after the lowest Colorado River flows on record. Arrows 1 and 2 point to the same bridge and former high-water mark shown in figure 9.3. Vegetation on the exposed delta consists of invasive Russian thistle and tamarisk, in many places more than eight feet tall and less than two years old. Photographs and scene description courtesy of John Dohrenwend.

The current drought is spotty, but the very dry years have lowered Colorado River reservoir levels so much that refilling to pre-1999 levels will require many higher than average rainfall years.[45] Yet this drought pales in comparison with earlier ones identified in tree ring records—some lasted two decades.[46] U.S. Geological Survey hydrologist Robert Webb warns, "What is unusual is not the drought periods, but the above average wet periods" of the late twentieth century. The wet decades coincided with an extraordinary era of western growth that still expands Arizona and Nevada populations. Trying to fulfill the Colorado and Rio Grande compacts' impossible promises, rashly made in rainy years, could prevent reservoirs and groundwater levels from recovering between droughts.

Political Climates

Time and tide wait for no man, and politics do not wait for scientific data. As John Wesley Powell found out, land-use decisions, in particular, never wait upon scientific evidence. The compacts, and their dams and reservoirs, were built to meet states' and interest groups' clamorous demands, well before scientific data could be gathered to show the impossibility of meeting those commitments. Once dams proved more costly and less beneficial than anticipated, public support for building them withered. By the 1980s, Congress was no longer funding new dams, although the BOR kept its jobs of maintaining the reservoirs behind the dams and controlling reservoir water allocations, while continuing to favor irrigation clients over wildlife.

Destroyed western landscapes and disappearing native species have decreased the quality of life in the arid southwestern United States, creating public support for environmental restoration. By the 1940s, Los Angeles had grown more rapidly than its local water supplies and was taking water from all along the eastern Sierra Nevada—first drying up Owens Lake and then draining Mono Lake.[47] In 1978, a national coalition of recreational interests, environmental scientists, local residents, and concerned citizens led a campaign to stop Los Angeles from totally draining Mono Lake, igniting a movement to return water to impoverished ecosystems.

After 16 years of research, formal hearings, and court battles, the State Water Resources Control Board issued an order protecting Mono Lake and its tributary streams. By 2002 the lake level had risen substantially. Under the Endangered Species Act, environmentalists also won river water allocations for fish habitat, and in the 1990s Los Angeles was forced to leave enough water in Owens Valley streams to support fish. More recent actions have forced Los Angeles to leave more water for reducing the choking dust storms on dried out Owens Lake bed— the U.S. source of world-class air pollution, comparable to the dying Aral Sea.[48] In 2006, water once again flowed through the Owens River channel and into upper Owens Lake.

Maintaining river flows limits the water that cities and farmers have come to depend on. And new water competition has arisen from Indian activism and the changing political climate, which promoted building casinos on Indian lands. Many tribes now are rich enough to hire top legal representation and are pursuing their prior rights to western rivers within or bordering their reservations.

Central Valley prescription

California takes the most water from rivers and aquifers of any state in the nation to supply its vast cities and vaster farms. In 1933, Congress approved the Central Valley Project (CVP) for irrigating California farms. The BOR dug a channel across the Sacramento River delta to divert Sacramento River waters, mingling the waters of many northern Sierra Nevada streams, to the thirsty south. At first, small canals carried CVP water to farms, but then CVP became part of the massive State Water Project, managed by southern California's powerful Metropolitan Water District.

By 1973, the California Aqueduct's transfer of northern water all the way to Los Angeles had damaged the San Francisco Bay and Sacramento Delta ecosystems. In 1982, California voters began resisting such massive transfers and defeated the proposed "Peripheral Canal," advertised as routing Sacramento River water around the upper delta to prevent salt water from invading the Bay.[49] Today, the Metropolitan Water District delivers an average of 1.7 billion gallons of water per day to a 5,200-square-mile service area.[50] In spite of some of the nation's best water conservation efforts, it still is not enough.

In 2000, the Planning and Conservation League challenged southern California's urban growth plans, and specifically its faith that the State Water Project someday would deliver 4.2 Mafy of water to Los Angeles and cities in Kern County—175% more than at present. The plans did not mention where additional water would come from, or the environmental impacts of taking it. State Appellate Court Judge Vance W. Raye detected an "aura of unreality" and ruled that urban expansion could not be based on "paper water." History tells us that even if deliveries could reach 4.2 Mafy, they eventually will fall short of demand.

For a decade, California's largest municipal and agricultural water districts hoped to expand their supplies through the collaborative state–federal CALFED "Bay-Delta" plan. In 2000, CALFED proposed new surface storage facilities for pumping additional Sacramento Delta water into CVP aqueducts. CALFED's stated purpose was to "provide good water quality for all uses, improve fish and wildlife habitat in the Sacramento Delta, *and reduce the gap between water supplies and projected demand*" (emphasis added). Ironically, CALFED's proposed water withdrawals had to be balanced by expensive ecosystem-protection programs, which agricultural interests opposed.[51]

Lacking consensus, fiscal shortfalls and organizational problems have put CALFED's projects on hold since 2004.[52] Another problem is the 2000 census, which showed slowing California population growth and undercut the Metropolitan Water District's demand projections. In 2005, the California Court of Appeals questioned the west's dogma of inevitable growth, finding CALFED's environmental review document "legally insufficient" in part because it is based on the "pre-conceived notion that inevitable growth in California necessitates...increased water delivery from north to south." The court also pointed out that CALFED "appears not to have considered...smaller water exports from the Bay-Delta region which might...lead to smaller population growth due to the unavailability of water to support such growth."[53] CALFED's appeal of this ruling has sent this attempt to shift California's water law toward sustainability to the state supreme court.

But as CALFED bites the dust, a half-decade of drought in the Colorado River compact states has raised new calls for constructing reservoirs and pipelines, particularly in California. To get around the court rulings and to keep developments booming, in 2007 reliable southern California advocates exhumed talk of a Peripheral Canal from the water project graveyard. Clearly, there will never be enough water to keep up with the requirement of population and development growth (see chapter 8).

Fish out of water

The abnormally dry to extreme drought conditions ruled California's and Oregon's still-rural Klamath Basin from 2001 through 2003,[54] pitting farmers against both Native Americans and wildlife.[55] The lower Klamath River Indian tribes still depend on salmon for their livelihood and main food supply, and U.S. treaties protect their fishing rights "for all time."[56] Before dams, the Klamath flowed heavily during normal winter rains and in late summer barely trickled or dried up, like all California's coastal rivers. Dams created reservoirs and wetlands that became

Lower Klamath and Tule Lake National Wildlife Refuges, sustaining "one of the mightiest concentrations of migratory birds on the planet." Lands within the refuges are leased for farming.[57] Klamath fish species include two endangered sucker fish[58] plus threatened anadromous salmon and steelhead trout.[59] National water policies futilely guarantee adequate water in the river to support both the salmon and the BOR's irrigation contracts.

Complying with long-standing treaties, policies, solicitor findings, and court rulings, the BOR shut off irrigation water to 90% of Klamath Basin's 200,000 acres in 2001. Ranchers and farmers went on a war footing. U.S. Interior Secretary Gale Norton approved 2001 water diversions to the irrigators, even though Congress would not waive protections for fish.[60] Secretary Norton also defeated the American Land Conservancy's eminently reasonable proposal to purchase and retire the water rights of lands belonging to willing farmers and ranchers, at high per-acre prices.

Some saw the farmers' apparent plight as a "wedge issue" aimed at weakening the Endangered Species Act or doing away with it.[61] Indians saw the government's help to irrigators as destroying the "backbone of the lower Klamath Indian tribes' economy, culture, and religion."[62] Absent peaceful solutions, threats and violence erupted. Tragically, thousands of Klamath salmon died in parched tributaries. Lawsuits from all sides charged variously that the DOI had failed to support endangered species, and that farmers dependent on federally subsidized irrigation water had suffered a "taking."[63] To reelect a pro-irrigation Senator from the region, the White House pressured its Department of the Interior appointees to make sure Klamath farmers got their 2002 water allocation.

Low stream flows in spring 2002 led the BOR to cut back water for fish and bird habitat during the fall bird migration. California Department of Fish and Game biologists rescued juvenile salmon stranded along the river channel, but diseases killed as many as 80,000 salmon in the Klamath mainstem.[64] "It felt like death, not just for the fish, but for the people," said a tribal member.[65] Agribusiness and BOR saw "no definitive proof" that agricultural diversions caused the fish kill, but biologists asserted that the unusually low river flows, carrying overly warm waters, probably had triggered the disease. "The flow dictates...how many parasites or bacteria the fish encounter. The warmer water temperature dictates how fast the diseases spread," said one.[66]

Eventually, Klamath farmers received $4 million for not using their water allocations in 2003,[67] but the long-term problem of too little water to meet both the demands of farmers and the needs of fish has not gone away.[68] "What happens in the Klamath is precedent setting," said Steve Pedery of the Oregon conservation group Water Watch. "Either we figure out how to deal with the fact we've promised too much water to too many people or we have train wrecks like this all over the West."[69]

All the Rio Grande, and then some

In the MRG, drought has extremely stressed the endangered Rio Grande silvery minnow and southwestern willow flycatcher.[70] Habitat-destroying river floodplain developments, levees, urban uses, water diversions for irrigation, and recreational and other uses have contributed to the minnow's decline. Albuquerque's over-pumping lowered the regional groundwater levels more than 100 feet along large stretches of the Rio Grande, draining soil water that had sustained the flycatch-er's bosque (cottonwood) woodland riparian habitat.[71] Urban wastewater treatment plants, agricultural return flows, grazing, and logging degrade the river water qual-ity, further threatening minnow and flycatcher survival (see chapters 1–3).[72] A 1996 minnow die-off prompted lawsuits, court decisions and appeals, angry outcries from water users, and political machinations.

One-seventh of the MRG's present water consumption goes to towns and cities. Crops and riparian habitat use well over half, 10–30% evaporates from reservoirs, and some 50,000 gallons of river water per year flow from the Rio Grande into depleted groundwater aquifers. The reservoir evaporation and flow to aquifers are unstoppable natural processes (figure 9.2). So the only options remaining for meet-ing lower Rio Grande water delivery agreements are lowering municipal and agri-cultural uses or limiting water for riparian vegetation even more.[73]

As drought deepened from 1999 through 2002, environmental groups fought to protect the minnow and its habitat against state, municipal, and federal water users.[74] In 2002, farmers received full water allocations from the Middle Rio Grande Conservancy District, which dried up a seven-mile stretch of the minnow's habitat.[75] As much as 95% of the district's irrigated crops are thirsty alfalfa and hay, much of it still watered by the notoriously wasteful and outmoded practice of flooding fields. In addition, the district's more than 1,200 miles of unlined water supply ditches leak badly.

Instead of negotiating with farmers to limit irrigation wastage, New Mexico looked at possible beneficial effects of emptying the river, reasoning that "the species evolved in a desert ecosystem and can survive one or more years of dry-ing." A U.S. Fish and Wildlife Service official concluded that "the state and their consultants are trying to…prove fish don't need water."[76] Meanwhile, the courts resolutely sided with the fish and habitat.[77] The usual political backlash displayed unusual bipartisan unanimity—Democratic Governor Bill Richardson denounced the Endangered Species Act, while Republican Congressman Steve Pearce decried "allowing the courts and extreme environmentalists to give a minnow more power than the Constitution gave states and private property owners."

Sleeping giants stir

Although the *Winters* decision did not miraculously restore water to Native Americans, the many gambling-enriched nations now are making water claims under their confirmed *Winters* decision treaty rights.[78] The results will have far-reaching consequences for all parts of the west and are being felt strongly in Colorado Compact states. One example is the 2004 water settlement between Phoenix and the Pima and Maricopa people of the southern Arizona Gila River Indian Community, who have farmed their lands with irrigation since long before Columbus. The settlement traded their claims on two million acre-feet of Gila and Salt river waters for a Colorado River water allotment, ending nearly 80 years of litigation and eight years of negotiation. The deal lets Phoenix continue taking most of the Gila and Salt rivers' water and gives the tribes "a huge slug of water from the Central Arizona Project...."[79]

This settlement translates to sending half of Arizona's annual Colorado River water allotment to the Pima, Maricopa, and other southern Arizona tribes with settled claims. Phoenix and central Arizona irrigators lose 267,000 acre-feet per year of federally subsidized water but can buy it back at something like 1,000 times the city's current water price. The potential profits for the Indians, from water sales and from greatly expanding their agricultural exports, innovatively avoid water-intensive casinos and resort developments.[80]

Eventually, the Diné and other Indian nations upstream from Arizona will get additional Colorado River water allotments, further reducing the water available for CAP to siphon. And that is not the end of the story. To get congressional votes for CAP in the 1970s, Arizona had guaranteed that California always could take its 4.4 Mafy water allotment—whether in high flow or dry times—before Phoenix or Tucson ever receives a drop. The U.S. Geological Survey estimate of the water reaching all the Lower Basin states averaged out at 5.4 Mafy in 2004, very nearly the amount that Arizona has guaranteed to California. Along with changing climate patterns, the mutual contradictions in all these settlements could turn CAP into a spectacular but dry channel.

Central New Mexico farmers also could lose existing water allocations to settle the water claims of the six independent Indian Pueblos.[81] The BOR has stored the Pueblos' unused irrigation water each year at no charge, reallocating the water to Middle Rio Grande Conservancy District irrigators along with Albuquerque's San Juan–Chama allotment but giving no credit to the Pueblos.[82] The Pueblo people now run burgeoning casinos and are negotiating for their water rights so that they can build hotels, golf courses, and other thirsty developments (see chapter 11). They are likely to demand the right to sell unused irrigation allocations like any other Rio Grande water user, carrying over unused and unsold water credits to future years.[83]

Water spreading

Times of seemingly abundant water supplies have created legal and illegal dependencies that cannot be honored in short supply times, adding to the burden of dry years. A history of illegal water spreading to unsuitable lands also broadened the political constituencies supporting unsound irrigation practices. In part, western water spreading began with old contracts that allotted enough water for flooding fields and other inefficient and now-outmoded irrigation methods. Newer methods use less water, leaving excess allotments that some water districts have sold to farmers on lands not authorized for irrigation.

The whistleblower protection organization Public Employees for Environmental Responsibility (PEER) has found instances of water spreading throughout the west. The BOR provides irrigation water to contractors and should be aware of how the water is used, but appears unwilling to publicly acknowledge the extent of water spreading. In 1994 the Department of the Interior Inspector General reported that at least two dozen irrigation districts in eight states were using federal water to irrigate more than 132,000 ineligible acres. The water deliveries to these districts between 1984 and 1992 amounted to an illegal subsidy of $37 million to $46 million. In a water-spreading suit against Oregon, Idaho, and Washington irrigators, a coalition of environmental and fishing groups charged that the BOR knowingly is "delivering irrigation water to unauthorized users" and has known of "widespread abuse for years."

In an even greater abuse of federally subsidized water, irrigation districts illegally profit from reselling federally subsidized agricultural water to cities and industries—a practice likely to expand in the future. The U.S. government established a Water Spreading Task Force in 1994 to assess the extent of losses to the federal treasury and initiate corrective action, but the incoming Congress ended the investigation's funding. The general lack of oversight for federally subsidized water seems likely to continue into the next decade.

Extended Credit

Water management is much like money management. For more than 80 years, the western states have lived on a water credit card funded by the U.S. Treasury, which built dams, reservoirs, aqueducts, and pipelines and provided abundant— seemingly endless—water supplies at low cost. The water credit system has allowed western populations to swell far beyond what arid regions can support during protracted droughts. Overall, populations in the 11 western states grew by a fifth between 1990 and 2000—Nevada's increased a whopping 66%, Arizona's 40%, and Colorado's and Utah's 30% each.[84]

The swelling populations are used to overconsuming and wasting water. But now the mighty water projects are running low, and many overdrafted aquifers no longer can supply the demand for both agricultural and municipal water supplies. Water costs are rising like the interest rates on overdrawn credit cards.

Most of the west's freshwater supplies go for agricultural irrigation, generally prorated on the basis of farm acreage. Paradoxically, Idaho, Wyoming, and Montana—cold states with small populations—have the highest levels of irrigation water use per capita in the west (table 9.1). Eliminating wasteful irrigation methods, cleaning up irrigator abuses such as water spreading, restricting the thirstiest crops, and fixing leaky pipelines would save huge amounts of irrigation water throughout the region.

Clearly, federal irrigation allotments are the largest remaining water supply that booming cities can try for. In 2000, the federal government kicked off this trend by encouraging southern California's Imperial Valley farmers to sell part of their Colorado River allotments to water-hungry San Diego and Los Angeles. The farmers resisted, but in a 2002 facedown Interior Secretary Gale Norton forced the BOR to finally account for wasted irrigation water, lowering water demand estimates. Native American nations also have obtained the rights to sell irrigation allotments. To keep Albuquerque sprawling, developers have put up highway billboards to attract water allotment holders who are willing to sell.

Trading water rights in this haphazard manner pushes up the price. And building the facilities for shipping northern water hundreds of miles to desert cities is hugely expensive, both in dollars and in damages to the ecosystems of origin. Substantial amounts of the water stored in reservoirs are lost through evaporation, especially in the hot southwestern states—and reservoirs progressively lose storage capacity as they fill with sediment.

Moving water from dams, through diversions, and miles and miles of aqueducts, canals, and irrigation plumbing also soaks up vast amounts of energy, and the energy costs of water are growing steadily. Pumping groundwater from deep wells also takes a lot of energy. Reservoirs and dams can produce hydroelectricity but have their own energy costs. The continual reservoir dredging uses enormous amounts of fuel during a dam's lifetime and is essentially an energy tax on the water supply. Rising energy costs inevitably will curtail the benefits of dredging and force the abandonment of smaller reservoirs.

Arizona's CAP, built after U.S. oil and natural gas production peaked and went into irreversible decline, should serve as a warning that the days of abundant water at low prices are over. World oil and gas production rates now are near or past their peaks, and the coming production declines are likely to send energy prices spiraling—including the prices of alternative fuels (see chapter 12 and appendix 9). Climbing energy costs will jack up the cost of technologies for water purification beyond what we can imagine today.

Wastewater disposal is another energy-intensive aspect of the water equation. Many western officials are hoping that recycling treated wastewater for many uses will help them avoid disposal problems and stave off growth limits. Tertiary sewage treatment with disinfection, which cleans water to current Clean Water Act and Safe Drinking Water Act standards, does not remove increasing amounts of human pathogens, heavy metals, toxic chemicals of all sorts, and unmetabolized medications. When wastewaters are dumped into surface streams or onto the ground, inevitably these substances will degrade ecosystems and find their way into drinking water supplies.[85] Recycling tertiary treated wastewater adds to the risk of contaminating aquifers and thus further reduces clean water supplies.[86] Recharging aquifers with chlorinated municipal drinking water also can contaminate groundwater with trihalomethane pollutants, which are potential health threats.[87]

Advanced wastewater treatment, in particular reverse osmosis (RO), forces wastewater through progressively finer filters, producing much cleaner water than does tertiary treatment.[88] But RO is highly energy intensive and will produce costly water. The impurities concentrated in filters may be toxic and must be disposed of as hazardous wastes (see chapter 10). Other schemes for dealing with limited western water supplies include desalinating ocean water, promoted as the ultimate solution to freshwater problems. Conventional desalination processes include RO, so are energy intensive.[89] But even with improved filter systems, building and maintaining desalination plants is costly, and logistical problems greatly restrict widespread application.

Desalinating seawater yields 40–60% fresh water, the remainder being saline wastewater; discharging the wastewater back into the ocean is likely to have devastating impacts on ocean ecosystems—especially near shore. Inland, desalinating brackish groundwater yields only 20–40% waste under ideal conditions. But pumping brackish groundwater costs more than getting saltwater from the ocean, while evaporating the waste slurry from the groundwater source yields "train loads" of waste salts needing disposal.[90]

Together, disposing of highly salty wastewater and the energy requirement represent 20–50% of desalination plants' operating costs. Absent additional costly pipelines and pumping facilities, this limits locations to coastal sites near power plants, greatly restricting cities' access to the water. The cost of desalinized water in California was $782 per acre-foot in 2002, and a Texas projected cost in 2010 is $978 per acre-foot, compared to subsidized irrigation water at $100–125 per acre-foot.[91]

Increasing the efficiency of water use is the best option for protecting clean water supplies, as Tucson, Los Angeles, San Diego, and even Albuquerque and Las Vegas have discovered already.[92] Probably a third (2.3 Mafy) of California's municipal water use can be saved annually if residences, industries, and farms would support widespread retrofits with water-saving appliances, irrigation devices, and industrial washing and cooling devices. Water suppliers also can save greatly by upgrading transmission and purification systems. Enhancing water efficiencies also requires keeping the infrastructure in repair and limiting evaporative losses. The main route to better water efficiency is price, and substantially increasing water efficiencies can eliminate the rising construction and operational costs of new water transmission project for the foreseeable future.[93]

A 2005 California Energy Commission report estimates that "water-related energy use consumes 19 percent of the state's electricity, 30 percent of its natural gas, and 88 billion gallons of diesel fuel every year, and the demand is growing."[94] Not surprisingly, California's state water plan aims to increase water use efficiencies as the single most effective way to increase energy supplies over the next 25 years. But few other western states appear to realize that lowering water use also reduces energy demand and saves money.

High water prices attract the interest of private enterprise, especially in the face of a shortage. The worldwide movement to sell water for profit is a sure sign that water scarcities are real.[95] Although private corporations long ago garnered large subsidized supplies of western water for farming, municipal water privatizers have not made many inroads. Exceptions are Stockton, California, and Laredo, Texas. As giant international conglomerates seek to take over water distribution worldwide, they undoubtedly will attempt to acquire cheap irrigation water allotments and resell them at great profit. But the reputation of profit-based water systems, so far, is mixed—in some places proving to be more expensive and unreliable than anticipated. Many fear that when water profits go out of state or out of country, these private water companies will not invest in water system improvements or water source protection for the rate payers.

Water Reality Bites

Many commentators have pointed out that water shortages are the greatest future challenge facing western civilization. This is particularly true for overdeveloped arid regions, such as the western United States. The main shortage will be one of clean fresh water due partly to population growth, which escalates demands for additional drinking and bathing water supplies, and partly to degraded water supplies from all accompanying human activities. The worldwide movement against water privatization has called access to clean water a critical human right. But that political position begs many important questions: Access to how much water? At what cost? At whose cost? Does this human right always negate the water rights of other species? What if exercising other human rights pollutes the water supply? And what if nature cannot meet human demands for water? All these questions also plague the west's past and current water policies.

Excepting the Carter administration era of defunding dam projects, Congresses and presidents have dodged politically hard decisions and reality-based policies for water supply and use in the western states. But the mismatch between demand (often called "need") and supplies is beginning to trouble the sleep of western planners. No state has abandoned the dream of supplying endless water to ever-growing populations, but some city and county planners are beginning to write about the possibility that growth can take their state "beyond the natural limits of available water resources."[96]

In fact, the entire west is heading toward the natural limits of available water resources. Establishing a more realistic system is unlikely, for that would require seeing far beyond technological improvements in water use efficiencies. It would demand tremendous changes in outlooks and practices, which would be even more politically divisive than the present status quo. But realistic policies for sustaining water supplies need to be supported nonetheless. If small changes could move the west and its governments toward greater responsibility, the eventual water shocks might not destroy all prospects for our unsuspecting descendants.

The most fundamental move toward reality is recognizing (officially) that natural precipitation is the source of water supplies, that neither legislation nor engineering approaches can create or control it, and that surface and groundwater are a single resource. Another important reality to incorporate into our outlook is the economic and resource-protective value of natural habitats—especially forested hillsides, riparian streamsides, river floodplains, and other groundwater recharge lands. All of them help the land resist erosion while yielding clean air and clean water supplies.

Just accepting those realities would lead to more honest evaluations of water supplies and of the western lands' capacity for supporting humans. They also foster the understanding that groundwater replenishment is a natural "use" of stream water that competes with human demands, and that groundwater pumping can deplete the rivers. Perhaps they would lead to laws that reduce per capita water uses and protect natural habitat remnants for the common good, keeping water supplies in their watersheds of origin and reducing wastes as much as possible. They should encourage ecological farming practices (see chapter 2) and reforms to many or most industrial processes—perhaps moving toward greater use of composting toilets or importing European systems that pipe household graywater directly to bathrooms for flushing toilets.[97]

Accepting reality would influence all western states to make groundwater conservation a critical priority. Several already promote groundwater management strategies, including recharging depleted aquifers with surface water, protecting undeveloped recharge areas, and using permeable paving materials. Groundwater management takes energy and raises the price of water but is essential for sustaining agriculture and urban areas that might otherwise collapse along with over-drafted aquifers.

Reality-based water planning would anticipate decreases in our primary supply as a direct consequence of human-accelerated global warming. The past climate record shows evidence of variations that brought decades-long droughts, which also may await us. Already, global warming has brought warmer winter and spring temperatures that dry out soils more quickly, which require more irrigation water than in the wetter past. Warmer winters mean smaller mountain snowfall totals—so less water flowing in the rivers, less water in dam reservoirs, and less seeping into aquifers. Warmer springs and hotter summers make mountain snowfields melt earlier in the year than "normal," releasing more snowmelt water in the winter and spring than current reservoirs can store, and reducing river flows at the driest times of year.[98]

Requiring that states, counties, and communities plan for drought and a hotter future would reduce politicians' temptation to call for dam building to add water storage capacity. In reality, all the good dam sites are already built on, and reservoirs already lose too much water through evaporation. In the warmest states, such as Arizona, very warm reservoir waters are incubating previously unknown threats to human health—one is the amoeba Naegleria fowleri, which can enter the head through a boisterous swimmer's nose and fatally attack the brain. Worse effects are coming as the climate warms. Early snowmelt water sits longer in reservoirs, losing even more to evaporation than in the cooler past. During long droughts, reliable snows and rains do not fill reservoirs, so new dams will only waste money and energy supplies.

In the current drought, even the Department of the Interior has recommended upgrading decrepit and wasteful water distribution systems and reducing wasteful irrigation practices over dam building, due to the associated costs of water importing and exporting, energy, maintaining infrastructure, and protecting the environment.[99] Imposing water efficiency programs on federally supplied water also would require changing antiquated water rights laws, which now allow each state to choose the water uses within its boundaries.

Sustainable policies would cut back water deliveries for all purposes, not just in times of drought. Reducing per capita water use reduces wastewater volumes and the costs of cleaning polluted water, while increasing the proportion of freshwater supplies. Instead of using the saved water for growth, urban sprawl, and pollution, reality-based policies would recycle it back to natural habitats.

We have already acknowledged that these realities are not likely to guide western water policies in the foreseeable future. But western states are being forced to recognize that water use efficiencies alone cannot solve the underlying problem of too little water to go around unless human populations slow, stop—and even reverse—their growth. Even now, population growth is outstripping California's water-saving efforts. If human ingenuity and resourcefulness cannot overcome our desires and limit human expansion, nature will do it for us, fulfilling New Testament prophecies of eventual famines and wars. That future will be ugly and grim for all species, including our own, both in our "Land of Little Rain"[100] and everywhere on Earth.

10   Garbage of the Golden West

It is folly to keep trying to throw this stuff away, because there is no "away."
Jim Hightower, *There's Nothing in the Middle of the Road but Yellow Stripes and Dead Armadillos*

In May 1970, *Look* magazine ran an International Paper Company advertisement, "The Story of the Disposable Environment," which envisioned a time when "the entire environment in which [we] live" would be discarded. "Colorful and sturdy" nursery furniture "will cost so little, you'll throw it away when [your child] outgrows it," the ad enthused, adding for the socially conscious, "experimental low-budget housing developments of this kind are already being tested." International Paper never addressed where the disposable housing, furniture, and hospital gowns, or the toxic chemicals used for processing raw materials and manufacturing products—or the fossil fuel emissions—would end up. More than 30 years later, we live with the consequences of that vision, which has transmuted the real environment that we depend on into a nightmarish one, dominated by colossal and increasingly hazardous wastes.

For nearly all of human history and prehistory, people dropped their wastes where they lived, expecting the discards would largely disappear. When wastes were relatively minor and all natural materials, many of them did disappear through "natural attenuation"—the diluting or neutralizing effects of natural processes. But even after tens of thousands of years, many items in ancient garbage remain recognizable, and poking through prehistoric dumps can reveal significant details about long-gone people and their ways of life.

History shows that soils and waters have limited capacities for processing even natural wastes. Garrett Hardin underscored these lessons in his 1968 essay "The Tragedy of the Commons." From Roman *urbs urbii* (cities) to nineteenth-century industrial complexes, the refuse dumped in and around larger population centers issued foul odors and helped spread diseases. Public health concerns eventually forced towns and cities to provide sewers, "sanitary" dumps, water treatment, and more recently, sewage treatment. Nowadays, however, our sewers and dumps receive a sizable proportion of synthetic chemicals with unknown properties as well as millions of tons of toxic wastes, hazardous to humans and other living things. Using our soils and common water and air resources in this manner is the surest way to destroy large sectors of our natural capital.

Unfortunately, our myriad toxic repositories still rely on natural attenuation.[1] It doesn't work. Nature constantly demonstrates that its processes cannot "treat" either the immense masses or the poisonous content of our wastes.[2] In fact, the reverse is true: Natural processes are very effective at spreading contaminants into soil, streams, and groundwater supplies. We may try dumping far out at sea, but the garbage comes floating back to our beaches. We may think that hazardous wastes buried in deep pits never will resurface, but so far contaminants that leak from landfills and radioactive waste dumps quickly get into water supplies. Winds pick up the dried sewage wastes spread on farms and spread them across suburban neighborhoods.

The sheer abundance of hazardous compounds in U.S. landfills and urban and industrial areas overwhelm even the capacity to study and understand their effects on the physical and biological environment—and on people, especially children and the elderly. The massive toxicity of our wastes thwarts nearly every scheme to reduce their health threats. Future archaeologists who poke through America's daunting waste piles, and future miners digging recyclable substances out of the pits, will not see our talents or native ingenuity. To them, we will be Earth's champion superconsumers, who cared little about what we threw away or where we put it.

Immense Tide

Each of us acquires and discards masses of goods every day, creating the nearly inconceivable waste volumes that the United States generates every year. A *small* catalog would include food, cosmetics, toiletries; food storage boxes, bags, films; clothing; electronic toys, lights, toothbrushes, knives, phones, iPods, cameras, tools, and the rechargeable or nonrechargeable batteries to run them; bathroom and kitchen soaps, detergents, scouring preparations, degreasers, and home air fresheners; barbeque starters; small hardware supplies; paints; and all kinds of tapes and disks. The immense trash component of packaging waste—cans, tubes, bottles and trays, boxes and bags, bubble paks, and cardboard containers; paper or plastic carrying bags; and shipping boxes plus packaging materials for catalog and Internet orders—is stupefying. Another huge component is the mass of newspapers, magazines, catalogs, and junk mail that would soon fill up all your living spaces if you didn't toss them.

The National Research Council (NRC) underestimates total annual wastes at six billion tons every year.[3] The U.S. Environmental Protection Agency (EPA) categorizes some 230 million tons of it as municipal solid waste (MSW), which goes into municipal landfills. The actual grand total is an astonishing 25 billion tons—*about a quarter of a ton (500 pounds) per person every day.*[4] This includes about 6.5 billion tons of gas emissions and 175 million tons of nongaseous hazardous pesticides, fertilizers, and car and road flotsam and jetsam. Another major contribution is the more than 18 billion tons of hard-rock mining waste (rock, tailings, and toxic water) and toxic waters from gas and oil production (see chapters 4, 12). Our yearly solid waste production alone could fill 90 million mine-hauling super dump trucks—the kind that tower three stories tall and have 12-foot-high tires. Placed nose to tail, they would girdle this planet more than seven times. Less than 2% of all these wastes are recycled.

These numbers still don't account for abandoned toxic debris, plus the polluted soil, streams and rivers, groundwater, and dust-laden winds on military training sites (see chapter 6). The numbers also omit nuclear weapons byproducts (see chapter 7), much of it not recognized or inventoried as waste,[5] or wastes generated elsewhere to provide Americans with foreign manufactured goods. The wastes from devastating fires, floods, and storms, such as the 2004 Florida hurricanes, and Katrina and Rita in 2005, add even more. The Army Corps of Engineers estimated Katrina's Mississippi wastes alone at 50 million cubic yards of debris—enough to fill Atlanta's Georgia Dome 13 times. As early as September 2005, politicians and business interests were demanding environmental regulation waivers to help them quickly move the mess out of camera range. With landfills nearing capacity, did you wonder where they went?

Waste roulette

The United States today harbors an estimated 90,000 hazardous waste dumps.[6] Our principal solid waste disposal methods include burial in landfills, surface dumping or spreading, and incineration. Some wastewaters and sludges are impounded temporarily in artificial ponds or lakes behind dams, but most get dumped directly into rivers, lakes, and oceans; onto convenient lands; or into shallow wells. We inject dangerous liquids into deep wells and vent gaseous wastes into the air. Government regulations require long-term storage in specially constructed vaults and repositories for only a few highly toxic waste materials, including radioactive wastes. But reclassification and lax oversight have sent many radioactive materials to landfills.[7]

The garbage in dumps doesn't just sit there benignly (figure 10.1). Pollutants escape from every kind of landfill—"open dumps," which stay uncovered until filled; "sanitary dumps," where a layer of soil gets laid over each day's trash; and "secure disposal units," which are underground vaults for packages of highly toxic substances.[8] Bacteria, nature's foremost waste-reduction agents, chew up garbage and excrete methane (CH_4) plus carbon dioxide (CO_2)—both potent greenhouse gases. Methane and CO_2 seep continuously out of old, unregulated, and abandoned landfills. Active modern landfills attempt to capture and use methane (the same substance as natural gas; see chapter 12) but still release CO_2. Both gases and liquids can escape from landfills whether capped and lined or not (figure 10.2).

Landfills also emit significant quantities of dimethyl mercury gas, which bacteria easily convert to methyl mercury. Tiny plants and animals concentrate mercury and other toxic contaminants in their organs and tissues (bioaccumulation). Larger fish eat large amounts of the smaller aquatic plants and animals, building toxic concentrations in their bodies that can be 170,000 times greater than in river water.

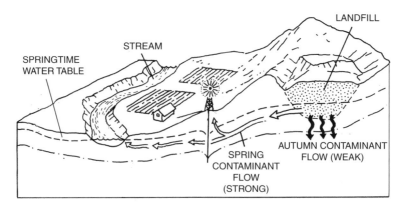

Figure 10.1 Leachates flowing from a landfill contaminate groundwater, ponds, and streams. Leachates enter groundwater directly in springtime (short dashed line, open arrows), contaminating shallow wells and surface waters. In autumn, the water table is lower (dot-dash line, dark arrows), affording less chance of contaminating the groundwater. The autumn contaminant flow is slowed by passing through an unsaturated zone before reaching groundwater. Modified from B. W. Pipkin et al. *Geology and the Environment*, 5th ed. (Pacific Grove, California: Brooks/Cole, 2008), figure 15.10, 455.

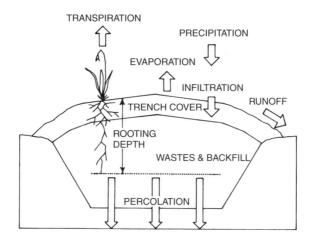

Figure 10.2 Exit routes for gaseous and liquid contaminants from a low-level radioactive waste repository. Modified from T. E. Hakonson et al. Trench-Cover Systems for Manipulating Water Balance on Low-Level Radioactive-Waste Repository Sites. In M. S. Bedinger and P. R. Stevens, eds. *Proceedings Safe Disposal of Radionuclides in Low-Level Radioactive-Waste Repository Sites—Low-Level Radioactive-Waste Disposal Workshop, U.S. Geological Survey, July 11–16, 1987, Big Bear Lake, California* (1987), figure 47.

This process of biomagnification concentrates contaminants more highly at each step up the food chain.[9] Readily bioaccumulating and biomagnifying methyl mercury, a powerful neurotoxin, is the most common fish contaminant in the U.S. human food chain. It particularly threatens the developing brains of infants and young children.[10]

Liquids dissolve (leach) toxic substances out of solid wastes and provide the major escape hatch for toxic pollutants. The leaching liquids may be waste liquids or rainwater seeping through landfills. Natural processes distribute the landfill leachates into the environment, and nature also determines how fast the toxic and pathogenic parts degrade (see chapter 13). To name only one threat, the versatile polybrominated diphenyl ether (PBDE) compounds, introduced in the late 1970s as flame retardants, now are found widely in human blood and nursing mothers' breast milk, even in remote rural districts. So PBDEs must be in the food we eat, the water we drink, and the air we breathe. Some are carcinogenic and cause organ malfunctions in laboratory animals.[11]

A complex web of federal, state, and local laws, regulations, and ordinances, burdened by overlapping and bewildering waste classifications, contradictory disposal-engineering schemes, and nearly incomprehensible jargon, regulates U.S. waste disposal to varying degrees (see appendix 2). By the time the 1976 Resource Conservation and Recovery Act (RCRA) set landfill design standards to protect the environment, large volumes of unidentified hazardous wastes—mostly household and industrial chemicals—already lurked in the nation's landfills and surface waters. RCRA requires landfills to prevent leaks with pit liners. Most liners have a design life of 30–100 years but manufacturers guarantee only 20 years, while regulations require them to last only 30 years. Closed landfills may be capped to keep the contents from eroding into streams, and/or surrounded by a "barrier" drainage system to prevent leaks from spreading.

Hydrologists know two kinds of dump liners—the ones that are leaking and the ones that are going to leak. Even the EPA's handbook admits, "Eventually liners will either degrade, tear, or crack and will allow liquids to migrate out of the [landfill] unit."[12] Wet–dry and freeze–thaw cycles also crack clay layers, and drainage barriers do not solve the leaking caps and liners problem. At best, all these methods offer a short-term solution and early warning of future problems.

Despite its shortcomings, RCRA and its subsequent amendments (see appendix 2) had the effect of substantially reducing hazardous-substance releases to the environment. A marked national decline in the number of MSW landfills, from more than 16,000 in 1987 to fewer than 2,000 in 2000,[13] represents the closing of many small county and municipal landfills that could not meet EPA's tightened 1992 standards or afford liners.

A comprehensive cleanup program, the so-called "Superfund" Law (see description of CERCLA in appendix 2), targeted the worst toxic dumps sites. Currently, there are 1,223 Superfund sites on the National Priority List (NPL). As many as 439,000 additional waste sites may qualify.[14] Four million people live within one mile of a designated Superfund site, while 40 million live within four miles of one.

To date, the "clean" Silicon Valley computer industry in Santa Clara County, California, has generated the most Superfund sites in any U.S. county. Leaks of volatile organic compounds (VOCs), xylene, trichloroethylene (TCE), trichloroethane

(TCA), and 1,1,1-TCA from underground waste storage tanks at Fairchild's, Intel's, and Raytheon's computer chip-making plants are leading examples.[15] Those leaks contaminated underground drinking water supplies and raised the rate of miscarriages and birth defects in nearby neighborhoods 2.5–3 times higher than in the general population. By 2001, the EPA had identified 29 additional Silicon Valley Superfund sites.

New York's Love Canal was the nation's first Superfund site. In March 2004, more than two decades after the contamination was discovered, the EPA declared the area safe for human habitation and took it off the NPL. The dioxin-laden wastes now are "contained" under a thick clay cap and high-density polyethylene liner. All must be closely monitored for the hazard life of the contaminants, which could range from decades to hundreds of years.

In Solid

Solid wastes include industrial wastes, household and other garbage, and sewage sludge. More than half of total solid wastes are agricultural and nearly 40% are from mining (figure 10.3).[16] Household garbage comes under the EPA category of MSW, consigned to leaky landfills. Municipal sewage goes to municipal treatment plants, while individual rural homes generally have individual septic-tank systems. A vast range of septic options serve everything else, from small developments to small towns.

Municipal sewage treatment plants yield solid and semisolid sludges, along with wastewaters. In 1998, the EPA estimated annual U.S. sewage sludge production at 6.9 million dry tons—28,000 super dump truck loads. About 60% is applied to agricultural and grazing lands, the cheapest option, while the remainder goes to landfills and incinerators.

MSW forever

When garbage makes the news, it is usually about municipal landfills and hazardous waste dumps, the largest potential sources of hazardous waste contamination close to large population centers. The EPA reports an astonishing array of hazardous and nonhazardous wastes as MSW: car batteries, furniture, appliances, tires, and hazardous household wastes, in addition to paper and other expected materials.[17] Rugs, furniture, and foam padding contain and emit hazardous PBDE flame retardants. All manner of electronic devices—an increasingly important category called "e-waste"[18]—also come under MSW (figure 10.3, table 10.1). PBDEs are used widely in electronics and in plastics, another dominant MSW category.[19]

Researcher Michael Dowling found little or no verifiable government or other monitoring of materials going to landfills, and no technical or scientific standards— only political ones—for distinguishing MSW from hazardous waste.[20] The EPA's completely artificial classification is designed for congressional reporting and public consumption. It does not reveal that construction and demolition debris, wastewater sludges, and nonhazardous industrial wastes make up two-thirds of

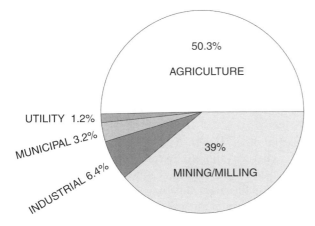

Figure 10.3 Pie diagram showing different proportions of solid waste types in the United States, 1992. From B. W. Pipkin et al. *Geology and the Environment*, 5th ed. (Pacific Grove, California: Brooks/Cole, 2008), figure 15.2, 449.

Table 10.1   Electronic Products Produced, Recovered, and Discarded in 2000 (Tons)

Product	Total Produced	Total Recovered	Percent Recovered	Discarded
Video	859,300	1,200	0.1	858,100
Audio	348,200	0	0	348,200
Information[a]	916,900	192,500	21	724,400
Total	2,124,400	193,700	9	1,930,700

U.S. Environmental Protection Agency. *Municipal Solid Waste in the United States: 2000 Facts and Figures* (EPA530R02001, 2002), table C5.

[a]Information products include all types of telephones, fax machines, word processors, personal computers, computer printers, monitors, and modems.

all MSW. Politically, terms such as "municipal landfill" or "sanitary landfill" are more comforting than "hazardous waste landfill" and meet less local community opposition.

The short initial use and reuse lifetimes of video (7–15 years), audio (3–15 years), and computers of all sorts (2–7 years)[21] guarantee a steady stream of e-waste to landfills. The metals and chemicals hidden in millions of short-lived electronic devices last forever and eventually will donate hazardous antimony, lead, cadmium, mercury, hexavalent chromium, cobalt, selenium, and other metals, as well as polyvinyl chlorides (PVCs), and PBDEs, to the soil and water around and beneath landfills.

About 350 million U.S. computers became obsolete between 1997 and 2004, representing two million tons of plastics and more than 500,000 tons of toxic metals—about 5,500 railroad car loads. The early desktop computers incorporated more than 11 pounds of plastics and about three pounds of lead; also 3.5 pounds of copper, half a pound each of nickel and tin, one pound of zinc, and many other noxious chemicals. Today's computers have smaller amounts of all the same toxic chemicals. Many millions of cell phones with average life times of 18 months go to landfills annually. Countless electronic toys—a child's dream but a dump night-mare—yield similar toxins. Both wet cell and dry cell batteries that run devices of all sorts contain toxic zinc, mercury, silver, cadmium, and lithium metals, toxic to aquatic and other wildlife.[22] All together, electronic devices contribute 70% of toxic metals in landfills.

In 2002, the California Department of Health Services sought to add some radioactive wastes to the list of materials allowed in ordinary California land-fills. In tests of 50 municipal and toxic-waste landfills, California's Environmental Protection Agency found 22 with unusually high radiation levels—6 of 14 unlined dumps had radioactive groundwater contamination exceeding safe levels, and one of the 26 lined dumps had groundwater contamination in excess of California radiation standards. Radioactivity in liner-trapped fluids exceeded maximum drink-ing water safety standards at 25 of the sites.

By any other name...

If you can't find places to bury wastes, should you spread them on the land? Since the 1970s, farmers have spread municipal sewage sludges on grazing and crop lands. The sludges contain toxic industrial and household chemicals and toxic heavy metals, viruses and bacterial pathogens from human wastes, unmetabolized antibiotics, and synthetic hormones and other medications. Wind and water can redistribute them, and plants can absorb toxic components and pass them on to animals higher in the food chain, whether wild, domestic, or human.

Not surprisingly, the arid west's cash-strapped cities and counties began running out of landfill and sewage disposal sites some decades ago. By 1987, more than 19,000 areas in 12 states had accepted sludge applications (table 10.1), with little attention to the potential for rains washing sludge into creeks, ponds, and lakes or for contaminants seeping into groundwater supplies. And no regulations prevent that from happening (see chapter 3).

People can ingest sewage sludge pathogens if they drink contaminated water, through skin contact with sludge or sludge-treated soil, or by inhaling minute bio-aerosol particles. Sludge-raised foods include grains that become breads, pastas, and breakfast cereals, and meat from animals raised on sludge-fertilized feeds.

In 1992, the EPA sponsored a public relations purification rite for sewage sludge, consisting of a new name, "biosolid" (eliciting the cynical question, "does it still stink?"); a changed characterization, from "polluted" to "clean"; and a redefinition as a "beneficial use" for cropland applications. A year later the EPA set chemical con-centration and land-loading rate limits and regulated pathogens and insect-, rodent-, and other disease-spreading species in sludge.[23] The EPA requires processing to reduce pathogen concentrations but still has not sufficiently addressed the inhaling issue, the potential for surface or groundwater contamination from sludge leachates, or potential health problems from antibiotics and hormones in treated sewage.[24]

Farmers and ranchers continually come in contact with pathogen-bearing sludge, but winds also waft the dust thousands of feet into the air and carry it to neighborhoods hundreds of miles downwind. Since 1994, whenever the east winds blow, residents in Denver, Colorado, have lived under bioaerosol showers from 51,000 easterly farmland acres spread with Denver Metro Wastewater District's sewage sludge. Describing a windy day near sludge-covered croplands, Deer Trail downwinder Sandy Turecek asked, "Did you ever see pictures of the [1930s] dust bowl?"

Critics disparaged the EPA's standards and poor implementation for sewage sludge applications, forcing the NAS to request an assessment. The 2002 NRC sewage sludge report tried to minimize problems, while suggesting that the EPA "(1) ensure that chemical and pathogen standards are supported by current scientific data and risk-assessment methods, and (2) demonstrate effective enforcement of [the regulations]...." The 2002 NRC report also urged the EPA to "validate the effectiveness of biosolids-management practices...to assure the public that sewage sludge is safe to use on cropland...."

A 1991 National Research Council report on environmental epidemiology had admitted that the complex pathogen mixtures in sludges are not known well enough to regulate on the basis of risk assessments. The 2002 NRC report reversed that logic, turning a *lack* of "documented scientific evidence that...[1993 regulations] failed to protect public health,"[25] into *support* for finding that the regulations do protect public health. The report also expressed confidence in risk assessments for mixtures of nine regulated chemicals, conveniently ignoring the thousands of unregulated chemicals and organic materials with unknown properties, which lurk in sewage sludges (see chapter 2).[26]

Conceding that more work is needed "to reduce the persistent uncertainty about the potential for adverse human health effects from exposure to biosolids," the 2002 NRC report also restated its recommendation for a new survey of sludge chemicals and pathogens, which the EPA had ignored for six years. Clearly, only public pressure will force EPA to fund public health studies on sewage sludges.

Waste to the Waters

Every year, the United States generates trillions of gallons of toxic waste liquids from urban stormwaters, sewage treatment, and nearly every industrial process—including agribusiness, chemicals, mining, petroleum, and weapons industries. A trillion gallons would fill 50,000 *Exxon Valdez*–sized tanker ships. Industries, vehicles, and homes in the United States also generate enough used oil-based lubricants to fill 1,250 Olympic swimming pools every year. Once upon a time, industry could legally dump petroleum-contaminated brines and waters from oil and gas operations (see chapter 12), as well as highly toxic and radioactive liquid wastes from weapons production, into streams, lakes, bays, or shallow unlined trenches (see chapter 7).

Ponds, pits, "lagoons," trenches, and dams still impound many hazardous liquid and semisolid sludge wastes (slurries), including drained agricultural irrigation water; acidic mine waters; mine tailings (figure 10.4; see also chapter 4); hydrocarbon and solvent vapors in oil and gas production wastes (see chapter 12); natural and industrial brines, some carrying heavy metals and organic compounds; fecal wastes from feedlots and other agricultural wastes (see chapter 2); and liquids from

Figure 10.4 Mine tailings impoundment, Mineral Park Mine, Cerbat Mountains, Arizona. Waters draining from the sulfide-rich copper mine's tailings impoundment are probably acidic. Discoloration in the washes can be traced for miles below the impoundment dam, showing that it overflows during wet periods. Contaminated water also likely seeps down to shallow groundwater. Photographed May 1982.

municipal sewage treatment. A 1987 inventory estimated that 190,000 impoundments store trillions of gallons of toxic liquid wastes nationwide, but probably there are many more than that.

Supposedly, the impoundments hold liquids until they dry up, but unlined impoundments and damaged impoundment linings regularly let contamination seep into the ground, potentially contaminating shallow domestic drinking and agricultural water-storage aquifers. The impoundment dams and other structures also are prone to failure, in some instances generating destructive mudflows, and generally releasing poisonous wastes into streams, killing any living thing in their paths (see chapter 4).

Tertiary sewage treatment with disinfection, the industry standard, does not remove viruses, antibiotics, chemotherapy drugs, birth control hormones, and other pharmaceuticals—not to mention illegal drugs. Municipal wastewaters also contain endocrine-system–disrupting phthalate plasticizers[27] and some volatile industrial compounds. Many of these substances can even survive more advanced and expensive ozone or filtration treatments.

Except for sewage outfalls, the United States banned solid and liquid waste discharges directly into the ocean in 1988—but from an old belief that diluting toxic materials reduces or eliminates their hazard, regulations still approve dumping wastes into large lakes. The same faith inspired federal regulations that allow many liquid wastes, including at least 40% of used oil-based lubricants, to be poured on the ground or down sewers. Again, natural processes don't fulfill the dilution fantasy. Instead, used motor oil contaminates groundwater with significant amounts of polluting gasoline oxidants, some carcinogenic, such as MTBE (methyl tertiary butyl ether), TAME (tertiary-amyl methyl ether), and BTEX (benzene, toluene, ethyl benzene, and xylene).[28] Sewage treatment plant wastewater components turn up in stream water and groundwater everywhere.

Instead of fading out like a Cheshire Cat smile, mercury and other toxic chemicals bioaccumulate in animals and in plant tissues and biomagnify at each higher food chain level (chapters 4, 7).[29] Many metals pass through the food chain to humans, causing lung and kidney damage. There are *no* safe concentrations for long-term exposures to any toxic chemicals that accumulate in living tissues. The Kesterson Wildlife Refuge in California, an artificial wetland created with selenium-laced agricultural irrigation wastewaters, is an ideal example. Selenium bioaccumulation and biomagnification yielded malformed bird hatchlings at Kesterson and sickened animals on nearby farms and ranches (see chapter 2).[30]

Down the hatches

The most hazardous liquid wastes cannot be dumped legally on the ground or into waters without treatment to remove most contaminants. Treatment costs are high, however, and some are considered prohibitive. So most toxic wastes go underground—either pumped to deep levels through high-pressure injection wells or drained a few feet underground in so-called "dry wells"—actually a name for cesspools and stormwater wells.

The Safe Drinking Water Act regulates five classes of wastewater injection wells (see appendix 3). Class I wells are deep "sophisticated, monitored wells," which must not let the waste return to the surface or affect underground drinking water sources for 10,000 years. Class I waste repositories also must be separated from overlying freshwater aquifers by low-permeability rocks, which will not transmit water easily. The most suitable underground rocks for disposal lie a few hundreds to more than 12,000 feet below the surface, have relatively high porosity due to a large proportion of open spaces between mineral grains or abundant fractures, and either are dry or contain saline or otherwise nondrinkable water.

Taken literally, the regulatory restrictions would outlaw class I injection wells because corrosion holes inevitably develop in well casings over time, and contaminants seep into adjacent rock layers and move through fractures in the confining rocks (figure 10.5), contaminating overlying freshwater aquifers.[31] Injecting fluids into very deep class I wells literally can shake the earth (box 10.1). And many class I wells leak. Yet more than 100,000 class I wells throughout the nation annually inject more than nine billion gallons of highly toxic hazardous waste into the ground without adequate public information or involvement, and without adequate financial coverage to properly seal them if they leak after closure.[32]

Class II wells are designated for reinjecting oil and gas brines instead of letting them collect on the ground (see chapter 12). Class III wells take solution-mining wastes—fluids injected into natural mineral deposits, including ordinary table salt

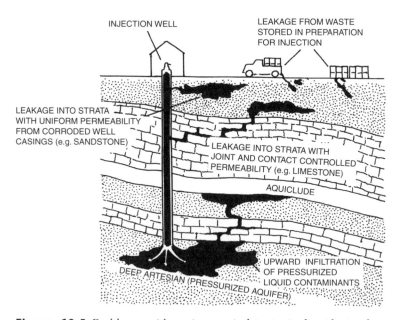

Figure 10.5 Problems with trying to isolate toxic liquids in deep rock layers. Stored liquids awaiting injection can spill accidentally (right and in center). Corroded well casings will leak into shallow and deep aquifers (left). Contaminants pumped into a deep aquifer can migrate upward from the injection zone through rock fractures under pressure. Modified from B. W. Pipkin et al. *Geology and the Environment*, 5th ed. (Pacific Grove, California: Brooks/Cole, 2008), figure 15.17, 460.

and uranium salts, and later pumped out to extract the dissolved minerals. Class IV wells once allowed shallow injection of hazardous and radioactive wastes, but the Underground Injection Control Program banned them in 1987. Previously, the Safe Drinking Water Act also permitted injecting wastes into or above potable groundwater aquifers as close as a quarter mile from the drinking water supply.

Class V wastewater injection wells are any holes deeper than their diameters.[33] A post hole meets the qualification. Class V disposal includes stormwater drainage wells; large-scale septic systems; state and municipal stormwater and sewage disposal sites; disposal wells for industrial byproducts or for injecting spent brines after minerals are extracted (apparently overlapping class III wells); backfilled mine wastes in mine shafts, pipelines, or other deep holes; injected wastewater from geothermal energy production; wastes from animal feedlots and other agricultural operations; and all of the Class V disposal methods, when carried out by a federal agency.

Class V wells have the potential to discharge contaminants directly into aquifers, or into soils above aquifers, and can pollute drinking water more directly than any of the other classes. For this reason, class V well regulations ought to protect

Box 10.1 Class I Earthquakes

In the early 1960s, the numbers of earthquakes recorded at Denver, Colorado, increased markedly. Analysis of numerous data records showed that the earthquake swarms followed closely the nearby Rocky Mountain Arsenal's schedule for injecting chemical warfare wastes to very great depths, about 2.3 miles beneath the surface. Starting up a short time after each injection episode, the earthquakes were a completely unanticipated side effect of deep fluid injection.

Geologic investigations revealed that the repository rocks were already fractured and under stress. The injected fluids lubricated the fractures, helping deep rock masses to shift position and relieve some of the stress. Every slight shift created an earthquake. Subsequent experimental tests pumped water into deep wells at the abandoned Rangely, Colorado, oil field, west of Denver, and confirmed that earthquake swarms follow episodes of deep injection.[a]

Similarly, Ashtabula, Ohio, had never experienced an earthquake until 1987, when millions of gallons of hazardous waste first were injected at high pressure to depths more than a mile beneath the town. Injections commenced in 1986, and earthquake swarms began after a short lag. The fluid waste injections apparently lubricated unknown ancient faults enough to let them slip suddenly generating multiple earthquakes. Earthquake swarms also accompanied the 2004 start of wastewater injections into The Geysers, California, geothermal field (see chapter 12).

Note
[a]B. W. Pipkin et al. *Geology and the Environment*, 4th ed. (Belmont, California: Brooks/Cole-Thompson Learning, 2005), 439–440.

the 86% of groundwater-derived public drinking water supplies. The EPA estimates there are more than 300,000 to more than 650,000 class V wells, giving little confidence that the agency even knows where the wells are, let alone what they may be adding to drinking water supplies.

Given hundreds of thousands of class V disposal "wells" still in use, and probably many thousands more abandoned without cleanup or containment oversight, the constant news reports of contaminated domestic water supplies are hardly surprising. For example, in 2000, VOCs perchloroethylene and TCE showed up in private wells near Santa Rosa, California, apparently leaked from dry cleaning and car repair shops. The businesses—some long gone—may have dumped those solvents in perfectly legal class V wells. Yet the EPA refused to revise class V well rules in 2002, because "the absence of frequent, widespread, or significant cases of actual contamination is good evidence of a low potential to endanger."[34]

Septic stews

Class V regulations do not address single-family or residential cesspools serving fewer than 20 persons daily. No existing class designation, regulation, or monitoring prevents residential and small-group "septic tank" sewage treatment systems from polluting groundwater. Septic systems consist of a tank and a system of leach, or absorption, fields. Solids settle out in the septic tank, in theory breaking down from bacterial action. Porous pipes in the leach fields distribute the liquid to let it sink into the ground. Natural attenuation then supposedly takes over, combining the liquids' organic materials with oxygen, so that soil bacteria, fungi, lichens, and other microorganisms will decompose them (see chapter 13).

Shallow groundwater contamination from the nation's more than 27 million operating septic systems is fairly common, and potentially a very great hazard.[35] Synthetic household chemicals, pathogens, personal care products, nitrogen compounds (nitrates and nitrites), and the hormones and other pharmaceuticals in human (and even in pet) wastes may break down extremely slowly under natural conditions, however, so many or most of them are not processed. Although widely sold for household use, such hazardous chemicals as naphthalene from in-tank toilet cleaners either do not break down in household septic systems or recombine into new toxic compounds.

Septic tanks may start leaking long before decomposition processes go to completion. Where a groundwater table is shallow or sloping, closely spaced septic leach fields are likely to contaminate groundwater or surface water with nitrates and unprocessed toxins. Unmetabolized antibiotics can kill beneficial soil microorganisms and reach deeper aquifers, along with pathogens and long-lived toxic substances.

Burning Problems

The potential for incineration to reduce hazardous waste volumes and toxicity, while generating electricity for industry, popularized large-scale waste burning from the 1950s to the later twentieth century. But incinerators emit gaseous wastes plus potentially harmful soot and other particulates, which unfortunately end up in the air we breathe. Burning wastes also release large amounts of greenhouse gases, including carbon dioxide (CO_2), carbon monoxide (CO), nitrogen and sulfur oxides, metals such as lead and mercury, and toxic VOCs and polycyclic aromatic hydrocarbons. The nation now contains more than 100 MSW incinerators, 1,600 medical waste incinerators, and hundreds of industrial incinerators, kiln facilities, industrial boilers, and furnaces, variously designed for burning solid and liquid, hazardous and nonhazardous wastes.[36]

Incineration does greatly reduce MSW quantities, but more than 10% is not burnable and either must be separated before ignition or removed afterward and disposed of somehow. The amounts of toxic substances that incinerators add to our air or that settle on land and water depend on the waste's composition and the heat and pressure of combustion. Like burning coal, burning MSW creates toxic incinerator residues that must be disposed of as hazardous waste. The potential adverse health effects of incinerator emissions on nearby neighborhoods have been long debated, but the U.S. government still has no epidemiological studies for assessing the impacts.

Breathing trash

Polluting gases are the greatest health concerns for incinerators. At moderate temperatures, hot plastic and various other organic substances emit toxic and carcinogenic dioxins and dioxin-like furans; deadly chlorine gas, hydrochloric acid, and cyanide; acidic sulfur dioxide; and abundant CO_2. Some of the poisonous organic contaminants break down at temperatures as high as 3,000°F, but those conditions maximize mercury and lead emissions.[37] In the current antiregulatory era, the EPA has not set or implemented standards or limits on particulate and dioxin and furan releases, or defined or rated toxic metal or other hazardous emissions.[38]

America's waste incineration fad peaked in the 1960s, when community resistance to proposed incinerator sites, increasingly stringent emission and residue disposal standards, and rising costs forced a substantial decline. New York City, once the nation's waste-burning leader, ended refuse incineration by the early 1990s. But a gigantic waste disposal problem persists. Some communities lack even the land for dumping incinerator residues—in desperation, Philadelphia loaded 16,000 tons of toxic ash on the steamship *Khian Sea* in the mid-1980s. Like the sailor in Coleridge's *Rime of the Ancient Mariner*, it sailed "alone, alone on the wide, wide sea" for two years, seeking a port. In 1988, an empty *Khian Sea* returned to its home berth,[39] the fate of its repulsive cargo a toxic mystery.

Burning bombs

For more than 50 years, the U.S. military generated formidable weaponry, while producing mind-boggling levels of highly toxic and radioactive wastes. Military wastes were handled so carelessly that vast tracts of land and water are off-limits to human uses essentially forever (see chapters 6, 7).

Probably the most contentious weapons disposal issue is the U.S. Army's plan to burn 31,496 tons of chemical weapons—primarily nerve agents and mustard gas—to satisfy the 1997 Chemical Weapons Convention Treaty.[40] The treaty obliged signatory nations to destroy aging weapons and all chemical weapons production facilities in 10 years, by April 2007. In spite of public opposition, NRC blessed an early plan to burn the weapons at seven continental U.S. storage sites, no matter how close to population centers—and also at two sites on Johnston Atoll in the Pacific Ocean.[41] In 2004, only 27% of chemical weapons had been incinerated at five of the sites, including both offshore locations.

Prior to any bomb disposal, chemical agents, propellants, and explosive materials must be separated from the metal shells, then from each other. Each part must be disposed of separately.[42] As usual, burning bomb chemicals releases toxic gases and particulates to the atmosphere, which can fall back onto populations, agricultural lands, and waters that people drink or use for recreation. Also usual are the problems of accidental toxic emissions and disposing of toxic ash, augmented by the Army's cavalier attitude toward the materials.[43]

The Johnston Atoll's chemical weapons stockpile has been incinerated already and plans to close the site are under way. Intense public and political opposition to incineration at Newport, Indiana, and Aberdeen, Maryland (close to Washington, D.C.) forced the government to try alternative disposal methods. Chemical weapons are being incinerated at Tooele, Utah, against ardent local opposition, while the Anniston, Alabama, and Umatilla, Oregon, sites plan to go ahead with burning.

In 1996 Congress instructed the Department of Defense to examine alternative disposal methods for the Blue Grass, Kentucky and Pueblo, Colorado, sites. One method that the NRC has deemed "mature and safe" is "hydrolysis," or dissolving the separated chemical agents and "energetic materials" (explosives and propellants) in water or caustic solutions. "Hydrolysate" products must be treated further to render them safe. Both hydrolysis and subsequent processing yield very large volumes of toxic "off-gases," including ammonia, cyanide, and nitrogen oxides, plus other toxic substances. The fate of lead in this process—and probably of many other toxic elements—is not known. In 2002 the Department of Defense decided to use hydrolysis at the Blue Grass and Pueblo sites, but disposal facilities remained incomplete as of late 2006.

The Forever Problem

In building its stockpile of 70,000 nuclear weapons, the United States also created about 1.3 billion cubic feet of radioactive wastes, by conservative estimate. That's enough to fill 140,000 train cars—laid end to end they would connect Los Angeles, California, to Tulsa, Oklahoma, standing still. Careless bomb making, multiple bomb detonations, and other bomb-related activities yielded more than double that amount of radioactive soil and rock, plus more than 63 billion cubic feet of contaminated water, mostly groundwater—enough to fill 9,500 Exxon Valdez tanker ships (see chapter 7). [44]

Military radioactive waste inventories list more than 900,000 tons still in storage, enough for 10,000 full railroad cars—a train nearly 100 miles long. The wastes include 645,000 tons of so-called "depleted uranium" (DU), mostly uranium-238 left over from producing bomb-grade uranium-235 from uranium ores. This DU is nearly as radioactive as bomb fuel and also is highly flammable upon impact. A dense, hard metal, perfect for piercing armored tanks and buildings, DU stockpiled by the military is being remanufactured into artillery shell and missile nose cones. [45] When the shells and bombs hit a target, everyone in the vicinity is exposed to dense clouds of radioactive smoke (see chapter 7).

Other nuclear weapons program wastes missing from official inventories include the long-lived radioactive substances in U.S. atmospheric bomb test fallout, which rained on seas, rivers, lakes, deserts, farms, forests, and neighborhoods all over the world. An estimated three million cubic feet of radioactive debris from the 1948 to 1958 Bikini and Enewetak Island bomb tests (see appendix 5) was excavated and dumped in the ocean. The military does list approximately 5,100 contaminated "surplus" facilities of unspecified volume or tonnage (see chapter 7).

The federal government has responsibility for keeping radioactive high-level wastes (HLW), transuranic wastes (TRU), uranium mill wastes, and low-level wastes (LLW)[46] out of the environment but has not demonstrated an ability to accomplish this daunting task. As usual, natural processes are not on our side.

- HLW materials contain long-lived radionuclides in used-up weapons and commercial nuclear reactor fuel rods, which must be "permanently" isolated in "geologic" repositories. HLWs also include highly radioactive liquids and sludges from plutonium production and processing, containing VOCs, toxic metals, and corrosive materials (see chapter 7).

- TRU wastes contain long-lived isotopes heavier than uranium, with relatively low total activities;[47] they include byproducts from fabricating plutonium weapons components and chemical plutonium separation (see chapter 7, box 7.1). TRU-contaminated materials include machinery, tools, filters, glassware, and scraps. Many TRU wastes also contain hazardous and toxic chemicals.

- Uranium mill wastes (see figure 10.6) are residues from uranium ore processing, which contain more than 85% of the ore's original radioactivity—mainly from radium-226

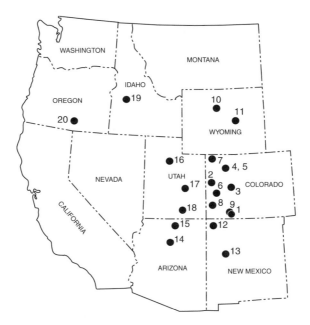

Figure 10.6 Uranium mill sites in the western United States. *Colorado*: 1, Durango; 2, Grand Junction; 3, Gunnison; 4, New Rifle; 5, Old Rifle; 6, Naturita; 7, Maybell; 8, Slick Rock (North Continent Site); 9, Slick Rock (Union Carbide Site). *Wyoming*: 10, Riverton; 11, Spook. *New Mexico*: 12, Shiprock; 13, Ambrosia Lake. *Arizona*: 14, Tuba City; 15, Monument Valley. *Utah*: 16. Salt lake City; 17, Green River; 18, Mexican Hat. *Others*: 19, Lowman, Idaho; 20, Lakeview, Oregon. Modified from U.S. Department of Energy. Integrated Data Base Report–1996: US Spent Nuclear Fuel and Radioactive Waste Inventories, Projections and Characteristics (1997), figures 5.1 and 6.2.

Table 10.2 Volume and Radioactivity of Commercial Low-Level Radioactive Wastes, 1986–1998

Generator	Volume Cubic Feet (× 1,000)	%	Radioactivity Curies	%
Academic[a]	420,000	3	9,000	0
Government[b]	1,210,000	8	329,000	5
Industry[c]	5,104,000	36	670,000	10
Medical[d]	215,000	1	20	0
Utility[e]	7,406,000	52	5,439,000	84
Total	14,356,000	100	6,448,000	99

Data from U.S. General Accounting Office. *Low-Level Radioactive Wastes: States Are Not Developing Disposal Facilities* (GAO/RCED99238, 1999), table 1.1.

[a]Includes university hospitals, medical, and nonmedical research; mostly isotopes with short decay times.
[b]State and federal agencies, such as the Army, licensed and regulated by the Nuclear Regulatory Commission.
[c]Private research and development companies and manufacturers; nondestructive testing, mining, fuel fabrication facilities, and radiopharmaceutical manufacturers.
[d]Hospitals and clinics, medical research facilities, and private medical offices; mostly short decay times.
[e]Nuclear power plants.

(half-life, 1,600 years) and thorium-230 (half-life, 75,400 years). Radon-222 gas (half-life, 3.8 days) and deadly polonium-210 (half-life, 138 days) are toxic radium-226 decay products.[48] The tailings also contain the toxic heavy metals and hazardous chemicals used in uranium-extracting processes.

- LLW is arbitrarily defined in the United States as all radioactive material except HLW, TRU, and mining/milling wastes. Commercial LLW comes mainly from industry, power plants, and government operations, with minor contributions from academic and medical sources (table 10.2). Mixed LLW consists of radioactive materials plus toxic organic and inorganic compounds, and non-radioactive heavy metals, mainly from nuclear weapons production.

Nuclear wastelands

Much of the HLW from Cold War military bomb production has been stored or disposed of at national nuclear reservations—principally Hanford Nuclear Reservation in Washington, the Idaho National Engineering and Environmental Laboratory (INEEL), and the Nevada Test Site (see chapter 7). Early on, HLW was deliberately dumped in unlined trenches, ponds, streams, and rivers. Some especially dangerous wastes were put in tanks, but many tanks leaked and may still be leaking. At Hanford and INEEL sites, all the waste components—long-lived radionuclides, corrosives, VOCs, and toxic metals—have appeared in groundwater (see chapter 7).

Nuclear power plant HLW is a growing radioactive disposal problem. As of 1996, the wastes from decommissioned power plant reactors amounted to approximately 14,000 tons of LLW waste per plant,[49] and combined military and commercial sources yielded more than 40,000 tons of HLW spent fuel. Although commercial reactors generate approximately 2,200 HLW tons per year, there is no permanent disposal site. As of 2006, about 55,000 tons of this increasing waste category was stored in pools at nuclear power reactor sites, leading to calls for reprocessing the wastes (see chapter 12).[50]

About 90% of Department of Energy spent fuel HLW is in aging pools at Hanford, INEEL, and Savannah River, South Carolina. To reduce disposal costs, in May 2004, the U.S. Senate Armed Services Committee relieved the Department of Energy of having to move radioactive HLW from tanks that might leak into the Savannah River to environmentally safer locations. Instead, the committee let the Department of Energy mix the wastes with mortar (grout) to solidify them. Grout is not suitable for long-term radioactive waste containment, however, so radioactive contamination still probably threatens the river and all who depend on it.[51] The department similarly avoided paying to remove HLW from leaking tanks at Hanford, simply by reclassifying the wastes as low level. This ploy could cause additional long-term public health and safety hazards in an already-afflicted region (see chapter 7).

Yucca Mountain solutions

In 2002, President George W. Bush recommended that Congress confirm the highly controversial Yucca Mountain, Nevada, as the first permanent national repository site for radioactive HLW disposal.[52] Currently, the repository consists of more than 100 miles of 1,000-foot-deep underground tunnels for storing specially designed HLW-filled containers. Neither Congress nor the responsible federal agencies ever promised a containment system that can be thoroughly evaluated, however. The law requires only "reasonable assurance" that the repository's operation will meet environmental regulations *for at least 10,000 years*.[53] The state of Nevada opposes the Yucca Mountain repository (chapter 7).

Studied for more than 20 years, Yucca Mountain has yielded a massive literature. The apparent geological stability, very dry climate, and very thick local unsaturated zone, 2,000 feet from the surface to the water table, initially recommended the site for an HLW repository.[54] In early years, most hydrologists thought these and other "geologic barriers" would provide dry, safe burial and prevent dangerous radioactive substances from leaking into groundwater virtually forever. The years of research provided an embarrassment of unforeseen problems, however—among them major fault zones, destructive earthquakes, unexpected water movement in the unsaturated zone, and previously unknown escape routes for radioactivity.

Most hydrologists had assumed that rainwater seeps extremely slowly—or not at all—through unsaturated zones in arid lands. But in 1996, researchers discovered radioactive chlorine-36 in rocks at the 1,000-foot depth of Yucca Mountain's repository. Atmospheric bomb tests from 1945 to 1963 had created the radioactive chlorine (appendix 5), which fell out onto ground surfaces worldwide. Its presence 1,000 feet underground meant that rainwater had moved the chlorine through the unsaturated zone in 51 years or less.[55] This discovery demonstrated that water moves through the unsaturated zone under Yucca Mountain far more rapidly than anyone had realized, probably along fractures, and challenged the site's assumed dryness. Other unresolved technical issues include potential future volcanic activity in the repository area and the ability of zeolite minerals, common in the repository rocks, to disperse uranium and plutonium far and wide.

Several decades ago, the regionally pervasive zeolite minerals in Yucca Mountain's volcanic rocks made the area seem particularly appropriate for retaining radioactive wastes. Zeolites absorb uranium and other radionuclides, so they were expected to bind radioactive material in surrounding rocks, limiting any unforeseen spread into the unsaturated zone. But radioactive plutonium from bomb tests has showed up on minuscule "colloid" particles in groundwater beneath the adjacent Nevada Test Site. As a result, we now know that altered volcanic rocks release zeolite colloids, which may carry the waste uranium and other radionuclides that they have absorbed into soils and groundwater (see chapter 7, box 7.2). The groundwater eventually will carry the radioactive colloids into populated areas outside the Nevada Test Site.[56]

As these problems surfaced, the Yucca Mountain repository's goals and standards shifted, reducing and finally eliminating reliance on geologic barriers to prevent offsite waste migration. The new "solutions" were untested—perhaps untestable—engineering designs, such as "antileak" storage canisters and "drip shields."[57] President Bush's recommendation to Congress was based on a "total system performance evaluation" of the proposed repository,[58] which applied current assumptions and theoretical models to future events, ignoring the massive uncertainties of estimating how engineered packaging, drip shields, or the geologic setting actually will perform over time.

The Secretary of Energy hailed the "sound science" of this performance evaluation. In September 2001, however, the National Research Council's Advisory Committee on Nuclear Waste concluded that the total performance evaluation "relies on modeling assumptions that mask a realistic assessment of risk," plus computations and analyses based on assumptions, not evidence.[59] That December, the U.S. General Accounting Office declared that "DOE will not be able to submit an acceptable application to the Nuclear Regulatory Commission...within the express statutory time frames...because it will take that long to resolve many technical issues." In January 2002, the Nuclear Waste Technical Review Board expressed "limited confidence in current performance estimates" and found "weak to moderate" technical bases for repository performance estimates.[60]

Official estimates assume that antileak canisters will fail after 5,000 years or so, largely from the corrosive effects of water dripping on them, rendering the engineered protections "akin to a torn wet blanket."[61] The exposed waste then will flood radionuclides into the surrounding environment, at rates that depend on the form and stability of the spent fuel and on the rates of zeolite colloid formation. This means that radioactive water someday might flow out of kitchen taps in Las Vegas.

This spent fuel will be mostly uranium, in the low-oxygen "reduced" UO_2 form. It unfortunately changes to highly mobile chemicals in the presence of even minor oxygen and moisture, however. Such "oxidizing conditions" prevail in Yucca Mountain's unsaturated zone, and oxidation likely will release radioactive materials relatively quickly. Concluded Yucca Mountain evaluators Rodney Ewing and Allison Macfarlane, "The concept of placing spent nuclear fuel in the unsaturated zone where it will experience oxidizing conditions is simply a poor strategy."[62] In June 2005, as the Department of Energy prepared to apply for a license to operate the Yucca Mountain facility, scientists still were discovering new facts about the fate of buried spent nuclear fuel when exposed to small amounts of water.

WIPPing TRU

From 1945 to 1970, government agencies dumped nuclear weapons program TRU wastes into unlined trenches, contaminating the soil. Once long-lived radionuclide concentrations in the trenches were found to be unsafe, the practice ended, but those dumps remain major environmental hazards. Unfortunately the contamination locations and amounts largely are undocumented and unaddressed. Next, TRU wastes were put in metal barrels (casks) for later retrieval and disposal at an engineered repository. In 2002, the Waste Isolation Pilot Plant (WIPP) opened in New Mexico and began receiving TRU waste shipments for burial in ancient salt beds, about 2,000 feet below the surface.

The WIPP consists of large caverns mined in highly soluble salt. In the 1950s, the mere existence of salt deposits seemed to prove their dryness, and rock salt beds appeared the best potential sites for waste repositories. The WIPP site's salt is brine-free and unwarped—but the same deposits a few miles from the repository contain brines and contorted layers. Over time, the salt-bed repository is expected to slowly deform, blocking and sealing it so that neither groundwater nor brines will penetrate to the wastes.

The natural sealing will make retrieval difficult or impossible if the caverns are ever breached by earthquakes or if the warming climate increases local rainfall. Both add to the chances for water to infiltrate deep enough into the ground to flood the repository. Radioactivity might escape if the waste casks corrode—a likely prospect because most of them contain corrosive and hazardous nonradioactive materials in addition to TRU wastes. Chemical interactions within waste casks could yield other additional corrosive—and even flammable—substances, abetting hazardous releases.[63]

True to its name, WIPP is a pilot project and much too small to accommodate even the currently retrievable TRU waste. The total requires four times the WIPP's disposal capacity. Even more sites will be needed for the nation's backlog of military and commercial TRU wastes. All the retrievable TRU poses fewer critical environmental threats than the unknown (very large) amounts of buried TRU waste at various weapons sites—not to mention those masses of TRU-contaminated soil (see chapter 7).

Tailings and leaks

Radioactive mill tailings from processing uranium ores have accumulated at 28 sites in seven western states. As of 1996, the radioactive tailings amount to more than four billion cubic feet—40,000 super dump truck loads. The tailings contain thorium-230 (half-life, 75,000 years) and uranium isotopes with even longer half-lives (see chapter 7, box 7.1), but regulations require them to be isolated from the environment for a mere 200–1,000 years. Any isolation would help; however, today the wastes mostly are exposed, "remediated" only with clay or wood chip caps.

Some controversial dump sites have not been treated, covered, or cleaned up. The most notorious is the Atlas Mine dump at Moab, Utah, where radioactive and otherwise toxic tailings remain piled on the Colorado River floodplain (see chapter 4). Comparable volumes of similarly contaminated, unregulated and untreated radioactive waste rock still are lying around other abandoned uranium mines. In Colorado, uranium mine tailings have been recycled into house foundations, sand trap fillings on golf courses, and even children's sandboxes, exposing families to high radioactivity levels in their daily lives (see chapters 4, 7).

In 1997, all U.S. military and commercial nuclear reactors had generated more than 180 million cubic feet of LLW, which the government thought safe enough to burn or bury in shallow trenches. The Nuclear Energy Institute estimates total power plant LLW averages about 250,000 cubic feet (27 railroad car loads) per plant (see chapter 12).[64] Real examples include the atypically small Fort St. Vrain, Colorado, power plant that generated nearly 143,000 cubic feet of wastes over its lifetime, and the large, modern Trojan plant in Oregon, which has yielded more than 181,000 cubic feet of wastes so far. The Trojan plant managers estimate a total of 400,000 cubic feet of wastes upon final disassembly.

Mixed LLW is variously regulated: the Department of Energy oversees federal wastes, while the Nuclear Regulatory Commission (successor to the Atomic Energy Commission) deals with wastes from commercial power plant, medical, and industrial reactor licensees. The Department of Energy disposes of some LLW at commercial facilities licensed by the Nuclear Regulatory Commission. The EPA abandoned efforts to regulate LLW but does oversee nonradioactive mixed-waste components under RCRA (see appendix 2). Although the National Research Council found that none of the sites is geologically satisfactory, all have been in use for decades.

Of the seven commercial LLW sites, only Richland, Washington (within the federal Hanford Nuclear Reservation); Envirocare, Utah; and Barnwell, South Carolina, still are open and operating. Contaminants that leaked beyond the site boundaries shut repositories at West Valley, New York; Maxey Flats, Kentucky; and Sheffield, Illinois. Two of them leaked before reaching capacity.[65] The governor of Nevada closed the Beatty site in 1992, following alarming discoveries of LLW buried outside the site boundary, large volumes of liquid LLW poured on the ground, a license violation, and dump workers' thievery of radioactive tools and other equipment for sale to unsuspecting customers.[66]

The Barnwell, Richland, and Beatty repositories all have leaked radioactive contamination. At Barnwell and Beatty, the contaminants have found their way to groundwater (box 10.2).[67] The depth of Richland's contamination is inadequately monitored and unknown—and is also masked by widespread weapons-processing contamination in the surrounding Hanford site (see chapter 7).

Although most western countries group radioactive wastes for disposal by the type, level of radioactivity, and length of radioactive half-lives, in the United States the Nuclear Regulatory Commission lumps a huge variety of radioactive materials

Box 10.2 Faith-Based Radioactive Waste Disposal

All but two of the National Research Council's Ward Valley review panel approved the largely uncharacterized site on the basis of beliefs about underground water movement, although virtually untested for arid unsaturated zones. As respected hydrologist Peter Wierenga put it,

> [If] the committee [panel] had to sanction the site just based on the data of the contractor...*there would have been no chance in the world that we would have had the majority of the committee do this.* As it is, there's all the additional information, some of it at Yucca Mountain, some of it at the Beatty site, some of it from New Mexico, and some of it from Texas.[a]

Wierenga meant that he and the others had relied on studies of Beatty and Yucca Mountain in Nevada, the Waste Isolation Pilot Plant (WIPP) in New Mexico, and Sierra Blanca in Texas.

But the information was incomplete. At the time, Yucca Mountain, WIPP, and Sierra Blanca were only proposed sites, also undergoing characterization studies. None had any radioactive wastes stored in the vicinity, and today only WIPP is receiving wastes. The closed Beatty site could provide information related to actual radioactive LLW disposal, however. Along with site proponents, the panel majority simply believed that the Beatty site's 300-foot-thick unsaturated zone had been a perfect geologic barrier to leaking radioactive contamination for 30 years. But no studies ever had tested that faith. The panel's belief that Beatty and Ward Valley geology are closely comparable also was unstudied and untested.[b]

In October 1995, a U.S. Geological Survey hydrologist revealed that 1994 tests had shown radioactive tritium at both shallow and deep unsaturated zone levels next to the Beatty dump.[c] The dump itself was the only source for the tritium, which had moved 300 feet deep in 30 years or less. The hydrologist had testified before the Nuclear Regulatory Commission review panel without revealing this information, which contradicted conclusions in his published papers.[d]

Even after the U.S. Geological Survey data's exposure,[e] Ward Valley proponents continued to push for site approval, but this time making every effort to *prevent* comparisons with Beatty.[f] The Beatty tritium data, and later discovery of bomb-test chlorine-36 deep in Yucca Mountain's unsaturated zone, revealed the frailty of untested scientific assumptions.[g] More recent tracer studies at the Idaho National Engineering and Environmental Laboratory reveal that water can move nearly a mile through the unsaturated zone in four months (see chapter 7).[h]

Notes
[a]Nuclear Regulatory Commission. *Letter Report to R. A. Meserve, Chairman, U.S. Nuclear Regulatory Commission Advisory Committee on Nuclear Waste* (18 September 2001, Official

continued

Box 10.2 Continued

Transcript of Proceedings), statement of Peter Wierenga (emphasis added). Wierenga mistakenly said that the National Research Council review panel was charged with recommending the Ward Valley site—in fact, it was charged only with reviewing the U.S. Geological Survey scientists' concerns.

[b]The site applicants and proponents for building the Ward Valley disposal facility were U.S. Ecology, the company licensed to operate the facility; the California Department of Health Services, which would oversee the operation; and the nuclear power industry. All promoted these untested beliefs.

[c]This revelation came as the Department of the Interior negotiated transfer of federal land for the repository site to the state of California. Agency spokespersons later called the tritium discovery a surprise and said the hydrologists had wanted to check that their samples had not become contaminated from an external source. Subsequent sampling showed that the tritium contamination is real and spreading.

[d]Panel members had asked U.S. Geological Survey hydrologist David Prudic and others for the most recent U.S. Geological Survey information at the first National Research Council panel meeting, 7 July 1994. Prudic's answer suggested that the U.S. Geological Survey had no data showing contamination in the unsaturated zone at Beatty. But responding to questions from California Congressman George Miller in August 1997, Prudic confirmed that he had received the tritium contamination data before that July meeting. Prudic also delivered a written report at the panel's second meeting, which concluded that water required tens of thousands of years to get below a depth of 30 feet in unsaturated zones at Beatty and Ward Valley, and that no rainwater could have penetrated deeper than 30 feet for thousands of years. Data Prudic already had in hand contradicted that conclusion (David Prudic. *Estimates of Percolation Rates and Ages of Water in Unsaturated Sediments at Two Mojave Desert Sites, California, Nevada* {U.S. Geological Survey Water-Resources Investigations Report 94–4160, 1994}).

[e]U.S. Geological Survey Beatty project hydrologists already had shared the data with the California Department of Health Services and Ward Valley licensee U.S. Ecology. But over the next 10 months while the panel completed its report, they did not enlighten the National Research Council panel or the U.S. Geological Survey director. An accidental revelation forced public disclosure after the report's May 1995 release. The selective data releases violated U.S. Geological Survey policy and regulations.

[f]Both the history of comparing the Beatty and Ward Valley sites and the importance of discovering tritium migration, both away from and deep into the ground beneath the Beatty site, are misrepresented in a report of the U.S. General Accounting Office (now the Government Accountability Office) (U.S. General Accounting Office. *Radioactive Waste, Interior's Continuing Review of the Proposed Transfer of the Ward Valley Waste Site* [GAO/RCED-97–184, 1997]). The General Accounting Office also omitted to mention that the Beatty discovery further undermined the review panel's principal findings that lateral and deep contaminated water movements in the unsaturated zone were unlikely.

[g]K. Campbell et al. Chlorine-36 Data at Yucca Mountain, Statistical Tests of Conceptual Models for Unsaturated-Zone Flow. *Journal of Contaminant Hydrology* 2003;62–63:43–61; for numerous records of bomb-pulse chlorine-36 in Yucca Mountain's unsaturated zone, documenting rapid groundwater percolation through fractured tuffs see J. T. Fabryka-Martin. *Analysis of Geochemical Data for the Unsaturated Zone* (Report ANL-NBS-GS-00004; Los Alamos, New Mexico: Los Alamos National Laboratory, 2000) and a report on bomb-pulse tritium and chlorine-36, indicating fast water flow through connected fracture networks throughout the unsaturated zone at Yucca Mountain (Marianne Guerin (Tritium and ^{36}Cl as Constraints on Fast Fracture Flow and Percolation Flux in the Unsaturated Zone at Yucca Mountain. *Journal of Contaminant Hydrology* 2001;51:257–288); see also, J. J. Hinds et al. Conceptual Evaluation of the Potential Role of Fractures in Unsaturated Processes at Yucca Mountain. *Journal of Contaminant Hydrology* 2003;62–63:111–132).

[h]Idaho National Engineering and Environmental Laboratory (INEEL) data show water moving 0.8–1.0 mile through interlayered volcanic lava flows and sediments in four months—probably through fractures, lava tubes, and lava rubble zones. Movement over horizontal distances

Box 10.2 Continued

within the unsaturated zone may exceed 46 feet per day (J. R. Nimmo et al. Kilometer-Scale Rapid Transport of Naphthalene Sulfonate Tracer in the Unsaturated Zone at the Idaho National Engineering and Environmental Laboratory. *Vadose Zone Journal* 2002;1:89–101). Large-scale 1999 tracer experiments showed that rapid lateral flow of groundwater in high-level aquifers, perched above stream valleys, which collect water seasonally is an important factor in spreading contamination at INEEL (K. S. Perkins. *Measurement of Sedimentary Interbed Hydraulic Properties and Their Hydrologic Influence Near the Idaho Nuclear Technology and Engineering Center at the Idaho National Engineering and Environmental Laboratory* [U.S. Geological Survey Water-Resources Investigations Report 03–4048, 2003]).

with starkly different characteristics into LLW. The list includes very short-lived radioactive substances and rather large amounts of long-lived radionuclides, such as plutonium, uranium, and thorium (see chapter 7, box 7.1). In fact, every defined HLW radionuclide also can be LLW (figure 10.7). As a result, U.S. LLW sites contain long-lived, highly toxic and flammable plutonium:

- About 450 pounds (>18,000 curies) at Richland, Washington

- About 140 pounds (~6,000 curies) at Maxey Flat, Kentucky

- About 47 pounds (~2,000 curies) at Beatty, Nevada

The Nuclear Regulatory Commission's regulations permit as much as 4,600 curies of cesium-137 per cubic meter of LLW—enough to produce a lethal radiation dose in about 20 minutes to a person standing three feet away.[68]

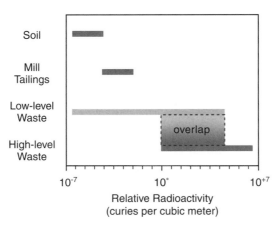

Figure 10.7 Graph showing overlapping U.S. definitions of low- and high-level radioactive waste categories (in curies per cubic meter), which govern the radioactivity of materials permitted in U.S. dumps, and ranges of radioactivity in soils and uranium mill tailings for comparison. Data from Robert Bernero, Nuclear Regulatory Commission (personal communication, 1996).

Preconceived notions

Added to poor site selection and characterization, the negative history of LLW repository leaks so far has blocked proposed LLW repositories at Sierra Blanca, Texas, and Ward Valley, California. Nuclear Regulatory Commission licensing regulations are partly responsible. For political reasons, the Nuclear Regulatory Commission requires that all essential characteristics of LLW disposal sites—geology, hydrology, biology, climate, and so on—be fully examined and evaluated in one year or less. But a year's study of previously unexamined arid western sites is grossly insufficient to fully determine whether or not they can isolate such hazardous materials.[69]

When Ward Valley was proposed for a state-run LLW site, only a single water well existed within 40 miles and no earth science or biological studies ever had been done there. During the single year of site characterization studies to support the license application, site safety concerns focused on whether contaminated radioactive water could get into the Colorado River and eventually into Los Angeles' drinking water (see chapter 9). To measure the depth to groundwater, as well as local surface and unsaturated zone water flows, consultants drilled five closely spaced wells in the proposed location of the repository's unlined burial trenches. They did not drill wells farther from the site to define the regional groundwater flow directions and sources, however. Nuclear Regulatory Commission regulations required that judgments about the site's safety and integrity had to be based on their large, but restricted data set.

Three U.S. Geological Survey geologists challenged the study's database, leading to a review by a National Research Council panel, under the NAS Radioactive Waste Management Board. After examining the site characterization field data—more than 300 bound volumes occupying 48 linear feet of library shelf—an expanded group of scientists with long experience of the Ward Valley area wrote formal reports detailing flaws in the study's data. The flaws proved so severe that they invalidated all the study's important conclusions. The important missing data rendered the study useless to even address the site's suitability for a radioactive waste repository.[70]

The National Research Council review panel held hearings in 1994 and released its final report in 1995. The panel majority disagreed with the geologists' concerns[71] but had to agree that the site characterization field data were significantly limited and that many were so poor that they could not reliably predict the repository's safety. Perhaps unconsciously responding to a political agenda, the panel scientists tried to base critical scientific conclusions on untested notions about geologic processes in arid lands, which turned out to be wrong (box 10.2). Appealing to preconceived notions about other arid sites (Beatty and Yucca Mountain in Nevada, WIPP in New Mexico, and Sierra Blanca in Texas), the majority concluded that Ward Valley would be a safe repository (box 10.2). Although only the Beatty site ever had stored radioactive waste, and no studies of Beatty had ever looked for leaks, the panel's report advocated moving ahead with site development while collecting more data. Two NAS panel members registered the first ever minority dissent in any report from the NAS's Radioactive Waste Management Board, however. The dissenters urged collecting more data checked for accuracy *before* approving the site.[72]

Just before the Ward Valley land could be transferred from federal to state control, a group of U.S. Geological Survey hydrologists revealed that they had withheld data from both their critical colleagues and the National Research Council review panel. The withheld data showed that radioactive contaminants had leaked from the Beatty dump, through the site's unsaturated zone, and all the way to groundwater (box 10.2). The government hydrologists had kept this critical information out of their review panel testimony and had not corrected the information as the panel wrote its report. They had testified to the panel that Beatty's unsaturated zone is leak-proof, misleading the willing majority to conclude that contaminants could not move away from the Ward Valley repository at shallow depth, or penetrate deeply into the unsaturated zone. In fact, *both had occurred at Beatty.*

The Ward Valley site never received approval for radioactive waste disposal, largely for political reasons rather than the flawed scientific evidence. Its actual suitability as an LLW repository may never be assessed. Contrary to the common lament that nuclear-waste decisions should be based on science and not politics, however, neither Ward Valley's shoddy site characterization data nor the information on water movement in Beatty's thick unsaturated zone would have come to light without pressure from key elected and appointed officials. These political figures listened to credible nonmainstream scientists and opposed both the George H. W. Bush and Clinton administrations' attempts to approve the Ward Valley site, no matter what the consequences.[73]

Wasted Outlook

The NAS has scored U.S. hazardous waste site management regulations as inconsistent with good public health policy on the bases of health threats, lack of national waste sites and contents inventories, and lack of data needed to study health effects and develop adequate public health protection programs. In spite of RCRA and the Superfund program, the United States has no comprehensive national program for identifying hazardous dumps, no protocol for managing hazardous wastes, no comprehensive national inventory of hazardous waste sites, no site discovery program, no minimum information on potential human exposures, and inadequate systems for identifying sites needing immediate action.[74] There are no rules that set technical criteria or procedures for assessing either a dump's long-term biological and chemical stability or the need for monitoring beyond 30 years.[75] Lax U.S. waste disposal oversight and enforcement limit the effect of even these protocols.

Even though landfills promise a future of toxic leaks and lawsuits, the space problem is the immediate challenge to governments at all levels. Cities and counties want to bury wastes far away from their population centers, but today few places are remote enough. And bad past performances have sensitized the public. Los Angeles has tried for years to site a landfill more than 100 miles away, next to the abandoned Eagle Mountain iron mine. The landfill would abut the southeastern edge of Joshua Tree National Park and an open section of the California Aqueduct, which carries drinking water to southern California towns and cities. Operating round the clock, the remote landfill would receive 20,000 tons of garbage every day, hauled by train and truck for 117 years. In addition to contamination worries, locals and national park aficionados charge that the dump would destroy the national park's natural values with noise and night floodlights, while filling the air and aqueduct water with hazardous dust.

After 30 years of applying sludges to agricultural land and 10 years of regulations, nearly a third of the country's 3,795 major municipal sewage treatment facilities still significantly fail to comply with Clean Water Act standards, while many lack valid permits.[76] The EPA's oversight of "biosolids-management practices" lacks coherence, and independent researchers continually find persistent toxins in "biosolids" after prescribed treatment and application to farm and ranch lands.[77] Worsening the regulatory problem, in 2002 the Army Corps of Engineers and EPA adopted new federal rules redefining "fill material" and "discharge of fill material" under the Clean Water Act, making it easier to dump toxic mining and landfill wastes into the nation's waters (chapter 4).

The United States pursues a "don't ask, don't tell" policy on chemical use and disposal as new chemicals, and chemical wastes, flood into the environment—more than 700 new chemicals per year.[78] The Toxic Substances Control Act does not require chemical companies to test new chemicals before the EPA reviews them, so we have no assessment of their health and environmental risks before they go into widespread use. The EPA tries to predict toxicity by comparing new chemicals to previously tested ones of similar type,[79] but this is just more faith-based "science." The toxic effects of most chemicals—including many long in use—still are unknown. Without testing, nobody knows if mixing new chemicals with thousands of other natural and synthetic chemicals already in the environment will enhance or broaden toxic effects (see chapters 2, 8).

Dumping treated wastewaters into ephemeral streams and low-flowing rivers poses another problem in the western states. Broad-based public opinion in Sonoma County, California, opposes the regional treatment plant discharging wastewater into wetlands, the ocean, or the Russian River. The cities face restrictions on surface drinking water resources and want to use wastewater for irrigating business and household landscapes and for groundwater recharge, to fulfill growth plans. But planned growth will increase wastewater so much that all these disposal options probably are insufficient. Like Banquo's ghost in Shakespeare's *Macbeth*, plans to dump tertiary treated wastewater into the Russian River keep coming back.

Fixes

A variety of technical fixes can neutralize or isolate hazardous wastes.[80] Some contamination can be simply filtered out of local water supplies, waste piles can be capped or otherwise covered to prevent contaminants from washing or blowing away, and acid mine waters can be treated with lime and other neutralizing substances. Various chemicals and bacteria can be injected into aquifer soils and rocks to turn hazardous compounds into nonhazardous ones. Waters contaminated with petroleum, petrochemicals, and VOCs either can be pumped to the surface and treated or the chemicals can be made into extractable oxides. Hydrolysis alters other kinds of substances for easier extraction.

Some of these techniques are in use, particularly at Superfund sites, but they are costly. Bottom lines are driven both by the immense sizes and complexity of contaminated sites and by the large amounts of energy consumed in mining chemicals used for treatments or synthesizing additives and treatment agents—and in transporting, pumping, or applying them. Completing Superfund cleanups took longer and cost more than anticipated,[81] and in some instances cleanup standards were trimmed to fit limited budgets. As stocks fell in 2000 and industrial interests gained greater political power, Superfund cleanup money dried up and may be a thing of the past. At today's prices, only the worst and most controversial sites will ever see any cleanup.

Poorly "remediated" sites remain vulnerable to destabilizing forces—something that nature can provide in abundance. For example, Hurricane Katrina inundated several Superfund sites in New Orleans—one, the Agriculture Street Landfill, about halfway between the French Quarter and Lake Pontchartrain—had been fenced, covered with matting, and topped with two feet of soil. The flood waters undoubtedly got through the soil and matting, and likely redistributed highly toxic wastes across the lower Mississippi delta.

A lot of radioactive waste is sitting around the United States with no place to go. Aging nuclear power plants soon will need to be decommissioned, requiring additional HLW and LLW disposal sites, which do not exist. In 2002, the U.S. government began planning new nuclear weapons programs, and current proposals to build more nuclear power plants inevitably will increase the need for more radioactive waste repositories. No scientists and no nations can yet predict the safety of radioactive repositories, but if American LLW were more correctly categorized and long-lived materials removed, the disposal problems—and costs—could be vastly reduced.

Some alkaline western soils can bind and hold toxic metals. For heavy metals and radioactive wastes, toxic oxides can be preserved as relatively inert compounds under antioxidizing ("reducing") conditions in bog soils, rock units under exhausted oil fields, and other appropriate sites. Injected into contaminated aquifers, hydrogen sulfide (H_2S) and some other reducing chemicals extract most heavy metals from groundwater. But so far, none of the nation's nuclear reservations or repositories has reducing characteristics. Perhaps as a result, all have leaked.

The unfortunate, unanticipated outcomes at radioactive waste dumps, along with antiparadigmatic discoveries over decades of nuclear waste site characterization studies, at least yield a better understanding of what we are up against. Instead of forcing approvals of remote but unknown sites, the United States should look for better ways to isolate such long-term hazardous wastes.

In-place treatment of contaminated water with nanoparticles, comparable to the size of atoms and molecules,[82] may lend itself to groundwater cleanup. The tiny particles may be extremely effective at binding contaminants, due to the very high surface area for small volume. But proposing to use nanoparticles at this stage is comparable to all the other untested and failed radioactive waste disposal methods. For example, the particles are tiny enough to integrate with living cells, and preliminary animal tests have shown them capable of entering and damaging lung, brain, and other organ tissues. Britain's Royal Society and Royal Academy of Engineering warned, "There is virtually no information…about the effect of nanoparticles on species other than humans or…how they behave in the air, water or soil, or…their ability to accumulate in food chains.…"[83]

Ironically, high-tech cleanup techniques also raise resource demand and generate additional wastes. When treating petroleum wastes, volatile noxious or greenhouse gas vapors are likely to escape into the atmosphere. Mining, processing, and producing cleanup agents just drive another cycle of waste generation.

The increasing volumes of hazardous modern garbage, and the costs of isolating them from human habitat, are beginning to force recycling programs upon counties and cities. But industry still is focused on dumping, even as recycling and reuse businesses can and do employ as many Americans as the auto industry. Recycling costs money, and with the exception of aluminum and e-wastes, most federal, state, and local tax laws paradoxically make even the easiest and cheapest recycling applications more expensive than mining, hauling, and processing new materials.

Separating hazardous from nonhazardous wastes for recycling is an important step, although early requirements that householders separate all categories before pickup hampered recycling programs for many years. Most successful modern programs sort non-yard wastes at transfer facilities or the dumps themselves. One benefit of RCRA is that the cost of maintaining landfills makes presorting garbage for recycling cost-effective.

Nearly all common products with hazardous components—vehicles (all kinds), batteries, computers and phones, even fluorescent lights—can be safely recycled or remanufactured. In 2002, the EPA proposed new rules for handling mercury-containing equipment and cathode ray tubes from electronic displays. Each ton of e-waste contains more gold than 17 tons of ore from mining, with much less toxic waste—so separating gold, silver, and platinum can pay for e-waste recycling.

Unfortunately, many of the wrong kinds of wastes are in the recycling stream. Electronic gear recycling cannot guarantee safe disposal, either. A high proportion of California's recycled computers go to Asia, where workers with no personal protection or environmental safeguards dismantle them to retrieve the poisonous metals. Massachusetts and a few other states have banned obsolete computers and TVs from dumps and landfills. In 2003, California even imposed a unit recycling fee on computers and televisions.

Consumption Trap

Americans are the world leaders in devouring resources. The avalanche of hazardous wastes that we produce represent the flip side of our glitteringly consumptive culture, however. In 1940 a tiny wastebasket ably handled the small volume of household trash it collected between pickups. By the 1980s, suburbanites were filling 30- or 50-gallon plastic dumpsters every week. If all other countries tried to emulate our lifestyle, their cumulative raw materials consumption, and need for waste disposal sites, would demolish the environments of four more planet Earths—even assuming no further growth of population or per capita demand.

The negative effects include massive wastes, fouled beaches, trashed lake and ocean waters, dumps seeping chemical pollution into soil and water supplies,

and dwindling disposal sites. Largely ignoring official denials, the public increasingly worries that higher levels of asthma, attention-deficit hyperactivity disorder
(ADHD), and late-onset autism in children; arthritis, lymphomas, reproductive
problems, and testicular and breast cancers, especially in younger men and women;
and Crohn's disease, lupus, and other autoimmune diseases are linked to low-level
but long-term chemical exposures.[84] Wastes, especially landfill and sewage contaminants found in soil and water everywhere, are the major source of Americans'
continual chemical exposures.

But even if we can suddenly stop consuming and generating toxic products
and toxic wastes, the United States will face increasing waste-derived pollution
for many generations to come. This is a largely unexamined public health threat
that water and sewage treatment policies do not adequately address. Federal environmental safeguards call for only 30 years of monitoring and treating landfill
leachates to be sure most wastes stay put. This is not enough, but the same rules
apply to groundwater and methane gas monitoring. Conversely, it is unlikely that
our current repository schemes will isolate the most hazardous wastes for 10,000
years. The likelihood that poisonous and disease-spreading materials will end up
in streams or groundwater should make us think about what we are willing to put
in the ground (figures 10.1, 10.2).

Expanded recycling is highly desirable but cannot provide the whole answer.
The hazardous chemicals in wastes ought to limit or prohibit recycling of chemically laden wastes, including mining, manufacturing, treated municipal sewage
solids and waters, and radioactive wastes. Sadly, the increasing costs of waste treatments and disposal, and the superabundance of nearly every waste type, has local,
state, and the federal government scheming to recycle hazardous substances back
into our daily lives—including into food.

Nobody wants hazardous materials sneaked into their lives. Just a mention on
television news of plans to recycle radioactive nickel into steel for ordinary household uses stopped the whole project in 2000 (see chapter 7). Apparently few
Americans favor radioactive hip implants, cooking pots, or car parts that increase
their daily radiation exposures. People do not like fertilizers made from arsenic-rich
mine wastes, either (see chapters 2, 4), and circulating e-mails suggest that gardeners want to avoid buying mulches containing contaminated Hurricane Katrina
wastes. Public relations campaigns intended to overcome some of these concerns
have not made much headway.

Slowly, public pressure is forcing public agencies to keep hazardous electronic
wastes, batteries, fluorescent light bulbs, and materials soaked in flame retardants
out of landfills. EPA is encouraging a new grassroots movement to return unused
medications instead of putting them into a waste system.

But whether the public knows and approves or not, they eat foods that have
been grown with sewage sludges, containing heavy metals, pathogens, pharmaceuticals, and chemical toxins. Antibiotics and estrogenic hormones from treated
wastewater discharged into rivers flow back to our kitchen faucets, increasing the
risk of antibiotic-resistant bacteria and hormone overdoses. An informed public
might reject these forms of recycling if scientific findings show that human food
sources concentrate the hazardous materials. An informed public probably will
have to demand to have that kind of research performed.

Before long, we will have to come to grips with consumption itself. Conservation writer Michael Frome has noted: "[T]he only possible and lasting solution to many of the world's environmental problems lies in reduced consumer demand." The simplest approach, lowering per capita consumption, is also the most difficult for a population that has never questioned a lifestyle of increasing demand and sees it as their birthright. But this lifestyle goes back little more than three generations. If Americans wanted to limit wastes and their hazards, they probably could change relatively quickly. But nobody expects Americans to take this path until forced.

Still, how could we lower consumption and waste? The easiest first step would be to reduce or eliminate packaging. Everybody already hates packages that can hardly be opened with regular scissors. We could ban them and return to selling retail food from all-bulk stocks, as do many supermarket butcher/bakery/vegetable sections, and greatly reduce individual packages for nonfood items. This would have the sea-change effect of lowering MSW. Concerned citizens might have to force the federal government to ban hazardous materials from landfills, and even start excavating and safely recycling older landfills. Public information campaigns would encourage enacting European-style product regulations, which shift the responsibility for recycling and disposal to manufacturers. All these measures would deflect our focus from consumption toward resource conservation and environmental cleanups.

Aroused public opinion could force a moratorium on new chemicals until the toxic effects of the current inventory, especially in combinations, can be defined. An appreciation of the health hazards in wastes could mobilize the nation to support penalties on industry to defray the financial burden of advanced treatments for Superfund sites and hazardous landfills.

Eventually natural resource depletion, especially petroleum depletion, will force the issue. Today's rising gas prices preview a coming era of limits. Many landfills already purify methane out of landfill gas and use it to generate electricity, and that sort of waste recycling already is growing in popularity. As repositories for most of the nation's, and many of the world's, resources, landfills and waste dumps inevitably will be mined. Our main concern is whether natural processes will saturate our environment with synthetic poisons before resource limitations take effect, visiting biblical plagues of human origin on us, our lifestyle, and our economic system.

11   Tragedy of the Playground

You call someplace Paradise—kiss it goodbye.
The Eagles, "Hotel California"

"Recreation" connotes revitalization, the re-creation of spirit. In an increasingly urbanized culture, people recreate in natural settings to lift their spirits and revitalize their outlook and motivation. Public lands in the western United States, which embrace much of the nation's remaining natural and wild areas, are especially attractive—and most are open for recreation. We authors certainly have found solace from camping, hiking, climbing, and skiing in backcountry areas. But late-twentieth-century American affluence has created a massive and unprecedented invasion of these lands, and particularly an invasion of motorized recreation.

All human uses of natural areas can, and generally do, degrade soils, kill plants, and increase erosion rates, with resultant water pollution and ecosystem damage.[1] In small numbers, and spread out widely, recreational disturbances can be minor, but millions of people regularly play on western public lands in mass gatherings that have large cumulative impacts. More now drive vehicles across forested or desert areas than pursue the less-damaging activities of hiking and small-group camping.

The Bureau of Land Management (BLM) and U.S. Department of Agriculture's Forest Service (USFS) oversee the largest amount of western land available for recreation. By law, the agencies must manage public lands for multiple uses and "sustained yield."[2] Instead, federal land-management agencies are partitioning them to separate incompatible pursuits, including many that consume land. For example, as logging, mining, and grazing pressures ease, recreational pressures are exploding in Colorado's White River National Forest, a short 50 miles west of Denver on Interstate Highway 70. Along with Denver's increasing population, snowmobile registrations jumped 70% in Colorado since 1985. Off-road vehicles (ORVs) are everywhere, and mountain bike use has jumped more than 200%. Between 1990 and 2004, all ORV registrations in Colorado increased *more than 650%*. Ski facilities also burgeoned, along with hiker and equestrian demands for greater backcountry access. The USFS's efforts to bring the conflicting uses under control is losing ground rapidly.[3]

Proliferating vehicular trail networks, temporary tent cities, and recreational vehicle encampments damage fragile ecosystems extensively and have become environmental threats. Some people feel they can do as they like in a recreational vehicle, no matter the cost to the land, others, and themselves. Dense ORV gatherings and encampments attract urban alcohol and drug problems, combined with uncontrolled firearms and reckless driving. And they often trash natural areas.

Luxurious resort-based recreation is increasingly popular, expanding urban–suburban sprawl across once-pristine mountain slopes and desert oases along with urban garbage and sewage disposal problems—not to mention air, water, noise, and light pollution. Remote and rugged western mountain ranges are scarred by ski runs more and more, and improbably green resorts and golf courses obliterate natural desert habitats. All consume huge amounts of water and add pollution to creeks and soil. Manufactured lakes have appeared where surface waters rarely flow, filled with groundwater that rapidly evaporates in the desert heat (figure 11.1).

In America's least justifiable tragedy of the commons, large-scale outdoor recreation threatens to further reduce natural ecosystems and leave few areas where larger native animals can survive (see chapters 5, 8). Reversing the impacts of play on our backcountry, the source of much remaining U.S. natural capital, may be impossible. Stopping them altogether would require un-American self-denial.

Figure 11.1 A Mojave Desert recreational lake east of Daggett, California, made with pumped groundwater—one of many created in the arid west since the 1980s. Open desert ponds and lakes lose about 10 feet of water to evaporation annually. Photographed April 2003.

But if Americans cannot agree to let government agencies effectively control recreation on natural western lands, they will surely destroy the paradises that offer respite from daily noise and stress.

Tracks in Paradise

All creatures leave marks on the land. Even a tiny beetle leaves tracks on the surface of a sliding sand dune. The unshod feet of early people carved tracks, trails, and roads into the soil, and those ancient trading routes still are visible in many parts of the world. Playing on the land also tramples it. Heavily used trails and camping sites suffer the same impacts as unpaved school playgrounds, which are notoriously barren areas of trampled soil. Many people fail to see the harm because soil is not well regarded—it is usually dismissed as "dirt." But soil is a critical resource and the foundation of most healthy ecosystems (see chapters 1–3, 5, 6, 13).

In deserts, foot trampling and vehicular impacts destroy soil-protecting crusts and pavements, along with biological soil components (see chapter 13).[4] Compaction degrades a soil's fertility because it significantly reduces the bacteria and fungi that help plants absorb nutrients (see chapter 2).[5] Soil compaction also reduces the amount of water that can seep into the soil.[6] On level ground, rainwater ponds in the compacted foot, hoof, or vehicle tracks, evaporating faster than absorbed water from uncompacted soils. On slopes, gravity forces rainwater to runoff downhill, but more water runs off of compacted than undisturbed slopes. The plants and microscopic life within a compacted soil can't obtain their normal moisture supply. Eventually, the basic functions of the interactive, life-supporting ecosystem close down.

Plants form an insulating cover that shields soil from the sun's heat in the daytime and slows heat loss at night. Both soil compaction and removing plant cover increase daytime soil temperatures by as much as 23°F and lower them at night on the order of 5°F (figure 11.2).[7] Exposure to more extreme temperatures slows new plant growth and has unknown effects on other aspects of soil health (see chapter 6).

Driving destruction

Recreational ORVs include mountain bikes, motorcycles, all-terrain vehicles (ATVs), four-wheel-drive vehicles (4WDs or "Jeeps"), and snowmobiles.[8] All are designed for operation on natural terrain and unimproved trails, and most of their play areas are on public lands. Like tank treads (see chapter 6), wheeled ORVs compact the soil underneath their tires and loosen soil beside the tracks (figure 11.3). The tires crush small plants in the tracks and uproot them in the loosened soil. Vehicular compaction also crushes animal burrows, often killing or maiming animals still in them.

The level of vehicle damage depends on the design and operation. A vehicle disturbs the ground least if driven slowly in a straight line on a dry, level surface. A typical two-wheeled motorcycle will damage one full acre in 16–20 miles of

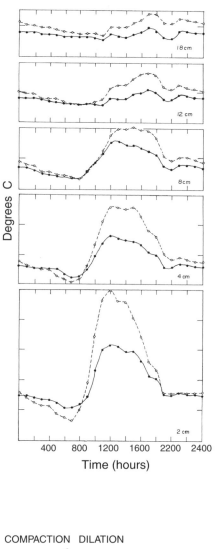

Figure 11.2 Graphs showing daily temperature changes at five different soil depths, measured at the same location, after vehicles had stripped vegetation and compacted the soils. Soil temperatures in the trails (dashed lines) are higher in daytime and lower at night than in natural, vegetated soils (solid lines).

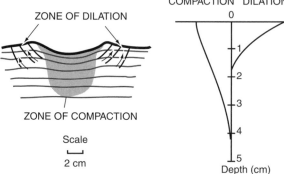

Figure 11.3 Diagram illustrating the zones of compression and dilation from a vehicle passing across a soil (left) and the subsurface effects (right). Soils are deformed and compacted in the zone of compression (left of zero), and loosened in the zones of dilation (right of zero).

travel if driven that conservatively. More wheels means more damage: three- and four-wheeled ATVs and 4WD vehicles will damage an acre in only six miles of travel. But ORV riders generally drive at high speeds over erratic courses and so tend to damage more land over shorter travel distances.

Studies of human trampling (see below) and military tank maneuvers (see chapter 6) show that they inflict long-term damage on vegetation.[9] The compaction and surface disturbances from one pass of a military tank still can be measured in 55-year-old tracks. Annual plants growing in the tracks are much smaller and provide less soil cover than do plants in untracked areas, and plants with laterally spreading roots grow better in the tracks than do plants with deeper tap roots (see chapter 6). Similar effects can be predicted for ORV areas.

A single ORV pass destroys desert biological soil crusts (figure 11.4), and a typical sandy loam desert soil is close to maximum compaction after only 10 motorcycle passes on a dry, level surface.[10] Desert race-staging sites and pits are completely denuded over large areas from trampling as well as driving. The soil and plant damage is more selective and less noticeable at low-use sites, because resilient creosote shrubs succumb more slowly. ORV trail networks become webs of bared tracks, which heavier use broadens into biologically barren areas at trail

Figure 11.4 Biological soil crust destroyed by a single pass of a four-wheel-drive vehicle, St. George, Utah. The tracks have channeled rainwater runoff, which also erodes them. Photographed May 1977.

Figure 11.5 Aerial view of Bureau of Land Management ORV Open Area, Dove Spring Canyon, California. Areas where denuded trail networks coalesce are completely bare. Extensive stretches of the canyon walls also are completely denuded. Photographed August 1977.

intersections (figure 11.5). In addition, weeds invade along access roads and ORV trails, eventually capturing most of the moisture and nutrients that native plants need, reducing native biodiversity in those areas even further.[11]

Abnormal erosion

Over time, ORV use and erosion can eliminate whole native habitats, starving, driving away, or outright killing the animals. ORV enthusiasts agree that riding across soils and bare rocks causes erosion—but erosion is natural, they argue, adding, "after all, the Grand Canyon was carved by erosion." Too true. Acting over exceedingly long time periods, natural erosion is the most potent of forces. Most people can scarcely guess at its power. Take the Grand Canyon, which the Colorado River eroded to a depth of about one mile and up to 18 miles in width over a period of about four million years. A conservative estimate of the natural erosion rate that created the Grand Canyon is about 620 feet per million years. Vehicular impacts on natural surfaces accelerate erosion, however. Measured rates in ORV trails are 10–20 times higher than natural ones,[12] even on moderate slopes.

Erosion on steep ORV trails commonly removes the entire soil mantle and exposes the underlying rock (figure 11.6), then begins eating at the rocks. Favored ORV recreation sites tend to be dunes, or hills carved in soft, poorly consolidated rock, which is highly susceptible to erosion.[13] Most ORV tires have knobs and

Figure 11.6 Motorcycle hillclimbs at Jawbone Canyon, a Bureau of Land Management Open Area in the northwestern Mojave Desert, California, have promoted soil erosion down to hard rock (light colored patches and strips). Soils will not regenerate on these slopes for many thousands of years. Photographed September 2003.

cup-shaped protrusions that mechanically erode soil and sediment as they help the vehicles climb steep slopes. The tires look and act like mini-ditch diggers, throwing up behind the vehicle a "rooster tail" of soil and small uprooted plants, which cascade downslope or drift in the breeze. On steep slopes of about 30% or more, mechanical erosion may displace as much as 50 tons of soft soil—enough to fill 10 dump trucks per mile of travel (figure 11.6).

ORV areas can yield as much as 610 pounds of eroded sediment per square foot.[14] At that rate, a motorcycle hillclimb area 150 feet long by 10 feet wide would shed 915,000 pounds of sediment, enough to fill five railroad cars or more than 90 dump trucks. If the level of ORV erosion were to extend over four million years, it would carve a supergrand Grand Canyon, 400 miles deep and up to 2,500 miles wide, large enough to swallow much of the North American continent.

Natural erosional processes always are present and inevitably spread the direct damages that vehicles inflict on the land. Even if a slope has relatively resilient soils, runoff from steep ORV trails will flow across nearby untracked areas, eroding deep gullies. The excessive runoff from compacted ORV trails at Hungry Valley, California, State Vehicular Recreation Area has enlarged nearby stream channels and undermined the roots of mature oak trees, which then fell into the channels. Even after partly denuded ORV areas are closed, the erosion continues to spread until the hills, valleys, and watercourses of the entire surrounding region finally adjust to the soil losses, increased runoff, and sediment volumes.

In *Deserts on the March*, Paul Sears declared that soil where it formed is our friend, but soil on the move is our enemy.[15] The oversupply of sand and other sediment eroded from ORV areas is one of those enemies, for it extends ORV damage to streams and lands both near and far. Runoff can carry the sediment away during rainstorms, choking and polluting streams and rivers and raising flood levels (see chapter 1). Eventually, the eroded soil from ORV areas is redeposited somewhere, likely burying the most fertile, upper soil layer on undamaged areas and smothering the plants growing in it (see chapter 2).

Loosened and mechanically eroded soil is especially vulnerable to wind erosion.[16] Winds can pick up loose sand and dust from bare hillclimb trails, and also from the deposits of eroded and redeposited sediment. Dust plumes originating from ORV areas actually are visible from space.[17] As in the 1930s Dustbowl era, windblown soil can sandblast plants, exposing and undercutting their roots and ultimately killing them (see chapter 2). Dust can cause or worsen severe respiratory ailments in people, domestic animals, and wildlife. Disease organisms that live in arid land soils, such as the valley fever fungus, *Coccidioides immitis*, spread widely in dust storms (see chapters 2, 5, 6).[18]

Vehicles take over

Improved state and local road systems have provided easy access to formerly remote forest and desert areas. In national forests, recreational vehicles follow mazes of logging roads and ways connecting vast "clearcut" gaps. John Muir would find little of nature remaining in these areas. But taking care of the land is not a characteristic of many ORV enthusiasts. More common is the worldview expressed by one rider: "All this talk about lizards and beetles. The world was made for people, not creepy-crawly things."[19]

Western deserts have few natural obstacles or regulatory enforcement to impede vehicles, and are even more overrun. ORVs have shredded desert lands for four decades, through BLM-permitted off-road motorcycle chases across California desert routes. These included the annual Thanksgiving weekend free-for-all motorcycle race from Barstow, California, to Las Vegas, Nevada, run from 1967 to 1974 and briefly resurrected in 1983, and long 4WD events through scenic Colorado River-side mountains. Annual Jeep jamborees on BLM areas in southern Utah and elsewhere exhibit the modern 4WD vehicles' awesome capabilities in rougher country.

ORV recreationists tend to gather in large groups. The Barstow to Las Vegas event often attracted more than 3,000 riders, plus families, pit crews, and spectators. Hundreds to thousands of vehicles drive through and around most authorized ORV areas on a yearly basis—at Algodones Dunes, near Glamis in southeastern California, nearly 200,000 have shown up on some holiday weekends. Tracked by both feet and ORVs, the encampments also are commonly trashed. By USFS report, as many as 800 people have camped all together at the Eldorado National Forest's scenic Spider Lake–Little Sluice area along the Rubicon Trail, a favorite ORV route for one-day events in California's Sierra Nevada. A huge litter of human feces and toilet paper, dotting "the landscape like daisies," forced its closure.

Some ORV riders express an affinity and yearning for natural areas, but participants at Utah Jeep jamborees drive through *tinajas* (also called tanks), which are shallow rock basins that collect winter and spring rainwater.[20] For millennia the tinajas have been precious, tenuous, wildlife watering places in an otherwise parched country. Now, the Jeeps leave black rubber stains on the tinajas' steep sides and leak toxic oil and other chemicals into the water. Another rare gem is beautiful Surprise Canyon, near Ridgecrest, California. The absence of a drivable trail in the canyon did not stop extreme "technical" four-wheeling enthusiasts from winching Jeeps and 4WD sport utility vehicles up a 7,000-foot climb, through seven waterfalls.[21] These events have smashed trees and shrubs, littering the canyon with tools, damaged car parts, and oil stains. ORVs also have damaged or destroyed archaeological sites—in particular, southeastern California's fragile ancient desert geoglyphs ("intaglios"), the patterns and figures that Native American ancestors' created by scraping lines into the soil's thin, darkly patinated, pebble-mosaic surface layer.

Misuse and rebellion

Vehicular recreation has become a major law enforcement problem on BLM land, especially on holidays. Safety is a significant issue, since unstable vehicle designs and rough terrain contribute to numerous adult and child injuries and death. The mayhem of uncontrolled, high-speed recreation, reckless driving, alcohol, and a disregard for safety helmets also plays a significant role. Insurance companies warn, "Crash tests...demonstrate that off-road vehicles and fun cars roll over more easily and are a bigger hazard to other road users than normal cars...passengers are exposed to a considerably stronger impact."[22] Drunken driving and attacks on enforcement officers are common, and so are injuries to children.[23] The high rate of ATV injuries promoted a design change from three to four wheels in the 1980s, but accidents increased as the sport boomed.

Scenes at some ORV areas can be so bizarre as to defy imagination. For example, rangers at California's Red Rock Canyon State Park told of three adults who fired up motorcycles soon after arriving on a Friday evening. One charged up a nearby hill and discovered a cliff just past the crest. Skidding to a halt, he turned and waited for the others. The second rider also managed to stop, but the unwarned third rider plunged over the cliff to his death. On weekends, the flat, barren BLM-managed El Mirage Dry Lake in California hosts a literal ORV circus of motorcycles, 4WD vehicles, ATVs, light aircraft, gliders, land-rocket cars, land-sailing craft, and untold novel devices. By 1983, deaths from ORV accidents at El Mirage averaged 26 per year. Particularly horrendous accidents included a young woman who was decapitated when she attempted to ride her motorcycle between a glider on takeoff and its tow plane.[24]

Poorly marked boundaries and a plethora of mining roads and utility corridors—and just plain bad attitudes—contribute to incidents of ORV trespass into private properties and motor-restricted public reserves. In the 1970s, one man told a CBS 60 *Minutes* interviewer of arriving at his weekend desert home to find dirt bikes driving through it, and the riders drinking beer from his refrigerator. Other desert residents described motorcycle riders chasing rabbits until their hearts burst, for "fun."

Some off-roaders bring their guns. Once a scientific colleague encountered a friendly couple in a desert canyon, driving a Jeep bristling with weapons, who lamented the disappearance of wildlife in the area. Some riders practice fast-draw and rifle shooting skills, often firing into vegetation where other off-roaders may be riding or relieving themselves.[25] Gun violence, commonly alcohol related, tends to flare in crowded recreational areas at wild ORV gatherings—such as holiday weekends at Algodones Dunes. Over the "really good" Halloween weekend at Algodones in 2000, "only" three people were killed, three others paralyzed, and one suffered critical head injuries.

Incompatible Uses

Because they destroy scenic and other recreational values, ORVs are incompatible with every other desert and forest land use except logging. Author David Sheridan wrote, "St. Francis of Assisi himself while driving an ORV on wild land could not avoid diminishing the recreational experience of many non-ORVers in the same area."[26] Few hikers, trail bikers, or equestrians willingly share trails or campsites with noisy vehicles and their polluting exhaust, especially since the riders churn up dust that rains down on nearby campsites. Today's recreational machines are bigger and faster than ever before, driving unmotorized recreationists off the land.

The grating cacophony of two-stroke engines, which power most off-road motorcycles and ATVs, combines the clamor of a chainsaw or jackhammer with the rasp of a machine gun, injecting the harshest of urban noises into remote backcountry. At a distance, the sound may be simply annoying, like a constantly buzzing fly—but close up the sounds are ear-splitting. At Algodones Dunes and other sites where thousands of ORVs play, the noise level is comparable to a factory floor or a ship's engine room. Per mile of travel, ORV engines emit 118 times as much "NOx, SOx, and rocks"—nitrogen and sulfur oxides, and particulates—as a modern automobile (see chapter 8). As much as one-third of the fuel passing through two-stroke engines, including the suspected carcinogen MTBE, comes out unburned.[27] Small particulates, both from ORV engines and kicked up from the surface, are in themselves health hazards.[28]

Uphill and down

The extent of vehicular damage to backcountry trails is another severe incompatibility, especially for hikers and photographers (figure 11.7). Even heavily used equestrian areas and foot trails can become severely compacted and hard to negotiate after several seasons of accelerated soil erosion. Foot trails along parts of the John Muir trail in California's Sierra Nevada are worn as much as two feet deep, for example. But tires strip soil surface much more efficiently than do feet and hooves. Hikers impair one acre of land in 40 miles of travel, so they accumulate negative effects at slower rates than either vehicles or horses. The John Muir trail

Figure 11.7 Motorcycle tracks on Wildhorse Mesa, southern Utah, a naturally barren area but once a major attraction for natural landscape photographers. Since 2003 changes in Department of the Interior policies, it is no longer a proposed wilderness study area, and widespread motorcycle tracking now mars the views. Photographed by Ray Bloxham, Southern Utah Wilderness Alliance, June 2000.

would be eroded much more deeply if motorcycles or ATVs had traveled along it over the same period.

Each type of recreation strips plant cover at a different rate, depending on slope and vegetation types.[29] Motorcycles traveling uphill on a 15-degree (26%) slope can completely denude a grassy hillside in only 100–200 passes, while going downhill they can completely eradicate the grass in 300 passes. In contrast, horses walking uphill must make 1,000 passes to denude a grassy slope (figure 11.8) but only 800 passes going downslope. Hikers walking uphill achieve only 30% denudation after 1,000 passes (figure 11.8) but 1,000 passes going downhill will fully denude it. Horse trampling reduces plant cover more rapidly than motorcycles on level grasslands, but motorcycles have greater effects on sloping grasslands (figure 11.9). In forested areas, horses and motorcycles destroy plants at comparable rates on both types of terrain (figure 11.9), but ORVs also can damage or destroy large plants, even trees.[30]

Erosion eventually destroys steeply sloping pathways. Having hiked the Grand Canyon and other trails used by both humans and pack animals, the authors would rather avoid horse and burro trails. Depending on the season, walkers must pick their way around deep hoof imprints and piles of droppings, or pools of ponded water plus horse or burro urine. And we walk past actual horse or burro trains

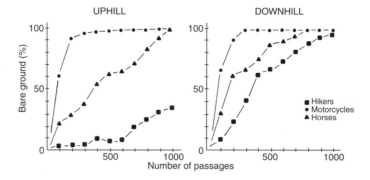

Figure 11.8 Hillslope denudation rates due to uphill and downhill travel by motorcycles, horses, and hikers. From T. Weaver and D. Dale. Trampling Effects of Hikers, Motorcycles and Horses in Meadows and Forests. *Journal of Applied Ecology* 1978;15:451–457. Copyright 1978, *Journal of Applied Ecology.* Reproduced with permission.

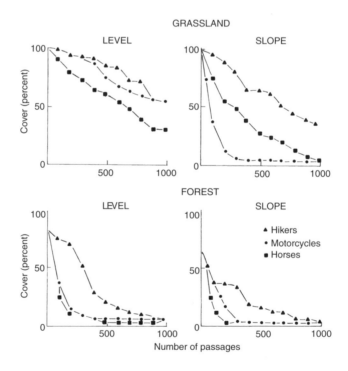

Figure 11.9 Reduced plant cover from the impacts of motorcycles, horses, and hikers on level and sloping grasslands and forest lands. From T. Weaver and D. Dale. Trampling Effects of Hikers, Motorcycles and Horses in Meadows and Forests. *Journal of Applied Ecology* 1978;15:451–457. Copyright 1978, *Journal of Applied Ecology.* Reproduced with permission.

with great caution on narrow trails. The deep ruts and slippery footing of dry compacted soil make deeply eroded ORV trails even more difficult going in summer and fall—and the extensive winter and spring mud generally is worse. ORV areas also are hazardous to hikers (not to mention smelly and noisy) when motorcycles or ATVs are playing recklessly.

Endangering others

Natural forest and desert settings are the life support for natural ecosystems. Outdoor recreational pursuits in natural areas are eradicating whole species—they are the main threat to 27% of the 1,207 species federally listed as endangered or threatened, or proposed for listing, nationwide. ORVs contribute half that impact.[31] The large scale of ORV soil compaction and denudation is especially destructive to desert wildlife.[32]

Naturalist Robert Stebbins observed that most desert animals protect themselves from both winter cold and summer heat in burrows below ground, or under or inside surface mounds or plant and rock debris.[33] On the order of 80–90% of total desert wildlife, including eggs in developmental stages, rely on fragile shelters and burrows—and some 75% of them are one foot or less below the surface. We don't know how many animals and their young are crushed beneath ORV tires, but the figure must be staggering.

ORV destruction of small and brittle shrubs, grasses, and annual plants undoubtedly impairs wildlife habitats, but the effects are not well defined. A BLM study of the next-to-last Barstow-to-Las Vegas race in 1974 hints at major animal and habitat destruction, however. Researchers found many fewer small mammals in the start area immediately after the race than in undisturbed areas nearby, and the number in the start area a year later was as much as eight times lower.

ORV noise can injure and even kill desert animals,[34] such as kangaroo rats. The rats have extremely sensitive hearing to protect them from predators, so motorcycle noises can damage their hearing for weeks, making them easy prey.[35] After the 1974 Barstow-to-Las Vegas race, normally nocturnal kangaroo rats were seen in the start area in daylight, acting disoriented and bleeding from the ears. Couch's spadefoot toad is another sound-sensitive animal,[36] which lives in burrows to a depth of about 20 inches. The toads must emerge during a thunderstorm to reproduce, and dune buggies sound enough like thunder—the signal of approaching rain—to bring the toads out of the ground, expecting to find rainwater pools and potential mates. If ORV noise brings them out in dry weather, their lack of body moisture and inability to get back in the burrow means certain death.[37]

Sensitivity to noises also may explain why most birds left Afton Canyon, a popular Mojave Desert ORV spot until the 1980s, as soon as weekend riders began to gather on Friday afternoons. The birds did not appear again until the following Thursday, so two days of ORV recreation drove away much of the bird population for most of a week.[38]

Hits Keep Coming

Unmotorized mountain biking is quiet, lacks exhaust, and may seem a relatively benign use of prepared trails and roads. But the bikes' knobby tires compact and

erode soils like motorcycles, with all the ancillary effects of changing soil temperatures, reducing infiltration, and accelerating hillside erosion. A mountain bike damages one acre of land in 44–50 miles of travel, comparable to a hiker's 40 miles.
The bikers cover ground much faster, however, so cumulatively they damage much
more land per outing. The tires also compress soil to a much greater degree than
do hikers' feet[39] and achieve maximum soil compaction with fewer passes.

Cross-country mountain bike travel also destroys and fragments habitat (see
chapter 1). Mountain bikes do not destroy larger plants, but the biker expression
"bring home a Christmas tree" represents a predilection for crashing through
underbrush, tearing branches from trees and shrubs. Fast bike speeds on mountain
and hill slopes and blind trail corners in dense forest vegetation make mountain
biking difficult to combine with hiking and horse riding. Opening remote areas to
mountain bikers to win support for wilderness designations threatens to heighten
this incompatibility—and to segment more habitat in and beside wilderness acres.
Recent studies suggest that avid male mountain bikers who ride as much as 3,000
miles per year may experience reduced fertility, however, and this could reduce
some of mountain biking's appeal.

Other nonmotorized recreational pursuits that damage arid lands include sand-
boarding (essentially snowboarding on sand dunes) and land sailing in wheeled
sailboats on dry desert lake basins. In 1999 The U.S. Fish and Wildlife Service
identified severe damage to an endangered dune grass species on California's
Eureka Dunes, under the National Park Service's jurisdiction, and recommended
an end to sand-boarding there. The sport also doubtless hurts animals that live in
shallow burrows in the sand.

Large parties or "raves" that intrude urban life into natural settings also trample
and inflict vehicular damage on natural soil and lake bed deposits. One prominent example is the convention-like, clothing-optional Burning Man Festival. Over
about a decade, it has attracted thousands of people to a dry lake bed in Nevada's
remote Black Rock Desert for one week out of the year. Due to the location on
public (BLM) land, tickets for the dense "temporary community based on radical
self expression and self-reliance" cost a minimal $225.[40]

Burning Man organizers believe that they leave no trace behind, but trampled
soils generally do not recover for decades or longer, especially in arid regions.
Since attendees arrive in vehicles and live largely in vehicles or tents, Burning
Man's impacts on the dry lake bed and surroundings resemble World War II military encampments in the Mojave and Arizona deserts. Although inhabited for only
two years, the military impacts still are visible and measurable (see chapter 6).
Nobody has measured the soil compaction in heavily used areas peripheral to the
Burning Man site, the animal and plant populations before and after the event, or
the volumes of dust produced before, during, and after the event.

Slippery slopes

Popular downhill and Nordic skiing and snowboarding sports, and motorized ski-
mobiling, have spread into national forests and other public lands. Between 1970
and the mid-1980s, downhill skiing's popularity burgeoned. The 11 western states
now accommodate at least 147 major ski resorts, 121 on national forest lands.

Some single resorts offer up to 4,640 skiable acres, as many as 31 ski lifts, and 174 cleared ski runs. Individual runs reach lengths of 38,000 feet and widths of 200 feet and more.[41] More and less luxurious resort facilities and services, and associated pollution, have come to public lands with them.

Ski resorts combine the environmental damages of roads, urban development (see chapters 5, 8), and foot traffic, magnified by concentrated heavy usage. Ski lifts run all year, in summer hauling mountain bikers upslope to careen down bared ski runs. The U.S. Environmental Protection Agency has noted that ski areas make up just 3% of the acreage in Colorado's White River National Forest, but

> no other land management prescription on the Forest directly results in more stream-water depletion, wetland impacts, air pollution, permanent vegetation change, or permanent habitat loss. [Since 1984], more wetland impacts and stream depletions resulted from ski area expansion and improvement than from all other Forest management activities combined, including many direct and indirect impacts that are permanent (irreversible and irretrievable).[42]

Together, ski runs and lifts and visitor accommodations have massively transfigured formerly remote mountainous areas. Roads snake into surrounding forests from the urbanized resorts, seriously fragmenting backcountry wildlife habitats (see chapters 1, 5). Snow compaction in the ski runs lengthens snowmelt, shifting sprouting times to later in the spring for plants up on the slopes. Ski developments and many majestic mountain vistas will remain marred with downhill ski run scars, resembling worm tracks under tree bark, for decades after they have been abandoned.

Including ski resorts, American taxpayers massively subsidize all commercial developments in national forests and other public lands. These multimillion dollar businesses hold 40-year term leases, paying only a small fraction of their take to the U.S. Treasury. Nine ski resorts, including Sugar Bowl at Lake Tahoe and Utah's Snowbasin, site of the 2002 Winter Olympics, paid back less than one cent for each dollar of sales. In 1998, all ski resorts on public lands in the nation returned only $18 million to U.S. coffers, barely the land value of even a small resort.

In a federal land swap, one politically connected Utah developer obtained 1,320 acres of public land for expanding Snowbasin's ski facilities to accommodate the 2002 Winter Olympic Games. The deal included additional public land for condos, a golf course, and other development. Congress also bestowed on the developer a $15 million, 3.5-mile paved road, courtesy of U.S. taxpayers, and exempted the first phase of new development—including lifts, a restaurant, snowmaking equipment, and other amenities—from environmental reviews. Without environmental precautions, biologists were unable to examine Snowbasin's Olympic site before the Olympics preparations wreaked widespread destruction on mountain wetlands.[43] Colorado's Breckenridge, Keystone, and Beaver Creek ski resorts also propose more expansions into additional public lands, and Vail wants to spread over another 885 acres into habitat that supports the threatened Canada lynx.

Too much popularity comes at a price. The high costs and long winter lift lines have begun to tarnish downhill skiing, sending many skiers to Nordic trails or motorized ski tracks, which also crisscross national forest areas and do double duty as summer hiking, equestrian, and mountain biking trails. As skier numbers declined from the 1980s through the 1990s, the resorts became year-round recreation meccas. That transformation fueled unprecedented building booms and urban sprawl in many mountain retreats, unchecked by rural county governments that never had established environmentally protective planning or project reviews. The initial resort developments were built on river bottomlands, supported with large water withdrawals from the streams, which decimated aquatic life. The withdrawals now exceed 160 million gallons per season—enough to supply nearly a thousand families for a year.

Ski resort proponents defend expanding resorts on public lands as a public convenience. Stacy Gardner, the National Ski Areas Association's Communications Director, said, "[I]t helps mitigate some of the long lines. Nobody wants to wait in line for a ski lift."[44]

The Ski Area Citizens' Coalition tries to promote environmental awareness at ski resorts. Since it can do little about the scarred hill slopes, the coalition focuses on expansion and development issues in the already urbanized zones and compiles an annual environmental policies report card. The coalition gave 35 of 51 resorts a C or below in 2000. Among the resorts graded B by the independent non-profit coalition is the Squaw Valley, California, former winter Olympics site, which is under investigation for water pollution due to erosion from its ski runs.

Power skis

Motorized snowmobiling recreation has effects similar to those of all vehicles. Snow insulates the ground from low winter air temperatures. The snowmobiles compact the snow, crushing plants and small animals that live beneath it, either on the ground surface or in shallow burrows. After compaction, the snow blanket loses much of its insulating effect, just as wetting a down sleeping bag lowers its insulating properties. The reduced insulation lowers ground temperatures by as much as 54°F. Hibernating animals must draw more warmth from internal energy sources than their bodies may be able to supply, and when the weather brings larger than normal temperature drops, they may freeze to death.[45]

Compacting snow may lower temperatures enough to freeze any plants that can survive beneath undisturbed snow. Perennial plants with large, fleshy roots are the most likely to freeze, and as many as 40% of them disappear from snowmobile areas. The plants that survive grow less vigorously and produce fewer flowers and seeds than they do in undisturbed areas.[46] The low ground-surface temperatures in snowmobile areas also delay spring plant growth.

Ground-level plants are similarly affected in downhill ski areas—but like wheeled ORVs, snowmobiles can go farther and spread the damage more widely. In the same way that wheeled vehicles disturb desert soil crusts and other biota, snowmobile traffic destroys soil fungi beneath the snow along with the bacteria that nourish tundra ecosystems.[47] Two-stroke snowmobile engine noises annoy nonmotorized recreationists and disturb both domestic animals and wildlife.[48] The problem of polluting snowmobile emissions has been reduced in Yellowstone National Park by restricting entry to four-stroke machines, however.[49] Snowmobile riders commonly violate closed areas, including designated wilderness areas, as do wheeled ORVs.[50]

The U.S. Consumer Product Safety Commission estimated that about 110 people die each year while riding snowmobiles, and hospital emergency departments treat about 13,400 snowmobile injuries. In a familiar litany, most accidents—many fatal—are related to lack of helmets, letting underage children drive, and risky behaviors related to alcohol and other drug use. In a 1996 accident on a national forest trail north of Yellowstone National Park, expert rider Brian Musselman sustained severe injuries that permanently incapacitated him. A court ruled that the USFS had failed to adequately warn snowmobilers about unsafe trail conditions and must pay damages of $4 million to Musselman and his family. This decision makes taxpayers liable for an individual's reckless snowmobile play on public lands.

The ski industry's future sustainability is predicated on water supply. Drought conditions since 1999 have reduced mountain snow levels in many states, shrinking drinking water supplies at ski resorts (see chapter 9). Rapidly climbing global temperatures likely will only worsen today's western snow and drinking water deficits.[51] Recent climate studies predict that warm winters may become more frequent in Washington's Cascade and Olympic ranges, for instance, affecting 19 ski resorts. Warm winters are likely for the many other western areas that support ski resorts, but with less frequency.[52] Long-lasting water pollution, a legacy of metals mining in many mountain areas, especially Colorado, restricts the clean water supply and worsens the problem (see chapters 4, 10).

Climatic changes limited snowmobiling in 2003, when unusually warm temperatures thawed the snow and delayed Yellowstone National Park's winter season. Snowmobilers clamored for snow to be trucked to the park in insulated semi trucks. One frustrated rider demanded, "We pay our taxes. The federal government should use some of that money to haul in snow from Alaska or northern Canada."[53] That demand extends to providing snow for year-round snowmobiling.

Good land spoiled

Mark Twain quipped that golf is a good walk spoiled. With apologies to golf-loving friends and relatives, we believe that converting natural western lands to golf courses is a case of good land spoiled, especially in desert areas. Golf courses carved out of arid ecosystems may shelter some bird and other wildlife species, but their construction destroys and fragments wildlife habitat, grossly alters the diversity of plants, and displaces most native animals that inhabited the areas for millennia. Many, if not most, golf courses import soils for their greens and tees. This eradicates natural soils, and any rare and endangered plant species that grew in them, at the same time killing off native animal species that fed on those native plants.

The 11 western states now harbor more than 2,000 golf courses, following a 95% increase between 1991 and 1999.[54] Golf course sizes vary from about 100 acres to more than 250 acres. In 2003, the United States supported nearly 15,000 golf courses, occupying more than 1.3 million acres, about the size of the state of Delaware. More and more courses spring up near vacation areas and proxy for open space in urban subdivisions.

A golf course's 35–67% of greens, tees, fairways, and roughs must be irrigated, fertilized and treated with pesticides, herbicides, and fungicides. The irrigation consumes huge volumes of scarce water in arid regions—especially drinkable groundwater. Both the irrigation drainage and rain runoff contaminate surface and groundwater supplies with fertilizers and pest control chemicals (see chapters 2,

9). Most courses store irrigation water in large artificial lakes or ponds, with large evaporation losses in desert areas (see chapter 9). The few golf courses built on landfills did not alter natural habitat further, but they do consume water and add to groundwater pollution.

The amount of water that golf courses consume depends on their location, the season, and weather patterns. Courses in temperate eastern U.S. climates use an average of 82,000–33,000 gallons per day.[55] In arid and semiarid regions, the average water consumption of a game invented in rainy Scotland likely amounts to much more. As of 1993, the area around Phoenix, Arizona, sported 108 golf courses "that collectively use enough water to replicate a rainforest."[56] Together, Palm Springs and Palm Desert, California, advertise 300 golf courses.

Golf courses tend to consort with casinos. In New Mexico, Native American rights to Rio Grande water support the current proliferation of posh resorts and golf courses on tribal lands (see chapter 9). The tribes' claims legally override urban and irrigation water allocations along the chronically stressed Middle Rio Grande, and all these conflicting human claims radically diminish protection for fish and other aquatic life in the river, as well as streamside habitat in the Middle Rio Grande Valley (see chapter 9). Elsewhere in the southwest, Native Americans are building casino resorts and golf courses on lands that depend on nonrenewable groundwater resources. While applauding the tribes' efforts to improve their lives, we have to point out that both natural and human ecosystems will collapse when overdrafted desert water supplies eventually run out.

Golf course grass is fed standard nitrate-phosphorus-potassium fertilizer mixtures (see chapter 2). As much as 180 tons of liquid fertilizers may be used on individual courses every year. To allay environmentalists' concerns, a few oversimplified studies have examined the fate of nitrogen and phosphorus applied to golf courses,[57] but no reliable data track the environmental effects either on or beyond the courses. In particular, the studies examine neither the complex migration paths that contaminated groundwater can follow nor the difficulties of predicting those paths (see chapter 13).

Pesticide applications to golf courses are as heavy as on agricultural crops[58] (see chapter 2) and similarly contaminate groundwater.[59] Tees and putting greens get the highest pesticide applications. The pesticide treatments combine insecticides, herbicides, fungicides, and worm-killing nematicides from a large and constantly changing list. Fungicides and herbicides essentially sterilize the soils to control fungi and weeds, greatly reducing the soils' capacity to retain or modify contaminants.[60] The long-lived herbicide clopyralid, used widely for eradicating broad-leaf weeds on golf courses and in city and domestic landscaping, has showed up in commercial composts and killed garden plants that the compost was meant to nurture. Both imported and original tee and green soils also commonly receive applications of substances that help water soak into the ground. Those treatments speed the movement of pesticides and other contaminants into groundwater.[61]

Regaining Paradise

Technology and affluence give people the time and funds to spend on play. People cannot help causing environmental damages wherever they gather in large numbers, and large numbers of people are choosing activities that cause severe harm to natural landscapes. Off-road recreational vehicles have a far greater capacity to damage natural lands than any other recreational choice. The motorized recreation industry promotes this wasteful use of declining fossil fuels, but gasoline prices are rising, and by 2010 they may soar. The question is—how much more damage will the environment sustain before petroleum becomes too expensive for towing several ATVs or motorcycles out to the country and back behind a huge mobile home?

Science still cannot predict if and how desert soil and vegetation recovers from ORV use. Estimates are based on other activities that cause similar impacts, and the processes and known rates of soil formation in arid lands. Severely compacted soils at 29 of 31 abandoned military bases and mining town sites have not recovered, even after 91 years without human occupation (see chapter 6). Different methods for calculating the time for soil recovery at those relatively level sites indicate periods of 92–124 years (see chapter 5, box 5.1).[62] Recovering the vegetation and animal species diversity of a healthy ecosystem is likely to take much longer, on the order of a millennium.[63]

As hiking's popularity receded, damaged mountain trails began to recover from the hordes of 1960s and 1970s hikers. Plants trampled under horse hooves or human feet can recover much faster than vegetation crushed by ORVs, probably because feet damage fewer root systems and areas of maximum soil compaction are less widespread. Biological soil-forming and transforming processes also resume more quickly in soils that are less severely compressed than in ORV areas (see chapter 13).

As yet, motorized recreation has not slowed noticeably. Most arid lands, plants, and animals are so sensitive and require such extremely long periods to recover after motorcycle, 4WD, and ATV damage that the only effective management under the Federal Land Policy and Management Act is to restrict access. Generally level areas can be barricaded; the soil can be ripped to eliminate compaction, revegetated to obscure tire marks, and then left to revegetate naturally. The BLM sponsors a program that obscures illegal desert roads and ORV tracks with dead plant material and artfully repositioned rocks. This tactic reduces or eliminates further traffic impacts but does not mitigate the effects of compacted soils or otherwise speed recovery (see chapter 5).

Relatively few programs to reclaim, or "assist," recovery have been attempted on ORV-damaged lands (figure 11.10). Projects to reclaim closed motorcycle hillclimbs in California State Vehicular Recreation Areas generally failed because the initial work was either inappropriate or not followed through properly.[64] Before hilly land can recover, new soil must form. Where soils are completely stripped from steep slopes and bedrock exposed, the rock must be deeply weathered before biological action can convert it to soil. This series of processes probably require many millennia. Unfortunately, many western desert soils are relics of past climates (see chapter 13). Most formed during past ice ages. Neither current nor near-future climatic conditions can restore them.

Restoring native plant associations is notoriously difficult and expensive in arid lands. The success or failure of complicated revegetation projects is unknown, because studies of replanted areas do not extend over long enough times to

Figure 11.10 A heavily ORV-impacted hill slope, Red Rock Canyon State Park, California, March 1998. Although closed to vehicles on Easter 1977, accelerated erosion continues despite efforts to revegetate damaged areas.

evaluate the results. No practical or economic method exists for reversing severe soil losses.

Closing ORV areas is politically difficult, however. Written restrictions have little effect, and enforcement is spotty or worse. As an example, motorcycles, ATVs, and other ORVs are entering wilderness areas, areas of critical environmental concern, and wilderness study areas, supposedly closed to them under BLM regulations. Such incursions spread severe damages and create routes where none existed previously, preempting congressional authority to protect roadless forest areas and other lands with wilderness values (see chapter 5).[65]

The national forests are only beginning to confront motorized recreation problems. By 1999, the demand for vehicular routes in national forests was 10 times greater than in 1950. Most of the 400,000 miles of USFS roads, some built for access but most for logging, should be closed and rehabilitated to reduce erosional problems and habitat segmentation (see chapter 5)—and to reduce the huge maintenance costs that get heaped on taxpayers (see chapter 1). Even national parks face a winter onslaught of tens of thousands of snowmobiles, radically changing the ground-surface temperature, extending the period of thaw, and leaving a trail of toxic pollutants in the snow—all of which damage animals and plants and pollute water.

Coalitions of wildlife defenders and nonmotorized winter recreationists have been unable to keep snowmobiles out of national parks. With their overwhelming support, in late 2000 the outgoing Clinton administration announced a limit on individual snowmobile operation in Yellowstone and Grand Teton national parks, aiming for a complete ban by 2003.[66] The limitation and ban affected the economies of local store and motel owners at the park's portals, who joined snowmobile manufacturers' and rider organizations' clamor against the limitations. Unfortunately, the George W. Bush administration reversed the ban nearly two years later.

All natural areas will be prey to vehicular destruction if the Department of the Interior's proposal to allow ORVs in wilderness areas stands. The George W. Bush administration is trying to remove the public's right to challenge such decisions.[67] A 2004 U.S. Supreme Court decision blocked an environmentalist bid to make the BLM protect wilderness study areas and other sensitive public lands from ORV damage, under the Department of Interior's wilderness study area nonimpairment regulations. The Court ruled unanimously that department can be sued only over affirmative actions and not for actions that it has not taken.[68]

Even the nonmotorized recreational pursuits have only limited compatibility and demand separate areas, increasing competition for public land. In effect, recreational lands are being sacrificed, with no pretense at protecting or sustaining their natural resources—including recreational uses—for future human or wildlife generations.

It is time to question if destructive recreational pursuits should invade natural ecosystems. Americans must also question whether the western states really can provide an endless supply of forest and desert to rip up, and of clean water to use up and contaminate. Can we pursue our every whim for "fun" and still provide quality recreation and natural solitude for future generations? The time to decide is now, before the few remaining areas of pristine beauty and solitude, which also provide us with the natural capital of clean air and water, disappear forever.

12   Driving to the End of America's Birthright

In the face of the basic fact that fossil fuel reserves are finite, the exact length of time these reserves will last is important....[T]he longer they last, the more time do we have, to invent ways of living off renewable or substitute energy sources and to adjust our economy to the vast changes which we can expect from such a shift.
Admiral Hyman Rickover, "Energy Resources and Our Future" (1957)

In a 1957 speech, the legendary and controversial scientist and submariner Admiral Hyman G. Rickover noted, "Our civilization rests upon a technological base which requires enormous quantities of fossil fuels."[1] Rickover understood that the United States was producing and using as much oil in the 1950s as we had in all our previous history and worried, "What assurance do we...have that our energy needs will continue to be supplied by fossil fuels: The answer is—in the long run—none." Rickover also warned that failing to conserve our oil wealth could leave us destitute. The United States doubled oil production and consumption again in the 1960s, and again in the 1970s—ignoring Rickover's appeal "to think soberly about our responsibilities to our descendents—those who will ring out the Fossil Fuel Age."

Unrestrained fossil fuel consumption has propelled the United States to a level of affluence previously unknown in human history. Fossil fuels, petroleum (oil and natural gas), and coal, represent the "stored sunlight of millions and millions of years deposited in an energy bank account in the Earth by geological processes." Since the early twentieth century, the whole world has been using up this inheritance "in a geological instant."[2] Cars and other transportation consume the major proportion of the world's oil, but petroleum also is the raw material for a wide variety of industrial products, fabrics, and medicines (see appendix 3). Without it, every facet of modern life would be less convenient, less comfortable, and far less mobile.

Massive energy consumption has addicted Americans to cheap fossil fuels. Energy addicts overheat the house and wear T-shirts all winter, tend to own two or more refrigerators, and maintain a vehicular fleet. Many believe that having and driving cars is a more important American liberty than voting (see chapter 5). Along with U.S. Senator Trent Lott, they feel that "the American people have a right to drive a great big road-hog SUV if they want to."[3]

Exercising our "right to drive" has a high price. By the 1970s, when petroleum prices first rose sharply, Americans knew that smokestacks belch unhealthy soot and

308

that smog comes out of tailpipes. Summer heat waves and killer storms finally are forcing us to understand that global fossil fuel burning is heating up the oceans and atmosphere much faster than natural forces could do. Far less well known is the long-lasting water and soil pollution in the 11 western states from mining, extracting, transporting, and manufacturing or refining coal and petroleum-based resources.

The 50 years since Rickover's speech covers most of the "long run" that he mentioned. Those descendants who will ring out the Fossil Fuel Age—perhaps more appropriately called the "Petroleum Interval"[4]—are living today. They probably include us and definitely include our children and grandchildren. Now, rising and volatile oil and natural gas prices face us with the addict's brutal quandary— can we wean ourselves from a terrible dependency before it kills us?

Optimists seek a methadone solution: "new" ways to feed our addiction without the bad effects. More realistically, cleaner but nonrenewable natural gas and highly polluting "renewable" energy sources might combine to bridge the gap between petroleum and some still unknown, hopefully cheap, replacement.[5] But finding the replacement will be difficult because no other known energy source combines oil's high per-volume energy content with convenient mobility and versatility as a raw material. And we probably have less than 50 years to accomplish the transition. Americans must recognize that our "birthright" demands a level of energy production so gigantic that renewable sources probably cannot meet it. And trying to maintain today's high level of energy consumption certainly will inflict severe environmental damage across the western states.

The Sunlight Store

Natural processes in the remote geologic past concentrated decayed plant and animal remains into petroleum pools or coal layers. Both are complex melds of carbon and hydrogen-based hydrocarbon (organic) compounds, the basis of all life. Oil formed from the remains of microscopic sea life, predominantly huge algae blooms that settle into ocean-floor sediments. Coal is concentrated land-based wood and leaves from ancient swamp-forest plants, deposited in sinking sedimentary basins. Rapid burial and gentle heating converted both oceanic and continental sediments into rocks, distilling the plant and animal matter into petroleum or coal, and releasing water and natural hydrocarbon gases (see chapter 13).[6]

The hydrocarbon gases are mostly methane (CH_4) with smaller amounts of heavier propane (ethane, butane), plus carbon dioxide (CO_2), hydrogen sulfide (H_2S), nitrogen (N), and helium (He). In deeper zones, high heat and pressure keep gases dissolved in petroleum, but at shallower depths the gases float on oil pools. Deeply buried oil may be hot enough to convert wholly to methane, which tends to disperse into and through adjacent rock masses.[7] Similarly, high natural pressures and temperatures yield hard anthracite coal, while the much softer and less pure bituminous coals form under lower heat and pressure. As coal hardens, methane ends up trapped in seams and pockets throughout the layers.

Gasoline is America's drug of choice, generally refined from formerly abundant "conventional oil" stocks (preferably "light, sweet crude" petroleum).[8] Within most of Earth's normal surface temperatures, conventional petroleum is a liquid, easy to transport and store. Once it also was very easy to recover—as soon as a drill hole penetrated a "conventional" oil pool, black gold rushed to the surface, driven by underground pressure or the dissolved gas, or both. Some dry but porous buried rock masses also produced conventional gas.

Most petroleum originated in two episodes, at about 90 and 145 million years ago, both intervals of highly concentrated greenhouse gases in the atmosphere and extremely warm global climates.[9] The thickest and hardest coal beds also formed in a greenhouse climate more than 300 million years ago, when all Earth's mobile continents formed a belt straddling the equator. Smaller and less pure coal beds and oil–gas pools developed more recently. Oil and coal may be developing from biomass even today—but the rates of formation are puny compared to our consumption. In truth, even if all the plants and animals living today went into making oil and coal, the result would not equal even one year of our consumption. During 1997 alone, the world burned up 400 times more organic energy in the form of fossil fuels than exists in all the land and sea biomass that Earth produces in a year. *This means that every year humans use up four centuries worth of ancient plants and animals.*[10]

From sunlight to pollution

Fossil fuels pollute air, land, and water at every stage of production, processing, transportation, and use. Refined gasoline is fortified with additional poisonous compounds (see box 12.1), so gasoline leaks and spills can highly contaminate soils, and surface and underground drinking water supplies. Leaking fuel storage tanks are commonplace throughout the United States. Any fuel made from plant or animal matter—including fossil fuels, wood, and biomass fuels and gasoline additives—emits toxic hydrocarbon compounds when burned, along with minor amounts of sulfur, nitrogen, and numerous other impurities, including radioactive substances such as uranium and thorium.

Emissions from fossil fuel burning include CO_2, which animals cannot breathe, as well as poisonous carbon monoxide (CO), nitrogen and sulfur oxides, and tiny soot particles (particulates) with attached impurities. Nitrogen gases interact with sunlight to create the ozone (O_3) in smog. Smog can reduce the quantity and quality of carbon available to soil, lessening forest and agricultural productivity.[11] Poorly ventilated underground mines expose coal miners to a cancer risk from breathing radon gas, a radioactive decay product of uranium. Polluting coal power plant emissions can be reduced, but in 1999 coal plants produced slightly more acidic sulfur and nitrogen oxide emissions per billion kilowatt-hours of electricity generated than did oil-fired plants, and much more than gas-fired power plants (table 12.1).

A single coal-fired power plant also produces more than a million tons of ash per year, made of tiny particles (particulates) that spread in dry air as well as in rainfall (see chapter 4). These particulates carry more than 10 tons of toxic arsenic, chromium, and lead and more than one ton of chromium and vanadium. When they fall to the ground, surface water readily draws (leaches) toxic substances out of the fine ash, and many enter food chains. Burning coal mined in the western

Table 12.1 1999 Electricity Generation and Emissions from Fossil Fuels

Energy Equivalents	Fossil Fuels		
	Coal	Oil	Gas
Net electricity generation[a]	1,965	109	596
Fossil fuel consumed[b]	952 million tons	196 million barrels	5,680 billion ft^3
Emissions[c] (tons per billion kilowatt-hours)			
Sulfur dioxide	6,034	6,028	20
Nitrogen oxides	2,511	1,853	646
Carbon monoxide	122	165	158
Total emissions[d] (thousands of tons)			
Sulfur dioxide	11,856	657	12
Nitrogen oxides	4,935	202	385
Carbon monoxide	239	18	94

[a]U.S. Energy Information Administration. *Annual Energy Review 2000* (2000), table 8.2. Units in billions of kilowatt-hours.
[b]U.S. Energy Information Administration. *Annual Energy Review 2000* (2000), table 8.8.
[c]U.S. Environmental Protection Agency. *National Pollutant Emission Estimates for 1999* (2001). Available: www.epa.gov/ttn/chief/trends/.
[d]U.S. Energy Information Administration. *Annual Energy Review 2000* (2000), table 8.2.

United States also puts about 51 tons of mercury vapor per year into the atmosphere—greater than any natural or other human source—plus selenium vapor and radioactive uranium gases.

Mercury and other toxic materials in coal-derived particulates both bioaccumulate, developing higher toxin concentrations in organisms than the surrounding environment, and biomagnify, progressively concentrating the toxins in animals higher up the food chain. Power plant scrubbers capture substantial amounts of toxic trace elements, but the scrubber materials then become another source of hazardous wastes.[12]

The sulfur in coal emissions is the major source of acid rain. Sulfur, nitrogen, and CO$_2$ gases can move thousands of miles, accumulating in the atmosphere before reacting with water vapor or rainwater to create sulfuric, nitric, and carbonic acid. Rainstorms flush the acids to ground, suddenly subjecting aquatic plants and animals to "acid shock." The same process can acidify snowmelt water. Acid waters leach nutrients out of soils, often poisoning vegetation and microscopic soil organisms. As long as we fill the air with fossil fuel combustion gases, acid rain and acid shock will degrade the environment.

Although the U.S. Geological Survey (USGS) has claimed that "modern coal-burning utilities do not cause [human] health problems,"[13] people in heavily polluted metropolitan areas have a 16% greater risk of dying from lung cancer than do rural populations. Toxic chemicals in the air give Americans a cancer risk at least 10 times higher than the U.S. Environmental Protection Agency's acceptable level, and 12 million Americans have cancer risks 100 times higher.[14] University of Southern California researchers have found that urban smog and particulates demonstrably worsen heart disease and contribute to the incidence of life-threatening childhood asthma, a disease reaching epidemic proportions.

Burning either fossil fuels or renewable biofuels contributes climate-warming greenhouse gases to the atmosphere. The immense volumes of greenhouse gases released by cars, industries, and power plants—especially coal-fired power plants—are mostly responsible for increasing global temperatures far above natural levels. Accelerated climate warming is far from coincidental—our addiction to fossil fuels is returning the Earth to greenhouse climates. such as the long past eras that favored petroleum and coal formation. Already the warming has melted Alaskan permafrost, and polar ice caps are rapidly disappearing. Rising sea levels are threatening low-lying coastal and island areas, and some whole communities are being forced to move. Americans spend vast sums attempting to clean up fossil fuel contamination and will pay even more in trying to adapt to a rapidly warming climate.

Bonanza's Past

In his 1957 speech, Admiral Rickover predicted that "total fossil fuel reserves recoverable at not over twice today's unit cost, are likely to run out at some time between the years 2000 and 2050, if present standards of living and population growth rates are taken into account." Confirming Rickover's prescience, by 2004, growing world populations had used up nearly half of the Earth's retrievable petroleum stores in about 150 years. Increasing numbers of oil geologists and economists now predict an end to the brief Petroleum Interval no later than 2050, and probably much sooner. The petroleum will not actually *run out* but will go into decline, sounding the death knell for our consumptive lifestyles.

The era of U.S. petroleum dominance passed three decades ago. New discoveries of oil fields on U.S. lands peaked out in 1930 and have diminished ever since (figure 12.1). From the peak-to-decline patterns of individual oil fields, in 1956 Shell Oil Company geologist M. King Hubbert predicted that U.S. petroleum production would peak in 1970, marking consumption of about half the nation's substantial oil endowment.[15] Hubbert was only one year off—U.S. petroleum production actually peaked in 1971 and now is well over the hill (figure 12.2).[16] Additional finds have had little effect. Even Prudhoe Bay oil flowing through the Alaskan pipeline merely slowed the rate of decrease.[17] Similarly, domestic natural gas discoveries peaked in the mid-1950s, and the production peak followed in 1971.[18]

While U.S. oil production plunged, Americans drove *more*, pushing gasoline consumption ever higher. Admiral Rickover advised President Jimmy Carter of the tremendous gap between new U.S. oil and gas discoveries and growing

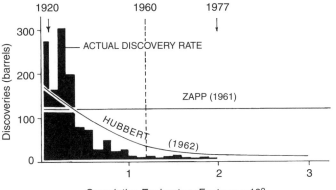

Figure 12.1 United States oil discoveries, 1910 to 1986. Total discoveries peaked in the early 1930s. The bars show actual discovery rates (in billions of barrels of oil), plotted against the total length (in feet) of all exploratory wells. Sloping line shows M. King Hubbert's 1962 prediction of declining future discoveries even as exploration increased. The flat line is the discovery rate predicted in a 1961 U.S. Geological Survey report (the Zapp report), which envisioned that increased exploration would yield new finds of more than 100 billion barrels of oil per year forever. Hubbert data and curve from L. F. Ivanhoe. Updated Hubbert Curves Analyze World Oil Supply. *World Oil* 1996;91; Zapp curve from H. W. Menard. Welcoming Remarks: Interdependence of Nations and the Influence of Resources Estimates on Government Policy. In F. C. Whitmore, Jr., and M. E. Williams, eds. *Resources for the Twenty-First Century, Proceedings of the International Centennial Symposium of the United States Geological Survey, Reston, Virginia, October 14–19, 1979* (U.S. Geological Survey Professional Paper 1193, 1982), 3–10.

consumption and Carter sounded the warning, naming our energy deficit "the moral equivalent of war." But American energy addicts voted instead for Ronald Reagan's impossible dream that more exploration and drilling would bring a future of energy independence.

In the 1990s, an average American used four tons of oil, more than two tons of natural gas, and 2.5 tons of coal each year, the energy equivalent of three gallons of oil per person per day. By 2000, Americans represented less than 5% of the world's population but consumed about one-third of its annual energy supplies. In 2006, each U.S. resident consumed his or her own weight in oil every seven days. More than 60% of that oil was, and is, imported. Reliable current estimates put proved U.S. oil supplies at only 2% of the world's petroleum reserves.

In 1973, domestic gas production peaked at 21.4 trillion cubic feet (Tcf). Currently the U.S. consumes about 23 Tcf of natural gas per year, and consumption is increasing by 2% per year. By 2003, "proven" U.S. gas reserves represented about 3% of world reserves, suggesting an eight-year supply. In 1987 Congress

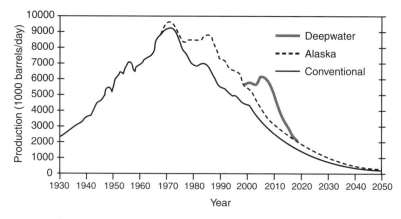

Figure 12.2 Oil depletion profile for the continental United States as well as Prudhoe Bay, Alaska, and deep-water areas currently being explored. Production in the lower 48 states peaked in 1970 at 9.4 million barrels per day, as M. K. Hubbert had predicted in 1956. The supergiant Alaskan field forms a secondary peak but did not prevent continued decline. Deep-water discoveries may yield another small satellite peak at lower production levels but cannot prevent further decline. Hubbert curve from M. K. Hubbert. *Nuclear Energy and Fossil Fuels.* Paper presented at the American Petroleum Institute, Proceedings of Spring Meeting, San Antonio (1956), 7–25; other data from C. J. Campbell. The Assessment and Importance of Oil Depletion. In K. Aleklett and C. J. Campbell, eds. *Proceedings of the First International Workshop on Oil Depletion* (Uppsala, Sweden, 2002), figure 8. Available: www.isv.uu.se.

repealed the Powerplant and Industrial Fuel Use Act that had prohibited using gas to generate electricity, helping increase demand by nearly 80%. Between 1995 and 2005, that increase drove a fourfold rise in gas prices.[19] Additional growth could reduce the domestic gas supply to many fewer than eight years.

Energy fantasies and realities

Predicting how much oil and gas can be produced domestically and globally, and for how long (reserves), can be perilous and politically slanted—even in the United States. Industry and government reports can be confusing, and sometimes purposely mix estimates for "technically recoverable" (theoretically producible with current technology) with "economically recoverable" amounts (producible, depending on a lot of factors). And reserve estimates may not be comparable between nations—production data are state secrets in some countries. The most credible reserve predictions are based on past production from all oil fields, reports of newly discovered fields, and guesses of how much undiscovered resource still is out there (see appendix 9). But newly discovered fields have poorly known potentials until actually drilled, and the guesses may be more hopeful than educated. According to one petroleum consultant, current and historical production levels are facts, reserves are opinion, and undiscovered resources are fantasy.[20]

In the case of undiscovered conventional oil and gas resources, usually reliable USGS reports attempt to quantify fantasies. Couching assessments in gambler's terms, the USGS assigns each oil field or region both a long-shot (5%) chance of eventually discovering and producing a large amount of oil and a high (95%) chance of discovering a minimum amount. But both the agency and news media typically report only *the average* of those high- and low-probability estimates, arbitrarily assigning them a 50% chance of discovery. The 95% estimates tend to be closest to actual production levels. Through 1999, USGS estimates of U.S. reserves varied from 200 to 220 billion barrels, but after 2000 they became wildly optimistic compared to most other predictions. That optimism is predicated on "reserve growth"—the assumption that future technological improvements will extract a higher proportion of oil and gas than now is possible (see appendix 9).

The truth is that drilling our lands cannot take us back to the halcyon days of cheap oil and gas. Finding new resources is energy intensive and the search has brought diminishing returns. New conventional oil fields generally are small or located in politically or physically hazardous and climatically extreme regions. They cost more than ever to produce, driving up fuel prices. For example, 2006 reports on new oil and gas discoveries in the Gulf of Mexico came with uncharacteristic industry warnings about the difficulty of producing oil from 29,000 feet—nearly as deep as six stacked Grand Canyons—below the hurricane-haunted Gulf. If the finds can be produced from this technically challenging area, they could increase U.S. oil reserves by half. Even so, raising U.S. reserves to 3% of the world's supply is no harbinger of energy independence.[21]

Only extremely large new discoveries of conventional oil and gas could raise reserves significantly and lower energy prices. The last big find dates to the 1980s. The rest of American fossil fuel resources are mostly "unconventional" and even more difficult and energy-intensive to extract and process than the conventional oil remaining in older fields. Although supposedly vast, the actual abundance and distribution of many unconventional deposits are relatively unknown.

Imports from other parts of the world may not supply American addicts for much longer, either. New oil field discoveries peaked for the rest of the world in about 1964, on average, and then declined (see appendix 9). World oil production began leveling off in 2004.[22] The trend seemed to affect even Saudi Arabia's most productive fields, the world's largest known oil reserves, in spite of the increasing numbers of new wells. Many observers expect world oil production to peak by 2010 and then decline over the following decades. When that happens, oil exporters will not be able to meet rising world demand.[23]

All told, the United States has produced nearly as much natural gas as all the rest of the world combined,[24] but much of it was wasted. For many decades, U.S. oilmen looked at the natural gas coming from their wells as a dangerous byproduct and burned it at wellheads or refineries, a procedure called flaring (figure 12.3). Flaring and venting still waste about 100 billion cubic feet of domestic gas each year, currently a sufficient supply for about 1.5 million households.[25] The United States now imports more than 19% of its annual domestic natural gas consumption, representing 24% of global production. World gas discoveries peaked in 1971,

Figure 12.3 Flaring (burning) of natural gas (methane) from an oil well near Maricopa, California. Photographed February 1980.

and we can expect the production peak around 2030.[26] Declining worldwide natural gas supplies and rising consumption will limit U.S. ability to sustain, let alone increase, dependence on foreign gas.

Unconventional Methadone

Oil and surrogate industries are developing alternative but nonrenewable fossil fuels to aid the transition from a total dependence on cheap conventional oil and natural gas to a more diversified energy future. These alternatives include "unconventional" petroleum and petroleum-like substances, and natural gas resources, in addition to coal products. "Unconventional" oil and gas (also called "continuous") hydrocarbon deposits are broadly disseminated throughout dense rock units or coal beds. Oil drops and films are tightly attached to sedimentary particles in these extensive deposits. Most unconventional oil sources are so viscous, or the rock so dense, or the binding forces so great, that the oil cannot flow toward a drill hole.

Unconventional natural gas (methane) deposits are disseminated through coal beds, other low-permeability rocks, or frozen watery sediments. U.S. natural gas resources are principally "coal-bed methane" and gas from "tight" (relatively nonporous) sedimentary rocks. Neither the USGS nor leading gas industry journals estimate recoverable gas from what some think are vast methane hydrate deposits— icy mixtures of methane and water, both in on-land permafrost and in undersea sediments at depths between about 1,000 and 3,500 feet.[27]

Like pouring oil from a bottle or squeezing toothpaste from a tube, extracting every bit of a free-flowing resource from the ground is impossible, but even less can be gotten out of unconventional sources. Also, developing these mostly potential resources is highly energy consuming and environmentally damaging. Burning them adds to air pollution while contributing vast amounts of greenhouse gases to the atmosphere.

Tarry sands and oily shales

The USGS did not include estimates of unconventional oil (or natural gas) in its U.S. resource assessments until 1995 because of the difficulties in developing them. Even then, the unconventional oil estimate amounted to only 1.5–2.7 billion barrels.[28] None of the estimates, nor the probabilities, is backed with credible information. Even the maximum figure represents less than half a year's supply at current U.S. petroleum consumption of 21 million barrels per day.

Other unconventional oil deposits include "tar sands" (oil sands)[29] and "oil shale." Tar sands are sandstone, siltstone, and claystone containing water-coated sediment particles surrounded by films of tarry oil. "Oil shale" really is a clay-rich limestone containing "kerogen"—waxy hydrocarbon solids that form in early petroleum development. Oil shales have been burned like coal since ancient times, and kerogen has been manufactured into petroleum for more than a century. Some parts of the world with little or no conventional oil or coal still process kerogen from oil shales on a small scale. Estimates put the amount of potential fuels in the world's shale oil deposits at trillions of barrels to 2.1 quadrillion barrels.[30]

The United States has small tar sand deposits, probably not exploitable until oil prices climb much higher. In contrast, the United States is well endowed with oil shales, in the five trillion barrel range.[31] Industry sources project that combined U.S. tar sand and oil shale deposits could supply oil for 100–1,000 years, but commercial production may never be high enough to support U.S. consumption.[32] Neither is a realistic major energy source until multiple problems can be overcome.

The world's largest tar sand resources are in nearby Canada and Venezuela. Simply by adding tar sands to their estimated oil reserves, in 2003 Canada magically increased its oil reserve total by 3,600%, vaulting to second place in world oil reserves (behind Saudi Arabia). But in 2004, Canadian tar sands produced only 1.1 million barrels per day, and by 2030 the industry projects five million barrels per day—just 24% of *current* U.S. consumption.[33]

Near-surface oil shale deposits are mined in open pits (see chapter 4), but deeper deposits may be dug from underground shafts, like coal. To turn solid kerogen into a liquid suitable for making gasoline, the mined rock must be pulverized and heated at the surface in special extraction vessels (retorts) at about 700°F for a long time, or 900–1,000°F for shorter times.[34] Hydrogen has to be added before refining the fuel and adding other chemical treatments. The processing also requires three to four barrels of water for each barrel of oil produced, which severely limits the feasibility of producing shale oil from U.S. arid lands.

Canada uses natural gas for adding hydrogen to tar sand fuel, which is rapidly depleting their gas fields. The gas will give out long before the tar sand deposits, and Canada will have to find other hydrogen sources to continue mining tar sands. Most of the known hydrogen-generating methods use up a lot of energy (see appendix 10).[35] Pumping the required water also consumes large amounts of energy. But using natural gas to produce tar sand oil is likened to changing gold into lead. Some Canadians are beginning to ask whether the gas is more valuable for domestic use than for tar sand oil production.

In-place oil shale extraction ("retorting") schemes are experimental and unproven. Current plans call for electrically heating 1,000-feet-thick columns of underground rock to 700°F for three to four *years*. The column first must be dewatered to prevent migrating groundwater from draining the heat away. This involves freezing groundwater at the rock column margins and then drilling wells and pumping it out of the central zone. Shell Oil Company reportedly has tried this method successfully for a very small experimental area in Utah.[36] Scaling it up to produce 100,000 barrels a day (0.4% of current U.S. consumption) will demand 1.2 gigawatts of electricity from a coal-fired power plant burning 14,000 tons of coal per day. Producing a million barrels of shale oil per day (a bit more than 4% of current consumption) would require 10 new gigawatt power plants and five new coal mines.[37]

Tight gas

A USGS report has stated that "increased use of energy gas…will be the solution we need to fulfill our industrial needs and maintain our standard of living…."[38] To reach this conclusion, USGS added an unsupported average chance of finding 617 Tcf of conventional and unconventional gas on land, plus 322 Tcf of conventional gas "reserve growth," to the 135 Tcf of proved reserves.[39] This theoretical reserve total represents 38 years' supply *at current gas consumption rates*, but 28 years at the U.S. current growth of 2% per year in gas use.[40] Even if industry finds that much additional methane, a large proportion may not be technically recoverable. To reach markets, it will have to be moved long distances across regions with no existing pipelines or pumping infrastructure. The Energy Information Administration estimates that the number of producing U.S. conventional gas wells increased 22% (to 385,000) from 1995 through 2004. But slow production from unconventional fields pushed average gas productivity 20% lower, so total production increased only 2.6%.[41] Slow production lowers gas depletion rates but cannot meet increasing demand.

Coal-bed methane production is expanding rapidly across nearly 200,000 square miles in Montana, Wyoming, Utah, Colorado, and New Mexico. The expected total U.S. production is somewhat less than two years' worth of current demand and probably will yield less than a year and a half, with predicted consumption

increases.[42] Gas may be more difficult and expensive to obtain from all other disseminated deposits, however. The USGS's 1995 report called for more than 1.5
million new wells[43] to extract "technically recoverable" natural gas across at least
11 million acres of disseminated deposits.[44]

The potential amount of gas hydrates under U.S. control is not certain. Actual
data are so poor that potential worldwide gas hydrate estimates range from 100,000
to 280 million Tcf.[45] Gas hydrate exploitation has not been economic to date, and
scientists still don't know whether most hydrates are widely dispersed or sufficiently
concentrated for commercial production.[46]

Extracting gas from watery sandstones and gravel beds may be economic,
however. A recent test in northern Canada did start natural gas flowing from
deep-water gravel layers, but the test well's production abruptly declined from
unknown causes. In contrast, geological hazards limit off-shore gas hydrate production. The deposits mostly are frozen into unstable, semisolid clay and claystone
deposits in settings prone to submarine landslides. Proposed recovery techniques
include injecting the sediments with steam, or fracturing them to separate the
gas. Both approaches promote slumping and landsliding in the unstable materials,
which could destroy or destabilize very costly deep-water drilling platforms and
pipelines.[47]

Oil or gas production from rocks deep beneath hydrate-bearing sediments could
incite similar hazards. The relatively hot oil flowing in pipes could trigger slumping in cold, watery overlying sediments, generating undersea equivalents of debris
flows (see chapter 13).[48] Land slumps and flows also could destabilize drill pads
built on permafrost.

Diminishing returns

The energy costs of finding additional conventional gas and oil deposits, and for
mining or retorting and processing unconventional resources, are complicated
and largely unexamined issues. Simply put, it takes energy to get energy. Few
reports examine *net energy* yield—the energy in produced fuels compared to
the energy used to get them (usually expressed in calories or British thermal
units, Btu, or a ratio called energy returned on energy invested [EROI]).

As a conventional oil or gas field depletes, more energy must be invested
to extract each barrel of oil or cubic foot of gas. The net energy production
declines steadily as fluid oil is pumped out and gas pressure drops. To maintain
pressure in oil reservoirs and enhance extraction, natural gas may be captured
and reinjected into wells. Groundwater seeps into the reservoir as the oil comes
out, continuously raising the proportion of extracted wastewater to as much
as 98%. Pumping the heavier wastewater uses more energy than pumping oil
alone. To extract stiffer residual oil and tars, some operations burn natural gas
to make and inject steam, which improves recovery by making the residual oil
flow better.

In the 1930s, oil wells were relatively shallow and could use the energy in only
one barrel of oil to produce 100 barrels. By the early 1970s, the U.S. ratio of oil
extracted to the energy used[49] began to decline. Today it lies somewhere between

11 to 1 and 17 to 1, and it likely will reach 1 to 1 in the next few decades. Clearly, mining and processing unconventional gas and oil resources have higher energy costs than does pumping conventional petroleum (table 12.2), but few reports examine whether the net energy yield is positive or negative.

Another critical issue is the varied energy equivalence of unconventional fossil fuel sources compared with what we are used to. For example, 5,600 cubic feet of natural gas contains the same amount of energy as a single barrel of oil. This means that the USGS's estimated total reserve of proved and undiscovered gas, including reserve growth, translates to only 192 billion barrels of oil. So even if we could substitute natural gas for all current U.S. petroleum consumption (7.3 billion

Table 12.2 Estimated Energy Returned on Investment (EROI)[a]

	EROI	Range
Fuel		
Crude oil (2000)	20	
Coal (2000)	80	
Gasoline	12	9–15
Corn ethanol	0.8	0.8–1.3
Oil shale[b]	6	9–15
Oil sand	2	
Coal liquefaction	4	<1–9
Electric Power Generators		
Nuclear	6	2–10
Coal	8	5–11
Hydroelectric	14	6–17
Geothermal	8	2–13
Wind (operational)	18	5–40
Solar thermal	4	1–7
Solar photovoltaic	8	1–15

Data for most fuel systems from C. J. Cleveland. *Energy Quality, Net Energy and the Coming Energy Transition.* Presented at the Association for Study of Peak Oil-USA conference, Boston, 25–27 October 2006; oil sand data from J. D. Hughes. *Unconventional Oil—Canada's Oil Sands and Their Role in the Global Context: Panacea or Pipe Dream?* Presented at the Association for Study of Peak Oil-USA conference, Boston, Massachusetts, 25–27 October 2006. Available: www.aspo-usa.com/. Data for electric power generators from Cutler Cleveland and Ida Kubiszewski. Energy Return on Investment (EROI) for Wind Energy. In Peter Saundry, ed. *Encyclopedia of Earth* (Washington, DC: Environmental Information Coalition, National Council for Science and the Environment, 2006).

[a]Values change over time as oil, coal, and natural gas become more difficult to find and produce and as quality diminishes. Oil EROI in 2006 was 11–15 but is projected to reach 1 in two decades or so.
[b]Oil shale data likely are overly optimistic.

barrels per year), the speculative resource provides just 26 years of supply. Factoring in the 1.5–2% annual growth in oil demand shrinks the estimated lifetime of our total gas resource to about 23 years (table 12.3).

Two tons of rock have to be mined to produce one barrel of oil from tar sand, and even more for oil shale. After mining, the rock may be heated to extract the tarry oil. But most require in-place extraction, using injected steam to heat underground masses for a week or so. The liquid flows to wells that pump it to the surface. Steam injection for this retorting process must repeat every month or two. Again, hydrogen is added, either from superheated steam or natural gas, to make the product refinable. The whole process is about equivalent to burning one to two barrels of oil for every three generated.

Oil shale contains only a sixth the energy of coal, pound for pound—comparable to the energy in a pound of baked potatoes (figure 12.4). All the energy costs for kerogen production—the mining, transport, crushing, heating, hydrogen, safe disposal of huge wastes, and other energy inputs—may yield negative net energy. We can only conclude that future production from oil shale deposits cannot materially help the United States continue consuming so much oil, let alone increase consumption.

Table 12.3 Lifetimes of Nonrenewable Resources, Related to Growth in Consumption

Growth Rate (%)	Lifetime (Years)[a]						
0	10	30	100	300	1,000	3,000	10,000
1	9.5	26	69	139	240	343	462
2	9.1	24	55	97	152	206	265
3	8.7	21	46	77	115	150	190
4	8.4	20	40	64	93	120	150
5	8.1	18	36	56	79	100	124
6	7.8	17	32	49	69	87	107
7	7.6	16	30	44	61	77	94
8	7.3	15	28	40	55	69	84
9	7.1	15	26	37	50	62	76
10	6.9	14	24	34	46	57	69

A. A. Bartlett. *Reflections on the 20th Anniversary of the Publication of the Paper, Forgotten Fundamentals of the Energy Crisis* (University of Nebraska, Physics Teachers' CD-ROM Tool Kit, 1998), 2.

[a]Lifetimes calculated from the equation: EET = $(1/k) \ln (kR/r_0 + 1)$, where expiration time, k = percentage rate of growth in consumption, and R/r_0 is the number of years the quantity R of the resource would last at the present rate of consumption, r_0.

How to use the table:
1. If a resource lasts 300 years at the present rate of consumption (0% growth), it will last 49 years at 6% growth per year.
2. If a resource lasts 18 years at 5% annual consumption growth, it will last 30 years with no growth.
3. If a resource lasts 55 years at 8% annual consumption growth, it will last 115 years at 3% annual growth.

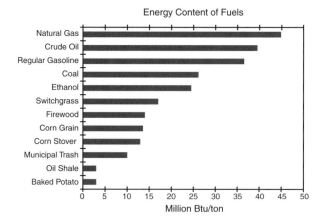

Figure 12.4 Energy contents of current fuels and other substances. The energy content of corn ethanol in this diagram is lowered by the amount of energy required to dry the corn. Corn stover is a term for corn stalks. Modified from Randy Udall and Steve Andrews. *The Illusive Bonanza: Oil Shale in Colorado* (Association for Study of Peak Oil-USA, 2005). Available: www.aspo-usa.com. Corn ethanol values are based on Professor Dennis Buffington, personal communication, 2 March 2006; switchgrass energy content is from S. B. McLaughlin et al. *Evaluating Physical, Chemical, and Energetic Properties of Perennial Grasses as Biofuels.* In Proceedings Bioenergy '96, Seventh National Bioenergy Conference, September 15–20, 1996, Nashville, Tennessee.

The Coal Train

The United States still generates much more electricity from coal-fired power plants than from oil- or natural gas-fired plants, in spite of its contributions to air pollution and global warming effects.[50] We have used a smaller portion of the nation's large coal reserves compared to oil and natural gas. *Assuming no change in demand*, the United States and particularly the western United States still has more than 275 billion tons of recoverable coal, a quarter of the world endowment. Supposedly, this is a 250-year supply. Only four decades ago, the U.S. coal supply was expected to last 500 years, however. The power of growing usage has halved America's coal resource in just 40 years.

From 1990 through 2000, U.S. coal use increased an average of 1.8% per year. In 2005 we burned a quarter of the coal consumed worldwide. Continuing that growth rate while also burning petroleum shrinks the putative supply to 95 years (table 12.3). Energy analysts posit that coal can replace declining petroleum supplies, even though it is poorly adapted to residential applications and does not provide the raw material for most petroleum-based products. Coal is more expensive and difficult to transport than is oil or gas, however, and must be transmuted into gas or liquid fuel for running vehicles. Even supposing it were possible to substitute coal for all U.S. fossil fuel consumption, the "250 year supply" reduces to a mere 70 years. Continuing the past decade's average annual 2% increase in total fossil energy consumption collapses the coal supply to 45 years (table 12.3).[51]

"Clean" coal gasification technology, used in several plants around the world, is widely believed to be a major stopgap for declining natural gas supplies that will lower CO_2 emissions. But the technology offers lower emissions than pulverized coal processes only when the CO_2 can be disposed of (sequestered). Turning coal into gas requires burning energy in some form to build pressure and produce a vapor, and then filtering out CO_2 and smog-causing pollutants. The result is a fuel with less than a third of the heat in an equal volume of natural gas. In fact, the processing uses 1.54 calories of natural gas to yield 1 calorie of liquefied coal fuel, and today liquefied coal is much pricier than natural gas.[52]

Burning coal-derived gas releases still more CO_2, while the filtered-out CO_2 and other polluting gases have to be sequestered somehow. The feasibility of sequestering CO_2 remains unproven, but even if feasible it will require building pipelines and roads, drilling wells, and pumping the gas—further reducing the net energy of gasified-coal fuels.[53]

Meanwhile, the highest quality coal deposits are largely used up. As in the declining oil and gas fields, mining coal from ever lower grade deposits yields lower and lower net energy per ton. By the late 1970s, the energy input for obtaining coal had increased from a fraction of a percent of the coal's energy content to 5%. Coal production could yield negative net energy by 2040.[54]

All Stages of Contamination

All stages of fossil fuel production can be destructive and contaminating (box 12.1). Proving and exploiting oil and natural gas deposits require drilling, both to discover the extent and to tap deposits. Drilling for conventional oil and gas involves pumping of drilling muds, cements, and lighter drilling fluids into the drill holes—a messy process. The largest aggregate oil spill in U.S. history is not the *Exxon Valdez*, but 20 million gallons of oil and chemicals that leaked, oozed, or splashed out of wells and pipelines over many years at Unocal Corporation's Guadalupe Dunes oil field, on California's magnificent southern coast.[55]

Coal beds must be drilled and then mined. Coal mining today commonly means stripping masses of surface sediment and rock from a hilltop to reach the coal layer, and then dumping the spoils into streambeds. This happens both in eastern and western U.S. coal fields.

Exploiting unconventional energy sources—especially tar sands and oil shales—has even worse environmental effects. To obtain one barrel of oil from mining tar sand requires removing up to 150 feet of barren rock, and then mining and processing the oil-rich rock, handling a minimum total of five tons of rock. Breaking up rock increases its edges and surfaces enough that not all of the wastes can be put back into the hole they came out of, so waste disposal is a significant problem. Extraction and processing release toxic heavy metals and other contaminants, and wastewaters carry large amounts of suspended sediment particles, so fine that they require decades of settling before the water can be released into streams or rivers. Canada's tar sand operations are scenes of "astounding" ruin: "As far as one can see, the landscape is dominated by giant pits, snaking pipes and refinery stacks."[56]

Box 12.1 Witches' Brew

Gasoline and engine lubricating oils are witches-brew blends of refined petroleum and 200 different poisonous "additive" substances, to variously improve combustion, enhance octane, control ignition, act as detergents, correct engine knocks, and control icing and rust. Additives include BTEX (benzene, toluene, ethyl benzene, and xylene), naphthalene, metals, solvents, and polycyclic aromatic hydrocarbons (PAHs). The Environmental Protection Agency still allows some lead additives. Methanol (methyl alcohol) is the most hazardous to handle of all the substances added to gasoline, although it is rarely used.

The most volatile additives evaporate into the air, and refining and transport spills and leaks from dumps, storage tanks, and fueling pumps put others into water and soils.[a] Some toxic components cling to soil particles—winds, or storm and flood waters carry them off. Others dissolve in water and can seep into underground drinking water supplies. Many have demonstrated and potential health impacts[b]—for example, premium unleaded gasolines may contain as much as 50% by weight benzene and related compounds, with both acute (immediate) and chronic (long-term) toxicity to humans.

Methyl tertiary-butyl ether (MTBE), formerly added to gasoline to improve combustion, has a very high octane value and also can substitute for lead compounds. MTBE can leak even from tanks designed to prevent it and moves rapidly through groundwater. MTBE is not a proven health hazard, but it imparts a strong and unpleasant taste and smell to water. A single spoonful is enough to foul an Olympic swimming pool. MTBE-contaminated groundwater has shut down numerous city and domestic water supply wells, leading to nationwide bans.[c]

In April 2002, Congress required that ethanol replace MTBE in gasoline, a move expected to boost ethanol consumption to five billion gallons annually by 2010, which mightily pleased the farm lobby. In January 2007, President George W. Bush supported ethanol production to replace gasoline. But ethanol is not a clean fuel—burning it in dilute concentrations in gasoline pollutes air with nitrous oxide and nitrous aldehydes. And the U.S. Environmental Protection Agency has discovered that corn ethanol manufacturing plants emit smog-producing volatile organic compounds, including aldehydes[d]—respiratory irritants that cause fatigue, skin rash, and allergic reactions and may be carcinogenic. High acetaldehyde concentrations can cause nausea, vomiting, and hallucinations.

Notes
[a]R. J. Irwin et al. *Environmental Contaminants Encyclopedia* (Fort Collins, Colo.: National Park Service, Water Resources Division, 1997).
[b]As of March 2005, the United States had more than 660,000 storage tanks for hazardous materials (mostly gasoline) in use, and about 1.6 million still in the ground but not in use.

Box 12.1 Continued

About 449,000 tank leaks were identified, and cleanups were initiated at about 416,000 sites. Only limited data are available for abandoned tanks (U.S. Government Accountability Office. *Environmental Protection: More Complete Data and Continued Emphasis on Leak Prevention Could Improve EPA's Underground Storage Tank Program* [GAO-06–45, 2005]).
[c]L. S. Dernbach. The Complicated Challenge of MTBE Cleanups. *Environmental Science and Technology News* 2000;34:516A–521A. Underground fuel tanks let aromatic gasoline hydrocarbons (BTEX) leak into the ground along with MTBE. The other gasoline components have much lower solubilities in water than MTBE and are much more biodegradable. Subsurface MTBE can contaminate community water supplies for tens to hundreds of years because groundwater migrates at varying rates from the leaking tanks to nearby wells. The dominant strategy for reducing MTBE contamination is diluting the contaminated well water with clean water. But wells being pumped at high rate of a million gallons per day may require large dilution factors to reach acceptable MTBE levels—as much as 10,000 gallons of uncontaminated water for every gallon of polluted well water.
[d]Unanticipated air pollutants coming from corn ethanol plants include acrolein, formaldehyde, acetaldehyde, and butanon. Cal Hodge (Ethanol Use in US Gasoline Should Be Banned, Not Expanded. *Oil and Gas Journal* 9 September 2002;21–22, 24–25, 28) provides a good discussion of ethanol's polluting effects.

The principal oil shale resources in the United States underlie more than 15,000 square miles of arid Wyoming, Colorado, and Utah, some of the most spectacular scenic lands in the western United States. Mining these scenic lands for energy will have the same effect as mining tar sands.[57] Producing oil shale by open-pit mining can disrupt as much as five times more land than mining a volume of coal with the same net energy content.[58] These wastes contain toxic metals that easily leach from tailings into groundwater supplies (see chapter 4). Surface retorting emits air pollution, and producing 100,000 barrels of shale oil a day would release more than 27,000 tons of greenhouse gases. The problem of putting the rock dug out of a hole back into the same hole is even worse for oil shale because retorting the rock expands it like popcorn. The oil shale waste disposal problem will become colossal, even in comparison to tar sands. Surface disposal of toxic waste rock damages unmined landscapes and destroys wildlife habitats.

Where the oil shale is deeply buried (as much as 1,000 feet), underground mining reduces surface damage but is much more expensive than open-pit mining, and so far uncommercial. The process involves excavating rooms in the oil shale deposit, and then collapsing the room walls. The excavated material is raised to the surface for retorting, while the rubble is retorted underground. The electricity use for in-place retorting is huge, so inventive minds propose pulverizing and retorting oil shale underground with nuclear dynamite (bombs) (see appendix 6). The resulting fuel would be radioactive.

Producing oil, natural gas, and coal-bed gas also yields salty wastewaters, commonly containing total dissolved solids twice the 500 ppm drinking water standard.[59] For a long time, U.S. oil producers simply dumped saline wastewaters on the ground, contaminating surface waters and leaving biologically dead zones where they ponded. Those zones remain biologically dead more than a century later. Today, petroleum wastewaters commonly are reinjected to maintain high oil

Figure 12.5 Crude oil seeps out of Lost Hills, California, oil field wells, impoundments, and pipelines and flows downhill into streams. Photographed March 1979.

reservoir pressures and keep oil flowing, an additional energy cost. In 1995, about 65% of wastewaters were reinjected, and that proportion may have increased.

Wells pump salty groundwater into coal-beds to release the methane from myriad small crevices and cracks, producing on average 38 times more wastewater per unit of produced energy than onshore conventional natural gas wells. The wastewater is injected into the ground in disposal wells, drained to surface evaporation ponds, or—the cheapest method—dumped into creeks and streams. Disposing of all the drilling fluids and materials can contaminate both underground and surface waters. Most coal is relatively hard and impermeable, and once the gas comes out, groundwater cannot quickly seep back into the fractures. So unlike oil fields, coal-bed wastewater volumes typically decrease as the gas extraction proceeds.

Deliberate waste dumping is a common practice at oil fields and coal-bed gas mines (figure 12.5). In 2003, a U.S. appeals court ruled that coal-bed methane wastewater is "industrial waste" to be regulated under the Clean Water Act. But illegal wastewater disposal behind in-stream dams is rampant in Wyoming gas fields, and in some cases has caused flooding.

The Last Rush

Ignoring the minimal chance of finding the equivalent of Saudi Arabia's oil reserves within U.S. borders, the George W. Bush administration opted for oil and gas exploration and development on national wildlife refuges and other protected lands in the 11 lower western states and Alaska, ostensibly to reduce U.S.

dependence on foreign oil. Bush's 2001 Executive Order 13212 spurred the federal land-managing agencies to speed reviews of energy-related permit applications. The Bureau of Land Management (BLM) staff assertively implemented the new policy, shifting from their old role of monitoring and mitigating energy exploration environmental impacts to issuing permits for energy development.[60] The highly visible fight over the Arctic National Wildlife Refuge in Alaska has deflected public attention away from the great scenic and wilderness values of other targeted lands— the last unspoiled remnants of roadless forest, canyon lands, and essential wildlife habitat in the lower western states.

Relaxed energy leasing rules and dramatic oil and gas price increases more than tripled the total number of BLM-approved drilling permits from 1999 to 2004, mostly in Colorado, Montana, New Mexico, Utah, and Wyoming. Lands the Clinton administration had protected as national monuments became particular targets.[61] Only a small fraction have the geological characteristics for oil, gas, and coal-bed methane discoveries, however, even in moderate amounts.[62] Qualitative USGS ratings of the lands' energy potentials point to the probable truth of a Wilderness Society finding that 15 western national monuments likely contain less than six days of economically recoverable natural gas and 15 days of economically recoverable oil at current U.S. consumption.[63]

Industry sources claim that oil reserves in and near Utah's new Grand Staircase-Escalante National Monument could be as high as four billion barrels,[64] even though the whole region contains only 24 geologically favorable oil and gas sites (figure 12.6)—all of which have been explored for many years. One site has

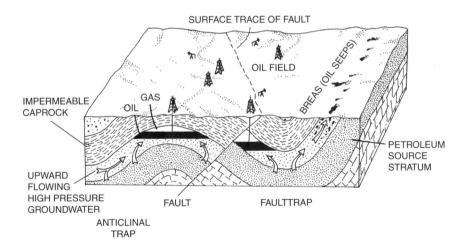

Figure 12.6 Diagram showing oil and gas trapped in a geological anticline (upward fold) and against a fault in subsurface rock layers. Arrows show groundwater moving upward under high pressure as wells withdraw gas and oil. Oil seeps appear where uncapped oil reservoir rock (stippled) comes to the surface. The seeps would naturally drain the reservoir over time. Modified from B. W. Pipkin et al. *Geology and the Environment*, 5th ed. (Pacific Grove, Calif.: Brooks/Cole, 2008), figure 14.2, 416.

trickled out 25 million barrels of oil since its 1964 discovery, barely more than one day of current consumption over 35 years.[65] In spite of the barely commercial kerogen fuel yields, in early 2007 the United States prepared to lease extensive areas of Colorado, Utah, and Wyoming for oil shale development.

The industry's National Petroleum Council (NPC) wants to look for 125 Tcf of natural gas under protected lands in the Rocky Mountains region—more than 2.5 times the mean USGS estimate and 4.5 times the 95% probability for *total* undiscovered gas in the entire United States (including CO_2 and other unburnable gases).[66] The USGS figure amounts to only a couple years' supply, and even the inflated NPC prediction represents less than six years at current consumption levels.

Contrary to the George W. Bush administration's belief, land-use restrictions have not prevented oil and gas exploration and exploitation. Most finds have been unconventional resources, including great expansion of coal-bed methane production over the past 15 years. Twenty thousand wells produce gas in the San Juan Basin of New Mexico and Colorado, and 15,000 more are planned. Wyoming's Powder River Basin already supports about 15,000 wells, with another 39,000–65,000 approved—all to be drilled before 2010. The BLM estimates that drilling the 39,000 wells will disturb 212,000 acres and generate 160 billion gallons of saline water.[67] An optimistic estimate of the total methane yield for all this disruption is about 246 days' supply, less than a year at current consumption.

Daily U.S. energy consumption totally dwarfs even exaggerated estimates of possible oil and gas production from these lands.[68] Further exploration and production in the western states are more likely to yield a high proportion of dry holes than significant additional resources, draining energy and actually worsening our dependence on imported energy.

Nuclear Methadone

As a nuclear engineer, Admiral Rickover saw nuclear fuels as the most promising alternative to fossil fuels. Admiral Lewis Strauss famously trumpeted that nuclear electricity would be too cheap to meter—as impossible a claim as energy independence. But cheaply built nuclear power plants have not proved to be safe enough. Fears of radiation-spreading accidents have lowered the U.S. nuclear power sector to about a fifth of current electricity demand, the same as in 1989. Significant problems include fuel supply limitations, nuclear weapons proliferation, and terrorists.[69] And we still have no way to safely dispose of radioactive wastes, which Rickover named the "problem which must be solved before there can be any widespread use of nuclear power" (see chapter 10).

Despite such daunting problems, the nuclear power industry, Nuclear Regulatory Commission, and media promote nuclear power as a safe and clean fossil fuel alternative, with no or very low CO_2 emissions.[70] The 2005 Energy Policy Act provided a $6 billion subsidy to boost incentives for building nuclear power plants in the United States, and some environmentalists support the nuclear option to reduce greenhouse emissions and global warming. But producing nuclear power requires reactor building, maintaining, securing, and decommissioning; waste repository construction, maintenance, and security; and shipping wastes long distances for disposal. All together, these processes and activities release enough greenhouse gas to cancel out the low-CO_2 benefit of nuclear-generated electricity.

In nuclear reactors, radioactive uranium atoms release neutrons to bombard neighboring uranium atoms, splitting them (fission) to release great heat (see chapter 7, box 7.1). The heat makes steam for generating electricity in turbines. Fuel elements must be packed into a reactor core close enough to promote fission reactions, but far enough apart to let cooling liquids circulate. Neutron-absorbing rods control the reactors' power levels—pushing rods farther into the core, or adding rods, lowers the power output. Inserting all control rods will shut a reactor down.

Today, all U.S. nuclear power plants are light-water reactors, fueled with a mix of around 95% uranium-238, plus no more than 5% uranium-235. The very low concentration of uranium-235 in reactor fuel prevents atom bomb-like explosions,[71] but the fuel is dangerously radioactive, and fission creates highly toxic and flammable plutonium-239. Also, large nuclear power plants contain large amounts of fuel—modern light-water reactors contain "several hundred" fuel channels, each holding 12 fuel bundles weighing 50 pounds.[72]

Maximum credible

Critically important nuclear plant cooling systems prevent uranium and plutonium from burning or melting. Loss of coolant leads to melting in the reactor core—the "maximum credible" nuclear power plant accident. The fuel melts at temperatures above 4,500°F, so once a fuel fire starts, control rods cannot quench it or stop it from burning through the heavy steel reactor vessel and four-foot-thick steel-lined concrete floor of the reactor containment building. If melted fuel meets reactor cooling water or groundwater, violent steam explosions can rupture the containment, releasing clouds of hot, highly radioactive iodine-131 and strontium-90. Releasing hot radiation over a populated area immediately threatens human health, and contamination from some of the component materials persists for centuries (for potential health effects, see chapter 7, box 7.1). Uranium or plutonium exploding out of a power plant would contaminate soil, surface water, and groundwater *essentially forever.*

The official cost estimates for such an accident pushed insurers' enthusiasm for nuclear power plants below zero (box 12.2). In 1957 Congress first passed the Price-Anderson Act, which caps the nuclear industry's liability for a serious accident at around $10 billion. Taxpayers get stuck for the rest of the anticipated $500 billion in damage (2006 dollars).

Many nuclear power plants in the United States, Canada, western Europe, and Asia have experienced overheating and fuel burning or melting (meltdowns), although none has had a maximum credible accident. Far more radiation has escaped from U.S. reactors, both into containments and outside plants, than had been considered possible (box 12.2), however.[73] In 1979, Pennsylvania's Three Mile Island nuclear power plant, located on an island in the Susquehanna River and close to the state capitol at Harrisburg, came the closest of any in North America to a maximum credible accident. Less than a year after Three Mile Island's startup, a pressure valve failed to close in the cooling system and water drained out of the reactor core—a classic loss-of-coolant accident. The core melted down, and close to 20 tons of melted uranium pooled in the lower reactor vessel. Contaminated

Box 12.2 Nuclear Safety Washout

Most nuclear power plants supply, and are located near, population centers. Largely for that reason, the Brookhaven National Laboratory's 1957 WASH-740 report estimated that a nuclear reactor core meltdown could kill more than 3,000 people outright and injure or irradiate 43,000 more.[a] The only foreseeable protection for local residents would be very fast evacuation before an accident could get out of hand. The price tag on so much destruction, plus long-term contamination of homes, businesses, and farms, made nuclear power plants commercially uninsurable. In 1957, Congress passed the Price-Anderson Act to secure investor confidence and limit industry liabilities.[b]

As power plants in the United States increased in size and number, most operating without remarkable incidents, a 1964 WASH-740 update predicted eight times more death and damage than the original.[c] It was never released. The nuclear industry pushed for safety estimates that emphasize the low probability of an ultimate power plant accident over possible effects. The 1975 WASH-1400 (Rasmussen) report followed that rationale[d]. Based on a nine-year, $3 million Massachusetts Institute of Technology study, the report assessed the chance of a power reactor accident harming 1,000 people at one in a million.

The very next year, a fire beneath reactor control rooms at the Brown's Ferry nuclear power plant in Alabama came very close to causing loss of coolant and a meltdown in one of the reactors.[e] The accident revealed bad plant design and maintenance practices, poor knowledge of backup and safety systems, poor firefighting techniques and command arrangements, and flawed evacuation plans. Both the Brown's Ferry and 1979 Three Mile Island accidents proved the Rasmussen report's conclusions to be overoptimistic and forced the governing Atomic Energy Commission to formally disown WASH-1400.

In 1982, Sandia National Laboratories prepared the CRAC-2 report for the Nuclear Regulatory Commission, which "estimated that damages from a severe nuclear accident could run as high as $314 billion"—or more than $560 billion in 2000 dollars. And improved modeling methods have more than doubled that estimate.[f] In 2005, Congress reauthorized the Price-Anderson Act for 20 more years, without significantly raising the liability cap. The investors remain protected, but Nuclear Regulatory Commission oversight of the 103 aging U.S. reactors is the public's only real insurance.[g]

Notes
[a] J. G. Fuller. *We Almost Lost Detroit*, 1st ed. (New York: Readers Digest Press, 1975), 56–59; U.S. Atomic Energy Commission. *WASH-740, Theoretical Possibilities and Consequences of Major Accidents in Large Nuclear Power Plants* (Brookhaven National Laboratory, 1957).
[b] Public Citizen. *Price-Anderson Act: The Billion Dollar Bailout for Nuclear Power Mishaps* (2004). Available: www.citizen.org
[c] Fuller. *We Almost Lost Detroit*, 131–164.

Box 12.2 Continued

[d]Norman Rasmussen et al. *WASH-1400, Reactor Safety Study* (Washington DC: U.S. Nuclear Regulatory Commission, 1975).
[e]The Browns Ferry fire knocked out power to the control and safety systems of two reactors, disabling the primary and backup coolant-water pumps, reactor shutdown systems, lights, and monitoring instruments for both reactors and their backup systems, plus the internal telephone system.
[f]Sandia National Laboratories. *Calculations of Reactor Accident Consequences, version 2, CRAC-2* (NUREG/CR-2326, U.S. Nuclear Regulatory Commission, 1982).
[g]U.S. Government Accountability Office. *Nuclear Regulatory Commission: Oversight of Nuclear Power Plant Safety Has Improved, but Refinements Are Needed* (GAO-06–1029, 2006).

water leaking from an open valve released radioactive gases throughout the plant and into nearby communities.[74] Eventually, the 1986 Chernobyl, Ukraine, accident hardened the American public's negative perception of nuclear power plant safety.

Shoddy or incorrect construction has characterized many nuclear power plants, and small accidents happen all the time. Part of California's Diablo Canyon nuclear plant was built inverse to the blueprint design. None of the three Brown's Ferry, Alabama reactors matched their blueprints (box 12.2), and all were shut down in 1985 for reconstruction. Nuclear power plant accidents have helped to improve other plants, and probably made reactors safer—but they resulted as much from human acts as technical problems. In the cases of Three Mile Island and Brown's Ferry, confusing instrument readings led to human errors that worsened the accidents.

Reactors in the United States are now aging, raising the likelihood that new vulnerabilities will develop and expose design or construction flaws. Radioactive tritium-water leaks have been discovered at more than a dozen nuclear power plants since 2000. The 2002 discovery of a pineapple-sized corrosion hole in a pressurized reactor vessel at the Davis-Besse plant near Toledo, Ohio, raised alarms about another 68 reactors. Only a skin-thin stainless steel liner prevented reactor rupture, which could jam or destroy the control rod mechanism and leave operators powerless to stop a meltdown.[75] In 2006, tritium contamination in underground drinking water supplies revealed a water leak at the San Onofre plant near San Clemente, California's largest. Dangerous tritium leaks are being discovered at other U.S. reactor sites.

A completely unanticipated coolant problem appeared in Europe in 2006, when a heat wave aggravated ongoing drought. Water levels dropped in lakes and rivers, limiting cooling waters and shutting down nuclear power plants in France, Spain, and Germany. Drought has plagued the western United States since about 1999, threatening to limit cooling water for all industrial uses. This problem may worsen as global warming intensifies.

Although well aware that a jealous maintenance staffer intentionally destroyed a small Idaho power reactor in 1961, killing himself and two others—one of them the object of his suspicions—industry and government officials do not publicly discuss the possibility of insider sabotage.[76] Even now, more than five years after terrorists brought down the World Trade Center in New York, nuclear power plants' defense arrangements remain minimal, on the unrealistic assumptions that assailants will consist of one external three-person team with one inside co-conspirator—all using hand-held automatic weapons, only moderate-sized truck bombs, and no boats or planes. Sadly, even though the five to nine armed defenders get six months advance warning of the test attack dates, *mock-terrorist teams succeed in reaching power plant control areas in half of all tests.*[77]

Reactor containments probably could not withstand a September 11th–type airliner crash, either, according to Jai Kim of Bucknell University's civil engineering department: "A 747, 767 would go through four feet of concrete easily....I don't think we have any defense against an attack like that." Critical cooling systems, electricity to run the reactor, and stored spent-fuel wastes all occupy unhardened, highly vulnerable buildings. The National Research Council agreed that "even an accidental plane crash was not factored into the designs of 96% of U.S. nuclear plants."

The fearful link

The United States mines uranium mainly from sedimentary deposits in 5 of the 11 western states.[78] Dangerous underground uranium mining (see chapter 7) ended in 1992, and ores now come out of open pits. The ores are either leached in place or in heaps to extract the uranium. The United States is estimated to have 112,000 tons of known recoverable uranium at $36 per pound—enough for seven to eight years *with no rise in consumption.* Increased prices and improved reactor efficiencies could substantially extend the resource's life, while building more nuclear power plants will shorten it.

Proven world uranium reserves could provide approximately a 52-year supply for today's conventional reactors, *assuming no growth in consumption.*[79] The richest uranium lodes have been found and exploited, and the lower grade ores now require much larger energy inputs per pound to produce and process (see chapter 4). Claims that a virtually limitless supply can be produced from ocean seawater and unmineralized rocks ignore the prohibitive energy costs of extracting uranium from common rocks and the vast environmental impacts of mining our back yards.[80]

India and Israel acquired their nuclear arsenals by reprocessing byproduct plutonium from power plants, and the more politically unstable Pakistan and North Korea, and perhaps Iran, have based or will base their nuclear weapons programs on highly enriched uranium from facilities that produce uranium power plant fuel. Until the 1980s, the United States had intended to extract plutonium from breeder reactors, but the power plant–to–atom bomb linkage led western nations to abandon that plan, hoping to restrain nuclear weapons proliferation.[81] In addition, the breeder process yields plutonium very slowly. It reaches the point of energy breakeven—when more fuel has been produced than consumed—only after decades of operation.[82]

Post-Cold War U.S.–Russian agreements to disassemble nuclear warheads could supply highly enriched uranium for power plant fuel. But instead of turning the bombs into peaceful fuels, the U.S. government decided instead to store decommissioned warheads. As a result, three of nine types of nuclear weapons, all designed during the Cold War, have had their useful lifetimes extended by 30 years (over the designed 20 years) at a very high price.[83]

Forever dangerous

In 2001, the United States chose radioactive waste "disposal" instead of reprocessing to discourage nuclear weapons' proliferation. But no country on Earth has solved the problem of safe disposal for highly water-leachable nuclear fuel rods. The wastes remain dangerous virtually forever—plutonium has a hazard life of 240,000 years (see chapter 10). Both uranium and plutonium are long-lived radioactive materials, which accumulate in organs of the body and have known human health hazards, particularly cancers, reproductive problems, and debilitating illnesses (see chapter 7).

All U.S. repositories for so-called "low-level" nuclear wastes have leaked. Sweden, Germany, and Switzerland have been unable to find geologically and politically acceptable storage sites for the most dangerous wastes, which strengthened their decisions to abandon nuclear power. In the United States, after decades of research and debate, only one high-level waste site, at Yucca Mountain, Nevada, is tentatively approved. As electricity consumption grows and nuclear power plants continue to operate, many more repositories for high-level, long-lived wastes will be needed—on the order of two or more every decade (see chapter 10 for a fuller discussion of radioactive waste disposal).

Starting in 1945, fallout from atmospheric atom bomb tests loaded soils and water with higher levels of long-lived radioactivity than any human populations ever have encountered. Increasing nuclear-powered electricity and weaponry promise a future of escalating human radiation exposures from long-lived wastes in soil and groundwater.[84] These doses add to human exposures already accumulating from high natural background radiation in areas of uranium-bearing rocks,[85] cosmic rays in airplanes, medical and airport X-rays, and therapeutic radiation treatments.

As of 2010, more than 70,000 tons of high-level wastes may be shipped to Yucca Mountain from all over the country.[86] California wastes may be loaded onto trucks and driven on busy highways through towns and countrysides. Wastes from the rest of the nation will ride trains on miles of elderly tracks susceptible to derailment, transferred to trucks somewhere far from Los Angeles, and driven to the repository on steep and winding central Nevada roads to avoid Las Vegas. These plans eerily recall the ineffective restriction of 1950s atom bomb tests to days when winds blew away from Los Angeles (see chapter 7).

Nuclear fiction

Science-fiction fans and energy optimists still expect that nuclear fusion power, the holy grail of huge research expenditures since the 1960s, eventually will replace fission reactors. Nuclear fusion is not to be confused with tabletop or "cold fusion"—a transitory phenomenon with little or no promise for producing usable energy.

Nuclear fusion welds nuclei of two radioactive forms of hydrogen, deuterium and tritium, into larger atomic nuclei, releasing the immense heat of suns and stars to boil water to make steam for generating electricity in turbines.[87] Fuel is no problem because ocean waters contain a practically limitless supply of deuterium, so fusion power's big problem is technical feasibility. Experimenters have fused atomic nuclei in labs at extremely high pressures and temperatures on the order of 18 *million* degrees Fahrenheit. But even small-scale fusion reactions still use hugely greater amounts of energy than can be derived from them. And contrary to press hype, fusion reactions do yield dangerous radioactive wastes.

"Renewable" Methadone

In 1957, Admiral Rickover came up with the same list of putative "renewable" energy sources that today many expect to replace fossil fuels and nuclear power, while solving pollution problems: "wood fuel, farm wastes, wind, water power, and solar heat." His "wood fuels and farm wastes" equate to ethyl alcohol (ethanol), methyl alcohol (methanol), other alcohols, and biodiesel. Rickover assumed "that the principal renewable fuel sources which we can expect to tap before fossil reserves run out will supply only 7 to 15% of future energy needs." They currently provide 6% of electricity consumption: 2.7% hydroelectric (dams); 2% from wood burning; and 1.4% all others (figure 12.7).

All the "renewable" energy sources have limited potential or limited supply capacities, and even in aggregate may lack the potential to substitute for even our present liquid fuel consumption. Unfortunately, some of these sources degrade, and ultimately destroy, the conditions required to keep them producing energy. The reservoirs behind hydroelectric dams eventually fill with sediment, for example. Both cutting trees and industrial farming promote soil erosion and lead to irreversible soil degradation and depleted soil nutrients, which diminish the potential for growing additional crops (see chapters 1, 2). Making fertilizer uses large amounts of energy, so fertilizing biofuel crops reduces their net energy yield even more (see chapter 2).

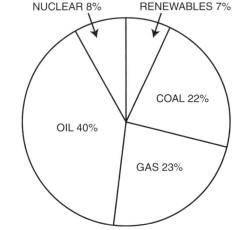

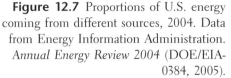

Figure 12.7 Proportions of U.S. energy coming from different sources, 2004. Data from Energy Information Administration. *Annual Energy Review 2004* (DOE/EIA-0384, 2005).

Cold and hot water power

Hydroelectric power from inland dams, and geothermal power from Earth's internal hot spots, probably will never contribute substantially to the national energy mix. Hydroelectric power from wave or tidal action still is minor in the United States but could expand along coastlines that see powerful and consistent wave action and tides.

Falling water generates hydroelectric power at inland dams, which collect river flows in reservoirs high above the original courses. When released, reservoir water falls to the river level through turbines, generating electricity. Within limits, power generation can be adjusted to meet peak electricity demand, although water supply can be a major limitation. A 1986–1988 dry spell noticeably lowered hydroelectric power production nationwide, showing the vulnerability of hydroelectric power to drought.

Hydroelectric power will remain important locally, but dams already occupy virtually all of the large hydropower sites, and many of the smaller ones, limiting future expansion. The fixed generating sites make hydro a relatively inflexible power source, a common problem for many alternative electricity-generating schemes. And dams have severe environmental impacts: They drown floodplain habitat under reservoirs, reduce downstream water flows, withhold nutrients from downstream areas, and trap sediment and starve coastlines (especially beaches), and the reservoirs serve as pollution-collecting sumps (see chapters 9, 10).[88] Catastrophic dam failures and other disasters—including dam-induced earthquakes from the weight of impounded water and landsliding into reservoirs, should be a cost balanced against the putative benefits of building dams.[89] In the long run, hydroelectric energy is nonrenewable because the reservoirs eventually will fill with sediment, making dams into artificial waterfalls.[90]

Geothermal energy uses natural underground heat sources, mostly from past and ongoing volcanic activity. The sources are either natural hot water reservoirs at shallow depth (several hundred to a few thousand feet) or a zone of hot dry rocks. In the most prized vapor-dominated reservoirs, hot-water–rich solutions fill spaces between rock particles, and steam can be piped directly from fractures to electricity-generating turbines. From liquid-dominated reservoirs, superheated water is pumped to the surface. The much lower surface pressure lets it flash to steam under carefully controlled conditions. Water also can be pumped into the rocks and heated, then pumped out again to extract the heat. In some locations geothermal steam or hot water is piped directly into buildings, but most sources generate electricity in steam turbines.[91]

Geothermal energy is relatively clean but even less flexible than hydroelectric power because it cannot be adjusted to meet variable demand. Favorable geothermal locations are very few, as indicated by the low levels of geothermal electricity generation in the United States (less than 3,000 megawatts), and the world (about 8,000 megawatts). The 1,000 megawatts produced annually at The Geysers geothermal power plant in California provides only enough electricity for a city the size of San Francisco.

This energy is renewable only to a limited degree. In fact, all U.S. electric power-generating geothermal plants are in decline because they have pumped heated groundwater out of the hot rock reservoirs so quickly that low natural rain levels cannot replenish them. The Geysers has tried to overcome this problem by injecting municipal wastewater from nearby towns and cities into the reservoir, a costly and energy-intensive solution. Around The Geysers, wastewater injection generates earthquake swarms that continually rattle local residents.

Deep scientific drill hole experiments in the 1970s and 1980s found hot, brine-filled fractured rock at great depth (6,000 feet to nearly three miles) at a few sites, stimulating additional ideas for generating geothermal power. Schemes include pairing a water-injection well with a hot-water recovery well, or drilling single wells into complex fracture zones. Both approaches require the expensive step of opening fractures by pumping fluids deep into the ground at high pressure (hydrofracturing) and are beset by serious problems, including the difficulties of drilling in hot rock, poor water flow through fractures, losing injected water, corrosive effects of brines on drilling equipment, and generating deep earthquakes.[92] After some 30 years of experiments, these ideas remain speculative.

Biofuel imbalances

We have seen that highly concentrated fossil energy originated as plant and animal materials that accumulated and became naturally refined over hundreds of millions of years. Attempting to replace them with biofuels assumes that we can harvest and process the unrefined source materials as they grow. Worldwide, that would put nearly all of Earth's agricultural land into fuel crops, with nothing left over for growing food.[93] The United States uses 85% more fossil fuel energy each year than the total energy annually stored in all plant biomass growing in the nation's soils. This fundamental energy imbalance rules out the possibility that domestic bioenergy can replace our fossil fuel consumption.[94]

Admiral Rickover's speech pointed to a similar problem: "Wood fuel and farm wastes are dubious as substitutes because of growing food requirements to be anticipated. Land is more likely to be used for food production than for tree crops; farm wastes may be more urgently needed to fertilize the soil than to fuel machines."

Wood burning has warmed people inside and outside of dwellings for millennia, and today industrial-scale wood chip burners make electricity on a large scale. But the energy costs of harvesting and transporting wood are high, while conversion efficiencies are low. To equal the electricity that a single modern wind turbine can generate each year with wood requires cutting and burning 400 acres of forest. A city of 100,000 people dependent on electricity from sustainable forest biomass would need nearly 500,000 woodland acres, producing more than a ton of wood per acre every year. To supply the whole United States with about 42% of its current electricity from woody biomass would require a sustainable forest of 250 *million* acres, the total of all forested land under U.S. Department of Agriculture Forest Service (USFS) management.[95] Unfortunately, USFS does not yet manage forests for sustainability (see chapter 1).

Throughout history, taking wood for fuel has resulted in major deforestation and devastating soil erosion worldwide (see chapter 1).[96] Turning to chip burners or liquid fuel from wood to increase U.S. electricity generation undoubtedly would accelerate U.S. deforestation and erosion.

Recalling that coal is the concentrated remains of trees explains why burning biomass or wood in fireplaces can cause air pollution and smog just like fossil fuels. Wood burning also releases greenhouse gases and more than 200 different chemical pollutants, 14 carcinogens, and at least 40 toxic wood smoke pollutants with negative human health impacts.[97] Just cutting forests releases immense amounts of greenhouse CO_2. Unfortunately, rising natural gas prices are raising the demand for firewood.

Growing crops for making ethanol or biodiesel fuels to run cars and farm machinery could offset some of our current and future oil demand. The corn ethanol industry is already a going concern in the United States, with more than 100 refineries in business, and 200 or so more planned or under construction (box 12.1). The surge in corn ethanol production is propelled by congressional requirements for ethanol additions to gasoline, a tariff on imported ethanol, and massive federal and state subsidies—much greater than those for oil and gas. The Global Subsidies Initiative, based in Switzerland, estimates that the subsidies cost U.S. taxpayers at least $5.1 billion and possibly as much as $6.8 billion in 2006. These costs are likely to increase with ethanol production.[98]

Ethanol is the fermentation product of simple sugars and has the same basic composition as vodka or gin. To make one gallon of ethanol takes the energy in one half-gallon of gasoline. But a gallon of ethanol contains only *two-thirds* of the energy in a gallon of gasoline. So to replace one gallon of gasoline you have to make six gallons of ethanol.[99] At that rate, devoting the entire 2005 corn crop (11.1 billion bushels) to ethanol would offset a bit more than 3% of current annual gasoline consumption. To offset the whole U.S. gasoline demand with corn ethanol, currently 140 billion gallons per year, corn crops would have to occupy all the land in the entire lower 48 states and Alaska.[100] Growing crops for ethanol would use a lot of water, of course (about four gallons of water to produce one gallon of ethanol), but even without fuel crops, agriculture already claims two-thirds of world water consumption. And water shortages are a growing problem. Other ethanol sources suffer from similar space and water problems, and none promises to yield significantly more energy than they consume in growing the crops and converting the biomass to fuel.[101]

Current ethanol research favors cellulosic sources, essentially the woody parts of plants, such as switchgrass, poplar and other trees, crop and feed residues (corn stalks, wheat stems, etc.), and urban, agricultural, and forestry wastes.[102] Professor Tad Patzek points out, "Close to one billion years of plant evolution have made cellulose very stable and resistant to biochemical attacks."[103] The problem of how to efficiently break down cellulose into simple sugars is still unsolved, so far preventing commercial conversion of cellulosic plants to ethanol. The same feedstock crops are in heavy demand for forest and paper products and for biochemicals, pharmaceuticals, and plastics. Along with the need for corn as food, these competing demands will drive hard choices and high prices.

Pushed by ethanol demand, corn prices already are rising, threatening to withdraw vast acreages from food crops, such as tomatoes. The effect is global—sugar has become scarcer and prices have risen since Brazil began converting sugar cane to ethanol, for instance.

Long the darling of environmentalists, biodiesel fuels have energy contents close to petroleum diesel and can be derived from recycled cooking oils and even chicken or cattle renderings. But the level of wastes generated today can supply only a small part of current diesel use (about 45 billion gallons per year), let alone projected future demands. Competing with ethanol for cropland, many U.S. farmers are growing soybeans for making biodiesel fuel. Running a car on pure biodiesel for a year requires anywhere from eight to 16.9 acres of soybeans, while a farmer cultivating 1,800 acres can use up 2,500 gallons of diesel fuel annually.[104]

Ethanol refineries are shifting from natural gas to cheaper but more polluting coal to power the conversion of corn to ethanol. The U.S. Environmental Protection Agency may soon seek to relax emission restrictions on all coal burning, not that it would matter much—biorefineries are ignoring even current restrictions. Some emit up to 430 times the allowed sulfur and nitrogen oxides; volatile organic compounds, including carcinogens; and greenhouse gases.[105] Far more common than the E85 blend (85% ethanol, 15% gasoline) will be cars run on gasoline with low ethanol concentrations added to promote complete burning. Both will add to ozone and smog.

Archer Daniels Midland Corporation spins ethanol as growing energy, not burning it. But vehicles do burn biofuels, and they emit more harmful emissions than anyone in the industry likes to admit. Total CO_2 emissions from making and burning corn ethanol are 50% higher than from energy-equivalent volumes of gasoline—and 100% higher if ethanol fermentation byproducts become cattle feed.[106] Soybeans fix atmospheric nitrogen, and the nitrous oxide emissions from soybean biodiesel may be comparable to CO_2 emissions from fossil fuels. Making and burning soybean biodiesel actually may emit more CO_2 than all aspects of producing and burning petroleum diesel.[107]

Growing biofuels adds negative environmental impacts. All current scenarios for growing energy presuppose continuous cropping—particularly of highly erosive row crops such as corn. Such intensive farming will degrade soil, eventually making corn-ethanol crops nonrenewable. As Admiral Rickover pointed out, turning crop and animal feed residues into ethanol eliminates the antierosion and nutrient-restoring benefits of leaving crop and feed residues on farmland.[108]

Some research on genetically modifying plants aims to make them generate more cellulose by shortening their span of winter dormancy, increasing growth rates, reducing stem-stiffening lignin, and preventing flowering.[109] Such schemes ignore potential negative ecosystem impacts from eliminating flowering and pollen, and soil nutrient depletion from extending growth cycles and plant yields (see chapters 1, 2).

Local opportunities for generating fuels abound—in particular, biogas, largely methane, can be harvested from dumps and farm animal wastes. Natural air-absent (anaerobic) processes yield methane and other gases from decomposing garbage, farm manure, and human wastes. The gas consists of about 60% methane and nearly 40% CO_2, with less than 1% of a water vapor–hydrogen sulfide mixture. The volume equivalent energy in biogas is about 60% of fossil natural gas and about one-quarter of propane. The highly corrosive water vapor and sulfide component degrades pipes and power-generating equipment, but removing those impurities lowers biogas's net energy yield. However, if the CO_2 is not captured and safely disposed or used, this fuel is not environmentally benign.

In the United States, biogas production has turned the methane hazard of large landfills into a source for generating electricity or running heating elements.

Sewage treatment plants and large cattle and swine operations (dairy farms, feed lots) also can generate biogas in various types of "digesters." Cattle and hog factory farms cause huge land and stream pollution and emit nauseating odors, but complying with environmental manure disposal regulations can be costly. Farmers can lessen those costs by decomposing the manure in digesters. With additional, but expensive, equipment, the farmer can balance out equipment maintenance and marginal economic returns by feeding biogas-generated electricity into local or regional electrical grids.[110] Digesters also yield odorless solids, usable as fertilizers, and even animal feed supplements. But CO_2 emissions remain a problem.

Sun and wind

Solar heat exchangers, including passive solar structures, collect solar-induced heat from structural walls, floors, or indoor spaces or in water heater tanks. Photovoltaic solar cells directly generate usable electricity on sunny to partly cloudy days. After dark, a building's structural thermal mass can retain heat, and solar-heated water can be stored in a heavily insulated tank. But solar electricity must be stored in batteries for use at night and during sunless intervals. Current solar devices use only a tenth to a fifth of incoming solar energy, but photovoltaic efficiencies can rise to nearly 40% for small business and domestic use. Solar panels or roof tiles can be installed on houses in urban areas.

The price for domestic solar electricity-generating systems is dropping but remains three to four times higher than the electricity from fossil fuel generators and some other renewable sources. The cost is prohibitive for most residential uses, but some state and federal incentive programs and rebates defray close to half of equipment and installation costs.

Large centralized solar power plants can feed electricity to regional energy grids where sunny days average about 300 per year. Only three now operate in southern California. Two of Israeli design comprise fluid-filled tubes nestled in trough-shaped mirrors, which focus the sun's rays on the tubes. The sun heats the fluid as it flows to a steam generator. A third plant in the Carrizo Plains generates electricity with photovoltaic cells.

Wind power generates electricity directly when the wind turns windmill blades (rotors), cranking a turbine. Windy days are intermittent, so wind-generated electricity must be stored in batteries to maximize its usefulness. Unlike solar power, wind turbines do not easily adapt to urban conditions, largely restricting individual wind generation to rural areas. In the 1980s, centralized "wind farm" or "wind park" developments began supplying electricity to regional power grids in the western United States, mainly California. The early turbines are primitive by today's standards, and some never generated a watt of energy,[111] but more efficient and less costly models now are operating. By 1999, wind-generated electricity at about five cents per kilowatt-hour was close to the average three cents per kilowatt-hour cost of fossil fuel-generated electricity, with far greater net energy return.[112]

Centralized wind farms must cover large tracts to ensure that each turbine achieves maximum efficiency. Turbines can be clustered but must be separated enough to minimize turbulence and the reduced wind speed effects from neighboring rotors—a distance of five to nine rotor diameters is common.[113] Where wind directions vary, the turbine rows must spaced more widely, using even more land, especially on slopes. For example, 6,500 turbines at the Altamont Pass, California, wind farm occupy about 47,000 acres.[114]

The United States has abundant sites where winds are strong enough for generating electricity.[115] Employing them to full potential, just a dozen or so states could supply all of the nation's current electricity consumption and perhaps more—estimates vary from 126% to 260% (table 12.4). In 2004, winds made enough electricity to power 1.6 million households, but wind farm developments were not urgently expanding. Rising petroleum prices after Hurricane Katrina have helped wind power costs catch up with fossil fuel-generated electricity, and Congress has reinstated a tax credit for wind producers. These factors could help expand the industry by as much as a third.

Table 12.4 High and Low Estimates of Annual U.S. Electricity Generation Potential from Wind Energy, 12 Top States (Billion Kilowatt-hours; Data Corrected for Land Use Exclusions)

State	High Estimate[a]	Percent Total United States, 2003[b]	Low Estimate[c]	Percent Total United States, 1999[d]
North Dakota	1,210	31.5	1,089	28.3
Texas	1,190	30.9	303	7.9
Kansas	1,070	27.8	363	9.4
South Dakota	1,030	26.8	726	18.9
Montana	1,020	26.5	454	11.8
Nebraska	868	22.6	212	5.5
Wyoming	747	19.4	454	11.8
Oklahoma	725	18.9	272	7.1
Minnesota	657	17.1	424	11.0
Iowa	551	14.3	151	3.9
Colorado	481	12.5	272	7.1
New Mexico	435	11.3	(121)[d]	(3.2)[d]
Totals	9,984	260	4,841	126

[a]Pacific Northwest National Laboratory. *An Assessment of the Available Windy Land Area and Wind Energy Potential in the Contiguous United States* (1991), cited in Science for Democratic Action. *Large-Scale Wind Energy Development in the United States* (2001), 9.
[b]Arjun Makhijani et al. *Cash Crop on the Wind Farm: A New Mexico Case Study of the Cost, Price, and Value of Wind-Generated Electricity* (Institute for Energy and Environmental Research, 2004), table 1.1. The figure used for total 2003 U.S. electricity generation is 3,846 billion kilowatt-hours (from Energy Information Administration., Available: www.eia.doe.gov).
[c]J. G. McGowan. Tilting Toward Windmills. *Technology Review* 1993;41, cited in C. W. Montgomery. *Environmental Geology*, 5th ed. (New York: WCB McGraw-Hill, 1997), 345. Montgomery's percentages are related to total 1990 U.S. electricity generation; these are recalculated to 1999.
[d]Assumed value.

Solar power and wind energy come the closest to being truly renewable large-scale energy sources, but together they supplied only 0.14% of the nation's total 2000 electricity consumption. Considering the alternatives, wind energy likely will capture a much greater share of the electricity market than at present.

Centralized solar and wind power generating developments are not as environmentally benign as are individual passive or photovoltaic solar installations, but in most respects they are far less damaging than other major electricity-generating sources. Their polluting effects primarily come from raw material mining, manufacturing, packaging, and transporting materials. Each emits some harmful substances and generates waste material, either through construction and maintenance, or in operation. Backup storage batteries, generally lead-acid based, also must be properly disposed of (see chapter 10).

Centralized solar power plants can disturb many acres of land, which cannot be used for any other purpose. Site construction for large solar developments plows up surface sediments that feed dust storms, and older designs have proved fatal to birds (figure 12.8).[116] To supply the nation's entire *current electricity consumption*, photovoltaic cell arrays operating at 10% efficiency would cover a land surface equivalent to the state of Maryland, about 10,000 square miles.

Figure 12.8 The Solar One experimental solar electricity-generating plant, Daggett, California (now defunct). Mirrors (heliostats) tracked the sun, focusing reflected sun rays on the central fluid-filled chamber (white hot in photograph). Any bird venturing into the focused rays was instantly vaporized. Photographed May 1982.

On a wind potential map, many of the most favorable western wind-generation sites are the crests of interior and coastal mountain ranges (figure 12.9), plus offshore along much of the west coast. Some Americans consider desert vistas monotonous and like seeing rotors turning against the horizon, unaware that large mountainside wind-power developments are major road-building projects that accelerate natural hillside erosional processes (figure 12.10; see chapter 5).

A 200-unit wind farm with 56-foot rotor diameters built on steep slopes, and including turbine pads, access roads, and transmission facilities, physically disturbs 210 acres of land. Side casting road construction techniques create even more damage (see chapter 5).[117] Wind energy proponents say correctly that erosion problems

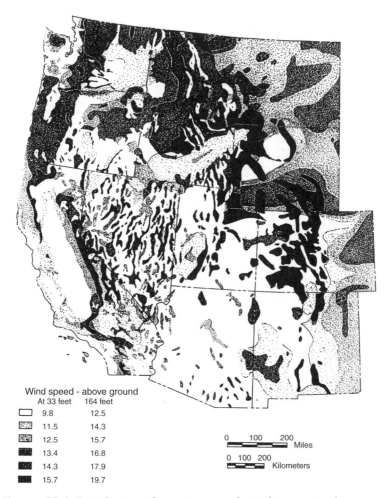

| Wind speed - above ground | |
At 33 feet	164 feet
9.8	12.5
11.5	14.3
12.5	15.7
13.4	16.8
14.3	17.9
15.7	19.7

0   100   200 Miles

0 100 200 Kilometers

Figure 12.9 Distribution of average annual wind power in the western United States. Highest wind-power areas are the northwestern coastal zone and crests of mountain ranges. Redrawn from Pacific Northwest Laboratory. *Wind Energy Resource Atlas of the United States*, 1986. Available: rredc.nrel.gov.

Figure 12.10 Erosion on a wind farm development, from rain running down roads and off the wind generator pads, Tehachapi Pass, California. Photographed March 1993.

can be mitigated with "proper" construction techniques, but wind-farm construction rarely uses them. The wind industry and even environmental advocates for wind power gloss over the erosional effects.

Even relatively low windmill densities do not reduce a wind farm's dominance on a mountain crest or hillside. Each latest generation turbine replaces seven older ones, so substituting four-footed lattice towers with single-footed tubular ones is supposed to reduce the number of towers and significantly reduce visual impacts. But the replacements are huge turbines on towers nearly 300 feet high, which may not improve visual effects. Abandoned wind farms degrade scenic values on public lands, leaving dysfunctional turbines and standing or felled towers in various states of disrepair, along with concrete foundations and other litter, plus accelerated road and pad erosion. To date, neither state, federal, nor industry funds are earmarked for cleaning them up.

Wind farms also threaten wildlife. Fences, installed for security and to avoid liabilities, segment wildlife habitat and cut off larger animals from food sources and home building or burrowing sites. Bird deaths are frequent at wind farms in migratory flyways and raptor habitat—especially for hawks and eagles, which tend to pursue prey on direct flight paths. A 2004 study of the Altamont Pass wind farm estimated as many as 4,721 bird deaths per year, affecting more than 40 different bird species, including large numbers of golden eagles, red-tailed hawks, American kestrels, and burrowing owls.[118] The only current proposal for protecting birds is to relocate turbines away from raptor habitats and build towers much higher to raise rotor blades above the flight paths.

The best generation sites for minimizing environmental damage are in potential high-wind areas of midwestern prairies and western plains, where wind farms can spread across flat or gently sloping farming and grazing lands. Farmers on the plains of eastern Washington and Oregon are finding wind farming compatible with grain farming, for instance. Except for the visual impacts, placing wind turbines offshore can avoid the environmental damages of land-based developments. In Europe, offshore wind turbines are successfully generating power on relatively broad, shallow ocean shelves, anchoring towers in water only 25–30 feet deep. Unfortunately perhaps, most of the western U.S. coastline has only narrow zones of shallow ocean floor, which drop off steeply at the continental rise.

The Silent Alternative

The various "renewable" energy sources provide options having lesser negative environmental effects than fossil fuels, but impacts grow with the use. If Americans want to avoid degrading more and more of our land, and the water and air that sustain us, they must start thinking more creatively, understand that there is no free lunch at the energy table, and examine the cradle-to-grave effects of always and forever expanding energy consumption. Rather than substituting new problems for old ones, the real solutions are to reduce energy demand and increase energy use efficiency.

Richard Cheney, vice president under George W. Bush and former leader of the National Petroleum Council while employed by Halliburton Energy Services, Inc., famously denigrated energy conservation as "a sign of personal virtue, but...not a sufficient basis for a sound, comprehensive energy policy." He was quite incorrect—in fact, cutting per capita energy use and conserving all fuel resources is the most reliable route to increasing energy supplies, lowering costs, and extending the time to petroleum exhaustion. Energy conservation in the United States has been increasing, but encouraging even greater conservation measures in homes, factories, offices, appliances, cars, municipal water systems, sewage treatment, and power plants can yield very large energy savings.

A ground-breaking 2000 report from five U.S. national laboratories, for example, proposed policies that could reduce electricity demand between 20% and 47%.[119] If adopted and vigorously pursued, that program could result in electricity savings equivalent to the power generated by 265–610 new 300-megawatt plants, reducing by as much as 59% the George W. Bush administration's proposal to build a total of 1,040 new 300-megawatt power plants—one each week for 20 years.

Many industrial nations, particularly in northern Europe, are far ahead of the United States in domestic energy conservation. Annual U.S. per capita energy consumption in 1998 was about 350 million Btus, and 356 million in 1999. Residents of Austria, Denmark, France, Italy, Germany, Switzerland, and the United Kingdom used less than half that much, on average. Only Canada and Norway, with substantially smaller populations, used more energy per capita than the United States. Polls showing the level of public satisfaction with their lifestyle are higher for some of those European nations than for America's population.

Lacking government support, the U.S. private sector is advancing designs and standards for new buildings and demanding more conserving vehicle designs to enhance energy conservation. Improved building designs will reduce energy consumption and greenhouse CO_2 emissions from U.S. buildings, which exceed even transportation and industrial contributions.[120] In 2007, President George W. Bush finally acknowledged the need for better fuel efficiency standards for cars, trucks, and sport utility vehicles. The costs of conservation programs are nearly canceled by overall economic benefits, including significant reductions in air pollution and CO_2 and other greenhouse gas emissions. They also reduce the petroleum dependence that makes us vulnerable to climate instabilities and political upheavals worldwide. Conservation also reduces the gross energy wastes from inefficient energy production and consumption. A new emphasis on conservation is the best—perhaps the only—way to help Americans and the world prepare for a new lifestyle over the Petroleum Interval's final two to three decades.

Power Down

Fifty years ago, Admiral Rickover saw that the United States should immediately begin to research and develop alternative energy sources to preserve its industrial economy. But until recently, America did not heed him or the many other energy Cassandras, including M. King Hubbert, who have been correct in their foresight but ignored. Now the time frame is very limited. Waiting until the world's cheap fossil energy supplies are about to peak, guaranteed to be a time of growing climatic and political instabilities, has limited our choices more than we can comprehend. One limit is the growing energy cost of developing alternatives.

Without an understanding of each energy source's limiting factors, the expectation that "renewable" resources can take over from petroleum may have lured us into a dangerous complacency. Hydroelectric and geothermal limitations will keep them at or near current levels, given adequate water supplies. Turning to nuclear power in the short run will not help us preserve cheap fuels for liquid-fuel cars or cut greenhouse gas emissions, while magnifying the unsolved problems of radioactive exposure and pollution.

There really is no methadone for our addiction. It is time to start listening to the Rickovers and other Cassandras. For one, Walter Youngquist warns, "The public in general seems of the view that the transition to alternatives will be simple. In my view it will be a quantum leap, and at best the resulting society will look vastly different (and probably in considerably reduced numbers) from what we see today." Since we have left serious experimentation with biofuels and other alternatives, and funding for such research, to a very late stage, we remain unaware of many, if not most, of their potential limitations.

Most critically, given the global competition for oil, when global petroleum production finally goes into decline, even a mix of alternatives may not be adequate to satisfy demand. Right now and in the foreseeable future, we do not have the capacity to install new generating systems fast enough to keep up with rising demand, and probably will not be able to replace every fossil-fuel use. We probably cannot fulfill current economic projections—just to replace a third of the projected 30 terawatt (one terawatt = one trillion watts) demand increase expected by 2050 would require a billion watt nuclear power plant to come on line *every other day* for the next half century.[121] Producing the 30-plus terawatts from biomass would leave the world with no agricultural land for growing food.

Solar and wind energy may well provide the best and most sustainable alternatives, helping to reduce attendant environmental problems. But relying a great deal on solar power would require major adjustments—it would mean shifting to an entirely different lifestyle than the one Americans take for granted. On that more serious note, Youngquist wrote, "Living on only what solar energy comes in each day, and surviving the cloudy days, will be a vastly different situation from doing as we are today....The challenge of living on current income, rather than our fossil energy inheritance, is very large, and not simple."[122] Observed Admiral Rickover: "Looking into the future, from the mid-20th century, we cannot feel overly confident that present high standards of living will of a certainty continue through the next century and beyond."

While Americans complain about higher gas prices and worry about global warming, we have not yet reached the level of pain that requires us to examine our present lifestyle. Therein lies a warning that the limitations of alternative energy resources in the post-world oil-production peak era can raise United States demands for alternative fossil energy resources, in spite of global warming. Developing them will exact a toll in ravaged land and water, dwarfing all previous disturbances and contamination in, on, and under energy-producing lands. Part of that toll could come from wars, which are hugely damaging to the environment that must sustain our descendants in a warmer, but less abundant future.

Competition for oil in a world of dwindling supply inevitably will challenge the American "birthright," enhancing the risk of global conflict over diminishing fossil energy and other resources.[123] Many believe that the time is already at hand—that serving America's oil addiction was the main motivation for the 2003 U.S. invasion of Iraq and also is driving colder relations with oil-producing Russia.

Energy analyst Robert Hirsch emphasizes the need to prepare: As peak world oil production approaches, "liquid fuel prices and price volatility will increase dramatically, and, without timely mitigation, the economic, social, and political costs will be unprecedented."[124] Weaning ourselves from oil will be a shift of monumental magnitude, needing time for careful thought and preplanning before we can plunge ahead without economic hitches or actual disruption. We are already very late to avoid a period of societal cold turkey, with agonizing and unpredictable social upheavals. To avoid it, we must start large conservation programs first, moving away from suburbs to smaller centralized communities, and away from cars and toward mass transit, while developing a broader energy mix for all energy needs with deliberation.

13   Nature's Way

Nature bats last and owns the stadium.
Paul Hawken and Amory and Hunter Lovins, *Natural Capitalism*

Humans have demonstrated the capacity to damage land and extinguish species from very early times (figure 13.1).[1] Our impacts have grown immensely as populations and technological prowess increased (figure 13.2), and by the twentieth century, they had begun changing the landscape in major ways.[2] To generations born since 1945, the pervasive human sculpting of natural landforms may even seem part of the natural scene. The effects certainly have reached a scale comparable to natural geological forces.

Humans directly displace approximately 35 billion tons of soil and rock per year worldwide, exceeding the work of rivers and streams and greatly surpassing natural erosion from glaciers or wind.[3] In the United States, road building, mining, construction, urban expansion, recreation, and military training and bomb testing move approximately 28 tons of earth per person each year—far outranking the world average of about six tons per person per year. Unintentional agricultural displacements are even greater—about 1,500 billion tons per year.

Natural processes obey the physical laws of motion and energy, which never take a break. In this book, we have tried to explain how human changes add to nature's effects (figure 13.3), in many cases multiplying the impacts of natural processes and causing severe environmental damage. Most people simply do not understand how the Earth works—but if nothing else, Hurricane Katrina's 2005 devastation of New Orleans made it clear that ignoring or underestimating the power of natural forces can severely imperil our present and future well-being.

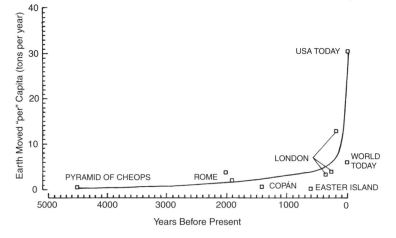

Figure 13.1 Estimates of per capita amounts of earth moved by selected relatively advanced societies in the past and today. From Roger L. Hooke. On the History of Humans as Geomorphic Agents. *Geology* 28 (2000), figure 4. Copyright 2000, Geological Society of America. Reproduced with permission.

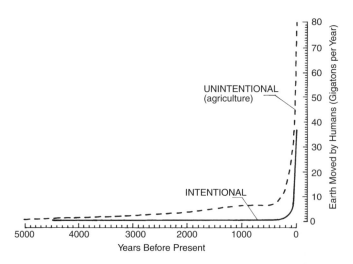

Figure 13.2 Estimate of the total amount of earth moved by humans today and at various times in the past, both intentionally (from mining, construction, etc.) and unintentionally (mainly due to agriculture). From Roger L. Hooke. On the History of Humans as Geomorphic Agents. *Geology* 28 (2000), figure 1. Copyright 2000, Geological Society of America. Reproduced with permission.

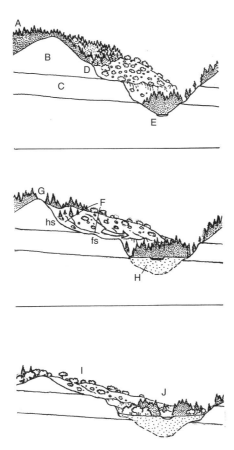

Figure 13.3 Roles of biology and geology in landscape evolution. (*Top*) Conifer forest (A) grows in soils developed on rock unit B on moderately steep slopes with low erosion rates. Open woodlands grow at lower elevations on soils developed mostly on underlying resistant rock unit C, which forms the steep canyon wall. The soils erode more easily below a sharp break in slope (D), due to the greater resistance of unit C. The canyon floor grows thick riparian vegetation on sediments deposited by the river cutting down into more easily eroded unit (E). (*Middle*) Erosion lowers the slopes, exposing rock unit B (G). Upland forests give way to more open lower elevation woodlands (F), accelerating mass wasting soil-erosion processes, which transfer sediments from unstable higher slopes (hs) to gentler foreslopes (fs). The canyon develops a broad flat floor, as erosion processes on higher slopes (arrow) and on the unit C cliff add sediment (H). (*Bottom*) Erosion has reduced the contrast between slopes on units B and C. Few areas are high enough to support conifer forest, so most woodlands are relatively open (I). The canyon has become a flat-floored meandering river floodplain (J) supporting mature riparian vegetation. The landscape's evolution to this older stage would take 10,000 years or more. Both fires and extensive logging on lower slope forests may accelerate these processes.

Irresistible Forces

Natural forces constantly are acting on and inside the Earth, building up land and tearing it down at the same time. Humans experience the Earth's internal heat engine through tectonic effects, which have segmented the surface along boundaries that define mobile tectonic plates. Plate interactions at the boundaries generate faults and earthquakes, raising hills and mountains or depressing basins and troughs.[4] Plate boundaries also contain most of the world's volcanoes, generated by the internal melting that creates igneous rocks.[5]

We think of eruptions and earthquakes as destructive, but tectonism and volcanism also are the main forces that add land masses. Over Earth's more than four-billion-year history, tectonism and rock-forming processes built the continents, and tectonic plate movements redistributed them around the globe. The 1980 eruptions of Washington's Mt. St. Helens demonstrate this destruction–construction balance of internal earth forces—they blew away one scenic mountain, destroying forests, lakes, and some of the people who clustered around it, and then began building a new one.

At Earth's surface, gravity and the sun's heat drive the surface forces that we contend with every day, everywhere. Solar heating, acting with Earth's rotation, creates the climatic belts and the oceanic and atmospheric circulations that run the weather machine—winds, ocean currents, and precipitation (snow, sleet, rain), or lack of them. Day–night and seasonal heating–cooling variations, due, respectively, to Earth's rotation and its axial tilt plus yearly revolution around the sun, are other important forces at the surface. The "natural systems"—hill slopes, rivulets and creeks, ponds, wetlands, rivers, lakes, bays and oceans, air and water currents, tides, and underground aquifer rocks—carry the energy flows and also are shaped by them.[6]

Like the front between two contending armies, the natural landscape that we live on represents a temporary compromise between building and destroying forces. Erosional forces are powerful enough to reduce all the continents to broad rolling plains, much like Australia, in only a few million years. In most places, tectonic and volcanic forces prevent that from happening.

Destructive surface forces break or degrade rocks into the rock debris called sediment. Internal earth forces help turn sediment into sedimentary rocks, under the right circumstances reforming organic materials in the sediment into petroleum and coal (see chapter 12). Both destructive and constructive forces also contribute to building soil, which sustains plants and the animals that eat them. For example, volcanoes created the attractive scenery of Mt. Rainier, Mt. Hood, and the other Cascade volcanoes over hundreds of thousands of years, but surface processes turned the volcanic lava and ash into rich soils.

Erosion and soils are like eggs and chickens. Erosion is the source of the sediment base for soil, but it also carries soil away. In the following, we discuss the processes of erosion first, and then the processes that build soils.

Erosion the destroyer

Gravity, running water, winds, and ice are the agents of erosion, which eat rocks away. Erosion lowers slopes and deposits sediment in valleys and other low spots (figure 13.3). Abrasive forces, including raindrop impacts, windstorm sand blasting, and rocks tumbling downhill and in water—all mechanically break rocks down into sediment (mechanical weathering). Also important are wedging effects along cracks, including the wedging of plant roots and water that freezes or salts that crystallize and expand in cracks. Natural chemicals partially dissolve the minerals in rocks (chemical weathering). The chemicals include slightly acid soil water and acids leached from mineral deposits, wildfire ashes, and decomposing plants and animals.

"Gravity," as the bumper sticker says, "is not just a good idea, it's the law." Gravity moves eroded sediment and soil masses from high elevations to lower ones. Water processes of sheetwash, rilling, gullying, flooding, and wind storms can transport sediments and dissolved chemicals long distances downslope and downwind.[7] All sediment-transporting systems simultaneously cause erosion.

Masses on the move

Primary gravity-driven "mass wasting" erosional processes embrace the many forms of landsliding—everything from rapid, chaotic snow or rock avalanches, mud flows, and debris flows, to the imperceptibly sluggish movements of loose wet soil and rocky surface rubble called "soil creep." Debris flows are among the most destructive of them all.[8]

Landslides and debris flows are common in coastal mountains and hillsides of California, Oregon, and Washington State, as well as in Rocky Mountain states. Debris flows originate in weak hillside soils. When water saturated and exposed to intense rainfall, they tend to give way suddenly and go careening downslope. The flows thicken as they take out plants—including mature trees—and strip slopes down to bare rock. The sounds of a debris flow charging relentlessly down a steep forest slope are as frightening as the roar of a tornado or the shock of a big earthquake, and can be as deadly.

In the wet winter of 1998, killer debris flows swept into mountainside settlements near Santa Cruz, California. Some of the residents had Katrina-like experiences. One man heard a roaring in the woods: "You couldn't miss that the mountainside was coming," he said. "The whole thing [was] just like a hundred locomotives coming down at us...limbs were going over our heads...." A "large, loud, scraping, moving-freight-train-type sound, then a giant explosion" heralded a debris flow that blasted off a steep hill slope and crashed through another family's back door. They found a stream of mud, rocks, logs, and shredded trees flowing in the back door, filling all corners of the house, and carrying their belongings out the front door. For over an hour they salvaged the pieces of their life from that river of mud. Another debris flow swept one house clean away. A woman watching television barely escaped from a downstairs room, while her husband, asleep upstairs, was swept away with the house and did not survive.[9]

Soil creep continuously affects virtually every soil-mantled slope. Because of its pervasiveness, soil creep destroys human structures more broadly than the more dramatic landslides and avalanches. Soil mantles vary from inches to many feet thick, averaging about 10 inches worldwide. Slopes with abundant plant roots have moderate to low soil moisture, so the soils rarely freeze in winter. Those relatively dry soils tend to be strong and have low creep rates of only fractions of an inch each year or less. But slopes with few plant roots tend to have higher soil moisture, and commonly the soils do freeze. When thawed, the soil may creep quite rapidly—as much as several inches or tens of inches a year. Creeping soil masses shift the bases of trees, telephone poles, and fence posts, progressively tilting and eventually toppling them.

Creeping soils are a large-scale problem in Alaska, where human-accelerated global warming has melted frozen subsoil (permafrost) into a thick soup. The soil's strength is so reduced that it no longer supports either trees or human structures. Rapid climate changes also are expected to increase the frequency of intense rainstorms and hurricanes, which will accelerate landsliding.

Rills and gullies

Gravity enhances the erosional force of running water, because raindrops hit bare soil with such great power that they dislodge particles and displace them downslope. In contrast, vegetated surfaces shield the soil from raindrop impacts, so allowing the water to collect and slowly seep into the ground.

Intense rainfall on hard soil will run off quickly downslope. During a downpour, runoff from steep to moderately sloping, unvegetated slopes initially flows in a generally continuous water film or layer, called sheetwash, which can scour sediment, sweeping up everything from gravel (pebble- to fist-sized rock fragments) to sand and smaller particles. Over a period of even a few hours, sheetwash erodes small channels, called rills, into the slope. Rilling initially resembles the coalescing raindrop rivulets on a car windshield, but later the rivulet channels deepen and larger rills merge, incorporating smaller channels. Sheetwash is the least erosive runoff effect, while rills yield 29 times more sediment from an equal area.

Any kind of land clearing activity and hill slope construction (see chapters 2, 4–5, 8, 12) can promote gully erosion, able to convert smooth, rolling hills to jagged wastelands. Gullies begin forming as rills below the top, or crest, of a slope. Eventually, the deep channel sides broaden and the bases deepen into full-fledged gullies.[10] Landowners trying to maintain hillside roads and structures commonly learn the hard way that gullies are destined to expand upslope. The rainwater flows downhill, but soil particles crumble off at the gully's top end, progressively extending the narrow channel toward the hillcrest.[11]

Rilling and gullying processes can be mitigated or even stopped but too often are allowed to progress unchecked. When unmitigated, they are especially destructive to agricultural and grazing lands in semiarid climates worldwide (see chapters 2, 3). Hoping to calculate the huge agricultural soil losses from sheetwash and rilling erosion, pioneering soil scientist W. H. Wischmeier and colleagues developed an empirical mathematical model, the Universal Soil Loss Equation.[12] Their idea also inspired a Wind Erosion Prediction System (box 13.1).[13]

Box 13.1 Universalizing Soil Loss Predictions

Both the Universal Soil Loss Equation (USLE) and Wind Erosion Prediction System (WEPS) are supposed to estimate the rate of erosional soil losses, to improve conservation planning and reduce air pollution. The Revised Universal Soil Loss Equation (RUSLE) improves estimates of erosive effects from grazing, military exercises, construction, mine spoils, logging, and recreational disturbances on nonagricultural forested and unforested lands.[a] But soil scientists still know little about how a soil is transported from the top of a hill to the base of the slope and into a river system.

The USLE cannot be universally applied to crop and grazing lands until it accurately incorporates gully erosion and the mechanics of sporadic or concentrated water flows. Neither USLE, RUSLE, nor WEPS yet take adequate account of interactions that detach soil particles from the surface,[b] the effects of the land's contours on overland water flow or blowing wind, or the erosive vulnerabilities of different terrains. They cannot predict whether each unit of eroded soil will come apart or remain intact, how it might be chemically or physically transformed, what characteristics it will acquire minute by minute, where any of it will be at any time, or where it may temporarily come to rest.[c]

As a result, the models yield annual estimates of soil losses from U.S. cropland erosion that vary from 2 to 6.8 billion tons. Model deficiencies can partly explain the large range of estimates but also represent controversy over the meaning of "loss." In some accounts "eroded" implies that soil is completely removed from the original site, that is, carried completely away in streams or rivers.[d] Others count soil stored near its source as a loss. Until the processes are better understood, the actual impacts of soil erosion will remain guesswork.

Notes
[a]U.S. Department of Agriculture, Agricultural Research Service. *Revised Universal Soil Loss Equation*, version 2 (RUSLE2 2005). Available: www.ars.usda.gov/Research/docs.htm?docid=6010; Natural Resources Conservation Service. *Wind Erosion Simulation Models, Wind Erosion Research Unit, Kansas State University* (2002). Available: www.weru.ksu.edu/weps.html. Both RUSLE and WEPS equations are given in the glossary.
[b]L. J. Lane et al. *The U.S. National Project to Develop Improved Erosion Prediction Technology to Replace the USLE, Sediment Budgets, Proceedings of the Porto Alegre Symposium, December 1988* (International Association of Hydrological Sciences Publication No. 174, 1988), 473–481; see also Rattan Lal, ed. *Soil Erosion Research Methods*, 2nd ed. (Delray Beach, Florida: St. Lucie Press, 1994), an excellent account of problems and approaches in soil erosion research.
[c]H. G. Wilshire et al. *Geologic Processes at the Land Surface* (U.S. Geological Survey Bulletin 2149, 1996), 17, 19.
[d]S. W. Trimble and Pierre Crosson. U.S. Soil Erosion Rates, Myth and Reality. *Science* 2000;289:248–250.

Dust in the wind

Even in arid and semiarid lands, wind erosion is subordinate to the effects of running water, but it still has significant impacts.[14] Strong winds gusting across a loose soil can dislodge and then move particles, which roll, tumble and bounce (saltate), similar to the way bed load sediment moves in a stream (described below). Bouncing sand grains kick more soil particles into motion with each skip and jump. The lightest dust-sized particles become suspended in turbulent air, developing dust clouds. High sustained wind speeds can turn the dust clouds into dust storms, which may travel long distances.

Wind-borne debris is a natural sandblaster, highly effective at etching, scouring, and pitting wood and glass, chipping surface paint and varnish, denuding plants of their leaves and stems, and burying animals and their burrows. In many windy mountain passes, intermittent dust and sand storms have grooved and scratched rocks on the ground for hundreds to thousands of years, creating streamlined "ventifact" (wind-fashioned) stones.

Dry lake beds, intermittently dry river courses, and associated dune fields are natural sources for sand and dust storms in the western United States. Plowing (chapter 2); heavy grazing (chapter 3); roads, especially dirt trails and tracks (chapters 5, 6, 11, 12); construction; and mining (chapter 4) all contribute major amounts of sand and dust that winds can carry. Sediment sources created by humans are largely responsible for the most destructive dust and sand storms recorded on the continent.[15]

The Delicate Organism

An intricate series of natural processes build the neglected resource called soil—a complexly modified mix of mineral and organic rubble. All living things must absorb nutrients to grow and thrive, and soils are nature's pre-eminent organ for conveying natural nutrients—nature's plant food—to the plants that higher animals eat. Healthy soils are the most vital long term life support for ecosystems, natural communities of interdependent plants and animals.

Soils probably develop in a series of discontinuous steps, which scientists still cannot define with any confidence. The basic constituent is sediment—the rock debris from erosion and weathering, described above. Chemical weathering also releases inorganic nutrients from the minerals that compose rocks, and converts what remains into clay minerals. Dead and decomposing plant and animal remains add critical organic carbon. The mixture is modified by air, rainwater, and chemical and biological processes that form new minerals, transport small clay mineral grains to deeper levels, and eventually create a layered soil structure (profile).[16]

Fertile, healthy soil is itself an ecosystem, with a rich and changing cast of living things (biota)—including microscopic bacteria, insects, earthworms and nematodes, fungi, and microscopic plants such as algae in its upper layer.[17] The biota decompose plant litter and maintain soil moisture, release soil-enriching or poisonous wastes, maintain open soil textures that roots can penetrate, and much more.[18] Particularly important are the bacteria, protozoans, microfungi, and algae, which variously fix mineral nutrients in their cells or release them. Some bacteria add oxygen to relatively insoluble natural chemicals, creating soluble forms—such as nitrates—that plants can absorb easily.[19] Other bacteria variously release hormones that enhance root growth; vitamins, proteins, and sugars that feed plants; and anti-biotics that protect them. The soil biota prey upon each other, balancing benefi-cial and harmful types. This balance preserves the fertility of undisturbed natural soils and the health of the plants growing in them.

Surface pebble accumulations, hardened mud, and biological soil crusts protect desert soils from erosion, helping them capture and store water and convey nutri-ents to plants (see chapters 3, 6, 11).[20] A natural soil's biotic balances also convey water-absorbing properties. For the most part, groundwater resupply, or recharge, largely takes place through soils (see chapter 9).

Climatic conditions, the type of parent material, and the shape of the land (topography) determine a soil's characteristics.[21] Compared to temperate climate soils (between the tropics and the Arctic or Antarctic circles), tropical soils are gen-erally nutrient poor because rainwater constantly flushes through them, dissolving nutrients and carrying them away. Temperate soils are exposed to less rain, so are more nutrient rich and more resistant to degradation and erosion. But most soils probably formed and became modified under varying climatic conditions, over long periods of time. Where soil is completely stripped to expose hard bedrock, a series of processes must break up enough rock, and weather it deeply, before biological processes can convert the debris into soil—especially on slopes (see chapters 6, 11). These processes probably require many millennia (figure 13.3).[22]

Many western U.S. desert soils probably are relics because they developed under past climate regimes. Some widespread southwestern soils formed tens of thousands of years ago when the climate was wetter than today, for example. Large tracts of the nation's most productive midwestern agricultural soils developed on loess, a fine rock flour that glaciers up to a mile thick scoured from North America's con-tinental bedrock over the past two million years of Earth history.[23]

Once scraped away, crushed under vehicles, or trampled upon, the delicate pro-tective biological soil crusts recover very slowly on desert soils (chapter 3, 5, 6, 11). They may regrow under shrubs in about 100 years, but in exposed areas the crusts' complex assemblage of organisms will not fully recover for many centuries or mil-lennia. Fossil soils cannot be replaced—in particular, the open textures and fertil-ity of loess soils cannot be repaired or restored.[24]

Ways of Water

Rain falling from the clouds may collect at the surface in ponds or lakes or flow from hillsides into large or small streams, including rivulets, creeks, and rivers. Eroded sediment, whether transported by mass wasting, washed down slopes in sheet floods or gullies, or carried in winds, eventually finds its way into stream channels. And rain and surface waters also can seep into soil from ponds, lakes, and streams to reach subsurface groundwater reservoirs, called aquifers (see chap-ter 9). In times of low stream flow, groundwater can drain from aquifers back into stream channels.

Of sediment and the river

The rainwater emerging from a gully carries loads of sediment into streambeds and onto valley floors.[25] The journey of a sediment particle depends on its size and weight. Larger gravel, cobbles, and boulders may be so heavy that ordinary stream flow cannot dislodge them, so they form gravel bar deposits that shift only in floods.[26] Very fine dust-sized particles normally flow with the water as suspended load, not touching the stream bottom. Sand-sized sediment particles can roll and tumble along the streambed (bed load) or may bounce (saltate) like windblown particles (see above).

Mid-nineteenth-century engineer James Buchanan Eads fashioned a primitive diving suit to experience the movement of sediment and water along the Mississippi River bed, reporting:

> The sand was drifting like a dense snowstorm at the bottom....At sixty-five feet below the surface I found the bed of the river, for at least three feet in depth, a moving mass so unstable that, in endeavoring to find a footing on it...my feet penetrated through it until I could feel...the sand rushing past my hands, driven by a current apparently as rapid as that on the surface. I could discover the sand in motion at least two feet below the surface of the bottom, and moving with a velocity diminishing in proportion to its depth.[27]

The stream's water volume and flow velocity determine its load capacity—the maximum amount of sediment a stream can carry. When a stream carries less sediment than its capacity, it will erode sediment from its banks or channel until the load reaches capacity. When the sediment load exceeds a stream's capacity, the excess begins to drop out, building sand bars, floodplains, gravel bars, and other sedimentary deposits. Floods cause a river to overflow its banks and carry sediment out of the stream channel and across adjacent land. As the waters subside, the sediment drops out. Flood-borne sediments build the river's floodplain on either side of its banks, piling up the greatest thickness right beside the river channel and forming natural levees.[28]

In tumbling streams on hillsides, sediment particles act like natural abrasives, scraping, scratching, and polishing the river bed rocks. Over time, flooding and abrasion lower the river channel. As the channel deepens, the banks on either side become higher and commonly steeper—and more susceptible to mass wasting and runoff erosion. This process sustains and enhances the natural sediment supply. Human impacts, such as mining gravels out of rivers, flooding sediment into them, or blocking the river flow with dams, refocus the rivers' energies to counter the particular disturbance. The river *has* to adjust, often by increasing erosion in some reaches and increasing flooding in others (see chapters 4, 8).

The lowest level to which a stream can erode its channel is its base level, and also its ultimate destination. The base level of most rivers worldwide corresponds to sea level, but it can also be a natural lake, such as Lake Tahoe in California; an artificial reservoir behind a dam, such as Lake Mead between Nevada and Arizona (see chapter 9); a saline interior lake, such as the Great Salt Lake, Utah; or a dry desert basin, such as Death Valley, California. River water flowing into the sea or a lake or reservoir loses energy to the larger body of sluggish water and drops its sediment, forming a delta.[29] Rivers flowing into a dry desert basin drop their sediments where the steep mountain front meets a flat basin floor, forming a moderately sloping alluvial fan that bridges the sharp transition.

Rivers flowing across broad coastal plains or other flat lands cannot erode more sediment and deepen their channels. Instead, they develop meandering channels that loop across the generally level landscape. Meanders let the river adjust its load to match its capacity by eroding sediment from its bank along the convex outer edge of meander bends and dropping sediment in shallow slackwater, so growing point bars along the meanders' inner edges. Sediment migrates downstream from bar to bar during periods of high water. Large floods disrupt this sediment transfer system altogether, many times completely reconfiguring the meander system. Tightly looped meanders may be cut off entirely, straightening the channel and leaving a U-shaped "ox-bow" lake. Many western rivers receive more sediment than the water can carry and drop the excess debris into the channel. Myriad streamlets meander around the debris piles of these "braided" rivers, joining each other, branching, and then joining again.

The water underground

Surface waters are seasonal and are relatively scarce in the arid west. Groundwater aquifers are replenished (recharged) from the surface episodically, and generally at a low rate of accumulation. To transmit water, soils and rocks must be permeable— but to store water they must contain open spaces (pores), such as the tiny openings between mineral grains in sedimentary rocks. Fractures and caves also provide openings. A rock may be highly porous or fractured, but unless the pores or fractures interconnect to provide throughgoing pathways, it can be relatively impermeable. Ideal aquifers are permeable *and* porous soils or rocks, which permit relatively rapid flow and hold abundant groundwater supplies. Stream sediments and sandstone rock layers make excellent aquifers. Clay soil, shale rock layers, and fractured igneous rocks are relatively to highly impermeable, and unfractured igneous rocks also pass and store little or no water.

In groundwater aquifers, the rock or soil pores and fractures are saturated, meaning that they are completely filled with water.[30] The top of a saturated zone is called the water table. In dry country, the surface water must seep through a zone of relatively dry soil or rock, called the unsaturated zone, vadose zone, or zone of aeration, to reach the water table (figure 13.4). In unsaturated zones, pore spaces and fractures may be barely moist or partly filled with water droplets or vapor. In both the groundwater and unsaturated zones, water either percolates—moving pore to pore along grain boundaries of the rock—or flows through interconnected fractures and other openings, called preferred pathways or macropores (see chapters 7, 10). Plants, leaf litter, and other organic debris have a large capacity to retain water and slowly release it into the soil or rock, so water seeps into the unsaturated zone more easily from vegetated than from denuded surfaces.

Unlike water stored in surface reservoirs, groundwater in aquifers continually moves from higher to lower levels on the subsurface water table.[31] In many cases, the groundwater and surface waters flow in the same direction, but faults and other underground conditions can deflect groundwater along different routes (see chapter 10). Where a water table comes to the surface on a hill slope or through valley or

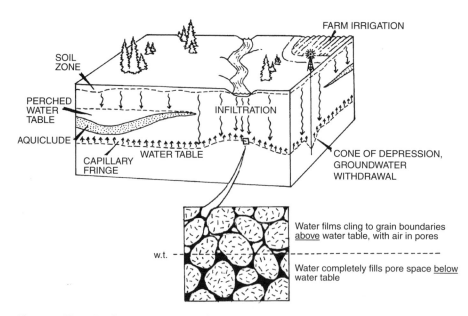

Figure 13.4 Surface water–groundwater connections. Groundwater comes from surface water that infiltrates (seeps) through the upper unsaturated zone (long dark arrows) to the water table (lower dashed line). At left and also below the irrigated farm, a relatively impermeable layer (aquiclude) intercepts water infiltrating from above it, creating a local (perched) water table. Surface water also infiltrates from the stream. The water table front is thickened by capillary pressures that force water upward an inch or so into the unsaturated zone (short dark arrows). Inset shows an idealized water table: Water forms films on individual sediment grains in the unsaturated zone above the water table, and fills spaces between grains in the saturated groundwater zone below it. Modified from B. W. Pipkin et al. *Geology and the Environment*, 5th ed. (Pacific Grove, California: Brooks/Cole, 2008), figure 8.10, 220.

canyon walls (perched water table), the groundwater flows or seeps out in unpressurized (effluent) springs. Some aquifers collect groundwater at high elevations and carry it to deep levels below valleys. This water may be under enough pressure to rise through cracks in rock and soil against the pull of gravity and emerge at the surface as bubbling artesian springs.

The time that surface water takes to get through an unsaturated zone to the water table and recharge an aquifer depends on the depth of the water table and the way the water moves. Recharging surface aquifers can occupy days or weeks, while refilling very deep ones might take millennia (see figure 13.5). The west's deepest aquifers hold (or held) water collected over thousands of years, which is considered "fossil" water. Water seeping through the body of a consolidated rock takes the longest time to move through the unsaturated zone. It moves much faster with pore-to-pore percolation along sinuous grain boundaries in unconsolidated sediments.

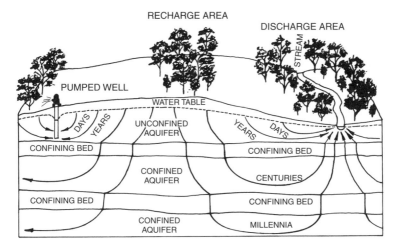

Figure 13.5 Time intervals generally required for recharging shallow and deep aquifers. Confining beds are relatively impermeable layers (aquitards) separating porous, permeable zones (aquifers). Arrows show directions that the groundwater moves, both naturally and in relation to a pumping well. From T. C. Winter et al. *Ground Water and Surface Water: A Single Resource* (U.S. Geological Survey Circular 1139, 2002), figure 3.

The former firmly held belief that rapid water movement through fractures can happen only in saturated zones no longer appears tenable.[32] Atom bomb tests and sloppy radioactive and other hazardous waste disposal practices have produced evidence of quicker water movement through arid unsaturated zones than previously believed (see chapters 7, 10).[33] The fast routes probably are preferred pathways, such as animal burrows, decayed plant root channels, and fractures.[34] Many pathway combinations are complex and unknown to scientists. Mused one, "The science of hydrology would be relatively simple if water were unable to penetrate below the earth's surface."[35]

The Human Touch

Flowing water and blowing wind disperse toxic materials through the environment,[36] commonly on sediment particles. Relatively insoluble pollutants, such as poisonous metal compounds, can be transported long distances when absorbed or adsorbed on sediment. Particularly during and after floods, stream flow increases from agricultural areas and mining districts (chapters 2, 4), military reservations (chapter 7), urban-industrial zones (chapter 8), and legal and illegal waste dumps (chapter 10), carrying sediments that are contaminated variously with pesticides, mine waste, cancer-causing organic solvents, radioactive substances, and other toxic compounds. Contaminated sediments get trapped in reservoirs and can become health hazards both for people who drink the water and for fish and other aquatic life (see chapter 4).

Whether diluted, transformed, or remediated, every substance that goes into surface waters or lodges in subsurface soil ultimately finds its way into groundwater as dissolved or suspended compounds. Water infiltrating an unsaturated zone can carry surface and buried pollutants to aquifers, but the rates of movement are not well known. Water may take unpredictable "chaotic" preferred pathways to the water table, which the mathematical models cannot emulate, so they require a host of simplifying assumptions. These assumptions may not be correct and so can weaken or invalidate the model results for determining how, and how fast, contamination can spread.[37]

Processes of fate

Wastes and other chemicals interact with a variety of natural processes that may change their forms and distribute them widely.[38] These so-called "fate" processes include the following:

- Volatilization: conversion from a solid or liquid state to a gaseous state

- Oxidation: increasing oxygen content or changing to a more positive ionic state, which can degrade many organic compounds under a range of environmental conditions

- Hydrolysis: interaction with or dissolution in water, mainly degrading organic compounds

- Sorption: attaching to the surfaces of dust particles in the air and other compounds, especially in mineral soils; includes ion exchange (water purifiers), exclusion and retardation, and dialysis

- Reaction: organic and inorganic interactions with aquifer constituents that may alter the character of both contaminant and aquifer materials[39]

- Biodegradation: biota-driven chemical transformations, mainly through the action of bacteria, fungi, and other microorganisms; some bacteria can digest petroleum and other hydrocarbons, converting them to water, hydrogen, and carbon dioxide

- Bioaccumulation: pollutant accumulation and concentration in plants and animals

Over time, bioaccumulation can build higher toxic chemical concentrations in an organism's tissue than are present in the environment, as exemplified by the DDT now in human breast milk.

The aggregate effect of fate processes on single contaminants is complex and depends on climate, the kinds of soils and contaminants, depth to the water table, and the kinds of rock materials in the unsaturated zone.[40] The dominant processes of volatilization and oxidation begin transforming organic and inorganic contaminants at the surface. In soils and the lower unsaturated zone, oxidation, hydrolysis, and many additional physical, chemical, and biological processes break down and dilute the toxic substances. The physical and chemical interactions are slow compared to microbial transformations, but below the soil layer microorganisms diminish in abundance and biological degradation processes become less effective.

Over short times and distances, natural chemical transformations may take some noxious chemicals out of the water. Natural processes cleanse most reliably when processing natural wastes. For example, we rely on the remediating interactions of soil bacteria to process septic tank effluents. Flowing surface and groundwater can disperse and dilute some pollutants, but the facile beliefs that flowing stream water can "purify" itself and that soil reliably cleanses percolating water of pollution over short distances have long been discredited. And large human populations overwhelm nature's abilities—too many septic fields commonly contaminate surface and groundwaters with septic wastes, for example (see chapter 10).

Current scientific research focuses on cleaning up industrial spills and dumps with natural remediation processes, such as using bacteria to decrease crude oil contamination from a spill in iron-rich soil.[41] Unfortunately, the cleansing effects are limited and also produce hazardous byproducts, such as methane, while increasing dissolved concentrations of potentially toxic metals. Thick unsaturated zones composed of unconsolidated sediment can accommodate contaminants in pore spaces and provide sorbing rock and mineral surfaces. Volatile materials trapped in pore spaces may convert to gaseous forms and migrate back to the surface, however.

But contaminants sorbed onto mineral surfaces are not necessarily stuck in the unsaturated zone, particularly if they are attached to substances that readily form "colloids"—solid particles so minute that their surfaces carry electric charges (see chapter 7, box 7.2).[42] Natural recharge processes flush contaminants of all sorts from the unsaturated zone downward into groundwater. Flushing may occur during periodic rainstorms, even in dry areas. Preferred pathways, such as fractures in rocks, can shorten the storage time of contaminants in the unsaturated zone by reducing the time it takes them to reach the water table (see chapter 10).[43]

Both inorganic and organic contaminants enter groundwater already dissolved in water, or as suspended particles or liquids that do not mix with water (figure 13.6). Fluids less dense than water will float on groundwater and can move through an aquifer separately. Fluids with similar densities to water may flow as separate layers within the groundwater, but a small proportion may dissolve in the water over time. Liquids more dense than water tend to sink, and part may dissolve and migrate with the water. Very heavy liquids will settle until they encounter a rock layer with reduced permeability (aquitard). After that, the movement is controlled by gravity and the direction of groundwater flow.

Many contaminants in flowing groundwater diffuse within the water, spreading from high to low concentration areas in the direction of groundwater flow, forming elongate plumes. Inorganic contaminants usually become sorbed onto aquifer materials, are broken down by oxidation, or are crystallized and coagulated. Soil and rock materials may pick up dissolved substances—including pollutants—from groundwater, exchanging them for other sorbed substances.

The different sorption and electrostatic interactions may separate a constellation of contaminants within a single plume (figure 13.7)—highly soluble substances travel more rapidly in the spreading plume than do less soluble ones, for instance.[44] Any change in the water composition can significantly alter the solubilities—and therefore the spreading potential—of dissolved substances.[45] The interactions can make the water more or less acidic (lower or higher pH), or richer or poorer in oxygen, either greatly expanding contaminant plumes or slowing their movement.

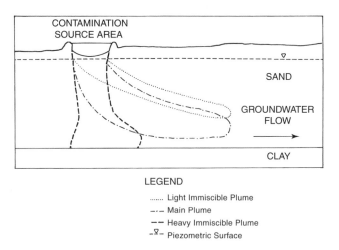

Figure 13.6 Schematic diagram showing migrating groundwater plumes carrying a number of different contaminants in a high level "unconfined" aquifer, having no overlying confining layer (aquitard or aquiclude). The horizontal dashed line (with caret symbol) is the water table (also called the "piezometric surface"). The clay layer at the base is an aquiclude. The density of the contaminating material determines where and how the plumes migrate. Modified from U.S. Environmental Protection Agency (1985). Available: www.epa.gov.

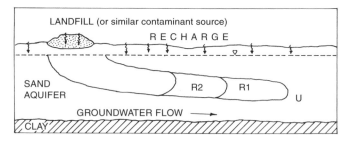

Figure 13.7 Continuous addition of leached landfill contaminants to groundwater, forming a contaminant plume. Some contaminants move in or along with the water (U), but others (R1, R2) interact with natural constituents in the water or the aquifer and may spread at different rates. In the diagram, R2 is more retarded than R1. Modified from MacKay et al. (1985), as cited in J. A. Fallin. *Hydrogeology, Environmental Contamination, Assessment, Remediation, and Effects of Superfund (CERCLA), and Other Hazardous Waste Sites in Western North America* (Boulder, Colorado: Janus Press International 1994).

The kind and extent of natural chemical exchanges depend on the type and abundance of the aquifer rock or soil and on the presence or absence of colloids. When waters from different sources mix, or when water is filtered or evaporates, colloids tend to coagulate into solid masses. Among the most common colloid types is silica (SiO$_2$), which can congeal as agate, chalcedony, or chert nodules, depending on the chemical characteristics of particular locales.[46] Rivers interacting with saltwater tend to deposit colloid-rich sediment in coastal deltas.

Colloids generally have the net effect of improving water quality because their electrically charged surfaces tend to grab up highly toxic contaminant ions, "scrubbing" or "depoisoning" groundwater, and most streams and rivers (see chapter 10). Without colloids, "a number of biologically damaging substances would [eventually] accumulate in ocean waters."[47] As a mass of water grows cleaner, its colloidal cargo may become quite poisonous, however. The contaminated particles may not congeal but spread widely in groundwater and into streams (see chapters 7, 10).[48] Intermittent desert streams may deposit colloids with attached toxic materials, including radioactive nuclides, on dry lake surfaces in arid basins. When dried, the powdery colloids and attached contaminants are vulnerable to strong winds that can distribute them far and wide.

Natural processes can neutralize acid mine waters, transforming toxic metals so that they drop out of the water and crystallize as metal salts on sediments. Again, these effects require particular rock compositions in contact with the acid water. The purifying effects can be reversed if the water flows into different rocks or if the amount of contamination overwhelms (saturates) the transforming rocks. Very little is yet known about interactions between soils and most toxic industrial organic compounds.

Erosion Future

Natural erosional processes reshape human excavations and construction in the same way they attack newly erupted volcanic deposits or a newly exposed fault face. Natural processes generally work slowly compared to human lifetimes, but the effects build up over time. Our huge construction projects, population growth, and resulting pollution expand the destructive aspects of natural erosion by orders of magnitude, very often mangling people's lives in the process.[49]

The impacts of human activities have grown as large as natural ones. Nature-amplified human disturbances in the arid western United States are a slower moving disaster than the killer hurricanes, landslides, and floods that grab our attention, but they have taken many thousands of acres out of productive use and caused widespread air and water pollution. In some places, those effects have become devastating for human health and economic life in the west.

Human-enhanced erosion preferentially removes soil faster than nature could possibly resupply it, along with whole plant and animal communities. If droughts become more frequent, soil losses likely will accelerate—especially where human activities have made them vulnerable. The multitude of contaminant sources, the multitude of chemicals introduced into the environment, and the ever-increasing population make a daunting task of protecting the vital resource of groundwater in a dry region. Soils provide primary treatment sites for many forms of pollution, so soil losses also reduce water-cleansing potentials. Active remediation of contaminated aquifers is a big business today, and much funded research aims to correct past mistakes—but the data have large gaps.[50] Preventing contamination is both cheaper and more certain.

The continuing and accelerating losses of fertile soils and clean water are the greatest long-term threats to our life support. Human activities that destroy more and more soils will add to the depletion of other resources to make us more and more vulnerable to future climatic variations and societal upheavals. The words of soil scientists Jacks and Whyte are worth bearing in mind: "Below that thin layer comprising the delicate organism known as soil is a planet as lifeless as the moon."[51]

Conclusion The Needs of Our Posterity

Our past history and security have given us the sentimental belief that the things we fear will never really happen—that everything turns out right in the end. But, prudent men will reject these tranquilizers and prefer to face the facts so that they can plan intelligently for the needs of their posterity.
Admiral Hyman Rickover, "Energy Resources and Our Future" (1957)

In 1957, Admiral Hyman G. Rickover gave a speech of breathtaking prescience, which warned that the United States would face a resource-depleted future unless it controlled energy consumption.[1] Speaking to an audience of American physicians, he advised that the nation needed to turn away from the allure of affluence and take a truly conservative path—that is, the path of conserving resources and carefully planning consumption levels, to keep itself rich in oil and other mineral wealth for many generations.

Rickover's warnings are especially relevant to the message that we authors hope to convey. The western states have contributed much of the nation's timber and mineral resources, and most of its major public works projects—including damming of the region's rivers to provide water supplies and power to western farmers and cities. The whole nation has benefited from exploiting these resources, but now the oil and mineral resources are highly depleted, and clean water also is in short supply. Rickover questioned some of the apparent benefits, however—"Much of the wilderness which nurtured what is most dynamic in the American character has now been buried under cities, factories and suburban developments where each picture window looks out on nothing more inspiring than the neighbor's back yard," he said. "The nation's resources—its lumber, mineral ores, and especially its petroleum—have been used for this remaking of a once-agrarian country into a relatively sterile urban–suburban landscape."

We can add only that this transition from natural lands to sterile urban–suburban and agricultural affluence is the force that has degraded our air, water, and soils. These are nature's gifts, open to everyone, which the lives and well-being of all creatures in the western states, including its people, depend upon.

Rickover feared mindless consumption and population growth, observing, "In the 8,000 years from the beginning of history to the year 2000 A.D. world population will have grown from 10 million to 4 billion, with 90% of that growth taking

365

place during the last five percent of that period....Calculations give us the astonishing estimate that one out of every 20 human beings born into this world is alive today." And that was 1957. Today, the United States holds only 5% of the world's people, but the largest population of affluent, acquisitive, natural-resource consumers on Earth. Building and supporting the suburban landscape, in particular, has gobbled up the lion's share of resources.

Admiral Rickover foresaw that supplying an ever-growing, ever more consumptive population could undercut America's ability to remain a powerful nation, noting,

> We began life in 1776 as a nation of less than four million people—spread over a vast continent—with seemingly inexhaustible riches of nature all about. We conserved what was scarce—human labor—and squandered what seemed abundant—natural resources—and we are still doing the same today....Merely to supply us with enough water and to carry away our waste products becomes more difficult and expensive daily.

He said a mouthful. The policies of extracting resources from abroad for U.S. consumption and exporting our lifestyle to other countries together have contributed to ravaging many parts of the world, in addition to our own back yard.

As we noted in chapter 13, humans now move nearly as much soil and rock as the world's rivers, glaciers, and winds can erode and carry away.[2] Agricultural disturbances unintentionally add a lot more, mostly in the form of valuable topsoil (see chapter 13, figure 13.2).[3] The United States alone intentionally displaces five times more soil and rock per capita than any other country in the world, and dramatically more than other developed countries (see figure 13.1). Every year, the American mining industry displaces about 3.8 billion tons of earth materials, house builders move nearly a billion tons, and road builders move three billion tons for paved roads. The total comes to 7.6 billion tons per year, without considering building and maintaining hundreds of thousands of miles of unpaved roads and nonresidential buildings; dredging rivers, harbors, and reservoirs; replenishing beaches; or landscaping.[4]

We authors have tried to point out that natural forces are ever present and respond to everything we do. Following quarrying, tilling, burning, and mechanical stripping of vegetative cover, natural erosion takes over at an accelerated pace, filling the reservoirs with more sediment from disturbed lands than from natural landscapes, spreading our wastes through the environment, and destroying more and more natural habitats. The effects of American agriculture alone have displaced more than a third of plant and animal species listed or proposed for listing under the Endangered Species Act, and urbanization threatens another third.[5]

This is all very bad news. All of these human actions have certainly put the west at risk and much of its potential for future viability has been lost. But losing is not the same as lost. In writing this book, we wanted to follow in Admiral Rickover's footsteps, telling the difficult truth to "prudent men" and women so that together we can start to "plan intelligently for the needs of [our] posterity." We have attempted to increase readers' scientific understanding of land-use issues and to document the physical and biological ravages from a century and a half of unsustainable development in the western states. We sincerely hope that this book will provide fuel for a more visible public debate and particularly more individual actions.

We hope that a better understanding of the Earth and its processes, the "natural capital" of Paul Hawken and collaborators—rich soils, clean air, and clean water resources and the processes that keep them rich or clean[6]—also will help to turn our society toward less destructive lifestyles. This is not just a touchy-feely goal—it includes a self-interested economic incentive. If Americans came to understand and value nature's processes enough to preserve chemical-free the critically important natural areas around our communities, and did not overwhelm them with our presence, soils would detoxify and decompose many of our wastes; forests would give us clean water; beneficial insects, bats, and birds would pollinate vegetation, including crops, and disperse seeds; soil microorganisms would recycle nutrients; naturally produced gases would screen out dangerous solar radiation—and much more—for *free*.[7]

We also aimed to dismantle some of the romantic myths that have driven resource depletion and destruction, especially in the west. Although rarely stated, one myth sets modern humans apart from nature, singularly entitled to corner and harness natural forces and dominate all other life forms without limit.[8] Oscar Wilde's definition of nature as "a damp place, above which fly thousands of ducks, uncooked," exemplifies that worldview. But we, like all living things, need clean water to drink and clean air to breathe. Indeed, we are what we eat, and healthy foods must grow in healthy soil and from clean water.[9]

Americans also believe in the "postmodern" myth that we authors call "technological optimism"—the belief that technologies yet to be developed can and will fix any problem, including resource depletion and environmental degradation. This myth sustains the belief that economic forces will painlessly and cheaply replace depleted resources with others that work just as well. Every proposed technological fix depends on a resource base, however, and will exact an environmental price. Unlike nature's processes, technological fixes are anything but free, and energy costs represent the major proportion. Remediations for the current level of human-caused erosion and pollution—and every invented process that tries to emulate nature's free services—consume ever more nonrenewable mineral and energy resources and generate ever more poisonous wastes.

From the standpoint of simple prudence, public policy discussions about the future should be based upon what is achievable now, rather than deferring mounting problems to solutions yet to be invented. But without lowering consumption and shifting away from toxic materials, technological optimism can only pile up costs and defer the payment for today's excesses to an uncertain future of scarce resources and gargantuan, poisonous waste dumps.

The storied "winning of the West" by stalwart individualists constitutes its own persistent mythology. But in truth, the west's economy grew at the public trough and never has left it. Massive investments of national treasure opened the frontier to settlement and then did away with it in a bare 50 years. Government surveys established road and railroad routes through unknown lands; the federal government subsidized railroads and then interstate highways. The government distributed nearly free homesteading lands, sent military forces and built forts to help settlers displace Native Americans, and supported state grammar schools, normal schools, colleges, and universities through federal land allocations and financial programs.

Industrial development in the west still benefits from cheap, unrestrictive permits for logging and grazing on public lands and nearly free access to public lands for prospecting, mining, and mineral processing. Federally built dams and irrigation projects opened arid western lands to agriculture, industry, and expanding cities—and still provide most of the west's water and hydroelectric power. There is much, much more. These public investments were not only crucial to "winning" (exploiting) the west—they remain fundamental to its vaunted independent lifestyle. Now the public needs to demand a return on its investment through reallocating government expenditures to better protect our *natural* capital.

Many ideas for resource-saving changes lurk within, and particularly at the ends, of every chapter in this book. They pose an overriding challenge, however, because most of these changes require more than altering daily habits—they require changing our *outlook* from always entitling the individual, to restricting individual choices *for the common good*. This is the focus required to negate Garrett Hardin's "Tragedy of the Commons"—the overexploitation of a finite resource to which everyone has free and unrestricted access.[10] There are many cooperative models in traditional cultures, and some modern societies still use them to reduce the potentially bad effects of farming, grazing, and logging on public or shared lands. In chapter 1, we cited the example of selective logging in Switzerland. Another is the system for sustainably sharing grazing space on traditional high-altitude summer pastures, such as a Norwegian "støl." If Americans can agree to use ingenuity similarly to improve both the quality of our lives and our environment, we truly will enrich both our communities and ourselves.

Some might characterize that idea as "socialism," which Americans tend to equate with dictatorship and tyranny. But that is a false correlation.[11] Democracy—the rule by popular will—should and can embody community values, as many European democracies demonstrate. Hawken and colleagues pointed out that our system of capitalism neither balances individual gain against the community needs nor correctly values natural services or resources.[12] If instead American capitalism factored in costs of resource depletion and degradation—including the costs of restoration and performing nature's functions ourselves (not to mention health costs)—and if commercial interests had to pay compensation for losses of natural capital, environmental destruction would decrease significantly.

Unfortunately, our society has failed to keep representative government "checks and balances" working for the common good. Only through legal challenges does America's democracy make the government enforce its own acts and policies, or follow the spirit of the laws' original intents. Both the environment and human populations have paid a heavy price for this failure. Abuses follow from the weakness of the ordinary citizen, or citizen groups, compared to the political influence of business lobbies and their sympathetic congressional allies, many of whom have grown rich on government contracts and subsidies. Ironically, this system mangles free markets.

Hardin pointed out that bureaucratic agencies are part of the problem:

> Since it is physically impossible to spell out all of the conditions under which it is safe to burn trash in the back yard or to run an automobile without smog control, we

delegate the details to agencies....The result is administrative law, which is rightly feared for…"Who shall watch the watchers themselves?"…[A]dministrators, trying to evaluate the morality of acts in the total system, are singularly liable to corruption, producing a government of men, not laws.

This is why polls that show a majority of citizens concerned about "the environment" do not protect environmental laws very effectively. We can make the system work for us only through concerted citizen actions, involving informed people who back achievable solutions to critical issues, with well-established goals.

In the absence of meaningful federal initiatives that address the west's and the nation's critical issues, many individuals, communities, private organizations, and local and state governments are filling the void. A multitude of committed people are addressing all of the major threats to our natural life support systems: Earth's climate, vulnerable food sources, clean air and water shortages, the growth and well-being of our children threatened by a synthetic environment, and the education of all citizens in how the Earth works. They exhibit the real ingenuity and progressive can-do attitudes that we once thought of as the mold of America and Americans. This book cites, and its Web site (www.losingthewest.com) provides links to, many of these organizations. They all need abundant support, no matter what political party is in power.

Concerned westerners can start by taking a few stands with the potential for making huge changes very quickly. Many public interest organizations already are campaigning for these changes, but they need the backing of a large community of informed and concerned citizens. The easiest and cheapest changes begin with conserving soil in several ways: first, by opposing campaigns to proliferate roads, and closing and repairing the unneeded old and newly blazed routes—especially across steep timberlands and through wilderness areas. The most effective ways to start conserving agricultural soils is to *stop* government subsidies for an unsustainable food production system and *start* supporting small farms that employ soil-conserving and soil-building practices. If the nation needs more utility lines, citizens can demand that permitting agencies require new lines to stay within old corridors and not allow additional routes. Grassroots organizations can concentrate on changing the public image of national forests to emphasize their natural gems, many equivalent to national parks. Citizen movements can and should insist on restricting the numbers of people who can gather on public lands so that trampled camp and play sites can recover. Innovative population convergences can take place just as well on private lands, letting the marketplace work its magic.

Campaigns also are under way to protect public grazing lands, by taking cattle and sheep off the lands entirely, by strictly regulating their numbers and grazing intervals, or by requiring adequate payment for grazing permits in the public interest. An imperative is to stop paying the externalized costs of businesses and industries with our taxes.

A public clamor could change the costs of grazing public lands to reflect the value of the land to all of its owners. Mining and logging permits ought to require the miners and loggers to clean up their own messes with their own money. And miners should pay significant royalties for extracting ores and commodities belonging to all the people. Mineral values on public lands belong to all Americans and should not be sold for a pittance to foreign investors with no stake in the land's long-term health. For example, efforts to change the ancient Mining Law for the common good arise nearly every year in Congress and require only support of a concerned and watchful citizenry. Loggers also should be obliged to deal with the reality of the marketplace by paying as much for permits to log on public as on private lands, and bear all of the costs of restoration, plus maintenance and fire protection.

Americans must weigh the monetary, energy-saving, and spiritual values of wilderness services against the clamor of anti-public land campaigners, who have misread Garrett Hardin's message about unregulated commons. Privatization is not the answer, because everyone may need access to those lands in the unknown future. We can join campaigns to elevate and expand wilderness preservation as part of our national treasury and strictly limit its uses. That change alone could refocus forest management practices along more ecological lines.

We also need to start valuing water as a scarce resource in a land of little rain, and finally give up the idea that we will find unclaimed supplies in some remote corner, to be bagged for transport, or piped far away at huge public expense. Examining the costs in terms of energy and other resource consumption will explain the reasons. Eventually, we will need to rework the old river water-distribution agreements to meet the realities of drought, depletion, and pollution. Each individual can start finding more efficient ways to use water and help develop programs that encourage putting water-saving devices and energy-saving retrofits into all buildings.

We must stop looking for new garbage dump sites and *start reducing the garbage to dump.* Locally and nationally, we can clamor for restarting the Superfund program and for cleaning up as many hazardous dumps as possible now, while fuel prices remain relatively low.

Cutting back on waste production means consuming less, of course—the easiest and cheapest approach. The biggest reductions could be made immediately by eliminating most packaging—especially plastic containers, with plastic shopping and storage bags, bubble-paks, and water bottles at the top of the list. This step would help reduce the burden of plastic on every part of our environment—along with wildlife–plastic encounters—while saving petroleum. Returning to using and reusing glass bottles would save even more.

Reducing per capita consumption is the key aspect needed to turn Americans toward true sustainability, which means living within the bounds of what nature provides. Other forms of sustainability, such as "sustainable development" and "sustainable growth," are less promising. The 1987 report to the United Nations World Commission on Environment and Development[13] defines sustainable development as meeting the needs of the present without compromising the ability of future generations to meet their own needs. The plight of third-world nations shows that population increases can lead to unsustainability, even at low levels of affluence and technology. The U.S. experience shows that growing affluence can lead to unsustainability even with a relatively stable population. This implies that sustaining a high-tech world has to mean shrinking both populations *and* affluence. The term "sustainable growth" is therefore an oxymoron. The potential for sustainable *development* is an unproven hypothesis.

So here is the rub: reaching for true sustainability represents a profound change to many aspects of current American culture and economic life. Yet it takes us back to the practices of only 60 or so years ago. Americans, particularly in the western states, have a long way to go before reduced per capita consumption takes them even to the level of most Europeans, who live quite well and tend to register higher satisfaction with their lives than Americans.

Whatever political choices or philosophical shifts the public and the nation is able to make, we soon will feel the rigor of nature's law in the form of diminishing energy supplies at much higher prices. At the same time, more-rapid-than-natural climate warming is driving more-rapid-than-anticipated melting of land ice, which is raising ocean levels faster than any current prediction. The encroaching seas will force many American populations to shift away from coastlines over the next 50 to 100 years, increasing urban population densities. Eventually, people and industry will be forced out of most of today's beachside towns and sites. For example, a very recent study foresees that by 2100 rising water levels in San Francisco Bay could combine with collapsed ground levels from overpumping California's deep Santa Clara Valley groundwater aquifers to drown Silicon Valley.

As nature forces these changes upon us, an unregulated and resentful population might make terrible choices, both politically and environmentally. Since Admiral Rickover could not persuade 1950s Americans to conserve U.S. oil, concerned citizens must now seriously plan to cope with this kind of future.

Admiral Rickover's foremost concern was that the United States maintain a dominant role in the world, knowing that energy and strategic mineral resources are critical elements for attaining that objective. Geographer Jared Diamond takes a broader view in his book *Collapse*, which explains the forces that brought whole civilizations and colonies to their knees. Diamond's work indicates that cultures were especially vulnerable if they could not support themselves from the local surroundings but came to rely on long-distance transport for many food supplies. Of course, those long-gone losers did obviously dumb things, "like cutting down their forests, overharvesting wild animal[s]...watching their topsoil erode away, and building cities in dry areas likely to run short of water." Also, "They had foolish leaders...who embroiled them in expensive and destabilizing wars, cared only about staying in power, and didn't pay attention to problems at home."[14]

In Professor Diamond's tales of unsuccessful past societies and governmental systems, we authors find echoes of our own concerns—such as the severe effects of clearcut logging, industrial farming, road building, overgrazing, and military and mechanized recreational activities on fragile western soils—not to mention sprawling populations living on an overstretched water credit card. Diamond contends that the United States is much better off because it can learn from the past, and from our own wisdom, which we interpret to mean scientific observations and common sense. But is he right?

A small population could have lived well for thousands of years on America's bounty, even at our current level of consumption. But the United States has grown to a population of 300 million, which is still growing and eating its way through the remaining forests, minerals, soils, and water supplies—the commons shared by all living things. Our increasingly hazardous wastes also are growing. The result is polluted land, air, and waters, which certainly affect our health and may be threatening our lives.

In 1957, the insightful Admiral Rickover noted, "We are rapidly approaching the time when exhaustion of better grade metals will force us to turn to poorer grades requiring in most cases greater expenditure of energy per unit of metal." Again he hit the nail on its head. Even as he spoke, the U.S. mining industry was exploiting quite low grade ore deposits. Today's mammoth western U.S. mines extract nothing that an old-time prospector would have called an ore. The gaping mine pits drain and despoil precious groundwater in an arid land.

Current drought highlights the west's looming problems of oversubscribed rivers and depleted groundwater aquifers, further limited by water pollution. This is a gap that cannot be filled. Our depleted water and energy supplies are obvious consequences of our long denial of the region's aridity and our heedless resource use. Industry now contemplates even vaster destruction of western lands and water, which will result from exploiting the very low-grade "oil shale" energy resource, with no guarantee of obtaining even as much energy as the production requires.

Admiral Rickover predicted that our consumptive lifestyle eventually would pit America's dwindling "conventional" petroleum production against spiraling oil consumption—another serious gap. "Looking into the future, from the mid-twentieth century," he said, "we cannot feel overly confident that present high standards of living will of a certainty continue through the next century and beyond." Our current reliance on foreign oil imports would have disturbed him greatly, because it signals a high level of vulnerability. And he foresaw the possibility of warfare over oil and other resources.

The soil losses, declining grades of metal ores, and even the water gap could be overlooked as long as energy prices remained low and seasonal storms kept the reservoirs filled. Even the petroleum gap could be overlooked as long as cheap imports flowed into it. But now all the problems that Admiral Rickover foresaw are coming together with the globally warming climate and driving us to a tipping point in our history—one that, in retrospect, a Babylonian, Greek, or an ancient southwestern U.S. Anasazi might recognize.

Here is the risk that we now face: if the United States as a whole cannot shift to a less Earth-degrading and more truly sustainable course, the last raids on the west's fossil energy resources will degrade its land and water resources to the point of no return. We are seeing this happen today on the lands ruined by coal-bed methane extraction and in the wildernesses slated for barely economic oil exploration. Keeping on this course will face us with an abrupt adjustment to a world of impoverished resources, lacking the means for peacefully transitioning to a different lifestyle. It very likely will erode American democracy and bring hardship, conflict, and strife of incalculable scale. So there is much to be done.

Before discussing the energy future, however, we have to cast aside a few more illusions. In presenting the scientific facts and limitations of energy transformations, we have tried to explain that *there is no free lunch with energy*. There isn't even a cheap one. The economist's technologically optimistic belief—that the higher prices of depleting gas and oil will make alternative fuels and electricity sources more cost-competitive—ignores a very significant energy fact: the rising price of petroleum will raise all other costs, including the prices of alternative fuels. The more energy required to produce those alternative fuels, the pricier they will be.

Everything will become hugely more expensive. Pick one—water? Pumping water *drinks* energy. It accounts for around 20% of California's energy use. For this reason, the California Energy Commission has named water conservation the main key to energy conservation. Agriculture? To grow each calorie of food product, today's agriculture spends *at least one and as much as 500 calories* of some energy resource. A significant proportion of that energy goes to pump water. Congressional support for corn ethanol, both as a gasoline additive and a fossil-fuel substitute, already is raising corn prices. Higher corn-ethanol prices will raise feed prices, seed prices, and food prices at the supermarket. In addition to higher market prices, the largest corn growers already get large taxpayer subsidies—and now they have an additional subsidy from growing corn *for ethanol*. This does not look like any form of capitalism—it seems more like Marc Reisner's characterization of the west as a durable welfare state.[15]

Rising oil prices could quickly balance out the taxpayer subsidies that keep conventionally grown food prices low and make organic food prices competitive. This would have the soil-friendly effect of squeezing today's industrial agriculture and supporting smaller organic farms, which use 30% less fuel, much less water, and few or no pesticides. When this happens, government policies will need to emulate the 1930s by supporting small- and medium-sized farmers who follow conservation rules for saving and rebuilding soils. We are going to need many small local farms, because long-distance food transport might become impossible by mid-century. For this reason, federal and state farm programs should have the intent of bringing Americans back to the land. Crop support should be limited and provide for payback—again along the lines of the New Deal crop loans. Government policies should give most support to crops and farming practices that do the least harm to our lands.

Large-scale animal rearing takes the most energy per calorie of food and is an enormous water and land polluter. Increasing energy costs will drive supermarket meat prices sky-high, therefore. The best model that we have encountered for raising animals for meat, while decreasing energy costs to very low levels, is the ecologically based system of management-intensive grazing.[16] Ecological limitations restrict the method to relatively small farms, which also raise some feed and market-oriented vegetable crops. Every population center eventually will need small meat-raising farms nearby, along with local slaughterhouses. The meat will probably cost more than now—but as fuel prices increase, locally grown meats probably will be competitive with supermarket meat from energy-guzzling factory farms.

The same issues apply to logging and mining. Tree cutting will have to shrink to smaller scales, and extremely low-grade mineralized materials will become too costly to mine. Mining industry newsletters now remark that the higher costs of diesel fuel, ultralarge tires, and other supplies squeeze potential profits. All are related to energy costs. To provide metals and other strategic materials, we suppose that cost balances will begin to favor large-scale recycling of wastes.

As energy supplies tighten and prices rise, all vehicle use will become more expensive, of course. Long-distance drives to vacation spots will be relegated to the pages of cultural-history books. Truck-transported goods will be much more expensive than those carried by rail, and urban-suburban zones with no passenger trains, trams, or streetcars will pose problems for commuters. Once energy prices have risen enough to severely decrease car traffic, we doubt that far-flung western populations now lacking trains and tramlines will have the financial resources to pay for them. Like Los Angeles and some other commuting centers, every county and city should be revitalizing train systems now.

Sprawling suburbs are likely to depopulate,[17] and large urban areas will be at risk of societal fragmentation. Like us, postpetroleum strategists foresee a need to relocalize populations in smaller towns, rebuild localized manufacturing, and get their food locally, including from their own gardens. As people require materials for building whatever new infrastructure our society can afford, they will recycle more and more, and develop end uses for the recycled substances. This reorientation has already begun. Now, if most Americans can become very conserving, the nation can go for a long time by dismantling the surplus of unused buildings and cars no longer in use.

The problem scenarios and possible mediations for a petroleum-poorer future are examined in many fine books and related Web sites.[18] In addition, many hundreds of organizations nationwide are working for positive changes to national policies that principally affect the western United States and that could help prepare us for a more resource-competitive future. Many of the names and Web addresses of these organizations are in chapter endnotes. The Web site for this book provides links to many of the sites and abundant other information sources.

To meet the coming "Long Emergency"[19] of high energy prices and general resource unavailability, people will have to help each other. We saw that spirit on display immediately after the 9/11 terrorist attacks on the streets of New York City and elsewhere. Sadly, it was later sullied by the federal Environmental Protection Agency's disregard for the health of workers exposed to toxic chemicals at ground zero. In 2005 the state and federal response to Hurricane Katrina showed that communities cannot rely on help from higher governmental levels during or after large emergencies without a radical change in national leadership. If this problem continues, small communities will have to become the lifeboats for keeping American values afloat.[20]

So this is the bottom line, and also the starting position: Americans have to start caring about the survival of small communities, their local towns, and their local resources. Towns and cities need to establish protective zoning policies for streams and riparian areas. School programs should connect every school child to a stream in their neighborhood (including inner-city neighborhoods), to ensure that they will understand the economic need for protecting and defending water courses and restoring riparian zones. All who can should support local farmers' markets. Or grow your own food if you can, join campaigns to save and restore land for growing food in your town or city, and help spread the idea. If common gardens will be watered with groundwater, urban governments will need to inventory contamination sources—including buried fuel, solvent, and leaking waste storage tanks. Protective zoning policies will be needed to keep future potentially polluting activities, including housing developments, far away from wellheads and groundwater aquifers' recharge areas.

To survive inevitable long droughts, every water supply district or agency will need to develop and implement programs for managing surface streams and groundwater aquifers as a single, interactive resource. Citizens in every western state must insist on protective rules, both governing groundwater withdrawals and limiting the import and export of local water supplies. The saved water must not go to expanding development, which would continue pushing our Land of Little Rain far past the point of sustainability. We may be past that point already—but we can still turn back.

Although the west may be severely at risk, the present authors do not believe that it will be lost. There is much room for optimism that western folks will find ways to extend what remains of our natural bounty farther into the future than our current course can take us. Americans are innovative and able to adjust to adversity as well as Europeans, and easily can adopt more conserving lifestyles. The challenges will be great for today's young people—tomorrow's engineers and entrepreneurs—but the world has many models for them to learn from. For example, future engineers face marvelous opportunities within the challenge of reshaping the industrial enterprise, which are being pursued by small institutions and some larger companies outside the United States. It is high time that Americans joined the effort and began reaping the profits.

The western United States has left its frontier days well behind. We can no longer pillage the land and move on because there is no other place to go. We cannot even get into space without support from Earth's bounty. Now is the time for learning how to live better with nature. If we cannot, our proud nation could go the way of many civilizations that preceded ours.

Appendix 1

Conserving U.S. Public Lands: A Chronology

Year(s)   Event or Law[1]

1781   The states cede former "Indian territories" of British Colonial adminis-
          tration, west of the Appalachian Mountains, to "public domain," making
          them the first public lands opened to settlement and development.

1803   The United States purchases the Louisiana Territory from France, more
          than 800,000 square miles, extending from the Mississippi River to the
          Rocky Mountains.

1862   Congress passes the Homestead Act, granting 160 acres of public land to
          any individual or family for grazing and agriculture.

1865   Yosemite Valley and a Hot Springs, Arkansas, site are removed from pub-
          lic domain to become national reserves, the seeds of the future national
          park system.

1872   Congress designates Yellowstone the country's first national park. The U.S.
          Army manages the park until formation of the National Park Service.

1872   Congress passes the General Mining Law, granting free access for min-
          erals prospecting on public domain lands to individuals and corporations
          and allowing for filing claims on any mineral deposit discovered. This law
          still governs mineral prospecting on extensive western tracts.

1878   Colorado River explorer John Wesley Powell publishes his landmark
          "Report on the Lands of the Arid Region of the United States," calling for
          careful resource management and prudent water conservation for proper
          settlement. The report does not envision massive reclamation projects to
          support large western populations.

1881   Congress establishes the Bureau of Forestry in the Department of
          Agriculture (USDA), starting management and protection for American
          forests.

1891   Concerned about logging-caused erosion and degradation in a number of western watersheds, the federal government establishes a forest reserve system, precursor to the national forest system.

1892   John Muir and a group of 26 friends establish the Sierra Club.

1896   In his report "The Significance of the Frontier in American History," Frederick Jackson Turner declares the American frontier is no more. He also discusses the importance of wilderness in supporting American democracy.

1897   Congress passes the Forest Management Act, authorizing logging in public forests and calling for both preservation and conservation of public forest lands as important management objectives.

1902   The Reclamation Act sets aside money from semiarid public land sales in 16 western states for building and maintaining irrigation projects. Money from sales of newly irrigated land is to be put into a revolving fund for more irrigation projects, which leads to the damming of nearly every western river.

1905   The USDA Division of Forestry (formerly the Bureau of Forestry) becomes the U.S. Forest Service (Department of Agriculture Forest Service). Then-Agriculture Secretary James Wilson lays out the "Land of Many Uses" forest management concept in a letter to conservation-minded Gifford Pinchot, first Chief Forester: "It will be clearly borne in mind that all land [in the national forests] is to be devoted to its most productive use for the permanent good of the whole people and not for the temporary benefit of individuals or companies."[2]

1908   The U.S. Supreme Court's Winters Doctrine states that federal law established Indian water rights, and therefore Indian tribes have "reserved rights" to water to fulfill the purposes of their reservations.

1916   Congress moves national park system management from the War Department to the Department of the Interior and establishes the National Park Service.

1919   Forest Service landscape architect Arthur Carhart initiates the concept of wilderness areas on public lands, successfully lobbying the government to prohibit summer homes around Trappers Lake, Colorado, still a pristine wilderness.

1922   Colorado River Compact (ratified 1928) allocates water to seven Colorado River basin states, based on a period of unusually high average river flows.

1924   Ecologist Aldo Leopold convinces the federal government to set aside over a half million acres of Gila National Forest (now the Aldo Leopold Wilderness Area) as a public lands wilderness for "recreational purposes."

1926   Under Chief Forester W. B. Greeley, the U.S. Forest Service conducts its first ever inventory of wilderness lands in the national forest system.

1928   The McSweeney-McNary Act sets up the Forest Research Service, today supporting a scientific staff of more than 700.[3]

1934   Congress passes the Taylor Grazing Act, which still governs managed grazing on Bureau of Land Management and other public lands, mostly in the western United States.

1935   Aldo Leopold, Bob Marshall, and others establish the Wilderness Society.

1939   The U.S. Forest Service reclassifies "primitive areas" (under the Shipton-Newton-Nolan Act) according to their sizes, as "wilderness," "wild," or "roadless" areas.

1940   Greatly appreciated by President Franklin Roosevelt, the professional nature photographs of Sierra Club member Ansel Adams are instrumental in establishing Kings Canyon National Park in California's central Sierra Nevada.

1946   The Grazing Service and General Land Office are combined to form the Bureau of Land Management, with oversight of most public lands in the western United States.

1950   Environmental activism prevents the Echo Park Dam in Dinosaur National Monument, the first great success for environmentalists challenging the western U.S. reclamation program.

1955   Wilderness Society Director Howard Zahniser prepares the first draft of the Wilderness Act, proposing "Wilderness Areas" on U.S. Forest Service, National Park Service, and Bureau of Land Management lands, to remain untouched by mining and logging.

1956   Senator Hubert Humphrey introduces the Wilderness Act in Congress. Congress simultaneously passes a bill prohibiting dams in any preexisting national parks or monuments. During the following decade, this law is challenged by proposals to put dams in the Grand Canyon.

1960   Congress extends the "Land of Many Uses" concept of national forests, passing the Multiple Use-Sustained Yield Act to allow logging, recreation, grazing, mining, oil and gas drilling, and related construction, but making wildlife protection and watershed conservation important objectives. Enforceable objectives are avoiding impairment of the land's productivity and environmental quality and managing for sustained yield of all resources.

1963   Congress passes the Wilderness Act, and President Johnson signs it into law in 1964.

1970   President Richard Nixon signs the National Environmental Policy Act, the most significant single federal environmental statute, requiring evaluation of any proposed federal action that could cause significant environmental damage (see appendix 2).

1970   Wisconsin Senator Gaylord Nelson spearheads the first Earth Day on April 22. Highlighted issues include burning rivers, health effects from polluted air and other wastes on inner-city neighborhoods, the effects of farm chemicals on field workers, and wildlands preservation. Congress adjourns for the day to listen to public concerns.

1973   Congress passes the Endangered Species Act, protecting threatened and endangered plants and animals and implicitly protecting lands providing habitat for those species.

1976   Congress passes the Federal Land Policy and Management Act, which repeals the Homestead Act and allows the Bureau of Land Management to manage its lands fully.

Appendix 2

Best Intentions: Federal Waste Disposal Laws

Year	Law[1]
1899	The Refuse Act/Rivers and Harbors Appropriations Act gives the U.S. Army Corps of Engineers (ACE) wide jurisdiction over activities that alter navigable waters. The most important subsection is the "Refuse Act," prohibiting discharge of "refuse matter of any kind or description whatsoever" into navigable waters, or other waters of the United States that feed into navigable waters, without a permit from the corps. For its first 60 years, the strong criminal sanctions are little used because the corps applies the law only to structures obstructing navigation. Eventually the corps expands its authority and begins setting permit requirements for pollutant discharges. In the late 1960s, environmentalists rediscover the act and put its enforcement provisions to work.
1947	The Federal Insecticide, Fungicide, and Rodenticide Act requires all pesticides distributed, sold, offered, or received to be registered with the U.S. Environmental Protection Agency (EPA), contingent on the EPA determining that the pesticides are effective as claimed, are labeled in conformance with federal standards, and will perform their functions without "unreasonable risks to humans and the environment, taking into account the economic, social, and environmental costs and benefits of the pesticides intended use"—as long as they are applied in accordance with instructions. The EPA must follow all registrations by establishing maximum permissible exposure "tolerance levels" for each chemical.
1960	The Federal Hazardous Substance Act gives the Consumer Product Safety Commission control over new toxins, hazardous substances, corrosives, and other substances in interstate commerce.

1965   The Solid Waste Disposal Act (SWDA) is the first in a series of federal initiatives for managing solid waste. Operative SWDA provisions are limited to research and grant programs, but the Act recognizes "a rising tide of scrap, discarded, and waste materials" as a major environmental problem.

1970   The Resource Recovery Act amends the Solid Waste Disposal Act, changing the emphasis from studying disposal practices to recovering materials and energy from the solid waste stream.

1970   Among the most significant achievements of the Richard Nixon administration, the National Environmental Policy Act (NEPA) provides "the environmental movement with a strategic statutory foundation in court because of the pervasive connections between government and business, and NEPA's quite accidental litigable requirement that all federal agencies must prepare an environmental impact statement (EIS)" before a federal agency undertakes any "major federal action significantly affecting the human environment."[2] Projects causing little environmental disturbance, and needing less rigorous analysis of impacts than an EIS, can be dealt with through Environmental Assessments (EA), but many EAs are written for projects with potentially significant impacts.

1970   The Clean Air Act of 1970 regulates air pollution emissions, requiring each state to draw up and enforce a "state implementation plan" and assigning permitted levels of emissions for all air pollution sources in the state to reduce the ambient level of each pollutant to its federal "primary standard." The Clean Air Act also requires the U.S. Environmental Protection Agency to regulate hazardous air pollutants (see chapters 4, 12).

1972   The Federal Water Pollution Control Act responds to the 1969 spectacle of Ohio's Cuyahoga River on fire and massive Santa Barbara, California, oil spills.[3] Federal water pollution regulatory statutes are not enforced until Amendments give them teeth (see Clean Water Act, below).

1974   To ensure public health, the Safe Drinking Water Act establishes water quality standards for drinking water suppliers, prohibits the use of lead pipes, solder, or flux in drinking water systems, and requires underground drinking water sources (aquifers) to be identified and protected. A less well-known component of the Act is the Underground Injection Control Program, which sets standards for, but does not oversee, underground injection well disposal of hazardous liquids in huge quantities to protect groundwater supplies for drinking. The states' direct regulatory programs have done no better with groundwater pollution than with surface water pollution.[4]

1975   The Hazardous Materials Transportation Act charges the U.S. Department of Transportation with regulating hazardous substances in transit, providing for spill control and prevention and central reporting of spill events.

1976   The Toxic Substances Control Act requires the chemical manufacturing industry to ensure that substances are tested for human health and environmental hazards before manufacture and sale. The U.S. Environmental Protection Agency can prohibit or limit uses of a substance based on negative tests, deny the permit by banning a substance, or grant a permit by not acting on test results.

1976   Resource Conservation and Recovery Act (RCRA) supplants and expands the 1965 Solid Waste Disposal Act, regulating all aspects of the solid waste "life cycle," from generation through treatment, storage, or disposal—inaccurately called "cradle to grave" regulation.[5] A landmark statute in its own right, 1984 RCRA amendments added leaking underground tanks and medical wastes to its scope.[6] Key RCRA provisions include treating hazardous wastes bound for land disposal to minimize environmental threats; design standards that certain facilities must meet, including installing double liners under landfills that treat, store, or dispose of hazardous waste; and requiring waste treatment, storage, or disposal facilities to obtain a permit before handling hazardous wastes.[7] RCRA's subtitle D gives municipal solid waste landfill owners wide latitude in managing wastes, as long as they monitor groundwater and control leachates, and provides funds for proper closure. But these regulations apply only to landfills created after 1991 that receive more than 20 tons of waste daily. The Act authorizes the EPA to seek substantial civil and criminal penalties for infractions.

1977   The Clean Water Act is a fusion of the 1972 Federal Water Pollution Control Act, with the Refuse Act and the Water Quality Improvement Act of 1970 added as amendments. Like the Clean Air Act, the Clean Water Act is a joint federal–state partnership "to restore and maintain the chemical, physical, and biological integrity of the nation's waters." Since 2001, new rules for implementing Sections 402 and 404 of the Clean Water Act make these goals much more difficult to attain.

1980   The Comprehensive Environmental Response, Compensation, and Liability Act (CERCLA)—a better known companion to the Resource Conservation and Recovery Act, CERCLA is the "Superfund Law" for disposing of toxic substances and remediating contaminated sites. Following CERCLA's 1986 amendment and reauthorization, Congress established a funding mechanism to make responsible parties clean up the most serious contamination sites. But much contamination results from anonymous illegal dumping or from multiple sources, so the responsible parties often cannot be identified easily. The simplest way to undermine this statute is simply to withhold or obstruct funding.

1986   Responding to the 1984 Bhopal, India, toxic gas release from a Union Carbide plant, the Emergency Planning and Community Right-to-Know Act establishes requirements for federal, state, and local governments, Indian tribes, and industry to do emergency planning and report to communities about the nature and characteristics of designated hazardous and toxic chemicals. States must identify facilities containing substantial amounts of hazardous materials, provide emergency actions if those materials are released, and report the nature and characteristics of certain hazardous materials to the public, so that officials are better prepared for hazardous-substance releases.

1995   Congress lets expire a special tax on industry to fund Superfund (under the Comprehensive Environmental Response, Compensation, and Liability Act), shifting the burden to taxpayers.

July 2002 | The George W. Bush administration cuts Superfund funding by more than $200 million, reducing or completely eliminating money for cleaning up 33 sites in 18 states. The 2004 budget of $450 million to deal with current obligations pales in comparison to the U.S. Environmental Protection Agency's estimate of $250 billion needed to clean up 355,000 hazardous waste sites in the next three decades.

September 2002 | President George W. Bush orders the Council on Environmental Quality to "review" National Environmental Policy Act (NEPA) requirements, with a view to "streamlining" the Act and removing "burdensome regulations." A 20-member NEPA task force completes its review in late 2005, showing a clear intent to raise economic considerations above environmental protection.[8] The review could be the first step in an attempt to gut NEPA.

Appendix 3

Everything Comes from the Earth

Everything that we use in daily life, even toothpaste, cosmetics and lotions, plastics and kitty litter—all are manufactured out of natural materials. Paint, paper, and plastics contain limestone, clay, and silica fillers. Water purification and many other industrial processes require filtration through limestone, salts, soda ash, and zeolite minerals. Abrasives for smoothing and scrubbing, in industry and at home, include volcanic pumice, silica, diatomite, garnet, and corundum—and diamond for polishing. The salt that we eat is a naturally occurring mineral, and medications and prepared foods commonly contain silica minerals (quartz, feldspar).

Petroleum is the residue of ancient sea life, preserved in rocks now found beneath continents, or along continent-ocean margins (see chapters 12, 13). Most petroleum is refined into fuel—but it is also the raw material for plastic, miracle drugs, and synthetic cloth. Drilling for oil (or water) requires tons of steel, abrasives, and fluids made of barite (a sulfide mineral), bentonite and other clays, mica, and perlite (a hydrated volcanic glass). Refining petroleum into gasoline employs clay and zeolite (filtering) minerals and platinum for catalyst.

Myriad common products are made from metals and other minerals mined from rocks, river beds, old river terraces, and old lake beds. We are using, and using up, many of these commodities at rapidly increasing rates (figure A3.1). Common products and source materials include the following:

- Agricultural fertilizer—phosphorus and potassium

- Air conditioning—galvanized ducts (iron and zinc): steel is galvanized by coating it with zinc

- Airplanes; soft drink cans; recreational equipment (bicycles, skateboards, tennis racquets, golf clubs, frame backpacks, fishing rods, skis, etc.)—aluminum (mined as bauxite or recycled)

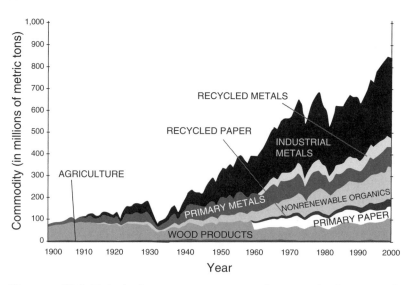

Figure A3.1 United States consumption of materials by general categories, 1900–2000. Data for 1900–1995 from Grecia Matos and Lorie Wagner. *Consumption of Materials in the United States, 1900 to 1995* (U.S. Geological Survey, 2000). Available: greenwood.cr.usgs.gov. Data for 1996–2000 are unpublished, courtesy of Grecia Matos, U.S. Geological Survey (personal communication).

- Automobiles—steel, aluminum (bauxite) or recycled, copper, zinc, magnesium
- Baby powder, cosmetics, lotions—talc, silica, titanium dioxide, mica, iron oxides, hectorite clay
- Batteries—lead, antimony, sulfur (in the form of sulfuric acid), lithium, cadmium
- Cat litter: clay, diatomite
- Cleaning abrasives—pumice, silica, calcium carbonate, diatomite, feldspar
- Ceiling tiles—clay, gypsum, perlite (hydrated volcanic glass), silica
- Doorknobs/locks/hinges—steel, iron, zinc or brass (copper plus zinc)
- Electronic equipment (home and industrial, including televisions, radios, stereos, telephones, computers, VCRs, DVDs, clocks, etc.)—aluminum, antimony, barite, beryllium, cobalt, columbium, copper, gallium, germanium, gold, indium, iron, lanthanides, lithium, manganese, mercury, mica, molybdenum, nickel, platinum, quartz crystals, rhenium, selenium, silica, silicon, silver, strontium, tantalum, tellurium, tin, tungsten, vanadium, yttrium, zinc, and zirconium
- Fertilizers—phosphate, potassium, sulfur, gypsum, limestone (calcium carbonate)
- Floor covering—limestone, clay, wollastonite
- Insulation for buildings—usually fiberglass (quartz, feldspar, trona) or vermiculite (mica mineral), increasingly replaced by products made of recycled magazines and cardboard (magazine paper contains clay)

- Interior wallboard—gypsum, clay, perlite, vermiculite (also a clay), borate minerals
- Jewelry—gold, sliver, platinum, diamond, corundum, beryl, quartz, and so forth
- Kitchen appliances—aluminum (bauxite); steel/stainless steel (alloyed with zinc, copper, lead, or other metals); motors contain copper
- Kitchen and dining utensils—clay in ceramic or china dishes, bowls, and cups; silica in glass vessels; steel or aluminum in pots and pans (some are ceramic coated); steel, silver, or aluminum in flatware
- Kitchen sinks—steel/stainless steel, or clay if ceramic, steel alloys in faucets
- Light bulbs—silica (quartz, feldspar, trona) glass, plus titanium, lithium in incandescent bulbs, neon gas in fluorescent bulbs, and mercury in mercury-vapor bulbs
- Metal roofing for buildings—steel, zinc, aluminum (bauxite)
- Money—zinc/copper alloys in pennies; copper/nickel alloys in nickels; dimes, quarters, half dollars, dollar coins formerly were silver, but now have a copper center sealed to outer layers of copper-nickel alloy
- Paint—clay, limestone, mica, talc, silica, fluorite, with colors from titanium oxide, copper oxides, iron oxides, zinc, and cadmium oxides
- Photographic black-and-white film—silver
- Plumbing and wiring—copper, aluminum, steel, iron, zinc, and clays
- Roads, paths, driveways, landscaping, gardening—concrete (limestone, sand, gravel), loose gravel, sand, landscaping rock
- Sawing/polishing abrasives—diamond, corundum, garnet (emery), silica, pumice
- Screws/nails—steel, iron, zinc (if galvanized)
- Siding/facing for buildings—aluminum, steel, zinc; field or quarried stone, or brick—clays, bauxite, feldspar, chromite, zircon, silica, graphite, kyanite and other silicate minerals
- Toilets, bathroom sinks, bathtubs, tiles, plumbing—ceramic (clay, feldspar, limestone); fittings that connect pipes (copper, iron, steel alloys)
- Tooth fillings—gold, amalgam (mercury and silver), clay (ceramic)
- Toothpaste—calcium carbonate and fluorite, or sodium carbonate and sodium fluoride
- Walls/foundations for buildings—concrete (limestone, sand, gravel), reinforced with steel rods; cement (limestone, bauxite, clay, shale, and gypsum); brick (sand and clay), plus decorative rock facings
- Window glass—silica (quartz, feldspar, trona), mixed with limestone

Products made from petroleum include the following:

- Plastics of all types
- Buildings—house siding, asphalt roofing, insulation, water and sewer pipes, air conditioning (heating and cooling) ducts, indoor/outdoor paints, carpet and vinyl flooring (made of petrochemicals filled with limestone)

- Clothing—vinyl fibers, spandex, polypropylene
- Fuels—gasoline, diesel fuel, lubricating and fuel oils for all kinds of civilian and military vehicles
- Furnishings—kitchen appliances contain various plastics reinforced with industrial minerals, kitchen and dining utensils, and tablecloths
- Vehicle parts—car dashboards, steering wheels, seat covers, frame liners, hubcaps (containing industrial minerals for reinforcement); boat, bus, train and airplane interior paneling, covers, seats, trays, and so on
- Sports equipment—skis, skateboards, tennis racquets

Appendix 4

Biochemical War and You

These substances, most having potentially debilitating or fatal human health effects, have been tested on American military personnel and civilians all across the United States.

Table A4.1 Substance Trials and Potential Effects

Agent/Simulant	No. of Trials[a]	Potential Health Effects
Bacillus subtilis var. *niger*	24	Bacterium; may cause unspecified disease
Betapropriolactone	3	Possible carcinogen
BHP[b]	3	Irritant of skin, eye, respiratory tract
Calcofluor	2	Limited evidence of health effects
Coliphage	1	Virus infects *E. coli*
Coxiella burnetii	1	Q fever bacterium; may cause hepatitis or pneumonia; life threatening; part of U.S. biological arsenal
Di(2-ethylhexyl)phthalate	1	Eye, skin, respiratory tract irritant; causes cancer in some test animals
Diethylphthlate	1	Eye irritant
Dimethyl methylphosphonate	2	Suspected carcinogen
Escherichia coli	5	Acute infections of urinary tract, lung, bloodstream, other organs; potentially fatal

continued

Table A4.1 Continued

Agent/Simulant	No. of Trials[a]	Potential Health Effects
Ester of benzilic acid	2	Incapacitation (temporary?)
Francisella tularensis (wet or dry)	2	Acute infections of urinary tract, lung, bloodstream, other organs; life threatening; dry variant has biological weapon potential
Methylacetoacetate	6	Eye, skin, respiratory tract irritant
Monoethanolamine	1	Harmful or fatal if ingested; lung damage if inhaled; eye, skin burns
Tear gas[b]	3	Eye, respiratory tract irritant
Pasteurella tularensis	1	Pneumonia, tularemia (infectious), complications of meningitis or peritonitis; about 6% mortality
Phosphorus-32	1	Radioactive; high-energy beta emitter; may cause cancer
Polymethyl methacrylate	2	No health effect information
Puccinia graminis tritici	1	Toxic plant fungus, not thought harmful to humans
Sarin nerve agent	13	Lethal
Serratia marcescens	6	Bacterium causes acute infections of urinary tract, lung, bloodstream, other organs
Uranine dye[b]	1	Asthma, eczema, severe allergic reactions
Soman nerve agent	4	Difficulty breathing, coma, death
Staphylococcal enterotoxin type B	1	Bacterium that can incapacitate for 2 weeks, "not generally thought of as a lethal agent"
Sulfur dioxide	1	Strong respiratory tract irritant; more than 100 ppm is immediately dangerous to life and health
Tabun	4	Nerve agent causes difficulty breathing, coma, death; little information on long-term health effects
Tiara	2	Luminescent gelatinous substance; no information on health effects
Tri(2-ethylhexyl) phosphate	2	Causes cancer in some animals, not a demonstrated human carcinogen

continued

Table A4.1   Continued

Agent/Simulant	No. of Trials[a]	Potential Health Effects
Trichloropropane	1	Eye, respiratory tract irritant
Trioctyl phosphate	2	Eye, respiratory tract irritant, causes cancer in some animals, not a demonstrated human carcinogen
VX	10	Nerve agent, extremely lethal
Zinc cadmium sulfide	11	Heavy metals toxic to humans

U.S. Department of Defense. *Project SHAD Fact Sheets* (U.S. Department of Defense, 9 October 2002). Available: deploymentlink.osd.mil/current_issues/shad/shad_intro.shtml.

[a]As many as five different agents may be listed under individual tests.
[b]Full chemical name: BHP, bis(2-ethylhexyl)hydrogen phosphite; tear gas, *ortho*-chlorobenzylidene malononitrile; uranine dye, sodium fluorescein.

Appendix 5

Destroyer of the Worlds

The United States exploded more than 1,000 individual nuclear devices in atmospheric, underwater, and underground tests at Pacific Ocean, Nevada, Alaska, South Atlantic Ocean, New Mexico, Colorado, and Mississippi sites. Atmospheric tests totaled 331.[1] The following is a brief annotated chronology of the testing program and related events.[2] Bomb yields are given in kilotons (Kt) or megatons (Mt), equivalent to 1,000 or one million tons of TNT, respectively.

1944–1945	Nonnuclear testing for the World War II Manhattan Project takes place at Salton Sea Test Base, Muroc Air Base, and China Lake Naval Ordnance Testing Station in California, and Wendover Field in Utah. The Manhattan Project scientists creating the bombs are stationed at Los Alamos, New Mexico.
16 July 1945	Trinity "event": First detonation of a nuclear bomb, yielding 21 Kt, at White Sands Missile Base, Alamagordo, New Mexico. "With the Trinity event man would unlock the demon from the very fabric of matter and plunge the world into the atomic age."[3] The blast instantly raises the ground temperature to levels that melt soil, vaporizes the tower, and kills all desert life within half a mile, reminding Manhattan Project leader J. Robert Oppenheimer of words from Hindu mythology—"I am the destroyer of the worlds."
6 August 1945	Hiroshima: Uranium fission bomb called "Little Boy," 15 Kt, kills 70,000 people. Of 90,000 buildings in the city, 60,000 are demolished.
9 August 1945	Nagasaki: Plutonium bomb called "Fat Man," 20 Kt, kills 42,000 people, injures 40,000, and destroys 39% of buildings.
July 1946	Operation Crossroads "tests": A series of bombs detonate at Bikini Atoll. Two blasts, "Able" and "Baker," test the impacts of atom bombs

on Navy ships. Able is a Trinity-type bomb, with a 21 Kt yield; Baker has a Fat Man design, with a 20-Kt yield. An aircraft drops Able on an armada of 185 ships, from landing craft to aircraft carriers; exploding at the surface, it sinks five ships and damages all others within one-half mile. Baker detonates 90 feet below 70 ships, some carrying animals, plants, and biological warfare agents, to test heat, blast, and radiation effects. The area remained highly radioactive and unsafe for some time. The documentary film Trinity and Beyond shows bomb-blasted sheep and Navy personnel working on radioactive ships with no protective clothing, never mentioning what has happened to the biological warfare agents. Scientists had not anticipated the ships' high radioactive contamination, leading President Truman to cancel Charlie, third in the Crossroads series. Charlie is detonated in 1955 (see below).

1946–1958	Bikini Atoll tests: 23 bombs are detonated.
1948–1958	Enewetak Atoll tests: 43 bombs are detonated, including the first thermonuclear fusion (hydrogen) bomb, in 1952.
April 1948	Operation Sandstone at Enewetak Atoll tests new weapons designs, including the 14 April X-Ray event, 37 Kt; and the 1 May Yoke event, 49 Kt. Islands are bulldozed to remove all vegetation. Three bombs detonate on 200-foot-high towers, using a new technique to double the explosive yield from the same amount of plutonium. The blasts send shock waves into the ocean. The first ground zero samples are collected remotely, but later workers at and near ground zero wear no protective clothing.
1949	Weapons construction shifts from Los Alamos to Sandia Lab, near Albuquerque, New Mexico, which becomes the production line for tactical and strategic weapons stockpiles.
August 1949	Soviets detonate their first atom bomb, five years earlier than anticipated, due to atomic secrets provided them by Manhattan Project physicist Klaus Fuchs. At the time of arrest, Fuchs is head of the theoretical division of the Harwell Atomic Research facility in England.
January 1951	The Nevada Proving Ground (later the Nevada Test Site) opens, because long-distance logistics for Pacific testing, and compensation for blowing up all the real estate on Bikini and Enewetak have become too messy and expensive. Except for atmospheric detonations of thermonuclear devices (H-bombs), which could contaminate far too much U.S. land and people, the U.S. atomic testing program now centers on the Nevada Test Site.
Winter, spring 1951	Operation Ranger: Comprises five bomb tests conducted at Nevada Test Site, including Ranger Able on January 27. A 1-Kt airplane drop. Spring 1951 sees four additional tests: For structural tests, Easy, 20 April, 47 Kt, detonates on a tower; Item burns tritium in the center of a bomb and is the first use of a boosting technique that increases the yield from about 20 Kt to 45.5 Kt; George, 225 Kt, burns a deuterium capsule and detonates on tower.
1953	Operation Ivy, Enewetak Islands: First full-scale hydrogen bomb tests at the atoll's north end. The Mike event bomb uses liquid hydrogen

isotopes, yields 10 Mt, and is too heavy for airplane delivery at 62 tons, but detonation details are not available.

Spring 1953   Upshot-Knothole series: The Atomic Energy Commission, with the Department of Defense, conducts 11 tests at Nevada Test Site to try out new weapons designs and study civilian needs during atomic attacks. Encore, 27 Kt, is air dropped, detonates 2,800 feet above ground. Grable, 15 Kt, is shot from a cannon and detonates 500 feet in the air near Encore; Grable causes much more damage. Lower elevation detonation produces a "precursor" ground disturbance—possibly a laterally moving blast cloud (base-surge), laden with dust, opening a whole new perspective on how to use nuclear weapons. The test towns were built for the Badger event, and after the blast, hundreds of Department of Defense personnel were exposed to radiation at the site.

Winter 1954   Castle Bravo: The largest atmospheric test of a nuclear device, a hydrogen bomb with designed yield 15 Mt, detonates 28 February at Enewetak. The blast's actual yield exceeds expectations by a factor of 2.5, affecting an area 66 miles in diameter. It strips the islands' vegetation, forms a crater 1.2 miles in diameter, and irradiates people who scientists thought would be out of harms way. An extremely "dirty" bomb, Castle Bravo contaminates Japanese on the Lucky Dragon—a fishing boat that wandered into the restricted zone, plus U.S. servicemen and native Marshall Islanders. This event clarified in the public mind the danger from nuclear weapon fallout.

1955   Soviets detonate their first hydrogen bomb.

1955   Operation Wigwam: The renamed Charlie event, holdover from Operation Crossroads in 1946 (see above), 30 Kt, detonates 500 miles off the San Diego coast and 2,000 feet below an unmanned barge towing three instrumented submarines, to determine the "fatal" submarine range of a deep underwater blast. Charlie had unknown environmental effects because wildlife and ecosystem effects were not observed.[4]

1956   Operation Redwing, Pacific Proving Ground (Enewetak and Bikini): 17 high-yield thermonuclear bomb performance tests. "Cherokee," 21 May, a 3.5-Mt air-delivered bomb, yields 3.8 Mt. Bomb craters have obliterated some of the atoll, and its rapidly disappearing surface causes experimenters to tow nuclear bombs out to open ocean on barges.

Fall 1957   Operation Plumbob: Some of the 24 tests take place at the Nevada Test Site to accelerate military training in nuclear warfare, while continuing studies of nuclear explosion effects. Suspended 1,500 feet above the surface under a balloon, the 74-Kt Hood event is the largest atmospheric test ever conducted in the continental United States. Pigs are placed in pens to test blast effects. Troops stationed in foxholes near ground zero and in low-flying helicopters wear no special protective clothing but have slightly more protection than the pigs. Scenes in Trinity and Beyond show researchers in protective clothing handling the dead pigs. The Rainier event, 3 Kt, the 21st Plumbob shot, is the first fully contained underground test, detonated in a shaft 790 feet below the surface.

1958   Operation Hardtack, Pacific Proving Ground (Enewetak, Bikini, and a new Johnston Island site): 35 tests, mounted to beat the impending atomic test moratorium and broadly perceived as saber rattling. The total in this series equaled all prior Pacific tests. All are intended as atmospheric tests, although one bomb explodes on the rocket before it can launch. The Teak and Orange events are high-altitude tests detonated in the upper atmosphere from missiles. Teak, 3.8 Mt, detonates 50 miles up, creating violent magnetic disturbances in the upper atmosphere, which silence radio transmissions from Hawaii to New Zealand for about eight hours. The Cactus event, 18 Kt, produces a crater 137 feet across and 37 feet deep. In 1980, the Cactus crater becomes the dump site for radioactive debris collected all over Enewetak atoll. Its concrete dome is not an effective barrier against rainwater infiltration and seepage for the long term, however.

1958   Operation Argus: Immediately following Teak and Orange, Argus consists of three 1-Kt airborne tests in the South Atlantic. Forerunners of Ronald Reagan's 1980s missile shield idea, the Argus tests detonated 300 miles above Earth, ostensibly to study trapped bomb radiation in the Van Allen belt, but really intended to create a radioactive shield that would impede a Soviet missile attack.

1958–1962   Twelve upper atmospheric or ocean tests in the Johnston Island area, 24 tests in the Christmas Island area, 4 tests in the Pacific Ocean, and 4 tests at high altitude over the South Atlantic Ocean. The Johnston Island tests were launched by missiles, but one exploded on the launch pad.

1957–1963   "Safety" or "equation-of-state" tests, Nellis Air Force Base, Nevada: High explosives used for detonating packages of plutonium and uranium to study the sizes and distribution of plutonium particles from nuclear weapons fires and accidents.

1961(?)   The Soviet Union detonates an air-deliverable monster 57-Mt bomb over the ocean, breaking the voluntary test moratorium. The United States follows with "massive retaliation" Operation Dominic, 36 tests at the "Pacific Ocean Proving Ground."

1962   The last U.S. atmospheric test before the 1963 Partial Test Ban Treaty prohibits further atmospheric testing.

1961–1973   Thirty-five Plowshare Program tests, Nevada Test Site, Colorado, New Mexico (see appendix 7).

1963–1971   Underground atomic tests at Fallon, Nevada; Hattiesburg, Mississippi; Amchitka, Alaska; and Nevada Test Site (Vela Uniform project): These bombs were detonated for seismologic studies to distinguish bomb blast seismic signals from such natural events as earthquakes, meteorite impacts, and volcanic eruptions, with the goal of verifying the test ban.

1965–1971   Three tests at Amchitka Island, western Aleutian Islands: The 1971 underground Cannikin test, 5 Mt, was deemed too large for the Nevada Test Site.[5]

1968   Faultless test, central Nevada: The spectacularly misnamed event is part of the U.S. antiballistic missile program, costing many billions of dollars even before "Star Wars" and missile shield programs are proposed. The explosion caused numerous fault movements that ruptured the desert surface.

1992   The last U.S. underground nuclear test (except for ongoing subcritical tests) is detonated at the Nevada Test Site in September.

Appendix 6

Plutonium Fields Forever

Six years before the 1963 Atmospheric Test Ban Treaty went into effect, a nuclear bomb test called "Project 57" took place at "Area 13" on Nellis Air Force Range, Nevada, near the Nevada Test Site. Area 13 is about 6.5 miles northwest of the center of Groom Dry Lake in Area 51, famous for highly publicized 1950s weird phenomena and extraterrestrial speculations.[1] Project 57 was one of about 30 so-called "safety" tests, ostensibly to develop monitoring and cleanup procedures, which purposely scattered plutonium particles to study the dispersal patterns.

Of all the "safety tests," only the Area 13 experiment involved a real nuclear warhead.[2] The bomb's detonation was not a nuclear reaction, however—instead, high explosives blew up and fragmented the warhead, scattering plutonium fragments across more than 1,000 acres (figure A6.1).[3] Unfortunately, follow-up studies on Area 13 were mostly preliminary, and they ended in the mid-1980s without examining the long-term consequences of plutonium contamination in the pulverized soils.

The Area 13 experiment was certainly an unqualified "success" from the perspective of dispersing highly toxic and flammable plutonium widely into the environment. The site still is highly contaminated and will remain so for some 240,000 years. Clearly, neither this project nor the other safety tests yielded practical plutonium cleanup methods, and any efforts to monitor the hazards fizzled. Because the Area 13 experiment emulated possible-to-likely terrorist attack scenarios using nuclear materials, it is of great interest in the aftermath of the 9/11 terrorist attacks on New York and Washington.

Soon after the test explosion, experimenters sequentially placed 70–80 caged dogs in the contaminated area over 4–161 days, to examine what would happen to them after inhaling plutonium in the dust.[4] The researchers then killed and dissected the dogs. Between 1970 and 1986, other Area 13 follow-up studies examined the sizes of plutonium fragments from the explosion; effects of

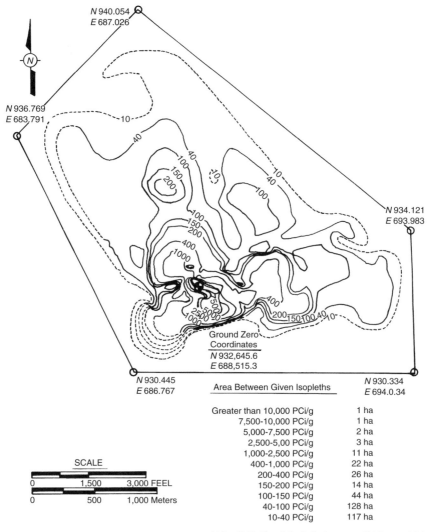

Figure A6.1 Map of plutonium contamination in Area 13 of the Nevada Test Site, contoured for 10–10,000 pCi/g (isopleths in picocuries per gram) contamination levels and showing the area (in hectares) for each contour interval. One hectare is 2.5 acres; see the glossary for explanation of picocuries. From Nevada Department of Conservation and Natural Resources, U.S. Department of Energy, and U.S. Department of Defense. *Federal Facility Agreement and Consent Order, Appendix I, Description of Facilities* (1999), figure I-6.

plutonium-contaminated vegetation and soil on cattle, kangaroo rats, and other small native mammals; how plant foliage traps plutonium; and other effects.[5] The later studies identified Area 13 vegetation as typical for the Great Basin Desert.[6] The site's soils contain about equal amounts of sand and silt, with some clay.[7]

Mounds occupy about 17% of the surface[8]—some are wind-blown coppice dunes of sand collected around shrubs, and the rest may be heaps created by colonies of small burrowing animals.[9]

Nevada Test Site Environmental Impact Statements, dated 1977 and 1996, contain striking contamination maps of Area 13 (figure A6.1), although neither record the data collection dates. Both maps display an unusual two-lobed pattern of plutonium contamination in Area 13 soils, but some minor details are puzzlingly discrepant. On both maps, two concentric "radiation fences" surround contaminated zones—the inner fence encircles the ground zero zone of highest contamination, and the outer fence encloses plutonium concentrations greater than 10 picocuries per gram (pCi/g)—a fatal human dose, equivalent to the radioactivity emissions of only a third of a trillionth of an ounce of radium in each ounce of soil. The total estimated inventory of plutonium-239 and plutonium-240 isotopes in the soils, sampled to a depth of two inches, is about 46 curies[10]—four times more than the average at nine other safety test sites.[11]

Plants grown in Area 13 have plutonium concentrations of 5.2–1,200 pCi/g (dry weight), in stems and in dust coatings on stems and leaves[12]—as much as a hundred times more than a fatal human dose. Laboratory experiments on untreated Area 13 soils[13] showed that plants absorb little plutonium through their roots, but plants significantly increased plutonium absorption after the soils were treated with agricultural-grade acetic acid and sulfur.[14] In addition, soil fungi and bacteria can modify plutonium compounds in ways that enhance the roots' ability to absorb them.[15]

On the presumption that plutonium transfers from a cow's blood system to its organs the same as it does in humans, despite differences in bovine and human metabolisms, one study grazed cattle, including some pregnant females, on Area 13.[16] The cattle ate mostly plants but could not avoid licking off soil particles from their snouts and swallowing them.[17] The ingested plants and soils contained as much as 400,000 pCi of plutonium-239 and 56,000 pCi of americium-241. The cattle ingested smaller amounts when grazing in the outer zone, and larger amounts when inside the inner fence, closer to ground zero.[18]

Eventually, the Area 13 cattle were butchered and examined for plutonium concentrations in stomach rumen and samples of gonad, bone, liver, and lung tissue.[19] If cattle really do model human systems, the results indicate that plutonium will concentrate in human gonads, bone, liver, and lungs. Gonads and upper leg bones (femurs) contained median levels of plutonium-239, 25 times higher than muscle tissue, while liver and lung tissues contained at least twice as much as gonads.[20] But the most important finding was that the Area 13 cows' uterine placentas transfer plutonium to their unborn calves. Femurs of calf fetuses contained one-fifth the plutonium-239 levels found in their dams' femurs.[21]

Rumen in the cattle's stomachs showed higher plutonium concentrations in later summer and fall than at other times of year. This difference probably reflects the proportion of winterfat in their late-seasonal diet—a fodder plant with pods that accumulate dust particles (including plutonium dust).[22] The seasonal difference continued for many years, showing that dust-sized plutonium kept on blowing around Area 13.

Area 13 researchers concluded that "resuspension"-prone fragments, small enough to be picked up by wind and water and redeposited elsewhere, compose up to 99% of Area 13's plutonium debris.[23] They did not monitor disturbance effects from raking the Area 13 ground surface, but raking experiments at other safety test sites put 27 times more plutonium-239 dust in the air, and wildfires produced 12 times more airborne plutonium than comes from undisturbed surfaces. These results suggest that the plutonium resuspension and dust inhalation exposure hazard may be 10 times greater for humans and other animals at Area 13 than at other safety test sites. Experimental "decontamination" efforts, ranging from irrigating to plowing, did disturb some Area 13 surfaces. All the methods disturb contaminants below the surface but would not remove them. This suggests that the primary aim of the decontamination efforts was to reduce resuspension.

Later studies apparently never rigorously addressed important questions of whether resuspension processes add contamination to Area 13, or where resuspended Area 13 material might go once it leaves the site. Resuspendable windborne contaminants could be transported a substantial distance to the northeast, while surface waters would carry contaminants to the southeast. High winds can, and probably do, carry radionuclides away from Area 13 and the other test sites— but apparently none of the studies monitored radionuclide movements.

Besides wind transport, plutonium also can escape from radioactive test sites in and on native wildlife, which put it into the broader ecosystem food chain. Potential food chain routes for plutonium are very complex, and depend on the animal and its lifestyle—whether burrowing or a surface dweller. Burrows have higher humidity and temperature levels than does surrounding soil, and food stored in burrows provides an ideal substrate for fungi and bacteria. Defecation in the burrows adds soluble nutrients and gases. Invertebrate animals may take portions of stored food from the burrows and distribute them through the soil and to the surface.

Birds, bees and other insects, rodents (mice, rats), and lizards are some of the animals that live on or pass through Area 13. Plutonium-239 and americium-241 studies on native animals are limited, however. Some focused on darkling beetles and kangaroo rats that lived on Area 13 for at least six months.[24] Soil fungi can alter plutonium oxides to more soluble compounds or organic complexes, and the fungus is a major food source for darkling beetles. Not surprisingly, Area 13 darkling beetles contain measurable plutonium (500 pCi) in their tissues. Darkling beetles are a food source for many small vertebrates, and Area 13 darkling-beetle predators certainly take in plutonium with their prey. Soil-eating animals such as earthworms add another dimension, because they have a broader range of movement than soil fungi or bacteria, and their predators are even more mobile. Adding birds to the range of possibilities only expands plutonium's potential for broad ecological contamination.

In the post-9/11 world, Russia and the United States joined forces to hunt down stray radioactive materials and thwart dirty bombers,[25] so leaving plutonium debris in the accessible environment could cause problems. Terrorists could fashion dirty bombs from extraordinarily contaminated soil in former Soviet atomic test areas, such as Kazakhstan. A large plutonium hot spot, probably resulting from dispersal tests, is being buried under a reinforced concrete cap six feet thick to prevent terrorists from getting the contaminated soil.[26] The Area 13 data show that the western United States harbors similarly radioactive soils.

Appendix 7

Bombs for Peace

Plowshare can help mankind reshape the earth into a Garden of Eden by overcoming the forces of nature.
Glenn Seaborg, *Man and Atom* (1971)

The first suggestion that nuclear bombs could be used as peaceful nuclear explosives (PNEs) came from John Von Neumann, a brilliant mathematician working on the World War II Manhattan Project. Since atomic (or fission) bombs produce deadly radioactive fallout, U.S. scientists ignored the idea until 1949, when the Soviet Union claimed their bombs were made for peaceful purposes, not war. At least theoretically, the Soviets had invented nuclear dynamite—using fission bombs for "razing mountains, irrigating deserts, cutting through the jungle, and spreading life, happiness, prosperity, and welfare."[1] American scientists found the Soviet statements unbelievable—but they did look into the feasibility of "moving mountains with nuclear explosions" and found it "an impractical idea."[2]

The first successful U.S. thermonuclear (hydrogen) bomb test, at Enewetak Atoll in the Pacific Ocean, obliterated Elugelab Island and dramatically demonstrated that nuclear fusion explosions had the power to excavate land. This lent credence to Soviet claims and suggested possible peaceful uses for nuclear explosions, capturing the imaginations of former Manhattan Project physicist Edward Teller ("father of the H-bomb") and Glenn Seaborg, head of the now-defunct Atomic Energy Commission (AEC).[3] The heavy hydrogen for bomb fuel would be cheaper and easier to produce than atomic bomb fuel, they thought, and fusion bomb explosions would yield lower radioactivity. Fusion bombs were going to be the highly touted "clean bombs."

Teller and Seaborg, the chief Plowshare proponents, could foresee no limitation on warhead sizes. They marketed all the positive features in Washington and oversaw the 1957 birth of Project Plowshare. A 1958 moratorium on nuclear testing, negotiated by the Eisenhower administration, put off the Plowshare program's start until 1961. The 1963 Partial Test Ban Treaty (PTBT), which prohibited radioactive "debris" releases beyond the borders of a testing nation, further circumscribed Plowshare—at least, in theory. Such obstructive developments climaxed in the late 1960s with the growth of environmental activism and passage of the National Environmental Policy Act in 1969. But in spite of a series of technological, political, and community failures, and with no notable successes, the project still persisted until 1973, through all the bomb test moratoriums and bans.

Nuclear Dynamite

Plowshare mounted a series of tests to determine the feasibility of excavating large holes or underground storage cavities and corridors—or craters for developing into harbors,[4] dams, highways, and canals—with atomic bombs.[5] Other possible projects included discovering additional heavy (transuranic) elements for the scientific record, developing new kinds of explosives, and improving oil and gas production. Teller and Seaborg could wax as poetic as the Soviets in their claims that PNE could be used to control weather and earthquakes, defend the Earth against comets and meteors, create vast reservoirs of water in the desert, produce diamonds, and divert rivers. It was all fantasy, of course.

The ideas on atoms-for-peace drawing boards included the following:

- Project Ketch, using a 24-kiloton bomb (equivalent to 24,000 tons of TNT) to make an underground cavity in north-central Pennsylvania, intended for natural gas storage. The principal beneficiary would be the private Columbia Gas Corporation. When Project Ketch attracted public attention, the possibility that it could lead to blasting thousands of other storage cavities with atom bombs probably was enough to incite fear. But news that the Nevada office of the AEC was in charge of public safety for the project probably didn't reassure everybody. Public protest defeated the project.[6]

- Blasting a river gorge on western Australia's Fortescue River with nuclear dynamite to "build" a dam. This Glenn Seaborg brainchild envisioned a bomb blast to cave in the sides of the gorge, at least partly blocking the river channel. The explosion actually would have created a loose pile of unconsolidated material varying in size from dust to large boulders. The bizarre scheme called for the pile to be injected with water and frozen to prevent the water from finding pathways through the debris. How the water would be frozen, or kept frozen in the dry and generally rather hot region, never was made clear—but seems to have involved another underground nuclear explosion.[7]

- Bomb excavation of canals and river diversions, for generating irrigation water and hydroelectric power nationwide, to be called the North American Water and Power Alliance.[8]

- Expanding North America's grain-growing region northward into the Yukon, by blasting the Arctic polar ice cap to bits with strategically placed nuclear bombs, thus warming the climate.[9] Little did the proponents know that high fossil fuel consumption could destroy the Arctic polar ice cap without the aid of bombs.

- Excavating the right-of-way for Interstate Highway 40 through the California Desert's Bristol Mountains. Plans called for simultaneous detonation of 23 nuclear devices with a total yield of 1,830 kilotons to create a linear excavation (figure A7.1).[10] The

Figure A7.1 Model of a proposed 1960s Project Carryall project, intended to "construct" part of the right-of-way for Interstate Highway 40 and a railroad through the Bristol Mountains, in southern California. Plans called for simultaneously detonating 23 nuclear bombs, totaling 1,830 kilotons, to create the trench (elongate dark area). A 100-kiloton detonation at a depth of 690 feet was to form the small oval crater (dark patch) at lower right, to catch Orange Blossom Wash runoff. From O. W. Perry et al. *Project Carryall, Feasibility Study. California State Division of Highways, the Atchison, Topeka and Santa Fe Railway Co.* (U.S. Atomic Energy Commission, 1963).

National Academy of Sciences, Army Corps of Engineers, and U.S. Geological Survey actually endorsed this project.[11]

- Blasting a sea-level canal across Nicaragua to avoid the tedious locks on the hillier Panama Canal route.[12]

- A series of water management projects for creating reservoirs, by either blowing a big hole in the ground surface or blowing a big hole underground, and letting blast debris subside into the explosion cavity. One project aimed to "conserve" surface waters along Arizona's Gila, Salt, and Little Colorado rivers by directing them into bomb craters. The craters could serve as storage reservoirs, or structures for replenishing (recharging) groundwater.[13] Another project sought to improve water quality in north-central Texas rivers by blasting cavities (termed "facilities") to intercept and prevent natural brine from seeping into the rivers.[14] Yet another project envisioned using nuclear blasts to tap very deep groundwater aquifers and develop new water supplies.[15]

- Producing petroleum-bearing hydrocarbons from oil shale in Colorado, either by removing a thick overlying layer of barren rock with rows of two-megaton (two million ton) nuclear explosives[16] or by creating underground "retorts"—fracturing rock in vertical zones (chimneys) with nuclear bombs to concentrate the hydrocarbons and let them come to the surface.[17]

- Propelling rockets, and generating heat and pressure for diamonds manufacture with nuclear explosions.[18]

- Controlling hurricanes by exploding a megaton-range nuclear bomb in the strongest wind zone of a mature hurricane.[19]

In all, the Plowshare program mounted 27 test nuclear bomb detonations: 10 to test PNE explosives design, 6 for determining excavation characteristics, 4 on heavy element production, 3 for natural gas stimulation, 2 on emplacement techniques, and 1 each for steam power generation and determining blast effects in limestone rocks.[20] An additional 21 weapons tests had Plowshare components.

Most of the Plowshare experiments were conducted on the Nevada Test Site near Las Vegas. Six nuclear tests elsewhere in the continental United States included four Plowshare Program events—the Gnome steam power-generating test 25 miles southeast of Carlsbad, New Mexico, and three attempts to stimulate natural gas production in New Mexico and Colorado.

Reality Tests and Fallout

The 1961 Gnome shot kicked off Plowshare experiments. Designed in part to generate steam power by detonating a nuclear bomb in underground salt deposits, Gnome was a technical failure. The blast caused an unanticipated above-ground release of highly radioactive steam, and then thousands of tons of salt collapsed into the cavity, ruining the experiment. Like its other goals, the plan for recovering rare isotopes from Gnome's nuclear reaction apparently did not work out.[21]

From the start, the AEC recognized that radioactivity releases would be a major public-perception problem for nuclear dynamite. Excavation experiments posed particular dilemmas because they necessarily disrupt the ground surface. One of the major Plowshare program goals, therefore, was reducing radioactive fallout. Early predictions of 95% radioactive fallout reductions proved wildly optimistic, however.

Following the March 1968 Buggy excavation experiment, proponents claimed a "*virtually* clean" explosive.[22] But radionuclide inventories recorded more than 20 curies of highly radioactive materials on 180 acres around the Buggy crater—the combined size of 160 football fields.[23] That represents 200 times the recommended lifetime radiation exposure for an individual, or a lifetime recommended radiation exposure for 200 million people.[24] Radioactivity spread even farther from ground zero as tiny suspended particles and gases in the main dust cloud that roared out of the crater.

Ideas about how to reduce radioactive fallout included reducing bomb sizes and finding an optimal burial depth to reduce the mass of blast debris. But Plowshare envisioned large to vast projects, requiring large bombs or many bombs. And fission bombs—regular atom bombs—would have to set off the thermonuclear bomb explosions, yielding significant radioactivity.[25] So Plowshare projects were designed for "containment"—essentially a hope that most of the radioactive debris would fall back into the bomb crater.

Even successful containment could not solve the radioactivity problem. For one thing, natural gas stimulation experiments had to be located where natural gas could be produced, not where radioactivity could be most easily contained. *New York Times* columnist H. Peter Metzger suspected that the radioactive contaminants would surely find paths into the water and from there would contaminate the biosphere—air, water, soil, plants, animals, and, not least, humans.[26]

Sedan was the first excavation test in support of the Panama Canal replacement project, conducted in the two-year gap between the nuclear test moratorium and PTBT. A 104-kiloton blast blew radioactive debris higher than 12,000 feet[27]—22 times higher than the Washington Monument. This was clear evidence that peaceful atom bombs might not meet PTBT conditions. The 1968 Schooner 31-kiloton excavation test lofted a radioactive cloud more than 13,700 feet into the atmosphere,[28] violating the PTBT when it drifted over Canada and Europe. The U.S. Public Health Service monitoring found that Schooner had produced the highest levels of radioactivity recorded in western U.S. cities since atmospheric nuclear testing stopped at the end 1962 (see appendix 5).

In 1970, the Atlantic Pacific Interoceanic Canal Commission report "confirmed the gut reaction of many lay people to the use of nuclear dynamite: namely that the imponderables involved in 'correcting natures mistakes' with large numbers of nuclear explosives are too terrifyingly great to contemplate."[29] The commission's report effectively ended Plowshare's excavation program. Undaunted, Plowshare turned to underground engineering projects.

The three most important underground tests were Gasbuggy (New Mexico, 1967), Rulison (Colorado, 1969), and Rio Blanco (Colorado, 1973), intended to stimulate natural gas production. Compared to the production records of older local gas wells, Gasbuggy successfully stimulated a six- to eightfold increase in gas production from wells drilled into the 26-kiloton blast cavity. But Gasbuggy also yielded an unexpectedly large amount of radioactive tritium. To reduce tritium, the Rulison test had used a 40-kiloton fission bomb. The results confounded experimenters by stimulating less natural gas than expected—and a lot more tritium.[30]

The Rio Blanco experiment was designed to simultaneously detonate three 30-kiloton bombs in the same drill hole, to minimize both tritium production and the size of the drill hole for setting the bombs. The detonations were expected to create a large chimney of fractured rock. Instead, they blasted three chimneys that did not connect and contaminated the gas with high levels of radioactive cesium-137 and strontium-90 (see chapter 7). The test also produced a lot more radioactive water than expected. The crowning insult was the minimal volume of natural gas that Rio Blanco produced.

All the gas stimulation results were unpredicted and still are unexplained. But even had the tests improved natural gas production, "the image of a woman in the kitchen with a baby on her knee and radioactive gas burning on the stove was a powerful deterrent to the commercial use of nuclear-stimulated gas in the home."[31]

The enormous number of nuclear detonations—as many as 1,000—required to fully develop small and relatively unproductive fields raised public antipathy against nuclear power as a tool for resource extraction. Propelled by high 2005 gas prices, the never-say-die industry went back into the Rulison district, to attempt releasing gas with hydrofracturing in a very tight formation—but hoping *not* to tap into the nuclear dynamite's residual radioactivity.

Plowshare mavens also wanted to try stimulating oil production, but the potential for radioactive contamination of the oil could not be assessed without a full-scale field test,[32] and that never happened. From "Project Grifon," an oil-stimulating experiment in Perm, near the town of Osi, the Soviets reported: "Immediately after the explosion, the level of radioactivity seemed normal, but since 1976 the number of wells that produce radioactively contaminated fuel has increased. No experts have been able to explain the reasons for an unexpected and unpredictable distribution of radioactive contamination in the oil fields."[33]

The Real Purpose

Edward Teller had urged President Dwight Eisenhower "to show the world that nuclear weapons, radioactivity and radiation were not harbingers of death but were in fact powerful, benign servants offering almost limitless benefits to humankind."[34] But the Plowshare Program's real legacy is radioactively polluted land and groundwater, which will plague us far into the future. That result is no different from the legacy of military bomb tests, although smaller in scope. It is a sorry inheritance under the name of peace—but as AEC Commissioner Lewis Strauss confessed, Plowshare was not really about peace. It was meant to "highlight the peaceful applications of nuclear explosive devices and thereby create a climate of world opinion that is more favorable to weapons development and tests."[35]

In addition to unwanted radiation, the Plowshare program's demise was largely determined by the *cost* of nuclear dynamite. Added to Plowshare project costs, the enormous price tag for nuclear research and development, plus the nuclear devices themselves, effectively eliminated commercial industry participation. So the costs were borne largely by the weapons side of the nuclear program.

Nuclear science's inability to either create a "clean" nuclear bomb, or to control release of radioactive materials to the atmosphere, also removed the potential for peaceful uses of nuclear explosives—the last substantive obstacle to a comprehensive nuclear weapons test ban. Now, only the U.S. government stands in the way of a world treaty to ban nuclear weapons testing.[36] But even though Plowshare was a complete technical failure, its positive outlook and promotion of nuclear weapons and tests remains "a powerful culture of denial [that has] sunk strong, deep roots into the heart of scientific and industrial America."[37]

Appendix 8

The Bunker Buster Fantasy

Amid all the talk about weapons of mass destruction, a curious bill passed through the U.S. Senate Armed Services Committee, repealing a ban on the research and development of low-yield nuclear weapons.
Scott Baldauf, "US May Stoke Asian Arms Race," *Christian Science Monitor*

When the United States went to war against Afghanistan in 2001, the military believed that terrorists were hiding chemical and biological weaponry—perhaps even nuclear weapons—in sophisticated, deeply buried bunkers. Iraq dictator Saddam Hussein also was suspected of hiding in deeply buried bunkers, and of having underground caches for weapons of mass destruction.[1]

Bombs cannot destroy a buried target unless they can penetrate deeply into the earth without either detonating too soon or destroying the payload before detonation. Arguing that nuclear weapons may offer the only way to destroy deeply buried weapons caches or weapon manufacturing sites, the Pentagon pushed to develop small nuclear bombs to be mounted on deeply penetrating missiles.[2] In 1997, the United States already had developed a so-called "bunker buster"—the "Low-Yield Earth-Penetrating Nuclear Weapon"—probably in contravention of U.S. policy to not develop new nuclear weapons.[3]

Bunker buster proponents explained that nuclear penetrator weapons could be used close to urban areas with minimal collateral civilian casualties, because the blast and radiation effects would be contained underground.[4] Tests of the 1997 prototype did not support these assumptions, however. Dropped from an altitude of 40,000 feet, it penetrated only about 20 feet into dry earth. The current (2004) B61-Mod 11 U.S. nuclear earth penetrator weapon, a massive bomb with estimated 300-kiloton (Kt) yield, can penetrate only about 10 feet into frozen tundra.[5] Penetration depths must be much greater to prevent an explosive fireball and deadly radiation clouds from reaching the surface.

Hundreds of underground nuclear weapons tests took place at the Nevada Test Site (NTS), near Las Vegas, between 1957 and 1992. They mostly detonated bombs in sealed excavations at depths scaled for containing varied bomb yields (see chapter 7, appendix 5). But many bomb test cavities failed to contain either the blast or the radiation. Figure A8.1 depicts the generalized sequence of events

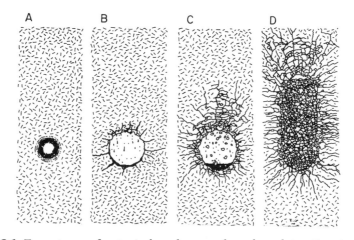

Figure A8.1 Four stages of a typical underground nuclear detonation in a cutaway view showing the ground surface (top) and detonation cavity at depth. Actual events depend on particular geologic and hydrologic conditions. (A) A spherical cavity forms in the first few microseconds and fills with hot gases at extremely high pressures, melting surrounding rocks (black zone). (B) The cavity expands, fracturing the surrounding rocks until the pressure of underground gases equals that of overlying rocks. (C) The cavity cools, some of the gases liquefy, and the molten rock runs to the cavity's floor. Within a few seconds the cavity's roof begins to collapse. (D) The collapse has propagated toward the surface, creating the chimney. At the stage shown, the chimney has reached the surface; next, the uppermost part of the chimney will collapse, forming a crater. At 200-foot depth, the chimney materials may push above the ground surface to form a dome, whereas with full containment the chimney does not reach the ground surface. Modified from Committee for Environmental Information. Underground Nuclear Testing. *Environment* 1969;11:6–7.

derived from these outcomes. An underground detonation creates a surface crater, and radioactive debris escapes into the air. If the detonation containment is only enough to prevent an explosive fireball from reaching the surface, a collapse crater is likely to form and leak radioactive gases to the air (see figure 7.7).[6] If the detonation is fully contained, radioactive rubble falls into an underground blast chamber above the point of detonation.

Figure A8.2 and table A8.1 show the least and most conservative formulations for calculating required detonation depths over a range of bomb yields. Least protective penetration depths may be calculated from the equation

$$D = 230(\mathrm{Kt}^{0.294}),$$

where D is the depth in feet and Kt is the weapon's TNT-equivalent yield in kilotons. Calculations of penetration depths that might fully contain a nuclear detonation are based on the NTS experiences, with added safety factors.[7] All calculations assume ideal conditions, including an ideal impact site in yielding, homogeneous, dry target material, and a perfectly vertical downward missile trajectory, impacting

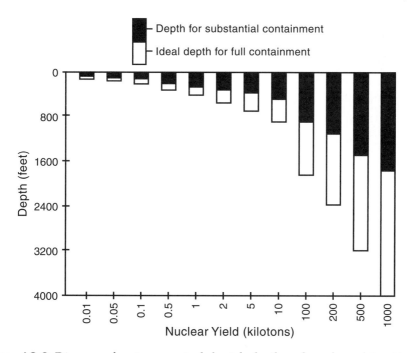

Figure A8.2 Diagram showing required burial depths of nuclear detonations to prevent the explosion fireball from breaking out at the surface (substantial containment, dark bars), scaled to $230(Kt^{0.294})$, and to fully contain radioactive emissions (full containment, white bars), scaled to $400(Kt^{0.333})$. Bomb yields vary from 0.01 to 1,000 Kt. If the explosion is not contained, a fireball breaks through the surface and carries large amounts of radioactive debris into the atmosphere. At shallower than required detonation depths, the bomb fractures the area it penetrates, throws radioactive materials into the air, and forms a crater. At greater depths, the bomb creates a chimney, leading to surface collapse, as in figure A8.1, with potential radiation release. The cloud of debris typically consists of a narrow vertical column and a broad, flowing debris cloud at ground level, called a base surge. From C. E. Paine et al. *Countering Proliferation, or Compounding It?* (Natural Resources Defense Council, 2003). Available: nrdc.org; R. W. Nelson. Low-Yield Earth-Penetrating Nuclear Weapons. *Journal of the Federation of American Scientists* 2001;54:4; R. W. Nelson. Low-Yield Earth-Penetrating Nuclear Weapons. *Science and Global Security* 2002;10:8.

the surface at a 90° angle. Lower entry angles require the penetrator and bomb to get through greater rock thicknesses if the payload is to get deep enough for blast containment. The results show that a 1-Kt bomb must detonate at least 230 feet underground (figure A8.2, table A8.1) to contain the blast. The NTS experiences prescribe that nuclear bombs should be placed a minimum depth of 600 feet below the surface to contain a blast.[8]

Forcing a bomb that deep into the earth requires a penetrating missile more than 22 feet long, slightly taller than a two-story building, and a lot of luck.[9]

Table A8.1 Calculated Detonation Depths Required to Contain Underground Nuclear Explosions

Bomb Yield (Kilotons, TNT Equivalent)	Substantial Containment Depth (Feet)	Ideal Full Containment Depth (Feet)
0.01	59	65
0.05	95	111
0.1	118	140
0.5	187	240
1	230	302
2	279	380
5	371	516
10	453	650
100	889	1,407
200	1,092	1,765
500	1,430	2,396
1000	1,752	3,018

C. E. Paine et al. *Countering Proliferation, or Compounding It?* (Natural Resources Defense Council, 2003), 39., Available: nrdc.org; R. W. Nelson. Low-Yield Earth-Penetrating Nuclear Weapons. *Science and Global Security* 2002;10.

If the blast is not completely contained, an area of about 1.8 square miles will be covered with radioactive debris. A cloud of radioactive gases and fine particulates ejected vertically from the crater will spread fallout over much larger areas downwind—possibly over civilian populations, or over U.S. and allied troops.

The law of frustrated intentions always operates: No matter "how perfect the process may be [in theory, the weapon] operates in an imperfect setting."[10] Far from perfect, the ground conditions at penetrator-weapon targets generally are known poorly, if at all. How the ground and the penetrating device interact depends on the strength, porosity, sealing strength, moisture content, position of the groundwater table[11]—and, importantly, on the abundance and orientations of natural fractures in the rock. Fracture systems may radiate in many directions or only a few directions, exerting nonsymmetrical forces that present formidable obstacles to the survival of penetrating weapons plunging into the earth, and to a great extent determine both the ultimate penetration depth and bomb blast effects.[12]

Penetrating only about 20 feet of hardened steel/concrete or hard rock targets—the current state of the art—will not contain even the smallest nuclear bomb envisioned for a bunker buster.[13] But what military weapons theorists seem to miss is that a missile penetrating most solid target materials, including earth materials, *creates an open passageway between the surface and the point of detonation.*[14] Clearly, even if penetrators *could* plunge bombs deep enough underground to avoid cratering, the blasts still would not be contained.

The very earliest underground nuclear bomb test, Pascal-A (26 July 1957), detonated in an unsealed shaft at a depth of about 500 feet, emulating a penetrator. The Pascal-A yield was expected to be equivalent to only one or two pounds of TNT, but actually was about 55 tons. Even so, the depth of burial was far greater than necessary to contain a sealed cavity blast. Detonated after dark, Pascal-A "was the world's finest Roman candle," according to two physicists who observed it:

> Blue fire shot hundreds of feet in the air. Everybody was down in the area, and they all jumped in their cars and drove like crazy....They were damn lucky they didn't go right through that [fallout] cloud...bad as it was, spectacular as it was, there was only a tenth of the radiation on the ground around there than there would have been if it had been done on the surface.[15]

Subsequent tests in unsealed shafts typically released 5–10% of their total radioactive products into the local environment, but even that amount can be deadly.[16]

Calculations show further that plans for megaton-yield (1,000 Kt) deep "robust penetrator" bombs are fantasies. To contain a one-megaton bomb blast, the missile must jam it more than 1,700 feet deep in the least conservative case, and 3,000 feet deep in the most conservative case (figure A8.2, table A8.1). No missile can reach such depths. Far from destroying chemical, biological, and nuclear weapons in deep bunkers, "robust penetrators" would only scatter toxic and radioactive debris far and wide across the countryside and into the atmosphere (see chapter 7).[17]

Nuclear weapons have not been tested—at least not openly—since Congress imposed a moratorium in 1992 and President Clinton signed the Comprehensive Test Ban Treaty in 1996.[18] Opening the way for resuming nuclear tests requires repeal of the 1993 Spratt-Furse Amendment, which prohibits nuclear laboratories from research and development that could lead to a nuclear weapon of less than 5-Kt yield, because "low-yield nuclear weapons blur the distinction between nuclear and conventional war."[19]

Congress accomplished the Spratt-Furse repeal in November 2003, and National Nuclear Security Administration chief Linto Brooks quickly instructed the weapons labs to start working on advanced designs for "agent defeat and reduced collateral damage" weapons (meaning bunker busters). Congress has oscillated between stripping this project of funding or funding it for research only.[20] U.S. development and testing of nuclear penetrator weapons certainly would abrogate the Partial Test Ban Treaty and eliminate any possibility for U.S. ratification of the Comprehensive Test Ban (see chapter 7).

Appendix 9

U.S. and Them: The United States and World Oil Reserves

Published geological and political estimates of undiscovered oil resources have no set time limits stated or implied for the postulated discoveries. Such open-ended estimates effectively imply that the volume of resources yet to be discovered will lie somewhere between zero and infinity and will be found sometime between now and eternity.
L. F. Ivanhoe, *World Oil*

Published estimates of ultimately recoverable world oil vary widely, from about 1,800 billion barrels to 4,820 billion barrels. Deciding which of the estimates are more nearly correct is critically important for accomplishing future energy transitions. Overly optimistic estimates have risen steadily since the 1980s on the basis of smoke-and-mirror accounting techniques, fooling the U.S. Energy Information Administration and others into concluding that oil production will continue rising indefinitely (see chapter 12). But in 2004, production levels at many of the world's principal oil fields showed advancing depletion, illustrating some basic facts: Oil is a nonrenewable resource, and world oil production is at or close to its peak. We soon will discover that the finite amounts of petroleum still in the ground cannot feed our demand for oil.[1]

Private and government petroleum experts formulate predictions of future global oil and gas discovery and production by estimating the world's total original oil endowment. The calculation requires adding together three factors: (1) cumulative production to date; (2) estimated "reserves" (remaining recoverable oil) in known producing and as-yet-undeveloped fields; and (3) more or less educated guesses of how many oil fields remain to be discovered.[2] The cumulative past production ought to be a simple matter of reading pipeline gauges, but the veracity of reports varies, and believable numbers are hard to come by.[3] A number of sources conclude that the world had produced about 800 billion barrels of oil by the mid-1990s. Although the U.S. Geological Survey (USGS) published substantially lower figures in 2000, the Energy Information Administration estimated that annual global oil production reached about 27 billion barrels in 2005, and in August 2006 the Association for the Study of Peak Oil estimated that oil-producing countries have extracted a total of about 970 billion barrels since the beginning of the world's brief "Petroleum Interval" (see chapter 12).[4]

Many different sources publish estimates of reserves plus undiscovered oil for both the United States and the world, but the term "reserves" has no universal definition. The U.S. Securities and Exchange Commission regulations restrict "reserves" to oil that is present and recoverable "with reasonable certainty" under current technological and economic conditions. Not only are "proved" and "undiscovered" reserve reports not comparable between many oil-producing countries, but they also represent disparate, and variably credible, assessment methods—some of which must be seriously in error.[5] The notorious unreliability of reserve estimates is responsible for part of the huge variation among total global oil estimates of conventional oil resources (table A9.1).

Table A9.1   Estimates of Ultimate World Oil Recovery and Year of Peak Production

Estimate Source, Year[a]	Ultimate Recovery (Billion Barrels)[b]	Peak Production[c] (Year)
Hubbert, 1969	2,100	2000
Moody, 1978	3,200	2004
Odell and Rosing, 1983	3,000	2025
Bookout, 1989	2,000	2010
Townes, 1993	3,000	2010
Campbell, 1994[d]	1,650	1997
Laherrère, 1994	1,750	2000
Ivanhoe, 1995	2,000	1996
	1,500	1988
MacKenzie, 1996	1,800	2007
	2,300	2014
	2,600	2019
Romm and Curtis, 1996	—	2030
Mabro, 1996	1,800	2000
Ivanhoe, 1997	—	2000–2010
Edwards, 1997	2,836	2020
Campbell, 1997	1,800	1998–2001
International Energy Agency, 1998	2,800	2010–2020
Energy Information Administration, 1998	4,700	2030
Campbell and Laherrère, 1998	—	2001–2010
Laherrère, 1999	2,700	2010
Duncan and Youngquist, 1999	—	2007

continued

Table A9.1 Continued

Estimate Source, Year[a]	Ultimate Recovery (Billion Barrels)[b]	Peak Production[c] (Year)
Duncan, 2000	—	2006
	2,000	2004
Bartlett, 2000	3,000	2019
	4,000	2030
U.S. Geological Survey, 2000[e]	3,021[f]	After 2025
Edwards, 2001	3,670[g]	2020–2030[h]
Deffeyes, 2001	2,000	2003–2007
Energy Information Administration, 2001	2,248	2021–2045
	3,003	2030–2075
	3,896	2037–2112
Campbell, 2004	1,850	2005
Zagar and Campbell, 2005	1,850	2010
Association for the Study of Peak		
Oil and Gas, 2006	1,900	2005
Cambridge Energy Research		
Associates, 2006	4,820[i]	2030+

[a]Most data sources are as cited in J. D. Edwards. Twenty-First Century Energy: Decline of Fossil Fuel, Increase of Renewable Nonpolluting Energy Sources. In M. W. Downey et al., eds. *Petroleum Provinces of the Twenty-First Century* (American Association of Petroleum Geologists Memoir 74, 2001), table 7; and Moujahed Al-Husseini. The Debate over Hubbert's Peak: A Review. *GeoArabia* 2006;11:181–209, table 1.
[b]Quantities represent conventional crude oil and exclude condensates, natural gas liquids, oil shales, tar sands, and very heavy oils.
[c]With the exception of Edwards (2001), no growth rate is specified.
[d]These estimates are continuously reevaluated as data are refined.
[e]Data from U.S. Geological Survey Assessment Team. *National Assessment of United States Oil and Gas Resources* (U.S. Geological Survey Circular 1118, 1995); U.S. Geological Survey World Energy Assessment Team. *World Petroleum Assessment 2000* (4 CD-ROM disks; Digital Data Series DDS60, 2000).
[f]This value is the mean of a 95% probability of 1,990 billion barrels of oil and a 5% probability of 3,843 billion barrels of oil. No estimates were given in the 1995 U.S. assessment for reserve growth, used for the 2000 world assessment.
[g]Includes unconventional deposits, Venezuelan heavy oil, and tar sands.
[h]Assumes 1.5% growth rate per year.
[i]Cambridge Energy Research Associates (Press Release, 14 November 2006; no supporting data provided); this assessment is widely discredited.

Growing evidence suggests that reported Middle East oil reserves serve commercial or political purposes rather than scientific ones,[6] leading to gross overestimates. For example, Iraq recorded no new discoveries from 1985 through 1998, yet reported the same oil-reserve figure of 100 billion barrels every one of those years while continuing to produce oil. In 2003, Iran also raised its reserve estimates from

80 to 126 billion barrels without explanation. Dozens of other countries engage in similar practices.[7] The Organization of Petroleum Exporting Countries (OPEC) added exaggeration incentives between 1987 and 1988, when it tied export quotas to reserves—the higher the reserves, the more oil OPEC permitted member countries to sell. Six OPEC members abruptly upped their reported reserves by a staggering 300 billion barrels, with no reports of new discoveries to support the increase. In 2004, the Royal Dutch/Shell Group disclosed that it also had greatly overstated the company's reserves.

The USGS is responsible for publishing official projections of U.S. and world reserves and undiscovered resources. Through 1999, most USGS estimates of U.S. reserves varied from 200 to 220 billion barrels, but recent reports have grown highly (some say incredibly) optimistic. The USGS now couches its assessments in gambler's terms—giving both a long-shot (5%) chance of discovering and producing a large amount of oil and a high (95%) chance of discovering a minimum amount. But the agency, and news media, typically emphasize the average of high- and low-probability estimates, arbitrarily assigned a 50% chance of discovery.

The high and low USGS reserve figures are not well founded, and the mean values are even less defensible. Oil expert Colin Campbell calls them "wild guesses that could as well have been half or double the true number, yet they influence the computation of the *Mean* values...." Campbell cites the revealing example of an undrilled (generally unknown) basin in East Greenland, which USGS assessed at a 95% chance of producing more than zero (i.e., at least one barrel of oil), and a 5% chance of more than 112 *billion* barrels of oil. The USGS then gave this basin a mean (50%) likelihood of producing 47 billion barrels.[8] Also based on mean values, in 2000 the USGS projected domestic reserves of 362 billion barrels, 65% more than it reported in 1995[9] and estimated a mean total world endowment of 3,021 billion barrels. The latter figure includes an estimate of around 2,000 billion barrels for the world's oil reserve, which is 50% more than other expert projections and also 50% higher than the USGS's 1995 estimate.[10]

Off-the-charts USGS optimism about remaining oil amounts is based in part on a "reserve growth" factor. In its *World Petroleum Assessment 2000*,[11] the USGS claims that world conventional oil reserves can increase by 730 billion barrels almost entirely from "growth" of reserves in existing fields. How can reserves grow? The term simply means that more oil comes out of the field than original estimates forecast.

According to the USGS, reserve growth is a combination of incomplete original reserve estimates and technological advances that extract oil more efficiently.[12] But Campbell and others conclude that reserve growth is only apparent—it's really just an artifact of poor data and bad reporting.[13] They point to the history of technological advances during the 1980s oil price boom, which mostly increased production rather than reserves. The net result for most oil fields was a temporary surge in production followed by more rapid decline after peaking (figure A9.1). Beginning in the early 1980s, consumption has outstripped discovery (figure A9.2).

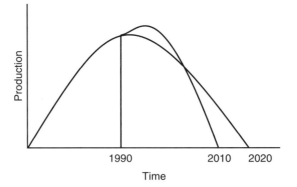

Figure A9.1 Graphs show how enhanced recovery techniques provide an immediate increase in oil production but sooner and faster depletion (highest peak line). In this case, 1990 is the time enhanced recovery begins, 2010 is the time that oil production will end as a result of enhanced recovery, and 2020 is the time of production exhaustion without enhancement. From John Gowdy and Roxana Julia. *Technology and Petroleum Exhaustion: Evidence from Two Mega-Oilfields* (Rensselar Working Papers in Economics, 2005), figure 2. Reproduced with permission.

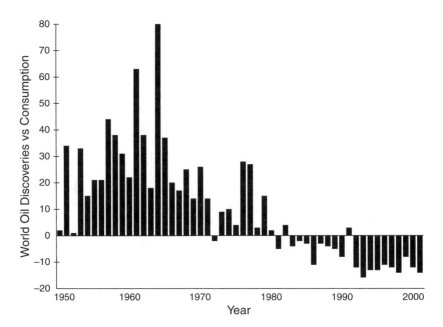

Figure A9.2 Plot of world oil discoveries minus consumption. The world produced more oil than it consumed until 1982, when it moved into deficit and began consuming much more oil than is being discovered. From C. J. Campbell. The Assessment and Importance of Oil Depletion. In K. Aleklett and C. J. Campbell, eds. *Proceedings of the First International Workshop on Oil Depletion* (Uppsala, Sweden, 2002), figure 2. Reproduced with permission.

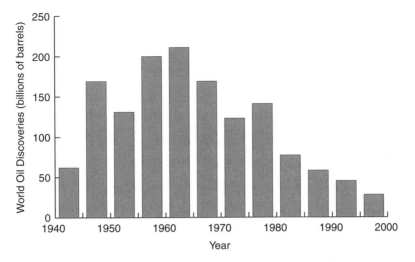

Figure A9.3 Graph showing the decline of worldwide oil discoveries, in five-year increments. Modified from L. B. Magoon. *Are We Running Out of Oil?* (poster; U.S. Geological Survey Open-File Report 00–320, 2000), figure 4.

Permanent decline is inevitable for oil, a finite resource of limited geological distribution. Inescapable facts include the mid-1960s peak of world oil discoveries (figure A9.3). That peak mirrors the lower 48 states' 1970 experience and tells us that the world's production peak is imminent.

When—not if—world production peaks and begins its inexorable decline is a subject of much debate.[14] The most highly inflated projections use the USGS's still-unrealistic 5% chance-of-discovery figure with the unlikely caveat of no growth in consumption—they give the world another 25–100 years to peak production.[15] But depletion profiles for various classes of petroleum liquids and natural gas suggest that the world peak and subsequent decline are much closer at hand.[16] The consensus of highly variable expert opinions place the peak of world oil production close to 2010, or perhaps sooner (table A9.1).[17]

The 2000 USGS report kindly provided a test for its optimistic estimates. Figure A9.4 shows the actual world oil discovery history from 1950 to 2005, compared with USGS estimates of potential new discoveries between 1995 and 2025, at each of the usual 5%, 50%, and 95% probability levels. The mean (50%) estimate for new discoveries and field expansion for the 30-year period was 1,420 billion barrels. Since larger fields tend to be discovered early on, attaining the mean prediction level requires average discoveries of more than 45 billion barrels of oil in each of the first eight years. But 1995–2003 discoveries averaged only 11 billion barrels per year, and the average from 1995 through 2005 was only 10 billion barrels per year.[18] The diagram shows that actual discoveries since 1995 are lower than the USGS's lowest level (95% probability) prediction, however.

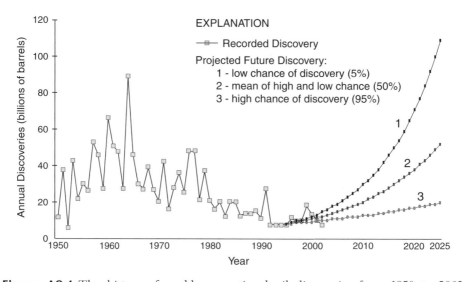

Figure A9.4 The history of world conventional oil discoveries from 1950 to 2003 (open data points), compared with the U.S. Geological Survey's highly optimistic 2000 estimates of the amounts still to be discovered between 1995 and 2025 (solid data points), rated at 1, low (5%); 2, intermediate (mean); and 3, high (95%) probability of the predicted outcome (the mean value is an average of the other two). The estimate given highest probability has the lowest predicted outcome, and vice versa. Curves 1, 2, and 3 prorate average discovery levels required to fulfill each predicted outcome. The actual discovery data show a rise in discoveries in only two of the years since 1995. The current falling trend since 1999 suggests that even prediction 3 is overly optimistic. Data for world discoveries through 2001 provided by C. J. Campbell (personal communications, November 2001 and December 2003); discoveries 2002–2005 from Rembrandt Koppelaar. USGS WPA [World Petroleum Assessment] 2000 part 1: A Look At Expected Oil Discoveries. *The Oil Drum* 27 November 2006; U.S. Geological Survey World Energy Assessment Team. *World Petroleum Assessment 2000* (4 CD-ROM disks; U.S. Geological Survey Digital Data Series DDS-60, 2000). Modified from C. J. Campbell. The Oil Peak: A Turning Point, *Solar Today* July/August 2001:41. Reproduced with permission.

The consensus expert opinion on world oil puts the ultimately recoverable total (past production, reserves, and undiscovered amounts) at about 2,000 billion barrels (table A9.1). The 2000 USGS prediction of new discoveries is off the mark so far, and in 2006 cumulative world production reached a level equal to half the level of reasonable estimates for ultimately recoverable total world petroleum. So the world peak is upon us. Since our own peak, the United States has grown largely dependent on imported oil resources. We must examine our near-future energy options now and take critical steps for coping with the energy-limited future.

Appendix 10

"Democratizing" Energy: Hydrogen Fuel Cells

For the first time in human history, we have within our grasp a ubiquitous form of energy, what proponents call "forever fuel."
Jeremy Rifkin, *The Hydrogen Economy*

Burning hydrogen in fuel cells briefly seemed the solution to environmental pollution from vehicles' fossil fuel emissions (see chapter 12) and the ideal replacement for declining petroleum supplies (see appendix 9). In 2002, the U.S. government abandoned research on high-mileage gasoline-conserving vehicles to embrace fuel cell research for running future vehicles.[1] Hydrogen is not a primary energy source like oil, however, but an energy carrier, like electricity. Hydrogen gas (H_2) is used to run fuel cells. It is the most common element in the universe, but relatively rare on Earth's surface because it is extremely light in weight and easily escapes into space.[2] So H_2 must be generated from hydrogen-bearing compounds such as water and organic hydrocarbons, including coal, petroleum, natural gas, wood, and grain. Unfortunately, the processes for extracting hydrogen are both energy inefficient and polluting.[3]

Hydrogen can be liberated from its compounds in three main ways, all requiring substantial energy input:[4]

- Heat and catalysts extract (reform) hydrogen from the hydrocarbons listed above.
- Electricity splits water molecules (electrolysis).
- A variety of experimental processes include sunlight, plasma discharges, and microorganisms, or splitting water with heat from very-high-temperature reactors.

None of these options will lessen the nation's dependence on imported fossil fuels.

Current annual U.S. hydrogen production is nine million tons.[5] The United States currently produces 97% of it by reforming natural gas or gasifying coal,[6] both fossil fuels. We would need more than 12 times that amount—110 million tons of hydrogen per year—just to run today's U.S. gasoline-burning fleet of light trucks and cars. Seawater could provide an essentially limitless hydrogen supply, but separating H_2 from water molecules requires immense amounts of electricity. The amount of electricity the United States now produces is but a small fraction of what would be needed to extract hydrogen from electrolysis of water enough to run all of today's cars and trucks. And we mostly generate that electricity from burning coal. Not least, extracting hydrogen gas from fossil fuels would consume even larger volumes than we burn directly in cars, so would use them up faster.

Pollution is another problem. On the face of it, an ordinary gasoline-powered car emits 12 ounces of CO_2 per mile, while a car powered with hydrogen emits none. But this nice picture ignores the energy input used for generating the hydrogen. On a source-to-wheel basis, getting the hydrogen from electrolysis of water is only 25% efficient. Using electricity from today's power grid for hydrogen production would generate emissions equivalent to 14 ounces of CO_2 per mile, which eliminates a hydrogen vehicle's perceived advantage of reducing smog and greenhouse gases. To lower pollutants, the electricity would have to come from hydroelectric, nuclear, biomass, or solar and wind power sources, most of which have severe environmental impacts (see chapter 12).[7] Gasifying coal for extracting hydrogen adds greenhouse gases to the atmosphere, particularly if the generators are small and dispersed.[8] The process for reforming natural gas to get hydrogen has an efficiency of 77%, but capturing and storing the CO_2 byproduct drop the overall efficiency to 58%.[9]

The vehicles that would consume this "pollution-free" energy don't come out of thin air, either. The metal ores mined for making vehicles originated in rare natural events over the length of geologic time (see chapter 4). They are as nonrenewable as fossil fuels and will not supply endlessly growing demands. Mining and refining the metal ores and other raw materials used in cars and trucks take enormous amounts of energy,[10] and regularly disposing of intentionally obsolete cars adds environmental damage, even with recycling.[11]

Aside from these issues, hydrogen has some advantages over gasoline but also significant disadvantages. Per unit of energy, hydrogen weighs 64% less than gasoline.[12] Considering only tank weights and volumes, a tank of liquid hydrogen ought to weigh less per mile than gasoline. But the *energy* in a gallon of hydrogen at atmospheric pressure amounts to only a tiny fraction of the energy in a gallon of gasoline: A car can go 3,000 times farther on gasoline than an equal volume of liquid hydrogen at atmospheric pressure. To be a useful transportation fuel, hydrogen must be somehow concentrated, and feasible methods must be found for moving the gas around and storing it. At present, the options include the following:[13]

- Liquefying H_2 by supercooling to about –423°F
- Compressing it under high pressure in tanks, like propane
- Binding very pure hydrogen to finely divided metals—iron, nickel, titanium, magnesium, manganese, or metal alloys—to form hydride compounds
- Storing hydrogen-rich gas at atmospheric pressure and pressurizing it when pumping to vehicle tanks

All options but the last require special arrangements for storing hydrogen—in particular, super-cooled hydrogen must be kept refrigerated, and highly pressured gas must be both carried and stored in special tanks. Compressing and liquefying hydrogen require additional large amounts of energy, and keeping it refrigerated takes even more.[14]

Because of the difference in energy content per volume, a hydrogen-propelled car must carry two to five times the amount of fuel a normal gasoline tank holds before it can travel as far as Americans now do between fuel fill-ups (table A10.1).[15] Storing the hydrogen in hydride compounds has some advantages, because the compounds are stable and the hydrogen can be released by reducing the container pressure or increasing the temperature. Metal hydrides also concentrate hydrogen as much 6.5 times more than an equal volume of highly compressed hydrogen gas, and about 1.5 times more than liquefied hydrogen.[16] Significant hydride disadvantages include the large amount of energy required to purify hydrogen[17] and the large container weight, especially compared to the weight of gasoline with equivalent energy. A car carrying the amount of energy equal to a normal tank of gasoline in the form of hydrogen absorbed on metal hydrides could be as heavy as an army tank and consume much more fuel per mile than either a Humvee or a house-sized camper vehicle.[18]

The metal powders contribute more than 90% of the metal hydride's weight. Iron and nickel have to be mined from hard rock, with attendant environmental impacts (see chapter 4). Light-weight magnesium is extracted electrically from magnesium chloride salt in a Utah desert basin, annually releasing tens of millions of pounds of toxic chlorine gas laced with dioxins, for example. Magnesium production tops the U.S. Environmental Protection Agency's Toxics Release Inventory for off-site pollution (see chapter 4). After a few hundred recharges, the powders become toxic wastes for disposal, driving the pollution spiral.

Energy conservation expert Amory Lovins sees no obstacles to developing light-weight hydrogen storage tanks for cars, which also allow an adequate driving range between refuelings. Lovins also thinks that on-board hydrogen is no more unsafe than other fuels.[19] But 22% of industrial hydrogen accidents currently arise from undetected leaks despite special training, standard operating procedures, protective clothing, and electronic flame and gas detectors available to the limited number of

Table A10.1 Equivalent Gasoline and Hydrogen Weights and Volumes

Fuel	Weight (Pounds)	Volume (Gallons)
Gasoline/combustion engine	110	18
Compressed H_2 (5,000 Psi[a])	198	85
Compressed H_2 (10,000 Psi)	~220	48
Liquefied H_2	99	50
H_2 on metal hydrides	440–1,320	48

Bjørnar Kruse et al. Hydrogen, Status and Possibilities, *Bellona Foundation Oslo*, Report 6–2002. Available: www.bellona.no.en/energy/hydrogen/report_6–2002/index.html.

[a]Psi, pounds per square inch pressure.

workers. This track record may suggest that the general public cannot safely manage refueling their own cars.

Developing a refueling infrastructure—thousands of hydrogen generators dispersed along U.S. highways and secondary roads—will be expensive,[20] and the safety precautions for controlling liabilities may be especially costly. But even if the hydrogen industry accepted regulation, they could not ensure that filling station managers and attendants would cooperate.[21] An unexplored problem is the potential for hydrogen leaks in commercial facilities and closed garages attached to homes.[22]

University of California professor Michael Prather warns that "the chemicals that we dispose of in the atmosphere often return as unexpected environmental problems."[23] Once it is used widely for fuel, hydrogen leaks to the atmosphere likely will increase greatly, with unknown consequences.[24] Even though switching to hydrogen would reduce nitrogen oxide emissions in smog, it also would increase methane, a much more potent greenhouse gas than CO_2. When highly concentrated, hydrogen might become an indirect greenhouse gas, affecting global atmospheric chemistry through photochemical transformations that create methane.[25] The likely climate impacts largely depend on the type and quantity of gases released in generating hydrogen.[26]

Clearly, the seductive "hydrogen economy" solves none of the problems associated with fossil fuels, especially the dwindling supply. The more conserving choice for reducing gasoline consumption is hybrid cars, but until Democrats recaptured Congress in 2006, both the George W. Bush administration and legislators had jettisoned that option in favor of continuing the oil guzzling habits of the United States for the next 10–20 years or so while focusing research on an affordable and practical hydrogen-powered car. Nobody in the government has made it clear that the hydrogen is coming from fossil fuels.

According to Jeremy Rifkin, all 750 million cars and other vehicles on the world's roads today could be supplanted with fuel cell vehicles running on hydrogen, "as cheap as personal computers, cell phones, and palm pilots"[27] opening "the possibility…to truly democratize energy, making it available to every human being on Earth." The facts suggest just the opposite—hydrogen from dwindling fossil fuel supplies is apt to follow gas and oil supplies in becoming decreasingly abundant and increasingly expensive. Each facet of the envisioned hydrogen economy—production, transport, storage, fuel cells, safety, and infrastructure—poses severe problems on its own. As writer Robert Service noted, "For a hydrogen economy to succeed…all of these challenges must be solved simultaneously."[28] Pumping large amounts of money into the hydrogen scheme instead of easing our demands on fossil fuels through proven technologies for improving fuel efficiency invites failure. Worse, it invites complacency and inaction.

GLOSSARY

Absorb, absorption A process that incorporates liquid molecules into the molecular structure of a solid or incorporates gas molecules into liquid.

Acid deposition Acids in the air falling onto the Earth's surface, including into surface waters and onto exposed soils, plant leaves, and all human structures. The acids come primarily from particulates emitted at sulfide ore smelters, coal-burning particulates at industrial sites and power plants, and sulfur, nitrogen, and carbon dioxide (CO_2) in smog. Sulfur and the other pollutants interact with both water in rain clouds and water particles in drier air, creating sulfuric acid, nitric acid, and carbonic acid. The gases can travel thousands of miles before converting to acids and falling to the ground out of dry air as dry acid deposition, or in rain as acid rain. Acids accumulated on fallen snow can acidify snowmelt waters.

Acid mine drainage Acid waters commonly arising from metal and coal mining. Many common copper, lead, iron, and zinc ores and related ore minerals are sulfide minerals, which break down to hydrogen sulfide, sulfuric acid (when sulfur reacts with water in an oxygen-rich environment), and other acid products. Waste rock from coal mining also contains sulfide minerals. Mine entrances and waste rock heaps promote and concentrate this process, feeding toxic acidified waters into neighboring streams and water bodies. Acidic water can leach, or dissolve, metal compounds from rocks and wastes and carry them elsewhere.

Acid rain Acid deposition in rainfall. Acid-forming pollutants accumulate in the atmosphere between storms and acids form when storm episodes add moisture to the air. Rainfall removes the acids from the air and deposits them at ground level. Each acid rain event suddenly raises surface water acidity, administering an "acid shock" to aquatic plants and animals. Both acid rain and acid snowmelt leach nutrients out of soils, often poisoning vegetation and soil microorganisms. *See* acid deposition.

Acre-foot A liquid measure, normally for water: the amount that can cover an acre to a depth of one foot (a football field is about 1.1 acre)—about 325,900 gallons. In the United States, an acre-foot generally corresponds to the annual water consumption of one to two 4-person households.

Adsorbed, adsorption Bonding of gas or liquid molecules to the surfaces of solids with which they are in contact. Soils hold adsorbed water so strongly that they resist both the pull of gravity and capillary action.

Aerosols Very tiny solid or liquid colloid particles, carried in air, smoke, or fog.

Agribusiness Large-scale industrial farming, which is highly mechanized and chemical dependent, primarily raising monocultures (only one or two livestock products or field crops).

Agroforesty Harvesting trees from human-planted and cultivated forests (plantations).

Albedo The fraction of light reflected from a surface, generally expressed as a percentage. Clouds and snow have high albedo, and surfaces covered by vegetation have low albedo (because of the light absorbed for photosynthesis).

Alcohol Organic compounds having a hydroxyl (OH) group bonded (attached) to carbon atoms.

Aldehyde Organic compounds having hydrogen atoms bonded (attached) to carbon-oxygen (carbonyl) molecules.

Alluvial fan Moderately sloping, wedge-shaped sediment deposits in arid regions, which connect steep mountain fronts with gently sloping desert basin floors. The fans start at the narrow mouths of mountain canyons and widen toward the basins. Alluvial fans emerging from adjacent canyons overlap to form alluvial aprons along relatively straight, fault-bounded mountain fronts.

Alluvium Sediments—mostly sand, silt, and clay—carried in creeks, steams, and rivers and deposited in their channels and floodplains.

Alpha particle, alpha radiation *See* radioactivity.

Alternative energy Energy sources other than the fossil fuels coal, oil, and natural gas. Solar, wind, and hydrothermal and geothermal energy sources produce electrical energy primarily and yield somewhat lower pollution than do fossil fuels.

Anadromous Describes fish that spend most of their lives in the ocean but migrate up freshwater rivers and streams to spawn and then die.

Anthropogenic Generated by humans and human activities. In describing substances, "anthropogenic" can be synonymous with synthetic.

Anticline Layers of rocks warped by natural process into an upright arch.

Aquatic Having to do with water (Latin *aque*). Aquatic life and aquatic organisms are plants and animals that must live in water, including in creeks, streams, lakes, ponds, and oceans.

Aquiclude An underground rock or soil layer with so few openings between mineral grains, and so lacking in connections between spaces, that it allows virtually no water flow into or through it. In practice, any such material will allow some water to seep into and through it, but generally very slowly, over very long time periods.

Aquifer Underground rock or soil able to hold water in open spaces between the constituent mineral grains, or in abundant cracks and fractures. Aquifers also contain connections between the water-filled spaces that help water to seep into and through them.

Aquitard An underground rock or soil layer with few openings between mineral grains, and relatively few connections between spaces, which retards water from seeping into and flowing through it.

Armoring A process of wind and water carrying away the fine sediment particles from an unpaved road surface or river channel below a dam, leaving only coarse fragments. Armoring forms a thin protective layer on roads. In rivers the particles are all larger than the minimum size that waters released from a reservoir can remove.

Arroyo *See* gully erosion.

Artesian, artesian spring Underground water under pressure (called hydrostatic pressure) because it is confined in an aquifer between aquitards, aquicludes, geologic faults, or some combination of these features. The artesian aquifer collects rainfall at a higher elevation than the pressured water's outlet, where it may bubble up through soil or rise above the ground surface in a fountaining artesian spring.

Atom Submicroscopic particle: one of the building blocks of all matter, whether animal, vegetable, or mineral. Atoms consist of even smaller particles: energetic forces pull negatively charged, virtually weightless, electron clouds to a central atomic nucleus. Nuclei of atoms of each of the 90 known naturally occurring chemical elements (hydrogen, oxygen, potassium, iron, etc.) have a different, fixed number of positively charged protons and a variable number of uncharged neutrons. Each element's atoms lack electric charge because they contain equal numbers of protons and electrons.

Atomic nucleus The central, massive core of an atom, packed tightly with protons and neutrons, which together determine an atom's weight.

Attenuation *See* natural attenuation.

Base level The elevation of a stream or river's lowest point. Sea level is the ultimate base level for flowing water, but rivers and streams can flow to local base levels at a lake or reservoir or a dry desert basin.

Beneficial, beneficial insects Natural processes and animal behaviors that contribute to human well-being, either directly through processing wastes or by performing economically important functions, such as pollination.

Benthic Animals and plants that live on sea floors or river beds.

Beta particle, beta radiation. *See* **radioactivity.**

Bioaccumulation A process that increases the concentration of a substance in an organism (plant or animal), compared to the same substance's concentration in its environment. Bioaccumulative substances are taken up and stored faster than the organism's system can break down (metabolize) and excrete them. Bioaccumulation provides plants and animals with life-sustaining levels of critical nutrients but also concentrates environmental contaminants. It can expose humans and other organisms to high levels of toxic chemicals and has become a critical consideration for regulating synthetic chemicals.

Bioaerosols Extremely minute airborne particles that transport pathogens, generally from feedlots and sewage treatment plant wastes.

Biocide A chemical agent capable of killing all life. The agricultural fumigant methyl bromide may be described as a biocide.

Biodegrade, biodegradation Natural recycling processes carried out by bacteria, fungi, insects, worms, and other organisms that eat wastes or dead organisms, breaking down the organic matter into nutrients that other organisms can use. Humans can harness natural biodegradation to reduce wastes and clean up some types of environmental contaminants. Composting and wastewater treatment accelerate biodegradation—one converts organic wastes to farm fertilizers; the other is used to prevent sewage pollution problems when the water is released to the environment. Microbes also are used to clean up oil spills and other types of organic pollution.

Biodiversity, biological diversity A blanket term denoting the natural biological variations, including the genetic variability within DNA and the species that it produces, that undergirds all life and sustains human life and well-being; also, the variety of organisms in an ecosystem (and throughout the world), including the range of competitive or collaborative species interactions that provide food and shelter and perform myriad other services necessary for life support.

Bioinvasion The spread of exotic (nonnative) animals and plants into natural ecosystems.

Biological soil crust A common protective surface layer on arid soils worldwide, from an association of bacteria, algae, lichens, mosses, and fungi, at or very close to the surface, which bind soil particles together and reduce water and wind erosion.

Biomagnification Progressively increasing concentrations of bioaccumulative substances, through dietary linkages between single-celled plants and the species higher in a food chain. In one food chain, a water flea eats algae, then is eaten by a minnow, which falls prey to a trout, consumed in turn by a bird or a person. Each higher organism eats large numbers of the lower ones, transferring ever greater amounts of bioaccumulative substances to each higher food-chain level. Even when contaminant levels are low in water, air, or soil, animals at the top of a food chain may receive high enough concentrations to kill them or adversely affect their behaviors, reproduction, or disease resistance, endangering them and even their species. One study reported soil-eating earthworms contained 141 parts per million (ppm) DDT, a

magnification of 14 times the amount in local soils, and robins that eat the earthworms contained 444 ppm DDT, a magnification of 44 times.

Biomass The combined mass (essentially weight) of vegetation and animal life in a life-supporting habitat; also designates an accumulation of organic material or a fuel from vegetable or animal sources. Stove or fireplace wood is a directly used biomass fuel, whereas manure, soybeans, wood wastes, waste cooking oils, corn, and many other organic materials can be distilled into combustible fuels. Woodlands are being cut to chip the trees for burning in biomass generators.

Bioremediation Biodegradation strategies for treating environmental pollution with microbes, such as bacteria, which can convert toxic to nontoxic compounds. Bioremediation is most useful for reducing sewage and transforming organic compounds to simple compounds of water (H_2O), carbon dioxide (CO_2), and nitrogen and sulfur oxides—but so far the conditions must generally be oxygen rich and the problem areas well defined.

Biosolid A public-relations name for sewage sludge, the solid residues produced from municipal wastewater treatment, for marketing them as agricultural fertilizers.

Biota The aggregate of plants and animals in a location or an ecosystem. Generally applied to nonhuman life but can include all living things.

Botanical pesticide Natural pesticidal substances derived from plants, including rotenone, pyrethrum, and *Bacillus thuringiensis*, or *Bt*.

Canopy The branches and leaves of tall forest trees, which shade the forest floor.

Chemical element A substance that cannot be subdivided or changed into another by ordinary chemical transformations (heating, burning, dissolving, absorbing, etc.), by natural interactions, or with catalysts or electrical currents, and so forth. An element may be changed into another by nuclear reactions or radioactive decay. All material substances in the universe are combinations of the 90 naturally occurring chemical elements. The most important elements for life on Earth are atmospheric oxygen, carbon, and nitrogen, hydrogen in water, and the magnesium, aluminum silicon, sulfur, calcium, and iron that make up rocks and soils.

Clay Sediment particles smaller than silt but generally too large to breathe into the lungs, and which collect in soils. Particles are generally less than 1/8,000 inch (0.004 mm) and more than 1/120,000 inch (0.00024 mm) in size. Most clay particles tend to be sheet-like minerals, composed mostly of aluminum, oxygen, and silicon layers, held together by charged potassium ions and water molecules, which result from chemical breakdown or weathering of many minerals in rocks. *See* ion.

Clearcutting A logging practice that cuts down every tree in a tract, regardless of size, along with all the understory plants, to maximize the commercial take in minimum time.

Climax association A (somewhat theoretical) association of plants and other biota that develop through the natural processes of plant succession, given relatively stable climate and few impacts from natural fire, landslides, floods, and the like.

Climax forest A forest that has developed a climax association of successor plants and other biota, under a stable climate regime. Climax forests require 100–500 years to establish in the Pacific Northwest, while Canadian and Alaskan boreal conifer and hardwood forests take 500–1,000 years. Many old-growth forests are climax associations.

Colloids Tiny dust-sized particles, so small that they resist settling out of water, causing a cloudy appearance. When waters from different sources mix, or when water is filtered or evaporates, the colloids tend to coagulate, or solidify.

Conventional oil and gas deposits Relatively fluid crude oil; when found in subterranean concentrations, called pools, commonly under enough pressure from overlying rocks and associated natural gas that it naturally flows toward and into a drill hole, and may even rise to the surface, sometimes fountaining out of the hole. A desirable grade of conventional oil, based on fluidity and low sulfur content, is called "light, sweet crude." Because of its fluidity, conventional oil has generally been cheap to extract and refine. Even after a reservoir's gas pressure drops, pumps can still raise conventional oils to the surface.

Coppicing A logging practice that leaves stumps behind to sprout and regrow timber.

Cover crops Plants seeded between rows of perennial crops, such as grapes and fruit or nut trees. The cover plants' roots hold the soil, and their leaves shelter the soil, lessening the impact of raindrops. Both functions reduce soil erosion from rainstorms and windstorms.

Cumulative production Total petroleum produced over the history of an oil field, an oil-producing basin, a country, or a region of the world. Cumulative production and the amount of petroleum reserves must be known before estimating undiscovered petroleum resources.

Curie (Ci) The amount of radioactivity emitted by one gram (1/28th of an ounce) of radium, per second. For other radioactive materials, a curie is equal to the amount that yields 3.7×10^{10} atomic nuclear disintegrations per second. Ingesting amounts of radium greater than 1.2 microcurie (a microcurie is 10^{-6}, or one-millionth, of a curie, equivalent to one-millionth of a gram of radium) can cause serious health effects. *See also* nanocurie, picocurie.

Cut slope Notch cut into a hillside to create a roadbed.

Deadfall Accumulations of leaf litter and dead wood from fallen shrubs and trees, which help to replenish forest soil nutrients.

Debris flow A primary erosional mass wasting process driven by wetting and gravity, in the form of rapid and devastating flows of soupy mud and rock, the consistency of runny concrete. Debris flows are able to root up and incorporate any vegetation or other debris in its path.

Depleted uranium (DU) *See* uranium.

Desert pavement A protective rock mosaic on desert soil surfaces of small to large, generally dark desert-varnish covered pebbles and larger cobbles.

Desert varnish A very thin dark layer of clay minerals and metal oxide chemicals, especially manganese oxides, from natural weathering processes, that accumulates on the exposed parts of pebbles and cobbles in the surface layer of desert soils and on rock outcrops.

Drainage The means by which waters flow across the land surface or are discharged into and through soils; the natural or artificial (man-made) conduits that direct the flow of water—a collective term for the connected systems of rills, rivulets, gullies, washes, streams, rivers, lakes, and other natural surface water bodies or of roadside ditches, culverts, or irrigation channels, and canals that collect and distribute surface or underground water.

Drainage basin *See* watershed.

Drainage channels Generally sinuous grooves or slots that serve to collect and direct water from rainstorms. Flowing rainwater erodes the grooves into the ground surface.

Drilling fluids Thick muds, cements, and thinner fluids pumped into a hole being drilled to find and extract petroleum or gas deposits. Some of the drilling fluids lubricate and cool the drill. Thick, heavy drilling muds and cements keep highly pressured oil (or gas) from rising up and perhaps exploding out of the hole. Other drilling fluids are thin "completion fluids," used to clean out the drill hole or to stabilize the gas or petroleum-producing horizon.

Dry hole An oil or gas well, or even water well, that does not encounter supplies of the targeted resource.

Dry ravel Streams of loosened mineral bits and other soil matter that move down relatively steep slopes, exceeding about 10% slope (greater than a $5°$ angle), which previously were logged or burned. Plant losses allowed the soils to dry out, reducing the internal cohesion between soil particles.

Ecosystem An interdependent community of plants and animals, and the soil that supports them. Living components include submicroscopic spores, fungi, and bacteria, as well as birds, reptiles, amphibians, myriad rodents, weasels, deer, and other large mammals, which interact with each other and depend upon each other for life support. The plants and animals, collectively called biota, also build soil, helping to absorb and retain surface water.

Ecosystem migration Vegetation changes driven by climate changes over relatively long time periods (compared to human life times), as shifting temperature and rainfall patterns exceed the tolerance of plants to cold or heat, or wetness or aridity. Pioneer plants that can thrive under the changed conditions spread their seeds and replace stressed and dying species. Eventually, the process of plant succession may create new, sometimes quite different ecosystems, perhaps changing scrub or grassland ecosystems into forest, or replacing trees with scrub or grass until a forest has changed to a savannah ecosystem.

Edge habitat A transitional zone between ecosystems. Edge habitats in forests tend to be sunnier, warmer, and drier than deeper woods and have a slightly different mix of animals and vegetation and greater exposure to predators, parasites, and disease. Many birds and other animals accustomed to living in darker, damper, and cooler interior habitats fare poorly in edge habitats.

Effluent spring *See* spring.

Electrolysis A method of breaking down chemicals by passing an electric current through masses of solid substances, or through solutions; a favored process for breaking water molecules into hydrogen and oxygen gases.

Electron Negatively charged particles that cluster around an atomic nucleus. An electron released from an atom is called a beta particle. Radioactive substances emit beta particles when neutrons in the nucleus fly apart spontaneously, leaving an extra positively charged proton in the nucleus and releasing the electron formerly embedded in the neutron. Substances bombarded with artificially generated atomic particles in nuclear reactors also can emit beta particles. *See* radioactivity.

Endocrine disruptor, hormone disruptor A chemical that disrupts animals' hormone-synthesizing endocrine systems, which develop sexual characteristics and control sexual reproduction. Many endocrine-disrupting chemicals act as female hormones (estrogens), accumulating in fat cells and substituting for them in reproductive functions. Many pesticides and other industrial chemicals, such as dioxins, act as estrogenic hormones. Feminized alligators and frogs have been found in polluted waters containing endocrine-disrupting chemicals.

Ephemeral A transitory phenomenon. In arid regions, many or most streams are ephemeral, flowing only in the wintertime or only during rainstorms.

Erosion Here defined as all destructive processes operating at the Earth's surface, driven by surface water, wind, ice, and gravity that mechanically wear away or dissolve and otherwise chemically transform rocks and soils, and simultaneously move them from one place to another. Erosional processes include weathering, solution, abrasion, and mass wasting. Water leaches (dissolves) natural chemicals from the minerals that make up rocks and soils. Mass wasting, water and wind, and locally ice, have the power to transport eroded sediment rubble downslope. Gravity is the principal cause of mass wasting, involving variable amounts of water. Particularly important erosional effects include sheetwash, rilling, gullying, and flooding due to running water on hill slopes, and blasting by sand particles carried in windstorms.

Eutrophication Oxygen starvation in a water body due to oversupply of nitrogen, phosphorus, and other similar chemicals in runoff from farms, homes, and industries. Under a variety of conditions, the chemicals can use up all the oxygen in streams and pond waters, suffocating fish and other oxygen-dependent aquatic life forms.

Evaporite A mineral deposit produced from the evaporation of a concentrated salt solution, especially in a restricted or enclosed body of seawater or in a desert salt lake. *See* salination.

Evapotranspiration, transpiration A feeding process of plants, especially trees, which draw water from the ground, pull it through their stems or trunks, and release it through leaves or needles. The water carries many soil nutrients and microorganisms that the trees need to grow and flourish.

Floodplain The relatively level, constantly flood-prone land beside a stream or river channel, constructed of silt and other sediment that the river water transports. During floods, silty river waters rise and cover the floodplains, leaving the silt when they retreat.

Fossil fuels Coal, petroleum substances, and natural gases, composed of complex mixtures of volatile carbon and hydrogen–based organic compounds (hydrocarbons). All represent ancient life forms of the geologic past, representing stored solar energy that can be released through burning. Coal began as buried deposits of land plants, including ferns and early trees. Petroleum generally formed in marine environments, from mixtures of sea creatures and organic debris from nearby land. Associated natural gas, mostly methane, is distilled as living matter decomposes (decaying buried organic materials in landfills also yield methane).

Fragmentation Habitat subdivision, which can have subtle to extreme effects on ecosystems. Forests and other habitats are fragmented by natural barriers and obstacles and by destructive natural events. All human clearings subdivide habitat. Logging is a major cause of forest fragmentation, but roads, urban developments, and clearings for every purpose, including reservoirs, grazing and other agriculture, and recreation, all contribute. Some woodland and shrub habitats have been subdivided into such extremely small patches that they can barely function as ecosystems. *See* plant succession.

Gamma particle or ray *See* radioactivity.

Glacier Compacted and recrystallized snow, grown into a large mass of ice that lasts from year to year. Formed on land, at least in part, active glaciers move slowly downslope or outward in all directions under their own weight.

Gravel Sediment particles created by erosion, typically consisting of rounded rock fragments greater than about 1/12 (0.08) inch across, including concentrations of large boulders, fist-sized cobbles, skipping-stone–sized pebbles, or smaller granules, plus combinations of many fragment sizes.

Greenhouse climate Globally warm climate conditions, due to enhanced greenhouse effects of very high atmospheric greenhouse gas concentrations. In several episodes of Earth's relatively distant geologic past, about 90 and 145 million years ago, atmospheric greenhouse gas concentrations were very much higher and the climates much warmer compared to conditions since humans appeared, about 200,000 years ago. The ancient greenhouse climates facilitated formation of coal and petroleum deposits.

Greenhouse effect The warming effect created by light-transmissive substances, such as glass in windows, and greenhouse gases in the atmosphere. Both let short-wavelength sunlight pass through but block longer wavelength infrared, or heat, radiation. Inside a greenhouse, light radiation becomes converted to infrared radiation (heat). As long as the sun is shining, greenhouse windows trap the heat-, which raises the inside temperature. The Earth's surface similarly converts sunlight to heat, which is trapped by greenhouse gases in the stratosphere.

Greenhouse gases Hydrocarbon gases that accumulate in the stratosphere and create the greenhouse effect. The gases include methane (CH_4) and related propane (C_3H_8) and butane (C_4H_{10}); carbon dioxide (CO_2); hydrogen sulfide (H_2S); nitrogen (N) and various nitrogen compounds; and helium (He). All are produced by natural processes, but human activities emit much larger volumes.

Green manure A crop that is plowed under to enrich soil; includes clovers, which add nutrients to soil by interactions between the plants' roots and soil microbes; or when stems or leaves are cut and allowed to decay on or in the soil.

Groundwater Water held in soil and buried rock, principally in porous and permeable materials, called aquifers. Below the water table in the groundwater zone, also called the saturated (or phreatic) zone, open spaces in the rock or soil are completely filled with water. Groundwater continuously flows from collection (recharge) sites toward ponds or lakes and rivers, leaks out at springs, or is pumped out at wells.

Groundwater mining Taking more water from aquifers in a groundwater basin than can be replenished by natural recharge from rainfall, for human purposes.

Groundwater table See water table.

Gully Generally steep-sided channels, ravines, or arroyos, cut into soil and soft rocks by running water; any erosional channel having a cross-sectional area greater than one square foot or, in practice, so deep that it cannot be crossed by a wheeled vehicle or eliminated by plowing.

Habitat The life support that ecosystems provide to wildlife, including both animal and plant life. To survive, every species needs a home in a supportive habitat, consisting of potential living sites and food sources, in a comfortable climate, to

which its forebears adapted. Some species can adapt easily to changed conditions, while others have a limited adaptation potential if their habitat disappears.

Half-life The time required for half of a given amount of radioactive substance to decay, or give off half of its radioactivity, changing into another form, which could be a different radioactive element or a nonradioactive substance.

Hardrock mining Mining metals from hard igneous or metamorphic rocks, which generally form under conditions of high heat and variable pressure. The hard rocks are relatively resistant to erosion compared to softer sedimentary rocks, and mining or quarrying them requires drilling and blasting.

Hazard life The length of time a radioactive material will remain hazardous, defined as about 10 or 20 times the half-life of the longest lived radioactive element in a material.

Hormone disruptor See endocrine disruptor.

Hybridization Creating plant or animal types with desired traits, by breeding individuals of different genetic backgrounds. Plant hybrids are bred from seed over many generations to obtain stable characteristics.

Hydraulic mining Using pressurized water to mine metals, in flecks to boulder-sized nuggets, from stream and lake or pond sediments, known as placers (rhymes with *mashers*). During and after the California Gold Rush, large water cannons flushed masses of sediment fills out of deep Sierra Nevada canyons. After processing to remove the gold nuggets from the lighter rock debris, rains washed much of the waste sediment onto agricultural lands downstream.

Hydrocarbons Compounds made of linked carbon atoms and attached hydrogen atoms. Hydrocarbon compounds are the basic material of organic compounds, including the tissues of living things.

Hydrolysis Chemical interaction between water and other chemicals, to turn them into insoluble compounds that can be extracted from soils and polluted waters.

Hypoxic Extremely oxygen-poor water due to eutrophication, containing less than two parts per million (ppm) dissolved oxygen.

Igneous rock Rocks formed from melting of largely silicate minerals at various depths within the Earth's crust and mantle. Granite forms from deep rock melts, called magma, that crystallize deep underground. Volcanic rocks crystallize after surface melts, now called lava, erupt at the Earth's surface.

Impoundment Man-made dams, pits, and other constructs designed to hold wastewaters and other mining or industrial liquids and slurries.

Individualistic community An alternate to the plant succession forest development hypothesis, developed by H. A. Gleason in 1939, which sees forests as collections of individual trees, plants, and animals, requiring similar climate and other habitat conditions, but without significant coevolutionary relationships.

Infiltrate, infiltration The process of fluids, principally water or water-rich solutions, seeping into and through soils and rocks through openings (pores) or fractures.

Infrared radiation Long-wavelength energy, invisible to the human eye but felt as heat. Infrared radiation is bracketed by shorter wavelength visible red light and by even longer wavelength far-infrared radiation.

Inorganic Nonliving substances such as rocks and metals, composed principally of the chemical elements silicon (Si), aluminum (Al), and oxygen (O), or metals bonded with oxygen or sulfur (S), composing the Earth's rocky exterior and rocky to metallic interior zones.

Ion An electrically charged atomic particle or compound group of atoms, able to cling to a variety of other ionic substances having opposite electrical charge; an atom that has either lost or gained an electron.

Ionization Processes that create electrically charged ions from atoms, including dissolution, atomic interactions, and collisions between atoms and short-wavelength electromagnetic radiation, such as X-rays and gamma rays. *See* radiation.

Isotopes Differently weighted atoms of a chemical element. Every atom of an element has an identical number of electrons and protons but can have various numbers of neutrons in the atomic nucleus. For example, all calcium atoms have 20 protons, but may contain 20, 22, 23, 26, or 28 neutrons. Not all isotopes are radioactive. An atom with too many neutrons in the nucleus for its size and energy balance loses some of them—the neutrons fly apart spontaneously, emitting energy and other atomic particles. Some atoms emit radioactivity after capturing ionic particles in the environment. Radioactive isotopes, also called radionuclides, can also be created in nuclear reactors. *See* radioactivity.

Keystone species An animal or plant species that performs functions critical to other species in the same ecosystem. The life support contributions of keystone species are so critical that many other species could not survive without them.

Knickpoint A step in a stream or river channel (often large enough to be a waterfall), locally caused by resistant rock layers and regionally by a variety of natural forces that raise the river's base level, giving it more potential energy so that it erodes deeper into its channel. As the stream erodes deeper, it also eats away at the channel step, pushing the knickpoint farther upstream. Human interventions, such as weirs and dams, also are knickpoints.

Land application A method of waste disposal, spreading substances such as sewage sludge on the land.

Landfill A pit for disposing of solid waste; a dry impoundment.

Landslide A mass of soil and rock moving downslope under the influence of gravity; also, a landform resulting from the downslope movement of a soil and rock mass.

Leach, leaching, leached The action of water and other liquids to dissolve or otherwise extract chemical constituents from rocks, soil, and any other substances; a process used to extract soluble metals or metal compounds from ore (usually rock), or soluble contaminants from wastes.

Leachate fluid Generally, rainwater containing contaminants leached out of solid waste; also can be industrial solvents used to leach target materials out of ores.

Lichen A plant organism composed of a fungus and an alga living interdependently, often growing on rocks and tree trunks or in biological soil crusts.

Load capacity The maximum amount of sediment that a stream can carry. The capacity at any particular time depends on the amount of water flowing in the river and the rate of flow, determined in part by the river channel's overall slope.

Lode seams Rock containing metal ores, for example, metal sulfide, carbonate, or oxide compounds, or pure native metal; most commonly found in igneous or metamorphic rocks.

Loess soil Soil formed on deposits composed largely of fine wind-blown dust derived from glacial outwash plains during the ice ages. Loess (pronounced *luss*) soils are common in the central United States, eastern China, and parts of Europe, and originally were very rich in nutrients.

Macropore *See* preferred pathway.

Mass wasting A general term for gravity-driven erosion processes that move masses of soil and rock, including slow soil creep and landslides, as well as rapid avalanches, mud flows, and debris flows. A common trigger for mass wasting is a rapid increase of moisture in the mass.

Metamorphic rock Any type of rock material that recrystallized under heat and pressure conditions different from those of its original formation. Metamorphic rocks generally are harder than the original rock types due to interactions between minerals or between minerals and fluids.

Microbe, microbial Microscopic organisms, especially bacteria, viruses, and other germ forms.

Microcurie *See* curie.

Micrometer One-millionth (10^{-6}, or 0.000001) of a meter, equivalent to 0.000039 of an inch.

Mill tailings Waste materials from processed mineralized rock (ore), generally containing high concentrations of toxic metals and the chemicals used for mineral extraction processes.

Molecule The smallest particle that exhibits a substance's chemical and physical attributes: formed of bonded (attached) atoms. The molecules of many substances, such as carbon dioxide (CO_2), have the ability to form more complex compounds, such as calcium carbonate ($CaCO_3$), while others are relatively inert and do not form other compounds.

Monoculture A vegetative community composed mostly of one species. Generally the preferred result of human activities, which replace diverse ecosystems with crop plants for agricultural or logging purposes.

Nanocurie (nCi) One-billionth (10^{-9}) of a curie, equivalent to the radiation emitted by one-billionth of a gram (1/28 of an ounce) of radium. *See* curie.

Nanocurie per gram (nCi/g) Equivalent to the radiation produced by 1/28,000,000,000 of an ounce of radium per 0.036 ounce of a different radioactive material.

Natural attenuation Natural processes that dilute, disperse, biodegrade, or irreversibly sorb contaminants in soils and groundwaters. Radioactive decay is another form of natural attenuation.

Net energy The amount of energy in a fuel (measured in calories, thousands of calories [Kcal], or British thermal units [Btu]), less the amount of energy used to produce or extract, process, and deliver the fuel. Energy inputs include building dams, making batteries, or drilling or energy-collecting equipment, running pumps, heating water and generating steam, and so forth; also called energy return on investment (EROI) or Energy Returned on Energy Invested (EROEI).

Neutron An electrically neutral atomic nucleus constituent, composed of a positively charged proton with a negatively charged (virtually weightless) electron embedded in it, and having the same weight as a nuclear proton. The numbers of neutrons in an element can vary, leading to differently weighted atoms of the same element, called isotopes.

Nuclear fission Processes that break atomic nuclei apart, releasing electrons and other nuclear particles, and energy. Atom bombs are based on splitting large, heavy uranium-235 atoms with bombarding neutrons, releasing huge amounts of energy and clouds of radioactive particles. Today's most common nuclear power reactors use atomic fission to heat water and generate steam.

Nuclear fusion Process that combines light atomic nuclei, principally hydrogen, to create the nuclei of heavier elements, which are slightly lighter than the sum of the small ones. The difference in mass (related to weight) is converted to energy. Nuclear fusion is the source of the sun's energy; it requires huge pressures and temperatures. Hydrogen fusion also is the basis of hydrogen bombs; the required pressure and temperature conditions are created from the explosion of an atom (fission) bomb, packaged with the hydrogen fuel.

Open dump A landfill that does not get covered with soil at intervals but remains uncovered until filled.

Ore Rock, sediment, or soil containing concentrations of commercially valuable, mineable minerals. Values depend on market prices and on energy and reclamation costs.

Organic Any compounds based on myriad chemical combinations of carbon (C) and hydrogen (H), allied with small amounts of nitrogen (N), phosphorus (P), and sulfur (S), or oxides or hydrides of those substances. Natural organic compounds form the basis of life, mostly decomposing to carbon dioxide CO_2 and water (H_2O). Synthetic (human-made) organic compounds, including pesticides, also are based on carbon and hydrogen but may contain such poisonous substances as chlorine. The term also refers to farming and gardening methods based on natural processes and biodiversity, applying only naturally produced substances.

Organism Any living thing, including animals, plants, and viruses and some other life forms that cannot be classified in one of those groups.

Overgrazing Heavy livestock (cattle and sheep) grazing that eliminates most shrubs and grasses, reducing biodiversity, opening land to accelerated rill and gully erosion; bare stretches of level ground produce dust clouds under even moderate breezes. Cowpaths, called "grazing-step terracettes," typically girdle hill slopes. Stream banks are broken down and streamside (riparian) vegetation largely eliminated or replaced by exotic vegetation. Ponds and springs dry up and disappear.

Oxidizing conditions, oxidized An environment promoting chemical interactions that combine metallic or submetallic elements with oxygen. Oxidation can also describe a type of reaction in which the atoms in an element lose electrons and become more positively charged.

Particulates Microscopic, breathable solid particles in dust, smoke, and exhaust emissions, produced from natural processes or human activities, such as burning coal and gasoline fossil fuels. Particulates include ash from coal, soot, and smoke, and from vehicle emissions— especially from diesel-powered vehicles— combined with many other hazardous air pollutants. Dust storms are composed of mineral and organic particulates. Particles less than 100 micrometers in size can be inhaled through the mouth and nose. Dust storms carry thoracic PM_{10} dust, 10 micrometers and less (1/2,500, or 0.0004, of an inch) in size, which can reach into the lungs. Respirable particles, less than 5 micrometers (1/5,000, or 0.0002, of an inch), certainly reach into the gas exchange region of the lungs, and are the most hazardous. Recent research shows that PM_{10} and smaller particulates cause heart attacks and respiratory illnesses.

Pathogen, pathogenic Any microbe that can cause disease.

Percolate *See* infiltrate.

Permeability, permeable The capacity of a porous rock, soil, or sediment to allow fluid flow through it, due to connections between the open spaces (pores) or cracks in the material. Rocks and soils with highly interconnected pore spaces allow water to move through. *See* porosity.

Petroleum Naturally occurring flammable liquid, including oil and liquid natural gases, composed of complex organic compounds (hydrocarbons), containing a variety of impurities, such as sulfur and nitrogen. Either natural or human-managed distillation yields a range of combustible fuels, petrochemicals, and lubricants.

Petroleum reserves Petroleum resources that have been discovered but not yet extracted. The true extent of a reserve is unknown before an oil field starts into production. Reserve estimates for a field or province tend to change as production progresses.

pH Measurable acidity or alkalinity of a solution, or substance such as soil, based on the concentration of hydrogen (H^+) ions relative to hydroxyl ions (OH^-) in the solution. pH arises from dissociating and reforming water molecules (H_2O = H^+ + OH^-), augmented by water interacting with other chemical substances, especially certain salts. An equal abundance of each ion yields a pH of 7, indicating neutral conditions. Values of pH < 7 indicate a greater abundance of hydrogen ions, yielding acid conditions, and pH values > 7 indicate a greater abundance of hydroxyl ions, yielding alkaline (also called base or basic) conditions. Acid waters associated with hardrock mines have the lowest pH values ever reported.

Photosynthesis The processes of plant metabolism and respiration, which use the energy in sunlight to convert water and nutrients from air and soil into chemical forms of energy that sustain plants and animals. Plants take in water (H_2O) and carbon dioxide (CO_2)

from the soil and air. The green-tinted chlorophyll and other pigments in leaves and stalks absorb sunlight and produce carbohydrates (sugars) plus oxygen gas (O_2), in a complex series of transformations. All the food that animals eat, including carnivores, are highly to fully dependent on photosynthesis. In addition, all petroleum, natural gas, wood, and biomass fuels, including biodiesel and ethanol, are either direct or indirect products of photosynthesis.

Phreatic zone *See* saturated zone; *see also* groundwater.

Phytoplankton Microscopic plants that live in the ocean, floating on currents. With zooplankton, they form the base of the oceanic food chain.

Phytoremediation Strategies for treating environmental pollution by using plants to take up toxic contaminants and change them to less toxic or nontoxic forms. Phytoremediation is used mainly for extracting heavy metals, such as cobalt, nickel, and zinc, by concentrating them in plant tissues. The remediation plants must be disposed of in some environmentally protective manner, as yet unknown.

Picocurie (pCi) One-trillionth (10^{-12}) of a curie, equivalent to the radiation emitted by one-trillionth of a gram of radium or, in English units, to the radiation emitted by 0.036 trillionths (or 1/28,000,000,000,000) of an ounce of radium. *See* curie.

Picocuries per gram (pCi/g) In English measurement, one pCi/g is equivalent to the radiation produced by 0.036 trillionths of an ounce of radium per ounce of substance. *See* curie.

Picocuries per liter (pCi/L) In English measurement, one pCi/L is equivalent to the radiation produced by 0.036 trillionths of an ounce of radium per 1.01 quart of substance. In 1941, the lifetime limit of allowable radiation

exposure for a human body was set at a tenth of a microcurie of radium, and 10 pCi of radon, per liter of air.

Pioneer plants Typically short-lived, rapidly growing, and early maturing species of grasses, shrubs, wildflowers, or trees, which produce numerous small, widely dispersing seeds and generally can enter and colonize an open or transitioning landscape before other species. Pioneers tend to have low-density wood and small leaves and need a fair degree of climatic stability. Dry interior western forests with high fire frequencies contain a particularly diverse array of pioneer plants. *See* plant succession.

Placer, terrace placer Heavy metal particles, generally nuggets to dust-sized gold, silver, and platinum fragments, concentrated in stream channels. The metal placers (rhymes with *mashers*) are eroded from ores in hill country or mountainous areas and become segregated by weight from lighter water-borne sediments.

Plankton Microscopic plants (phytoplankton) and microscopic to very small animals (zooplankton), which float in ocean currents. *See* phytoplankton.

Plantation forestry Controlled, intensive cultivation of trees that are cloned or hybridized to yield predictable volumes of lumber per acre.

Plant succession Hypothesis of Frank E. Clements, first published in 1916, generally accepted as the normal sequence of natural forest and ecosystem development. Plant succession starts with pioneer plants initially colonizing unvegetated land, and leads to the final climax stage in which the plant and animal species reach a stable interdependence. At climax the biomass produced balances the total nutrient and water uptake. Natural processes, such as floods, high winds, landslides, and lightning-set fires, frequently reopen forested landscapes to pioneer species invasions, but given a stable climate and few other disturbances, pioneer plants are replaced by stands of larger successor plants that cast more shade, in many cases suppressing understory plant growth and thinning the forest's density. Over the long term, the growth of successors yields a diverse "climax" association of plants and animals that will not vary while the climate remains within the species' tolerance limits, or until another disturbance occurs. *See* individualistic community.

Plutonium An unstable alpha-particle–emitting radioactive metal, synthesized in nuclear reactors by bombarding uranium-235 with neutrons. Plutonium-239 is fissionable, burns easily if exposed to oxygen, and is one of the most toxic substances known, with a radioactive hazard life of 240,000 years. It can sustain chain reactions and so can be used as bomb fuel. The production of plutonium from uranium is the basis for the breeder power reactor concept, which apparently is the main source of current worldwide nuclear weapons proliferation.

Porosity, porous The degree of open spaces in a rock due to openings (pores) between particles of soil or rock or from cracks in the rock, per unit volume. Highly porous rocks may have the capacity to store groundwater and thus can be aquifers. *See* permeability.

Ppb (parts per billion) A very minor proportion, particularly of a dissolved substance, in relation to a billion parts of another substance, such as water (calculated as a percentage, which expresses the proportion of a substance in relation to a hundred parts of another). Parts per billion is most commonly expressed in metric measure, as micrograms (one millionth of a gram) per liter (mg/L)—equivalent in English units to 0.000000035 (3.5×10^{-8}) of

an ounce (barely a drop of liquid) in 1.01 quarts, or two ounces of a liquid substance distributed evenly through three filled tanker ships with the capacity of the *Exxon Valdez.*

Ppm (parts per million) A minor proportion, particularly of a dissolved substance, in relation to a million parts of another substance. Parts per million is most commonly expressed as milligrams (1/1,000th of a gram) per liter (mg/L)—equivalent in English units to 0.000035 (3.5×10^{-5}) of an ounce in 1.01 quarts—or two ounces of salt in a railroad tank car full of water.

Preferred pathways, macropores Interconnected fractures, or systems of fractures and other openings larger than the pores between particles in a rock or soil, including joints and geologic faults (offset fractures), which provide fast flow pathways for groundwater, and for water migrating through the unsaturated zone. *See* porosity.

Produced water Generally salty, petroleum-contaminated water, produced from oil and gas wells along with the oil or gas. When produced from coal beds as a byproduct of coal-bed methane production, the water is also contaminated with inorganic dissolved solids, such as sodium, chlorine, calcium, and bicarbonate, and such organic substances as toxic phenols, benzene, and toluene. Oil fields generate increasingly larger amounts of produced water as the oil is extracted, whereas produced water drops off during production of coal-bed methane.

Proton Primary constituent of an atomic nucleus, having a positive charge equal to the negative charge of an electron, and a weight (defined as one gram-atomic unit) nearly 2,000 times the weight of an electron. The number of protons in an atomic nucleus defines the character and contributes to the weight of each chemical element, whether naturally occurring or man-made.

Radiation, electromagnetic radiation Energy, including sound and visible light, created by atomic to molecular motions and transitions, transmitted as waves (the shape of ocean waves). Each form of radiant energy occupies a characteristic wavelength range (the inverse of frequency) or spectrum. Radio energy is transmitted at very long wavelengths, microwaves are shorter, and progressively shorter wavelengths are associated with heat (or infrared) energy, visible light, ultraviolet radiation, X-rays, gamma rays, and cosmic rays. Red colors are longer wavelengths, and violet is the shortest wavelength in the visible light spectrum. The longer wavelength energies of heat and visible light radiation can damage skin and eyes, but shorter wavelength, nonsensory radiation, such as ultraviolet radiation in sunlight, and higher energy X-rays, gamma rays, and cosmic rays, have much greater capacities for causing skin damage and for penetrating skin and damaging internal organs. Even small doses of ultraviolet A, B, and C wavelengths can damage skin, and high doses can cause skin cancers. The higher energy radiation is even more threatening. The effects of long-wavelength radiation on living tissue are still relatively unknown, but microwaves can interfere with electronic devices (e.g., heart pacemakers) and the radiation sources in televisions, computers, and cell phones may damage brain tissues.

Radioactive decay The process of an atomic nucleus undergoing spontaneous disintegration, in which it ejects a nuclear particle and energy. In other decay processes, an atom captures an electron or nuclear particle. *See* radioactivity.

Radioactivity, radioactive The process and the emissions resulting from radioactive decay. Emissions consist of positively charged alpha particles, containing two protons and two neutrons (equivalent to the nucleus of a helium atom); negatively charged beta particles, the same as electrons; or gamma rays, which are X-ray-like bursts of energy. Depending on the process causing the nucleus to break up, emissions may include several or all of the types. Also depending on the chemical element and the process, radioactive decay can release large amounts of heat. Collisions between neutrons and heavy atomic nuclei in nuclear reactors can eject neutrons from the nucleus of uranium and plutonium atoms.

Radionuclide, radioisotope An atom of any element having more neutrons than can be stabilized in its nucleus, which is therefore prone to radioactive decay. Some radioisotopes capture a particle before becoming destabilized enough to decay. Many light element radioisotopes are naturally occurring but radionuclides heavier than uranium are not found in nature and have been synthesized in nuclear reactors or atom bomb explosions, along with large amounts of fast-decaying light radioisotopes. *See* isotope.

Recharge The process of rainwater infiltration into the ground to replenish the groundwater stored in aquifers.

Reducing conditions, reduced An environment promoting chemical interactions that separate metal oxide compounds into the metallic element plus oxygen gas. Reduction can also describe a type of reaction in which the atoms in an element gain electrons and become more negatively charged.

Renewable resources, renewable energy Materials and energy sources that cannot be used up or that can be used up but somehow restored from natural processes or regrown. Solar and wind energy cannot be used up, but harvestable plants, such as trees, corn, or other biomass crops, can be regrown, and some consider them renewable resources. However, plants need to grow in soil—perhaps the most important nonrenewable resource. Biomass fuel crops will quickly deplete soils if used at today's levels of fuel consumption.

Retort, retorting Process of purifying liquid or extracting a liquid or vapor from a solid substance by heating them in a vessel (also called a "retort"). The retorting vessel often is equipped with a tube for conveying the liquid or vapor to another vessel for further processing. Distinctive glass retorts, with spherical body and long tapering tube, have been used for distilling liquids since medieval times. In-place retorting is a relatively new process that takes place in a body of rock, yielding fluids when heated.

Rill erosion, rills, rilling Erosion generally due to sheetwash, or sheets of rainwater running off a hill slope, eroding into softer material and producing very small, closely spaced channels, called rills.

Riparian Pertaining to streams and streamside (riverside) areas. Riparian zones make up less than 1% of arid western public lands but provide the habitat for most of the wildlife in arid regions.

Runoff The proportion of rainwater or snowmelt that is not soaked up by the soil but runs on the surface and drains into stream channels, or directly into ponds, lakes, or the ocean.

RUSLE Revised Universal Soil Loss Equation. *See* USLE.

Salination Processes that concentrate salts in water and soils. Rainwater can leach nutrients, such as sodium, calcium, potassium, chlorine, carbonate, nitrates,

sulfates, and others, out of soil. Interior desert lake basins can develop high concentrations of these materials, because the lake water evaporates during hot summers. Utah's Great Salt Lake is the largest U.S. example. Some desert lakes have dried entirely, leaving behind layered salt deposits, called evaporites, formed from combinations of ions leached out of rocks and soils (common table salt is sodium chloride, NaCl). When groundwater is used for irrigation, salts from both the water and the soil concentrate on the ground surface as the water evaporates, leaving salt on and in the soil.

Saltate, saltation A mode of sediment transportation that propels particles forward (downwind, or in the direction of stream flow) in a series of intermittent leaps, jumps, hops, and bounces across a surface. In water, sediment particles bounce and hop along the bottom of a stream channel, and on dry land high winds bounce particles off the ground surface. With each bounce or leap, saltating sediment particles impact other rock masses in the river or on the ground, abrading them and reducing them in size.

Sand Sediment particles larger than a coarse silt grain, less than 1/9 inch (3 mm) and more than 1/800 inch (0.03 mm) in size. Any mineral can be reduced to particles of this size, but quartz, the most resistant mineral to weathering, generally makes up the largest fraction of sand.

Sanitary landfill A landfill or dump that covers every day's addition of garbage, trash, and hazardous materials with a layer of soil.

Saturated zone Rock or soil below the groundwater table, where the open pore spaces are completely filled with water.

Savannah An ecosystem composed mostly of grasses and shrubs, with scattered individual trees or tree clusters.

Secure disposal A waste dump where highly toxic substances are packaged and placed in underground vaults.

Sediment Broken rock fragments of varying sizes, from gigantic boulders to near-microscopic clay particles and dust, produced by natural erosional processes—including everything from chemical solution to mechanical abrasion. From coarsest to finest, the major sediment categories are gravel, sand, silt, clay, and dust. *See* erosion, particulates.

Sedimentary rocks Rocks precipitated chemically from water or made of the debris from weathering and erosion of preexisting rocks at Earth's surface. Sediment becomes rock when the debris particles are pressed together and chemically cemented, generally under the pressure of overlying sediment deposits, which can accumulate to depths of many thousands of feet.

Seed tree harvesting A modified form of logging by clearcutting that typically leaves 5–12 mature trees standing per acre, to help reseed desired trees in the logged area.

Seismic Relating to earth vibrations and earthquakes.

Selective cutting Logging by periodically taking small numbers of mature trees from a forest, leaving the character and ecological stability of the surrounding forest intact. This approach to logging can be the least impacting on a forest but does not deliver large lumber volumes quickly. Loggers and supportive agencies argue that selective cutting by itself does not foster the healthy development of some forests and can promote degradation and replacement of commercially-desirable trees by succession processes.

Sewage sludge Solid or semisolid residues left from treating sewage to remove various contaminating substances.

Sheetwash An overland flow or downslope movement of rainwater, taking the form of a thin, continuous film over relatively smooth soil or rock surfaces and not concentrated into channels larger than rills.

Shelter wood harvesting A logging practice that removes only mature trees from a tract but leaves a portion of the larger trees to protect and encourage regenerating sprouts or seedlings. The larger trees are later removed in successive stages to foster growth of the sprouts and saplings.

Side-casting A technique for building roads on sloping terrain, by cutting a notch into the slope and pushing the cut material aside. The debris usually falls downslope.

Silt Sediment particles smaller than sand size, meaning greater than 1/6,500 inch (0.004 mm) and less than 1/800 inch (0.03 mm). Clay and silt particles are the primary constituents of most soils.

Skidding In logging, removing logs by dragging them across the land.

Slash residue The trimmings from logs and cut plants left over after removing the largest, commercial trees in a logging operation; includes shrubs and countless logs of unmarketable small trees.

Sludge Any thick, messy wet-mud-like substances, such as oil and gas drilling fluids; the mostly-solid residue from sewage treatment. .

Slurry, slurries A thin fluid suspension of solids in water or other fluid.

Smelt, smelter, smelting The process of burning or roasting ores to extract and purify the desired minerals. A smelter is the site of smelting and is usually located close to the mines that produce the ore.

Smog, photochemical smog Ozone (O_3) created from waste gases, including nitrogen dioxide, particulates, sulfur, and carbon monoxide (CO) in vehicle exhaust and industrial emissions, largely from burning fossil fuels. The ozone forms when ultraviolet radiation in sunlight activates the nitrogen oxides and volatile organic compounds produced from burning oil, coal, and gas. A myriad other chemical reactions take place between atmospheric gases and all the tailpipe pollutants. Since sunlight creates the smog, it generally dissipates after dark and builds up again the next day.

Soil The complexly modified mix of mineral rubble and decayed and transformed organic material that mantles most of Earth's dry land surface. Soil evolves continuously from weather- and climate-driven chemical and mechanical processes and biological action to form a nutritive blanket in which plants grow. Plants sustain all levels of life on Earth, so natural soils are critical natural resources.

Soil creep The slow, gravity-induced, downslope movement of soil masses on steep slopes, likely a very widespread phenomenon on both cleared and uncleared forest slopes.

Soil erosion Processes by which soil moves or is displaced from its place or origin, generally by wind or water. Some forms of soil removal, such as recreational vehicles' displacement of soil, is called mechanical soil erosion. Other human earth-moving activities generally are not included.

Soil profile, soil structure The sequence of soil layers, or horizons, with depth, developed by the actions of air, rainwater, and chemical and biological agents on weathered rock materials. Circulating water leaches the materials and transports the leached components to deeper levels. New clay and iron-rich minerals form, and smaller grains—especially clay—also move deeper, eventually creating the layers. The principal soil horizons are designated A, B, and C with depth. The uppermost A horizon varies from 1–2 to 20 or so inches thick and contains most

of the soil's organic materials and small clay minerals that may be transported to lower levels, along with leached iron and aluminum. The concentrated plant nutrients, and their role in controlling water, heat, and air balances, make A horizons critical to plant growth. Thicker B (subsoil) horizons consist of weathered rock materials and are the maximum accumulation zone for clay minerals and iron and aluminum oxides. Because of clay concentrations—and in arid regions, calcium carbonate (caliche) concentrations—B horizons of many soils are unfavorable to plant root growth, and their high clay contents impede infiltrating water. C horizons are rock materials, affected little by soil-forming processes.

Sorbed, sorption See absorption, adsorption.

Sound science Supposedly reliable, peer-reviewed scientific interpretations based on validly performed experiments and collected data, but today a political term, even oxymoronic, used by financially interested parties to publicly debunk actual peer-reviewed scientific studies on land uses, erosion, habitat reduction, water and air pollution, the effects of pesticides, and other industrial environmental impacts.

Species A group of living things, either plant or animal, that can interbreed and create fertile offspring having similar form, habits, and functions.

Species-area effect The effect of constricting habitat size, which reduces the numbers of animal and plant species the habitat can support.

Spring, effluent spring A site where groundwater flows naturally out of rock or soil, onto the land surface, or into a stream or other body of water.

Stratosphere The layer of the atmosphere overlying the troposphere (where clouds form), and extending from

6.2 miles to 31 miles above Earth's surface; characterized by a region of constant temperature in the lowermost few miles.

Stream channel The elongate depression that contains the main flow of either a permanent or intermittent stream or river. The channel may continuously contain water (permanent stream) or periodically (intermittent stream), and it is shaped by the periods of concentrated stream flow.

Succession *See* plant succession.

Successor plants In the process of natural plant succession, the successor plants tend to replace pioneer plants because they produce large, fertile seeds, grow larger leaves and denser wood, and have greater tolerance for climate variations. Successors tend to require fewer resources than the pioneers and eventually may outcompete them, to produce a climax association, or climax forest. If the climate remains stable over many centuries, climax forests may become old-growth forests. *See* plant succession.

Suspended load The part of the load of sediment, carried either in stream water or by high winds, that has no contact with the streambed or ground surface for an extended period of time. Suspended loads in streams consist of clay, silt, and sand. Suspended loads carried in high winds tend to be clay and dust (particulates), but rarely silt.

Sustainable, sustainability A term generally applied to industrial and agricultural and water-use practices and modes that can be sustained over long time periods, balancing resource depletion with resource renewal, recycling, and using natural systems for water purification. True sustainability consists of taking nothing from the Earth that cannot be replaced and returning no wastes that cannot be broken down by natural systems to provide nutrients for other organisms.

Sustainable growth An oxymoronic concept, used to promote continued development using "green" building or other supposedly Earth-friendly precepts. Natural processes and natural resources cannot sustain endless human population growth and development, however.

Synergistic, synergy Combined or coordinated action or force; processes or chemicals interacting and producing effects greater than the effects of any one by itself.

Tailings Mining wastes consisting of ore materials that have been ground finely and treated, either by heating (smelting) or with chemicals, to extract the target minerals.

Talus, talus slope Angular rock fragments, varying in size from car or house dimensions to cobbles or pebbles, formed mostly by mechanical and chemical erosional processes. The rock masses split along internal expansion joints or become broken up by freezing and thawing or from chemical weathering dissolving the rock's minerals. The rocks may also break from impacting each other. The fragments fall off cliffs or exposed rock outcrops and form piles (also called talus cones), which may coalesce to form talus aprons. Talus piles generally mantle steep hill slopes.

Tectonic, tectonism Processes originating deep within the Earth, which create ocean basins, and break up and rearrange the positions and orientations of land masses, in the process creating mountain ranges, with accompanying earthquakes and volcanic eruptions.

Temperate climate zone The portion of the northern hemisphere between the tropics of Cancer and the Arctic Circle; and in the southern hemisphere, between the Tropic of Capricorn and the Antarctic Circle.

Terawatt (Tw) A trillion watts. In 2001, the average total U.S. power consumption was 3.3 Tw.

Tight formations, tight sand Porous rocks in which the pore spaces are few and not well connected, but are filled with finer sediment or mineral cement; used in petroleum geology to indicate a rock that may contain petroleum but will not easily release it. *See* porosity, permeability.

Total dissolved solids The quantity of soluble substances in stream water, extracted from rocks and soil by weathering and leaching processes.

Transpiration *See* evapotranspiration.

Transuranic Refers to chemical elements heavier than uranium. Nearly all are products of nuclear bomb explosions or nuclear reactors.

Unconventional oil and gas deposits All petroleum deposits that are either so viscous or the rocks containing them are so dense, or the forces binding the petroleum to the reservoir are so great, that the oil cannot flow toward or into a drill hole. Unconventional deposits include "continuous" (or disseminated) deposits (tight formations), in which oil and gas are broadly distributed through dense rock units, and rock units having the petroleum tightly bound to particles within the rocks of the reservoir. Methane in coal beds is included in unconventional deposits, and some experts add deep-water deposits and those in remote and difficult terrains.

Understory, undergrowth The low-growing plants in a forested or savannah area, growing at ground level beneath a canopy of large trees and shrubs, especially in contrast with towering older trees.

Undiscovered petroleum resources Petroleum that is potentially exploitable but not yet discovered; a theoretical category, represented by estimates based both on informed and uninformed speculations.

Unsaturated zone, vadose zone The interval of soil or rock between the ground surface and the water table, where pore spaces may be completely dry or may contain water vapor or discontinuous films of liquid water. Rainwater must sink through the unsaturated zone to reach the water table and pass into the saturated zone (phreatic) of groundwater.

Unsustainable Practices that consume nonrenewable resources or that use renewable resources faster than they are replenished, for example, soil and groundwater.

Uranium A naturally occurring alpha-particle–emitting very hard and dense radioactive metal. Unprocessed uranium ore is more than 99% uranium-238, and less than 1% highly fissionable uranium-235 used in atom bombs. Nuclear power plant fuels are 95–97% uranium-238 and 3–5% uranium-235. Purifying uranium-235 for bomb fuel yields a large amount of residual uranium-238, called "depleted uranium" (DU). The United States is recycling DU into penetrators for nonnuclear shells and bombs, and has deployed them in a number of regional conflicts. Uranium is pyrophoric, meaning that it burns on impact, releasing dense clouds of breathable radioactive particulates, with a hazard life comparable to the age of the universe.

USLE Universal Soil Loss Equation, an empirical method for predicting erosion from sheetwash and rilling in agricultural soils. The Revised Universal Soil Loss Equation (RUSLE) uses the same approach: $A = R \times K \times LS \times C \times P$, where A = annual soil loss from sheet and rill erosion in tons per acre, R = rainfall erosivity factor, K = soil erodibility factor, LS = slope length and steepness factor, C = cover and management factor, and P = support practice factor. The factors have been revised with updated information, and new factor relationships have been derived from modern erosion theory and data. The intent of the revisions is to make the predictive capability applicable not only to tilled land but also to untilled lands such as grazing lands, those used for military exercises, construction sites, mine spoils, and disturbed forest lands.

Vadose zone *See* unsaturated zone.

Valley fever A severe respiratory disease, caused by the fungus *Coccidioides immitis* or *C. posadasii*, which lives in arid soils in central California, Arizona, and New Mexico.

Volcanism Processes deep in Earth that melt rocks and facilitate the eruption of those melts at Earth's surface; also, the kinds and characteristics of erupting volcanoes.

Watershed, drainage basin A local to regional network of drainage channels, including creeks, streams, and rivers. The watershed may connect small upland drainage channels to a larger lowland river, flowing directly to a lake or ocean. Watersheds also can consist entirely of small streams that drain from mountain ranges into dry intermountain basins, or inland marshes or lakes (e.g., Utah's Great Salt Lake). The watershed of any large river includes smaller watersheds of tributary rivers, streams, and creeks. Watershed lands are 90% uplands and valley bottom, and 10% narrow, vegetated "riparian" zones along streams and creeks. The vegetated creek zones are inseparable from the uplands and other parts of a watershed—whatever impacts one part impacts the whole.

Water table, groundwater table Top of the saturated zone, also called the phreatic zone, of underground water (groundwater). Above the water table, crevices and openings in soil and rocks may contain water vapor or water droplets but are not filled with water.

Below the water table, openings are completely filled with water.

Weathering Processes that weaken and corrode rock materials at the Earth's surface, from exposure to atmospheric conditions. The rocks become changed in color, texture, composition, cohesion, or form. Weathering implies little or no transportation of the loosened or altered sediment generated, so that the processes can produce a waste mantle (which may become soil) or prepare the disintegrated and chemically altered material for transportation by wind, water, or even ice sheets (glaciers). Weathering processes include mechanical breaking of rock from ice freezing and expanding in cracks and between mineral grains, or from salts, carried in rainwater, crystallizing in cracks. Chemical weathering starts with naturally slightly acid rainwater and organic acids from naturally decomposed plants and animals leaching the minerals in rocks, especially under humid conditions. Acid rain and other forms of natural and human-produced acid deposition, such as acid mine drainage, accelerate chemical dissolution of rocks.

WEPS Wind Erosion Prediction System, a parallel equation to USLE, developed for predicting wind erosion: $E = I \times K \times C \times L \times V$, where E = average annual soil loss in tons per acre, I = soil erodibility factor, K = soil roughness factor, C = climate factor, L = unsheltered distance across an agricultural field, and V = vegetative cover factor. Like USLE, WEPS is undergoing continuous refinement to improve conservation planning and reduce air pollution.

Wetland Ecosystem that is water covered at shallow depth for part or all of the time. Wetlands include coastal and interior marshes and swamps, and associated pools, sloughs, and bayous, but also can include vernal pools, which are wet only in spring and support amphibians that need pools or puddles to reproduce but can exist under drier conditions the rest of the year.

Zeolite minerals Natural minerals, composed largely of aluminum and silica, some calcium and sodium, and water molecules. Zeolites form when a rock's other minerals are altered, often by hot fluids from volcanic activity. They readily lose ionic constituents and take up others, so they are commonly used as industrial ion exchangers for water purification.

NOTES

Introduction

1. Garrett Hardin. The Tragedy of the Commons. *Science* 1968;162:1243–1248. Available: www.garretthardinsociety.org/articles/art_tragedy_of_the_commons.html. Also see Wikipedia's thorough discussion of the implications and misinterpretations of Hardin's essay (Available: en.wikipedia.org/wiki/Tragedy_of_the_commons).

2. Paul Hawken, Amory Lovins, and L. Hunter Lovins. *Natural Capitalism: Creating the Next Industrial Revolution* (Boston: Little, Brown, 1999).

Chapter 1

1. M. Williams. *Deforesting the Earth: From Prehistory to Global Crisis* (Chicago: University of Chicago Press, 2003).

2. Other than Christmas trees and mistletoe, spreading Christianity suffocated most tree-worshipping rituals in northern Europe during the "Dark Ages" and made tree cutting more acceptable. Christian doctrines also converted formerly beneficent tree spirits into bad or mischievous elves, gnomes, and fairies (Sir James George Frazer. *The Golden Bough* (London: Touchstone Books, 1995; first published 1922)).

3. John Perlin. *A Forest Journey* (New York: W.W. Norton, 1989), 25–68. Since the dawn of civilization, roughly 20% of primeval forest cover has been removed overall. The largest amount of removal has occurred in temperate zones in both northern and southern hemispheres (World Resources Institute. *World Resources 1990–1991* (New York: Oxford University Press, 1992). N. Myers defines deforestation as

> the complete destruction of forest cover through clearing for agriculture [so]...that not a tree remains and the land is given over to non-forest purposes...[and where] very heavy and unduly negligent logging...[results in a]...decline of biomass and depletion

444

of ecosystem services…so severe that the residual forest can no longer qualify as forest in any practical sense of the word. (quoted in Williams, *Deforesting the Earth*, 452) Worldwide, net deforestation estimates vary due to poor historical records, semantic arguments, and the vague definitions given forests or forested areas ("closed canopy" or "open woodland," etc.). United Nations Food and Agriculture Organization and European Economic Commission data show a 5% drop in world forest cover, from 18 million square miles in 1980 to 17 million square miles in 1990—almost entirely concentrated in the tropics (Williams. *Deforesting the Earth*, table 13.3, 451).

4. Williams. *Deforesting the Earth*.

5. The U.S. Department of Agriculture Forest Service estimates that forested land in 1630 covered 1,045 million acres of what eventually became the United States, declining to 759 million acres by 1907. In 1997, forests covered 747 million acres, or 33% of U.S. land area (U.S. Department of Agriculture Forest Service. *U.S. Forest Facts and Historical Trends* (2001). Available: fia.fs.fed.us).

6. From the diary of a Mayflower passenger (Perlin. *A Forest Journey*, 270).

7. Many cut-over woodlands have grown back, some two or more times. As crop yields soared, competition for markets forced farmers to cultivate only the most productive lands, allowing many cleared areas to revert back to woodlands (W. L. Thomas, Jr., ed. *Man's Role in Changing the Face of the Earth* (Chicago: University of Chicago Press, 1956), 417).

8. U.S. Department of Agriculture. *U.S. Forest Facts and Historical Trends*.

9. Notable timber alarmists included President Theodore Roosevelt. In his address "The Forest in the Life of the Nation" to the National Forestry Congress, Roosevelt warned of "inevitable timber famine," perhaps within a few decades (*Proceedings of the American Forest Congress* (Washington, DC: H. M. Suter, 1905), 3–12). Others included Frederick Starr, who thought that the timber "day of reckoning" might set in as early as 1895, and Gifford Pinchot, first U.S. Chief Forester, whose zealous concern eventually undermined his influence in government (Frederick Starr. *American Forests: Their Destruction and Preservation* (U.S. Department of Agriculture Annual Report, 1865), 210–234; Raphael Zon and William Sparhawk. *Forest Resources of the World* (New York: McGraw-Hill, 1923)).

10. U.S. Department of Agriculture. *U.S. Forest Facts and Historical Trends*.

11. In 1997, 7% of all U.S. forest lands had been reserved, either as national or state parks, or under other protected status. Of forested lands open to logging, only 6% were more than 175 years old. The trend of reserving old forests in the western United States, effectively removing stands containing trees more than 29 inches in diameter from logging, accelerated in the 1970s, but policy shifts in the 1990s curtailed harvests in stands with larger trees (U.S. Department of Agriculture. *U.S. Forest Facts and Historical Trends*).

12. G. A. Giusti and A. M. Merelender. *Inconsistent Application of Environmental Laws and Policies to California's Oak Woodlands* (Forest Service General Technical Report PSW-GTR-184–2002, U.S. Department of Agriculture, 2002).

13. From an interview with Mike Dombeck, chief of the U.S. Forest Service from 1997 to 2001, aired on National Public Radio's *Living on Earth*, March 2001.

14. J. H. Connell and R. O. Slayter. Mechanisms of Succession in Natural Communities and Their Role in Community Stability and Organization. *American Naturalist* 1997;111:1119–1144; M. Rees et al. Long-Term Studies of Vegetation Dynamics. *Science* 2001;293:650–655.

15. Unfortunately, practices that lead to long-term rainforest destruction in the tropics are also widespread in America's western temperate zone.

16. Frank Clements introduced the theory of plant succession and climax community in 1916, laying the foundation for contemporary thinking about the natural dynamics of forests. He greatly influenced E. P. Odum, whose textbooks and papers in turn influenced countless university biology students in the later twentieth century (F. E. Clements. *Plant Succession: An Analysis of the Development of Vegetation* (Washington, DC: Carnegie Institute Publication 242, 1916); also F. E. Clements. *Plant Succession and Indicators* (New York: Wilson, 1928); E. P. Odum. *Fundamentals of Ecology* (Philadelphia: W. B. Saunders, 1974); E. P. Odum. Energy Flow in Ecosystems: A Historical Review. *American Zoologist* 1968;8:11–18; and E. P. Odum. The Strategy of Ecosystem Development. *Science* 1969;164:262–270).

17. A Pacific Northwest Douglas fir forest may need 500–1,000 years to reach maturity (Jerry F. Franklin and Thomas A. Spies, Composition, Function, and Structure of Old-Growth Douglas-Fir Forests. In L. F. Ruggiero et al., eds. *Wildlife and Vegetation of Un-managed Douglas-fir Forests*. USDA Forest Service General Technical Report PNW-GTR-285, 1991, 71–80. Available: www.humboldt.edu/~storage/pdfmill/Batch%205/unmanaged. pdf; see table 1.1). Washington State's Department of Natural Resources uses growth times, determined by counting growth rings, to define forest succession stages (see endnote 62 of this chapter). The *National Forest Policy Statement of Canada* (Canadian Council of Forest Ministers. National Forest Strategy, 2003) defines old growth forest as an ecologically mature forest "that is and has been subject to negligible, unnatural disturbances such as logging, road development, or clearing, or a forest in which the upper stratum or over story is in the late mature to overmature growth phase." U.S. Forest Service geologist Fred Swanson (personal communication, 2003) gives a more general definition of late-stage (or "mature") growth as 80–200 years and "old growth" as forest more than two centuries old. In practice, foresters and loggers may define "old growth" by the amount of biomass as logs, downed broken limbs, etc.; the number of trees per acre of particular size; and the number of large tree species per acre. Some "old-growth" areas actually may have been logged repeatedly. The specific criteria for minimum size, downfall biomass, etc., are subject to preservationist interests or economic pressures.

18. A. S. Waitt. Pattern and Process in the Plant Community. *Journal of Ecology* 1947;3:1–22.

19. H. A. Gleason. The Individualistic Concept of Plant Association. *American Midland Naturalist* 1939;21:92–110.

20. A good review of the state of forest ecology theory is provided in H. H. Shugart, A *Theory of Forest Dynamics: The Ecological Implications of Forest Succession Models* (Caldwell, N.J.: Blackburn Press, 2003).

21. The Chicxulub meteorite impact destroyed virtually all life in the Caribbean and Central American region, and severely reduced North American biodiversity, long before humans came to North America. Recovery from that great disaster required about 10 million years (Tim Flannery. *The Eternal Frontier—an Ecological History of North America and Its Peoples* (New York: Atlantic Monthly Press, 2001), 36).

22. The pioneer white spruce species spread northward at a mean rate of about two miles per year into western Canada—now high plains, but then open tundra. Seeds spread on persistent, strong southeasterly winds from the retreating ice front. In the adjoining

Rocky Mountains, the northern limit of lodgepole pine forest started a 12,000-year migration across an irregular landscape from about the position of the U.S.–Canadian border to the latitude of today's Anchorage, Alaska. White spruce now dominates northern Canada and Alaska, while lodgepoles still are migrating northward at more than 600 feet per year (E. C. Pielou, *After the Ice Age: The Return of Life to Glaciated North America* (Chicago: University of Chicago Press, 1991; M. B. Davis and R. G. Shaw. Range Shifts and Adaptive Responses to Quaternary Climate Change. *Science* 2001;292:673–679; M. Davis and S. Sugita. Reevaluating the Pollen Record of Holocene Tree Migration. In B. Huntley, ed. *Past and Future Environmental Changes: The Spatial and Evolutionary Responses of Terrestrial Biota* (Proceedings of a NATO Advanced Research Workshop, 1995).

23. T. Vale, ed. *Fire, Native Peoples and the Natural Landscape* (Washington, DC: Island Press, 2002). Vale goes on to say:

> Settlements, agricultural landforms, and agricultural fields represent major changes in the landscape, although at modest spatial scales. Of these three impacts, moreover, some dwellings and constructed landforms have persisted long after they were no longer used (today we admire them in some of our National Parks). Nonagricultural plant manipulations illustrate more modest alterations of nature, and at limited spatial and temporal scales.... Wood cutting seems similarly constrained in magnitude...variable in spatial scale, and rather ephemeral in temporal terms. The reduction in numbers of grazing ungulates by excessive hunting not only represents an impact unto itself but also potentially caused vegetation change...over large areas, although the temporal persistence of such modifications without continued ungulate killing may have been fleeting. (p. 32)

24. N. H. Lopinot and W. I. Woods. Wood Over-exploitation and the Collapse of Cahokia. In M. Scarry, ed. *Foraging and Farming in the Eastern Woodlands* (Gainesville, Florida: University of Florida Press, 1993), 206–231.

25. Flannery, *The Eternal Frontier*, 218–229. Native American population pressures on forests certainly were muted compared to today. Highest estimates number the Native North American population at around 20 million. Diseases introduced by Europeans severely reduced them, perhaps more than 75%.

26. O. C. Stewart. Fire as the First Great Force Employed by Man. In W. L. Thomas, Jr., ed. *Man's Role in Changing the Face of the Earth* (Chicago: University of Chicago Press, 1956), 115–133; W. Doolittle. *Cultivated Landscapes of Native North America* (New York: Oxford University Press, 2000)).

27. Urban–suburban developments commonly replace natural grasslands and deserts with urban woodlands, uniquely human-engineered landscapes, which also replace native tree species with exotics. The scale is much smaller than replanting in logged areas.

28. U.S. Department of Agriculture. *U.S. Forest Facts and Historical Trends*. In the nation as a whole, only 17% of commercial timberlands lie within national forests. "Commercial" timberland includes all unprotected forest lands capable of growing trees at the rate of 20 cubic feet per acre per year (Forest Guardians. Available: www.forestguardians.org/).

29. Most private forests are in western Washington State, northwestern Oregon, and the redwood belt of northern California.

30. The world faces potentially catastrophic declines in fossil fuels, but depleting wood products would become a similarly serious limit to population and economic growth (J. N. Abramovitz

and A. T. Mattoon. Reorienting the Forest Products Economy. In *State of The World 1999: A Worldwatch Institute Report on Progress Toward a Sustainable Society* (New York: W. W. Norton, 1999), 60–77).

31. World Resources Institute. *1996–1997 Biannual Report of the World Resources Institute* (United Nations Environment Programme, United Nations Development Bank, and World Bank; New York: Oxford University Press, 1996).

32. Wood Energy Council. *Survey of Energy Resources: Wood (including charcoal).* Available: www.worldenergy.org/wec-geis/publications/reports/ser/wood/wood.

33. Long-term experiments in the pinyon–juniper (*Juniperus* sp.–*Pinus edulis*) woodland at Sunset Crater National Monument in Arizona showed the influence of local soil conditions on both tree species. Pinyons and junipers grow both in 800-year-old, nitrogen-poor soil developed on dry volcanic ash, cinder, and lava, and in an adjacent older, nutrient-rich, moist sandy soil with higher organic content. All pinyon trees growing in the sandy loam look nearly alike, but the volcanic soil supports two types—a group of fully developed specimens and another distinct population of shrubby trees stunted by *Dioryctria albovitella*, a stem-boring moth that selectively destroys terminal shoots. The stunted trees use about two-thirds less nutrients to make resin than the moth-resistant group. Providing more water and nutrients to susceptible trees tripled resin production and reduced moth attacks (N. S. Cobb et al. Increased Moth Herbivory Associated with Environmental Stress of Pinyon Pine at Local and Regional Levels. *Oecologie* 1997;109:389–397).

34. U.S. Department of Agriculture. *U.S. Forest Facts and Historical Trends.*

35. Replanting nursery-grown seedlings in clearcuts and protecting them physically and chemically from insect predators, weeds, fungi, and seed-eating rodents and birds can shorten the intervals between cuts. Before transplanting seedlings, forestry workers typically clear debris from the clearcut area, then plow and furrow it to improve drainage, eliminate competitive plants, and enhance access to nutrients. Some operations inoculate seedlings with microorganisms to help them extract nutrients, and may plant soil-nitrogen–enhancing legumes between rows. Pesticides and herbicides commonly are applied to protect saplings until they mature. Sampling and testing the downslope water quality is the most practical way to measure the off-site spread of these chemicals.

36. R. C. Sidle et al. Hillslope Stability and Land Use. *American Geophysical Union Water Resources Monograph Series* 1985;11:72–119.

37. F. J. Swanson and C. T. Dyrness. Impact of Clear-Cutting and Road Construction on Soil Erosion by Landslides in the Western Cascade Range, Oregon. *Geology* 1975;3:393–396.

38. D. R. Montgomery et al. Forest Clearing and Regional Landsliding. *Geology* 2000;28:311–314.

39. Rates of soil mass movements were 2.5–5.6 times higher in clearcuts than in surrounding forests (F. J. Swanson and G. Grant. *Rates of Soil Erosion by Surface and Mass Erosion Processes in the Willamette National Forest* (Report prepared for Willamette National Forest, Eugene, Oregon, 1982; on file at Pacific Northwest Forestry Sciences Laboratory, 3200 Jefferson Way, Corvallis, Oreg., 97331)). A 1985–1991 study of clearcuts in the landslide-prone Caspar Creek area of northern California reported no apparent increase in landslides up to six years after logging began, yet the suspended sediment load in local stream waters showed an unanticipated 89% increase. The actual total, including bed load sediment, may be much higher (P. H. Cafferata and T. E. Spittler. *Logging Impacts of the 1970s vs. the 1990s in the*

Caspar Creek Watershed (U.S. Department of Agriculture Forest Service, General Technical Report PSW-GTR-168, 1998), 103–115).

40. S. Gresswell et al. *Mass Movement Response to Forest Management in the Central Oregon Coast Ranges* (Resource Bulletin PNW-84, U.S. Department of Agriculture Forest Service, Pacific Northwest Forest and Range Experiment Station, 1979). Landsliding in the Pacific Northwest and northern Rocky Mountain clearcuts appears to increase for 4–10 years after logging, and 20 years may pass before landslide frequency returns to precut levels. This corresponds roughly to the interval between cutting in many areas (W. F. Megahan et al. Landslide Occurrence in the Western and Central Northern Rocky Mountain Physiographic Province in Idaho. In *Proceedings of the Fifth North American Forest Soils Conference* (Fort Collins: Colorado State University, 1978), 116–139). The frequency of debris avalanches and debris flows increased considerably for nine years after clearcutting on slopes in Maybeso Creek Valley, Alaska. During that interval, postcut landsliding covered a region five times greater than had been landslide prone the whole preceding century. Significantly, one major storm, which occurred six years after the logging, triggered more than half the observed landslides (D. M. Bishop and M. E. Stevens. *Landslides on Logged Areas in Southeast Alaska* (U.S. Forest Service Northern Forest Experiment Station Research Paper NOR-1, 1964)).

41. In 2003, Humboldt County sued Pacific Lumber, previous owner of the newly created Headwaters Forest Reserve, alleging that the company had altered a 1999 report required for state and federal environmental reviews. The altered report falsely stated that only a small percentage of that drainage's landslides were in recently logged areas. A day before California's Department of Forestry approved the environmental review, Pacific Lumber issued a corrected version of the report. The corrected report was released only locally to prevent its reaching the department's Sacramento headquarters in time to influence a timber harvest plan approval process. A former top California forestry official filed a declaration with the court supporting the county's allegation. In 2004, Pacific Lumber initiated and funded a recall campaign against the county's district attorney, Paul Gallegos. The recall failed.

42. W. E. Weaver et al. Magnitude and Causes of Gully Erosion in the Lower Redwood Creek Basin, Northwestern California. In K. M. Nolan et al., eds. *Geomorphic Processes and Aquatic Habitat in the Redwood Creek Basin, Northwestern California* (U.S. Geological Survey Professional Paper 1454, 1995), I1–I12.

43. S. E. Ryan and G. E. Grant. Downstream Effects of Timber Harvesting on Channel Morphology in Elk River Basin, Oregon. *Journal of Environmental Quality* 1991;20:60–72.

44. Swanson and Dyrness, Impact of Clear-Cutting; L. H. MacDonald and J. D. Stednick. *Forests and Water: A State-of-the-Art Review for Colorado* (CWRRI Completion Report No. 196, Colorado Water Resources Research Institute, 2003).

45. Sixty-five percent of the 45–117 cubic yards of sediment eroded per acre was mostly in the form of dry ravel (K. A. Bennett. *Effects of Slash Burning on Surface Erosion Rates in the Oregon Coast Range* (M.S. dissertation, Oregon State University, 1982)).

46. J. S. Krammes. *Seasonal Debris Movement from Steep Mountain Side Slopes in Southern California* (Miscellaneous Publication 970, U.S. Department of Agriculture, 1965), 85–88. Total erosional yield from unburned south-facing chaparral was 1.3 cubic yards per acre, mostly in the form of dry ravel. In burned areas less than a year old, the eroded sediment yield increased to 21.5 cubic yards per acre, more than a dump truck load, tapering off to 3.4 cubic yards per acre the next year, and 5.5 the year after that.

47. Maintaining roads on U.S. Forest Service lands cost more than $8 billion annually, subsidized by taxpayers for the benefit of private companies (MacDonald and Stednick, *Forests and Water*).

48. Sidle et al., *Hillslope Stability*.

49. Cafferata and Spittler, *Logging Impacts*, 103–115.

50. R. L. Everett and S. H. Sharrow. *Soil and Water Temperature in Harvested and Nonharvested Pinyon-Juniper Stands* (Intermountain Research Station Research Paper INT-342, U.S. Department of Agriculture Forest Service, 1985).

51. Megahan et al., *Landslide Occurrence*.

52. W. W. Bourgeois. 1978. Timber Harvesting Activities on Steep Vancouver Island Terrain. In *Proceedings of the Fifth North American Forest Soils Conference* (Fort Collins: Colorado State University), 393–409.

53. C. W. Montgomery. *Environmental Geology*, 5th ed. (Boston: McGraw-Hill, 1997), 400.

54. Public Employees for Environmental Responsibility. Big Apple Water Cops Under Siege. *PEEReview* Summer 2000; *New York City's Water Supply System, History*. Available: www.ci.nyc.ny.us/html/dep/html/history.html; *Watershed Protection Plan Summary*. Available: www.nyc.gov/html/dep/html/fadplan.html. Accessed October 2000.

55. One example of a keystone species is the banner-tail kangaroo rat (*Dipodomys spectabilis*) in the southwestern United States. It digs large burrow mounds that provide vital habitat for rattlesnakes, burrowing owls, arthropods, and other organisms. As the rat burrows, it brings up soil nutrients, which support plants that thrive in disturbed areas. The disappearance of banner-tail kangaroo rats has led to a systematic decline of many coexisting species (J. H. Brown et al. Complex Species Interactions and the Dynamics of Ecological Systems— Long-Term Experiments. *Science* 2001;293:643–650).

56. For example, the artificially isolated ponderosa pine–antelope bitterbrush–Indian rice grass woodland ecosystem now occupies only a single square mile in central Oregon (Bourgeois, Timber Harvesting Activities).

57. P. S. Levin and D. A. Levin. The Real Biodiversity Crisis. *American Scientist* 2002;90:6–15.

58. J. H. Cissel et al. Landscape Management Using Historical Fire Regimes: Blue River, Oregon. *Ecological Applications* 1999;9:1217–1231. Cissel et al. suggest that each public forest management unit establish a reserve system, designating areas by landscape types. Timber harvests would be prescribed for each area to approximate the frequency, severity, and spatial extents of past fires. Harvest blocks would be mapped as a basis for projecting forest patterns 200 years forward and anticipating the effects. This rotational scheme substitutes tree cutting for uncontained natural fires. To the extent currently possible, it is intended to tie-in with the natural processes that topple trees, sometimes clear large forest segments, and change landscapes.

59. In his classic study of fire threat and national fire policy development, focused particularly on fire in the American West, S. J. Pyne wrote:

As a fire breaks down available fuel, it releases heat and nutrients. The heat may kill many organisms, consume others, and reshape a microclimate by allowing more sunlight, wind, and so forth. Many organisms adapt against this wave of heat by developing thick bark, storing food in tuberous roots, or resprouting soon after a fire passes. Others, like certain insects with infrared sensors, seek out the heat. Some plants seem to encourage properties that promote…fire, thereby driving off less tolerant competitors; a number of trees, for example, have seritonous cones that open to release seeds only after violent heating. In a similar manner, the breakdown of biochemical compounds causes some organisms to die out or depart from a burn, while others seize on the simpler compounds for the promotion of their own growth. (S. J. Pyne, *Fire in America: A Cultural History of Wildland and Rural Fire* (Seattle: Weyerhaeuser Environmental Books, University of Washington Press, 1997; originally printed 1982), 35–36)

60. F. J. Swanson. Geomorphology and Ecosystems. In R. W. Waring, ed. *Proceedings of the 40th Annual Biology Colloquium 1979* (Corvallis: Oregon State University, 1980), 159–170.

Management use of natural variability relies on two concepts: that past conditions and processes provide context and guidance for managing ecological systems today, and that disturbance-driven spatial and temporal variability is a vital attribute of nearly all ecological systems.…Understanding the history of ecological systems (their past variability, and the principal processes that influenced them) helps managers set goals that are more likely to maintain and protect ecological systems and meet the social values desired for an area. Until we significantly improve our ecological understanding of systems, this knowledge of past ecosystem functioning is also one of the best means for predicting impacts to ecological systems today. (P. Landres et al. Overview of the Use of Natural Variability Concepts in Managing Ecological Systems. *Ecological Applications* 1999;9:1179–1188)

61. The evaporation of water through a tree's leaves is the force that draws water up from the roots all the way to the uppermost branches of the tallest tree—a pump driving its life functions. The circulating water feeds the tree, carrying dissolved nutrients to all the leaves or needles, which manufacture sugars and other plant foods through the process of photosynthesis. Crown fires that kill most of the leaves will kill the tree.

62. Trees continue to widen their trunks as long as they are alive, each year adding a thin layer of wood under the bark at the trunk's outer margin. Each layer is usually a complete ring, called a tree ring. The new wood is lighter in appearance in the early growing season and darker later in the season. The ages of trees can be determined by harmlessly taking cores from many trees in a forested area, counting the rings, and correlating their number and appearance to rings in the cut stumps of older trees in logged areas, or with the rings in cores from older deadfall trees. The cores taken from live trees penetrate only to the center of the trunk. Rings have been correlated between forests to create a standard for dating any tree or woodland (see H. D. Grissino-Mayer. *Ultimate Tree-Ring*. Available: web.utk.edu/~grissino).

63. The estimated possible range of variation is 50–400 years (F. J. Swanson, personal communication, 2003; R. E. Martin et al. Fire in the Pacific Northwest: Perspectives and Problems. In W. F. Megahan and D. C. Molitor, eds. *Proceedings of the Tall Timbers Fire Ecology Conference* (1976), 1–17).

64. The Peshtigo Blow Up fire burned nearly four million acres of heavily logged land in Wisconsin and Michigan, and killed 1,500 persons. It was a catastrophe that might be better known today if a lesser conflagration had not infamously burned the City of Chicago to the ground at about the same time.

65. H. Gannett. *The Forests of the United States* (U.S. Geological Survey Annual Report, 20 Part 5(2), 1899), 1–37.

66. G. Gruell. *Fire in Sierra Nevada Forests: A Photographic Interpretation of Ecological Change Since 1849* (Missoula, Montana: Mountain Press Publishing, 2001). The images do not compare a "healthy forest" state with one of poorly managed forest lands, however. During the nineteenth century, western forests were assaulted by unsustainable logging, grazing, and mining interests, especially in the Sierra Nevada. At best, the photographs can be used to compare a state of overexploitation with one of misguided or ill-informed fire-suppression policy.

67. D. M. Graber. *Coevolution of National Park Service Fire Policy and the Role of the National Parks* (U.S. Department of Agriculture General Technical Report INT-182, 1985).

68. Study of prescribed burnings, repeated every 15 years in Los Angeles brushland, showed almost a threefold increase in annual erosion rates—to about 55 cubic yards per acre, nearly three dump truck loads every year. The problem of burns in this kind of ecosystem is that slopes cannot develop enough root strength between burns to hold the slopes (R. M. Rice et al. *Slope Stability Effects of Fuel Management Strategies: Inferences from Monte Carlo Simulations* (General Technical Report PSW-58, U.S. Department of Agriculture Forest Service, 1982), 365–371).

69. Long-term fire retardants contain ammonium phosphate and sulfate, fertilizer salts that persist until erosion removes them. Short-term foams contain surfactants, or wetting agents, that dry out quickly. These chemicals pose threats to vegetation, enhance weed invasion, and reduce species richness. They are harmful to terrestrial and aquatic organisms and impair water quality (Robyn Adams and Dianne Simmons. *Ecological Effects of Fire Fighting Foams and Retardants.* Presented at the Australian Bushfire Conference, Albury, July 1999 (School of Environmental and Information Sciences, Charles Sturt University). Available: life.csu.edu.au/bushfire99/papers/adams/index.htm; see also S. F. McDonald et al. Acute Toxicity of Fire-Retardant and Foam-Suppressant Chemicals to *Hyalella azteca* (Saussure). *Environmental Toxicology and Chemistry* 1997;16:1370–1376; D. L. Larson et al. Effects of Fire Retardant Chemicals and Fire Suppressant Foam on Shrub Steppe Vegetation in Northern Nevada. *International Journal of Wildland Fire* 1999; 9(2):115–127; L. A. Norris and W. L. Webb. *Effects of Fire Retardant on Water Quality* (General Technical Report PSW-109, U.S. Department of Agriculture Forest Service, 1989), 79–86).

70. Northern Arizona University forestry professors Wallace Covington and Margaret Moore first mapped trees by size. They cored both standing trees and large stumps and then counted and correlated tree rings to find the age of most trees present before 1890. Using computers, they could eliminate trees that grew after settlement and add back the big trees that were cut after 1890 (W. W. Covington and M. M. Moore. Postsettlement Changes in Natural Fire Regimes and Forest Structure: Ecological Restoration of Old-Growth Ponderosa Pine Forests. In R. N. Sampson and D. L. Adams, eds. *Assessing Forest Ecosystem Health in the Inland West* (New York: Food Products Press, 1994), 153–181).

71. Covington and Moore, Postsettlement Changes.

72. Covington and Moore's 1990s experiment cut out 50% of trees in a mixed ponderosa pine forest measuring less than five inches diameter at breast height, but left a few thickets for wildlife cover. They burned the cut forest a few years later. Another, partly inadvertent, U.S. Department of Agriculture Forest Service restoration experiment used professional loggers

to cut 90% of the trees on a 90-acre test plot in the Kaibab National Forest, Arizona. A later (2000) wildfire burned through the plot at ground level, leaving the remaining older trees unscathed (J. S. MacNeil. Forest Fire Plan Kindles Debate. *Science* 2000;289:1448–1449). Follow-up studies examined only the fire potential and apparently did not gauge impacts on biodiversity or any other forest values.

73. Rick Keister. Salvage Logging Reborn. *High Country News* 1995; 27(14). The emergency salvage legislation was a rider to a 1995 Congressional bill.

74. D. C. Donato et al. Post-wildfire Logging Hinders Regeneration and Increases Fire Risk. *Science* 2006;311:352; M. P. Amaranthus et al. Decaying Logs as Moisture Reservoirs After Drought and Wildfire. In E. B. Alexander, ed. *Proceedings of Watershed 1989 Conference of the Stewardship of Soil, Air, and Water Resources* (U.S. Department of Agriculture Forest Service RIO-MB-77, 1989), 191. Logging slash from timber operations left on the ground promotes rapid spread of flames and releases the greatest amount of heat per unit area as it burns (S. L. Stephens. Evaluation of the Effects of Silvicultural and Fuels Treatments on Potential Fire Behavior in Sierra Nevada Mixed-Conifer Forests. *Forest Ecology and Management* 1989;105:31).

75. For example, fire-restoration cutting has been proposed for old stands of pinyon-juniper woodlands with trees greater than 400 years old, which remain in about the same condition as they were at the time of settlement, and do not need restoration (W. H. Romme et al. *Ancient Piñon-Juniper Forests of Mesa Verde and the West: A Cautionary Note for Forest Restoration Programs.* Presented at the Conference on Fire, Fuel Treatments, and Ecological Restoration: Proper Place, Appropriate Time, Fort Collins, Colorado [16–18 April 2002]).

76. The National Environmental Policy Act and the Appeals Reform Act both give citizens the right to read and critique specific plans for projects that could have negative environmental impacts. Proposed revisions to these laws restrict citizen involvement in environmental protection and limit appeals of U.S. Department of Agriculture Forest Service timber sales and other land management decisions.

77. According to initial information released by the White House's Healthy Forests Initiative. Available: www.whitehouse.gov/. The concentration of large ponderosa pines may have increased 15-fold, from 25 in a typical six-acre area of northern Rocky Mountains forest a century ago, to between 35 and 500 today. Given the dynamic gap-mosaic structure of natural forests, it is nearly impossible to evaluate this assertion.

78. The proposed rule exempts restoration, called "hazardous fuels reduction projects" from environmental assessments under National Environmental Policy Act. Most such projects would become "categorical exemptions," which allow local managers to determine that cutting for hazardous fuels reduction will have no significant impacts, and eliminate the need for monitoring or mitigations while also obstructing public input. No data are required to support the "no significant impact" finding. Projects to be allowed under categorical exemptions have no size limits or restrictions on harvest type.

79. Taking advantage of a rider slipped into the massive domestic funding bill in February 2003, the Bush administration endorsed 10-year "stewardship" contracts for fire-protection clearance projects in national forests. The contracts allow timber contractors to cut large commercially valuable trees in exchange for clearing smaller trees and undergrowth, with no limits on the size of trees to be cut or the acreage cleared.

80. The U.S. Department of Agriculture Forest Service brochure urged more logging to prevent wildfires in California's Sierra Nevada, displaying a 1909 photograph of the

Lick Creek area in Montana as a previously uncut forest. Large tree stumps are very easily identified in the scene, which was easily traced to a classic collection of early woodland photography.

81. R. W. Gorte. *Forest Service Timber Sale Practices and Procedures: Analysis of Alternative Systems* (Congressional Research Service Report 95–1077 ENR, Washington, DC: U.S. Library of Congress, 1995), 3–4.

82. R. W. Gorte. *Below-Cost Timber Sales: Overview* (Congressional Research Service Report 95–15 ENR, Washington, DC: U.S. Library of Congress, 1994), 20 p.

83. Wild Rockies Salvage Campaign. *The Economic Reality of Salvage Logging.* Available: www.wildrockies.org/Talus/Campaign/Salvage/EconReal.html.

84. The Red Star burn looks like it will become another candidate for a hot wildfire in 30–40 years. To clean up debris, the U.S. government ideally would subsidize a modern equivalent of the Civilian Conservation Corps working on a massive scale in coordination with a well-regulated commercial timber harvest. Ecologially informed forest managers would supervise. Enormous federal budget deficits make this an unlikely innovation.

85. J. D. McIver and L. Starr, eds. *Environmental Effects of Post-fire Logging: Literature Review and Annotated Bibliography* (General Technical Report PNW-GTR-486, U.S. Department of Agriculture Forest Service Pacific Northwest Research Station, 2000). Another constant scapegoat is the race against time for completing bidding and environmental impact studies of salvaging operations before the quality of burned wood deteriorates, significantly reducing its market value. The requirement of removing flammable debris after completing operations, complicated by the need to leave enough deadfall on the ground to protect against soil degradation and erosion, also reduces salvage revenue. Since no two locales are alike, the treatment has to be prescribed on a case-by-case basis. U.S. Department of Agriculture was able to sell only 69% of its 2002 postfire timber, lowering many minimum bids by as much as 80%.

86. Most parts of the western United States now have too few mills and operating lumber companies to support the legally required competitive bidding process for contracting large-scale timber harvests. In 1992, California had 56 operational timber mills, while the number today is less than 30—all dominated by Sierra Pacific Industries. Western Lumber Industry data (Available: www.wwpa.org) show that the total number of sawmills in the West declined from about 500 in 1991 to only about 300 in 1999, when the industry employed 192,573 persons. Also in that interval, the per-mill lumber production increased on average from around 40 million board feet per year to about 60 million, while imports of wood into the United States from abroad increased from nearly zero to more than 250 million board feet.

87. U.S. General Accounting Office. *Forest Service: Distribution of Timber Sales Receipts Fiscal Years 1992–1994* (RCED-95–237FS, 1995).

88. R. W. Gorte. *National Forest Receipts: Sources and Dispositions* (Congressional Research Service Report No. 89–284 ENR, Washington, DC: U.S. Library of Congress, 1989).

89. Cascade Holistic Economic Consultants, Testimony of Randal O'Toole on Problems of Forest Service Accountability. In U.S. House, Committee on Government Operations, Subcommittee on Environment, Energy, and Natural Resources. *Review of the Forest Service's Timber Sales Program Hearing, 31 March, 1992. 102nd Cong., 2nd Session.* (Washington, DC: U.S. Government Printing Office, 1993), 93–104.

90. The Forest Service's internal Knutson-Vandenberg (K-V) trust fund account receives an unlimited portion of timber sale receipts, earmarked for reforestation, timber stand improvements, and other resource mitigation and enhancement activities in timber sale areas. Forest Service sources argue that applying timber receipts to these projects is an appropriate reinvestment in the national forests (Gorte. *Forest Service Timber Sale Practices*, 4).

91. R. W. Gorte. *Salvage Timber Sales and Forest Health* (Congressional Research Service Report No. 95–364 ENR, Washington, DC: U.S. Library of Congress, 1996).

92. In many areas, competition for national forest timber is strong, and when bidding raises the prices for timber sale permits, many purchasers anticipate problems with recovering costs. The president's Private Sector Survey on Cost Control (Peter W. Grace Commission. *Task Force Report on the Department of Agriculture (Draft Report)* (U.S. Senate Committee on Agriculture, Nutrition, and Forestry, S.Prt. 98–76. 98th Congress, 1st Session; Washington, DC: U.S. Government Printing Office, 1983), 170–183).

93. Gorte, *Forest Service Timber Sale Practices*, 3–4.

94. R. W. Gorte. *Forest Health: Overview* (Congressional Research Service Report No. 95–548 ENR, Washington, DC: U.S. Library of Congress, 1995).

95. T. Ingalsbee. *Restoration or Exploitation? Post-fire Salvage Logging in America's National Forests* (American Lands Alliance, 2003). Near Phoenix, Arizona, USFS proposed logging 2,259 acres of burned slopes steeper than 40% near the head of the Salt River in October 2003. More than 30% (1,221 acres) of salvage logging projects near Durango, Colorado, were in high to severe landslide hazard areas that had burned in 2002.

96. Overgrazing and fire suppression also contributed to the growing inventory of firs in western forests (Gorte, *Below-Cost Timber Sales*). Succession processes are also culpable, since Douglas fir is a successional climax species.

97. Regeneration costs from 1977–1994 came to $2,449,900,000 (Wild Rockies Salvage Campaign. *Salvage Background Information*. Available: www.wildrockies.org/Talus/Campaign/Salvage/EconReal.html).

98. Wild Rockies Salvage Campaign. *The Economic Reality of Salvage Logging*.

99. U.S. Department of Agriculture. *U.S. Forest Facts and Historical Trends*.

100. Agroforestry practices began in the fifteenth century from experiments of Scot William Blair. Today, all five million acres of British woodlands are human-planted. In the southeastern United States, a trend toward converting natural forests to plantations is under way, while Washington and Oregon are seeing the most intensive plantation expansion in western states, mostly in logged-over natural forests. According to USFS statistics, 10.4% of eastern lands, but only 3.6% of western forests, are human-planted (U.S. Department of Agriculture, *U.S. Forest Facts and Historical Trends*).

101. The state of California had designated spotted owl habitat areas since 1989 but failed to stem the species' decline. The northern spotted owl was listed as a federal Endangered Species in 1990, and the U.S. Forest Service issued guidelines for protecting the California spotted owl in 1993 (U.S. Department of Agriculture Forest Service. *The California Spotted Owl: A Technical Assessment of Its Current Status* (1992)).

102. V. L. Parker. Update on the Sierra Nevada: Still in Peril. *Fremontia* 2003;31:5–13. Other species warranting careful monitoring and protected breeding areas include the great grey owl, reduced to fewer than a hundred birds; the willow flycatcher, with only 82 known

nesting sites; and the northern goshawk; plus nonbird species such as the Pacific fisher, Sierra Nevada red fox, mountain yellow-legged frog, Yosemite toad; and many species of now-rare or unusual plants.

103. The framework classified about 40% of Sierran national forests in the protected category of Old Forest Emphasis Areas (OFEAs) and proposed managing numerous Protected Activity Centers (PACs) to protect endangered or sensitive species against grazing, logging, mining, and recreation. Commercial activities were to be restricted or prohibited altogether in some PACs. Around towns and other human developments, the plan designated Urban Wildland Intermix Zones (UWIZs), consisting of quarter-mile-wide Defense Zones next to built-up areas, plus outer mile-and-a-quarter-wide Threat Zones. Supposing that "treated" (logged) areas scattered through the forests slow and moderate wildfire intensities and reduce burn acreage, it adopted the Finney model of a patchwork of fuel-impoverished Strategically Placed Area Treatment areas, through-out all forest lands except OFEAs, PACs, and UWIZs (M. A. Finney. *FARSIDE—Fire Area Simulator-Model Development and Evaluation* (U.S. Department of Agriculture Forest Service Research Paper RMRS-RP-4, 1998)). In a compromise between wildlife protection and severe fire prevention, the Framework restricted mechanical tree cutting and log removal to Defense Zones near developed areas, and in PACs only permitted prescribed burning. It preferred burns in OFEAs but allowed tree cutting and slash-pile burning in zones of high fire risk.

104. The Framework Review Report (Constance Millar et al. *Sierra Nevada Science Review*, Report of the Science Review Team [U. S. Department of Agriculture Forest Service, 1998]) assailed data suggesting that willow flycatcher and great grey owl habitat cannot include lim-ited grazing, and alleged a critical lack of evidence and statistical precision in evaluating spotted owl populations. Subsequent study suggests that owl populations are declining in one of the four critical research areas underpinning the Framework plan, however. Assuming that the Framework would provide adequate protection, the U.S. Fish and Wildlife Service had proposed removing the California spotted owl from the endangered species list, at the same time warning that changing the plan to conform with the Management Review and Recommendations Report (U.S. Department of Agriculture Forest Service. *Sierra Nevada Forest Plan Amendment: Management Review and Recommendations* [Pacific Southwest Region R5-MB-012, 2003]) could threaten the owl's survival.

105. The Management Review and Recommendations Report consulted only district rangers, sidestepping the USFS's resource specialists who had worked on the Framework. The Review and Recommendations Report team interviewed and took field trips with all but one of 32 Sierra Nevada district rangers. A California Native Plant Society author has alleged that protest-ing USFS scientific staff were transferred or silenced (Parker, Update on the Sierra Nevada).

106. U.S. Department of Agriculture Forest Service. *Sierra Forest Plan Amendment: Management Review and Recommendations*, 17–18. Objecting to the restriction on clearing, some rangers noted that prescribed burning would be overly hazardous without extensive pre-paratory clearing in the densest of the Sierran forests.

107. Under the Healthy Forests Initiative, in March 2003, the George W. Bush Administration announced plans to raise the Sierra's logging take limit to 450 million board feet per year (Environment News Service. *Forest Service to Triple Logging in Sierra Nevada* (2004). Available: forests.org/articles/reader.asp?linkid=28758).

108. Blackwell also replaced specific guidelines for local officials with vague suggestions. For example, where the Framework specified meadow vegetation 12 inches high to support

great grey owls, Blackwell allows a height "commensurate with site capability and habitat needs of prey species."

109. Potlatch Forest Industries is headquartered in Spokane, Washington. Plantation forestry remains largely a future prospect because, so far, scientists have been unable to propagate most trees by cloning. Poplars, cottonwoods, and aspens are the exceptions (C. C. Mann and M. L. Plummer. Forest Biotech Edges Out of the Lab. *Science* 2002;295:1626–1629).

110. While a clearcut Douglas fir grove in the Western Cascades may take 60–100 years to recover, GM Douglas fir stands may be cut over once every 35 years (Professor R. J. Carson, Whitman College, personal communication, 2003).

111. Natalie Goldstein. *Earth Almanac: An Annual Geophysical Review of the State of the Planet*, 2nd ed. (Westport, Conn.: Oryx Press, 2002).

112. The *Bt*-producing gene comes from plant bacteria that secrete it as a natural pesticide. Preparations of *Bt* bacteria have long been available as commercial pesticides for topical application, and organic farmers use them widely but sparingly.

> In the USA alone more than 120 field trials (including fruit trees) have been approved over the past decade, the majority of them in the last two years. Even though commercial GM plantation trees are not likely to be seen for at least two years, field trials are underway in a variety of countries including Chile, Indonesia, South Africa, New Zealand, and China, as well as several European countries. Currently the majority of the trials focus on three major plantation species: poplar, *radiata* [Monterey] pine, and eucalyptus. (M. Rautner. Designer Trees. *Biotechnology and Development Monitor* 2001;No. 44:2–7)

Other *Bt* tree species also are planned.

113. Proposed GM projects pose economic problems for the forest industry, because of the short-term cost of the research and the years to decades required for evaluating results. Traditional tree breeding by conventional hybridization techniques, now in only the second or third generation, also can increase the lumber or paper yield, with predictable returns and outcomes that continue to be more attractive to the economically conservative forest industry than engineering approaches (Mann and Plummer, Forest Biotech).

114. Rainforest Action Network. *Weyerhauser and GE Trees*. Available: ran.org/fileadmin/materials/old_growth/GE_Trees.pdf.

115. Pielou, *After the Ice Age*.

116. A. Goudie. *The Human Impact on the Natural Environment*, 5th ed. (Cambridge, Mass.: MIT Press, 2000), 42–43. The worst 2003 fires were set by humans, including a forest ranger.

117. Alexander von Humboldt (1852), quoted in C. J. Glacken, Changing Ideas of the Habitable World. In W. L. Thomas, Jr., ed. *Man's Role in Changing the Face of the Earth* (Chicago: University of Chicago Press, 1956), 78.

118. Intergovernmental Panel on Climate Change. Summary for Policymakers. In *Climate Change 2001: The Scientific Basis* (Working Group I of the Intergovernmental Panel on Climate Change, 2001). Available: www.ipcc.ch/.

119. The work of Margaret B. Davis of the University of Minnesota, in particular, suggests that particular eastern hardwoods will perish across whole regions in the next couple of centuries owing to rapid climate change.

120. Parker, Update on the Sierra Nevada.

121. "If wood use accelerates to the point where everyone consumes as much as the average person in an industrial country does today, by 2010 the world would consume more than twice as much wood as it does today. And if by 2010 everyone in the world consumed as much paper as the average American does today, total paper consumption would be more than eight times current world consumption." World wood consumption is likely to reach more than 77 billion cubic feet per year by 2010. (Abramovitz and Mattoon, Reorienting the Forest Products Economy). For other statistics, see World Resources Institute (*1996–1997 Biannual Report*).

Chapter 2

1. Michael Pollan. Playing God in the Garden. *New York Times* 25 October 1998. This article provides fascinating insights into genetic modification of crops and pesticide demands.

2. World Resources Institute. *World Resources 2000–2001: People and Ecosystems* (Washington, DC: World Resources Institute, 2001), 53–68.

3. David Pimentel et al. Environmental and Economic Costs of Soil Erosion and Conservation Benefits. *Science* 1995;267:1117–1123. Between 1982 and 2002, the total cropland in the contiguous United States decreased 12%, to 368 million acres, and cultivated cropland decreased 17%, but noncultivated cropland increased 27% (U.S. Department of Agriculture Natural Resources Conservation Service. *Natural Resources Inventory* (2002)).

4. Agriculture has displaced 38% of plant and animal species listed or proposed for listing under the Endangered Species Act, and urbanization threatens 35% (D. S. Wilcove et al. Quantifying Threats to Imperiled Species in the United States. *BioScience* 1998;48:607–615; see also R. F. Noss and R. L. Peters. Endangered Ecosystems: A Status Report on America's Vanishing Habitat and Wildlife. *Defenders of Wildlife* December 1995).

5. "In 1940 the average farm in the United States produced 2.3 calories of food energy for every calorie of fossil energy it used. By 1974 (the last year in which anyone looked closely at this issue), that ratio was 1:1" (Richard Manning. The Oil We Eat: Following the Food Chain Back to Iraq. *Harpers Monthly* 24 February 2004). The ratio probably is lower now.

6. Quoted in Pollan, Playing God in the Garden.

7. Some differences in the 1975 and 1997 vitamin and mineral contents were strikingly large. Vitamin C in cauliflower went down 40%, calcium in broccoli dropped by half, and iron in watercress was 88% lower. This article also reports survey results of what Americans eat (Vegetables Without Vitamins. *Life Extension Magazine* March 2001. Available: www.lef.org/magazine/mag2001/mar2001_report_vegetables.html).

8. This discussion borrows heavily from Chip Ward (*Canaries on the Rim: Living Downwind in the West* [New York: Verso Press, 1999], 147).

9. Startling results from a recent study of vinclozolin, a fungicide used in the wine industry, and methoxychlor, an insecticide chemically related to DDT, showed transmission of reproductive abnormalities through four generations of male lab animals (M. D. Anway et al. Epigenetic Transgenerational Actions of Endocrine Disruptors and Male Fertility. *Science* 2005;308:1466–1469). Toxins' poorly understood effects may support the controversial idea that

endocrine disruptors cause populationwide reproductive problems, such as lower sperm counts in men. See also Caroline Cox (Two Pesticides Cause Infertility Lasting Four Generations. *Journal of Pesticide Reform* 2005;25:5).

10. Michael Pollan. *The Omnivore's Dilemma: A Natural History of Four Meals* (New York: Penguin Press, 2006), 49–53.

11. The 2002 Farm Bill allocated 90% of farm subsidies to producers of wheat, corn, cotton, rice, and soybeans, although the single largest beneficiary is a sugar producer. Other crops allotted government subsidies until 2012 include sugar beets, apples, dairy, livestock, wool and mohair, peanuts, honey, and other grains (B. M. Riedl. Top 10 Reasons to Veto the Farm Bill (Heritage Foundation Backgrounder No. 1538, 2002). Available: www.heritage. org/Research/Agriculture/BG1538.cfm).

12. U.S. Department of Agriculture. Taking Stock for the New Century (p. 6), summarized in Environmental Media Services. *Fast Facts* (17 May 2002 update). Available: www. ems.org/farm_bill/facts.html. The subsidy system changes yielded to a series of farm emergencies, but even more to political pressure.

13. Houses built on former rice-growing lands come with perpetual taxpayer support for nonfarmed suburban yards. Realtors advertise the agriculture support payments in their listings (Dan Morgan et al. Farm Program Pays $1.3 Billion to People Who Don't Farm. *Washington Post* 2 July 2006).

14. Environmental Working Group. *Summary Farm Payments Database* (2002). Available: www.ewg.org/farm/subsidies/whogets.php. Many expect that the conservation funds will be inadequate to meet the heavy demand—the U.S. Department of Agriculture already has a growing backlog of farmland conservation assistance applications totaling $2.5 billion.

15. The final legislation added $83 billion to farm support payments over 10 years, a 78% increase. Ninety percent of the total payout goes to a few large growers. The Environmental Working Group's analysis of U.S. Department of Agriculture records suggests that 85% of the $2.2 billion certificate exchange program (a system of taxpayer-funded price guarantees in the 1996 farm bill) went to farmers and large agribusinesses in just four states: Arkansas, Mississippi, Texas, and California (Environmental Working Group, *Summary Farm Payments Database*).

16. The figure of $32,652 per year for a family of four represents 185% of the federally defined poverty line. Projecting past trends forward from 2002, the Heritage Foundation predicted that the 2002 farm bill's final cost over 10 years could increase the Congressional Budget Office's 2002 estimated total of $171 billion to payouts of $342 billion or more (Riedl, Top 10 Reasons).

17. Kiley Russell. Small Farmers Warn of Looming Economic Doom *Associated Press* (21 December, 2000).

18. Peter Rosset. The Multiple Functions and Benefits of Small Farm Agriculture in the Context of Global Trade Negotiations, *The Institute for Food and Development Policy/Food First and the Transnational Institute* (23 November, 1999). Available: www.foodfirst.org/pubs/policybs/pb4.html.

19. Michael Olson. *Metrofarm* (Santa Cruz, California: TS Books, 1994), iii.

20. U.S. Environmental Protection Agency. *Atlas of America's Polluted Waters* (EPA 840-B-00–0002, 2000). The U.S. Geological Survey has developed a useful classification scheme

for distinguishing the water pollution potential from row crops and orchards, vineyards, and nurseries (R. J. Gilliom and G. P. Thelin. *Classification and Mapping of Agricultural Land for National Water-Quality Assessment* [U.S. Geological Survey Circular 1131, 1997]). The growing problem of air pollution from factory farms is addressed by the National Research Council (*Air Emissions from Animal Feeding Operations: Current Knowledge, Future Needs* [Washington, DC: National Academies Press, 2003]) and V. P. Aneja et al. (Emerging National Research Needs for Agricultural Air Quality. *EOS, Transactions of the American Geophysical Union* 2006;87:25, 29).

21. Douglas Tompkins. Prologue. In Andrew Kimbrell, ed. *Fatal Harvest: The Tragedy of Industrial Agriculture* (Covelo, Calif.: Island Press, 2002), xi.

22. The Conservation Reserve Program assists conversions of highly erodible cropland to vegetative cover for 10-year intervals (U.S. General Accounting Office. *Agricultural Conservation: USDA Needs to Better Ensure Protection of Highly Erodible Cropland and Wetlands* [GAO-03-418, 2003]).

23. Row crop plants do not provide sufficient soil-protecting leaf canopy cover between rows to soften raindrop impacts. Planting crop rows straight up and down hill slopes expands erosional potentials.

24. Donald Worster. *Dust Bowl: The Southern Plains in the 1930s* (New York: Oxford University Press, 1979); Timothy Egan. *The Worst Hard Time: The Untold Story of Those Who Survived the Great American Dust Bowl* (New York: A Mariner Book, 2006).

25. Dust storms can carry a large inventory of soil bacteria, viruses, and fungi. About 25% cause plant diseases and about 10% cause human diseases. Typical soils contain about one million bacteria per gram. Using conservative estimates of 10,000 bacteria per gram in airborne soil, and one billion metric tons of sediment in the atmosphere, the number of bacteria carried with dust annually is about one quintillion (10^{18})—enough to form a microbial bridge between Earth and Jupiter. The bacteria can include human pathogens that attack the young and people with previous illnesses, systemic weaknesses, malfunctioning immune systems, or other forms of lowered resistance (J. K. M. Brown and M. S. Hovmøller. Aerial Dispersal of Pathogens on the Global and Continental Scales and Its Impact on Plant Disease. *Science* 2002;297:537–541; D. W. Griffin et al. The Global Transport of Dust. *American Scientist* 2002;90:228–235; Hajime Akimoto. Global Air Quality and Pollution. *Science* 2003;302:1716–1719). Dust storms also carry pesticides, fertilizers, and other pollution. In 2001, a dust cloud as long and wide as the Japanese island chain carried degraded Mongolian soils, and a large concentration of industrial and automotive carbon monoxide pollution, across the western United States.

26. A. D. Hyers and M. G. Marcus. Land Use and Desert Dust Hazards in Central Arizona. In T. L. Péwé, ed. *Desert Dust: Origin, Characteristics, and Effect on Man* (Geological Society of America Special Paper 186, 1981), 267–280; C. R. Leathers. Plant Components of Desert Dust in Arizona and Their Significance for Man. In T. L. Péwé, ed. *Desert Dust: Origin, Characteristics, and Effect on Man* (Geological Society of America Special Paper 186, 1981), 191–206; H. G. Wilshire, Human Causes of Accelerated Wind Erosion in California's Deserts. In D. R. Coates and J. B. Vitek, eds. *Thresholds in Geomorphology* (London: George Allen and Unwin, 1980), 415–433; R. T. Ervin and J. A. Lee. Impact of Conservation Practices on Airborne Dust in the Southern High Plains of Texas. *Journal of Soil and Water Conservation* 1994;49:430–437.

27. H. G. Wilshire et al. Field Observations of the December 1977 Wind Storm, San Joaquin Valley, California. In T. L. Péwé, ed. *Desert Dust: Origin, Characteristics, and Effect on Man* (Geological Society of America Special Paper 186, 1981), 233–251; H. G. Wilshire et al. *Geologic Processes at the Land Surface* (U.S. Geological Survey Bulletin 2149, 1996), 19, 25–26. Another major windstorm originated in New Mexico the same year (J. F. McCauley et al. The U.S. Dust Storm of February 1977. In T. L. Péwé, ed. *Desert Dust: Origin, Characteristics, and Effect on Man* [Geological Society of America Special Paper 186, 1981], 123–147).

28. Nitrogen, phosphorus, and potassium are the major essential nutrients for plant growth. Nitrogen is "fixed" when inert N_2 gas in air becomes reactive ammonia (NH_3), ammonium ion (NH_4^+), or nitrogen oxides. A number of plants and animals, even humans, fix nitrogen naturally. Other important minor nutrients come from rock debris and decayed plant and animal soil components, including sulfur, magnesium, and trace nutrients sodium, cobalt, iron, zinc, calcium, boron, chlorine, manganese, and copper (U.S. Geological Survey. *Fertilizers: Sustaining Global Food Supplies* [Fact Sheet FS-155–99, 1999]). The world's total yearly anthropogenic fixed nitrogen production now is 1.5 times more than the natural level. Fertilizer is the principal synthetic source; legumes and rice yield about 25%, and nitrogen oxide emissions from burning fossil fuels about 17%. Reactive nitrogen drives numerous environmental problems, including acid rain and urban smog formation, water pollution, and stratospheric ozone destruction. It is also a potent greenhouse gas. The great excess of synthetic fixed nitrogen feeds all these problems. University of Virginia biogeochemist James Galloway notes: "Once you break that [inert N_2] triple bond, that N atom stays reactive for a very long time and then cascades through the environment" before microbes finally convert it back to N_2. Quoted by Jocelyn Kaiser. The Other Global Pollutant: Nitrogen Proves Tough to Curb. *Science* 2001;294: 1268–1269.

29. The industrial process combines inert nitrogen (N_2) from the atmosphere with hydrogen from natural gas (CH_4) to form ammonia (NH_3) at 660°F and 3,000 psi pressure. The ammonia is converted to ammonium ion (NH_4^+) for synthesizing urea (NH_2CONH_2), ammonium nitrate (NH_4NO_3), and ammonium sulfate [$(NH_4)_2SO_4$], the other nitrogen compounds that go into fertilizers. Nitric acid (HNO_3) is a byproduct of all the processes.

30. Vaclav Smil. Global Population and the Nitrogen Cycle. *Scientific American* 1997;277:76–81.

31. David Pimentel. Environmental and Economic Benefits of Sustainable Agriculture. In M. G. Paoletti et al., eds. *Socio-economic and Policy Issues for Sustainable Farming Systems* (Padova, Italy: Cooperativa Amicizia, 1993), 10. Nitrous oxide has 320 times the warming potential of carbon dioxide (E. A. Holland et al. U.S. Nitrogen Science Plan Focuses Collaborative Efforts. *EOS, Transactions of the American Geophysical Union* 2005;86[27]:253, 256).

32. The United States applies about five million tons of phosphate fertilizers annually. The mined phosphate rock is first treated with sulfuric acid (H_2SO_4), to make phosphoric acid (H_3PO_4). Either mixing phosphoric acid with ammonia or treating phosphate rock with phosphoric acid produces fertilizers. Potassium for fertilizers is either mined from ancient lake and subsurface salt deposits, or manufactured as potassium sulfate and potassium nitrate (saltpeter).

33. M. J. McLaughlin et al. (Review: The Behavior and Environmental Impact of Contaminants in Fertilizers. *Australian Journal of Soil Research* 1996;34:1–54) provide a comprehensive review of fertilizer contaminants in soils. The study paid particular attention to

arsenic, cadmium, fluorine, lead, and mercury contaminants. Cadmium and fluorine accumulate in fertilized soils faster than the other elements studied. Cadmium is of greatest concern because it ends up in edible parts of food crops.

34. The term "pesticide" includes insecticides, herbicides, rodenticides, fungicides, nematicides (worm killers), and acaracides (ticks and mite killers) as well as disinfectants, fumigants, wood preservatives, and plant growth regulators. "Conventional pesticides" are chemicals produced exclusively or primarily for pesticides. Others have multiple uses, for example, petroleum products, sulfur, wood preservatives, sanitizers, specialty biocides for use as disinfectants in swimming pools and spas, and chlorine/hypochlorites used for potable and wastewater purification. "Active ingredients" are intended to kill or control the targeted pest, must be registered with the U.S. Environmental Protection Agency, and must undergo a series of tests to assess their toxicity, environmental fate, and effects on nontarget organisms. Over the past decade, new active ingredients were registered at generally increasing rates; 164 came into use between 1990 and 1997 (A. L. Aspelin and A. H. Grube. *Pesticides Industry Sales and Usage—1996 and 1997 Market Estimates* (U.S. Environmental Protection Agency, 1999). Available: www.epa.gov).

35. Pimentel et al., Environmental and Economic Costs.

36. Leo Horrigan et al. How Sustainable Agriculture Can Address the Environmental and Human Health Harms of Industrial Agriculture. *Environmental Health Perspectives* 2002;110:445–456; Pimentel, Environmental and Economic Benefits, 4–20.

37. M. I. Mattina et al. Plant Uptake and Translocation of Highly Weathered, Soil-Bound Technical Chlordane Residues: Data from Field and Rhizotron Studies. *Environmental Toxicology and Chemistry* 2004;23:2756–2762.

38. In addition to DDT, the common 1960s organochlorines included aldrin, dieldrin, and chlordane. As DDT's effectiveness faded, it was replaced by toxaphene, a particularly intractable complex organochlorine compound. Although generally degrading more rapidly than DDT, toxaphene's most highly toxic components degrade most slowly. By 1966, 20,000 tons of organochlorine insecticides were in use on agricultural lands annually, with toxaphene leading at 14,000 tons. Cotton-destroying insects eventually developed toxaphene resistance and reduced its use well before the registration was canceled in 1983 (C. J. Schmitt. Environmental Contaminants. In M. J. Mac et al., eds. *Status and Trends of the Nation's Biological Resources* [U.S. Department of the Interior, U.S. Geological Survey, 1998], 131–165; this excellent report provides an interesting history of pesticide use and impacts).

39. Theo Colborn et al. Developmental Effects of Endocrine-Disrupting Chemicals in Wildlife and Humans. *Environmental Health Perspectives* 1993;101:378–384; M. J. Mac. Endocrine-Disrupting Compounds in the Environment. In M. J. Mac et al., eds. *Status and Trends of the Nation's Biological Resources* (U.S. Department of the Interior, U.S. Geological Survey, 1998), 148; a persistent DDT metabolite is the likely cause of abnormal sex development in rats (W. R. Kelce et al. Persistent DDT metabolite p-p.-DDE Is a Potent Androgen Receptor Antagonist. *Nature* 1995;375:581–585).

40. Between 1971 and 1988, organochlorine insecticide use dropped from 43,000 to 3,600 pounds while organophosphate insecticides applications rose from 43,500 to 65,000 pounds (J. E. Barbash and E. A. Resek. *Pesticides in Ground Water: Distribution, Trends, and Governing Factors* [Chelsea, Michigan: Ann Arbor Press, 1996], table 3.6, 172–173). Several organochlorines still are in home and garden use, including dicofol, chlordane, endosulfan, heptachlor,

lindane, and methoxychlor (L. H. Nowell et al. *Pesticides in Stream Sediment and Aquatic Biota: Distribution, Trends, and Governing Factors* [New York: Lewis Publishers, 1999], 471).

41. Four potentially long-lived and bioaccumulative pesticides in table 2.4 are organophosphate compounds. These pesticides are detected in soils years after application, possibly due to sorption onto soil particles. The organophosphates generally are highly toxic, and some evidence suggests they are also mutagenic and teratogenic. Organophosphate pesticide exposures may cause a large number of modern nervous and immune system diseases in mammals (K. V. Ragnarsdottir. Environmental Fate and Toxicology of Organophosphate Pesticides. *Journal of the Geological Society, London* 2000;157:859–876); resistance to these pesticides also is developing (I. Denholm et al. Insecticide Resistance on the Move. *Science* 2002;297:2222–2223).

42. U.S. Geological Survey. *Pesticides in Stream Sediment and Aquatic Biota* [Fact Sheet 092–00, 2000], table 1. Estimates of the potential for a pesticide's long-term environmental damage are based on graphs of its soil half-life, the time for half of a pesticide dose to disappear from soil, plotted against the pesticide's solubility in water.

43. Schmitt, Environmental Contaminants; David Pimentel et al. Environmental and Economic Costs of Pesticide Use. *BioScience* 1992;42:750–760.

44. Schmitt, Environmental Contaminants, 151–152. Heavily used herbicides include atrazine, alachlor, metolachlor, trifluralin, cyanazine, metribuzin, glyphosate, and 2,4-D.

45. R. A. Relyea. The Impact of Insecticides and Herbicides on the Biodiversity and Productivity of Aquatic Communities. *Ecological Applications* 2005;15:618–627; R. A. Relyea. The Lethal Impact of Roundup on Aquatic and Terrestrial Amphibians. *Ecological Applications* 2005;15:1118–1124; Tyrone Hayes et al. Feminization of Male Frogs in the Wild. *Nature* 2002;419:895–896; T. B. Hayes et al. Hermaphroditic, Demasculinized Frogs After Exposure to the Herbicide Atrazine at Low Ecologically Relevant Doses. *Proceedings of the National Academy of Sciences of the USA* 2002;99:5476–5480. Studies published in 2003 showed that pesticide mixtures used on corn crops, including atrazine, metolachlor, and alachlor, have adverse effects on frogs far exceeding the effect of any one of the chemicals (Rebecca Renner. Pesticide Mixture Enhances Frog Abnormalities, *Environmental Science and Technology* 2003;37:52A; S. H. Swan et al. Semen Quality in Relation to Biomarkers of Pesticide Exposure. *Environmental Health Perspectives* 2003;111:1478–1484.

46. By 1996, more than 500 insect and mite species, 150 plant pathogen species, and 270 weed species had developed resistance to a pesticide (C. M. Benbrook et al. *Pest Management at the Crossroads* (Yonkers, New York: Consumer's Union, 1996), 2; see also National Research Council. *Pesticide Resistance: Strategies and Tactics for Management* (Washington, DC: National Academies Press, 1986), 16–17; Jennifer Curtis et al. *After Silent Spring, the Unsolved Problems of Pesticide Use in the United States* (New York: NRDC Publications, 1993), 28–29.

47. V. M. D'Costa et al. Sampling the Antibiotic Resistone. *Science* 2006;311:374–377; Alexander Tomasz. Weapons of Microbial Drug Resistance Abound in Soil Flora. *Science* 2006;311:342–343.

48. U.S. General Accounting Office. *Food Safety: The Agricultural Use of Antibiotics and Its Implications for Human Health* [GAO/RCED-99-74, 1999], 1.

49. Dan Ferber. Livestock Feed Ban Preserves Drugs' Power. *Science* 2002;295:27–28. Note also that, in addition to the antibiotics we eat in meat from treated animals, some are

excreted as urine and feces that end up in manure fertilizers. Crop plants may absorb these antibiotics and convey them to humans (K. Kumar et al. Antibiotic Uptake by Plants from Soil Fertilized with Animal Manure. *Journal of Environmental Quality* 2005;34:2082–2085).

50. Interview with University of California–Berkeley journalism professor Michael Pollan, in B. A. Powell. Is McDonald's New Policy on Antibiotics Simply Greenwash? (Organic Consumers Association, 1993). Available: www.organicconsumers.org/Toxic/mcdonalds_antibiotics.cfm. Since McDonald announced its policy, antibiotics use for chickens declined "a bit," then went back up. Cattle are fed antibiotics to keep them alive on the factory-farm diet of government-subsidized corn. Cow digestive systems contain bacteria that break down grass and cannot digest the acids that corn produces in the gut. The acids dissolve stomach linings, letting some of the bacteria escape into the cow's bloodstream, often causing liver abscesses. Antibiotics control the level of liver infections. Abscessed livers must be discarded, so feeding corn to cows limits U.S. beef liver production.

51. U.S. Geological Survey. *The USGS Role in TMDL Assessments* (Fact Sheet FS-130–01, 2002). In 2000, the General Accounting Office found that only six states reported having enough data for fully assessing their water quality, and only three states had a majority of the data needed for developing Total Maximum Daily [pollutant] Load (TMDL) standards for logging, agriculture, and urban development nonpoint pollution sources (U.S. General Accounting Office. *Water Quality: Identification and Remediation of Polluted Waters Impeded by Data Gaps* [GAO/T-RCED-00–88, 2000]). Other important nonpoint pollution sources include grazing lands, rights-of-way, and turf grass.

52. W. B. Solley et al. 1998. *Estimated Use of Water in the United States in 1995* (U.S. Geological Survey Circular 1200), table 31; see also U.S. General Accounting Office. *Freshwater Supply: States' Views of How Federal Agencies Could Help Them Meet the Challenges of Expected Shortages* (GAO-03-514, 2003), 47.

53. In aquatic environments, organic pesticides can dissolve in water, form colloids (very small particles that resist settling out of water, causing a cloudy appearance), or attach to soil fragments that travel suspended in the water or move along the channel bottom. The pesticides are redistributed by fish and bottom-dwelling organisms that feed in sediments, and this is how they enter the lower part of human food chains. For example, Yakima River studies identified surrounding agricultural areas as the principal sources of nutrients, suspended sediments, pesticides, and fecal indicator bacteria (including *E. coli*) in the river. DDT and its metabolites, and chlorinated compounds dieldrin, *cis*-chlordane, *trans*-nonachlor, dicofol, PCBs, and toxaphene all were present somewhere in the sampled environment, along with organophosphates chlorpyrifos, diazinon, dimethoate, malathion, parathion, phorate, and phosphamidon; thiocarbamate and sulfite compounds (EPTC and propargite); acetamide compounds (alachlor and metolachlor); triazine compounds (atrazine, prometon, and simazine); and chlorophenoxyacetic acid and benzoic compounds (2,4-D and dicamba). Other identified compounds included polycyclic aromatic hydrocarbons (PAHs), including two carcinogenic types called chrysene and 1,2-benzanthracene. Most frequently, dieldrin, diazinon, parathion, and DDT plus its metabolites (DDE + DDD) concentrations exceeded U.S. Environmental Protection Agency chronic toxicity water-quality criteria, as well as exceeding criteria for protecting aquatic life. At 18 of the sites, meeting the standards required 29–99% reductions in DDT + DDE + DDD concentrations. Other requirements included reducing dieldrin by 47–95% at seven sites, diazinon by 31–98% at 14 sites, and parathion by 19–93% at four sites (J. L. Morace et al. *Surface-Water-Quality Assessment of the Yakima River Basin in Washington, Overview of Major Findings, 1987–91*

[U.S. Geological Survey, Water-Resources Investigations Report 98–4113, 1999], 62, 74–76; J. F. Rinella et al. *Surface-Water-Quality Assessment of the Yakima River Basin, Washington, Distribution of Pesticides and Other Organic Compounds in Water, Sediment, and Aquatic Biota, 1987–91* [U.S. Geological Survey, Water-Supply Paper 2354-B, 1999], 1–2).

54. Although not as severe in the west as in the southeastern United States, DDT and its breakdown products still are present in all western states except New Mexico, Arizona, Nevada, and Wyoming. The organochlorines show up in many rivers and streams, aquatic animals, and agricultural soils (U.S. Geological Survey, Pesticides in Stream Sediment). Large DDT and PCB deposits lie shallowly buried in sediments across nearly 11,000 undersea acres on the continental shelf at Palos Verdes, California. About 220 tons of DDT contamination came from the Montrose Company, the world's largest DDT producer, which had discharged treated but still-contaminated wastewater into the ocean for five decades, until the early 1970s. More recently, deposited DDT-poor sediment lies on top of the contaminated materials. But animals live in the sediments to depths greater than 10 inches, stirring them up and mixing DDT into the cleaner surface sediment. Storms also redistribute contaminants to the surface, so they remain a major environmental hazard, both to organisms living in the sediments and to fish, humans, and other organisms higher on the food chain via bioaccumulation and biomagnification (J. K. Stull et al. Contaminant Dispersal on the Palos Verdes Continental Margin: I. Sediments and Biota Near a Major California Wastewater Discharge. *Science of the Total Environment* 1996;179:73–90). Hoping to prevent the DDT from getting into the food chain in the future, a U.S. Environmental Protection Agency pilot project will test a scheme for burying the DDT deposits under hundreds of thousands of tons of sand, in 180 acres of ocean floor near the contamination sources. If the project succeeds, an additional 2,000–2,500 acres will be capped in this way.

55. Schmitt. Environmental Contaminants, 142; see also Nowell et al. Pesticides in Stream Sediment, 469.

56. More dead zones are appearing worldwide, totaling about 200 in 2006 according to the United Nations. Agricultural runoff is only one of multiple causes, including sewage, animal waste, atmospheric pollution, sediment remobilization, and marine litter. Estuaries throughout the United States have recurring problems from harmful algal blooms. For example, in 1999, Washington State closed Puget Sound shellfishing because of an algal bloom of *Alexandrium* species, which causes paralytic shellfish poisoning (PSP). PSP toxin produces neurological effects in humans who eat the shellfish—including numbness, loss of balance, ataxia, and fever that may be life-threatening. There is no antidote; in severe cases, respiratory arrest may occur within 24 hours (Natalie Goldstein. *Earth Almanac*, 2nd ed. [Westport, Connecticut: Oryx Press, 2002], 256–259; this book provides a valuable comprehensive treatment of a wide assortment of environmental issues).

57. In fall, lower river flows decrease the nutrient supply and storm mixing replenishes oxygen in the bottom waters, so the hypoxic zone diminishes or disappears, to form again the next spring (D. A. Goolsby. Mississippi Basin Nitrogen Flux Believed to Cause Gulf Hypoxia. *EOS, Transactions of the American Geophysical Union* 2000;81:321, 326–327; T. M. Kennedy and T. W. Lyons. Hypoxia in the Gulf of Mexico—Causes, Consequences, and Political Considerations. *GSA Today, Geological Society of America* 2000;10:36–38).

58. Excessive irrigation water applications quickly get back into shallow groundwater supplies. Intensive irrigation in the central Columbia Plateau of Washington and Idaho takes 2.5 million acre-feet (800 billion gallons) of water from the Columbia River, in a region where

shallow groundwater supplies 84% of the drinking water (A. K. Williamson et al. *Water Quality in the Central Columbia Plateau, Washington and Idaho, 1992–95* [U.S. Geological Survey Circular 1144, 1998], 5–6).

59. Solley et al. Estimated Use of Water, tables 2, 4, and 8. Agricultural uses include crop and pasture irrigation and livestock waters.

60. P. F. Brussard and D. S. Dobkin. Great Basin: Mojave Desert Region. In M. J. Mac et al., eds. *Status and Trends of the Nation's Biological Resources* (U.S. Geological Survey, 1998), 510, 532; M. A. Bogan et al. Southwest. In M. J. Mac et al., eds. *Status and Trends of the Nation's Biological Resources* (U.S. Geological Survey, 1998), 548; Michael Collier et al. *Dams and Rivers* (U.S. Geological Survey Circular 1126, 1996); C. S. Crawford et al. *Middle Rio Grande Ecosystem—Bosque Biological Management Plan, Middle Rio Grande Biological Interagency Team* (U.S. Fish and Wildlife Service, 1993). The conflict between irrigated agriculture and wildlife is a global problem (A. D. Lemly et al. Irrigated Agriculture and Wildlife Conservation: Conflict on a Global Scale. *Environmental Management* 2000;25:485–512).

61. Daniel Hillel. *Out of the Earth: Civilization and the Life of the Soil* (Berkeley: University of California Press, 1991), 145–147.

62. John Letey. Soil Salinity Poses Challenges for Sustainable Agriculture and Wildlife. *California Agriculture* 2000;54:43–48; T. Jacobsen and R. M. Adams. Salt and Silt in Ancient Mesopotamian Agriculture. *Science* 1958;128:1251–1258.

63. T. S. Presser. The Kesterson Effect. *Environmental Management* 1994;18:437–454; T. S. Presser et al. Bioaccumulation of Selenium from Natural Geologic Sources in Western States and Its Potential Consequences. *Environmental Management* 1994;18:423–436; U.S. Geological Survey. *Unforeseen Consequences of Irrigation Drainage* (Fact Sheet FS-038-97, 1997).

64. Letey, Soil Salinity Poses Challenges. Letey also discusses various schemes for draining or sequestering the drain water with less ecological damage than occurred at Kesterson. The groundwater pumping approach eventually destroys high-quality groundwater supplies. Reverse osmosis, in particular, is costly in terms of money and energy. Any solution will be financed from public coffers.

65. L. H. Creekmore et al. Impacts of Animal Feeding Operations on Wildlife Health. In F. D. Wilde et al., eds. *Effects of Animal Feeding Operations on Water Resources and the Environment, Proceedings of the Technical Meeting, Fort Collins, Colorado, August 30-September 1, 1999* (U.S. Geological Survey Open-File Report 00–204, 2000), 69. The growth of factory farms for meat production has steadily increased the demand for agricultural land for growing feed crops, leading to massive conversions of Brazilian grasslands and rain forest to soybean crops. While westerners have long led the world in per capita meat consumption, the competition is growing. If world consumption met western levels, land requirements for feed crops would expand to about 6.2 billion acres, two-thirds more than is presently used for all agriculture (Rosamond Naylor et al. Losing the Links Between Livestock and Land. *Science* 2005;310:1621–1622). More serious than the pervasive stench of factory farms (also called confined animal feeding operations, or CAFOs) is the lack of knowledge about the fate and transport of gaseous and particulate air emissions and their impact on human and wildlife health (National Research Council. *Air Transmissions from Animal Feeding Operation: Current Knowledge, Future Needs* [Washington DC, National Academies Press, 2003]). Feeding soybeans to livestock will become a lot more expensive as soybean crops are converted to biofuel production.

66. Gerd Hamscher et al. Antibiotics in Dust Originating from a Pig-Fattening Farm—a New Source of Health Hazard for Farmers? *Environmental Health Perspectives* 2003;111:1590–1594; S. G. Gibbs et al. Isolation of Antibiotic-Resistant Bacteria from the Air Plume Downwind of a Swine Confined or Concentrated Animal Feeding Operation. *Environmental Health Perspectives* 2006;114:1032–1037.

67. *Salmonella* bacteria species cause salmonellosis, avian cholera is caused by *Pastuerella multocida*, and avian botulism by *Clostridium botulinum* (Creekmore et al. Impacts of Animal Feeding Operations, 69).

68. Three independent studies tested 37 fishmeal samples from six countries and found PCB contamination in nearly every sample (Environmental Working Group. PCBs in Farmed Salmon: Factory Methods, Unnatural Results [2003]. Available: www.ewg.org/reports/farmedP-CBs/part2.php). Farmed salmon tissues contain significantly higher concentrations of 151 out of 158 chemical contaminants than does wild salmon. In addition to 110 different PCBs, the contaminants included brominated flame retardants, organochlorine pesticides such as DDT and dieldrin, and carcinogenic combustion products (PAHs—polycyclic aromatic hydrocarbons) (M. D. Easton et al. Preliminary Examination of Contaminant Loadings in Farmed Salmon, Wild Salmon and Commercial Salmon Feed. *Chemosphere* 2002;46:1053–1074). A National Research Council panel found that the aquaculture industry's "intensive management approach" promotes PCB accumulation in fish tissues. The fishmeal feed is made out of wild fish taken from known high-pollution areas. Wild Alaskan salmon eat Pacific Ocean fish that are naturally lower in persistent pollutants (National Research Council. *Dioxins and Dioxin-like Compounds in the Food Supply: Strategies to Decrease Exposure* (Washington, DC: National Academies Press, 2003; see also R. A. Hites et al. Global Assessment of Organic Contaminants in Farmed Salmon. *Science* 2004;303:226–229).

69. Oliver Houck. Tales from a Troubled Marriage, Science and Law in Environmental Policy. *Science* 2003;302:1926–1929. This is an excellent, readable account of the problems in applying science to environmental policy.

70. O.W.L. Foundation. Available: www.owlfoundation.net/IRWP_wastewater_forum-Corson.html.

71. Kris Christen. Synergistic Effects of Chemical Mixtures and Degradation Byproducts Not Reflected in Water Quality Standards, USGS Finds. *Environmental Science and Technology* 1999;33:230A.

72. D. O. Carpenter et al. Understanding the Human Health Effects of Chemical Mixtures. *Environmental Health Perspectives* 2002;110:25–42.

73. Hayes et al. Feminization of Male Frogs.

74. Renner. Pesticide Mixture Enhances Frog Abnormalities, 52A; Elisabete Silva et al. Something from "Nothing," Eight Weak Estrogenic Chemicals Combined at Concentrations Below NOECs Produce Significant Mixture Effects. *Environmental Science and Technology* 2002;36:1751–1756; M. F. Cavieres et al. Developmental Toxicity of a Commercial Herbicide Mixture in Mice, I. Effects on Embryo Implantation and Litter Size. *Environmental Health Perspectives* 2002;110:1081–1085.

75. Holly Knight and Caroline Cox. *Worst Kept Secrets, Toxic Inert Ingredients in Pesticides* (Eugene, Oregon: Northwest Coalition for Alternatives to Pesticides, 1998), a publication that packs a wealth of information in its 15 pages.

76. Emily Green. Washington Apple Study Finds Organic Growing is Best. *Los Angeles Times* (19 April 2001).

77. Pollan. The Omnivore's Dilemma. 134–184.

78. Benbrook et al. *Pest Management at the Crossroads.*

79. David Pimentel et al. Organic and Conventional Farming Systems: Environmental and Economic Issues. *BioScience* 2005;55(7).

80. Paul Mäder et al. Soil Fertility and Biodiversity in Organic Farming. *Science* 2002;296:1694–1697.

81. Virginia Worthington. Nutritional Quality of Organic Versus Conventional Fruits, Vegetables, and Grains. *Journal of Alternative and Complementary Medicine* 2001;7:161–173.

82. J. P. Reganold et al. Sustainability of Three Apple Production Systems. *Nature* 2001;410:926–930.

83. B. P. Baker et al. Pesticide Residues in Conventional, IPM-Grown, and Organic Foods. *Food Additives and Contaminants* 2002;19:427–446. 73% of the 12 conventional vegetable crops had measurable pesticide residues. 47% of integrated pest management crops and 23% of organic crops had pesticide residues. Not counting the persistent chemicals, 71% of conventional crops, 46% of integrated pest management crops, and 13% of organic crops had chemical residues.

84. U.S. Department of Agriculture. *Pesticide Data Program, Annual Summary Calendar Year 2003* (2005). Available: www.ams.usda.gov/science/pdp; see also Jen Miller. Troubling Statistics about What's in Your Food. *Journal of Pesticide Reform* 2005;25:4.

85. E. O. Wilson. *The Future of Life* (New York: Adolph A. Knopf, 2002).

86. The Green Revolution developed traditionally hybridized grain varieties to increase crop yields and feed growing populations in developing countries. One was a short-statured rice that yielded dramatically increased harvest with increased fertilization. These varieties also had short growing seasons, allowing two or three crops per year. Multiple crops required additional water, mostly coming from groundwater and causing water tables to drop. Pest problems increased, but attempts to control them with increased pesticide applications also reduced the pests' natural predators and triggered even worse pest problems. Increased fertilizer and pesticide applications have degraded wetlands and threaten drinking water supplies. Long-term studies show that declining yields of 1.4–2% per year from 1979 through 1991 have reduced yields by 38–58% compared to 1968. The reasons for these declines are not understood (National Research Council. *Environmental Effects of Transgenic Plants: The Scope and Adequacy of Regulation* (Washington, DC: National Academies Press, 2002), 34–35).

87. C. M. Benbrook. *Genetically Engineered Crops and Pesticide Use in the United States: The First Nine Years* (BioTech InfoNet Technical Paper No. 7, 2004).

88. Jane Nielson and Howard Wilshire. *Sowing Pandora's Fields* (18 August 2006). Available: www.losingthewest.com/sowing_pandoras_fields.pdf.

89. Edward Groth et al. *Do You Know What You're Eating? An Analysis of U.S. Government Data on Pesticide Residues* (Consumers Union of the United States Inc., 1999), 20.

90. Lennart Hardell and Mikael Eriksson. A Case-Control Study of Non-Hodgkin Lymphoma and Exposure to Pesticides. *Cancer* 1999;8:1353–1360; Swan et al., Semen Quality. Diazinon is an organophosphate insecticide.

91. Quoted in Pollan, *Playing God in the Garden.*

92. E. A. Guillette et al. An Anthropological Approach to the Evaluation of Children Exposed to Pesticides in Mexico. *Environmental Health Perspectives* 1998;106. Available: www.americaspolicy.org/pdf/bios/Yaqui_Valley_eng.pdf.

93. R. F. Service. As the West Goes Dry. *Science* 2004;303:1124–1127.

94. Whether increasing CO_2 helps or hinders farmers, industrial agriculture clearly contributes substantially to increasing the atmospheric greenhouse gases, CO_2, nitrous oxide, and methane. The net global warming potential of conventionally tilled lands is about +110, in CO_2 equivalents per square meter per year, a net release of greenhouse gases. In contrast, the potential of nearby unmanaged early successional ecosystems is –211, indicating net greenhouse gas uptake. No systems for growing annual crops appear to provide net uptake, although no-till practices come closest (G. P. Robertson et al. Greenhouse Gases in Intensive Agriculture, Contributions of Individual Gases to the Radiative Forcing of the Atmosphere. *Science* 2000;289:1922–1925; Janet Pelley. Cropland Net Emitter of Greenhouse Gases. *Environmental Science and Technology* 2000;34:455A–456A). Suggestions that increased crop production in marginally farmable semiarid lands might increase carbon storage in soils ignore critical side effects. Generally, increased crop production in semiarid regions requires irrigation, adding substantial CO_2 emissions from fossil fuels that drive irrigation pumps. In addition, groundwater commonly contains as much as 1% dissolved bicarbonate (HCO_3). The drying irrigation waters crystallize calcium carbonate ($CaCO_3$) and release CO_2 to the atmosphere. Applying manure fertilizers to marginal lands also is counterproductive, and growing the animals' feeds takes larger acreages than the manure can fertilize (W. H. Schlesinger. Carbon Sequestration in Soils. *Science* 1999;284:2095; this elegant one-page paper succinctly points out the pitfalls of some commonly suggested schemes for countering atmospheric buildup of CO_2).

95. Cynthia Rosenzweig and Daniel Hillel. *Climate Change and the Global Harvest, Potential Impacts of the Greenhouse Effect on Agriculture* (New York: Oxford University Press, 1998); for a review of this excellent book, see P. D. Moore. Climate Change and the Global Harvest, Potential Impacts of the Greenhouse Effect on Agriculture. *Nature* 1998;393:33–34. Increasing temperature parallels the increase of atmospheric CO_2. One consequence for crops is reduced soil moisture, which may counterbalance crop benefits, such as increased CO_2, or enhance negative effects.

96. A plant's identification as type C3 or C4 depends on whether its first-stage photosynthesis reactions are based on three or four carbon atoms. Less than 1% of Earth's plants are type C4. C3 plants typically show higher productivity increases than do C4 plants in enriched CO_2 conditions (see www.c02science.org/dictionary; www.c02science.org).

97. Rosenzweig and Hillel (*Climate Change and the Global Harvest*) list C3 and C4 weeds in table 4.1.

98. S. P. Long et al. Food for Thought: Lower-Than-Expected Crop Yield Stimulation with Rising CO_2 Concentrations. *Science* 2006;312:1918–1921; David Schimel. Climate Change and Crop Yields: Beyond Cassandra. *Science* 2006;312:1889–1890.

99. L. H. Ziska. Evaluation of Yield Loss in Field-Grown Sorghum from a C3 and C4 Weed as a Function of Increasing Atmospheric Carbon Dioxide. *Weed Science* 2003;51:914–918; L. H. Ziska and K. George. Rising Carbon Dioxide and Invasive Noxious Plants: Potential Threats and Consequences. *World Resource Review* 2004;16:427–447; L. H. Ziska et al. Changes in Biomass and Root:Shoot Ratio in a Field-Grown, Noxious Perennial Weed, Canada Thistle (*Cirsium Arvense* L. SCOP.) with Elevated CO_2: Implications for Chemical Control by Glyphosate. *Weed Science* 2004;52:584–588; L. H. Ziska et al. The Impact of Recent Increases in Atmospheric CO_2 on Biomass Production and Vegetative Retention of

Cheatgrass (*Bromus tectorum*): Implications for Fire Disturbance. *Global Change Biology* 2005;11:1325–1332.

100. B. L. Gardner. *American Agriculture in the Twentieth Century, How It Flourished and What It Cost* (Cambridge, Mass.: Harvard University Press, 2002); C. P. Trimmer. Unbalanced Bounty from American Farms. *Science* 2002;298:1339–1340, a review of Gardner. *American Agriculture*; D. D. Richter, Jr., and Daniel Markewitz. *Understanding Soil Change, Soil Sustainability Over Millennia, Centuries, and Decades* (Cambridge: Cambridge University Press, 2001), 5; World Resources Institute. *World Resources 2000–2001*, 53–68; R. E. Evenson and D. Gollin. Assessing the Impact of the Green Revolution, 1960–2000. *Science* 2003;300:758–762.

101. Mark Vellend. Habitat Loss Inhibits Recovery of Plant Diversity as Forests Regrow. *Ecology* 2003;84:1158–1164; Horrigan et al. How Sustainable Agriculture Can Address the Environmental and Human Health Harms; Gardner. *American Agriculture*; Pimentel. Environmental and Economic Benefits, 4–20.

102. Edward Thompson, Jr., and T. W. Warman. Meeting the Challenge of Farmland Protection in the 21st Century. *American Farmland Trust Magazine* Summer 2000.

103. Pollan, The Omnivore's Dilemma, 186–225.

104. In 2000, when U.S. Department of Agriculture began formulating organic standards for certifying farmers and organic products, the agribusiness industry made numerous attempts to define genetic engineering, food irradiation, and sewage sludge fertilizers as organic practices. Public outcry killed all those proposals. U.S. Department of Agriculture's draft 2004 guidelines represented the most recent attempt to allow pesticides and antibiotics in organic farming. They were quickly dropped after yet another storm of protest from public and organic farmers, but assaults on the integrity of organic standards continue.

Chapter 3

1. J. A. Ludwig. Primary Productivity in Arid Lands—Myths and Realities. *Journal of Arid Environments* 1987;13:1–7.

2. A comprehensive review of the large scientific literature on grazing effects concluded that livestock grazing "negatively affect[s] water quality and seasonal quantity, stream channel morphology, hydrology, riparian zone soils, in-stream and stream bank vegetation, and aquatic and riparian wildlife. No positive environmental impacts were found" (A. J. Belsky et al. Survey of Livestock Influences on Stream and Riparian Ecosystems in the Western United States. *Journal of Soil and Water Conservation* 1999;54:419–431). The U.S. Geological Survey (M. A. Bogan et al. Regional Trends of Biological Resources—Southwest. In M. J. Mac et al., eds. *Status and Trends of the Nation's Biological Resources*, vol. 2 (U.S. Geological Survey, 1998), 560–561) similarly concluded that heavy grazing in riparian areas deteriorates soil-stabilizing vegetation and causes stream bank erosion, reduced water quality, reduced water storage capacity, streambed widening, and reduced water depth. Most of the effects increase water temperature and reduce aquatic habitat quality. Another detailed study (Allison Jones. Effects of Cattle Grazing on North American Arid Ecosystems: A Quantitative Review. *Western North American Naturalist* 2000;60:155–164) gives quantitative measures of grazing impacts on 16 factors needed to preserve wildlife, vegetation, and healthy soils in western

U.S. arid lands. Eleven of the 16 analyses showed significant detrimental effects of cattle grazing on arid rangeland (see also T. L. Fleischner. Ecological Costs of Livestock Grazing in Western North America. *Conservation Biology* 1994;8:629–644).

3. R. F. Noss and A. Y. Cooperrider. *Saving Nature's Legacy* (Covelo, Calif.: Island Press, 1994), 240–241.

4. Of the 1,207 plant and animal species listed as endangered, threatened, or proposed for listing under the Endangered Species Act, 22% are at the brink of extinction (D. S. Wilcove et al. Quantifying Threats to Imperiled Species in the United States. *BioScience* 1998;48:607–615; report assesses sources of threats to 2,490 species of plants and animals, of which 700 are not [yet] listed). Brian Czech et al. (Economic Associations Among Causes of Species Endangerment in the United States. *BioScience* 2000;50:593–601) assessed inter-actions among the various pressures causing species deterioration (see also R. F. Noss and R. L. Peters. Endangered Ecosystems—a Status Report on America's Vanishing Habitat and Wildlife. *Defenders of Wildlife* December 1995; D. J. Tazik and C. O. Martin. Threatened and Endangered Species on U.S. Department of Defense Lands in the Arid West, USA. *Arid Land Research and Management* 2002;16:259–276).

5. Spanish settlers first brought cattle to the south Texas plains, then a part of Mexico. After the 1836 Texas revolution, cattle grew wild, rapidly multiplying to an estimated 300,000 head by 1850 and more than 3.5 million by 1860. Sporadic cattle drives to markets in Ohio, California, and Illinois (Chicago) apparently began in 1846. The Civil War interrupted cattle drives from 1860 to 1865, but feeding Confederate troops on Texas cattle did not stop herds from growing in size. The fabled cattle drives to notorious boomtown railhead cowtowns of Abilene, Wichita, and Dodge did not begin until the war ended. In 1880, newly invented barbed wire fenced off huge prairie acreages for farming, while railroads expanded, mak-ing cattle drives obsolete. In 1885, the cattle boom busted, with effects that continued to the end of the century (Walter Prescott Webb. *The Great Plains* [New York: Grosset and Dunlap, Grosset's Universal Library Edition, 1931], 207–237; John Rolfe Burroughs. *Where the Old West Stayed Young* [New York: Bonanza Books, 1962], 3–5; see also George Wuerthner. Beef, Cowboys, the West, American Icons. In George Wuerthner and Mollie Matteson, eds. *Welfare Ranching* [Covelo, California: Island Press, 2002], 27–30).

6. Paul Rogers and Jennifer LaFleur. Cash Cows, The Giveaway of the West. *San Jose Mercury News* 7 November 1999.

7. Jobs related to grazing on public lands constitute only 0.1% of western U.S. employ-ment (U.S. General Accounting Office. *Rangeland Management: BLM's Hot Desert Grazing Program Merits Reconsideration* [GAO/RCED-92–12, 1992]; U.S. Department of the Interior Bureau of Land Management, and U.S. Department of Agriculture Forest Service. *Rangeland Reform '94* [Final Environmental Impact Statement, 1994], 4–122).

8. Bureau of Land Management regulations initially required grazing permittees to con-trol "base property"; that is, they had to own some land in the vicinity of the federal grazing allotment. A common illegal dodge was to sublet the public lands' grazing right to a second party who does not control the base property and may not even own land in the vicinity (U.S. Department of the Interior. *Rangeland Reform '94*, 9).

9. In the 11 western states, 260 million acres of public lands are needed to support a mere 3% of all of the nation's livestock (D. L. Donahue. *The Western Range Revisited: Removing Livestock from Public Lands to Conserve Native Biodiversity* [Norman, Oklahoma: University of

Oklahoma Press, 1999], 252; see also Lynn Jacobs. *Waste of the West: Public Lands Ranching* [Tucson, Ariz.: Lynn Jacobs, 1991], 23—this report is a thoroughly referenced, comprehensive account of grazing abuses in the western United States). The U.S. General Accounting Office found that public lands in the southwestern U.S. "hot desert" region (Mojave, Sonoran, and Chihuahuan deserts) provided forage for no more than 1.6% of the nation's cattle and 3% of its sheep (U.S. General Accounting Office. *Rangeland Management*).

10. U.S. Government Accountability Office. *Livestock Grazing: Federal Expenditures and Receipts Vary, Depending on the Agency and the Purpose of the Fee Charged* (GAO-05–869, 2005).

11. G. F. Gifford and R. H. Hawkins. Hydrological Impact of Grazing on Infiltration: A Critical Review. *Water Resources Research* 1978;14:305–313; G. F. Gifford and R. H. Hawkins. Deterministic Hydrologic Modeling of Grazing System Impacts on Infiltration Rates. *Water Resources Bulletin* 1979;15:924–933. Measurements on large Nevada and Arizona experimental arid rangeland plots show that water infiltration decreases as both vegetative canopy cover and rock and gravel cover decrease (L. J. Lane et al. Semiarid Rangeland Areas of the Southwestern U.S.A. In Yu-Si Fok, ed. *International Conference on Infiltration Development and Application, Pre-conference Proceedings* [University of Hawaii at Manoa, U.S. Department of Agriculture, 1987], 365–376).

12. C. M. Rostagno. Infiltration and Sediment Production as Affected by Soil Surface Conditions in a Shrubland of Patagonia, Argentina. *Journal of Range Management* 1989;42:382–385.

13. Conservative erosion estimates indicate that loss of less than four inches of arid soils reduces nitrogen by 27%, phosphorus by 21%, and organic material by 38% (J. L. Charley and S. W. Cowling. Changes in Soil Nutrients Resulting from Overgrazing and Their Consequences in Plant Communities of Semi-Arid Areas. *Proceedings of the Ecological Society of Australia* 1968;3:28–38).

14. R. C. Balling, Jr. The Climatic Impact of a Sonoran Vegetation Discontinuity. *Climatic Change* 1988;13:99–109; R. C. Balling, Jr. The Impact of Summer Rainfall on the Temperature Gradient Along the United States-Mexico Border. *Journal of Applied Meteorology* 1989;28:304–308; O. E. Sala and J. M. Paruelo. Ecosystem Services in Grasslands. In G. C. Daily, ed. *Nature's Services—Societal Dependence on Natural Ecosystems* (Covelo, Calif.: Island Press, 1995), 237–252; Jennifer Couzin. Landscape Changes Make Regional Climate Run Hot and Cold. *Science* 1999;283:317, 319.

15. Jacobs. *Waste of the West*, 94; see also R. D. Ohmart. Historical and Present Impacts of Livestock Grazing on Fish and Wildlife Resources in Western Riparian Habitats. In P. Kraussman, ed. *Rangeland Wildlife* (Denver: Society for Range Management, 1996), 246–279; Belsky et al. Survey of Livestock Influences; U.S. General Accounting Office. *Public Rangelands: Some Riparian Areas Restored but Widespread Improvement Will Be Slow* (GAO/RCED-88–105, 1988).

16. C. F. Wilkinson. *Crossing the Next Meridian* (Covelo, Calif.: Island Press, 1992), 78.

17. See extensive references in Allison Jones. *Review and Analysis of Cattle Grazing Effects in the Arid West, with Implications for U.S. Bureau of Land Management Grazing Management in Southern Utah* (Southern Utah Landscape Restoration Project, 2001); Belsky et al. Survey of Livestock Influences; Andrew Goudie. *The Human Impact on the Natural Environment*, 5th ed. (Cambridge, Mass.: MIT Press, 2000), 307; Donahue. *The Western*

Range Revisited, chapter 5; Jacobs. *Waste of the West*, 82–110; Fleischner. Ecological Costs of Livestock Grazing; Bogan et al. Regional Trends of Biological Resources, 560–561.

18. Soil deforms under the weight of grazing animals. On heavily grazed slopes, deformed soils form the conspicuous grazing-step terracettes along cow trails. The animals continue to graze as they follow a trail; after browsing all of the plants within reach, they move farther along the terracette tread to find unbrowsed forage. Once they run out of fodder at a terracette level, they move higher or lower on the slope to find more. Studies of terracette spacings show dependence on animal sizes and numbers, slope steepness, and obstacles on the slopes (J. K. Howard and C. G. Higgins. Dimensions of Grazing-Step Terracettes and Their Significance. In V. Gardiner, ed. *International Geomorphology 1986, Proceedings of the First International Conference on Geomorphology* [London: John Wiley and Sons, 1987], 545–568).

19. J. Belnap and O. L. Lange, eds. *Biological Soil Crusts: Structure, Function, and Management* (Ecological Studies 150; New York: Springer-Verlag, 2001); H. G. Wilshire. The Impact of Vehicles on Desert Soil Stabilizers. In R. H. Webb and H. G. Wilshire, eds. *Environmental Effects of Off-Road Vehicles* (New York: Springer-Verlag, 1983), 31–50; Jayne Belnap and D. A. Gillette. Vulnerability of Desert Biological Soil Crusts to Wind Erosion: The Influences of Crust Development, Soil Texture, and Disturbance. *Journal of Arid Environments* 1998;39:133–142.

20. R. P. Beasley et al. *Erosion and Sediment Pollution Control*, 2nd ed. (Ames, Iowa: Iowa State University Press, 1984); Jayne Belnap and D. A. Gillett. Disturbance of Biological Soil Crusts, Impacts on Potential Wind Erodibility of Sandy Desert Soils in SE Utah. *Land Degradation and Development* 1998;8:355–362. Studies suggest that the lichen component of biological crusts can recover in 45 years (Jayne Belnap. Recovery Rates of Cryptobiotic Crusts—Inoculant Use and Assessment Methods. *Great Basin Naturalist* 1993;53:89–95), but the moss component may require another 200 years to become reestablished. Even incomplete recovery restores some functions of the crust, such as reducing susceptibility to wind erosion.

21. Allan Savory. *Holistic Resource Management* (Washington, DC: Island Press, 1988). Holistic management supporters promote the belief that crust-destroying livestock trampling is good for the land, but the adverse effects are difficult to defend.

22. W. J. Foreyt. Pneumonia in Bighorn Sheep, Effects of *Pasteurella haemolytica* from Domestic Sheep and Effects on Survival and Long-Term Reproduction. *Biennial Symposium of the Northern Wild Sheep and Goat Council* 1990;7:92–101; R. J. Monello et al. Ecological Correlates of Pneumonia in Bighorn Sheep Herds. *Canadian Journal of Zoology* 2001;79:1423–1432.

23. Goudie. *The Human Impact*, 90. Exotic Johnson grass (*Sorghum halepense*) and kudzu (*Pueraria lobata*) were introduced as crops but became invasive weed pests. Most exotic invader plants, however, were introduced accidentally along with crop seeds, soil ballast, various imported plants, and foodstuffs (David Pimentel et al. Environmental and Economic Costs of Nonindigenous Species in the United States. *BioScience* 2000;50:58) and on wind-blown materials (J. K. M. Brown and M. S. Hovmøller. Aerial Dispersal of Pathogens on the Global and Continental Scales and Its Impact on Plant Disease. *Science* 2002;297:537–541).

24. J. D. Williams and G. K. Meffe. Nonindigenous Species. In M. J. Mac et al., eds. *Status and Trends of the Nation's Biological Resources*, part 1 (U.S. Geological Survey, 1998), 117–129, quotation 117; see also U.S. General Accounting Office. *Invasive Species: Obstacles Hinder Federal Rapid Response to Growing Threat* (GAO-01 724, 2001), 7.

25. J. Lacey et al. Observations on Spotted and Diffuse Knapweed Invasion into Ungrazed Bunchgrass Communities in Western Montana. *Rangelands* 1990;12:8–13; R. J. Tausch et al. Patterns of Annual Grass Dominance on Anaho Island: Implications for Great Basin Vegetation Management. In S. B. Monsen and S. G. Kitchen, eds. *Proceedings—Ecology and Management of Annual Rangelands* (General Technical Report INT-GTR-313, U.S. Department of Agriculture Forest Service, Intermountain Research Station, 1994), 120–125; T. J. Stohlgren et al. Exotic Plant Species Invade Hot Spots of Native Plant Diversity. *Ecological Monograph* 1999;69:25–46.

26. E. L. Painter and A. J. Belsky. Application of Herbivore Optimization Theory to Rangelands of the Western United States. *Ecological Applications* 1993;3:2–9; A. J. Belsky. Overcompensation—Herbivore Optimization or Red Herring? *Evolutionary Ecology* 1993;7:109–121. In places, a single exotic plant species competitively overruns an entire ecosystem; for example, nearly 10 million acres of northern California native grasslands have been lost to the yellow star thistle (*Centaurea solstitalis*). Pimentel et al. (Environmental and Economic Costs, 53–65) give a comprehensive analysis of all dominant groups of invasive nonindigenous plants and animals. In addition to huge economic costs of fighting invasive species and their spread, on natural areas "they have strangled native plants, taken over wetland habitats, crowded out native species, and deprived waterfowl and other species of food sources" (U.S. General Accounting Office. *Perspectives on Invasive Species* [GAO-03–1089R, 2003]).

27. W. A. Dick-Peddie. *New Mexico Vegetation: Past, Present, and Future* (Albuquerque: University of New Mexico Press, 1993); Bogan et al. Regional Trends of Biological Resources.

28. Dick-Peddie. *New Mexico Vegetation;* Bogan et al. Regional Trends of Biological Resources; C. J. Bahre and M. L. Shelton. Historic Vegetation Change, Mesquite Increases, and Climate in Southeastern Arizona. *Journal of Biogeography* 1993;20:489–504; S. C. Goslee et al. High-Resolution Images Reveal Rate and Pattern of Shrub Encroachment over Six Decades in New Mexico, U.S.A. *Journal of Arid Environments* 2003;54:755–767; A. Joy Belsky. Viewpoint: Western Juniper Expansion—Is It a Threat to Arid Northwestern Ecosystems? *Journal of Range Management* 1996;49:53–59.

29. Medusahead wild rye is *Taeniatherum asperum.* Cheatgrass (*Bromus tectorum*) monoculture covers an estimated 1.1 million acres of southwestern rangelands and is the dominant understory vegetation on 1.8 million acres. Another 1.3 million acres has a strong vulnerability to cheatgrass invasion (W. D. Billings. *Bromus tectorum,* a Biotic Cause of Ecosystem Impoverishment in the Great Basin. In G. M. Woodwell, ed. *The Earth in Transition—Patterns and Processes of Biotic Impoverishment* [Cambridge: Cambridge University Press, 1990]). Cheatgrass also has invaded large areas of Great Basin shrubby grassland—in all of Nevada, plus adjacent parts of Utah, Arizona, California, Oregon, and Idaho—displacing perennial bunch grasses and greatly increasing fire frequencies, from once every 60–110 years to once every three to five years.

30. D. E. Busch et al. Water Uptake in Woody Riparian Phreatophytes of the Southwestern United States: A Stable Isotope Study. *Ecological Applications* 1992;2:450–459. Saltcedar (*Tamarix* spp.) now largely dominates extensive riparian zones of the Mojave, Virgin, Humboldt, and Walker rivers, as well as many lesser streams of the West (J. E. Lovich et al. Tamarisk Control on Public Lands in the Desert of Southern California: Two Case Studies. In *46th Annual California Weed Conference* [California Weed Science Society, 1994], 166–177).

31. S. B. Monsen. The Competitive Influences of Cheatgrass (*Bromus tectorum*) on Site Restoration. In S. B. Monsen and S. G. Kitchen, eds. *Proceedings: Ecology and Management*

of Annual Rangelands (General Technical Report INT-GTR-313, U.S. Department of Agriculture Forest Service, Intermountain Research Station, 1994), 43–50. Monsen reported that native species still had not recovered on south- and west-facing slopes 58 years after grazing ceased.

32. C. A. Brandt and W. H. Rickard. Alien Taxa in the North American Shrub-Steppe Four Decades After Cessation of Livestock Grazing and Cultivation Agriculture. *Biological Conservation* 1994;68:95–105; Billings. *Bromus tectorum*; T. P. Yorks et al. Vegetation Differences in Desert Shrublands in Western Utah's Pine Valley Between 1933 and 1989. *Journal of Range Management* 1992;45:569–578.

33. L. H. Ziska and K. George. Rising Carbon Dioxide and Invasive, Noxious Plants: Potential Threats and Consequences. *World Resource Review* 2004;16:427–447; L. H. Ziska et al. The Impact of Recent Increases in Atmospheric CO_2 on Biomass Production and Vegetative Retention of Cheatgrass (*Bromus tectorum*): Implications for Fire Disturbance. *Global Change Biology* 2005;11:1325–1332. Another study of an especially troublesome grassland invader—yellow starthistle—showed a 70% increase in above-ground biomass and 132% increase in mid-day photosynthesis in an enhanced CO_2 atmosphere in monoculture. The weed enjoyed 69% increase in above-ground biomass in competition with native species in comparison to a total polyculture increase of 28% above-ground biomass (J. S. Dukes. Comparison of the Effect of Elevated CO_2 on an Invasive Species [*Centaurea solstitialis*] in Monoculture and Community Settings. *Plant Ecology* 2002;160:225–234; Cynthia Rosenzweig and Daniel Hillel. *Climate Change and the Global Harvest* [New York: Oxford University Press, 1998], chapter 4).

34. M. J. Trlica and L. R. Rittenhouse. Grazing and Plant Performance. *Ecological Applications* 1993;3:21–23.

35. R. L. Everett and S. H. Sharrow. *Soil and Water Temperature in Harvested and Nonharvested Pinyon-Juniper Stands* (Intermountain Research Station Research Paper INT-342, U.S. Department of Agriculture Forest Service, 1985).

36. Belsky. Viewpoint: Western Juniper Expansion; A. R. Hibbert. Water Yield Improvement Potential by Vegetation Management on Western Rangelands. *Water Resources Bulletin* 1983;19:375–381. Some studies do suggest that pinyon–juniper expansion really may be bad for the range—that the species actually do enhance erosion and reduce livestock forage. But plants and shrubs that replace junipers also intercept and transpire moisture (J. W. Doughty. The Problems with Custodial Management of Pinyon-Juniper Woodlands. In R. L. Everett, comp. *Proceedings, Pinyon-Juniper Conference* (General Technical Report INT-215, U.S. Department of Agriculture Forest Service, 1987), 29–33; R. J. Tausch and P. T. Tueller. Foliage Biomass and Cover Relationship Between Tree- and Shrub-Dominated Communities in Pinyon-Juniper Woodlands. *Great Basin Naturalist* 1990;50:121–134.

37. A draft programmatic environmental report for a vegetation treatment program, including pinyon–juniper eradication, in 17 western states was released November 2005 (U.S. Department of the Interior Bureau of Land Management. *Vegetation Treatments on Bureau of Land Management Lands in 17 Western States* [2005]). This report updates and replaces four earlier environmental impact statements on similar projects completed by the Bureau of Land Management between 1986 and 1992. The proposed program includes "treating" (i.e., clearing) up to six million acres per year, using prescribed burns, herbicides, biological control agents, and mechanical and manual methods. Assessment of the effects of the various clearing methods was to be part of a 10-year project for pinyon–juniper eradication in 13 western states begun in 1991, but repeated Freedom of Information Act requests for the monitoring

data over three years yielded only a list of herbicides used and evasive responses to inquiries, raising the suspicion that no monitoring has been done.

38. U.S. Department of the Interior, Bureau of Land Management. *Record of Decision, Vegetation Treatment on BLM Lands in Thirteen Western States* (1991). The data shown in table 3.2 were obtained by Freedom of Information Act Request.

39. Caroline Cox. 2,4-D: Ecological Effects. *Journal of Pesticide Reform* 1999;19:14–19. Fish showed no harm from either of the pesticides alone but were damaged when exposed to the two in combination. In general, projects such as the Bureau of Land Management's do not examine potential effects of combined pesticide use or of single pesticides applied to areas that had other pesticide residues from previous applications.

40. Tyrone Hayes et al. Feminization of Male Frogs in the Wild. *Nature* 2002;419:895–896; T. B. Hayes et al. Hermaphroditic, Demasculinized Frogs After Exposure to the Herbicide Atrazine at Low Ecologically Relevant Doses. *Proceedings of the National Academy of Sciences* 2002;99:5476–5480.

41. T. N. Johnsen, Jr. Using Herbicides for Pinyon-Juniper Control in the Southwest. In R. L. Everett, ed. *Proceedings, Pinyon-Juniper Conference* (Intermountain Research Station General Technical Report INT-215, U.S. Department of Agriculture Forest Service, 1987), 330–342.

42. Caroline Cox. Sulfometuron Methyl (Oust). *Journal of Pesticide Reform* 2002;22:15–20.

43. The war against pinyon–juniper forests raises substantial questions about its usefulness, economic benefit, and environmental consequences (Bogan et al. Regional Trends of Biological Resources, 555). Supposed "social benefits" of pinyon–juniper destruction include maintaining traditional lifestyles, jobs, community stability, and stability of ranch operations that depend on public lands (R. S. Dalen and W. R. Snyder. Economic and Social Aspects of Pinyon-Juniper Treatment, Then and Now. In R. L. Everett, comp. *Proceedings, Pinyon-Juniper Conference* [Intermountain Research Station General Technical Report INT-215, U.S. Department of Agriculture Forest Service, 1987], 343–350).

44. G. F. Gifford. Myths and Fables and the Pinyon-Juniper Type. In R. L. Everett, comp. *Proceedings, Pinyon-Juniper Conference* (Intermountain Research Station, General Technical Report INT-215, U.S. Department of Agriculture Forest Service, 1987), 34–37. A 10-year Bureau of Land Management plan to "control" pinyon–juniper woodlands and sage shrub communities in 13 western states (U.S. Department of the Interior Bureau of Land Management. *Final Environmental Impact Statement, Vegetation Treatment on BLM Lands in Thirteen Western States*, 1988) calls for clearing 372,000 acres annually, 3.8% manually, 15.6% mechanically, 16.2% biologically (using some insects and pathogens, but mostly [93%] through cattle, sheep, or goat grazing), 26.3% by burning, and 38.1% by chemicals (i.e., herbicides).

45. The Bureau of Land Management assessed 47% of their lands in fair to poor condition, but using the same data Natural Resources Defense Council found 68% in fair to poor condition (Johanna Wald and David Alberswerth. *Our Ailing Public Rangelands* [National Wildlife Federation and the Natural Resources Defense Council, 1985]; Johanna Wald and David Alberswerth. *Our Ailing Public Rangelands—Still Ailing!* [National Wildlife Federation and the Natural Resources Defense Council, 1989]).

46. The 50 million acres in unknown condition constitute 19% of lands managed by the Bureau of Land Management. In 1987, the General Accounting Office assessed 54% of U.S. Department of Agriculture Forest Service rangelands to be in unsatisfactory condition, while

Forest Service managers estimated more than 40% to be in unsatisfactory condition but did not know the condition of 24 million acres (23%) (U.S. General Accounting Office. *Rangeland Management*). In 1994, the Bureau of Land Management pronounced 57.8% of rangelands static (neither improving nor declining), 10.5% declining, and 13.9% with unknown direction of change (U.S. Department of the Interior. *Rangeland Reform '94*, 3–27, table 3–7).

47. U.S. Department of the Interior Bureau of Land Management. *State of the Public Rangelands* (1990); E. Chaney et al. *Livestock Grazing on Western Riparian Areas* (Eagle, Idaho: Northwest Resource Information Center, Inc., 1990).

48. A National Research Council Committee and the Department of the Interior simultaneously attempted to put rangeland condition estimates on a solid scientific footing (F. E. Busby et al. *Rangeland Health: New Methods to Classify, Inventory, and Monitor Rangelands* (Washington, DC: National Academies Press, 1994); Department of the Interior. *Rangeland Reform '94*; see also L. F. James et al. A New Approach to Monitoring Rangelands. *Arid Land Research and Management* 2003;17:319–328. In 1995 the Society for Range Management called for a focus on maintaining soils to ensure rangeland health (Committee on Unity in Concepts and Terminology. *Society for Range Management* [1995]). In 2000, the Bureau of Land Management attempted to integrate and elaborate the suggestions of the National Research Council, earlier Bureau of Land Management reports, and the concepts of the Society for Range Management (U.S. Department of the Interior Bureau of Land Management. *Interpreting Indicators of Rangeland Health*, version 3 [Technical Reference 1734–6, 2000]).

49. S. Archer and F. E. Smeins. Non-linear Dynamics in Grazed Ecosystems, Thresholds, Multiple Steady States and Positive Feedbacks. In *Is the Range Condition Concept Compatible with Ecosystem Dynamics?* (Spokane, Washington: Society for Range Management, 1992); R. J. Tausch et al. Viewpoint: Plant Community Thresholds, Multiple Steady States, and Multiple Successional Pathways, Legacy of the Quaternary? *Journal of Range Management* 1993;46:439–447. Archer and Smeins cite three major destructive regional ecosystem shifts in recent history: perennial bunchgrass and open sagebrush stands in the Great Basin changed to dense sagebrush and exotic annual grasses, such as cheatgrass; tobosa and black grama grasslands in southwestern desert areas changed to creosote, tarbrush, or mesquite shrublands; and California's Mediterranean grasslands changed from perennial bunchgrasses to annual grasses.

50. Fully characterizing an ecosystem is such a major undertaking that all of the rangeland health evaluation systems, new and old, employ "indicators" of particular land properties, shortcuts thought to sufficiently define ecosystem attributes for management purposes. For example, instead of a complete soil survey, a soil's stability is evaluated through the presence or absence of rills and wind scour features. In all evaluation systems, the observed status of indicators must be compared with the same indicators in a reference area or with an independent standard for validity. The National Research Council's evaluation scheme is a matrix of 12 indicators, which may be rated "healthy," "at risk," or "unhealthy," divided among three "phases," or attributes of the natural systems, such as soil stability and watershed function; distribution of nutrient cycling and energy flow; and recovery mechanisms (Busby et al. *Rangeland Health*, table 4–8). The Bureau of Land Management scheme is a matrix of 17 indicators for soil, vegetation, and some hydrologic factors, evaluated as "extreme," "moderate to extreme," "moderate," "slight to moderate," and "none to slight" (U.S. Department of the Interior Bureau of Land Management. *Interpreting Indicators of Rangeland Health*, appendix 6).

51. A. G. de Soyza et al. Indicators of Great Basin Rangeland Health. *Journal of Arid Environments* 2000;45:289–304.

52. Busby et al. *Rangeland Health*, 13. Suggested alternatives for replacing plant succession as the basis for monitoring rangeland conditions include risk assessment, sustainability, and desertification (N. E. West. Theoretical Underpinnings of Rangeland Monitoring. *Arid Land Research and Management* 2003;17:333–346), but these are at least as vague and unevenly used as succession.

53. The Interior Secretary developed the Bureau of Land Management's Proper Functioning Condition assessment policy in 1995, now expressed in regulations. Important problems include failure to identify and analyze reference (control) sites for comparing natural riparian system functions, lack of a wildlife element, focus on flooding as the only major event affecting riparian habitat, and ignoring insect outbreaks, fire, windstorms, and introduced exotic species. No account is taken of exotic invasion threats to native plant communities or of exotic species distributions. The assessment protocol insufficiently emphasized human impacts and impacts of grazing and trampling on riparian vegetation and stream banks (James Catlin et al. *Critique of BLM's Method to Assess the Health of Riparian and Wetland Areas* [Wild Utah Project, 2001]).

54. Catlin et al. Critique of BLM's Method; L. E. Stevens et al. *Refining Southwestern Riparian Ecosystem Evaluation: A Review and Test of BLM's Proper Functioning Condition Assessment Guidelines* (Draft Report Submitted to Mark Miller, U.S. Bureau of Land Management, Kanab, Utah, 18 April 2001). This report provides a much tighter evaluation system for riparian areas than Bureau of Land Management's and is far better for eliminating bias and making evaluations verifiable, and therefore defensible.

55. Copper in soils varied from 2.10 to 29.78 parts per million (recorded as mg/kg), and cadmium from 0.01 to 0.20 ppm (Richard Aguilar et al. Desertification of Southwestern Rangelands and Rehabilitation Using Municipal Sewage Sludge. In *Making Sustainability Operational—Fourth Mexico/U.S. Symposium* [U.S. Department of Agriculture Forest Service General Technical Report RM 240, 1993], 28–35).

56. U.S. Department of Agriculture. *The New American Farmer* (U.S. Department of Agriculture, 2001), 122–124.

57. Ian McMillan. How the Experts Teach Ranchers to Overgraze. *Earth's Advocate* (Environmental Center of San Luis Obispo County, Calif., August 1976).

58. M. I. Dyer et al. Herbivory and Its Consequences. *Ecological Applications* 1993;3:10–16; S. J. McNaughton. Grasses and Grazers, Science and Management. *Ecological Applications* 1993;3:17–20; Painter and Belsky. Application of Herbivore Optimization Theory, 2–9; D. T. Patten. Herbivore Optimization and Overcompensation, Does Native Herbivory in Western Rangelands Support These Theories? *Ecological Applications* 1993;3:35–36; Noss and Cooperrider. *Saving Nature's Legacy*, 225–226; a comprehensive review is provided in A. J. Belsky. Does Herbivory Benefit Plants? A Review of the Evidence. *American Naturalist* 1986;127:870–892.

59. American bison were rare in the arid western United States, which therefore did not have a history of large herbivore grazing (J. Berger and C. Cunningham. *Bison: Mating and Conservation in Small Populations* [New York: Columbia University Press, 1994]; R. N. Mack and J. N. Thompson. Evolution in Steppe with Few Large, Hoofed Mammals. *American Naturalist* 1982;119:757–773; Noss and Cooperrider. *Saving Nature's Legacy*, 226). Noss and

Cooperrider (p. 220) observe that the traditional song, *Home on the Range*, contains nary a word about livestock.

60. Donahue. *The Western Range Revisited*, chapter 8.

61. U.S. General Accounting Office. *Rangeland Management*, 47.

62. Barry Lopez. *Of Wolves and Men* (New York: Charles Scribner's Sons, 1978). It is difficult to see how "rangeland health" can be sustained, or made sustainable, given the wholesale slaughter of native ecosystem-adapted predators, along with literally millions of other living creatures who happened to be in the way, solely to assure the welfare of cows and sheep, not to mention gross ecosystem changes from introduced exotic animals and plants, and whole-ecosystem conversions to exotic grasslands (chapter 1).

63. U.S. General Accounting Office. Public Rangelands: Some Riparian Areas Restored but Widespread Improvement Will Be Slow (GAO/RCED-88–105,1988).

64. Claire Vitucci. Lawmakers Back Rancher in Grazing-Rights Dispute. *The Riverside Press-Enterprise* (2 May 2001).

65. Recovering altered channel dimensions and sediment distributions is likely to take longer. Lawsuits against the U.S. Forest Service in 1998, which led to removal of 15,000 cattle from 230 miles of rivers and streams in Arizona and New Mexico, will provide good tests of the potential for full recovery.

66. The Bureau of Land Management found 46% of land grazed by its 20 largest permit holders in unsatisfactory condition, contrasting with 27% of all grazing permit lands in unsatisfactory condition (U.S. Department of the Interior Office of Inspector General. *Audit Report: Selected Grazing Lease Activities* [Report No. 92-I-1364, Bureau of Land Management, 1992]). Ten percent of Bureau of Land Management permit-holders control 65% of the livestock on Bureau of Land Management public lands, and 10% of Forest Service permit holders control 49% of cattle on Forest Service public lands (Rogers and LaFleur. Cash Cows). For other assessments of public lands cattle ranching subsidies, see Patricia Wolff [*The Taxpayer's Guide to Subsidized Ranching in the Southwest* (Center for Biological Diversity, 1999]. Available: www.biologicaldiversity.org/swcbd/Programs/grazing/taxguide.html).

67. Wallace Stegner. *Beyond the Hundreth Meridian* (Boston: Houghton Mifflin, 1962; Sentry Edition), 338.

Chapter 4

1. Lewis Mumford. *Technics and Civilization* (New York: Harcourt Brace Jovanovich, 1963).

2. Mineral Information Institute, 475 17th Street, Denver, CO 80202 USA. Available: www.sciencemaster.com/earth/item/mii.php.

3. Stanford's fortune also built the Leland Stanford Junior University on his horse farm near Palo Alto, California. The university memorializes his only child, who died before the age of twenty (Rossiter Johnson, ed. *The Twentieth Century Biographical Dictionary of Notable Americans*, Vol. 9 (Boston: Biographical Society, 1904)).

4. J. E. Young. *Mining the Earth* (Worldwatch Institute Paper 109, 1992); C. W. Montgomery. *Environmental Geology*, 5th ed. (San Francisco: WCB McGraw-Hill, 1997), 287.

5. The Glamis project, California, a highly controversial proposed open-pit gold mine, will have an estimated yield of only 0.002% gold, producing one ounce of gold from 422 tons of mined rock (John Leshy. *Regulation of Hardrock Mining* [Memorandum from the Department of the Interior Solicitor to the Secretary of the Department of the Interior and Acting Director of the Bureau of Land Management, 1999]).

6. There are perhaps 100 large open-pit gold, silver, and copper mines in the western United States. The Kennecott copper mine in Utah is a pit 2.5 miles in diameter and more than 2,600 feet deep (Kennecott Utah Copper. *Bingham Canyon Mine*(April 1997).) When mining is terminated in a few years, the Betze-Post gold pit in Nevada will be nearly two miles long, three-quarters of a mile wide, and 1,800 feet deep; the companion Gold Quarry pit is 1.4 × 1.2 miles across and proposes to reach a depth of 1,800 feet. The Berkeley copper pit in Montana is 1.5 miles long, 1 mile wide, and 2,000 feet deep (C. F. Wilkinson. *Crossing the Next Meridian* [Covelo, California: Island Press, 1992], 28–31).

7. J. L. Sinclair. *The Town That Vanished into Thin Air* (2001). Available: www.zianet.com/snm/santarit.htm.

8. Walter Youngquist. *GeoDestinies* (Portland, Oreg.: National Book Company, 1997), 21. Modern mining, and digging large-scale open-pit mines, in particular, is one of the nation's most energy-intensive industries. Besides fueling ore trucks, major electricity sinks include running pumps for dewatering mines, giant electric shovels, drag lines, and large electric drills (Kennecott Utah Copper. *A World Class Resource, Educator and Tour Guide Information Booklet* [Communication Department, Bingham Canyon Mine, 1997], 4–5).

9. The report's introduction recognizes that "impacts on water quality, vegetation, and aquatic biota often extend beyond the immediate area of a mine site" (National Research Council. *Hardrock Mining on Federal Lands* (Washington, DC: National Academies Press, 1999)). The report also concludes that existing regulations are working, but at the same time cites numerous "regulatory gaps" (regulations that are not working or that need strengthening).

10. At least 557,650 abandoned mines pockmark the United States (Burden of Gilt. *Clementine* [Mineral Policy Center] 1993;1–2). Mines "reclaimed" under present laws add a good many more to the tally. In 1975, the EPA counted 16,188 abandoned or inactive underground mines in the 11 western states: 9% coal, 88% metals, 3% nonmetals. Western national forest lands alone harbor nearly 117,000 abandoned and inactive mines (D. J. Shields et al. *Distribution of Abandoned and Inactive Mines on National Forest System Lands* [General Technical Report RM-GTR-260, U.S. Department of Agriculture Forest Service, Rocky Mountain Forest and Range Experiment Station, 1995]). The nation also contains tens of thousands of miles of abandoned and inactive mine and exploration roads. The National Park Service counts 4,000 abandoned mines and more than 5,000 miles of abandoned mine roads, mostly in the West (Michael Baker, Jr. *Inactive and Abandoned Underground Mines, Water Pollution Prevention and Control* [U.S. Environmental Protection Agency, 1975]).

11. Lizards and beetles apparently came in through ground-level perforations. Warping from the force of driving the pipe into the soil lowered the inside ground level, so smaller animals dropped below the entry hole and could not get out (L. F. LaPre. Wildlife Deathtraps. *Clementine* [Mineral Policy Center] 1990;3–5).

12. L. A. James. Tailings Fans and Valley-Spur Cutoffs Created by Hydraulic Mining. *Earth Surface Processes and Landforms* 2004;29:869–882.

13. John McPhee, cited by Rebecca Solnit. The New Gold Rush. *Sierra Magazine* (July–August 2000) 50–57, 86.

14. Estimates project that every American will consume 3.75 million pounds (1,875 tons) of minerals and fuels in a lifetime, including 1,900 pounds of copper, 23,400 pounds of clay, 30,400 pounds of salt, 1,000 pounds of zinc, and 1.7 million pounds of stone, sand, and gravel, 83,900 gallons of petroleum, 68,900 pounds of cement, 69,100 pounds of other minerals, 6 million cubic feet of natural gas, 42,600 pounds of iron ore, 1,100 pounds of lead, 5,900 pounds of aluminum, 27,800 pounds of phosphate, 589,000 pounds of coal, and 1.8 troy ounces of gold (numbers rounded) (B. W. Pipkin et al. *Geology and the Environment*, 4th ed. [Belmont, California: Brooks/Cole–Thompson Learning, 2005], 345–346; the Mineral Information Institute. Available: www.mii.org/).

15. Thirty years of mining borax from underground workings created more than 200 miles of tunnels. Open-pit mining began in 1957. The pit is now below the water table and pumping relatively small amounts of groundwater. The pit will expand eastward, engulfing the underground workings, for approximately another 1.5 miles and reach depths of about 1,300 feet, severely affecting groundwater.

16. Phosphates in the northwestern United States are within folded and faulted sedimentary rocks of Permian age (260 million years old). Mines following the relatively thin inclined ore beds tend to be long, deep, and narrow (P. J. Lamothe and J. R. Herring. *Selenium and Other Trace Elements in Air Samples Collected Near the Wooley Valley Phosphate Mine Waste Pile, Angus Creek and Little Long Valley, Caribou County, Idaho* [U.S. Geological Survey Open-File Report 00–514, 2000]).

17. The "cone" of depression would be truly conical in shape in homogeneous aquifers. Water movement is controlled by complex fractures, mineralized zones, and the variable character of the rock in the deep aquifers being dewatered; thus, the actual drawdown characteristics are complex (D. K. Maurer et al. *Water Resources and Effects of Changes in Groundwater Use Along the Carlin Trend, North-Central Nevada* [U.S. Geological Survey Water Resources Investigations Report 96–4134, 1996]). The Gold Quarry mining operations are expected to create a five million acre-foot (1.6 trillion gallon) groundwater deficit in the Humboldt Basin, equivalent to 25 years of average flow in the Humboldt River (Maurer et al. *Water Resources and Effects*).

18. The groundwater pumped from deep bedrock aquifers is good quality, but once at the surface it is pumped into infiltration pits, about five to nine acres in size in soils with high soluble salt concentrations. The salts accumulated incrementally over thousands of years, due to insufficient rainfall to flush them below the evapotranspiration soil zone. The large pumped water volume dissolves highly concentrated salts in the soil and carries them into shallow groundwater aquifers. Total dissolved solids have increased more than 10 times in the shallow aquifers, making that water unsuitable for drinking and irrigation. The shallow aquifers have no proven connection with the deep bedrock aquifers (Great Basin Mine Watch. NDEP Renews the Permit for Pipeline Infiltration Basins. *Bristlecone* 2001;2:8–9; Great Basin Mine Watch. Dewatering the Humboldt River Basin. *Bristlecone* 2001;2:10–11.

19. The Humboldt River drains into Humboldt Sink, an area of low elevation with no outlets. Much of the sink contains ephemeral wetlands, a very small remnant of the huge Pleistocene Lake Lahontan (M. C. Reheis et al. Pliocene to Middle Pleistocene Lakes in the Western Great Basin, Ages and Connections. In Robert Hershler et al., eds. *Great*

Basin Aquatic Systems History (Smithsonian Contributions to the Earth Sciences No. 33; Washington, DC: Smithsonian Institution Press, 2002), 53–108.

20. Quoted by Wilkinson, *Crossing the Next Meridian*, 32.

21. W. H. Langer and V. M. Glanzman. *Natural Aggregate, Building America's Future* (U.S. Geological Survey Circular 1110, 1993).

22. U.S. Department of the Interior, Office of Inspector General. *Audit Report, Sand and Gravel Operations, Las Vegas District, Bureau of Land Management* (Report No. 92-I-1347, 1992).

23. J. F. Mount. *California Rivers and Streams: The Conflict Between Fluvial Processes and Land Use* (Berkeley: University of California Press, 1995), 216–225; an elegant book giving lucid, authoritative accounts of myriad problems from conflicts between land use and natural river systems.

24. In the year 1999 alone, the Bureau of Land Management reported more than 290 million cubic feet of materials, excavated under nearly 3,000 permits issued in the 11 western states (U.S. Department of the Interior Bureau of Land Management. *Public Rewards from Public Lands* [2000]).

25. U.S. Environmental Protection Agency. *2000 Toxics Release Inventory* (Public Data Release, EPA 260-R-02–003, Office of Environmental Information 2002), tables 4–3, 4–5, 4–8, and 5–3. Available: www.epa.gov/tri.

26. Caroline Ash and Richard Stone. A Question of Dose. *Science* 2003;300:925; and additional papers, *Science* 2003;300:926–947.

27. J. N. Moore and S. N. Luoma. Hazardous Wastes from Large-Scale Metal Extraction. *Environmental Science and Technology* 1990;24:1279–1285; a brief report, with excellent overview of many problems associated with toxic element releases during metal extraction.

28. Moore and Luoma. Hazardous Wastes; E. V. Axtmann and S. N. Luoma. Large-Scale Distribution of Metal Contamination in the Fine-Grained Sediments of the Clark Fork River, Montana, U.S.A. *Applied Geochemistry* 1991;6:75–88. The most common metal mineral is iron sulfide, also called iron pyrite and "fools gold." The sulfide ores occupy relatively soft seams in hard igneous and metamorphic rocks, surrounded by relatively barren "wall" rocks, which still can contain 30% and more sulfide minerals.

29. The "fate" of a chemical in the environment (chapter 13) depends on natural chemical transformations that release it from minerals or rock material, how it is transported, where it is deposited, and what happens to it during and after deposition (J. N. Moore et al. Partitioning of Arsenic and Metals in Reducing Sulfidic Sediments. *Environmental Science and Technology* 1988;22:432–437).

30. Natural oxidizing processes produce acid waters where sulfide-bearing ore veins cross the surface of a mountainside, for example. When iron-sulfide–rich mineralized rock interacts with oxygen or water (oxidation), it becomes streaked and mottled with red and yellow minerals, the "colors" that prospectors look for. Bacterial action facilitates the interactions between water and sulfide ores, yielding sulfate and metal and hydrogen ions. In water, the sulfate forms highly corrosive hydrogen sulfate (sulfuric) acid (Moore and Luoma. Hazardous Wastes, 1280). Coal deposits also contain iron sulfides (mainly pyrite and marcasite, a ferric iron sulfide) and yield acid mine drainage.

31. D. Z. Piper et al. *The Phosphoria Formation at the Hot Springs Mine in Southeast Idaho, a Source of Selenium and Other Trace Elements to Surface Water, Ground Water, Vegetation,*

and Biota (U.S. Geological Survey Open-File Report 00–050, 2000). Experiments with mosses (bryophytes) exposed to water seeping from a mine dump for 10 days indicated that selenium increased by three orders of magnitude (J. R. Herring et al. *Chemical Composition of Deployed and Indigenous Aquatic Bryophytes in a Seep Flowing from a Phosphate Mine Waste Pile and in the Associated Angus Creek Drainage, Caribou County, Southeast Idaho* [U.S. Geological Survey Open-File Report 01–0026, 2001]). Phosphoria Formation mine dumps also release other potentially toxic elements, including cadmium, chromium, copper, molybdenum, vanadium, uranium, and zinc (U.S. Geological Survey. *Western Phosphate Field, U.S.A., Science in Support of Land Management* [Fact Sheet FS-100–02, 2002]).

32. Moore and Luoma. *Hazardous Wastes from Large-Scale Metal Extraction.*

33. The exposed surface area of broken rock is much greater than the same volume of unbroken rock still in the ground. Potentially toxic materials are disseminated throughout the rock materials, so breaking rocks exposes them more to leaching by rainwater. For this reason, broken rocks produce far greater amounts of acid mine drainage than the same volume of rock still in the ground. Tailings are ore material wastes, originally more concentrated in toxic materials than mine waste rock, and they also contain toxic mercury, sulfuric acid, or cyanide chemicals used for extracting the target minerals. Therefore, the amounts of toxic chemicals in tailings are very much greater than in broken waste rock. Tailings were ground to small sizes for treating and extracting the ore minerals, hugely increasing the surface area and making them even more vulnerable to leaching and dispersal in the environment than crudely broken waste rocks.

34. Francine Madden and Bettina Camcigil. *TRI Toolkit, Using the Toxics Release Inventory to Promote Environmentally Responsible Mining in Your Community* (Mineral Policy Center, 2000), 10. The Environmental Protection Agency requires annual reports on potential nonmetal contaminants used in ore processing, as follows: acrylamide, ammonia, benzene, bromine, bromoform, chlorine, cresols, cyanide compounds, cyclohexane, ethyl benzene, formaldehyde, glycol ethers, hydrazine, hydrochloric acid, naphthalene, nitric acid, phenol, phosphoric acid, propylene, sulfuric acid, thiourea, toluene, and xylene.

35. Mike Medberry. As the Tailings Tumble. *Clementine* (Mineral Policy Center) 1997;8–9. Near Atlanta, Idaho, the Talache Mine's 80-foot-tall earthen tailings impoundment collapsed in 1997, sending a wall of toxic mud and water nearly a mile downstream into Montezuma Creek and the Middle Fork of the Boise River. The EPA estimates that the slide released 70,000 cubic yards of contaminated mud, which invaded a pine forest and inundated a unique 60-acre wetland meadow. The long-abandoned mine is one of many that threaten landslides and pollution hazards (Luoma et al. Mining Impacts, 1279–1285; National Research Council. *Coal Waste Impoundments, Risks, Responses, and Alternatives* [Washington, DC: National Academies Press, 2002]).

36. U.S. Geological Survey. *Monitoring the Effects of Ground-Water Withdrawals from the N Aquifer in the Black Mesa Area, Northeastern Arizona* (Fact Sheet 064–99, 1999); Margot Truini et al. *Ground-Water, Surface-Water, and Water-Chemistry Data, Black Mesa Area, Northeastern Arizona: 2003–04* (U.S. Geological Survey Open-File Report 2005–1080, 2005). From 2001 to 2002, slurries dominated the uses of the 4,530 acre-feet (about 1.5 billion gallons) of industrial water withdrawn. The water withdrawals ceased at the end of 2005 with closure of the Laughlin power plant.

37. Pipkin et al. *Geology and the Environment,* 401–402.

38. Mineral Policy Center. *Six Mines, Six Mishaps* (Mineral Policy Center Report No. 1, Regulatory Reform, 1999).

39. D. J. Cain and S. N. Luoma. Benthic Insects as Indicators of Large-Scale Trace Metal Contamination in the Clark Fork River, Montana. In G. E. Mallard and D. A. Aronson, eds. *U.S. Geological Survey Toxic Substances Hydrology Program, Proceedings of the Technical Meeting, Monterey, California, March 11–15* (U.S. Geological Survey Water-Resources Investigations Report 91–4034, 1991), 525–529.

40. T. J. Coulthard and M. G. Macklin. Modeling Long-Term Contamination in River Systems from Historical Metal Mining. *Geology* 2003;31:451–454.

41. J. N. Moore et al. Downstream Effects of Mine Effluent on an Intermontane Riparian System. *Canadian Journal of Fish and Aquatic Sciences* 1991;48:222–232.

42. J. E. Gray, ed. *Geologic Studies of Mercury by the U.S. Geological Survey* (U.S. Geological Survey Circular 1248, 2003); C. N. Alpers and M. P. Hunerlach. *Mercury Contamination from Historic Gold Mining in California* (U.S. Geological Survey Fact Sheet FS-061–00, 2000); R. K. R. Ambers and B. N. Hygelund. Contamination of Two Oregon Reservoirs by Cinnabar Mining and Mercury Amalgamation. *Environmental Geology* 2001;40:699–707; B. N. Hygelund et al. Tracing the Source of Mercury Contamination in the Dorena Lake Watershed, Western Oregon. *Environmental Geology* 2001;40:853–859.

43. Pipkin et al. *Geology and the Environment*, 366–368.

44. S. A. Brackett. Bittersweet Victory. *Clementine* (Mineral Policy Center) 1996;4–5.

45. "Hard" rocks are igneous (originally molten) rocks, such as granite or volcanic rhyolite, and metamorphic rocks, which have been affected by changed conditions of heat or pressure.

46. The courts found the 1872 Mining Law extraordinarily difficult to interpret and over nearly 100 years repeatedly recommended that amending it would be beneficial to the mining industry (J. D. Leshy. *The Mining Law: A Study in Perpetual Motion* [Washington, DC: Resources for the Future, 1987] chapter 14.; U.S. General Accounting Office. *Mining Law Reform and Balanced Resource Management* [EMD-78–93, 1979]). The 2002 Mineral Exploration and Development Act represents a new effort to reform the law, in effect seeking to reinstate the Clinton administration regulation reforms under section 43 CFR 3809 (commonly referred to as the 3809 regulations), which the George W. Bush administration largely nullified (Mineral Policy Center. *Two New Bills in Congress Will Protect Water and Public Lands* [2002], 1, 3).

47. Leshy. *The Mining Law*, 287.

48. The numerous giveaways include 2,701 acres of public lands, containing more than $15.8 billion worth of minerals, sold for $13,095 between 1994 and 1996 (S. A. Brackett. Rock-Bottom Prices. *Clementine* [Mineral Policy Center] 1996;1–3). The General Mining Law of 1872 allows individuals or companies that discover valuable minerals on public lands to purchase (patent) those lands for $2.50–5.00 per acre. Since 1872, mining interests have patented a land area equivalent in size to the state of Connecticut, containing minerals worth more than $245 billion. Congress passed a moratorium on patenting in 1994, which has been renewed annually since 1994. About 55 patent claims were grandfathered during the moratorium. The George W. Bush administration appears intent on granting these claims— the most recent (June 2004) is sale of 155 acres on Mt. Emmons, three miles from the ski resort town of Crested Butte, Colorado, to multinational Phelps Dodge Corporation for $775. The land in that part of the state goes for about $1,000,000 per acre. In an annual exercise

in futility, a new mining law reform bill, the Mineral Exploration and Development Act of 2003, called for a permanent ban on patents, and again met formidable resistance from western Congressional representatives (Mineral Policy Center. *The Rahall-Shays-Inslee Mining Reform Bill* (2003). Available: www.mineralpolicy.org).

49. Wilkinson. *Crossing the Next Meridian*, 33; U.S. General Accounting Office. *Modernization of 1872 Mining Law Needed to Encourage Domestic Mineral Production, Protect the Environment, and Improve Public Land Management* (B-118678, 1974); U.S. General Accounting Office. *Federal Land Management, Unauthorized Activities Occurring on Hardrock Mining Claims* (GAO/RCED-90–111, 1990). Dislodging illegal squatters is very difficult and expensive.

50. David Sheridan. *Hard Rock Mining on the Public Land* (Washington, DC: Council on Environmental Quality, 1977), 11. Families can collect up to 25 pounds of pinyon nuts annually without a permit on Bureau of Land Management lands; the 30 tons or so gathered commercially in good years require a permit (Jim Baca. Our Public Lands, the National Perspective. *Newsbeat* [Bureau of Land Management] October 1993, 1–8).

51. Leshy. *The Mining Law*, 85. The American Mining Congress, National Mining Association, and other industry lobbies' outcries against any change in the Mining Law—"This way of life must be preserved," "It is the quintessential American West," "Reform will ruin local economies and destroy communities," "Small miners are the nation's backbone"—are nearly identical to protests against grazing reform. While admitting that past mining (grazing) practices damaged land, they claim that miners (graziers) now are good land stewards.

52. Keith Knoblock. The Mining Law Must Be Allowed to Work for a Long Time to Come. *Journal of the American Mining Congress* 1989;75:6–8. In 1976, the American Mining Congress surveyed 41 large mining companies to try to demonstrate the important role of small miners in domestic mineral exploration. The survey credited a small miner for one mineable discovery, but later inquiries showed that the prospector had merely staked speculative paper claims between two known deposits without further exploration. A larger company obtained the right to drill these claims and eventually discovered the mineable resource (Leshy. *The Mining Law*, 85).

53. The 1970s estimates suggest that small miners provide less than 1% of Nevada's annual mineral production, and a small miner's chance of selling his claim to a large company, capable of modern mining techniques, was about one in 2,000 (Sheridan. *Hard Rock Mining*).

54. Environmental Working Group. *Who Owns the West? Section 3. Foreign Control of Mining Rights on Public Lands*. Available: www.ewg.org/mining/report/index.php (accessed May 2004).

55. House Subcommittee on Energy and Mineral Resources, *Oversight Hearing on Mining Regulatory Issues and Improving the General Mining Laws*, 106th Congress (August 1999).

56. Leshy. *The Mining Law*, chapter 4. Principal restraints on free access to federal lands include the court-supported executive privilege, allowing U.S. presidents to withdraw lands from mining entry; the 1920 Mineral Leasing Act that removed some minerals, including oil and gas, from the Mining Law; and the passage of the Federal Land Management and Policy Act in 1976, giving the Bureau of Land Management authority to deny applications for mining to protect other resources, never exercised until 2002.

57. In 1992 Congress enacted an annual maintenance, or holding, fee of $100 for unpatented mining claims. Earlier, claim holders merely had to show $100 worth of

improvement or exploration on a claim each year, which led to rampant land fraud. When first enacted, the maintenance fee cut the total number of mine claims in half. The fiscal year interior appropriation bill renewed the $100 per year claim maintenance fee, through 2003 (Mineral Policy Center. Claim Maintenance. *MineWire* 2002;5[2]).

58. The 3809 regulations govern mining on lands under Bureau of Land Management jurisdiction. Established in 1981 before onset of heap-leach gold mining, by 2000 the regulations needed much revision. Ironically, the Clinton administration revisions finally went into effect on 20 January, 2001, George W. Bush's inauguration day. Among the new regulations' many reforms is a requirement for bonding all mine operators at a sufficient level to pay for real reclamation. The Bureau of Land Management has long held the authority to require bonds for reclamation but generally has done so only for operators with a record of non-compliance (U.S. General Accounting Office. *Public Lands: Interior Should Ensure Against Abuses from Hardrock Mining* [GAO-RCED-86–48, 1986]; U.S. Department of the Interior Bureau of Land Management. *Final Environmental Impact Statement, Surface Management Regulations for Locatable Mineral Operations* [2000]; Jim Baca. 1872 Mining Law, Time for Reform. *Geotimes* 1991;6).

59. Glamis Gold Ltd. filed more than 5,000 claims on 10,500 acres in western Imperial County, of southeastern California, including many traditional Quechan (*kay-shann*) tribe sites of religious instruction and retreat. The mining claims encompass southern parts of the Trail of Dreams, connecting lower Colorado River Quechan lands with northerly Spirit Mountain, which Quechan, Chemehuevi, Cocopah, and Mojave tradition identifies as those peoples' place of origin. The most critical sites for tribal rites since time immemorial provide excellent views of Spirit Mountain. In 2003 California's governor signed into law a requirement that any mined site on Quechan lands must be fully restored, making the Glamis project potentially uneconomic. Subsequently, Glamis Gold submitted a claim to arbitration under NAFTA agreements claiming that federal government actions and the California law effectively expropriate its investments (U.S. State Department. Available: www.state.gov/s/l/c10986.htm).

60. U.S. Department of the Interior Bureau of Land Management. *Record of Decision for the Imperial Project Gold Mine Proposal, Imperial County, California* (Case File No. 670–41027, 2001). The 1976 Federal Land Management and Policy Act gave the Bureau of Land Management explicit authority to weigh other land values against mineral extraction, and unambiguously identifies minerals as one of the resources for consideration "so that resources will be utilized in the combination that will best meet the present and future needs of the American people." The U.S. Forest Service has much the same authority under the 1974 Forest and Rangeland Renewable Resources Planning Act (Leshy. The Mining Law, 199–205). Careful consideration and conservative decisions seem little to ask for when exploiting one resource is likely to destroy another.

61. Mineral Policy Center. *The Norton Mining Rule* (Mineral Policy Center Fact Sheet, 2001). A federal court ruled in 2003 that the Bureau of Land Management does have the authority to balance other interests against the inevitable destruction caused by mining, but by 2004 the George W. Bush administration was not doing more to protect the public's interest (U.S. Department of the Interior Bureau of Land Management. 2001. Proposed Rule 43 CFR Part 3800. *Federal Register* 66(210):54863–54870; U.S. Department of the Interior Bureau of Land Management. Final Rule 43 CFR Part 3800. *Federal Register* 2001;66[210]:54834–54862).

62. The Bureau of Land Management claimed that the 2001 Final Rule left the 2000 rules "essentially intact" and left the 2000 rule on bonding requirements unchanged, requiring that the bond had to be sufficient for the Bureau of Land Management to hire an independent contractor for reclaiming disturbed land. In actuality, the Final Rule ties the bonding level to performance standards and substantially waters down performance requirements—so the bond is not very likely to pay for as much reclamation as the 2000 rule. The new Final Rule also abandoned the 2000 rule provision that anyone standing to financially benefit from a mining claim could be held directly responsible for environmental damages from mining that claim, to protect taxpayers from having to pay big cleanup costs. Those 2000 rule provisions also would have obstructed the widespread practice of creating a "sub-corporation" for a variety of operations, which could declare bankruptcy to protect the "parent corporation" if the costs of cleanup became too burdensome. The 2001 Final Rule even eliminated the fines established under civil penalties to punish really irresponsible operators (Mineral Policy Center, *The Norton Mining Rule*).

63. Concluded Judge Henry H. Kennedy, "It is clear that mining operations have highly significant, and sometimes devastating, environmental consequences.…[T]he 2001 regulations prioritize the interests of miners over the public interest.…[S]uch prioritization may well constitute unwise and unsustainable policy." His decision is unlikely to affect what happens on the ground until administrative attitudes change.

64. J. R. Kuipers. *Hardrock Reclamation Bonding Practices in the Western United States* (Boulder, Colorado: National Wildlife Federation, 2000). This excellent report is the most comprehensive one available on the inadequacies of bonding regulations for reclaiming western hardrock mines. Under the Surface Mining and Reclamation Act, local governments generally approved reclamation plans and environmental review of the plans. This proved very unsatisfactory. In 2005, the California Supreme Court granted the state authority to oversee local decisions on the adequacy of reclamation plans.

65. Baca. *1872 Mining Law: Time for Reform*, 6.

66. V. L. Ketellapper et al. The Mining History and Environmental Clean-up at the Summitville Mine. In *Hydrogeology of the San Luis Valley and Environmental Issues Downstream from the Summitville Mine* (Geological Society of America Field Guide, 1996); J. E. Gray and G. S. Plumlee. Geologic Characteristics of the Summitville Mine and Their Environmental Implications. In *Hydrogeology of the San Luis Valley and Environmental Issues Downstream from the Summitville Mine* (Geological Society of America Field Guide, 1996). The compositions of these mineralized rocks have little capacity to moderate (buffer) the acid levels in the mine waters.

67. Grand Canyon Trust Projects. 2000. *Atlas Mine Tailings, Analysis of Recent Studies*. Available: beta3.c-t-g.com/science.htm; National Research Council. *Remedial Action at the Moab Site, Now and for the Long Term* (Washington, DC: National Academies Press, 2002); Environment News Service. Moab Mine Dump Cleanup Plan Gets Reassessed. *Environment* 2000. Available: ens.lycos.com.

68. P. E. Mariner et al. Fingerprinting Arsenic Contamination in the Sediments of the Hylebos Waterway, Commencement Bay Superfund Site, Tacoma, Washington. *Environmental and Engineering Geoscience* 1997;3:359–368.

69. H. P. Metzger. *The Atomic Establishment* (New York: Simon and Schuster, 1972), 171–194.

70. Extensive western U.S. coal deposits are flat-lying, appropriate for area stripping, but most are deeply buried and will have to be mined by underground methods, if at all.

71. Public Employees for Environmental Responsibility. *Empty Promise* (White Paper No. 17, 1997). This report was written with direct assistance from employees of the Office of Surface Mining.

72. Nothing in the publicly accessible record indicates whether an "inspection" was only a drive-by or other abbreviated approach, or whether the mine operator was notified in advance of the inspection (e.g., U.S. Department of the Interior, Office of Surface Mining. *20th Anniversary Surface Mining Control and Reclamation Act, a Report on the Protection and Restoration of the Nation's Land and Water Resources Under the Surface Mining Law, Part 2: Statistical Information* [1999]). The fine print for the fiscal year 2002 budget also showed significant curtailments in U.S. Environmental Protection Agency inspections, investigations, and enforcement actions under the Surface Mining and Reclamation Act.

73. Final rules for Section 404 of the Clean Water Act, both for the Corps of Engineers and the EPA, were published 9 May 2002, as "clarifications" with far-reaching implications. The rules appeared to be the industry-friendly George W. Bush administration's response to lawsuits against mountain-top coal removal practices charging that they violate both the Clean Water Act and mining regulations prohibiting unnecessary and undue degradation of the public's land and water. The new rules appear to violate both the Clean Water Act's stated goals and mining regulations. Although Army Corps of Engineers permits formerly forbade using mining and other waste materials for fill, the so-called clarifications allow mining overburden, slurry, or tailings disposal in United States waters. The April 2000 proposed rules referred only to coal mine wastes, but the Final Rules extend to all mining wastes, including known hazards to the environment and human health, and permit radical modification of landscapes and waterways. Many sites that would qualify under the new disposal rules are Superfund sites (U.S. Department of the Army, Corps of Engineers and U.S. Environmental Protection Agency. Final Revisions to the Clean Water Act, Regulatory Definitions of "Fill Material" and "Discharge of Fill Material." 33 CFR Part 323 and 40 CFR Part 232. *Federal Register* 2002;67:31129–31143).

74. Montgomery. *Environmental Geology*, 286; Grecia Matos and Lorie Wagner. *Consumption of Materials in the United States, 1900–1995* [U.S. Geological Survey]1999. Available: pubs.usgs.gov/annrev/ar-23–107/.

75. R. B. Gordon et al. Metal Stocks and Sustainability. *Proceedings of the National Academy of Sciences of the USA* 2006;103:1209–1214.

76. Youngquist. *GeoDestinies*, 21.

77. S. N. Luoma. Bioavailability of Trace Metals to Aquatic Organisms, a Review. *Science of the Total Environment* 1983;28:1–22; D. J. Beltman et al. Benthic Invertebrate Metals Exposure, Accumulation, and Community-Level Effects Downstream from a Hard-Rock Mine Site. *Environmental Toxicology and Chemistry* 1999;18:299–307.

78. Solnit. The New Gold Rush, 50–57, 86.

79. The mining industry succeeded in removing these wastes from the Toxics Release Inventory by convincing a judge to rule for them. A change back to listing the mining waste rocks might be sought through legal action or political pressure.

80. In fighting off backfilling requirements, mining companies argue that slightly mineralized pit walls should be preserved for a future time when they could be mined economically, and that the waste volumes are greater than needed for backfilling.

81. Eighty-four percent of world gold serves nonessential purposes (manufacturing jewelry). In addition, central banks and international institutions hold more than 33,000 tons of gold reserves (8,600 tons in the United States). These gold reserves represent more than 13 times the annual mine production worldwide and could satisfy gold demand for eight years, including for coins and jewelry (U.S. Geological Survey. *Mineral Commodity Summaries 2001*, 71). Holding all this gold off the market promotes more gold mining and environmental damage (J. E. Young. *Gold, at What Price?* [Mineral Policy Center, February 2000], 5. Available: www.mpc@mineralpolicy.org. Mineral Policy Center now is called Earthworks.

82. Solnit. *The New Gold Rush*, 57.

Chapter 5

1. U.S. Department of Agriculture Forest Service. *1998 Report of the Forest Service: Performance Highlights of the Natural Resource Agenda* (1999). Available: www.fs.fed.us/pl/pdb/98report/06_resource_agenda_highlights.html).

2. Harvard Sociology Professor Orlando Patterson surveyed 1,500 ordinary Americans in 2001 to find out how they defined "freedom". Overwhelmingly, those polled think of freedom as doing what you want, and especially moving about freely. The most frequent examples of experiencing freedom were expressed in terms of driving a car or moving from state to state, rather than speech or civic functions (Orlando Patterson. A "Sick" Democracy, John Harvard's Journal. *Harvard Magazine* 2001;103:70).

3. Pipeline corridors include 325,000 miles of mainly interstate natural gas transmission lines, 1.7 million miles of mainly intrastate natural gas distribution lines, and 156,000 miles of mainly interstate hazardous liquid pipelines (U.S. General Accounting Office. *Pipeline Safety* [GAO-01–1075, 2001]; U.S. General Accounting Office. *Pipeline Safety and Security* [GAO-02–785, 2002]); American Petroleum Institute. *Basic Petroleum Data Book* (Petroleum Industry Statistics, 1999); U.S. Energy Information Administration. *Deliverability on the Interstate Natural Gas Pipeline System, Appendix B, Natural Gas Pipeline and System Expansions, 1997–2000* (1999), 119–130), electrical transmission lines (U.S. Department of Agriculture, Rural Utilities Services. *1997 Statistical Report, Rural Electric Borrowers* [Informational Publication 201–1, 1998], 14, 37), and canals (C. F. Wilkinson. West's Grand Old Water Doctrine Dies. In Char Miller, ed. *Water in the West* [reprint; Corvallis Oregon: Oregon State University Press, 2000], 16–27). As of November 2006, a crash program was under way to designate new and wider utility corridors in the 11 western states (West-wide Energy Corridor Programmatic Environmental Impact Statement). Available: corridoreis.anl.gov/.

4. Estimates suggest that U.S. public roads in the 48 contiguous states directly and indirectly affect as much as 20% of the land area (about 590,000 square miles) (R. T. T. Foreman. Estimates of the Area Affected Ecologically by the Road System in the United States. *Conservation Biology* 2000;14:31–35; M. C. Larsen and J. E. Parks. How Wide Is a Road? The Association of Roads and Mass-Wasting in a Forested Montane Environment. *Earth Surface Processes and Landforms* 1997;22:835–848; Chuck Cottrell. Roads and Habitat Fragmentation. *Road-RIPorter* [Wildlands Center for Preventing Roads] 1997;2:12–13; R. A. Reed et al. Contributions of Roads to Forest Fragmentation in the Rocky Mountains. *Conservation Biology* 1996;10:1098–1106).

5. R. T. T. Foreman et al. Ecological Effects of Roads, Toward Three Summary Indices and an Overview for North America. In K. Canters, ed. *Habitat Fragmentation and Infrastructure* (Delft, The Netherlands: Ministry of Transport, Public Works, and Water Management, 1997), 40–54. This estimate is prorated from the national figure. Western roads likely do not carry the level of traffic that Foreman et al. assumed for eastern roads.

6. The U.S. Forest Service estimates it has jurisdiction over 60,450 miles of "unclassified" roads (U.S. Department of Agriculture Forest Service. *Forest Service Roadless Area Conservation Final Environmental Impact Statement* [2000], table 3–5). Off-road recreational vehicles continually create "ghost roads," so this figure is likely to be substantially underestimated. David Havlick summarizes road types and agency jurisdictions (D. G. Havlick. *No Place Distant* [Covelo, California: Island Press, 2002], table 1–1). This excellent book contains a wealth of information of road impacts on public lands.

7. S. C. Trombulak and C. A. Frissell. Review of Ecological Effects of Roads on Terrestrial and Aquatic Communities. *Conservation Biology* 2000;14:18–30.

8. U.S. Department of Agriculture. *Forest Service Roads, a Synthesis of Scientific Information* (1999). This report provides a valuable summary and comprehensive references to the literature on both benefits and environmental effects of forest roads (see also L. H. MacDonald and J. D. Stednick. *Forests and Water: A State-of-the-Art Review for Colorado* (CWRRI Completion Report No. 196, Colorado Water Resources Research Institute, 2003).

9. U.S. Department of Agriculture. *Forest Service Roads*, 24; W. E. Weaver et al. Magnitude and Causes of Gully Erosion in the Lower Redwood Creek Basin, Northwestern California. In K. M. Nolan et al., eds. *Geomorphic Processes and Aquatic Habitat in the Redwood Creek Basin, Northwestern California* (U.S. Geological Survey Professional Paper 1454, 1995), I1–I21; B. C. Wemple et al. Channel Network Extension by Logging Roads in Two Basins, Western Cascades, Oregon. *Water Resources Bulletin* 1996;32:1195–1207; J. G. King and L. C. Tennyson. Alteration of Streamflow Characteristics Following Road Construction in North Central Idaho. *Water Resources Research* 1984;20:1159–1163; D. R. Montgomery. Road Surface Drainage, Channel Initiation, and Slope Instability. *Water Resources Research* 1994;30:1925–1932.

10. Jacky Croke and Simon Mockler. Gully Initiation and Road-to-Stream Linkage in a Forested Catchment, Southeastern Australia. *Earth Surface Processes and Landforms* 2001;26:205–217.

11. J. L. La Marche and D. P. Lettenmaier. Effects of Forest Roads on Flood Flows in the Deschutes River, Washington. *Earth Surface Processes and Landforms* 2001;26:115–134.

12. R. D. Harr and R. A. Nichols. Stabilizing Forest Roads to Help Restore Fish Habitats, a Northwest Washington Example. *Fisheries* 1993;18:18–22.

13. G. S. Plumlee and T. L. Ziegler. The Medical Geochemistry of Dusts, Soils, and Other Earth Materials. In B.S. Lollar, ed. *Treatise on Geochemistry*, vol. 4 (New York: Elsevier, 2003), 263–310; M. W. Bultman et al. An Overview of the Ecology of Soil-Borne Human Pathogens. In O. Selinus, ed. *Medical Geology: Earth Science in Support of Public Health Protection* (New York: Elsevier, Academic Press, 2003). C. R. Leathers. Plant Components of Desert Dust in Arizona and Their Significance for Man. In T. L. Péwé, ed. *Desert Dust, Origin, Characteristics, and Effect on Man* (Geological Society of America Special Paper 186, 1981), 191–206.

14. Dust suppressants include, among others, surfactants, which are wetting agents that last a short time and require frequent applications; adhesives, such as tree sap; petroleum products; chloride salts; and electrochemical stabilizers made from sulfonated petroleum.

All seal the road surface to some degree, which increases runoff from slopes and therefore increases erosion downslope (Lance Frazer. Down with Road Dust. *Environmental Health Perspectives* 2003;111:A892–A895).

15. B. C. Wemple et al. Forest Roads and Geomorphic Process Interactions, Cascade Range, Oregon. *Earth Surface Processes and Landforms* 2001;26:191–204.

16. John McPhee. Los Angeles Against the Mountains. In *The Control of Nature* (New York: Noonday Press, 1989), 185–191; R. L. Beschta. Long-Term Patterns of Sediment Production Following Road Construction and Logging in the Oregon Coast Range. *Water Resources Research* 1978;14:1011–1016.

17. P. J. B. Fransen et al. Forest Road Erosion in New Zealand, Overview. *Earth Surface Processes and Landforms* 2001;26:165–174; D. E. McClelland et al. *Assessment of the 1995 and 1996 Flood and Landslides on the Clearwater National Forest, Part I, Landslide Assessment* (U.S. Department of Agriculture Forest Service, 1997); C. M. Falter and Craig Rabe. *Assessment of the 1995 and 1996 Flood and Landslides on the Clearwater National Forest, Part II, Stream Response* (U.S. Department of Agriculture Forest Service, 1997).

18. W. M. Brown. Historical Setting of the Storm, Perspectives on Population, Development, and Damaging Rainstorms in the San Francisco Bay Region. In S. D. Ellen and G. F. Wieczorek, eds. *Landslides, Floods, and Marine Effects of the Storm of January 3–5, 1982, in the San Francisco Bay Region, California* (U.S. Geological Survey Professional Paper 1434, 1988), 12.

19. R. C. Sidle et al. Hillslope Stability and Land Use. *American Geophysical Union, Water Resources Monograph Series* 1985;11:72–119; F. J. Swanson and C. T. Dyrness. Impact of Clear-Cutting and Road Construction on Soil Erosion by Landslides in the Western Cascade Range, Oregon. *Geology* 1975;3:393–396; D. N. Swanston and F. J. Swanson, Timber Harvesting, Mass Erosion, and Steepland Forest Geomorphology in the Pacific Northwest. In D. R. Coates, ed. *Geomorphology and Engineering* (Stroudsburg, Pennsylvania: Dowden, Hutchinson, and Ross, 1976), 199–221.

20. Swanson and Dyrness. Impact of Clear-Cutting, 393–396; H. G. Wilshire et al. *Geologic Processes at the Land Surface*. U.S. Geological Survey Bulletin 2149, 1996, 22.

21. U.S. Department of Agriculture Forest Service. Forest Service Roads: A Synthesis of Scientific Information (8 March 1999) 22.

22. T. A. Black and C. H. Luce. Changes in Erosion from Gravel Surfaced Forest Roads Through Time. *Proceedings of the International Mountain Logging and 10th Pacific Northwest Skyline Symposium, March 28–April 7, Corvallis, Oregon* (1999), 204–218; C. H. Luce and T. A. Black. Spatial and Temporal Patterns in Erosion from Forest Roads. *Water Science and Application* 2001;2:165–178.

23. C. H. Luce and T. A. Black. Sediment Production from Forest Roads in Western Oregon. *Water Resources Research* 1999;35:2561–2570; U.S. General Accounting Office. *Oregon Watersheds, Many Activities Contribute to Increased Turbidity During Large Storms* (GAO/RCED-98-220, 1998).

24. J. L. Florsheim et al. Effect of Baselevel Change on Floodplain and Fan Sediment Storage and Ephemeral Tributary Channel Morphology, Navarro River, California. *Earth Surface Processes and Landforms* 2001;26:219–232.

25. T. M. Wood. *Herbicide Use in the Management of Roadside Vegetation, Western Oregon, 1999–2000, Effects on the Water Quality of Nearby Streams* (Water-Resources Investigations Report 01–406, U.S. Geological Survey, 2001).

26. T. A. Zink et al. The Effect of a Disturbance Corridor on an Ecological Reserve. *Restoration Ecology* 1995;3:304–310; C. H. Greenberg et al. Roadside Soils, a Corridor for Invasion of Xeric Scrub by Nonindigenous Plants. *Natural Areas Journal* 1997;17:99–109; J. L. Gelbard and Jayne Belnap. Roads as Conduits for Exotic Plant Invasions in a Semiarid Landscape. *Conservation Biology* 2003;17:420–432.

27. Karen Wood. Roads and Toxic Pollutants. *Road-RIPorter* (Wildlands Center for Preventing Roads) 1998;3:10–11; J. D. Balades et al. Chronic Pollution of Intercity Motorway Runoff Water. *Water Science and Technology* 1985;17:1165–1174.

28. G. M. Filippelli et al. Urban Lead Poisoning and Medical Geology: An Unfinished Story. *Geological Society of America, GSA Today* 2005;15:4–11.

29. Vehicle-deposited platinum-group elements (rhodium, platinum, and palladium) are found with nickel, copper, zinc, and lead and reach concentrations of 64–73 parts per billion in roadside soils. Platinum levels remain higher than background 150 feet from the roadways. Platinum is the only one of the group that plants can absorb through their roots (J. C. Ely et al. Implications of Platinum-Group Element Accumulation Along U.S. Roads from Catalytic-Converter Attrition. *Environmental Science and Technology* 2001;35:3816–3822).

30. Department of Transportation Office of Pipeline Safety. *Hazardous Liquid Pipeline Operators Accident Summary Statistics by Year, 1/1/1986–12/31/2002.* Available: http://ops.dot. gov; Environmental Defense. *Increase in Oil Pipeline Accidents Noted* (1999). Available: www. environmentaldefense.org; U.S. General Accounting Office. *Pipeline Safety* (2001), 3.

31. F. C. Vasek et al. Effects of Pipeline Construction on Creosote Bush Scrub Vegetation in the Mojave Desert. *Madroño* 1975;23:1–13.

32. H. B. Johnson et al. Productivity, Diversity and Stability Relationships in Mojave Desert Roadside Vegetation. *Bulletin of the Torrey Botanical Club* 1975;102:106–115. The enhanced vegetation attracts larger insect populations than the plants farther away from the roadways and corridors (D. C. Lightfoot and W. G. Whitford. Productivity of Creosote Bush Foliage and Associated Canopy Arthropods Along a Desert Roadside. *American Midland Naturalist* 1991;125:310–322).

33. Christina Tague and Larry Band. Simulating the Impact of Road Construction and Forest Harvesting on Hydrologic Response. *Earth Surface Processes and Landforms* 2001;26:135–151.

34. W. H. Schlesinger and C. S. Jones. The Comparative Importance of Overland Runoff and Mean Annual Rainfall to Shrub Communities of the Mojave Desert. *Botanical Gazetteer* 1984;145:116–124; W. H. Schlesinger et al. Biological Feedbacks in Global Desertification. *Science* 1990;247:1043–1048.

35. Trombulak and Frissell. Review of Ecological Effects of Roads, 18–30; H. J. Mader. Animal Habitat Isolation by Roads and Agricultural Fields. *Biological Conservation* 1984;29:81–96; D. J. Levey et al. Effects of Landscape Corridors on Seed Dispersal by Birds. *Science* 2005;309:146–148; G. C. Daily, ed. *Nature's Services, Societal Dependence on Natural Ecosystems* (Washington, DC: Island Press, 1997), 35–36, 317–322.

36. R. F. Noss and A. Y. Cooperrider. *Saving Nature's Legacy* (Covelo, California: Island Press, 1994), 55–57; Reed Noss. *The Ecological Effects of Roads, Missoula, Montana* (Wildlands Center for Preventing Roads, 1999). Available: www.wildrockies.org/WildCPR/.

37. E. W. Lathrop and E. F. Archbold. Plant Response to Los Angeles Aqueduct Construction in the Mojave Desert. *Environmental Management* 1980;4:137–148; E. W. Lathrop and E. F. Archbold. Plant Response to Utility Right of Way Construction in the Mojave Desert. *Environmental Management* 1980;4:215–226; G. D. Brum et al. Recovery Rates and Rehabilitation of Powerline Corridors. In R. H. Webb and H. G. Wilshire, eds. *Environmental Effects of Off-Road Vehicles* (New York: Springer-Verlag, 1983), 303–314; F. C. Vasek et al. Effects of Power Transmission Lines on Vegetation of the Mojave Desert. *Madroño* 1975;23:114–131.

38. H. G. Wilshire. Environmental Impacts of Oil and Gas Pipelines. In *Energy and the Environment, Application of Geosciences to Decision-Making* (U.S. Geological Survey Circular 1108, 1995), 117–118.

39. E. A. van der Grift. The Impacts of Railroads on Wildlife. *Road-RIPorter* (Wildlands Center for Preventing Roads) 2001;6:8–10.

40. R. W. Gorte. *Forest Health: Overview* (Washington, DC: U.S. Library of Congress), Congressional Research Service Report No. 95–548 ENR (28 April 1995).

41. Robert Perks et al. *The Bush Administration's Assault on the Environment* (Natural Resources Defense Council, 2002), 41. Available: www.nrdc.org.

42. *Forest Service Manual 7700, Final Directive* (Transportation System, Amendment No. 7700–2003-2).

43. The proposed rule continued the Clinton Roadless Area Conservation Rule for 60 days, at which time it would be no longer applicable to public forests (U.S. Department of Agriculture Forest Service. *Notice of Proposed Rulemaking, State Petitions for Inventoried Roadless Area Management* (36 CFR Part 294, RIN 0596-AC10, 2004)).

44. This account is modified from an original draft provided by Robert Wyigul, Earthjustice Legal Defense Fund. R.S. 2477 was repealed by enactment of the Federal Land Management and Policy Act (FLPMA) of 21 October 1976, while preserving valid existing rights-of-way as of the date of FLPMA's approval. FLPMA reserved lands under Bureau of Land Management's jurisdiction for other uses, thus withdrawing any routes of travel created after 1976 from R.S. 2477 claims. Less commonly, R.S. 2477 claims are being lodged for routes on forest lands under U.S. Forest Service jurisdiction. Unlike Bureau of Land Management lands, passage of the 1891 National Forest System Enabling Act reserved many forest lands for other uses. To satisfy R.S. 2477 criteria, routes on National Forest lands must have been created before 1891 or before the date of creation of newer national forests (Ronni Flannery. Federal Court Closes the Gate on R.S. 2477 Claims. *Road-RIPorter* (Wildlands Center for Preventing Roads) 2001;6[4]:10–11).

45. Havlick. *No Place Distant*, 70–73. The U.S. General Accounting Office (*Recognition of R.S. 2477 Rights-of-Way Under the Department of the Interior's FLPMA Disclaimer Rules and Its Memorandum of Understanding with the State of Utah* [B-300912. Enclosure, 2–21, 2001]) lays out the legal and administrative history of R.S. 2477. Other R.S. 2477 claims include a map submitted by the state of Utah to the DOI claiming 100,000 miles of routes as highways under R.S. 2477, including national park trails and routes across every designated wilderness in the state; Moffat County, Colorado, claimed 240 Dinosaur National Monument trails, and the state of Alaska claimed 164 separate routes through 14 Alaska National Parks, totaling nearly 3,000 miles, as highways under R.S. 2477.

46. See note 45.

47. The lawsuit respondents included the Bureau of Land Management, and southern Utah San Juan, Kane, and Garfield Counties. The state of Utah tried to intervene, but the court found that the state lacked sufficient interest in the disputed roads, as the counties long had claimed. In 2000, the state threatened a lawsuit against the Department of the Interior to establish its own right to make R.S. 2477 claims.

48. In 2001, the court ruled against the Bureau of Land Management and Utah counties, finding that a valid right-of-way under R.S. 2477 must be constructed by "some form of purposeful, physical building or improving, and…cannot be achieved solely by the effects of haphazard use." The land was reserved for other purposes before the 1976 enactment of the Federal Land Policy and Management Act (FLPMA). The counties appealed these decisions to the 10th Circuit Court of Appeals, but the appeal was dismissed in 2003 (U.S. Court for the District of Utah, Central Division, Order, Case No. 2:96-CV-836C [25 June, 2001], Tena Campbell, United States District Judge).

49. The Department of the Interior also reformed regulations to help the state and other parties qualify for R.S. 2477 claims (U.S. Department of the Interior, Bureau of Land Management. Conveyances, Disclaimers and Correction Documents 43 CFR Part 1860. *Federal Register* 2003;68:494–503).

50. Wildlands Center For Preventing Roads. *Investing in Communities, Investing in the Land, Summary Report* (2003). Available: www.wildlandscpr.org. This excellent and hopeful report is adapted from a study by the Center for Environmental Economic Development, assessing the benefits and costs of a national program for road removal on U.S. Forest Service Lands.

Chapter 6

1. The term "soil" in the literature on contaminant migration is used loosely to include all unconsolidated sediments lying on basement rock (of any age), in addition to weathered (or unweathered) sediment mantles affected by soil-forming processes.

2. J. S. Lynch et al. Patton's Desert Training Center. *Journal of the Council on America's Military Past* December 1982; reprinted as a pamphlet (date unknown).

3. The numerous camp and maneuver locations are isolated enough to determine how desert ecosystems respond naturally after a variety of light to heavy human impacts (D. V. Prose. *Land Disturbances from Military Training Operations, Mojave Desert, California* (Miscellaneous Field Studies Map MF-1855, U.S. Geological Survey, 1986).

4. D. V. Prose et al. Effects of Substrate Disturbance on Secondary Plant Succession; Mojave Desert, California. *Journal of Applied Ecology* 1987;24:305–313; complete data sets in D. V. Prose and S. K. Metzger. *Recovery of Soils and Vegetation in World War II Base Camps, Mojave Desert* (U.S. Geological Survey Open-File Report 85–234, 1985).

5. K. K. Nichols and P. R. Bierman. Fifty-Four Years of Ephemeral Channel Response to Two Years of Intense World War II Military Activity, Camp Iron Mountain, Mojave Desert, California. *Geological Society of America, Reviews of Engineering Geology* 2001;14:123–136.

6. Biological soil crusts also stabilize and insulate desert soils from extreme temperatures, and preserve much of their fertility by fixing nitrogen and retaining moisture (J. Belnap and

O. L. Lange, eds. *Biological Soil Crusts: Structure, Function, and Management* [Ecological Studies 150; New York: Springer-Verlag, 2001]).

7. The species growing in the control area but not in disturbed areas include ratany (*Krameria parvifolia*), silver cholla (*Opuntia echinocarpa*), pencil cactus (*Opuntia ramosissima*), and Mojave yucca (*Yucca schidigera*).

8. Creosote (*Larrea tridentate*) grows only sparsely in the disturbed areas, while burroweed (*Ambrosia dumosa*) seems particularly adaptable and has regrown abundantly. Creosote generally regrows slightly better in many former tent areas where the plants were hand cleared, leaving some root crowns intact and able to resprout, than in graded roadways.

9. Notable exceptions include a few individuals of brittle bush (*Encelia farinosa*), cheesebush, and *Porophyllum gracile* that were not present originally but grow in disturbed areas. Like *Ambrosia*, these short-lived plants are specially adapted to germinating and growing in naturally disturbed areas, such as rocky stream courses, and also do well in areas of human disturbance.

10. Anja Kade and S. D. Warren. Soils and Plant Recovery After Historic Military Disturbances in the Sonoran Desert, USA. *Arid Land Research and Management* 2002;16:231–243. Two out of four perennial plant species had recovered at the motor pool site. Note that many disturbed areas in the Mojave Desert show strong returns of burroweed (*Ambrosia dumosa*) whether the soil was compacted or not. In places essentially a monoculture of *Ambrosia* has replaced a creosote (*Larrea*)–*Yucca* community.

11. The recovery rates were calculated using both linear and logarithmic models, and follow-up studies of many abandoned sites, which showed that the most severely compacted soils in abandoned mining towns had not recovered in 91 years (R. H. Webb. Recovery of Severely Compacted Soils in the Mojave Desert, California, USA. *Arid Land Research and Management* 2002;16:291–305).

12. Jayne Belnap and Steve Warren. Patton's Tracks in the Mojave Desert, USA, an Ecological Legacy. *Arid Land Research and Management* 2002;16:245–259; this study showed more rapid recovery of biological crusts beneath perennial plant cover (estimated to require about 100 years), but much slower recovery between shrubs (estimated to require more than 1,000 years and perhaps as much as 2,000 years). The rates of recovery vary with lichen species, but a biological crust community is "recovered" only when the most sensitive species has recovered under the harshest climatic conditions for the habitat.

13. H. G. Wilshire. The Impact of Vehicles on Desert Soil Stabilizers. In R. H. Webb and H. G. Wilshire, eds. *Environmental Effects of Off-Road Vehicles, Impacts and Management in Arid Regions* (New York: Springer-Verlag, 1983), 31–50; Jayne Belnap and D. A. Gillette. Vulnerability of Desert Biological Soil Crusts to Wind Erosion, the Influences of Crust Development, Soil Texture, and Disturbance. *Journal of Arid Environments* 1998;39:133–142; Jayne Belnap. Surface Disturbances, Their Role in Accelerating Desertification. *Environmental Monitoring and Assessment* 1985;37:39–57.

14. D. V. Prose and H. G. Wilshire. *The Lasting Effects of Tank Maneuvers on Desert Soils and Intershrub Flora* (Open-File Report 00–512; U.S. Geological Survey, 2000); D. V. Prose. Persisting Effects of Armored Military Maneuvers on Some Soils of the Mojave Desert. *Environmental Geology and Water Science* 1985;7:163–170.

15. Desert varnish is deposited as dust on the surface of pebbles and cobbles exposed to the weather (Tanzhuo Liu and W. S. Broecker. How Fast Does Rock Varnish Grow? *Geology*

2000;28:183–186). Aboriginal people, both in the California deserts and on the Nazca Plain of Peru, constructed human, animal, and geometric figures most visible from the air, by scraping aside the surface layer of dark desert-varnished pebbles and exposing the light-colored silt beneath. Native people also scraped surface varnishes off of hard rock surfaces to create petroglyphs.

16. Prose and Wilshire. *The Lasting Effects of Tank Maneuvers.*

17. J. W. Steiger and R. H. Webb. *Recovery of Perennial Vegetation in Military Target Sites in the Eastern Mojave Desert, Arizona* (U.S. Geological Survey Open-File Report 00–355, 2000). Available: geopubs.wr.usgs.gov/open-file/of00–355/. The degree of recovery is measured by comparing plant cover, density, and volume of each species growing within and outside of each target site. Percentage-similarity and correlation coefficient indices vary from 22.7 to 95.1.

18. Joshua tree is *Yucca brevefolia*; ocotillo is *Fouquieria splendens*; paloverde is *Cercidium floridum*. The Dutch Flat strafing target berms are relatively fine-grained sediment, smoothed by erosion to a greater extent than the nearby upland. Disturbance-tolerant burroweed (*Ambrosia dumosa*) dominates the plant community, as usual.

19. U.S. Department of Defense. *FY96 Worldwide List of Military Installations* (1996); U.S. Departments of the Air Force, Navy, and Interior. *Special Nevada Report, Submitted in Accordance with Public Law 99–606* (DE-AC08–88NV10715, prepared by Science Applications International Corporation and Desert Research Institute, 1991). Available: ludb.clui.org/tag/Military/Training+%3B2F+Testing+%3B2F+Bombing+Range/.

20. U.S. Army. *Training Land, Unit Training Land Requirements* (U.S. Army Training and Doctrine Command, Fort Monroe, Virginia, 1978), 11. Asserting in 1996 that "the increasing sophistication and speed of modern weapons…require greater time and distance factors," the U.S. Army requested a 50% increase in the National Training Center's size (U.S. Department of the Interior Bureau of Land Management. *Draft Environmental Impact Statement for the Army's Land Acquisition Project for the National Training Center, Fort Irwin, California, and Proposed Amendment to the California Desert Conservation Area Plan* (prepared by U.S. Army Corps of Engineers with the Chambers Group, Inc., 1996), ES-1).

21. J. E. Lovich and David Bainbridge. Anthropogenic Degradation of the Southern California Desert Ecosystem and Prospects for Natural Recovery and Restoration. *Environmental Management* 1999;24:309–326.

22. D. C. Anderson and W. K. Ostler. Revegetation of Degraded Lands at U.S. Department of Energy and U.S. Department of Defense Installations, Strategies and Successes. *Arid Land Research and Management* 2002;16:197–212.

23. "Unexploded ordnance" is primed and fired ordnance that failed to explode. Problem military munitions also include discarded (not primed and fired) ordnance and constituents closely associated with military munitions. The military's euphemism for unexploded ordnance is "nonstockpile chemical munitions" (Chip Ward. *Canaries on the Rim: Living Downwind in the West* [New York: Verso Press, 1999], 101; U.S. General Accounting Office. *Military Munitions: Department of Defense Needs to Develop a Comprehensive Approach for Cleaning Up Contaminated Sites* [GAO-04–147, 2003]).

24. Center for Public Environmental Oversight. *Toxic Ranges, a National Problem* (2002). Available: www.cpeo.org. This report provides comprehensive information about contamination on military test and training ranges and is an excellent citizens' guide to ensuring thorough assessment and cleanup of formerly utilized defense sites released for public use.

There are more than 200 chemicals associated with munitions (U.S. General Accounting Office. *Department of Defense Operational Ranges: More Reliable Cleanup Cost Estimates and a Proactive Approach to Identifying Contamination Are Needed* [GAO-04–601, 2004], 14). The most common contaminating agents in practice munitions are TNT, RDX (both possible human carcinogens), HMX (damaging to liver and central nervous systems), perchlorate (thyroid disorders), and white phosphorus (reproductive, liver, heart, and kidney damage)—see table 6.1.

25. R. Misrach and M. W. Misrach. *Bravo 20, the Bombing of the American West* (Baltimore: Johns Hopkins University Press, 1990); U.S. Department of the Navy. *Renewal of the B-20 Land Withdrawal, Naval Air Station, Fallon, Nevada* (Draft Legislative Environmental Impact Statement, U.S. Department of the Navy, 1998).

26. J. A. MacDonald. Cleaning Up Unexploded Ordnance. *Environmental Science and Technology* 2001;35:372A–376A; U.S. General Accounting Office. *Department of Defense Operational Ranges.* The Department of Defense has no coherent policy for inventorying the more than 200 munitions-associated chemicals; the lists of existing active sites vary from year to year, as do estimated cleanup costs. In 2001, the Department of Defense estimated $14 billion would be required for cleanup, whereas the General Accounting Office estimated costs in excess of $100 billion. MacDonald gives Department of Defense estimates ranging from $107 to $392 billion. Remediation costs through mid-2005 amount to about $20 billion (U.S. General Accounting Office. *Groundwater Contamination: DOD Uses and Develops a Range of Remediation Technologies to Clean Up Military Sites* [GAO-05–666, 2005]). These estimates do not include cleanup costs of nontraining range sites such as manufacturing facilities, munitions burial pits, or open-burn and detonation sites; formerly used defense sites since returned to civilian use; and contamination that is not federally regulated, such as perchlorate.

27. U.S. General Accounting Office. *Environmental Liabilities: DOD Training Range Cleanup Cost Estimates Are Likely Understated* (GAO-01–479, 2001), 18. A good example of accidental unexploded ordnance hazards discovery was the 1993 find of buried ordnance at a long-forgotten World War I site in the Spring Valley area of Washington, D.C. (U.S. General Accounting Office. *Environmental Contamination, Many Uncertainties Affect the Progress of the Spring Valley Cleanup* [GAO-02–556, 2002]). Also, the Army Corps of Engineers estimates that unexploded 61-mm mortar shells, 155-mm artillery rounds, and 2,000-pound bombs may lie buried deeper than 10 feet in the earth.

28. U.S. General Accounting Office. *Environmental Contamination: Corps Needs to Reassess Its Determinations That Many Former Defense Sites Do Not Need Cleanup* (GAO-02–658, 2002), 29–30.

29. U.S. General Accounting Office. *Environmental Contamination, Cleanup Actions at Formerly Used Defense Sites* (GAO-01–557, 2001), 2. Substantial confusion plagues accounts of the numbers of contaminated sites and acreages. Depending on the meaning of the word "sites" in various reports, listed acreages may represent either the sizes of entire bases or training ranges or of potentially contaminated areas within them.

30. D. G. Milchunas et al. Plant Community Responses to Disturbance by Mechanized Military Maneuvers. *Journal of Environmental Quality* 1999;28:1533–1547; D. G. Milchunas et al. Plant Community Structure in Relation to Long-Term Disturbance by Mechanized Military Maneuvers in a Semiarid Region. *Environmental Management* 2000;25:525–539. One

surprising ecological benefit from the early stage of military maneuvers at Fort Carson has been to displace cattle grazing. The grazing effects on plant diversity and ground coverings were pervasive (see chapter 3), while the initial phases of military training affected a much smaller proportion of the land. Eventually, the impacts of military vehicles are bound to spread, however, and will catch up to and probably surpass those of cattle. Longer term training at Fort Carson has had varying effects on open woodlands, grasslands, and shrublands, but all build up faster than the land can recover. Unfortunately, increases in both the diversity of plant species and plant cover are due to weeds and exotic plants that the U.S. military brought with it.

31. A. J. Krzysik. *Ecological Assessment of the Effects of Army Training Activities on a Desert Ecosystem, National Training Center, Fort Irwin, California* (USA-CERL Technical Report N-85/13, U.S. Army Corps of Engineers, 1985).

32. Krzysik. Ecological Assessment of the Effects of Army Training Activities.

33. U.S. Department of the Army. *Record of Decision, Supplemental Final Environmental Impact Statement for Proposed Addition of Maneuver Training Land at Fort Irwin, California* (3 March, 2006). Significant impacts that cannot be mitigated include vegetation losses of 52% to 100% on low-use to high-use lands, respectively, including loss of 25% of Lane Mountain milk-vetch habitat; and severe impacts to as much as 140,000 acres of desert tortoise and wildlife habitat (about 84,000 acres of tortoise critical habitat), and "taking" (killing) up to 1,000 tortoises. The *Record of Decision* states: "Impacts to the biological environment…and visual resources are not considered an irreversible or irretrievable commitment as experience has shown that the desert landscape will recover to its natural condition if military use is halted." No credible studies support this finding. The Army proposes to "mitigate" impacts by purchasing 99,000 private acres elsewhere, removing grazing rights from them, and designating 17,000 acres for tortoise conservation and management.

34. The language in a 2002 Department of Defense plea for exemption from environmental laws reads:

> Federal departments and agencies shall not place the conservation of public lands, or the preservation or recovery of endangered, threatened, or other protected species found on military lands, above the need to ensure that soldiers…receive the greatest possible preparation for, and protection from, the hazards and rigor of combat through realistic training on military lands and in military air space.

35. In what conservationists called "the biggest rollback of the Marine Mammal Protection Act since it was enacted 30 years ago," the exemptions allow the Navy to "redefine activities considered harassment." Passed out of Congress on 17 November 2003, the bill also exempted Department of Defense sites from critical habitat designations if an "adequate natural resources plan is in place."

36. U.S. Department of Defense. *Defense Environmental Restoration Program, 10th Annual Report to Congress for Fiscal Year 1995* (1996).

37. Aimee Houghton and Lenny Siegel. *Military Contamination and Cleanup Atlas for the United States* (San Francisco: Pacific Studies Center and Career/Pro, San Francisco State University, 1995). Toxic substances that contaminate soil and water include battery electrolytes, BTEX (benzene, toluene, ethyl benzene, xylene), cadmium, chromium, lead, zinc, other heavy metals, chlorinated solvents, corrosives, chloroform, cyanide, dichlorodiphenyl trichloroethane (DDT), dioxins, ethylene glycol, fly ash, furan organic solvents, herbicides,

hydrazine and JP-5 jet fuels, lead-based paint, methylene chloride, naphthalene, nitrates, pentachlorophenol, perchloroethylene, pesticides, phenols, polyaromatic hydrocarbons, polychlorinated biphenyls (PCBs), radon, refrigerants, rocket fuel, royal demolition explosive (RDX), silver nitrate, tetrachloroethene, and trichloroethylene (U.S. Department of Defense. *Defense Environmental Restoration Program*).

38. Old, unfounded beliefs are hard to kill, and the incorrect assumption that soil can "clean" any contaminants dumped on it remains a widely held belief. This assumption is risky for long-lived contaminants, which may move slowly through relatively dry soil until they reach groundwater.

39. U.S. Department of Defense. *Defense Environmental Restoration Program, Annual Report to Congress, Fiscal Year 2002* (2003), chapter 2, figure 3. In 1995: the DOD reported that 12,185 U.S. Army sites, 4,433 U.S. Navy sites, 5,595 U.S. Air Force sites, 634 Defense Logistics Agency sites, 36 Defense Special Weapons sites, and 4,049 former military sites required remediation, for a total of 26,932 sites. Cleanups supposedly are complete at 13,013 sites (48%) (U.S. Department of Defense. *Defense Environmental Restoration Program* [1996], appendix B, table B-7).

40. U.S. Department of Energy. *Accelerating Cleanup, Paths to Closure* (DOE/EM-0362, Nevada Operations Office, 1998), 1–6 to 1–7. Notably, other documents cited by DOE claim that contaminated groundwater currently is being cleaned up (U.S. Department of Energy. *Accelerating Cleanup, Paths to Closure*, 3–21). More recent documents that apparently supplant the *Accelerating Cleanup* series explain why many major contamination problems are unsolved and how current remediation methods could cause more ecological damage than they correct (U.S. Department of Energy. *From Cleanup to Stewardship* [DOE/EM-0466, 1999]). The large list of sites requiring long-term management, the nature and complexity of many of the sites, and limited knowledge of underground geology make institutional management challenging, if not impossible. Past failed efforts at long-term hazardous materials management inspire little confidence that the materials can be isolated from the environment and human contact (National Research Council. *Long-Term Institutional Management of U.S. Department of Energy Legacy Waste Sites* [Washington, DC: National Academies Press, 2000], 2–7).

41. The Department of Defense estimated that 16 million acres of potentially contaminated training ranges have been transferred to the public or other agencies, and many more acres are in the transfer process (U.S. General Accounting Office. *Environmental Liabilities*, 11; W. Weber and R. Moore. Study on Feasibility of Mass Area Ordnance Decontamination [Technical Report TR-161; Indian Head, Md.: Naval Explosive Ordnance Disposal Facility, 1974], part II).

42. D. J. Hempel, ed. *Unexploded Ordnance on Lands Managed by the Department of the Interior* (U.S. Department of the Interior Bureau of Land Management, 1994). Estimates include 115 sites in 5,358,000 Bureau of Land Management acres, 22 sites in 1,214,000 U.S. Fish and Wildlife Service acres, 18 sites in 338,000 National Park Service acres, 12 Bureau of Indian Affairs sites, and two sites in 121,000 Bureau of Reclamation acres.

43. U.S. General Accounting Office. *Environmental Liabilities*, 4.

44. U.S. Environmental Protection Agency. *Used or Fired Munitions and Unexploded Ordnance at Closed, Transferred, and Transferring Military Ranges* (EPA 505-R-00–01, 2000); U.S. General Accounting Office. *Environmental Liabilities*, 12. We could not find

an authoritative inventory of the civilian or military unexploded ordnance casualties, despite those recounted here and despite the explosion of an old shell in a housing development on former Camp Elliot near San Diego that killed two boys (Lenny Siegel. Unexploded Ordnance and Explosive Wastes in California, a Silent, Threatening Giant. Center for Public Environmental Oversight, Newsletter 2000;7[2]:1–7).

45. Department of Defense's most recent estimate of full unexploded ordnance cleanup costs varies from $16 billion to $165 billion, requiring 150 to more than 1,500 years to complete at current funding rates (U.S. General Accounting Office. *Department of Defense Operational Ranges,* 1).

46. The balance of available funds in the Superfund trust has decreased significantly since 1996 (U.S. General Accounting Office. *Superfund Program, Current Status and Future Fiscal Challenges* [GAO-03–850, 2003]).

47. Lenny Siegel. *Risk-Based End State Policy at Energy Department* (Center for Public Environmental Oversight, 2003). Available: www.cpeo.org. Siegel notes that it can make sense to begin with the end in mind, but

> massive, complex, and secretive nuclear weapons plants are not ideal candidates for risk-based cleanup. They are not like gas stations, plating shops, or drum collection sites. Remedies that focus on interrupting pathways tend to be successful where the risk is minor in the first place, migration is unlikely, or the hazard can be expected to attenuate on its own.

Neither major weapons manufacturing or testing areas nor unexploded ordnance sites, where the hazards are long-lived and contaminant migration is likely, meet these criteria (see also U.S. General Accounting Office. *Nuclear Waste, Challenges to Achieving Potential Savings in DOE's High-Level Waste Cleanup Program* [GAO-03–593, 2003]; U.S. General Accounting Office. *Nuclear Waste, Challenges and Savings Opportunities in DOE's High-Level Waste Cleanup Program* [GAO-03–930T, 2003]).

48. Biological and chemical weapons have been used for military purposes for millennia. Assyrians are said to have poisoned enemy wells with rye ergot, a grain infected with fungal disease in the sixth century B.C., and in 1346, the Tartar army besieging the Crimean city of Kaffa, catapulted bubonic plague–infected soldiers' corpses into the city (E. K. Noji. Biological Agents as Natural Hazards and Bioterrorism as a "New" Natural Disaster Threat. *Natural Hazards Observer* 2000;25:1–3; S. M. Block. The Growing Threat of Biological Weapons. *American Scientist* 2001;89:2–11). Advanced technology has greatly "improved" our ability to spread fatal diseases.

49. Carol Gallagher. *American Ground Zero, The Secret Nuclear War* (Cambridge, Massachusetts: MIT Press, 1993), xxiii; Ward. *Canaries on the Rim,* 114.

50. Ward. *Canaries on the Rim,* 101, 105. Bacteria types used or tested in bioweapons include anthrax, cholera, and plague; the toxins include botulinum, various mycotoxins (molds), and staphylococcus (food contaminants). Viral agents include Ebola, hantaviruses, dengue, smallpox, malaria, and Rift Valley and yellow fevers. The carriers for these agents include such things as contaminated water or soil, rodents, and mosquitoes. Vaccines and treatments are available for the bacterial agents, but many of the viruses and toxins have no known successful treatments and no vaccines (Arthur Getis. Understanding Biological Warfare. In S. L. Cutter et al., eds. *The Geographical Dimensions of Terrorism* [New York: Rutledge, 2003], 181–185). Japan stockpiled tularemia (*Francisella tularensis*), the bacterium that causes rabbit fever, in World War II, as did both the United States and Soviet Union

during the Cold War, and it is now under hurried examination as a bioweapon. Tularemia is one of the most infectious organisms known: inhaling as few as 10 of the microbes can cause debilitating illness. Military personnel were used for two trials of the tularemia bacterium, each consisting of an unspecified number of tests (see appendix 7) (Gretchen Vogel. An Obscure Weapon of the Cold War Edges into the Limelight. *Science* 2003;302:222–223).

51. Virginia Brodine et al. The Wind from Dugway. *Environment* 1969;11:2–9, 40–45.

52. Ward. *Canaries on the Rim*, 103–104.

53. In April 1997, the U.S. Senate ratified the Convention on the Prohibition of the Development, Production, Stockpiling, and Use of Chemical Weapons and on Their Destruction (known as the Chemical Weapons Convention) (U.S. General Accounting Office. *Chemical Weapons, Better Management Tools Needed to Guide DOD's Stockpile Destruction Program* [GAO-04–221T, 2003]). Chemical weapons disposal under the Chemical Demilitarization Program is fragmented and disorganized, jeopardizing public safety and prolonging hazardous materials releases to the environment. And costs are rising—2001 estimates of $24 billion represent an increase of 60% since 1998 (U.S. General Accounting Office. *Chemical Weapons, Sustained Leadership, Along with Key Strategic Management Tools, Is Needed to Guide DOD's Destruction Program* [GAO-03–1031, 2003]).

54. Anthrax is a particularly deadly pathogen. The fatality rate from inhaling anthrax spores is more than 80%, while the smallpox fatality rate is only 30% (D. A. Henderson. The Looming Threat of Bioterrorism. *Science* 1999;283:1279–1282; D. A. Henderson. Bioterrorism as a Public Health Threat. *Emerging Infectious Diseases* 1999;Special Issue:488–492). Unlike smallpox and plague, also used in biological weapons, anthrax spores can stick around for decades (John Bohannon. From Bioweapons Backwater to Main Attraction. *Science* 2003;300:414–415). Q fever causes fever, and chills, severe headaches, eye pain, chest pain, cough, sore throat, weight loss, nausea, vomiting, and neurological problems such as visual and auditory hallucinations. Unlike anthrax, Q fever is short-lived and only about 4% lethal (Ed Regis. *The Biology of Doom* [New York: Henry Holt, 1999], 49).

55. Ward. *Canaries on the Rim*, 101.

56. The U.S. Environmental Protection Agency (E. T. Urbansky et al. Survey of Fertilizers and Related Materials for Perchlorate [ClO_4^-] (U.S. Environmental Protection Agency, Final Report, EPA/600/R-01/1ba, 2001) detected perchlorate only in samples of Chilean caliche. Only 0.14% of U.S. fertilizers come from that source. Furthermore, no perchlorate has been found in any agriculture-grade potassium chloride. The report concludes that "evidence to date largely argues against fertilizers as sources of environmental perchlorate" (p. 11). The level of potential groundwater contamination from the perchlorate in incompletely burned flares (commonly stubbed out and left along roads after an emergency), as well as from unburned flares and matches used at illegal methamphetamine labs, is unknown (T. K. G. Mohr and Jim Crowley. Perchlorate, Is It All Rocket Science? *HydroVisions* (Groundwater Resources Association of California) 2002;11[4]:7, 23).

57. One of the perchlorate plants was destroyed in a 1988 explosion and relocated to Cedar City, Utah. Perchlorate forms as salts of ammonium (NH_4^+), magnesium, potassium, and sodium. It does not bond readily to soil particles but is highly soluble in water and travels much faster in water than do other dissolved contaminants (Environmental Working Group. *Rocket Science, Perchlorate and the Toxic Legacy of the Cold War* [2002]. Available: www.ewg.org).

58. April 2003 analyses of 22 lettuce samples in northern California markets revealed per-chlorate contamination levels in four samples. The Centers for Disease Control and Prevention reported in October 2006 that perchlorate may suppress thyroid function in women, with greatest threats to unborn fetuses or conveyed to infants in breast milk. Studies indicate that much lower levels of perchlorate drinking water contamination than the U.S. Environmental Protection Agency permits can interfere with thyroid hormones crucial to developing many organ systems, including nervous and reproductive systems. Possible developmental effects include mental retardation; vision, speech, and hearing impairment; and abnormal testicular development in males (Environmental Working Group. *Rocket Science*, chapter 2).

59. In 2006, the U.S. Environmental Protection Agency adopted an interim perchlorate goal of 24.5 ppb, to strong criticism from California's Office of Environmental Health Hazard Assessment. California proposed a 6 ppb standard in March 2006, and Massachusetts has adopted a standard of 2 ppb. The non-profit Environmental Working Group's publication *Toxic Rocket Fuel Found in Milk Samples from Texas Supermarkets* [2003] reported that all of seven cow's milk samples had perchlorate levels between 1.7 and 6.4 ppb. In subsequent Environmental Working Group studies all but one of 33 milk samples purchased in California had perchlorate contamination. At present, perchlorate is known to contaminate more than 500 drinking water supplies in at least 20 states, serving well more than 20 million people.

60. Some explain the perchlorate contamination in groundwater both upgradient (against the direction of groundwater flow) and downgradient (in the direction of groundwater flow) from the flare factory as due to multiple sources, with the upgradient contamination coming from fertilizers. The fertilizers are not a likely source of significant perchlorate contamination, however (Rebecca Renner. Fertilizers Not a Source of Perchlorate. *Environmental Science and Technology* 2001;35:359A; see also n. 56).

61. Howard Wilshire. Geologic Features and Their Potential Effects on Contaminant Migration, Santa Susana Field Laboratory. In *Independent Scientific Studies of Potential Community Impacts from Field Lab Nuclear Meltdown and Other Contamination Released* (Santa Susana Field Laboratory Panel Studies and Findings, October 2006). Available: www.ssflpanel.org; Ali Tabidian. Migration of SSFL Perchlorate Contamination Offsite. In *Independent Scientific Studies of Potential Community Impacts from Field Lab Nuclear Meltdown and Other Contamination Released* (Santa Susana Field Laboratory Panel Studies and Findings, October 2006). Available: www.ssflpanel.org.

62. Lenny Siegel. *Perchlorate Summary* Center for Public Environmental Oversight, 2002;9:1–4; U.S. General Accounting Office. *Department of Defense Operational Ranges*, 5. Siegel reports that no effort is being made to clean up perchlorate contamination on military sites because "DOD policy did not require that they do so." See also U.S. Government Accountability Office. *Environmental Cleanup: Transfer of Contaminated Federal Property and Recovery of Cleanup Costs* (GAO-05–1011R, 2005). In 2005, the U.S. Government Accountability Office noted that perchlorate contamination has been found in water and soil at almost 400 sites in the United States where the concentration levels ranged from a minimum reporting level of 4 ppb to millions of ppb. Approximately two-thirds of sites had concentration levels at or below the U.S. Environmental Protection Agency's provisional cleanup standard of 18 ppb (U.S. Government Accountability Office. *Perchlorate: A System to Track Sampling and Cleanup Results Is Needed* (GAO-05–462, 2005).

63. National Research Council. *Health Implications of Perchlorate Ingestion* (Washington, DC: National Academies Press, 2005).

64. Martin Enserink. Secret Weapons Tests' Details Revealed. *Science* 2002;298:513–514.

65. Only 57 of the tests actually took place, but each consisted of an unspecified number of "trials" on test subjects. As many as five different agents are listed under individual tests, and presumably represent different trials. If this is correct, 112 trials are reported through 9 October 2002. Project SHAD (Shipboard Hazard and Defense) was part of Project 112 (U.S. Department of Defense. *Project SHAD Fact Sheets* [2002]. Available: deploymentlink.osd.mil/current_issues/shad/shad_intro.shtml). The U.S. General Accounting Office reports 50 tests, but their list shows as many as 52 tests, omitting test 66–3, Swamp Oak I, conducted March–April 1966; test 69–14, conducted July–November 1971; and test 74–10 Phase II. The GAO also lists test 68–13, Rapid Tan, as a single test, even though it actually was three separate tests (Rapid Tan I, II, and III) run at different times and at two different locations (U.S. General Accounting Office. *Chemical and Biological Defense, DOD Needs to Continue to Collect and Provide Information on Tests and Potentially Exposed Personnel* [GAO-04–410, 2004]).

66. U.S. Department of Defense. *DoD Releases Information on 1960 Tests* (2002). Available: www.defenselink.mil/news/.

67. U.S. Department of Defense. *Project SHAD Fact Sheets.*

68. "Simulants" used in open-air testing through 1969 included the fungus *Aspergillus fumigatus*, bacteria *Serratia marcescens* and *Bacillus subtili*, and the inorganic chemical zinc cadmium sulfide (L. A. Cole, *The Eleventh Plague, the Politics of Biological and Chemical Warfare* [New York: W. H. Freeman, 1997], 18–19, 61–66). The biological simulants are potentially harmful to infants and the elderly. Zinc cadmium sulfide continued to be used for years after its listing as a carcinogen (Ward. *Canaries on the Rim*, 106–108).

69. Regis. *The Biology of Doom*, 197. The Aum Shinrikyo sect's release of sarin gas in the Tokyo subway was only the second toxic chemical release aimed at a modern urban population. Operation LAC was the first.

70. U.S. General Accounting Office. *Chemical and Biological Defense*, 19–20.

71. Gallagher. *American Ground Zero*, xvii.

72. Robin Herbert et al. The World Trade Center Disaster and the Health of Workers: Five-Year Assessment of a Unique Medical Screening Program. *Environmental Health Perspectives* 2006;114:1853–1858. Of nearly 10,000 9/11 workers, 69% reported new or worsened respiratory ailments, and symptoms persisted in 59% of these workers to the time of examination.

73. In 1969 President Nixon first banned "biological and bacteriological" weapons, and in 1970 extended the ban to chemical weapons (Regis. *The Biology of Doom*, 210).

74. The United States signed the Biological Weapons Convention in 1972 and ratified it in 1975. It came into force after ratification by the United Kingdom and the USSR in 1975 and now has 137 parties to the ratification. The convention prohibits the development, stockpiling, and acquisition of biological agents and toxins "of types and in quantities that have no justification for prophylactic, protective, or other peaceful purposes." Unlike the Chemical Weapons Convention, the Biological Weapons Convention has no verification or enforcement provisions (Martin Enserink. On Biowarfare's Frontline. *Science* 2002;296:1954–1956; Martin Enserink. Biodefense Hampered by Inadequate Tests. *Science* 2001;294:1266–1267).

75. United States chemical and biological weapons research is carried out under the Department of Defense's Chemical and Biological Defense Program, the Defense Advanced

Research Projects' Biological Warfare Defense Program, the Department of Energy's Chemical and Biological Nonproliferation Program, and the Technical Support Working Group's Counterterror Technical Support Program (U.S. General Accounting Office. *Chemical and Biological Defense: Coordination of Nonmedical Chemical and Biological R&D Programs* [GAO/NSIAD-99–160, 1999]). Varied goals of the four agencies include "denying military advantage" and "allowing U.S. forces to operate largely unimpeded by chemical and biological attacks" (U.S. General Accounting Office. *Chemical and Biological Defense, Program Planning and Evaluation Should Follow Results Act Framework* (GAO/NSIAD-99–159, 1999))

76. David Malakoff. U.S. Biodefense Boom, Eight New Study Centers. *Science* 2003;301:1450–1451; Martin Enserink. New Biodefense Splurge Creates Hotbeds, Shatters Dreams. *Science* 2003;302:206–207. The Biological Weapons Convention treaty bans nations from developing, producing, acquiring, retaining, or stockpiling a weapon or other means of delivering germ agents "for hostile purposes or in armed conflict." Any research on potentially dangerous biological substances is perched at the edge of tight government controls to the dismay of researchers, who prefer to self-regulate. Tighter experimental reviews have been called for that would demonstrate how to render a vaccine ineffective, confer resistance to antibiotics or antivirals, enhance a pathogen's virulence or render a nonpathogen virulent, increase a pathogen's transmissibility, alter a pathogen's host range, enable evasion of diagnostic tests, and enable weaponization of pathogens and toxins (National Research Council, *Biotechnology Research in an Age of Terrorism, Confronting the Dual Use Dilemma* (Washington, DC: National Academies Press, 2004), 5; see also National Academy of Sciences, *Globalization, Biosecurity, and the Future of the Life Sciences* (Washington, DC: National Academies Press, 2006)).

77. Erika Check. U.S. Army Attacked over Published Patent for "Bioweapons Grenade." *Nature* 2003;423:789.

78. Mousepox kills all mice, even those supposedly protected by vaccines. The reason for the noncontagious character of rabbitpox and cowpox is not understood. When October 2003 biosecurity conference attendees in Geneva questioned the need for these experiments, "an American voice in the back boomed out: 'Nine-eleven,'" amid "murmurs of agreement." We know of no way that weapons developed in these experiments could protect Americans against such decidedly low-tech terrorist attacks. They serve instead to broaden the world's dangers.

79. Project Jefferson, to make a vaccine-resistant anthrax, is another of the many current secret projects to develop genetically modified pathogens. All are widely viewed as potential Biological Weapons Convention violations. When made public in 2001, the Pentagon announced its intent to complete the project and classify it. John D. Steinbruner, director of the Center for International and Security Studies at the University of Maryland, greeted that alarming news with the comment, "In their hands, this technology is potentially extremely dangerous" (quoted in Michael Scherer. The Next Worst Thing. *Mother Jones* March/April 2004;17–19).

80. Year 2000 recommendations to the Marine Corps (National Research Council. *An Assessment of Non-lethal Weapons Science and Technology* (Washington, DC: National Academies Press, 2003)). Perusal of the NAS' list of nonlethal weapons (appendix 2) suggests that Dr. Strangelove produced offspring who now work on U.S. weapons program projects.

81. Quote from Jonathan Tucker of the Monterey Institute of International Studies, quoted in Alexander Stone. U.S. Research on Sedatives in Combat Sets Off Alarms. *Science* 2002;297:764.

82. U.S. General Accounting Office. *Nonproliferation: Delays in Implementing the Chemical Weapons Convention Raise Concerns About Proliferation* (GAO-04–361, 2004). This report focuses on the unlikely prospect for achieving the convention's requirement for all chemical weapons stockpiles to be destroyed by 2007, with possible extensions to 2012. It does not address the potential violations of the convention by U.S. research, which also will exacerbate proliferation problems.

83. T. H. Nguyen. Microchallenges of Chemical Weapons Proliferation. *Science* 2005;309:1021.

84. U.S. General Accounting Office. *Bioterrorism, a Threat to Agriculture and the Food Supply* (GAO-04–259T, 2003).

85. Quoted in Scherer. The Next Worst Thing.

86. U.S. General Accounting Office. *Issues Facing the Army's Future Combat Systems Program* (GAO-03–1010R, 2003); U.S. General Accounting Office. *Defense Acquisitions, the Army's Future Combat Systems' Features, Risks, and Alternatives* (GAO-04–635T, 2004). The 2003 Iraq invasion demonstrated that superior information either may not be available or not generally believed.

87. Martin Enserink. This Time It Was Real: Knowledge of Anthrax Put to the Test. *Science* 2001;294:490–491; U.S. General Accounting Office. *Bioterrorism, Public Health Response to Anthrax Incidents of 2001* (GAO-02–365, 2003). There is some hope that mail delivery of anthrax and other such substances can be defeated (U.S. General Accounting Office. *Diffuse Security Threats: Technologies for Mail Sanitization Exist, but Challenges Remain* [GAO-04–152, 2003]; see also J. P. Fitch et al. Technology Challenges in Responding to Biological or Chemical Attacks in the Civilian Sector. *Science* 2003;302:1350–1354; Gary Matsumoto. Anthrax Powder, State of the Art? *Science* 2003;302:1492–1497).

Chapter 7

1. Contaminants include both radioactive and highly toxic nonradioactive wastes (U.S. Departments of the Air Force, Navy, and Interior. *Special Nevada Report, Submitted in Accordance with Public Law 99–606* (DE-AC08–88NV10715, prepared by Science Applications International Corporation and Desert Research Institute, 1991); U.S. Department of Energy. *Estimating the Cold War Mortgage*: Vol. 2. *Site Summaries* (1995); U.S. Department of Energy. *Implementation Plan for the Nevada Test Site Environmental Impact Statement* (DOE/NV-390, UC-700, Revision O, 1995); U.S. Department of Energy. *The 1996 Baseline Environmental Management Report*, vols. 2 and 3 (1996); U.S. Department of Energy. *Focused Evaluation of Selected Remedial Alternatives for the Underground Test Area* (DOE/NV-465, 1997)).

2. A more scientifically literate American public might have insisted on limiting nuclear weapons and power plants long ago, as have citizens in many parts of Europe. Nuclear power plant meltdowns at Three Mile Island, Pennsylvania, in 1979, and Chernobyl, Ukraine, in 1986, and the 1999 near disaster at the Tokaimura, Japan, nuclear fuel reprocessing plant have frustrated government, industry, and media efforts to build benign images for both forms of atomic energy. Recent spates of Middle Eastern and Asian nuclear weapons proliferation have exposed the major role of atomic power reactors in spreading atomic weaponry

(J. O. Lubenau and D. J. Strom. Safety and Security of Radiation Sources in the Aftermath of 11 September 2001. *Health Physics* 2002;83:155–164).

3. Catherine Caufield. *Multiple Exposures, Chronicles of the Radiation Age* (Chicago: University of Chicago Press, 1990), 34–35.

4. Principal natural and military radiation sources are defined in box 7.1. Other radiation exposures come from adding radium to luminous paint and pottery glazes. Other sources of damaging electromagnetic radiation are medical and dental X-rays. A major beta-particle radiation source is tritium (hydrogen-3), a heavy isotope of hydrogen. Tritium has a weight of 3 atomic mass units, while natural hydrogen, hydrogen-1, has an atomic weight of 1. Tritium occurs naturally in very low concentrations, but atmospheric bomb tests inserted large quantities of man-made tritium into the global atmosphere. It is abundant in bomb and power-plant wastes, especially reactor wastewater.

5. Gordon Edwards. *Health and Environmental Issues Linked to the Nuclear Fuel Chain*, Section B: Health Effects, 1996. Available: www.ccnr.org/ceac_B.html.

6. Radiation poisoning killed Edwin Lehman in 1925. He had inadvertently breathed in low-radioactivity radium dust while working in his laboratory. Lehman's body contained about 2.3×10^{-7} ounces, equivalent to 8.1×10^{-9} (about eight billionths) of a gram of radium (Caufield. *Multiple Exposures*, 34).

7. M. C. Olson. *Unacceptable Risk: The Nuclear Power Controversy* (New York: Bantam Books, 1976), 92.

8. U.S. General Accounting Office. *Nuclear Health and Safety: Consensus on Acceptable Radiation Risk to the Public Is Lacking* (GAO/RCED-94-190, 1994).

9. Natural background radiation comes from radioactive elements in natural soil and rocks, notably potassium-40, uranium-235, uranium-238, and thorium-232, and radioactive decay products, such as radium-226. In rocks and soil, uranium is present as uranium-238, with only tiny amounts of the bomb-making isotope, uranium-235. The amount of natural radiation exposure depends on local rock and soil compositions. Radium contents are significant in some rocks, but radium decays to radioactive radon gas (radon-222), which can collect in tunnels and basements, decaying relatively quickly to form molecule-sized radioactive dust particles that can be breathed into the lungs. We also receive natural gamma radiation from the sun's cosmic rays.

10. National Research Council. *Health Effects of Exposure to Low Levels of Ionizing Radiation: BEIR V* (Washington, DC: National Academies Press, 1990). Higher estimates come from studies on the effects of low-level exposures of radiation plant workers (Beatie Ritz. Radiation Exposure and Cancer Mortality in Uranium Processing Workers. *Epidemiology* 1999;10:531–538; Steven Wing et al. Mortality Among Workers at Oak Ridge National Laboratory, Evidence of Radiation Effects in Follow-Up Through 1984. *Journal of the American Medical Association* 1991;265:1397–1402).

11. Harvey Wasserman and Norman Solomon. *Killing Our Own: The Disaster of America's Experience with Atomic Radiation* (New York: Dell Publishing Co., 1982), part II, chapter 6. Available: www.ratical.org/radiation/KillingOurOwn.

12. Producing 1,500 short tons of plutonium also yielded about four billion curies each of strontium-90 and cesium-137, both toxic and both with hazard lives of about 300 years (see n. 35). The annual plutonium output (77 tons) produces about 210 million curies of

each. Cesium-137 is both an external and internal radiation hazard because it emits both beta and gamma radiation. Beta emitter strontium-90 is particularly hazardous when ingested because it can substitute for calcium in bone (Arjun Makhijani and Scott Saleska. The Production of Nuclear Weapons and Environmental Hazards. In A. Makhijani et al., eds. *Nuclear Wastelands, A Global Guide to Nuclear Weapons Production and Its Health and Environmental Effects* (Cambridge, Mass.: MIT Press, 1995), 23–63; R. C. Ewing. Nuclear Waste Forms for Actinides. In J. V. Smith, ed. *Colloquium on Geology, Mineralogy, and Human Welfare* (Washington, DC: National Academies Press, 1999), 3432–3439.

13. H. P. Metzger. *The Atomic Establishment* (New York: Simon and Schuster, 1972), 80–81. Radon gas is always present in uranium mines (Gunter Faure. *Principles of Isotope Geology*, 2nd ed. [New York: John Wiley and Sons, 1986]). Radon decay products include highly toxic polonium-210, the substance that killed former Russian intelligence agent Alexander Litvinenko in 2006. In unventilated spaces, the submicroscopic gas-derived polonium-210 dust particles hang in the air and are inhaled.

14. Deborah Hastings. Navajo Uranium Miners Never Warned Work Could Kill Them (Associated Press, 30 July 2000); see Arjun Makhijani and S. I. Schwartz, Victims of the Bomb. In S. I. Schwartz, ed. *Atomic Audit* (Washington, DC: Brookings Institution Press, 1998), 401–404.

15. Alex Shoumatoff. *Legends of the American Desert, Sojourns in the Greater Southwest* (New York: Harper Perennial, 1999), 477.

16. Michael D'Antonio. *Atomic Harvest* (New York: Crown Publishers, 1993), 102.

17. D'Antonio. *Atomic Harvest*, 102; Robert Alvarez. Energy in Decay. *Bulletin of the Atomic Scientists* May/June 2000;25–35.

18. Marie Curie (1867–1934) died of aplastic anemia, a form of leukemia, probably initiated during her long-term work with uranium- and-radium-bearing pitchblende (Caufield, *Multiple Exposures*, 23). Enrico Fermi (1901–1953) died of stomach cancer at the young age of 53 (S. E. Atkins. *Historical Encyclopedia of Atomic Energy* [Westport, Connecticut: Greenwood Press, 2000], 130–131). Richard Feynman (1918–1988), Nobel Prize–winning physicist and former Manhattan Project scientist, lived to the extended age of 70, fighting off a series of very rare cancers for nearly a decade. They included abdominal myxoid liposarcoma, a cancer of soft fat and connective tissue discovered in 1978, which recurred in 1981; Waldenström's macroglobulinemia, a rare bone marrow cancer discovered in 1986; and in 1987 the rare abdominal cancer that killed him (James Gleick. *Genius, the Life and Science of Richard Feynman* [New York: Pantheon Books, 1992], 401–404, 417, 437).

19. Advisory Committee on Human Radiation Experiments. *Final Report* (U.S. Department of Energy Document O-16–063416-4, 1995), 937. Available: tis.eh.doe.gov/ohre/roadmap/achre/report.html.

20. Carole Gallagher. *American Ground Zero* (Cambridge, Mass.: MIT Press, 1993), 1–108. This volume gives names and faces to many of the U.S. citizens whose lives were sacrificed in the development, testing, and experimentation with atomic bombs. "Atom bombs," such as the ones developed and used in World War II, are fission bombs because they contain a core of fissionable material such as uranium-235 or plutonium-239.

21. Wayne Brittenden. *The Dragon That Slew St. George* (British Broadcasting Corporation, 2001; U.S. distribution through radio station WAMU, American University Soundprint Media Center [available: www.soundprint.org).

22. William Burr et al. *The Costs and Consequences of Nuclear Secrecy*. In S. I. Schwartz, ed. *Atomic Audit* (Washington, DC: Brookings Institution Press, 1998), box 8–3.

23. Michele Stenehjem. Indecent Exposure. *Natural History* 1990;9:6–21. Even in the 1980s, Dr. Herbert Cahn, agricultural commissioner of Benton County, Washington, location of the Hanford site, "nearly lost his job…when he suggested handing out potassium iodide pills that would protect the public from an accidental release of radioactive iodine" (D'Antonio. *Atomic Harvest*, 65).

24. D'Antonio. *Atomic Harvest*, 66. The study, by Ernest Sternglass, compared infant mortality between 1943 and 1945. The 1945 levels increased by 50% in Franklin County, 60% in Umatilla County to the south of Hanford, and 160% in Benton County, closest to Hanford operations.

25. Burr et al. *The Costs and Consequences of Nuclear Secrecy*.

26. The Nuclear Regulatory Commission's "Rogovin Report" on the Three Mile Island nuclear power plant serial meltdowns estimated that the plant had released 15 curies of radioactive iodine isotopes, or "radioiodine," a small fraction of the total release of about 13 million curies (Mitchell Rogovin and G. T. Frampton. *Three Mile Island: A Report to the Commissioners and to the Public* (NUREG/CR-1250, vols. 1 and 2, 1980).

27. D'Antonio. *Atomic Harvest*, 72–85.

28. June Stark Casey learned on Mother's Day of 1986 that she was one of 270,000 Washington State residents who had been exposed continually and secretly to 1.1 million curies of radioactive iodine-131, a known carcinogen, released from Hanford Nuclear Reservation in the 1940s and 1950s. June was adversely affected by "Green Run," a secret experiment conducted on 2 December 1949 by GE's Nucleonics Department, which intentionally released over 11,000 curies of radioactive iodine and 20,000 curies of xenon into the atmosphere. (June Stark Casey, first-person account in a speech to the Vision of Culture in the 21st Century—Regeneration or Degeneration? conference [Cairo, Egypt, November 2000]).

29. The main purpose of bomb "tests" may have been to keep antagonists notified of U.S. power. Detonating atom bombs so far from the mainland required complex transportation and logistics, and the costs climbed when some Pacific island populations had to be resettled.

30. Rebecca Solnit. *Savage Dreams* (paperback edition; Berkeley: University of California Press, 1999), 153–154. The first decades of contact with whites drastically reduced the population of Western Shoshone, but the survivors adapted settler lifestyles and stayed. They are still there and still fighting the government for title to their old homeland, in spite of continuing explosive tests and radioactive waste disposal on that land.

31. Gallagher. *American Ground Zero*, xxiii.

32. Quoted in Solnit. *Savage Dreams*, 151.

33. Brittenden. *The Dragon that Slew St. George*; Gallagher. *American Ground Zero*, 110–311; John G. Fuller. *The Day We Bombed Utah* (Signet edition; New York: New American Library, 1985), 152–154.

34. Fuller. *The Day We Bombed Utah*, 21; D'Antonio. *Atomic Harvest*, 20, 72–75. A one-kiloton above-ground explosion releases about 10 million curies (see n. 35). Total radioactivity releases from Nevada Test Site atmospheric tests amounted to about 12 billion curies by 1963, greatly exceeding the Chernobyl accident's underestimated release of 100 million curies (U.S.

Congress, Office of Technology Assessment. *Complex Cleanup, the Environmental Legacy of Nuclear Weapons Production* (OTA-0–484; Washington, DC: U.S. Government Printing Office, 1991)). Such large amounts of radioactivity are dominated by short-lived radionuclides, which decrease very rapidly. Those tests also released long-lived radionuclides such as plutonium-239 to the atmosphere, which remain toxic for hundreds of thousands of years.

35. A complex nomenclature defines the radiation levels of various radioactive materials' emissions and express radiation dose and hazard levels, in cgs (centimeter gram second) units. All are difficult to translate into simple English units of measurement (David Close and Lisa Ledwidge. Measuring Radiation: Devices and Methods. *Science for Democratic Action* 2000;8:11–14). Radioactive releases often are expressed in curies, and for simplicity we have followed the older practice of expressing human radiation exposure standards and emission levels in curies. Named for the pioneer radiation physicist Marie Curie, the curie is the amount of radioactivity emitted per second by one gram (1/28th of an ounce) of radium (see n. 37). For other radioactive materials, a curie is equal to the amount that produces 3.7×10^{10} nuclear disintegrations per second. The amount of radioactivity declines as radioactive materials decay, so the total number of curies changes with time. The values given are based on the amounts remaining as of 1 January 1994 (U.S. Department of Energy. *Linking Legacies, Connecting the Cold War Nuclear Weapons Production Processes to Their Environmental Consequences* (DOE/EM-0319, 1997), 58).

36. This estimate of the Chernobyl release is based on the commonly cited low-end figure of 100 million curies of radiation released over 10 days, apparently based on an official Soviet government underestimate of 80 million curies, adjusted to 10 days after the accident, by which time some of the shortest lived isotopes had decayed to nothing. The high estimate is 200 million (Richard Stone. Living in the Shadow of Chernobyl. *Science* 2001;295:420–421). An independent estimate of the initial Chernobyl release is 240 million curies of radioactive iodine and cesium (Z. A. Medvedev. *The Legacy of Chernobyl* [New York: W. W. Norton, 1990], 78).

37. Short-term observations of the radium workers suggested that ingesting amounts greater than 1.2 microcurie (one-millionth of a curie, equivalent to one-millionth of a gram of radium) can cause serious health effects. (A gram is 1/28 of an ounce). In 1941, the lifetime limit of allowable radiation exposure for a human body was set at a tenth of a microcurie of radium and 10 picocuries (pCi) of radon per liter (L; 1.01 quart) of air (Caufield. *Multiple Exposures*, 39–40). A picocurie is one-trillionth (10^{-12}) of a curie, equivalent to the radiation emitted by one-trillionth of a gram (0.035 trillionth of an ounce) of radium. In English measurement, 1 pCi/L is about equivalent to the radiation produced by 0.035-trillionths of an ounce of radium per quart of liquid.

38. U.S. Department of Energy. *Closing the Circle on the Splitting of the Atom* (DOE/EM-0266, 1996), 74; U.S. Government Accountability Office. *Department of Energy: Preliminary Information on the Potential for Columbia River Contamination from the Hanford Site* (GAO-06–77R, 2005).

39. Milled uranium-ore tailings consist of finely ground rock materials with substantial radioactivity levels, from alpha-emitting uranium, thorium-230, radium-226, and the products of radium-226 decay. The total radioactivity in the tailings can exceed 1,000 picocuries per gram (pCi/g) (see n. 37). Liquids drained from the tailings can evaporate or infiltrate the ground from ponds. These liquids contain 7,500 pCi/L of radium-226, 22,000 pCi/L of thorium-230, and 0.01% uranium (U.S. Department of Energy. *Integrated Data Base Report, 1996,*

U.S. Spent Nuclear Fuel and Radioactive Waste Inventories, Projections, and Characteristics [DOE/RW-0006, revision 13, 1997], table 5.2).

40. The more than 900,000 tons stored nationally include 645,000 tons of depleted uranium hexafluoride, 4,500 tons of lead, 45,800 tons of lithium, more than 19,000 tons of natural uranium, more than 8,100 tons of low-enriched uranium, 192 tons of highly enriched uranium, 58 tons of plutonium, more than 1,900 tons of "Nuclear Materials Management and Safeguard System" materials, 170,000 tons of scrap metal, 20,800 cubic feet of sodium, 2,900 tons of spent nuclear fuel, 1,450 tons of chemicals, and about 10 million "pieces" of weapons components (U.S. Department of Energy. *Taking Stock, a Look at the Opportunities and Challenges Posed by Inventories from the Cold War Era,* vol. 1 [DOE/EM-0275, 1996], table 2–1).

41. U.S. Department of Energy. *Linking Legacies,* 58.

42. U.S. General Accounting Office. *Nuclear Waste: Understanding of Waste Migration at Hanford Is Inadequate for Key Decisions* (GAO/RCED-98–80, 1998); U.S. Department of Energy. *Accelerating Cleanup, Paths to Closure* (DOE/EM-0362, 1998); Stenehjem. Indecent Exposure.

43. J. L. Waite. *Tank Wastes Discharged Directly to the Soil at the Hanford Site* (WHC-MR-0227, Westinghouse Hanford Co., 1991). This amount of radioactivity is about 500 times less than estimates for the total release during the Three Mile Island power plant reactor meltdowns, but the Hanford discharges contained larger amounts of health-threatening components (Burr et al. *The Costs and Consequences of Nuclear Secrecy*). The corrosive materials are fluorides, ferrocyanide, sodium aluminate, sodium oxalate, ammonium nitrate, nitrate, phosphates, and sulfates.

44. U.S. Department of Energy. *Linking Legacies,* table 4–5.

45. Thousands of cubic meters of water were pumped per minute through each of the reactors for cooling purposes (R. C. Newcomb et al. *Geology and Ground-Water Characteristics of the Hanford Reservation of the U.S. Atomic Energy Commission* [U.S. Geological Survey Professional Paper 717, 1972]). This report was completed in 1953, then classified for no apparent reason except for its prediction that the tanks would eventually leak. It was declassified in 1960 but not released by the U.S. Geological Survey until 1972.

46. Roots of bean plants that grew in reactor effluent accumulated 33% of available neptunium-239; also, Columbia River green algae, sponges, and insect larvae showed neptunium-239 concentration factors of 280, 40, and 30 times the concentration in river water (J. J. Davis et al. Radioactive Materials in Aquatic and Terrestrial Organisms Exposed to Reactor Effluent Water. In *Proceedings of the Second United Nations Conference on Peaceful Uses of Atomic Energy,* vol. 18 [Geneva, 1958]).

47. Stenehjem. Indecent Exposure, 8.

48. Makhijani and Schwartz. Victims of the Bomb, 410–411.

49. Pat Lavelle. *Facing Reality at Hanford* (Government Accountability Project Report, 2000). Available: www.whistleblower.org.

50. Liquids and sludges contaminated by a wide range of hazardous and radioactive materials were discharged in "swamps," or ponds of waste liquid and slurry, from which contaminants could percolate into the ground; deep "reverse wells" that pumped liquid wastes directly into the subsurface, commonly to groundwater; and "cribs," or trenches that mostly

covered about 900 square feet of surface area, up to 1,400 feet long and 30 feet deep. The trenches are filled with rock and covered with rock and soil. Contaminated liquids were fed into the trenches from perforated pipes and allowed to percolate into the ground.

51. U.S. Department of Energy. *Hanford Tank Farms Vadose Zone, Baseline Characterization Current Status and Issues Briefing* (GJO­55­ TAR/GJO­ HAN­22; prepared by MACTEC-ERS for Department of Energy, Albuquerque Operations Office, Grand Junction Office, Grand Junction, Colorado, 1998).

52. U.S. Department of Energy. *Closing the Circle*; U.S. Department of Energy. *Linking Legacies*, 168–169.

53. Michael Balter. Filtering a River of Cancer. *Science* 1995;267:1084–1086. An informative discussion of issues affecting migration of radium and plutonium in the environment is presented by Brice Smith (*The Environmental Transport of Radium and Plutonium: A Review* (Takoma Park, Maryland: Institute for Energy and Environmental Research, 2006). Available: www.ieer.org/reports/envtransport/).

54. U.S. General Accounting Office. *Nuclear Waste: Understanding of Waste Migration at Hanford Inadequate for Key Decisions*, 5.

55. In late 2006, the Department of Energy still had few data on the abundance and character of contaminants temporarily resident in the unsaturated zone, or on plans for treating contaminated areas to prevent the substances from reaching the Columbia River. This is a huge and expensive problem (U.S. Government Accountability Office. *Nuclear Waste: DOE's Efforts to Protect the Columbia River from Contamination Could Be Further Strengthened* [GAO-06–1018, 2006]).

56. The scientists drilled only shallow monitoring wells, on the untested assumption that leaked contaminants would not spread far or penetrate very deep beneath the surface.

57. Waite. *Tank Wastes Discharged Directly to the Soil.* "Specific retention trenches" were assumed capable of retaining liquid radioactive wastes for disposal. Calculations based on assumed values for the moisture-retention capacity of soils above the water table set waste volume limits at 6–10% of the soil volume between the trench floor and the water table. This limit was supposed to allow soils to absorb the wastes well above the water table. The concept ignored faults and other fractures, which provide preferred pathways for concentrated flow. Operators exceeded the limits from time to time.

58. U.S. General Accounting Office. *Nuclear Waste: Understanding of Waste Migration at Hanford Is Inadequate for Key Decisions*, 5; U.S. Department of Energy. *Closing the Circle*, 73.

59. A single tritium atom can combine with two oxygen atoms to make "tritiated water." Tritium cannot be detected by gamma-radiation counters.

60. D. J. Brown and W. A. Haney. *The Movement of Contaminated Ground Water from the 200 Areas to the Columbia River* (HW-80909, General Electric Co., Hanford Atomic Products Operation, 1964). Monitoring wells close to the Columbia River in 1963 showed tritium concentrations of 40,000 pCi/L. Between 1975 and 1982, the concentrations gradually increased to near 250,000 pCi/L and stabilized at that level through 1985. The national safe drinking water standard is 20,000 pCi/L (see n. 37) (U.S. Geological Survey. *Subsurface Transport of Radionuclides in Shallow Deposits of the Hanford Nuclear Reservation, Washington, Review of Selected Previous Work and Suggestions for Further Study* (Open-File Report 87–222, 1987)).

61. Brown and Haney. *The Movement of Contaminated Ground Water.*

62. The analysts looked for only these contaminants, but others may be present (S. J. Trent. *Hydrogeologic Model for the 200 West Ground Water Aggregate Area* [WHC-SD-EN-TI-014, Westinghouse Hanford Company, 1992]; M. P. Connelly et al. *Hydrogeologic Model for the 200 East Groundwater Aggregate Area* [WHC-SD-EN-TI-019, Westinghouse Hanford Company, 1992]).

63. These figures are almost certainly gross underestimates, because they do not account for contaminants moving through both the unsaturated zone and in groundwater. Of course, little information is available (see Gregory Nimz and J. L. Thompson. *Underground Radionuclide Migration at the Nevada Test Site* [U.S. Department of Energy Report DOE/NV-346, 1992]).

64. U.S. Department of Energy. *Implementation Plan for the Nevada Test Site Environmental Impact Statement* (DOE/NV-390, UC-700, Revision O, 1995), 3–9; U.S. Department of Energy. *Geology, Soils, Water Resources, Radionuclide Inventory* (Technical Resource Report for the Final Environmental Impact Statement for the Nevada Test Site and Off-Site Locations in the State of Nevada, Nevada Operations Office, Las Vegas, Nev., 1996). 82; U.S. Department of Energy. *The 1996 Baseline*, 11.

65. See n. 37.

66. U.S. Energy Research and Development Administration. *Final Environmental Impact Statement, Nevada Test Site, Nye County, Nevada* (ERDA-1551, 1977), 2–88.

67. Hapless beagles still are "advancing knowledge" about radioactivity affects on animals. Train loads of contaminated beagle carcasses from experimental stations, including the University of California–Davis, are disposed of as radioactive waste at the Hanford Site.

68. R. J. Laczniak et al. *Summary of Hydrogeologic Controls on Ground-Water Flow at the Nevada Test Site, Nye County, Nevada* (U.S. Geological Survey Water Resources Investigations Report 96–4109, 1996), 42. The figures do not include recent subcritical tests.

69. "Subcritical" tests disperse plutonium by detonating high explosives, so do not create fission and activation radionuclides. The tests were conducted on the ground surface until 1963, and after that in tunnels and shallow boreholes. Current subcritical "hydrodynamic" tests supposedly assess the behavior of plutonium devices when exposed to explosive shocks (U.S. Government Accountability Office. *Nuclear Weapons: Views on Proposals to Transform the Nuclear Weapons Complex* [GAO-06–606T, 2006]).

70. Nevada Test Site radioactive concentrations register greater than 40 pCi/g (see n. 37) (U.S. Department of Energy. *The 1996 Baseline*, 14).

71. U.S. Department of Energy. *The 1996 Baseline*, 14. These estimates exclude uranium. Reported inventories include 35 curies of cobalt-60, 330 of strontium-90, 310 of cesium-137, 130 of europium-152, 20 of europium-154 and europium-155, 160 of plutonium-238, 910 of plutonium-239 and plutonium-240, and 150 of americium-241. Inadequate sampling makes the accuracy of these data questionable, however.

72. The success of attempts to clean up some of these sites, including Double Tracks and Clean Slate, is not known, and the cleanup standard level of 40 of pCi/g (see n. 36) indicates the sites will remain unsafe for human uses.

73. U.S. Energy Research and Development Administration. *Final Environmental Impact Statement, Nevada Test Site*.

74. One hundred eleven tests were detonated below the water table, 71 within a one-crater-radius distance from the water table, 39 within one to two crater radii from the water

table, and 95 within two to five crater radii. Bombs that exploded within two crater radii above the water table are probable sources of groundwater contamination, and those within two to five crater radii above the water table may have directly contaminated groundwater.

75. Studies on long-distance radionuclide transport within the unsaturated zone have invalidated the common assumption that discharging radioactive materials into thick unsaturated zones does not threaten groundwater (J. R. Nimmo et al. Kilometer-Scale Rapid Transport of Naphthalene Sulfonate Tracer in the Unsaturated Zone at the Idaho National Engineering and Environmental Laboratory. *Vadose Zone Journal* 2002;1:89–101).

76. Sorption processes include absorption of a material in the holes within a mineral's atomic structure and adsorption on mineral surfaces.

77. Laczniak et al. *Summary of Hydrogeologic Controls.* The preponderance of evidence favors subsurface groundwater transmission independent of surface barriers (S. T. Nelson and A. L. Mayo. Testing the Interbasin Flow Hypothesis at Death Valley, California. *EOS, Transactions of the American Geophysical Union* 2004;85[37]:349, 355–356; I. J. Winograd et al. Comment on "Testing the Interbasin Flow Hypothesis at Death Valley, California." *EOS, Transactions of the American Geophysical Union* 2005;86[32]:295).

78. G. M. Russell and G. L. Locke. *Summary of Data Concerning Radiological Contamination at Well PM-2, Nevada Test Site, Nye County, Nevada* (U.S. Geological Survey Open-File Report 96–599, 1997).

79. E. A. Bryant and June Fabryka-Martin. *Survey of Hazardous Materials Used in Nuclear Testing* (LA-12014-MS, Los Alamos National Laboratory, 1991).

80. U.S. Department of Energy. *Integrated Data Base*, table 0–6.

81. Transuranic wastes are alpha-emitting elements heavier than uranium, with half-lives greater than 20 years and a combined activity level of at least 100 nanocuries per gram (one-billionth of a curie in each 0.036 trillionth ounce of waste.) (U.S. Department of Energy. *Linking Legacies*, 40).

82. Lockheed Martin Idaho Technologies Company. *A Comprehensive Inventory of Radiological and Nonradiological Contaminants in Waste Buried in the Subsurface Disposal Area of the INEL RWMC During the Years 1952–1983. Idaho National Engineering Laboratory*, INEL-95/0310, Rev. 1 (1995).

83. Michele Boyd and Arjun Makhijani. Poison in the Vadose Zone, Threats to the Snake River Plain Aquifer from Migrating Nuclear Waste. *Science for Democratic Action* 2001;10(1):1–10; Nimmo et al. Kilometer-Scale Rapid Transport.

84. Comment made by an Idaho National Engineering and Environmental Laboratory representative at the U.S. Geological Survey 1999 Unsaturated Zone Interest Group Meeting, 13–15 January 1999. An important study examined the redistribution of plutonium wastes in a Los Alamos canyon dump site after 55 years. The plutonium concentrations in alluvium had highly irregular variations—over five orders of magnitude. Concentrations are highest in fine-grained sediment near the source and lower in more distant coarser grained deposits (S. L. Reneau et al. Geomorphic Controls on Contaminant Distribution Along an Ephemeral Stream. *Earth Surface Processes and Landforms* 2004;29:1209–1223; E. A. Martell et al. Fire Damage. *Environment* 1970;12:14, 21; U.S. General Accounting Office. *Fire Protection, Barriers to Effective Implementation of NRC's Safety Oversight Process* [GAO/RCED-00–39, 2000]).

85. C. J. Johnson et al. Plutonium Hazard in Respirable Dust on the Surface of Soil. *Science* 1976;193:488–490; J. P. Kaszuba et al. Modeling of Actinide Geochemistry for Reactive Transport and Waste Isolation. *Geological Society of America, Abstracts with Programs* 1998;30:A-87. The samples taken from this area were collected at depths as great as eight inches. Respirable particles are less than about 1/20,000 of an inch in diameter.

86. Dr. Brown taught at Brigham Young University briefly in the 1950s. He recalled, "Dr. Joseph 'Lynn' Lyon, Chief of the Division of Epidemiology of the Department of Family and Community Medicine at the University of Utah and codirector of the Utah Cancer Registry, headed cancer studies of southern Utah residents that, to his surprise, showed that there was a three-fold increase in leukemia."

87. U.S. Department of Energy. *Closing the Circle.*

88. The panel consisted of H. H. Hess (chairman), M. King Hubbert, Richard J. Russell, John N. Adkins, William E. Benson, John C. Frye, William B. Heroy, and Charles V. Theis. Hess, Russell, and Adkins had served as chairmen of the National Academy of Science's Earth Sciences Division, and Hubbert later served in that capacity.

89. P. M. Boffey. *The Brain Bank of America, an Inquiry into the Politics of Science* (New York: McGraw-Hill, 1975), 90.

90. H. H. Hess et al. *The Disposal of Radioactive Waste on Land, Report of the Committee on Waste Disposal of the Division of Earth Sciences* (Publication 519, National Academy of Science, 1957). The discussion is an edited condensation from the record of the stenotype reporter of a meeting on 10–12 September 1955 (appendix B). The following excerpts are found on pp. 41–45. "Cribs" are unlined excavations 30–40 feet deep, filled with broken rock. For explanation of adsorption, see the glossary at the end of the book.

Mr. Piper (Atomic Energy Commission): "A few areas adjacent to old cribs…have been tested by sinking holes at intervals around them, sampling the earth material, and analyzing it. This has delineated, beneath the cribs, roughly pear-shaped zones in which [radioactive] fission [atom bomb] products have been adsorbed by the earth materials. *I know of no case where the total quantity of fission products fixed in the pear-shaped zone can be shown to be a major part of the products that were in the total volume of fluid discharged into the overlying crib.*" (p. 41, emphasis added)

Dr. Gilluly (U.S. Geological Survey): "You said this pear-shaped area of poisoned soil doesn't contain anywhere near a major fraction of the material that was fed into it. What happened to the rest of it? Where did it go?"

Mr. Piper: "Some…may be tied up in sludge in the bottom of the crib…but even making allowance for that, we haven't demonstrated at Hanford that there is anything like complete interception by adsorption."

Dr. Griggs (University of California): "I asked a question as to whether contamination occurs in ground water."

Mr. Piper: "Not directly under the areas of absorption that have been tested by drilling."

Dr. Griggs: "How do you reconcile the fact that there is no contamination beneath this pear-shaped area?"

Mr. Piper: "I am not sure that any [test wells] reached ground-water level immediately beneath any of the absorption zones that were drilled out. I don't think we prove so from [our] sampling."

Dr. Gilluly: "What happened to the stuff then?"

Mr. Piper: "Some could have gone down to the water table. We can't prove that it didn't." (pp. 44–45)

91. Casey Ruud. *Briefing on Single Shell Tank Vadose Zone Issues* (U.S. Department of Energy-Richland Office, Tank Farm Operations, Vadose Zone Characterization Project, 1996); J. G. Conaway et al. *Tank Waste Remediation System, Vadose Zone Contamination Issue, Independent Expert Panel Status Report* (U.S. Department of Energy DOE/RL-97–49, revision 0, 1997).

92. The report also shows the Atomic Energy Commission presumed that mobile components (e.g., ruthenium) might reach groundwater—but, having a short half-life (about a year), would decay to a harmless state before it could reach the Columbia River. The committee responded that water movement through the unsaturated zone is poorly understood, and to continue dumping large low-level waste volumes in the unsaturated zone "is of limited application and probably involves unacceptable long term risks" (Hess et al. *The Disposal of Radioactive Waste on Land*, 7).

93. "[T]he larger the masses of water, the more unpredictable the mechanism of dilution becomes. Those who are familiar with the dilution of industrial waste in larger rivers and harbors know the tendency of these wastes to move in narrow, uncontrolled streams....It takes a special circumstance to make available for dilution the full volume of a large mass of water" (Hess et al. *The Disposal of Radioactive Waste on Land*, 19).

94. J. E. Galley et al. *Report of the Committee on Geologic Aspects of Radioactive Waste Disposal to the Division of Reactor Development and Technology* (U.S. Atomic Energy Commission, National Academy of Sciences, Division of Earth Sciences, 1966).

95. Hess et al. *The Disposal of Radioactive Waste on Land*, 3.

96. Galley et al. *Report of the Committee on Geologic Aspects of Radioactive Waste*, 2.

97. Boffey. *The Brain Bank of America*, 89–111.

98. The next, long-delayed, 1970 National Academy of Sciences committee report was only nine pages long (double-spaced) and mostly provided superficial observations about disposal sites. The only value judgment praised "the extensiveness and care in waste management at each site visited. The Committee is gratified by the quality and scope of the R&D program sponsored by the [Atomic Energy Commission] in radioactive waste management" (National Academy of Sciences. *Radioactive Waste Management, An Interim Report of the Committee on Radioactive Waste Management* [Washington, DC: National Academies Press. 1970]).

99. U.S. General Accounting Office. *Nuclear Waste: Understanding of Waste Migration at Hanford Is Inadequate for Key Decisions* (GAO/RCED-98–80, 1998).

100. In 2002, the Department of Energy's original $192 billion cost estimate, exclusive of prior expenditures and time required for completion of cleanup (until 2070), were revised downward to $142 billion and completion time of 2035 under the Accelerated Cleanup Program. Neither goal is likely to be met (U.S. Government Accountability Office. *Nuclear Waste: Better Performance Reporting Needed to Assess DOE's Ability to Achieve the Goals of the Accelerated Cleanup Program* [GAO-05–764, 2005]).

101. U.S. Government Accountability Office. *Hanford Waste Treatment Plant: Contractor and DOE Management Problems Have Led to Higher Costs, Construction Delays, and Safety Concerns* [GAO-06–602T, 2006]).

102. An important exception is the excellent report by the National Research Council, *Long-Term Institutional Management of U.S. Department of Energy Legacy Waste Sites* (Washington, DC: National Academies Press, 2000).

103. National Research Council. *Ward Valley: An Examination of Seven Issues in Earth Sciences and Ecology* (Washington, DC: National Academies Press, 1995); H. G. Wilshire. Is Ward Valley Safe? Serious Issues Remain. *Geotimes* June 1997;18–20.104. A. B. Kersting. *The Role of Colloids in the Transport of Plutonium and Americium: Implications for Rocky Flats Environmental Technology Site* (University of California Lawrence Livermore National Laboratory UCRL-TR-200099, 2003).

105. U.S. Government Accountability Office. *Securing U.S. Nuclear Materials: DOE Needs to Take Action to Safely Consolidate Plutonium* (GAO-05–665, 2005).

106. V. J. Brechin. Nevada Test Site National Sacrifice Zone, $7.3 Trillion Betrayal. *Geological Society of America Abstracts with Programs* 2000;32:A-500; U.S. Department of Energy. *Nevada Environmental Restoration Project, Focused Evaluation of Selected Remedial Alternatives for the Underground Test Area* (DOE/NV-465 UC-700, 1997).

107. U.S. Department of Energy. *Accelerating Cleanup*, 1–6.

108. U.S. Department of Energy. *Linking Legacies*, table 4–5. Plans in 1989 called for vitrifying some 53 million gallons of highly radioactive tank sludge, to reduce the problems of safe disposal. After spending $4 billion to develop vitrification facilities (melting the material and then chilling it to a glass), not a gallon of sludge has been vitrified as of 2006.

109. U.S. Department of Energy. Proposal to Downgrade INEL High Level Waste Using 2005 Reagan Act (Document E6–20107). *Federal Register* 712006;68814–68815.

110. Frank Clifford. Widespread Use of Radioactive Scrap Assailed. *Los Angeles Times* 12 June 2000.

111. "Depleted uranium" consists of less-radioactive uranium-238 waste from the process of separating highly radioactive uranium-235 for bomb fuel. Uranium is enriched as uranium hexafluoride (UF_6), a gas. Depleted uranium for munitions is obtained by converting the UF_6 gas, containing mostly uranium-238, to metal. This process is too expensive to use for reducing the huge amounts of UF_6 gas stored in corroding cylinders at Hermiston, Oregon (Ernest Goitein, personal communication, 2000).

112. U.S. Department of the Air Force. *Final Environmental Assessment for Resumption of Use of Depleted Uranium Rounds at Nellis Air Force Range Target 63–10* (1998). The United States fired American-manufactured tank-piercing shells with exploding depleted uranium (DU) tips in the Gulf War of 1991, as did NATO forces in Bosnia-Herzegovina (10,000-plus rounds) in 1996. U.S. troops acting as part of NATO forces deployed in Kosovo in 1999. American forces probably also used DU-tipped munitions in Afghanistan in 2001 and more recently, and in the 2003 American Iraq invasion and its aftermath. The case of a DU-contaminated Iraq War veteran, and his child's deformity, is described by Dave Lindorff (Radioactive Wounds of War. *In These Times* 19 September 2005;6–7).

113. The $5.8 trillion figure includes building the bomb (7.0%), deploying the bomb (55.7%), targeting and controlling the bomb (14.3%), defending against the bomb (16.1%), dismantling the bomb (0.5%), nuclear waste management and environmental remediation (6.3%), victims of U.S. nuclear weapons (0.04%), nuclear secrecy (0.05%), and congressional oversight of nuclear weapons programs (0.02%) (S. I. Schwartz. Introduction. In S. I. Schwartz, ed. *Atomic Audit* [Washington, DC: Brookings Institution Press, 1988], figure 1).

114. Future costs also may have to include whatever comes of the U.S. government's 2002 plan to resume nuclear weapons production and testing. New nuclear weapons programs involve huge costs for developing and testing new delivery systems.

115. M. Koide et al. Records of Plutonium Fallout in Marine and Terrestrial Samples. *Journal of Geophysical Research* 1975;80:4153–4162. The radionuclides in terrestrial and marine sediments are principally plutonium and strontium-90. This study revealed the puzzling fact that strontium-90 fallout is higher over sea—double the amount over land (H. L. Volchok et al. Oceanic Distributions of Radionuclides from Nuclear Explosions. In National Research Council, *Radioactivity in the Marine Environment* (Washington, DC: National Academies Press, 1971), 42–89).

116. Some of the radioactivity may have natural oceanic sources, but radioactive concentrations in bird droppings never were reported before 2003 (Andy Coghlan. Sea Birds Drop Radioactivity on Land. *New Scientist* [2003]. Available: www.newscientist.com/news/news. jsp?ID=NS99993220).

117. National Research Council. *Long-Term Institutional Management of the U.S. Department of Energy Legacy Waste Sites* (Washington, DC: National Academies Press, 2000), 3–7.

118. The bold plan to destroy substantial numbers of nuclear warheads turned into an agreement between Russia and the United States to warehouse them instead. "Destruction" and "liquidation" seem somehow incompatible with retention by both countries of their full in-hand nuclear arsenals, but President George W. Bush promised that "this treaty will liquidate the legacy of the Cold War" (D. E. Sanger. Russians, U.S. Agree to Nuke Arms Cuts. *New York Times* 14 May, 2002). The 2002 Moscow treaty called for reducing the number of U.S.-deployed warheads to between 1,700 and 2,200 by 2012 (U.S. Government Accountability Office. *Nuclear WeaponsViews on Proposals to Transform the Nuclear Weapons Complex*). The United States has about 10,100 nuclear warheads, of which 5,735 are operational.

119. The United States boycotted the November 2001 United Nations conference to encourage support for the Comprehensive Test Ban Treaty (D. G. Kimball. CTBT Rogue State? *Arms Control Today* 1 December 2001). The National Academy of Sciences has discredited objections of 2001 opponents to the Comprehensive Nuclear Test Ban Treaty, which led to the U.S. failure to ratify it (National Academy of Sciences, *Committee On Technical Issues Related to the Comprehensive Nuclear Test Ban Treaty* (Washington, DC: National Academies Press, 2002); L. R. Sykes. Four Decades of Progress in Seismic Identification Help Verify the CTBT. *EOS, Transactions of the American Geophysical Union* 2002;83:497, 500).

120. Excerpts from the Nuclear Posture Review submitted to Congress 31 December 2001 indicated a desire to develop a "modern, responsive nuclear weapons sector of the infrastructure," including expanded capacity to assemble nuclear warheads, produce and certify nuclear triggers (plutonium pits), and resume tritium production. Among other things, the posture would provide the flexibility to allow nuclear destruction of hard and deeply buried targets; develop "a revitalized nuclear weapons complex that will…be able, if directed, to design, develop, manufacture, and certify [i.e., test] new warheads in response to new national requirements; and maintain readiness to resume underground nuclear testing if required." As for The Comprehensive Test Ban, this Nuclear Posture Review asserts that

[w]hile the United States is making every effort to maintain the [nuclear bomb] stockpile without additional nuclear testing, this may not be possible for the indefinite

future. Each year DOD and DOE will reassess the need to resume nuclear testing and will make recommendations to the President. Nuclear nations have a responsibility to assure the safety and reliability of their own nuclear weapons. (Global Security. *Nuclear Posture Review* (2002). Available: www.globalsecurity.org) See also excellent publications by Brice Smith (The "Usable" Nuke Strikes Back. *Science for Democratic Action* 2003;11:1–8), Arjun Makhijani and Lisa Ledwidge (Back to the Bad Old Days. *Science for Democratic Action* 2003;11:1, 9–13, 16), Julian Borger (Dr. Strangeloves Meet to Plan New Nuclear Era. *The Guardian*, 7 August 2003), David Malakoff (New Nukes Revive Old Debate. *Science* 2003;301:32–34), and the Natural Resources Defense Council (*Faking Nuclear Restraint* (2002). Available: nrdc.org/nuclear/restraint/asp; see also Natural Resources Defense Council. *The Moscow Treaty's Hidden Flaws* [2003]).

121. Arjun Makhijani, Brice Smith, and Michael C. Thorne. Science for the Vulnerable: Setting Radiation and Multiple Exposure Environmental Standards to Protect Those Most at Risk. *Institute for Energy and Environmental Research* (19 October 2006). Available: www.ieer.org/campaign/report.pdf.

122. Martin Enserink. Bracing for Gulf War Syndrome II. *Science* 2003;299:1966–1967.

123. Corbin Harney. Available: www.shundahai.org.

Chapter 8

1. A mega-conurbation is a huge melding of separate communities that have grown together, eliminating rural land that once separated them (Lewis Mumford. The Natural History of Urbanization. In W. L. Thomas, Jr., ed. *Man's Role in Changing the Face of the Earth* [Chicago: University of Chicago Press, 1956], 382–398). Between 1982 and 2002, the amount of developed land in the contiguous United States increased more than 47%, to 107 million acres—an area larger than the State of California. The pace of development increased 57% from the 1980s to the 1990s (U.S. Department of Agriculture Natural Resources Conservation Service. *Natural Resources Inventory* 2002 [2002]).

2. James Howard Kunstler. *The Geography of Nowhere* (New York: Simon and Schuster, 1993); this book is rich with insights into what went wrong in the way America developed.

3. The 51 million acres figure—equivalent to 2.7% of contiguous U.S. surface area under urban developments (including water)—is based on nighttime satellite images, with "city lights" showing up urban areas (M. L. Imhoff et al. Assessing the Impact of Urban Sprawl on Soil Resources in the United States Using Nighttime "City Lights" Satellite Images and Digital Soil Maps. In T. D. Sisk, ed. *Biological Science Report* [USGS/BRD/BSR 1998–0003, U.S. Geological Survey, Biological Resources Division, revised September 1999]). The larger figure represents a survey of developments on nonfederal land and includes outlying built-up areas of 10 acres or more, plus rural transportation corridors linking cities and suburbs. Most federal lands are undeveloped in the west, so the acreage may be slightly underestimated but probably is more correct than the 51 million acres figure (U.S. Census Bureau. *Statistical Abstract of the United States* [2001], table 383. Available: www.census.gov/statab/www/).

4. The top five states for population growth between 1990 and 2000 are Nevada, 66.3%; Arizona, 40.0%; Colorado, 30.6%; Utah, 29.6%; and Idaho, 28.5%. In 1990–2000, the 11 western states included nearly half (15) of the 37 fastest-growing metropolitan areas in the

United States, having 25% or greater growth—with Las Vegas far in the lead (83.3% population growth) (Giantglossary. Available: www.giantglossary.com/).

5. Leon Kolankiewicz and Roy Beck. *Weighing Sprawl Factors in Large U.S. Cities* (2001). Available: NumbersUSA.com.

6. Kolankiewicz and Beck. *Weighing Sprawl Factors.*

7. U.S. Census Bureau. *Statistical Abstract of the United States*, table 16.

8. These 8,000-square-foot houses are pipsqueaks compared to Bill Gates's compound along the shores of Lake Washington, totaling some 66,000 square feet and costing $97 million. Clustered developments of huge single-family residences constitute "mansionization."

9. U.S. Geological Survey. *Effects of Urbanization on Stream Ecosystems* (Fact Sheet FS-042–02, 2002). Clearing forests commonly starts the early, rapid degradation of stream ecosystems, altering the erosional and stream-flow characteristics, the stream temperature, and habitat. These physical processes may severely degrade plant and animal associations well before chemical contaminants from the urbanized areas can further degrade the ecosystems.

10. Margaret Maizel et al. Historical Interrelationships Between Population Settlement and Farmland in the Coterminous United States, 1790–1992. In T. D. Sisk, ed. *Biological Science Report* (USGS/BRD/BSR 1998–0003, U.S. Geological Survey, Biological Resources Division, revised September 1999).

11. Thirty-eight percent of the 1,801 plant and animal species listed, or proposed for listing, under the Endangered Species Act are threatened by agriculture and 35% by urbanization (D. S. Wilcove et al. Quantifying Threats to Imperiled Species in the United States. *BioScience* 1998;48:607–615). The proportion of species affected by different sources of habitat destruction are agriculture, 38%; urbanization, 35%; reservoirs and other water developments, 30%; outdoor recreation, 27%; livestock grazing, 22%; infrastructure, including roads, 17%; modified fire ecology, 14%; logging, 12%; mining and oil/gas development, 11%. The percentages add to more than 100 because different causes may combine to threaten more than one species (D. J. Tazik and C. O. Martin. Threatened and Endangered Species on U.S. Department of Defense Lands in the Arid West, USA. *Arid Land Research and Management* 2002;16:259–276).

12. The largest 12 sprawlers in the western United States consumed the following additional acreage between 1970 and 1990: Los Angeles, Calif., 252,000; Phoenix, Ariz., 227,000; San Diego, Calif., 198,000; San Francisco, Calif., 124,000; Seattle, Wash., 112,000; Denver, Colo., 106,000; Riverside, Calif., 96,000; Tucson, Ariz., 91,000; Portland, Oreg., 77,000; Albuquerque, N.M., 71,000; Las Vegas, Nev., 70,000; Tacoma, Wash., 67,000. The change in per capita land consumption for these cities varied from a 35% decline to a 21% increase, reflecting a variety of land use policies (Kolankiewicz and Beck. *Weighing Sprawl Factors*).

13. Marc Reisner. *Cadillac Desert* (New York: Penguin Books, 1986), chapter 4. The dewatering and destruction of the Colorado River's delta, and the impact on fishermen's livelihoods, also is covered in the 1997 *Cadillac Desert* TV mini-series, based on Reisner's book.

14. As the U.S. Fish and Wildlife Service explains it, "The adverse modification standard may be reached closer to the recovery end of the survival continuum, whereas the jeopardy standard traditionally has been applied nearer to the extinction end of the continuum" (Kassie Siegel. *A Synopsis of the "Adverse Modification" Standard Governing the Protection of Critical Habitat: The Failure of the U.S. Fish and Wildlife Service and the National Marine*

Fisheries Service to Uphold That Standard (Center for Biological Diversity, 6 December2000); see also Kassie Siegel. A *Review of Recent Court Cases Upholding the Importance of Critical Habitat to the Conservation of Threatened and Endangered Species* [Center for Biological Diversity, 6 December, 2000]).

15. Martin Taylor et al. *Critical Habitat Significantly Enhances Endangered Species Recovery* (Center for Biological Diversity, 2003). Available: www.biologicaldiversity.org/swcbd/programs/policy/ch/index.html. Critical habitat areas have been designated for only 34% of the species under recovery plans. In April 2004, the George W. Bush administration announced new procedures that would further curtail critical habitat designation if it is not supported by "sound science," plus economic analysis, and will be applied only in limited areas vital to species conservation. This use of "sound science" tends to give it a political rather than a scientific meaning.

16. Federal regulations allow wetlands to be filled for development, mitigated by replacing every destroyed wetlands acre with 1.78 acres of manufactured wetlands (National Research Council. *Compensating for Wetland Losses Under the Clean Water Act* [Washington, DC: National Academies Press, 2001]). Replacing wetlands is costly and has a high percentage of failures (87% in the state of Washington), due to poor designs, poor construction, and lack of (costly) maintenance.

17. Habitat for the bay checkerspot butterfly (*Euphydryas editha bayensis*), listed as threatened in 1987, is widely scattered outcrops of serpentinite (altered rocks from deep in the earth) in the San Francisco Bay area. Serpentinite soils support a unique array of plants, which in turn support unusual animals, all adapted to grow in very nonnutritious soils. The sizes of individual serpentinite outcrops are so small that checkerspots may die out entirely in some favorable areas from time to time, and surviving populations eventually reoccupy the vacated habitats. For this reason, critical habitat for the species should allow both occupancy and suitability of serpentinite sites, and the ability to migrate among those sites. Rapid Bay Area urbanization, accompanied by exotic plant invasions, constitutes the principal threat to the bay checkerspot's survival and the reason for listing (D. D. Murphy and S. B. Weiss. Ecological Studies and the Conservation of the Bay Checkerspot butterfly, *Euphydryas editha bayensis. Biological Conservation* 1988;46:183–200). In 2001, nearly 24,000 acres in San Mateo and Santa Clara counties in California were designated critical habitat for the bay checkerspot; the critical habitat lies within the first-of-its-kind HCP adopted in 1983 for a number of plant and animal species. A lawsuit brought by the Pacific Legal Foundation, settled in March 2006, requires reevaluation of critical habitat of the bay checkerspot to ensure "a balanced approach to environmental policy that takes into account the well-being of humans and the economy as well as the status of at-risk species."

18. California City has a population density of 11.4 acres per person, and Salton City, 128 acres per person. Compare with Los Angeles's 0.12 acres/person and about 0.20 acres/person in Bakersfield and Fresno (Kolankiewicz and Beck. *Weighing Sprawl Factors*, appendix B).

19. Howard Wilshire. Human Causes of Wind Erosion in California's Deserts. In D. R. Coates and J. D. Vitek, eds. *Thresholds in Geomorphology* (London: George Allen and Unwin, 1980), 415–433; J. K. Nakata et al. Origin of Mojave Desert Dust Storms Photographed From Space on 1 January, 1973. In T. L. Péwé, ed. *Desert Dust: Origin, Characteristics, and Effect on Man* (Special Paper 186, Geological Society of America, 1981), 223–232. Water filled the Salton basin before and during the ice ages, but it dried up long before Europeans arrived in North America. Today's Salton Sea accidentally developed in 1905, when the Colorado River

breached an irrigation dike. River water flowed into the Salton Sink for two years before it could be diverted back to the main channel.

20. In about 1982 a tourist stopped along Arizona highway 95 where some geologists (including Jane Nielson) were working. "So where is this damn bridge?" He asked. "It's that one in town with all the flagpoles," said one of the geologists. "That one?" said the tourist. "That's just a plain stone bridge." Agreed the geologist, "Yeah, just a plain stone bridge."

21. Wilcove et al. Quantifying Threats to Imperiled Species. Assessing interactions between the various causes of habitat loss yields the same results (Brian Czech et al. Economic Associations Among Causes of Species Endangerment in the United States. *BioScience* 2000;50:593–601).

22. J. M. Shepherd and S. J. Burian. Detection of Urban-Induced Rainfall Anomalies in a Major Coastal City. *Earth Interactions* 2003;7; Daniel Rosenfeld. Suppression of Rain and Snow by Urban and Industrial Air Pollution. *Science* 2000;287:1793–1796; O. B. Toon. How Pollution Suppresses Rain. *Science* 2000;287:1763–1764.

23. National Aeronautics and Space Administration. *Here Comes Urban Heat* (2000). Available: science.nasa.gov. Satellite imagery of Georgia shows that Atlanta's urban sprawl consumed 380,000 acres of trees at an average rate of 55 acres per day from the 1970s through the 1990s.

24. Amir Givati and Daniel Rosenfeld. Quantifying Precipitation Suppression Due to Air Pollution. *Journal of Applied Meteorology* 2004;43:1038–1056; Daniel Rosenfeld and Amir Givati. Evidence of Orographic Precipitation Suppression by Air Pollution-Induced Aerosols in the Western United States. *Journal of Applied Meteorology and Climatology* 2006;45:893–911; see also Amir Givati and Daniel Rosenfeld. Separation Between Cloud-Seeding and Air-Pollution. *Journal of Applied Meteorology* 2005;44:1298–1314.

25. Despite their concentration in chaparral scrub lands, and not forests, and the fact that 65% of the burned areas were private or state lands, the 2003 fires are likely to become a battering ram for pushing President George W. Bush's Healthy Forests initiative through Congress (see chapter 1).

26. In the 1950s, a series of freshwater injection wells were drilled to create a freshwater "wall," preventing further saltwater infiltration. They proved ineffective barriers because they prevent direct inflow of seawater only through aquifers exposed at the sea floor. The seawater finds other pathways, including through buried ancient stream channels and fault zones filled with crushed rock (B. D. Edwards and K. R. Evans. *Saltwater Intrusion in Los Angeles Area Coastal Aquifers, the Marine Connection* (U.S. Geological Survey Fact Sheet 030–02, 2002); J. A. Izbicki. *Seawater Intrusion in a Coastal California Aquifer* (U.S. Geological Survey Fact Sheet 125–96, 1996)).

27. D. B. Levy et al. The Shallow Ground Water Chemistry of Arsenic, Fluorine, and Minor Elements, Eastern Owens Lake, California. *Applied Geochemistry* 1999;14:53–65; G. S. Plumlee and T. L. Ziegler. The Medical Geochemistry of Dusts, Soils, and Other Earth Materials. In B. S. Lollar, ed. *Treatise on Geochemistry*, vol. 4 (New York: Elsevier, 2003), 263–310.

28. Nicholas Pinter. One Step Forward, Two Steps Back on U.S. Floodplains. *Science* 2005;308:207–208.

29. D. W. Kolpin et al. Pharmaceuticals, Hormones, and Other Organic Wastewater Contaminants in U.S. Streams, 1999–2000, a National Reconnaissance. *Environmental Science and Technology* 2002;36:1202–1211. Reports of recent research at Oregon State University

state that a city's treated wastewater can be used to show the range of drugs, both therapeutic and recreational, being used in American communities. (Wastewater Sampling Provides Community Drug Test. Associated Press (22 August 2007). Available: www.cbc.ca/new/story/2007/08/22/tech-drugs.html.)

30. Jan Svejkovsky and Burton Jones. Satellite Imagery Detects Coastal Stormwater and Sewage Runoff. *EOS, Transactions American Geophysical Union* 2001;82:621–630.

31. J. G. B. Derraik. The Pollution of the Marine Environment by Plastic Debris, a Review. *Marine Pollution Bulletin* 2002;44:842–852. This review found that plastic debris has had adverse effects on 86% of all sea turtle species, 44% of all seabird species, and 43% of all marine mammal species (Kevin Krajick. Message in a Bottle. *Smithsonian* July2001;36–47).

32. The U.S. Environmental Protection Agency listed a perfluorinated chemical (perfluorooctanoic acid) as a probable carcinogen (Lancaster Centre for Chemicals Management [10 December 2006]. Available: www.lec.lancs.ac.uk/ccm/research/perfluorinated/index.htm).

33. Kristina Thayer et al. *Toxic to Animals and People, Persistent Forever, Pervasive in Human Blood: PFCs, a Family of Chemicals That Contaminate the Planet* (Environmental Working Group, 2003).

34. In 1996, the U.S. General Accounting Office reported wide state-to-state variations in the controls imposed on pollutant discharges into the nation's waters, even from pollutant point sources (U.S. General Accounting Office. *Water Pollution, Differences Among the States in Issuing Permits Limiting the Discharge of Pollutants* [GAO/RCED-96–42, 1996]). For example, 1,407 permits examined for municipal discharges containing cadmium, copper, lead, mercury, and zinc, showed that only 12–14% of the permits imposed discharge limits, whereas 67–75% imposed no controls, and the rest (14–16%) required only monitoring. In 2000, the General Accounting Office reported a still more dismal situation for logging, agricultural, urban development, and other nonpoint pollutant sources, finding that only 6 of 50 states had most of the data needed to fully assess their state's water quality, and only three states reported having a majority of the data needed to develop total maximum daily loads for nonpoint pollutant sources (U.S. General Accounting Office. *Water Quality, Identification and Remediation of Polluted Waters Impeded by Data Gaps* [GAO/T-RCED-00–88, 2000]).

35. Richard Caplan. *Permit to Pollute, How the Government's Lax Enforcement of the Clean Water Act Is Poisoning Our Waters* (U.S. Public Interest Research Group, 2002).

36. Toxicity is defined as the dose level that kills. As the early chemist/toxicologist known as Paracelsus (1493–1541) put it, "The dose makes the poison." Chemical company representatives like to point out that essential foods such as table salt, and even water, can be poisonous if administered in large enough doses. However, even very small exposures to chemicals that have no role in body or organ functions can build up if they persist over long times, because they may accumulate in living tissues (bioaccumulation).

37. James Howard Kunstler. *The Long Emergency: Surviving the Converging Catastrophes of the Twenty-First Century* (New York: Atlantic Monthly Press, 2005), chapter 7; Richard Heinberg. *Power Down: Options and Actions for a Post-carbon World* (Gabriola Island, Canada: New Society Publishers, 2004), chapter 5. The unsustainability of our conventional food supply is underscored by the estimate that an average food item on an American dinner plate travels more than 1,500 miles by truck from source to supermarket (Rich Pirog et al. *Food, Fuel, and Freeways* [Leopold Center for Sustainable Agriculture, Iowa State University, 2001]).

Chapter 9

1. Ruth Langridge. Changing Legal Regimes and the Allocation of Water Between Two Rivers. *Natural Resources Journal* 2002;42:283–330.

2. John Wesley Powell was the first white man to lead an exploration party through the Grand Canyon. He later helped establish and direct the U.S. Bureau of Ethnology (now part of the Smithsonian Institution) and the U.S. Geological Survey. The son of a midwestern farmer, Powell took a truly conservative position on agricultural water supplies as second U.S. Geological Survey Director. He delivered his proposal on restricting western irrigation to booing delegates at a 1893 International Irrigation Congress (Wallace Stegner. *Beyond the Hundredth Meridian* [Sentry edition, Boston: Houghton Mifflin, 1966; first published 1953, Riverside Press, Cambridge, Mass.], 336–343).

3. Langridge. Changing Legal Regimes.

4. Washington Alliance for a Competitive Economy. *Water Case Study: Water for a Growing Economy* (2003). Available: www.awb.org/otherissues/competitiveness/2003reportwaterB.asp.

5. Stegner. *Beyond the Hundredth Meridian*, 336–343.

6. Marc Reisner. *Cadillac Desert* (London: Penguin Books, 1986), 73–93, 131–133, 168–175, 181–189, 202–209. Some of America's richest agribusinesses are doubly subsidized. Many Central Valley Project farms use federally subsidized irrigation water to grow federally subsidized crops, especially rice. Some California dairy operations receive the subsidized water to grow subsidized corn for cattle feed, producing cheese, milk, and other subsidized dairy products (Environmental Working Group. *Double Dippers: How Big Ag Taps into Taxpayers Pockets* [2005]. Available: www.ewg.org).

7. Figures are based on 2000 U.S. Census Bureau data (Available: www.census.gov), and 2000 U.S. Geological Survey consumption data for publicly supplied water (S. S. Hutson et al. *Estimated Use of Water in the United States in 2000* [U.S. Geological Survey Circular 1268, revised February 2005], table 2. Available: pubs.usgs.gov/circ/2004/circ1268/htdocs/text-total.html). In 2006 Arizona's growth exceeded Nevada's.

8. California Energy Commission Staff. *California's Water-Energy Relationship* (prepared in support of the 2005 Integrated Energy Policy Proceeding; 04-IEPR-01E, 2005). Available: www.energy.ca.gov/2005publications/CEC-700–2005-011/CEC-700–2005-011-SF.PDF.

9. Michael Collier et al. *Dams and Rivers: A Primer on the Downstream Effects of Dams* (U.S. Geological Survey Circular 1126, 1996), 2; this is an excellent, very readable account of the downstream effects of dams.

10. Collier et al. *Dams and Rivers*; R. M. Hirsch et al. The Influence of Man on Hydrologic Systems. In *The Geology of North America*, vol. 0–1, *Surface Water Hydrology* (Geological Society of America Special Publication, 1990), 329–359; Christer Nilsson et al. Fragmentation and Flow Regulation of the World's Large River Systems. *Science* 2005;308:405–408.

11. Inundating land under reservoirs is most harmful on prime agricultural areas in the nation's temperate zones. In the arid west, inundating critically important riparian habitat—only 1% of western arid lands—is especially harmful.

12. P. E. Grams. *Degradation of Alluvial Sand Bars Along the Snake River Below Hells Canyon Dam, Hells Canyon National Recreation Area, Idaho* (senior dissertation, Middlebury College, 1991), cited in Collier et al. *Dams and Rivers*, n. 2.

13. G. P. Williams and M. G. Wolman. *Downstream Effects of Dams on Alluvial Rivers* (U.S. Geological Survey Professional Paper 1286, 1984).

14. Collier et al. *Dams and Rivers*, n. 2.

15. Las Vegas dumps 170 million gallons of treated effluent into Las Vegas Wash daily, an arm of Lake Mead reservoir behind Hoover Dam, which also supplies the city's drinking water. Dumping the effluent has deeply eroded the wash and wiped out wetlands, while adding sediment and other pollutants to the lake. Drought dropped the level of Lake Mead, letting polluted water from Las Vegas Wash move closer to its drinking water intakes. In 2004, Las Vegas extended one of the intakes by 50 feet at a cost of $6.5 million, and is planning to build a third intake, 200 feet deeper than existing ones, at a cost of $650 million. Another water-protective proposal is to pipe the effluent into deeper water close to Hoover Dam, which could pass the problem downstream. Rio Grande fish are polluted with pesticides of both urban and agricultural origin, and toxic metals from mining further impair water quality. Six of ten sites sampled by U.S. Geological Survey show significant habitat degradation from pollution, bank erosion, and loss of riparian vegetation. Will Rogers once described the Rio Grande as "the only river I ever saw that needed irrigation" (G. W. Levings et al. *Water Quality in the Rio Grande Valley, Colorado, New Mexico, and Texas, 1992–95.* [U.S. Geological Survey Circular 1162, 1998]).

16. At least 2,000 irrigation dams in the U.S. have filled with sediment already (Wes Jackson. *Man and the Environment* [Dubuque, Iowa: Wm. C. Brown Company, 1971]), and many thousands of small stock ponds likely are filled. Examples of the useful remaining life for western U.S. dams as of 2002 (based on the capacity loss rate at time of measurement) are Guernsey, North Platte River Wyo., 27 years; Alamogordo, Pecos River, N.M., 59 years; Black Butte, Stony Creek, Calif., 96 years; Conchas, Canadian River, N.M., 226 years; Elephant Butte, Rio Grande, N.M., 251 years; San Carlos, Gila River, Ariz., 439 years; Hoover Dam, Colorado River, Ariz.-Nev., 648 years (Reisner. *Cadillac Desert*, 491–492).

17. T. C. Winter et al. *Ground Water and Surface Water, a Single Resource* (U.S. Geological Survey Circular 1139, 1999). An excellent report, readable and understandable by the lay public. Another very useful, much-reprinted report is R. C. Heath (*Basic Ground-Water Hydrology* [U.S. Geological Survey Water-Supply Paper 2220, 1983]).

18. Fossil water accumulated over the millions of years when nobody was around to extract it from wells. Some of the water was "connate," that is, present in the rock when it formed.

19. H. G. Wilshire et al. *Geologic Processes at the Land Surface* (U.S. Geological Survey Bulletin 2149, 1996), 5. Declining groundwater levels and their effects are widespread in the western United States (S. A. Leake et al. *Desert Basins of the Southwest* [U.S. Geological Survey Fact Sheet 086–00, 2000]).

20. Devin Galloway et al., eds. *Land Subsidence in the United States* (U.S. Geological Survey Circular 1182, 1999).

21. Water spreading is the unauthorized use of federally subsidized water on lands not suitable for irrigation, and/or on lands beyond authorized project boundaries. The Bureau of Reclamation classifies lands as irrigable or nonirrigable based on their suitability for sustainable agriculture, but the agency has not yet classified large tracts of the western U.S. (Public Employees for Environmental Responsibility. Water Rustling in the West. *PEEReview* Winter 1999).

22. Environmental Working Group. *Soaking Uncle Sam: Why Westlands Water District's New Contract Is All Wet* (2005). Available: www.ewg.org. Westlands pays the Bureau of Reclamation as little as $31 an acre-foot for federal water, compared to $200 per acre-foot that southern California cities pay for state project water, and $500 per acre-foot that northern California's Marin County pays for Russian River Water. The Westlands U.S. Bureau of Reclamation contract is only one of about 200 that give California's Central Valley agribusiness control over the same amount of water they received 40 years ago, despite reducing agricultural demand. U.S. Congressman George Miller said, "This isn't about farming. It's about building an annuity for people who want to sell government-subsidized water to Southern California or whoever needs it" (Bettina Boxall, Water Pacts Give State's Growers New Profit Stream. *Los Angeles Times* [16 February, 2005].)

23. Reisner (*Cadillac Desert*, 129–131, chapter 8) gives a fascinating account of the water allocation misconceptions and political machinations among the seven states containing the Colorado River and its major tributaries, the Green and San Juan rivers. See also Glenn Totten. On the Edge: Defusing Tensions on the Colorado River. *Western Water* July/August 2005;4–13; Sue McClurg. Facing the Future: Modifying Management of the Colorado River. *Western Water* January/February 2006;4–13.

24. The other states feared that large and politically powerful California would grab most of the water, so wrangling over allocations went on for decades. Only six states ratified the Compact in 1928, because Arizona held out for concessions. For a detailed breakdown of the various compacts, laws, and agreements affecting distribution of Colorado River waters, see G. A. Mueller and P. C. Marsh (*Lost, a Desert River and Its Native Fishes, a Historical Perspective of the Lower Colorado River* [U.S. Geological Survey Information Technology Report USGS/BRD/ITR 2002–0010, 2002], 22–23).

25. The original allocations were based on the U.S. Bureau of Reclamation's river flow measurements between 1905 and 1922, yielding an average 17.5 million acre-feet per year (Mafy). Some flows during that period, averaging 16.1 million acre-feet at Lee's Ferry, turned out to be the highest of the twentieth century (R. H. Webb et al. *Climatic Fluctuations, Drought, and Flow in the Colorado River* [U.S. Geological Survey Fact Sheet 3062–04, 2004]. Available: water.usgs.gov/preview/pubs/fs/2004/3062). Accepted "normal" western rainfall levels are based on abnormally high precipitation between 1925 and 2000. The mismatch of actual to theoretical river flows, and the relatively few federally funded projects in Colorado and Utah, limited the total Upper Basin take. By the 1960s, the allocations clearly could not be satisfied unless Colorado River flows were at least 2.5 Mafy more. The agreement compels Upper Basin states to deliver 8.23 Mafy (on a 10-year moving average) to cover state appropriations for the Lower Basin states, reserving 1.5 Mafy for Mexico. Proposals for augmenting the Colorado River included building dams in the Grand Canyon and generating high-priced peak power to finance water imports from northern California, Washington, or Canada. Huge public opposition to dams in the Grand Canyon, and little northern interest in selling their water, eventually squelched the schemes. Studies indicate that the Upper Basin states may suffer much more than Lower Basin states in an extended drought (L. J. MacDonnell et al. The Law of the Colorado River: Coping with Severe Sustained Drought. *Water Resources Bulletin* 1995;31:825–836; B. L. Harding, et al. Impacts of a Severe Sustained Drought on Colorado River Water Resources. *Water Resources Bulletin* 1995;31:815–824).

26. The Arizona Department of Water Resources Web site provides a short description of all Colorado Compact agreements and amendments (Available: www.azwater.gov/dwr/ColoradoRiverManagement/Law_of_River.html).

27. Langridge. Changing Legal Regimes and the Allocation of Water Between Two Rivers.

28. Water downstream from Elephant Butte Reservoir is mostly allocated to Texas irrigators, but New Mexico irrigators can extract a small percentage. Flow measurements that determine the level of Rio Grande Compact water allocations, meant to preserve documented early twentieth-century water consumption patterns, are measured at Santa Fe, Socorro, and other river gages and on principal tributaries (*Rio Grande Compact* (Fort Collins, Colorado: Colorado State University, n.d.)). The six Pueblos—Cochiti, Santo Domingo, San Felipe, Santa Ana, Sandia, and Isleta—all are on the Rio Grande mainstem in New Mexico. The Pueblos were promised sufficient water to irrigate 8,847 acres for domestic and livestock needs, and for all "newly reclaimed lands," in total giving the six Pueblos sufficient water for more than 20,000 acres (Peter Chestnut. *A Pueblo Perspective on the Rio Grande Project* [Water Resources Research Institute Conference Proceedings, Colorado State University, Cooperative Extension Southwest Region, 1999]).

29. "[L]ike much western water development, [the San Juan–Chama Project] is an unsettling story that speaks directly to the limits of technical expertise, the treacherous allure of the grandiose solution, and the consequences that can unfold when the government opens its checkbook and closes its regulation manual" (Charles F. Wilkinson. *Crossing the Next Meridian* [Covelo, California: Island Press, 1992], 219–231; see also National Research Council. *Water Transfers in the West: Efficiency, Equity, and the Environment* [Washington, DC: National Academies Press, 1992], chapter 7).

30. The silvery minnow (*Hybognathus amarus*) was listed as endangered in 1994 and the southwestern willow flycatcher (*Empidonax traillii extimus*) was listed in 1995. The threatened Rio Grande bosque habitat consists of cottonwood and willow ecosystems.

31. Albuquerque's San Juan–Chama allotment is 48,200 acre-feet per year. Critical Rio Grande and Pecos River water shortages started with a six-year drought in the 1950s, when deliveries below Elephant Butte Dam fell more than 500,000 acre-feet short of contracted amounts, violating the Rio Grande Compact prohibition against deficits above 200,000 acre-feet. To meet the deliveries, the state increased groundwater mining at Albuquerque and built the "low-flow conveyance channel" to Elephant Butte Dam, which drained about 200,000 acre-feet of groundwater into the river channel (New Mexico Bureau of Geology and Mineral Resources. *Managing Our Most Important Resource*. New Mexico Earth Matters. Summer 2003;1–4).

32. J. M. Kernodle et al. *Simulation of Groundwater-Flow in the Albuquerque Basin, Central New Mexico, 1901–1994, with Projections to 2020* (U.S. Geological Survey Water-Resources Investigations Report 94–4251, 1995); see also D. P. McAda and Peggy Barroll. *Simulation of Ground-Water Flow in the Middle Rio Grande Basin Between Cochiti and San Acacia, New Mexico* (U.S. Geological Survey Water-Resources Investigations Report 02–4200, 2002); J. R. Bartolino and J. C. Cole. *Ground-Water Resources of the Middle Rio Grande Basin, New Mexico* (U.S. Geological Survey Circular 1222, 2002).

33. Albuquerque Official City Web site. *Water Information for Albuquerque*. Available: www.cabq.gov/waterresources/sjc.html. Albuquerque now pumps an unsustainable 110,000 acre-feet per year (Afy) of groundwater, about half of which (60,000 Afy) naturally flows back to the aquifer from the Rio Grande. Albuquerque also returns 55,000 Afy to the river as treated wastewater. The city has rights to 48,200 Afy of San Juan–Chama water and plans to divert 94,000 Afy from the enhanced Rio Grande—nearly twice the allotment—and return 47,000 Afy of treated wastewater.

34. U.S. Geological Survey. *Estimating Evaporative Losses Incurred by Conveyance of City of Albuquerque San Juan-Chama Water: Jemez to Albuquerque* (2006). Available: nm.water. usgs.gov/EvapLoss.htm.

35. If the city discharged part of the groundwater back into the river, it would increase the flow over the short term because water moves very slowly back into a depleting aquifer. But even if groundwater pumping ceased immediately, the aquifers will continue steadily withdrawing river water to reestablish equilibrium (S. S. Papadopulos and Associates, Inc. *Middle Rio Grande Water Supply Study* (prepared for the U.S. Army Corps of Engineers and the New Mexico Interstate Stream Commission, Executive Summary, 2000), ES-1–ES-5).

36. Wikipedia. *Navajo Indian Irrigation Project*. Available: en.wikipedia.org/wiki/Navajo_indian_ irrigation_project. The Jicarilla Apache Nation reservation lies entirely within New Mexico.

37. Wikipedia. *Navajo Indian Irrigation Project*:

Composed of over 110,000 acres the project is currently 70% complete (with about 68,000 acres in production). In many respects, this development is controversial because the project has had numerous economic and structural problems throughout its history. One of the main complaints…is that it has taken over 40 years to construct and still remains unfinished. High federal budget deficits are sometimes blamed.…Some critics also contend this project does not rate high budgetary priority because it benefits only Native Americans.

Ironically, the San Juan–Chama could never have made it through Congress without the NIIP tie.

38. Wilkinson. *Crossing the Next Meridian*, 229.

39. Current flows likely are lower, since the Rio Grande and San Juan River headwaters were depleted by extreme to exceptional drought from May 2002 through August 2004.

40. In 1952, Arizona filed a lawsuit to stop California's expanding piracy and claim its full share of Colorado River water. After many years of legal wrangling, the U.S. Supreme Court resolved the lawsuit in 1963, ruling in favor of Arizona on all important issues, much to California's surprise. That decision finally allowed Arizona congressmen to authorize the Central Arizona Project (Reisner. *Cadillac Desert*, 266–271).

41. Reisner. *Cadillac Desert*, 310–311.

42. K. L. Jacobs and J. M. Holway. Managing for Sustainability in an Arid Climate: Lessons Learned from 20 Years of Groundwater Management in Arizona, USA. *Hydrogeology Journal* 2004;12:52–65.

43. Webb et al. *Climatic Fluctuations*. The average Colorado River flow was 10.2 million acre-feet per year (Mafy) during the 1930s drought. The Colorado River's average annual volume, subtracting consumptive uses in the Upper Basin states, was 12.4 Mafy from 1895 through 2003. River inflow to the Lower Basin states at Lee's Ferry, Utah, declined 0.5 million acre-feet (Maf) per decade from 1895 through 2003. The lowest annual flow on record is 3.8 Maf in 2002, followed by 3.9 Maf in 1934 and 4.8 Maf in 1977.

44. Lake Powell reservoir is named after John Wesley Powell, who probably would have opposed the dam as hotly as the Sierra Club's David Brower. To satisfy Upper Basin water rights, Glen Canyon Dam must release 8.23 million acre-feet of water a year from Lake Powell, but river inflow in the late 1990s did not match the required release (Collier et al. *Dams and Rivers*, 67). For water level and capacity data for Lake Powell and Lake Mead, see

U.S. Department of the Interior (*Reclamation: Managing Water in the West: Upper Colorado Region*. Available: www.usbr.gov/uc/).

45. While the Pacific Northwest remained in severe drought, the winter storms in water-year 2004–2005 increased Rocky Mountains snowpacks and the Colorado River's flow. By late July 2005, Lake Powell storage had risen to 52% of capacity, and Lake Mead to 59%. In November 2006, Lake Powell stood at 51% and Lake Mead at 54% of capacity.

46. Thomas Piechota et al. The Western U.S. Drought: How Bad Is It? *EOS, Transactions of the American Geophysical Union* 2004;85(32):301, 304; E. R. Cook et al. Long-Term Aridity Changes in the Western United States. *Science* 2004;306:1015–1018.

47. In acquiring Owens Valley lands and water and diverting Owens River water to the Los Angeles aqueduct, Los Angeles employed "chicanery, subterfuge, spies, bribery, a campaign of divide and conquer, and a strategy of lies" (Reisner. *Cadillac Desert*, 65).

48. Reisner. *Cadillac Desert*, 65. The drying of Owens Lake resulted in massive dust storms from the dry lake bed. Geochemical evidence shows that dust from Owens Dry Lake is being transported at least 250 miles east of the dust source areas (M. C. Reheis et al. Geochemical Evidence for Diversity of Dust Sources in the Southwestern United States. *Geochimica et Cosmochimica Acta* 2002;66:1569–1587; see also M. C. Reheis. A 16-Year Record of Eolian Dust in Southern Nevada and California, USA: Controls on Dust Generation and Accumulation. *Journal of Arid Environments* 2006;67:487–520). In 2005, after 30 years of litigation, Los Angeles was forced to restore portions of the Owens River and undertake rehabilitation of the greatest dust-yielding Owens Lake bed areas. Methods for reducing dust include shallow flooding, restoration of native salt-tolerant vegetation, and covering the surface with gravel.

49. The state completed investigations of the Sacramento delta's freshwater supply in 1930. In 1940, the Bureau of Reclamation completed the Central Valley Project's first canal and began exporting delta water for irrigation. In 1944, the Shasta Dam and reservoir added water to the Sacramento River and delta during low-flow periods, limiting saline water invasion into the upper delta, and by 1951 the Delta Cross Channel was efficiently moving water to the south-flowing Delta Mendota Canal. In 1959, the state legislature passed the Delta Protection and Burns-Porter acts, to assist State Water Project financing and increase delta exports again. The State Water Project built the Oroville Dam on the Feather River and the California Aqueduct to carry the water south. By 1965, Southern Californians wanted more water, but the delta was suffering from saltwater intrusion. Voters could see that the Peripheral Canal would be more of a threat than a remedy, and defeated it and the package of statewide facilities for other water transfers by a three to two margin.

50. Wilkinson. *Crossing the Next Meridian*, 252; Reisner. *Cadillac Desert*, 356–357. Southern California's Metropolitan Water District (MWD) is a consortium of 26 cities and water districts, providing drinking water to nearly 18 million people in parts of Los Angeles, Orange, San Diego, Riverside, San Bernardino, and Ventura counties. The MWD's stated mission is to provide its service area with adequate and reliable supplies of high quality water to meet present and future needs, "in an environmentally and economically responsible way" (Metropolitan Water District of Southern California Web Site. Available: www.mwdoc.com/metropolitan_water_district.htm).

51. Gary Pitzer, CALFED at a Crossroads: A Decade of the Bay-Delta Program. *Western Water* (March-April, 2005).

52. California Legislative Analyst Web Site. Available: www.lao.ca.gov/laoapp/main.aspx.

53. Gary Pitzer. State Seeks Rehearing of CALFED Environmental Ruling. *Western Water* (November-December 2005;3).

54. Drought level classifications are normal, abnormally dry, moderate, severe, extreme, and exceptional (*U.S. Drought Monitor.* Available: www.drought.unl.edu). In the summer of 2001, and again in mid-2002, extreme to exceptional drought conditions spread from Montana to New Mexico. The Klamath River drainage basin was in severe to extreme drought from July to December 2001.

55. Rising in south-central Oregon, the middle and lower Klamath River crosses relatively dry parts of northern Oregon and California on its way to the wetter coast.

56. A 1995 federal solicitor's opinion found that endangered Klamath River species have the highest water rights priority and that the fish-dependent Indian tribes have a higher right than irrigators (Langridge. Changing Legal Regimes, 291–293, 302).

57. Lower Klamath and Tule Lake National Wildlife Refuges provide critical Pacific flyway stopovers. The Clear Lake Reservoir was designed to prevent flooding the farmlands by increasing evaporative water losses and storing water that otherwise would reach the refuges (K. A. Rykbost and Rodney Todd. An Overview of the Klamath Reclamation Project and Related Upper Klamath Basin Hydrology. In *Water Allocation in the Klamath Reclamation Project, 2001, an Assessment of Natural Resource, Economic, Social, and Institutional Issues with a Focus on the Upper Klamath Basin* [Corvallis, Oregon: Oregon State University and University of California, 2002], 52). In addition to migratory birds, the refuge wetlands provide habitat for bald eagles (Jeff Manning and W. D. Edge. Relationship Between Bald Eagle Biology and Federal Environmental Decisions on the Klamath Reclamation Project. In *Water Allocation in the Klamath Reclamation Project, 2001, an Assessment of Natural Resource, Economic, Social, and Institutional Issues with a Focus on the Upper Klamath Basin* (Corvallis, Oregon: Oregon State University and University of California, 2002), chapter 15). Low water has reduced migratory bird numbers in the refuge from seven million to two million.

58. The Endangered Species Act enjoins federal agencies to protect and preserve endangered and threatened species. The Upper Klamath Lake reservoir provides habitat for the Lost River sucker (*Deltistes luxatus*) and shortnose sucker (*Chasmistes brevirostris*), listed as endangered in 1988. The suckers spawn in the lake's Williamson and Sprague River tributaries and populate the reservoirs (D. F. Markle, personal communication, 2004). Human alterations have profoundly altered the lake's ecosystems, and diking and draining of adjacent wetlands for farming reduced it in size by about one-third. Upper Klamath Lake is very shallow, about seven feet deep in summer. Bureau of Reclamation began regulating water releases for irrigation and other purposes after 1919 construction of the Link River Dam (Rykbost and Todd, An Overview of the Klamath Reclamation Project, 47–49).

59. Anadromous fish live in ocean waters but swim up mainstem rivers and reproduce in freshwater tributary streams before dying. Coho salmon (*Oncorhynchus kisutch*) were listed as threatened in 1992.

60. On 25 July 2001, Interior Secretary Gale Norton ordered Upper Klamath Lake water released to ranchers for greening up their pastures. Although the water had been retained in the lake to benefit endangered fish, none of it was expected to reach the Lower Klamath National Wildlife Refuge. On a Fox News Sunday program, Secretary Norton cited the plight

of Klamath Basin farmers as "Exhibit A for why the [Endangered Species Act] needs an overhaul" (Endangered Species Coalition. Klamath "Exhibit A" for ESA Overhaul, *GREENLines* Issue 1439 (7 August 2001).

61. In contrast, Montana ranchers voluntarily have reduced irrigation diversions to leave water in the upper Big Hole River and other areas critical to dwindling populations of native Arctic grayling, a candidate for Endangered Species Act protection. In 2004 a $1 million federal fund was established to pay Montana ranchers to shorten their irrigation season and keep water flowing in the river.

62. Elizabeth Furse. High Price for Klamath Irrigation. *Seattle Times* 12 July 2001.

63. The alleged taking was not of irrigators' water, which does not belong to the land, but a taking of the *use* of the lands (George Woodward and Jeff Romm. A Policy Assessment of the 2001 Klamath Reclamation Project Water Allocation Decisions. In *Water Allocation in the Klamath Reclamation Project, 2001, an Assessment of Natural Resource, Economic, Social, and Institutional Issues with a Focus on the Upper Klamath Basin* [Corvallis, Oregon: Oregon State University and University of California, 2002], 349–351). In 2004, the George W. Bush administration settled a California Central Valley farmer and irrigation district lawsuit over early 1990s water allocation reductions for $16.7 million to protect endangered fish. The settlement was intended as a precedent for crippling endangered species protection without changing the Endangered Species Act.

64. U.S. Fish and Wildlife Service. *Klamath River Fish Die-off September 2002* (AFWO-F-02–03, 2003), 40–42.

65. Added Troy Fletcher, Klamath River Indians Tribal Executive Director, "We never in our worst nightmare imagined this would happen. This is a terrible I-told-you-so." Congressman Mike Thompson remarked, "We are six months into the administration's 10-year water plan and the result is over 30,000 dead salmon" (Public Employees for Environmental Responsibility. Political Pressure Lowers Klamath Water Levels. *PEEReview* Winter 2003;3). On 17 July 2003, a federal district judge ruled the 10-year plan in violation of the Endangered Species Act, and an unprotective National Marine Fisheries Service Biological Opinion (BiOp) "arbitrary and capricious." The judge did not limit irrigation water deliveries to farmers, but ruled that the conflicting and uncertain nature of scientific opinion meant that the Bureau of Reclamation did not violate its own duty to protect endangered species. On 18 October 2005, the Ninth Circuit Court of Appeals remanded the case to the district court for injunctive relief to prevent harm to coho salmon in the Klamath River. On 27 March 2006, the district court granted an injunction and ordered defendants National Marine Fisheries Service and Bureau of Reclamation to reinitiate consultation on the Klamath Irrigation Project and issue a new BiOp based on the current scientific evidence and the full risks to threatened coho salmon. The court also directed the Bureau of Reclamation to limit Klamath Project irrigation deliveries if they would cause water flows in the Klamath River to fall below levels the National Marine Fisheries Service's BiOp specifically identified as necessary to prevent jeopardy, until a new Klamath Irrigation Project consultation is completed and reviewed by the district court.

66. September 2002 stream flows were among the four lowest recorded for that month on the Klamath mainstem since 1960, and water temperatures exceeded 65°F for nearly all of that month. Multiple days at or above 65°F can increase the likelihood of diseases (D. D. Lynch and J. C. Risley. *Klamath River Basin Hydrologic Conditions Prior to the*

September 2002 Die-off of Salmon and Steelhead [U.S. Geological Survey Water-Resources Investigations Report 03–4099, 2003]).

67. A National Research Council committee suggested paying farmers to reduce irrigation or not irrigate at all. In 2003 the Bureau of Reclamation established a water bank to enhance river flows for endangered fish, and may pay farmers $65 million in compensation through 2011. The water bank depends on groundwater pumping from private wells and has increased groundwater demand eightfold, causing serious water level decline and adversely affecting other wells. This is too much of a rob-Peter-to-pay-Paul program for a long-term solution to the Klamath Basin water competition (U.S. Government Accountability Office. *Klamath River Basin: Reclamation Met Its Water Bank Obligations, but Information Provided to Water Bank Stakeholders Could Be Improved* (GAO-05–283, 2005)).

68. Woodward and Romm. A Policy Assessment, 352; an excellent account of the tangled social and legal issues arising from Endangered Species Act implementation in the Klamath River Basin. Questioning the burden on farmers, farmworkers, and farm communities of withdrawing irrigation water for a broader public purpose (protecting endangered species), the authors also say, "Klamath Basin interests [need] to coalesce and build a social fabric that can shape how the [Endangered Species Act] is implemented, and how the benefits and burdens of satisfying specific public interests are distributed, within a context of diverse and important basin interests," 352 In contrast, Rykbost and Todd (An Overview of the Klamath Reclamation Project, 47, 61–62) attribute the "sudden" burden in part to 100 years of pampering farmers, who received water no matter what the supply limitations while the government trampled Native Americans' prior rights and neglected aquatic life and migratory birds. Water preemption fostered agribusiness interests, which also receive other heavy subsidies.

69. In 2006, a federal court ruled that the Bureau of Reclamation cannot continue irrigating at levels that strangle Klamath River flows in dry years, and the National Marine Fisheries Service recommended drastic salmon fishing cutbacks. PacifiCorp, owner of four hydroelectric dams on the lower Klamath River, appeared ready to remove the dams so that salmon can pass upriver without encountering shallow reservoirs, high summer water temperatures, and toxic algae blooms.

70. Once found along nearly 2,500 miles of the Rio Grande and the Pecos River, the silvery minnow survives tenuously only below Cochiti Dam and above Elephant Butte Reservoir in the middle Rio Grande, about 5% of its historical range.

71. Leake et al. *Desert Basins of the Southwest*; J. R. Bartolino and W. L. Cunningham. *Ground-Water Depletion Across the Nation* (U.S. Geological Survey Fact Sheet-103–03, 2003).

72. U.S. Fish and Wildlife Service. *Biological and Conference Opinions on the Effects of the Actions Associated with the Programmatic Biological Assessment of Bureau of Reclamation's Water and River Maintenance Operations, Army Corps of Engineers' Flood Control Operation, and Related Non-federal Actions on the Middle Rio Grande, New Mexico* (17 March 2003), 46–49, 56–57, 83–84. Available: southwest.fws.gov.

73. Papadopulos and Associates. *Middle Rio Grande Water Supply Study.* Towns and cities use about 14% of middle Rio Grande water, crops and river bank vegetation take a total of 67%, and about 19% evaporates from reservoirs, but this loss varies the most. Evaporation rates depend on reservoir surface areas, so the only possible control is lowering reservoir levels to reduce the surface area. Removing invasive water-hungry exotic plants, such as tamarisk, or salt cedar, could save substantial water volumes and would help restore Rio Grande

bosque habitat, now only 10–20% of its former extent. The preferred removal method, aerial spraying of Arsenal, a potent broad-spectrum herbicide (the active ingredient is imazapyr), may also kill nontarget crops and native cottonwoods. Imazapyr persists in soils for more than a year and is very mobile in water, so it could pollute groundwater.

74. The U.S. district court ordered parties to reach an agreement that let the Bureau of Reclamation provide water for silvery minnow habitat. After that agreement fell apart, in 2002, the court allowed the Bureau of Reclamation to use Albuquerque's and the Middle Rio Grande Conservancy District's San Juan–Chama allotments for the minnow. The state and other parties immediately appealed.

75. New Mexico established the Middle Rio Grande Conservancy District to drain marshlands for farming. From Bureau of Reclamation and U.S. Army Corps of Engineers reservoirs, it now contracts for more than 300,000 acre-feet of federally subsidized water, supplied to farmers in four counties on either side of the river (U.S. Fish and Wildlife Service. *Rio Grande Basin* Web site. Updated August 17, 2004. Available: www.fws.gov/southwest/mrgbi/maps/Rio_Grande_Basin/index.html).

76. Laura Paskus. Southwest Drought Desiccates Fish Before Farmers. *High Country News* 2002;34:5.

77. On 23 September 2002, the district court found the U.S. Fish and Wildlife Service's (USFWS) "Reasonable and Prudent Alternative" to be arbitrary and capricious (U.S. Fish and Wildlife Service. *Biological and Conference Opinions*, 4). It ordered the Bureau of Reclamation to release San Juan–Chama waters to maintain river flows and to renew formal consultation with the USFWS and the water-managing agencies. This decision was considered so important to western states' water policies that New Mexico appealed, supported by five other states with similar water problems. The 10th Circuit Court of Appeals upheld the district court's ruling on 12 June 2003. The Department of the Interior opined that the appeals court decision "represents a significant setback for collaborative efforts and makes a difficult situation more challenging."

78. Reisner. *Cadillac Desert*, 270–271. Indian water rights are based on "practicably irrigable acreage," which can yield huge water allotments. In the 1960s, the Colorado River Indian Tribes Reservation in Arizona won rights to 717,000 acre-feet of Colorado River water. If delivered, the Indians would be wading in water two and a half feet deep. The Navajo Nation is 63 times larger and could claim a share exceeding Arizona's entire 2.8 million acre-foot Colorado River allocation. The water is eminently salable (Matt Jenkins. The Colorado River's Sleeping Giant Stirs. *High Country News* 2003;35(8):11).

79. Daniel Kraker, The New Water Czars, *High Country News* 2004;36 (5):7–12.

80. Kraker. The New Water Czars. The tribes receive about 200,000 acre-feet of water that formerly went to non-Indian farmers, mostly central Arizona cotton growers. Another 67,000 acre-feet goes to an "account" for resolving future Indian water claims. In exchange, the cotton farmers receive a waiver of more than $73 million in debt for constructing Central Arizona Project (CAP) water distribution systems. U.S. taxpayers are paying off the Arizona cotton farmers' unpaid debt for CAP.

81. U.S. Fish and Wildlife Service. *Programmatic Biological Opinion on the Effects of Actions Associated with the U.S. Bureau of Reclamation's, U.S. Army Corps of Engineers, and Non-federal Entities' Discretionary Actions Related to Water Management on the Middle Rio Grande, New Mexico* (2001), 71. Available: ifw2es.fws.gov.

82. Using the Bureau of Reclamation's Imperial Irrigation District (California) estimate of 3.9 acre-feet per acre per year for crop evapotranspiration, the Middle Rio Grande Conservancy District (MRGCD) determined an appropriate irrigation allocation to the six Pueblos would be more than 77,500 acre-feet per year, exclusive of water needed to flush salt from soil (U.S. Bureau of Reclamation. Part 417: Determinations and Recommendations, Imperial Irrigation District, Calendar Year 2003. *U.S. Bureau of Reclamation* (2003), 47). The MRGCD must also supply the Pueblos' domestic and livestock water. The Indians expect to be a party to Bureau of Reclamation and MRGCD contract negotiations, observing that the twenty-first century challenge for lawyers and other water people is "to arrive at solutions that are fair and appropriately respectful of Indian Pueblo water rights and social needs" (Chestnut. *A Pueblo Perspective*).

83. Paul Krza. Indian Power. *High Country News* 2003;35(8):1, 8–10, 12.

84. U.S. Census Bureau. (2000). Available: www.census.gov.

85. The Environmental Working Group determined that tap water serving 195 million people in 42 states is contaminated with more than 140 unregulated chemicals that lack safe drinking water standards. Compliance with U.S. Environmental Protection Agency (EPA) standards for other chemicals is nearly 100%, so it is essential for the EPA to keep current with setting safety standards (Environmental Working Group. *A National Assessment of Tap Water Quality* [20 December 2005]. Available: www.ewg.org/).

86. Takashi Asano and J. A. Cotruvo. Groundwater Recharge with Reclaimed Municipal Wastewater: Health and Regulatory Considerations. *Water Research* 2004;38:1941–1951.

87. M. S. Fram et al. Processes Affecting the Trihalomethane Concentrations Associated with the Injection, Storage, and Recovery Tests at Lancaster, Antelope Valley, California. In G. R. Aiken and E. L. Kuniansky, eds. *U.S. Geological Survey Artificial Recharge Workshop Proceedings, Sacramento, California, 2–4 April, 2002* (U.S. Geological Survey Open-File Report 02–89, 2002).

88. Reverse osmosis involves driving water through a semipermeable membrane that blocks passage of dissolved salts. The two dominant conventional methods for desalination are evaporation and reverse osmosis.

89. Typical seawater reverse osmosis requires 1.5–2.5 kilowatt hours of electricity to produce 265 gallons of water, and thermal distillation plants take 10 times as much (R. F. Service. Desalination Freshens Up. *Science* 2006;313:1088–1090; Heather Cooley et al. *Desalination with a Grain of Salt: A California Perspective* [Oakland, California: Pacific Institutes for Studies in Development, Environment, and Security, 2006]).

90. Capital investments are high for desalination plants because they require corrosion-resistant pipes and equipment, plus special filters and cleaning membranes that need frequent backwashing and replacement. Adding water delivery pumps and pipeline infrastructure and operating costs, the consumer faces exorbitant prices, particularly compared to the costs of subsidized irrigation water transfers. For a thorough look at desalination's prospects, see Gary Pitzer (Tapping the World's Largest Reservoir, Desalination. *Western Water* (January-February, 2003;4–13) and W. M. Alley (*Desalination of Ground Water, Earth Science Perspectives* [U.S. Geological Survey Fact Sheet 075–03, 2003]).

91. Cooley et al. *Desalination with a Grain of Salt*, 44. Both the California and Texas figures represent drops from first-year costs of nearly $2,000 per acre-foot. Inasmuch as energy

represents one-third to one-half of desalination production costs, prices are likely to go up, not down.

92. In July 2003, the Las Vegas Valley Water District approved an average 29% increase in their customers' water bills, hoping to cut usage about 8% in what is believed to be the worst drought in 500 years. In 2004, the proposed water restrictions were overwhelmingly approved by residents of Clark County, Nevada. The $12.42 Las Vegas price for 7,480 gallons of water compares to $15.54 in Tucson, $21.20 in Los Angeles, and $23.41 in San Diego.

93. P. H. Gleick. *Waste Not, Want Not, the Potential for Urban Water Conservation in California* (Pacific Institute Executive Summary, 2003); Paul Rogers. Study: Water Supply Adequate, State Is Urged to Conserve More to Meet Its Needs. *San Jose Mercury News* 19 November 2003.

94. California Energy Commission Staff. *California's Water-Energy Relationship* (2005).

95. Elizabeth Mygatt. *World's Water Resources Face Mounting Pressure* (Earth Policy Institute, 26 July 2006). Available: www.earth-policy.org/. Irrigated agriculture uses 70% of western freshwater resources; 20% goes to industrial and 10% to municipal uses.

96. Susanna Eden and S. B. Megdall. Water and Growth. In *Arizona's Rapid Growth and Development: Natural Resources and Infrastructure* (Background report prepared for the 88th Arizona Town Hall, April 2006), chapter 4.

97. Michael Pollan. *The Omnivore's Dilemma: A Natural History of Four Meals* (Penguin Press, 2006), 185–273.

98. Thomas Pagano et al. Water Year 2004: Western Water Managers Feel the Heat. *EOS, Transactions of the American Geophysical Union* 2004;85(40):385, 392–393; R. F. Service. As the West Goes Dry. *Science* 2004;303:1124–1127; A. W. Nolin and Christopher Daly. Mapping "At-Risk" Snow in the Pacific Northwest, U.S.A. *Journal of Hydrometeorology* 2006;7:1164–1171.

99. U.S. Department of the Interior. *Water 2025, Preventing Crises and Conflict in the West* (2003).

100. Mary Austin. *The Land of Little Rain* (Garden City, New York: Doubleday, 1962; first published by Houghton Mifflin, 1903).

Chapter 10

1. Estimates for U.S. licensed sites alone put hazardous wastes at more than 300 million tons (National Research Council. *Environmental Epidemiology*: Vol. 1. *Public Health and Hazardous Wastes* (Washington, DC: National Academies Press, 1991), 1).

2. B. A. Bekins et al. Capabilities and Challenges of Natural Attenuation in the Subsurface, Lessons from the U.S. Geological Survey Toxic Substances Hydrology Program. In D. W. Morganwalp and H. T. Buxton, eds. *U.S. Geological Survey Toxic Substances Hydrology Program, Proceedings of the Technical Meeting, Charleston, South Carolina, March 8–12*, vol. 3 (U.S. Geological Survey Water-Resources Investigations Report 99–4018C, 1999), 37–56.

3. National Research Council. *Environmental Epidemiology*, 1–2. As of 2002, the U.S. population had increased by about 40 million people; if waste accumulated at the same rate,

the total at the end of 2002 would have reached 6.9 billion tons per year, but due to growing consumption, it probably increases faster.

4. Paul Hawken et al. *Natural Capitalism* (Boston: Little, Brown and Company, 1999), 52. Hawken and colleagues factored an additional 2,000 pounds per day per person of wastewater into total per capita waste, which after use "is sufficiently contaminated that it cannot be reintroduced into marine or riparian systems." This yields 2,500 pounds of daily American waste per capita, some 21 times the estimate of the National Research Council. In contrast, the U.S. Environmental Protection Agency estimates that treatment facilities received a total of 33,100 wastewater gallons per day in 1998, approximately 960 pounds per person per day—less than half of Hawken et al.'s estimate (U.S. Environmental Protection Agency. *Biosolids Generation, Use, and Disposal in the United States* [EPA530-R-99–009, 1999], table A-2). To the best of our knowledge, all wastewater goes into streams, rivers, and the oceans, except for a proportion that evaporates during treatment or is injected into the ground for disposal. If not stored in tanks or injected, we have no idea where water too contaminated to release into marine or riparian systems is discharged.

5. U.S. Department of Energy, Office of Environmental Management. *Linking Legacies, Connecting the Cold War Nuclear Weapons Production Processes to their Environmental Consequences* (DOE/EM-0319, 1997), 58, 77. Also omitted from the waste estimates are about 1.8 billion pounds of unclassified chemicals, lead, lithium, scrap, and other materials, including radioactive uranium (both low enriched and highly enriched), depleted uranium, and plutonium. The total amount constitutes 2,255 full loads for a Boeing 747 aircraft (U.S. Department of Energy. *Taking Stock, a Look at the Opportunities and Challenges Posed by Inventories from the Cold War Era* [DOE/EM-0275, 1996], 19–21). The amount of materials still classified and unknown likely is large.

6. Paul Hawken. *The Ecology of Commerce* (New York: Harper Collins Publishers, 1993), 45; U.S. General Accounting Office. *Hazardous Waste, Remediation Waste Requirements Can Increase the Time and Cost of Cleanups* (GAO/RCED-98–4, 1998). Hawken cites tens of thousands of sites that contain hazardous waste from past and current industrial activities. The Bureau of Land Management maintains a database on hazardous waste sites that listed 250 sites in 1992, mostly in the western states, undoubtedly a significant underestimate (National Research Council. *Hazardous Materials on the Public Lands* [Washington, DC: National Academies Press, 1992], 27).

7. S. E. Hasan. *Geology and Hazardous Waste Management* (Upper Saddle River, N.J.: Prentice Hall, 1996), 193.

8. B. W. Pipkin et al. *Geology and the Environment*, 4th ed. (Belmont, Calif.: Brooks/ Cole-Thompson Learning, 2005), 419–435; J. F. Barker et al. The Organic Geochemistry of a Sanitary Landfill Leachate Plume. *Journal of Contaminant Hydrology* 1986;1:171–189. Barker et al. found aromatic hydrocarbon contamination (mainly benzenes) throughout a 2,300-foot-long plume from a sanitary landfill to the groundwater discharge point at a creek; see also D. M. Mackay et al. *Transport of Organic Solutes, Proceedings of National Conference on Environmental Engineering* (American Society of Civil Engineering, 1983), 24–31.

9. Kris Christen. Tracking Mercury in Landfill Gas. *Environmental Science and Technology* 2001;35:354A.

10. Leonardo Transande et al. Public Health and Economic Consequences of Methyl Mercury Toxicity to the Developing Brain. *Environmental Health Perspectives* 2005;113:590–596.

11. Arnold Schecter et al. Polybrominated Diphenyl Ethers (PBDEs) in U.S. Mothers' Milk. *Environmental Health Perspectives* 2003;111:1723–1729; Kim Hooper and T. A. McDonald. The PBDEs, an Emerging Environmental Challenge and Another Reason for Breast-Milk Monitoring Programs. *Environmental Health Perspectives* 2000;108:387–392; L. S. Birnbaum and D. F. Staskal. Brominated Flame Retardants, Cause for Concern? *Environmental Health Perspectives* 2004;112:9–17.

12. U.S. Environmental Protection Agency. *Lining of Waste Impoundments and Waste Disposal Facilities (SW-870).* (Springfield, Va.: National Technical Information Service, Office of Research and Development, March, 1983) NTIS publication no. PB86–192796, 1.

13. The accuracy of these numbers is very much in question. Tables of state landfills show no uniform standards even for what constitutes a municipal solid waste landfill, and no agency is responsible for verifying the numbers (U.S. Environmental Protection Agency. *Municipal Solid Waste Landfills,* 2007 Available: www.epa.gov/msw/).

14. National Research Council. *Environmental Epidemiology,* 9. The number of possible National Priority List (NPL) sites is from an Office of Technology Assessment report. Candidate NPL sites include federal facilities, leaking underground storage tanks, underground injection wells, municipal gas facilities, wood-preserving facilities, and mine waste, pesticide-contaminated, and radioactive release sites.

15. Impurities and additives in trichloroethylene products significantly complicate characterization studies on contaminated sites and cleanup protocols (Peter Strauss. TCE Additives and Impurities Newsletter, *Center for Public Environmental Oversight,* 2001;4–5).

16. Pipkin et al. *Geology and the Environment,* 418–419; U.S. Environmental Protection Agency. *Municipal Solid Waste in the United States, 2000 Facts and Figures* (EPA530-R-02–001, 2002), 25.

17. Household hazardous wastes include paints, solvents, pest poisons, oven cleaners, pharmaceuticals, hormones, steroids, a galaxy of "personal care" substances, all manner of batteries containing such toxics as mercury and cadmium, and a host of other hazardous substances.

18. E-waste products that wind up in municipal solid waste dumps include televisions of all varieties, VCR decks, camcorders, laserdisc players, DVD players, and personal computers; also audio and information products such as radios, telephones of all sorts, fax machines, printers, word processors, monitors, modems, and other miscellaneous items (Beverly Howell. *Electronic Waste, the Dangers* (Hazardous Technical Information Services Bulletin 11, 2001); Beverly Howell. *Electronic Waste, Personal Computers, What's Inside?* (Hazardous Technical Information Services Bulletin 12, 2002).

19. The United States produces tens of millions of tons of polyvinyl chloride (PVC) annually, containing up to 40% plasticizers. Accordingly, plasticizers are found throughout the environment. Recent studies show that plasticizer metabolites (breakdown products) persist in the environment, and these are acutely toxic to fish, unlike the plasticizers they come from (Owen Horn et al. Plasticizer Metabolites in the Environment. *Water Research* 2004;38:3693–3698). The chemical bisphenol A (BPA), widely used in plastics manufacturing, remains in plastic products. In tests, BPA disrupts lab animals' reproductive systems and can disrupt the transfer of genetic material from sperm and egg to an embryo (Jon Cohen. Lab Accident Reveals Potential Health Risks of a Common Compound. *Science* 2003;300:31–32). High levels of brominated fire retardants such as polybrominated diphenyl ethers (PBDEs) have

been found in the breast milk of American women. In laboratory animals, they impair attention, learning, memory, and behavior at low levels. They are persistent in the environment
and bioaccumulate, building up in peoples' bodies over a lifetime (Environmental Working
Group. *Study Finds Record High Levels of Toxic Fire Retardants in Breast Milk from American
Mothers* [2003]. Available: www.ewg.org/reports/mothersmilk/es.php).

20. Michael Dowling. Defining and Classifying Hazardous Wastes. *Environment*
1985;27:36–41.

21. U.S. Environmental Protection Agency, *Municipal Solid Waste in the United States*,
147, 151.

22. Bette Fishbein. *Industry Program to Collect Nickel-Cadmium (Ni-Cd) Batteries* (2006).
Available: www.informinc.org/recyclenicd.php. The RCRA exempts household waste from hazardous waste regulations, so even though classified as hazardous under federal regulations,
batteries are exempted residential waste. Lead-acid car batteries are dominant wet-cell types,
designated hazardous waste, and should be barred from disposal in municipal landfills. The
wet-cell battery recycling record is good, 97% in 2000, but in 1994 about 1.7 million tons
went to landfills, contributing about 65% of the lead in MSW. Dry-cell batteries are either
primary or rechargeable; about 3.5 billion, weighing some 291 million pounds, were sold for
household use in 1992 and approximately 145,000 tons are discarded each year as MSW.
The batteries contain zinc, mercury, silver, cadmium, and lithium metals, toxic to aquatic
and other wildlife. The burgeoning numbers of dry-cell batteries in the United States come
with rising sales of battery-operated toys and tools, appliances such as toothbrushes and shav-
ers, video cameras, cellular phones, and laptop computers.

23. Regulations codified in CFR 40 (part 503) of the Clean Water Act defines sewage
sludge as the solid, semisolid, or liquid residue generated during the treatment of domestic
sewage in a treatment works. The term "biosolids" does not appear in the law. In 1995, the
U.S. Environmental Protection Agency (EPA) defined biosolids as "the primarily organic solid
product yielded by municipal wastewater treatment process that can be beneficially recycled,"
excluding the 40% of sewage sludge that goes to landfills for burial and incinerators for
burning. Nine inorganic chemicals are now regulated, and in 2002 the EPA was considering
adding dioxins and other organic chemicals to the list (National Research Council. *Biosolids
Applied to Land: Advancing Standards and Practices* (Washington, DC: National Academies
Press, 2002), 1). In late October 2003, model risk analysis prompted an EPA decision not
to regulate dioxin disposal in land-applied sewage sludge (U.S. Environmental Protection
Agency. *Agency Final Action Not to Regulate Dioxins in Land-Applied Sewage Sludge* (EPA-
822-F-03–007, 2003)).

24. Citing budgetary constraints, since 2002 the U.S. Environmental Protection Agency
(EPA) has decided not to regulate dioxins in sludge for land disposal. After Bakersfield,
California, residents halted dumping of sewage sludge from Los Angeles on nearby farmland,
in late 2006 the EPA approved Los Angeles's application for deep injection of a slurry made
from sludge plus treated wastewater into an abandoned oil field. This disposal method is
expected to provide methane as the sludge decomposes and also sequester CO_2 that would
otherwise escape to the atmosphere.

25. National Research Council. *Biosolids Applied to Land*, 1–3, 6.

26. For example, the toxic effects of zinc and copper mixtures are substantially higher
than the sum of their individual toxicities (V. P. Utgikar et al. Toxicity of Metals and Metal

Mixtures—Analysis of Concentration and Time Dependence for Zinc and Copper. *Water Research* 2004;38:3651–3658).

27. See n. 19.

28. C. W. Montgomery. *Environmental Geology*, 5th ed. (San Francisco: WCB McGraw-Hill, 1997), 363–364. Unused oil does not contain the gasoline components (R. J. Baker et al. Used Motor Oil as a Source of MTBE, TAME, and BTEX to Ground Water. *Ground Water Monitoring and Remediation* 2002;22:46–51). Baker and coauthors found levels of MTBE of about 100 milligrams per liter (mg/L), the same as parts per million, plus 2.2–87 mg/L TAME, 29–66 mg/L benzene, and about 500–2,000 mg/L other BTEX components. For the year 2000, the U.S. Environmental Protection Agency reported more than 6,000 tons of persistent toxic chemical releases. These bioaccumulative substances last a long time in the environment and accumulate in body tissues of animals and plants. Organisms higher on the food chain (us) may be severely harmed or poisoned by consuming plants and animals that accumulate and concentrate the toxins (U.S. Environmental Protection Agency. *2000 Toxics Release Inventory, Public Data Release* [EPA 260-R-02–003 (2002)]. Available: www.epa.gov/tri; see also U.S. General Accounting Office. *Environmental Protection: Information on the Purchase, Use, and Disposal of Engine Lubricating Oil* [GAO-03–340, 2003]).

29. See n. 28.

30. H. G. Wilshire et al. *Geologic Processes at the Land Surface* (U.S. Geological Survey Bulletin 2149, 1996); Tom Harris. *Death in the Marsh* (Covelo, California: Island Press, 1991).

31. U.S. Environmental Protection Agency. *Classes of Injection Wells*. Available: www.epa.gov/safewater/uic/classes.html.

32. U.S. General Accounting Office. *Deep Injection Wells: EPA Needs to Involve Communities Earlier and Ensure That Financial Assurance Requirements Are Adequate* (GAO-03–761, 2003).

33. Note that "injection wells" is a misnomer for most Class V wells, which are only "wells" in the sense that they are holes with depths greater than their widths. In most, "injection" is by gravity drainage.

34. U.S. Environmental Protection Agency. *Underground Injection Control Program, Notice of Final Determination for Class V Wells*, 40 CFR Part 144, Federal Register 67 (7 June, 2002): 39587.

35. The septic system numbers are 1987 figures (P. J. Squillace et al. VOCs, Pesticides, Nitrate, and Their Mixtures in Groundwater Used for Drinking Water in the United States. *Environmental Science and Technology* 2002;36:1923–1930).

36. National Research Council. *Waste Incineration and Public Health* (Washington, DC: National Academies Press, 2000).

37. B. W. Pipkin and D. D. Trent. *Geology and the Environment*, 3rd ed. (Pacific Grove, Calif.: Brooks/Cole, 2001), 506–507. Significant amounts of dioxins come from burning plastics, from newspaper, and from pulp contaminated with sodium, potassium, or calcium chloride compounds (Akio Yasuhara et al. Role of Inorganic Chlorides in Formation of PCDDs, PCDFs and Coplanar PCBs from Combustion of Plastics, Newspaper, and Pulp in an Incinerator. *Environmental Science and Technology* 2002;36:3924–3927).

38. Several hundred dioxins are known and 29 are considered toxic. Both international and national agencies have examined dioxin's toxicity and human health impacts. Exposure

standards have not been specified, and contradictory findings have yet to be reconciled. Dioxins persist for long periods in the environment because they are insoluble in water and relatively immobile in soil and sediment. They travel far in the atmosphere and in running water and have dispersed widely in the environment. Attached to fine particulates, they can fall from the air in dust and rain and flood into streams and standing bodies of water attached to sediment. Dioxins bioaccumulate in the fatty tissue of farmed fish and land animals that eat dioxin-contaminated plants or commercial feed or drink dioxin-laced water. They concentrate further in humans' fatty tissue. Dioxins are the leading industrial pollutant found in human blood samples and in mothers' milk, apparently from meats, fish, and dairy products (U.S. General Accounting Office. *Environmental Health Risks: Information on EPA's Draft Reassessment of Dioxins* [GAO-02–515, 2002], 1–4; National Research Council. *Dioxins and Dioxin-like Compounds in the Food Supply: Strategies to Decrease Exposure* [Washington, DC: National Academies Press, 2003], ES-1, 45–55). During breast feeding, human infants are exposed to higher concentrations of dioxins than at any later time in their lives. In only six months, a breast-feeding baby in the United States gets the maximum recommended lifetime dose of dioxin (Theo Colborn et al. *Our Stolen Future* [New York: Plume of Penguin Books USA, 1997], 107).

39. Pipkin et al. *Geology and the Environment*, 421.

40. The military's original disposal method for chemical armaments, called the "CHASE" ("cut holes and sink 'em") scheme, loaded the weapons onto old ships, towed them out to sea, and sank them. The game plan turned to incineration when CHASE was publicly unmasked (Chip Ward. *Canaries on the Rim: Living Downwind in the West* [New York: Verso Press, 1999], 150). Weapons waste incineration is not new—the nation's researchers never will be able to assess the human health costs of burning 370,000 gallons of plutonium-laced solvents in open, smoky fires at the Savannah River nuclear weapons site in the 1950s and 1960s (Arjun Makhijani and Marc Fioravanti. Cleaning Up the Cold War Mess. *Science for Democratic Action* 1999;7:3).

41. National Research Council. *Analysis of Engineering Design Studies for Demilitarization of Assembled Chemical Weapons at Pueblo Chemical Depot* (Washington, DC: National Academies Press, 2001); National Research Council. *Analysis of Engineering Design Studies for Demilitarization of Assembled Chemical Weapons at Blue Grass Army Depot* (Washington, DC: National Academies Press, 2002). The seven continental U.S. incineration sites are Anniston, Ala.; Aberdeen, Md.; Newport, Ind.; Blue Grass, Ky.; Pueblo, Colorado; Tooele, Utah; and Umatilla, Oreg.

42. U.S. Congress, Office of Technology Assessment. *Dismantling the Bomb and Managing the Nuclear Materials* (OTA-0–572; Washington, DC: U.S. Government Printing Office, September 1993); National Research Council. *Analysis of Engineering Design Studies*, 2001.

43. David W. Hall, chemist and former employee at Utah's Dugway Proving Ground, brought a whistleblower lawsuit against the Department of the Army for harassment after he raised safety concerns about the Army's handling of hazardous wastes, chemical warfare agents, and other toxic substances. The suit accused the Army of dumping toxic waste down a drain that flows to the community water supply, faking agent penetration tests for gas masks like those used in the Gulf War, exposing humans to chemical warfare agents and other toxic chemicals, and ignoring employees who took hazardous chemicals home with them. Dr. Hall was awarded $1.5 million in 2002 (Chemical Weapons Working Group. Available: www.cwwg.org). Whether the lessons of the past have been learned remains to be seen,

but the U.S. General Accounting Office appears hopeful (U.S. General Accounting Office. *Chemical Weapons, Lessons Learned Program Generally Effective but Could Be Improved and Expanded* [GAO-02–890, 2002]).

44. The volume of radioactive contaminated soil and rock is estimated at 2.8 billion cubic feet (U.S. Department of Energy. *Linking Legacies*, 7, 84–87, 105).

45. Burning depleted uranium (DU) gives off dense clouds of radioactive particles that bystanders can breathe into their lungs, where it does the most harm. The 1990 Persian Gulf War exposed many U.S. military personnel to DU-laden smoke—even though the Pentagon "had internal reports…stating unequivocally that DU can damage lungs, kidneys, and other organs" (Akira Tashiro. *Discounted Casualties: The Human Cost of Depleted Uranium* [Hiroshima, Japan: Chugoku Shimbun; English translation by Transnet, 2001], 37). Many believe that the effects of DU could explain some "Gulf War syndrome" illnesses. Pentagon officials have admitted that they were not thinking much about the danger from DU before using significant quantities of DU weaponry in combat, yet Kosovo bombings employed DU weapons, and probably so do ongoing actions in Afghanistan and Iraq.

46. Information on waste types largely comes from Marc Fioravanti and Arjun Makhijani. *Containing the Cold War Mess, Restructuring the Environmental Management of the U.S. Nuclear Weapons Complex* (Takoma Park, Maryland; Institute for Energy and Environmental Research, 1997), appendix B; Arjun Makhijani et al. Nuclear Waste Management and Environmental Remediation. In S. I. Schwartz, ed. *Atomic Audit* (Washington, DC: Brookings Institution Press, 1998), 353–394; U.S. Department of Energy, *Linking Legacies*, chapter 3.

47. Transuranic elements emit large, slow-moving alpha particles, equivalent to a helium nucleus, which cannot penetrate skin but can cause severe internal damage if ingested. The wastes have combined radioactivity of at least 100 nanocuries (100×10^{-9} curies) from each gram, or 1/28 (0.04) ounce, of waste. A curie is the amount of radioactivity emitted by one gram of radium. A nanocurie is one billionth of a curie, equivalent to the radiation emitted by one-billionth of a gram (0.04×10^{-9} ounces) of radium. In 1941, the lifetime limit of allowable radiation exposure for a human body was set at a tenth of a microcurie (0.1×10^{-6} curie) of radium (see glossary) (Catherine Caufield. *Multiple Exposures, Chronicles of the Radiation Age* [Chicago: University of Chicago Press, 1990], 39–40).

48. Arjun Makhijani and Scott Saleska. The Production of Nuclear Weapons and Environmental Hazards. In Arjun Makhijani et al., eds. *Nuclear Wastelands, a Guide to Nuclear Weapons Production and Its Health and Environmental Effects* (Cambridge, Mass.: MIT Press, 1995), 30, 34.

49. Diane D'Arrigo. *Energy Fact Sheet, "Low-Level" Radioactive Waste* (Nuclear Information and Resource Service, 1992). Ultimately, the 104 nuclear power plants currently licensed to operate in the United States will have to be dismantled and somehow discarded. Already, 17 others have been permanently shut down and decommissioned or are being prepared for decommissioning. Only one low-level radioactive waste site is licensed to receive the types of waste generated by 115 of these plants, including the 17 now shut down (U.S. General Accounting Office. *Low-Level Radioactive Wastes: States Are Not Developing Disposal Facilities* [GAO/RCED-99–238, 1999], 43). Building more nuclear power plants to meet growing energy demand as petroleum supplies decline will exacerbate our radioactive waste problem.

50. One-third of power plant reactor fuel rods are replaced annually, continuously creating "high-level" spent fuel wastes. The 7% generated by military reactors are counted as

inventory, not waste. The rest of spent fuel wastes come from commercial reactors and still produce approximately 30 billion curies of radioactivity. The level of radioemissions from military wastes is not available (Makhijani and Fioravanti. Cleaning Up the Cold War Mess, 13; Eli Kintisch. DOE Outlines Two Roads to Recycling Spent Fuel. *Science* 2006;313:746).

51. Arjun Makhijani. *Nuclear Dumps by the Riverside* (Takoma Park, Md.: Institute for Energy and Environmental Research, 2003); see also Arjun Makhijani. *The Savannah River at Grievous Risk* (Institute for Energy and Environmental Research, 2004). Available: www.ieer.org/.

52. R. C. Ewing and Allison Macfarlane. Yucca Mountain. *Science* 2002;296:659–660. A National Research Council report identified seven barriers (out of many) to applying valid scientific approaches and analyses to Yucca Mountain and other waste disposal issues: (1) planning driven by existing organizational structures rather than by problems to be solved; (2) commitments made without adequately considering technical feasibility, cost, and schedule; (3) inability to look at more than one alternative at a time; (4) priorities driven by interpreting regulations narrowly; (5) producing documents as ends in themselves; (6) lack of organizational coordination; and (7) a "[no good if] not-invented-here" syndrome at particular labs (National Research Council. *Barriers to Science: Technical Management of the Department of Energy Environmental Remediation Program* [Washington, DC: National Academies Press, 1996]).

53. L. J. Carter and T. H. Pigford. Getting Yucca Mountain Right. *Bulletin of the Atomic Scientists* (1 March 1998);56–61. For an excellent description of the geologic setting of Yucca Mountain and of the character of the wastes to be disposed there, see A. M. Macfarlane and R. C. Ewing (Introduction. In A. M. Macfarlane and R. C. Ewing, eds. *Uncertainty Underground: Yucca Mountain and the Nation's High-Level Nuclear Waste* [Cambridge, Mass.: MIT Press, 2006], 1–26).

54. The earliest concept for the Yucca Mountain repository was to bury the waste below deep water tables, but this idea was abandoned in the early 1980s because of high fracture transmissivity and high groundwater temperature (T. C. Hanks et al. *Yucca Mountain as a Radioactive-Waste Repository* [U.S. Geological Survey Circular 1184, 1999]).

55. Small amounts of the radioactive chlorine are everywhere. It is useful as a tracer to track the distribution of radionuclides but is not present in sufficient quantities to be harmful (K. Campbell et al. Chlorine-36 Data at Yucca Mountain, Statistical Tests of Conceptual Models for Unsaturated-Zone Flow. *Journal of Contaminant Hydrology* 2003;62–63:43–61; June Fabryka-Martin et al. Water and Radionuclide Transport in the Unsaturated Zone. In A. M. Macfarlane and R. C. Ewing, eds. *Uncertainty Underground: Yucca Mountain and the Nation's High-Level Nuclear Waste* [Cambridge, Mass.: MIT Press, 2006], 179–197).

56. A. B. Kersting et al. Migration of Plutonium in Groundwater at the Nevada Test Site. *Nature* 1999;397:56–59. Kersting and colleagues found plutonium in groundwater less than a mile from its source (identified by the ratio of plutonium-240 to plutonium-239) at underground nuclear bomb test sites. Previously, plutonium was assumed to be relatively immobile because it has low solubility in groundwater and strongly sorbs onto rock materials. Researchers now know that relatively insoluble plutonium is carried into groundwater exactly *because* it is easily sorbed by colloids derived from zeolites and other minerals. The repository area's geology consists largely of volcanic rocks with abundant zeolite alteration.

57. R. C. Ewing. Less Geology in the Geological Disposal of Nuclear Waste. *Science* 1999;286:415–417.

58. A number of official reports dispute that claim. The dispute is moot once it becomes clear that "sound science" is actually a political term for "a technical-sounding smokescreen that supports the currently favored policy decision" (Ewing and Macfarlane. Yucca Mountain, 659–660).

59. Quotation cited by Ewing and Macfarlane. Yucca Mountain, 659–660. Commenting on the evaluation's assumption that the climate will remain unchanged for the duration of regulatory control—10,000 years by law, but hundreds of thousands of years from the viewpoint of radioactive hazards, Ewing and Macfarlane noted that the percolation flux of water through the unsaturated zone at Yucca Mountain is uncertain " becasue it will depend on how the climate changes. Ewing and MacFarlane concluded, "It is a poor design strategy that relies on assumed boundary conditions, rather than the properties of the repository itself" (for discussion of Yucca Mountain climate issues see MaryLynn Musgrove and D. P. Schrag. Climate Change at Yucca Mountain: Lessons from Earth History. In A. M. Macfarlane and R. C. Ewing, eds. *Uncertainty Underground: Yucca Mountain and the Nation's High-Level Nuclear Waste* [Cambridge, Mass.: MIT Press, 2006]: 149–162).

60. Quotations cited in Ewing and Macfarlane, Yucca Mountain, 659–660; Advisory Committee on Nuclear Waste. *Letter Report to R. A. Meserve, Chairman, U.S. Nuclear Regulatory Commission, Advisory Committee on Nuclear Waste* (18 September 2001); U.S. General Accounting Office. *Nuclear Waste: Technical, Schedule and Cost Uncertainties of the Yucca Mountain Repository Project* (GAO-02–191, 2001), 3; Nuclear Waste Technical Review Board. *Letter Report to Congress and the Department of Energy* (24 January 2002).

61. Ewing and Macfarlane. Yucca Mountain, 659–660.

62. Ewing and Macfarlane. Yucca Mountain, 659–660.

63. When irradiated, waste materials such as plastics and rubber may undergo chemical changes, or "radiolysis reactions," that can make waste more hazardous over time or can even create hazardous waste from originally nonhazardous substances. For example, alpha-particle irradiation of PVC plastic creates toxic benzene and vinyl chloride, plus flammable acetone and hydrogen gas, and corrosive hydrogen chloride. Alpha irradiation of rubber forms corrosive hydrogen chloride, and flammable acetone and hydrogen gas. Depending on the half-life of the radionuclides and their decay products in the waste, this process may continue for immensely long (thousands to billions of years) periods of time (Makhijani and Fioravanti. Cleaning Up the Cold War Mess, 8, 21). See glossary for atomic radiation definitions.

64. Nuclear Energy Institute. Available: www.nei.org.

65. D. J. Goode. Mixed-Waste Leachates in Ground Water at Low-Level Radioactive-Waste Repository Sites. In M. S. Bedinger and P. R. Stevens, eds. *Proceedings Safe Disposal of Radionuclides in Low-Level Radioactive-Waste Repository Sites, Low-Level Radioactive-Waste Disposal Workshop July 11–16, Big Bear Lake, California* (U.S. Geological Survey, 1990), 49–57; B. J. Ryan. *Results of Hydrologic Research at a Low-Level Radioactive-Waste Disposal Site Near Sheffield, Illinois* (U.S. Geological Survey Water-Supply Paper 2367, 1991); J. M. Cleveland and T. F. Rees. Characterization of Plutonium in Maxey Flats Radioactive Trench Leachates. *Science* 1981;212:1506–1509; M. A. Lyverse. Movement of Tritiated Leachate Through Fractured Rock at a Low-Level Radioactive Waste Disposal Site Near Morehead, Kentucky. In *Third Annual Eastern Regional Ground Water Conference, July 28–30, Springfield, Massachusetts* (1986), 465–471.

66. U.S. General Accounting Office. *Radioactive Waste: Answers to Questions Related to the Proposed Ward Valley Low-Level Radioactive Waste Disposal Facility* (GAO/RCED-98–40R, 1998), 34, 49.

67. H. G. Wilshire and Irving Friedman. *Data Related to Contaminant Distribution at the Beatty, Nevada and Richland, Washington Low-Level Radioactive Waste Disposal Sites* (U.S. Geological Survey Open-File Report 97–865, 1997); H. G. Wilshire and Irving Friedman. Contaminant Migration at Two Low-Level Radioactive Waste Sites in Arid Western United States, a Review. *Environmental Geology* 1999;37:112–123; D. E. Prudic et al. Tritium and [14]C Concentrations in Unsaturated Zone Gases at Test Hole UZB-2, Amargosa Desert Research Site, 1994–98. In D. W. Morganwalp and H. T. Buxton, eds. *U.S. Geological Survey Toxic Substances Hydrology Program, Proceedings of the Technical Meeting, Charleston, South Carolina, March 8–12*, vol. 3 (U.S. Geological Survey Water-Resources Investigations Report 99–4018C, 1999), 475–483; R. W. Healy et al. Tritium in Water Vapor in the Shallow Unsaturated Zone at the Amargosa Desert Research Site. In D. W. Morganwalp and H. T. Buxton, eds. *U.S. Geological Survey Toxic Substances Hydrology Program, Proceedings of the Technical Meeting, Charleston, South Carolina, March 8–12*, vol. 3 (U.S. Geological Survey Water-Resources Investigations Report 99–4018C, 1999), 485–490. We have been unable to find out whether or not the Envirocare facility leaks.

68. U.S. General Accounting Office. *Radioactive Waste: Answers to Questions*, 50–53. The curies of radiation are calculated from average proportions of plutonium-239 (76.6%), plutonium-240 (18.8%), and plutonium-242 (4.6%) in low-level waste disposal sites (U.S. Department of Energy. Integrated Data Base Report,1996: U.S. Spent Nuclear Fuel and Radioactive Waste Inventories, Projections, and Characteristics (1997), table A-3). See n. 47 and glossary for definitions.

69. More likely, five years of study will be necessary for sites where little information is available at the outset of site characterization (K. L. Kipp and R. W. Healy. Characterizing and Monitoring Low-Level Radioactive-Waste Repository Sites. In M. S. Bedinger and P. R. Stevens, eds. *Proceedings Safe Disposal of Radionuclides in Low-Level Radioactive-Waste Repository Sites, Low-Level Radioactive-Waste Disposal Workshop July 11–16, Big Bear Lake, California* [U.S. Geological Survey, 1990], 6–8).

70. Howard Wilshire et al. *Ward Valley, Proposed Low-Level Radioactive Waste Site: A Report to the National Academy of Science* (unpublished, 1994; available in U.S. Geological Survey library, Menlo Park, California). Wilshire et al. concluded that the Ward Valley repository site scientific data were so poor that they could not be used to evaluate the site's unsaturated zone as a principal barrier against environmental contamination from radioactive waters leaking out of unlined waste trenches. Serious deficiencies in the data included inadequate sampling at the site, based on the mistaken notion that Ward Valley's physically and chemically complex unsaturated zone was actually homogeneous; too few data collected over only one year, far too short for an arid region lacking any previous detailed geologic and hydrologic investigations at the site; flawed measurements of such site properties as water potential, soil and groundwater temperatures and composition, and critical properties for characterizing the hydrologic behavior of the unsaturated and saturated zones; and inadequate investigation of the regional groundwater and geologic framework for modeling groundwater pathways to the Colorado River. The data were too flawed and limited for interpreting groundwater age or compositional variations; groundwater flow directions and rates; or unsaturated-zone tritium and chloride contents, among other characteristics. Most interpretations of the unsaturated

and saturated zone hydrology were theoretical (model driven), and the actual data used were internally contradictory and incomplete, often unrepresentative of the bulk of observations.

71. National Research Council. *Ward Valley: An Examination of Seven Issues in Earth Sciences and Ecology* (Washington, DC: National Academies Press, 1995).

72. The data showed tritium 100 feet deep in the unsaturated zone at the proposed Ward Valley low-level waste dump. Tritium is a radioactive isotope of hydrogen not found in nature, but produced in atmospheric nuclear bomb tests from the 1940s through 1963. The finding contradicted widely accepted models that water migrates exceedingly slowly through unsaturated zones—so slowly that tritium should not move 100 feet below the surface in just 30–50 years. The National Research Council panel majority preferred not to question their untested assumptions about unsaturated zone water migration rates and so concluded that the 100-foot samples somehow had become contaminated during sampling. Despite the panel's assertion that it did not *judge* the site as suitable, the majority concluded that preparations for building the repository could commence while the tritium findings were checked, clearly *implying* that they found the site suitable.

73. The principal political figures who opposed the George H. W. Bush and Clinton administrations' attempts to approve the Ward Valley repository included California Senator Barbara Boxer, Congressman George Miller, and then-Assistant Interior Secretary (and Californian) John Garamendi—all Democrats (H. G. Wilshire. Is Ward Valley Safe? Serious Issues Remain. *Geotimes* June, 1997;18–20).

74. The 1991 National Research Council report on environmental epidemiology noted also that neither evaluation processes nor the process components for selecting new dump sites are ever evaluated, reviewed, or validated, adding,

> The current regulatory system has failed to devise a protocol for managing hazardous waste sites that incorporates the essential components of public health policy. Not only is it possible that the public residing in some of these neighborhoods is imperiled, but the conditions for development of environmental epidemiology programs and methods are so adverse as to impede useful scientific investigations of many important questions. (National Research Council. *Environmental Epidemiology,* 7)

75. M. A. Barlaz et al. Critical Evaluation Factors Required to Terminate the Postclosure Monitoring Period at Solid Waste Landfills. *Environmental Science and Technology* 2002;36:3457–3464.

76. Richard Caplan. *Permit to Pollute, How the Government's Lax Enforcement of the Clean Water Act Is Poisoning Our Waters* (Washington, DC: U.S. Public Interest Research Group, 2002). A new Public Interest Research Group (PIRG) report provides an in-depth look at violations of Clean Water Act Permits by industrial and municipal facilities (Alison Cassady. *In Gross Violation, How Polluters Are Flooding America's Waterways with Toxic Chemicals* [Washington, DC: U.S. Public Interest Research Group, 2002]). This report finds that 81% of major facilities (5,116 total) exceeded their Clean Water Act effluent permits at least once in the period 1 January 1999 to 31 December 2001. Violations included 1,768 facilities (28%) that discharged high-hazard, known and suspected human carcinogens, known developmental or reproductive toxicants, and toxicants suspected to cause one or more serious noncancer health effects. In 10 western states (excluding California because it did not provide reliable data), 327 major facilities were responsible for 3,133 violations in the two-year period; 608 were high-hazard chemical violations. These findings are bolstered by an

EPA inspector general's report that charged the agency with mismanagement of the National Pollutant Discharge Elimination System, using antiquated computer systems. The George W. Bush administration suspended a Clinton administration proposal to increase controls on pollution from overflowing sewers and has deferred permit requirements for two years.

77. R. C. Hale et al. Persistent Pollutants in Land-Applied Sludges. *Nature* 2001;412:140–146; S. W. Pryor et al. Nonylphenol in Anaerobically Digested Sewage Sludge from New York State. *Environmental Science and Technology* 2002;36:3678–3682. These studies found high concentrations of nonylphenols, identified endocrine disrupters, in New York municipal sewage sludge samples, treated with anaerobic digestion. The nonylphenols stick on the surfaces of organic substances in the waste stream and are removed with the solid sludge. Concentrations of nonylphenols in five New York sludge samples ranged from 1,130 to 1,840 parts per billion (ppb). Samples reported from Los Angeles sludge averaged 754 ppb in five samples. The United States has no standards for nonylphenols in sludges. Denmark sets a 10 ppb standard for these contaminants in sludge used for agriculture. The authors suggest that more information is needed on the presence, toxicity, fate, and potential for bioaccumulation in U.S. agricultural soils. Compounds that yield nonylphenols are found in many domestic and industrial products. About 175 million pounds are used annually in the United States.

78. The new chemicals add to 7.3 million listed in the 1985 issue of *Chemical Abstracts* (Dowling. Defining and Classifying Hazardous Wastes). Dowling estimates that 1.5 million chemicals had been identified before registration began and about 70,000 of them were on the market in 1985.

79. There are more than 82,000 chemicals currently in the Toxic Substances Control Act Inventory, which excludes pesticides, food additives, drugs, and cosmetics, among other things. About 62,000 of these were in commercial use before the EPA began reviewing chemicals in 1979. Even the information obtained from EPA review may not be available to state and local governments or to the public, if the producer successfully claims confidentiality of trade secrets (U.S. Government Accountability Office. *Chemical Regulation: Options Exist to Improve EPA's Ability to Assess Health Risks and Manage Its Chemical Review Program* [GAO-05–458, 2005]).

80. We thank J. A. Fallin for his generous contributions to this section on waste remediation. They came in the form of many letters featuring his typical elegant and detailed diagrams and line drawings. Also thanks for the tasty dried fungi.

81. The estimated cost of cleaning up 156 hard-rock mining superfund sites is $7 billion to $24 billion. Sites now more than 20 years into cleanup will require many more years to complete (U.S. Environmental Protection Agency, Office of the Inspector General. *Nationwide Identification of Hardrock Mining Sites* (31 March 2004). Available: www.epa.gov/oig/reports/2004/20040331–2004-p-00005.pdf; see also Sierra Club. *Communities at Risk: How the Bush Administration Is Failing to Protect People's Health at Superfund Sites* (2004). Available: www.sierraclub.org/toxics/superfund/report04/report.pdf).

82. Nanoparticles vary from about the size of atoms—about 0.2 nanometers (nm) to 100 nm. For comparison, a single human hair is about 80,000 nm wide, a red blood cell about 7,000 nm, and a water molecule about 0.3 nm across.

83. D. W. Elliott and Wei-Xian Zhang. Field Assessment of Nanoscale Bimetallic Particles for Groundwater Treatment. *Environmental Science and Technology* 2001;35:4922–4926; Eva Oberdörster (Manufactured Nanomaterials [Fullerenes, C60] Induce Oxidative Stress in the

Brain of Juvenile Largemouth Bass. *Environmental Health Perspectives* 2004;112:1058) first demonstrated the adverse effect of nanoparticles on an aquatic species. Nanoparticles also pose regulatory problems because any two molecules of traditional chemical toxins with the same chemical formula behave alike, whereas two nanoparticles with the same composition but of differing size may have radically different chemical properties (R. F. Service. EPA Ponders Voluntary Nanotechnology Regulations. *Science* 2005;309:36). Britain's Royal Society and Royal Academy of Engineering recommended banning nanoparticles in cosmetics until fully tested and is critical of U.S. experiments that distribute nanoparticles in the environment (Royal Society and Royal Academy of Engineering. *Nanoscience and Nanotechnologies: Opportunities and Uncertainties* (London, 2004); see also, J. T. Nurmi et al. Characterization and Properties of Metallic Iron Nanoparticles: Spectroscopy, Electrochemistry, and Kinetics. *Environmental Science and Technology* 2005;39:1221–1230; R. F. Service. Calls Rise for More Research on Toxicology of Nanomaterials. *Science* 2005310:1609). A thorough review of risks and regulatory issues is provided by Karen Florini et al. (Nanotechnology: Getting It Right the First Time. *Sustainable Development Law and Policy* 2006;6(3)). As usual, U.S. cosmetics and numerous other commodities on the market already contain nanoparticles without adequate testing.

84. So many sources report links between pollutants and disease statistics that the following can only be a sample: plastic-making chemicals in breast milk and possible effects on children (Sandra Steingraber, Why the Precautionary Principle? A Meditation on Polyvinyl Chloride [PVC] and the Breasts of Mothers. *Rachel's Environment and Health Weekly* 1999;No. 658); pesticides and lymphomas (Lennart Hardell and Mikael Eriksson. A Case-Control Study of Non-Hodgkins Lymphoma and Exposure to Pesticides. *Cancer* 1999;8:1353–1360; Susan Osburn. *Research Report: Do Pesticides Cause Lymphoma?* [2000]. Available: www.lymphomahelp.org/docs/-research/researchreport/rr_2000.pdf), chemicals in plastic medical equipment linked to cancer, testicular atrophy, and cardiac toxicity (*Rachel's Environment and Health Weekly*. Precaution and PVC in Medicine. 1999;No. 661 [29 July 1999]); prenatal exposure to PCB hormone disruptors linked to infant cognitive development [J. Fialka. More Clinical Tests of Humans Exposed to Chemicals are Urged in US Study. *The Wall Street Journal* 4 August 1999], pesticides and other toxics linked to ADHD [Ted Schettler et al. *In Harms Way: Toxic Threats to Child Development* (Cambridge, Mass.: Greater Boston Physicians for Social Responsibility, May 2000)]. Available: www.igc.org.psr), and arthritis linked to widespread fluoridation (Paul Connett, Ellen Connett, and Michael Connett. Fluoridation: Time for a Second Look? *Rachel's Environment and Health Weekly* 2001;No. 724).

Chapter 11

1. R. L. Knight and D. N. Cole. Wildlife Responses to Recreationists. In R. L. Knight and K. J. Gutzwiller, eds. *Wildlife and Recreationists, Coexistence Through Management and Research* (Washington, DC: Island Press, 1995), 51–69.

2. Section 103 of the Federal Land Policy and Management Act (FLPMA) of 1976, which established the Bureau of Land Management, defines "multiple use" somewhat confusingly as

the management of the public lands and their various resource values so that they are utilized in the combination that will best meet the present and future needs of the

American people; making the most judicious use of the land for all of these resources; a combination of balanced and diverse resource uses that takes into account the long term needs of future generations for renewable and nonrenewable resources, including, but not limited to, recreation, range, timber, minerals, watershed, wildlife and fish, and natural scenic, scientific and historical values; and harmonious and coordinated management of the various resources without permanent impairment of the productivity of the land and the quality of the environment with consideration being given to the relative values of the resources and not necessarily to the combination of uses that will give the greatest economic return or the greatest unit output. Section 103 of the FLPMA defines "sustained yield" as the achievement and maintenance in perpetuity of a high-level annual or regular periodic output of the various renewable resources of the public lands consistent with multiple use.

3. Denver's population is now 2.3 million. Mountain bike use has jumped 214%, and Colorado off-road vehicle registrations have increased 658%.

4. Michael Liddle. *Recreation Ecology* (London: Chapman and Hall, 1997), 44–51. Biological crusts, a complex mosaic of unrelated cyanobacteria, green algae, lichens, mosses, microfungi, and other bacteria, are widespread in western U.S. arid lands. The crusts both stabilize the soils and insulate them from extreme temperatures; they also preserve much of the fertility in desert soils by fixing nitrogen and contributing carbon to soil, and retaining moisture. Off-road vehicle destruction of lichens growing on rock surfaces in the Slickrock country of Utah is another rapidly growing problem. Lichen are important food sources for invertebrates and vertebrates and excrete weak acids that help break rock surfaces down to soil-size particles (J. Belnap and O. L. Lange, eds. *Biological Soil Crusts, Structure, Function, and Management* [Ecological Studies 150; New York: Springer-Verlag, 2001]; H. G. Wilshire, The Impact of Vehicles on Desert Soil Stabilizers. In R. H. Webb and H. G. Wilshire, eds. *Environmental Effects of Off-Road Vehicles* [New York: Springer-Verlag, 1983], 31–50; S. E. Campbell et al. Desert Crust Formation and Soil Stabilization. *Arid Soil Research and Rehabilitation* 1989;3:217–228).

5. Bacterial oxidation creates soluble nitrates out of less soluble compounds, forms that make the nitrates easier for plants to feed on. Compaction reduces bacteria and fungi concentrations by 85%, from about 130,000 to 2,000 per ounce (Liddle. *Recreation Ecology*, 269–271).

6. R. H. Webb et al. Environmental Effects of Soil Property Changes with Off-Road Vehicle Use. *Environmental Management* 1978;2:219–233.

7. H. G. Wilshire et al. 1978. Impacts of Vehicles on Natural Terrain at Seven Sites in the San Francisco Bay Area. *Environmental Geology* 2:295–319

8. "Motorized recreation" is broadly classified as on-road, off-road or off-highway, and personal watercraft (Friends of the Earth. *Trails of Destruction, How Off-Road Vehicles Gain Access and Funding on Public Lands* [1998], 1), but in its broadest sense includes aircraft tourism/sightseeing, helicopter picnicking, helicopter skiing, and making the backcountry more hospitable to recreational homes, as well as dirt bikes, all-terrain vehicles, four-wheel drive vehicles, snowmobiles, and the like. Expanded environmental threats are posed by the Hummer, the civilian version of the military Humvee, although 2007's rising gas prices may curtail its popularity and damage its public image.

9. D. V. Prose and H. G. Wilshire. *The Lasting Effects of Tank Maneuvers on Desert Soils and Intershrub Flora* (U.S. Geological Survey Open-File Report 00–512, 2000); E. W. Lathrop

and P. G. Rowlands. Plant Ecology in Deserts, an Overview. In R. H. Webb and H. G. Wilshire, eds. *Environmental Effects of Off-Road Vehicles* (New York: Springer-Verlag, 1983), 139–140; Jayne Belnap and Steve Warren. Patton's Tracks in the Mojave Desert, USA, an Ecological Legacy. *Arid Land Research and Management* 2002;16:245–259.

10. R. H. Webb. Compaction of Desert Soils by Off-Road Vehicles. In R. H. Webb and H. G. Wilshire, eds. *Environmental Effects of Off-Road Vehicles* (New York: Springer-Verlag, 1983), 51–79.

11. M. Kummerow. Weeds in Wilderness, a Threat to Biodiversity. *Western Wildlands* 1992;18:12–17; R. W. Tyser and C. A. Worley. Alien Flora in Grasslands Adjacent to Road and Trail Corridors in Glacier National Park, Montana (USA). *Conservation Biology* 1992;6:253–262.

12. B. S. Hinckley et al. Accelerated Water Erosion in ORV-Use Areas. In R. H. Webb and H. G. Wilshire, eds. *Environmental Effects of Off-Road Vehicles* (New York: Springer-Verlag, 1983), 81–96; R. M. Iverson et al. Physical Effects of Vehicular Disturbance on Arid Landscapes. *Science* 1981;212:915–917.

13. H. G. Wilshire. *Study Results of 9 Sites Used by Off-Road Vehicles That Illustrate Land Modifications* (U.S. Geological Survey Open-File Report 77–601, 1977), ii; H. G. Wilshire. The Wheeled Locusts. *Wild Earth* 1992;2:27–31; W. J. Kockelman, Introduction. In R. H. Webb and H. G. Wilshire, eds. *Environmental Effects of Off-Road Vehicles* (New York: Springer-Verlag, 1983), 1–11.

14. Hinckley et al. Accelerated Water Erosion in ORV-use Areas, 81–96; Iverson et al., Physical Effects of Vehicular Disturbance, 915–917.

15. Paul Sears. *Deserts on the March*, 2nd ed. (Norman, Oklahoma: Oklahoma University Press, 1947).

16. H. G. Wilshire. Human Causes of Accelerated Wind Erosion in California's Deserts. In D. R. Coates and J. B. Vitek, eds. *Thresholds in Geomorphology* (London: George Allen and Unwin, 1980), 415–433.

17. J. K. Nakata et al. Origin of Mojave Desert Dust Plumes Photographed from Space, 1 January, 1973. *Geology* 1976;4:644–648.

18. California Dust Storm Spreads Fungus Infection. *Medical World News* 1978;19:44; C. R. Leathers. Plant Components of Desert Dust in Arizona and Their Significance for Man. In T. L. Péwé, ed. *Desert Dust, Origin, Characteristics, and Effect on Man* (Geological Society of America Special Paper 186, 1981), 191–206; H. G. Wilshire et al. *Geologic Processes at the Land Surface* (U.S. Geological Survey Bulletin 2149, 1996), 41.

19. Barbara Ransehousen, quoted in John Glionna. Sand Dunes Park Bracing for Invasion of Off-Roaders. *Los Angeles Times* 26 November 1998.

20. In Spanish, *tinaja* means earthen jar.

21. The Center for Biological Diversity successfully sued for closure of Surprise Canyon to off-road vehicles in 2001, allowing substantial riparian vegetation recovery through spring 2006.

22. Winterthur Insurance. *Accident Research*. Available: www.winterthur-insurance.ch/.

23. Especially in small communities near public lands, off-road vehicles have become the "outdoor television baby sitter." Inexperienced children and early teens who are allowed or encouraged to drive high-performance vehicles commonly suffer serious injury and even

death. CBS News reported that 17,900 children under the age of 16 were injured while driving all-terrain vehicles in 1973, and 37,000 in 2002. Driving or riding four-wheel all-terrain vehicles injured more than twice the number of children younger than 16 in 2002 than in 1993, with 14% of those accidents resulting in the death of a child under 12. Only half the states have set minimum ages for all-terrain vehicle operators—in Oregon the minimum age is 12, but in Utah a child of 8 can legally operate a powerful adult-sized vehicle.

24. As a gubernatorial appointee to the California Off-Highway Motor Vehicle Recreation Commission in 1983, Howard Wilshire and other commissioners heard this information from local residents. Some have wondered if taking a high school physics course might benefit off-road vehicle riders.

25. Activities observed by Howard Wilshire.

26. David Sheridan. *Off-Road Vehicles on Public Land* (Washington, DC: Council of Environmental Quality, 1979), 30.

27. MTBE (methyl tertiary butyl ether) is a gasoline additive designed to improve combustion. It now causes widespread pollution in the nation's water and is a suspected carcinogen (Karen Schambach. *California Off-Highway Vehicles, In the Money and Out of Control* (Report of a coalition of eight conservation groups, California Wilderness Coalition, Center for Sierra Nevada Conservation, and Friends of the River, 1999).

28. Melissa Kasnitz and Ed Masche. *Back Country Giveaways, How Bureaucratic Confusion Subsidizes Off-Highway Vehicle Harm* (Santa Barbara, Calif.: California Public Interest Research Group, 1996).

29. T. Weaver and D. Dale. Trampling Effects of Hikers, Motorcycles and Horses in Meadows and Forests. *Journal of Applied Ecology* 1978;15:451–457.

30. H. G. Wilshire et al. *Impacts of Off-Road Vehicles on Vegetation, Transactions 43rd North American Wildlife and Natural Resources Conference* (Wildlife Management Institute, 1978), 131–139.

31. D. S. Wilcove et al. Quantifying Threats to Imperiled Species in the United States. *BioScience* 1998;48:607–615. This report assessed the sources of peril for 2,490 species of plants and animals, 700 of which are not (yet) listed as threatened or endangered; Brian Czech and coworkers assessed the interactions of various activities causing species habitat deterioration (Brian Czech et al. Economic Associations Among Causes of Species Endangerment in the United States. *BioScience* 2000;50:593–601; see also R. F. Noss and R. L. Peters. 1995. *Endangered Ecosystems: A Status Report on America's Vanishing Habitat and Wildlife* [Defenders of Wildlife, December 1995]).

32. The California desert tortoise is *Gopherus agassizii*. The Mojave desert tortoise was officially listed as a "threatened" species in 1990, but by then a combination of human activities, including grazing, disease, off-road vehicles, urbanization, target shooting, poaching, and roadkills, had severely reduced its numbers.

33. Desert animals that live in burrows include mice, kangaroo rats, ground squirrels, lizards, snakes, desert tortoises, amphibians, insects, spiders, and others.

34. B. H. Brattstrom and M. C. Bondello. Effects of Off-Road Vehicle Noise on Desert Vertebrates. In R. H. Webb and H. G. Wilshire, eds. *Environmental Effects of Off-Road Vehicles* (New York: Springer-Verlag, 1983), 167–206; A. E. Bowles. Responses of Wildlife to Noise. In R. L. Knight and K. J. Gutzwiller, eds. *Wildlife and Recreationists* (Washington,

DC: Island Press, 1995), 109–156; D. J. Schubert and Jacob Smith. The Impacts of Off-Road Vehicle Noise on Wildlife. *Wildlands Center for Preventing Roads* 2000;5:12–14; N. C. Nicolai and J. E. Lovich. Preliminary Observations of the Behavior of Male, Flat-Tailed Horned Lizards Before and After an Off-Highway Vehicle Race in California. *California Fish and Game* 2000;86:208–212.

35. Kangaroo rats are *Dipodomys deserti* and *D. merriami*.

36. Couch's spadefoot toad is *Scaphiopus couchi*.

37. Brattstrom and Bondello. Effects of Off-Road Vehicle Noise.

38. Michael Weinstein. *Impact of Off-Road Vehicles on the Avifauna of Afton Canyon, California* (Report for the Bureau of Land Management, Contract No. CA-060-CT7–2734, 1978).

39. The distance of travel required to affect an acre is based solely on the size of the contact area between soil and traveler. Hence, a typical hiker mounted on a mountain bike would have his/her weight distributed over a smaller contact area, and so compresses the soil more.

40. Burning Man briefly becomes the most populous town in a low-resident county, so organizers refer to it as "Black Rock City."

41. SkiTown.com. *Ski Guide.* Available: www.ski-guide.com/. Additional data were obtained from individual ski resorts by e-mail communication; the U.S. Forest Service provided information on the number of ski resorts on public lands in the 11 western states by written communication, 19 January 2001, under a Freedom of Information Act request.

42. Nancy Watzman. Playground or Preserve? How the Recreation Industry Has Become the Newest Threat to Our Public Lands. *Washington Monthly* May 2001.

43. Todd Wilkinson. *Science Under Siege, The Politicians' War on Nature and Truth* (Boulder, Colorado: Johnson Books, 1998), 113–157.

44. Quoted by Watzman. Playground or Preserve?

45. W. D. Schmid. Modification of the Subnivian Microclimate by Snowmobiles. In A. O. Haugen, ed. *Symposium on Snow and Ice in Relation to Wildlife* (Iowa Cooperative Wildlife Research Unit, Iowa State University, 1970), 251–257. Snowmobile test runs indicated an increase in snow density 1.7 times greater than undisturbed snow, reaching a maximum in only four passes. The compacted snow areas transmitted heat upward from the ground surface five times faster than does undisturbed snow. Soil temperatures under the tracks also were sharply lower, and frost penetration effects went nearly two feet deep (A. R. Pesant. Snowmobiling Impact on Snow and Soil Properties and on Winter Cereal Crops. *Canadian Field-Naturalist* 1987;101:22–32).

46. W. J. Wanek. The Ecological Impact of Snowmobiling in Northern Minnesota. In D. F. Holecek, ed. *Symposium on Snowmobile and Off the Road Vehicle Research* (Technical Report No. 9, College of Agriculture and Natural Resources, Michigan State University, 1974), 57–76; H. G. Wilshire et al. *Impacts and Management of Off-Road Vehicles* (Report of the Committee on Environment and Public Policy, Geological Society of America, 1977).

47. C. W. Schadt et al. Seasonal Dynamics of Previously Unknown Fungal Lineages in Tundra Soils. *Science* 2003;301:1359–1361. The proportion of fungi to bacteria in Colorado tundra soils varies seasonally. The total mass of fungi, or biomass, is about 15 times the mass of bacteria in winter, and about six times as much in summer.

48. Studies of the stress-level indicator glucocorticoid in elk and wolf feces show a clear correlation between the intensity of snowmobile use and heightened stress levels in elk (*Cervus elaphus*) in Yellowstone National Park and in wolves (*Canis lupus*) in Yellowstone, Voyageurs, and Isle Royale National Parks (Scott Creel et al. Snowmobile Activity and Glucocorticoid Stress in Wolves and Elk. *Conservation Biology* 2002;16:809–814). Other studies showed that mammals and birds move burrow and nesting sites to avoid snowmobile areas (Jay Withgott. Signs of Stress Seen in Snowmobile Season. *Science* 2002;296:1784–1785). A recent National Park Conservation Association study in Yellowstone National Park (Yellowstone Sound Survey, President's Day Weekend, 2000; available: www.npca.org/media_center/reports/yellowstone.html) found that "Visitors in the most famous areas are inundated with snowmobile noise 90% of the time," and concluded "it is virtually impossible to escape snowmobile noise" anywhere in the park.

49. G. A. Bishop et al. Winter Motor-Vehicle Emissions in Yellowstone National Park. *Environmental Science and Technology* 2006;40:2505–2510.

50. In general, the low numbers of enforcement officials are too few to prevent illegal snowmobile incursions into wilderness (U.S. General Accounting Office. *Federal Lands: Agencies Need to Assess the Impact of Personal Watercraft and Snowmobile Use* (GAO/RCED-00–243, 2000)).

51. Andrew Todd and Diane McKnight. Abandoned Mines, Mountain Sports, and Climate Variability, Implications for the Colorado Tourism Economy. *EOS, Transactions of the American Geophysical Union* 2003;84:377, 386. Presently observable global climate changes led the United Nations Environment Program to predict that downhill skiing could disappear altogether at some U.S. and European resorts, and the retreating snowline will cut off base villages from their ski runs as soon as 2030.

52. Warmer winters cause more of the winter precipitation to come as rain and force earlier snowmelt. In addition, risks of fire are likely to increase (A. W. Nolin and Christopher Daly. Mapping "At-Risk" Snow in the Pacific Northwest, U.S.A. *Journal of Hydrometeorology* 2006;7:1164–1171). At least for a while, diminishing snowfall will be replaced by man-made snow—until the cheap energy and water required to manufacture snow runs out.

53. Bill Meelater. Motorheads Sue Over Park's Seasons. *High Country News* 35 (6)(2003): 3.

54. *The Official 1991 United States Golf Course Directory and Guide* (Ill.: Kayar Co., 1991) lists 1,048 golf courses by town or city. Figures for 1999 come from SkiTown.com (*Ski Guide*. Available: ski-guide.com/), which also lists golf courses under towns and cities. We searched each town or city listed on the Web site for public and private golf courses. Caroline Cox (Coming Soon, Roundup Ready Grass on Golf Courses. *Journal of Pesticide Reform* 2004;24:5) estimates 17,000 golf courses in the United States.

55. S. D. Merrigan et al. *Arizona Golf Course Pesticide Use Survey* (Publication No. 895024, University of Arizona, College of Agriculture, 1996); these authors determined an average use of 147,428,000 gallons per year on Arizona golf courses; Peter Levy. *The Practical Applications of Computerized Weather Watching* (2004). Available: www.weathermetrics. com; this report cites an average use for golf courses in all the western United States of 116,583,000 gallons per year.

56. Alex Shoumatoff. *Legends of the American Desert* (New York: HarperCollins , 1997), 368.

57. M. P. Kenna. What Happens to Pesticides Applied to Golf Courses? *U.S. Golf Association, Green Section Record* 1995;33:1–9; S. K. Starrett and N. E. Christians. Nitrogen

and Phosphorus Fate When Applied to Turfgrass in Golf Course Fairway Condition. *United States Golf Association, Green Section Record* 1995;33:23–25.

58. J. E. Barbash and E. A. Resek. *Pesticides in Ground Water, Distribution, Trends, and Governing Factors* (Chelsea, Mich.: Ann Arbor Press, 1996), 117–118. This immensely valuable book is the most comprehensive available reference to the sources and nature of pesticides in groundwater. Golf courses apply an average of more than 3.5 pounds of herbicides per acre per year, a similar amount of fungicides, and about 2.5 pounds of insecticides per acre per year, according to a national U.S. Environmental Protection Agency survey of golf course pesticide use. The total pesticide use on golf courses is more than nine pounds per acre. In some places pesticides are applied even more intensively than is typical for agriculture (Caroline Cox. Pesticides on Golf Courses, Mixing Toxins with Play? *Journal of Pesticide Reform* 1991;11:2–7).

59. U.S. Golf Association, Pesticide Runoff Model for Turfgrass: Development, Testing and Application (1999). Available: www.usga.com/green/. Similar information is no longer available from the U.S. Geological Survey (*Pesticides Used on and Detected in Ground Water Beneath Golf Courses* [1999]) as of December 2003. Many chemicals applied to golf courses, whether soluble in water or not, have contaminated groundwater beneath the golf courses. In arid settings, the greater distances from the golf courses to the water table may reduce groundwater contamination, but data are lacking, and other similar speculations have proved incorrect.

60. A. E. Smith et al. A Greenhouse System for Determining Pesticide Movement from Golf Course Greens. *Journal of Environmental Quality* 1993;22:864–867.

61. A. E. Smith and W. R. Tillotson. Potential Leaching of Herbicides Applied to Golf Course Greens. *American Chemical Society Symposium Series* 1993;52:168–181.

62. Recovery times can be calculated with a linear model (assuming that disturbed desert sites recover at a constant rate) or a logarithmic model (assuming that the processes promoting recovery act most quickly immediately after occupation and become slower with time). The reported results used both models (R. H. Webb. Recovery of Severely Compacted Soils in the Mojave Desert, California, USA. *Arid Lands Research and Management* 2002;16:291–305; see also R. H. Webb and H. G. Wilshire. Recovery of Soils and Vegetation in a Mojave Desert Ghost Town, Nevada. *Journal of Arid Environments* 1980;3:291–303; P. A. Knapp. Soil Loosening Processes Following the Abandonment of Two Arid Western Nevada Townsites. *Great Basin Naturalist* 1992;52:149–154).

63. R. H. Webb et al. *Perennial Vegetation Data from Permanent Plots on the Nevada Test Site, Nye County, Nevada* (U.S. Geological Survey Open-File Report 03–336, 2003), 12; see also U.S. Geological Survey. *Recoverability and Vulnerability of Desert Ecosystems* (Fact Sheet 058–03, 2003).

64. One project to reclaim deeply eroded, steep hillclimb trails on highly erodible soils at the Hollister Hills State Vehicular Recreation Area included cutting water bars across the trails, spreading biodegradable netting on lower slopes, and seeding the area. Without any follow-up work, winter runoff eroded through the water bars and stripped much of the netting, carrying away soil and seed, thus limited plant regrowth. Seedlings transplanted on other nearby trails all died from lack of care. At Hungry Valley State Vehicular Recreation Area, bulldozers leveled off and filled the slots that eroded motorcycle tire ruts had cut into steep slopes and then put hydromulch on the slopes and laid wattles of straw rolled in plastic netting across them to control erosion. Unfortunately, the bulldozed trail fills were not

engineered and are sliding off the slopes. The straw wattles have not prevented runoff channels from undercutting and outflanking them.

65. Wild Utah Project. *2000 Report* (Salt Lake City: Wild Utah Project, 2000), 17.

66. The proposed limit and ban addressed multiple studies on snowmobiling's adverse environmental effects and the incompatibilities between Nordic skiers and power skis (U.S. Department of the Interior, National Park Service. *Press Release: Winter Use Decision for Yellowstone and Grand Teton* (22 November 2000); Robert Perks et al. *Rewriting the Rules* (Natural Resources Defense Council, 2002), 40.

67. Brad Knickerbocker. Bush Takes Quiet Aim at "Green" Laws. *Christian Science Monitor* 8 November 2003.

68. *Norton, Secretary of the Interior, et al. v. Southern Utah Wilderness Alliance, et al.*, Certiorari to the United States Court of Appeals for the Tenth Circuit. No. 03–101. Argued 29 March 2004, decided 14 June 2004. Available: www.supremecourtus.gov/opinions/03pdf/03–101.pdf. The Southern Utah Wilderness Alliance now files individual emergency closure petitions, requesting immediate closure to motorized use on lands where the use is out of control. The Southern Utah Wilderness Alliance's petition for Utah's Factory Butte, supported by affidavits from recognized experts on archaeology, riparian environments, and sensitive soils, forced the Bureau of Land Management to take protective measures in September 2006. A petition also has been filed for the Vermillion Cliffs area (Southern Utah Wilderness Alliance. Enough Is Enough: Citizens Fight Back Against Off-Road Vehicles. *Redrock Wilderness* 2005;22[1]:5–9).

Chapter 12

1. Admiral Hyman G. Rickover often is called the father of the nuclear submarine. He obtained a master of science degree in electrical engineering from Columbia University and trained in nuclear power at Oak Ridge, Tennessee. Assigned to the U.S. Atomic Energy Commission (later the Department of Energy) Division of Reactor Development, and serving as director of the Naval Reactors Branch in the Naval Bureau of Ships, Rickover assumed control of the Navy's program for nuclear ship propulsion, directing the planning and construction of *Nautilus*, the world's first atomic-powered submarine, and eventually the nuclear-powered U.S. fleet. He also played a role in developing the U.S. civilian nuclear power industry and worked closely with President Jimmy Carter on energy issues in the 1970s. Admiral Rickover delivered the quoted remarks in a speech titled "Energy Resources and Our Future" at a banquet of the Annual Scientific Assembly of the Minnesota State Medical Association, St. Paul, Minnesota, 14 May 1957. Available: www.energybulletin.net/newswire.php?id=23151.

2. Walter Youngquist. *GeoDestinies* (Portland, Oreg.: National Book Company, 1997), 252. Since the historical period of abundant cheap oil will be less than two centuries, far too short to be called an "age" equivalent to the Stone Age or Bronze Age, Youngquist suggests "Petroleum Interval" as a more appropriate label.

3. Roll Call. Energy Policy Briefing, Roll Call Q & A, Lott Lights up the Democrats (12 March 2001).

4. Youngquist, GeoDestinies, 163.

5. Hydrogen is an energy carrier, like electricity, not an energy source. The idea of running vehicles on hydrogen-powered fuel cells as a transition to future energy sources, as yet unknown, had a short heyday. (see appendix 10).

6. Petroleum forms in subsiding basins, as sediment particles and associated organic matter become highly compressed under the weight of constantly accumulating overlying deposits. Water is all but incompressible and tends to flow out. Films of silica or calcium carbonate precipitate out of any remaining water, binding discrete sand and clay particles together as pressure makes them into rock (lithification). The subsiding basin takes the sediments into deeper and warmer earth zones, where organic matter within the mass slowly "cooks" and distills. Raw hydrocarbons first turn into a substance called "kerogen," and then into oil plus gas, or just gas (explanation adapted from Byron W. King. There's a Hole in the Bottom of the Sea. *Whiskey and Gunpowder/Energy Bulletin* 13 September 2006. Available: www.energy-bulletin.net/20455.html).

7. T. S. Dyman et al. Geologic Studies of Deep Natural-Gas Resources in the United States. In D. G. Howell, ed. *The Future of Energy Gases* (U.S. Geological Survey Professional Paper 1570, 1993), 171–203; R. E. Wyman. Challenges of Ultradeep Drilling. In D. G. Howell, ed. The *Future of Energy Gases* (U.S. Geological Survey Professional Paper 1570, 1993), 205–215; Youngquist, *GeoDestinies*, 178. The oil window varies with the geological setting, but in general, oil plus methane is found between 7,500 and 15,000 feet below Earth's surface, and only the gas below a depth of about 16,000 feet. Methane can persist to much greater depths but generally cannot be extracted. The heavier gases, propane, ethane, and butane, process easily into "natural gas liquids."

8. U.S. Geological Survey Energy Resource Surveys Program. *Describing Petroleum Reservoirs of the Future* (USGS Fact Sheet FS-020–97, 1997). C. J. Campbell uses the term "regular" instead of "conventional" for easily exploited deposits and includes oil in hostile environments, such as deep-water and polar deposits in unconventional categories (Klaus Illum. *Oil-Based Technology and Economy: A Short Introduction to Basic Issues and a Review of Oil Depletion Projections Derived from Different Theories and Methods* [Copenhagen, The Netherlands, 2004], 30).

9. Most petroleum extracted from the U.S. Gulf Coast and northern South America originated 90 million years ago at a time of extremely warm global climates, the effects of highly concentrated atmospheric greenhouse gases. Britain's North Sea oil and much of Middle Eastern and Russian petroleum come from an even earlier warm greenhouse period at 145 million years ago.

10. J. S. Dukes. Burning Buried Sunshine, Human Consumption of Ancient Solar Energy. *Climatic Change* 2003;61:31–44. The natural processes that created fossil fuels were very inefficient for preserving the carbon contents. Estimates put coal formation from plants at less than 10% efficiency, and oil and gas formation from phytoplankton at less than 0.01% efficiency. From these estimates, the mass of ancient organic matter in fossil fuels that the world consumed in 1997 was more than 400 times the current global net primary productivity of Earth biota.

11. Tropospheric ozone levels have increased 35% in the last century, causing adverse effects on forest and crop productivity, even though increased CO_2 levels can stimulate forest productivity (W. M. Loya et al. Reduction of Soil Carbon Formation by Tropospheric Ozone Under Increased Carbon Dioxide Levels. *Nature* 2003;425:705–707).

12. Ironically, scrubbers that capture sulfur from flu gas may themselves contain significant amounts of arsenic, antimony, selenium, lead, and other toxic elements. Limestone (calcium carbonate) from the scrubbers of one 600-megawatt plant carried most of the polluting trace elements coming from that plant.

13. U.S. Geological Survey. *Health Impacts of Coal Combustion* (Fact Sheet FS-094–00, 2000).

14. C. A. Pope III et al. Lung Cancer, Cardiopulmonary Mortality, and Long-Term Exposure to Fine Particulate Air Pollution. *JAMA* 2002;264:1132–1141. Aerosols soak up toxic materials like tiny sponges, and because of their small sizes can be breathed into the lungs and penetrate human tissue. Soot and other fine particulates may be major contributors to approximately one million premature deaths per year (James Hansen and Larissa Nazarenko. Soot Climate Forcing via Snow and Ice Albedos. *Proceedings of the National Academy of Sciences of the USA* 2004;101:423–428; André Nel. Air Pollution—Related Illness: Effects of Particles. *Science* 2005;308:804–806). Another study shows that "coarse" particulates (2.5–10 micrometers in diameter, 0.0001–0.0004 inches) are harmful to children (Mei Liu et al. The Influence of Ambient Coarse Particulate Matter on Asthma Hospitalization in Children, Case-Crossover and Time-Series Analyses. *Environmental Health Perspectives* 2002;110:575–581).

15. Other models besides Hubbert's conclude that peak production correlates to about 50% production of ultimately recoverable oil (Illum. *Oil-Based Technology and Economy*, 8–9).

16. Hubbert knew from experience that a graph of averaged unrestrained oil production levels for a field or region against time describes a roughly symmetrical bell-shaped curve. The production curve rises steeply until about half of the most easily recoverable resource is extracted, and then declines until production is uneconomic. In 1956, Hubbert modeled the rapid 1950s U.S. oil production rate as a similar bell curve, and accurately predicted that domestic oil production would peak in 1970 (see figure 12.2) (M. K. Hubbert. The Energy Resources of the Earth. *Scientific American* September 1971;60–70). If production is somehow controlled, averaged production curves may not have symmetrical bell shapes. "The [shape of the] ascending curve depends on skill/luck of the explorationists, while the descending side may fall off more rapidly due to the public's acquired taste for petroleum products, or more slowly due to government controls to reduce consumption" (L. F. Ivanhoe. Updated Hubbert Curves Analyze World Oil Supply. *World Oil* November 1996;91–94).

17. The decline of new oil discoveries on U.S. territories after 1930 meant that U.S. petroleum production also would reach a peak and then decline. The peak event depended on the rate of consumption. U.S. exploitation rates were restrained only by "driller's luck" and market forces. We now know that rate of consumption produced a 40-year gap between peak oil discoveries and peak production (C. J. Campbell. The Oil Peak, a Turning Point. *Solar Today* July/August 2001;42). Political decisions, recession, war, and corporate manipulations can slow production rates, lower peak production levels, and extend the time to peak production. They also can turn peaks into plateaus of long duration and slow the rates of decline. Very low consumption rates could extend the life of an oil resource indefinitely (Richard Heinberg. *The Party's Over: Oil, War and the Fate of Industrial Societies* [Gabriola Island, Canada: New Society Publishers, 2003], 34, 88–92).

18. Average U.S. gas productivity peaked at 435 thousand cubic feet (Mcf) per day per well in 1971, plunged 61% to about 170 Mcf per day in 1983, and then leveled off at about 160 Mcf per day until 1999, when it declined again, reaching 126 Mcf per day in 2004. As

exploration finds ever smaller gas deposits, new wells experience steeper and steeper depletion rates. Natural gas wells drilled between 1990 and 2004 show faster rates of declining
productivity than do older wells (25% overall). The reason is not fully understood (K. S.
Deffeyes. *Beyond Oil: The View from Hubbert's Peak* [New York: Hill and Wang, 2005], 80;
this is a very readable and informative account of petroleum issues).

19. U.S. Government Accountability Office. *Meeting Energy Demand in the 21st Century:
Many Challenges and Key Questions* (GAO-05-414T, 2005), 14–15. From 1995 to May 2005,
the price of natural gas rose $1.55 to $6.54 per Mcf.

20. Greg Croft, Inc. *Peak Oil: Fact and Fiction.* Available: www.gregcroft.com/peakoil.ivnu.
The U.S. government's Energy Information Administration (EIA) continues to project U.S. oil
consumption from a flat-earth platform: EIA's latest estimate suggests that U.S. oil consumption will rise from 21 million barrels per day in 2006 to 27 million barrels per day in 2030,
about 9.9 billion barrels per year (Energy Information Administration. *Annual Energy Outlook
2007* (December 2006)).

21. The discovery of the "Jack 2" oil field in the Gulf of Mexico, announced in 2006,
does not materially add to U.S. oil reserves. Oil analyst Jeff Vail notes that the touted three
billion barrels of oil, if actually produced, will fuel the world for no more than 35 days at
current consumption levels and is less than half current U.S. annual consumption. Chevron
had first confirmed the Gulf find (Jack field) in 2004. "They've known all along the expected
size of the field's reserves, and this figure has long been part of Peak Oil calculations. In
fact, the 6000 barrel per day flow rate of the test well was finalized and published in May
of this year (2006). There is absolutely no new information about this event to surface in the
past three months" (Jeff Vail. *EXTRA: Oil Discovery Saves Civilization: Theory of Power* [5
September 2006]. Available: www.jeffvail.net/2006/09/extra-oil-discovery-saves-civilization.html).
Speculating on the reason for delaying the find's announcement until 2006, Vail suggested
that the November 2006 mid-term congressional election might be one explanation.

22. Energy Information Administration. *Annual Energy Outlook 2007.*

23. The International Energy Agency forecasts that fossil fuels will meet only 80% of the
world's energy demand by 2030 are likely to prove optimistic.

24. In 120 years, the United States has produced about 839 trillion cubic feet (Tcf) of
gas while the rest of the world produced about 898 Tcf (U.S. Geological Survey. *Natural Gas
Production in the United States* (Fact Sheet FS-113–01, 2001); U.S. Geological Survey World
Energy Assessment Team. *World Petroleum Assessment 2000* (2000), Executive Summary, table 1).

25. U.S. Government Accountability Office. *Natural Gas Flaring and Venting: Opportunities
to Improve Data and Reduce Emissions* (GAO-04–809, 2004). Vented gas contributes more
greenhouse gases than flared gas. Not much of the methane likely reaches the upper atmosphere because most of it converts to CO_2 plus water (K. A. Kvenvolden. Potential Effects
of Gas Hydrate on Human Welfare. In J. V. Smith, ed. *Colloquium on Geology, Mineralogy,
and Human Welfare* (Washington, DC: National Academy of Sciences, 1999), 3420–3426).

26. Julian Darley. *High Noon for Natural Gas: The New Energy Crisis* (White River
Junction, Vermont: Chelsea Green Publishing Company, 2004), 92–97; *Association for Study
of Peak Oil Newsletter* 2004;No. 44:387. Natural gas is a land-locked resource because it cannot be easily or cheaply transported between continents. Liquifying natural gas to transport
it without pipelines entails a 20–30% loss of energy for liquefaction, transport, and regasifica-
tion (J. D. Hughes. *Natural Gas in North America: Should We Be Worried?* Paper presented

at the Association for Study of Peak Oil-USA, Boston, Mass., 25–27 October 2006. Available: www.aspo-usa.com/).

27. E. D. Sloan, Jr. Fundamental Principles and Applications of Natural Gas Hydrates. *Nature* 2003;426:353–359; Jack Dvorkin et al. Rock Physics of a Gas Hydrate Reservoir. *Leading Edge* September 2003;842–847; K. A. Kvenvolden and T. D. Lorenson. The Global Occurrence of Natural Gas Hydrate. *Geophysical Monograph* 2001;124:3–18. Note, however, that gas hydrates have recently been found at depths of only 165–400 feet on the northern Cascadia margin (M. Riedel et al. Gas Hydrate Transect Across Northern Cascadia Margin. *EOS, Transactions of the American Geophysical Union* 2006;87:325, 330, 332). Depending on local geothermal and surface temperatures, gas hydrates in Arctic permafrost may occur to about 1,500-foot depths, in amounts ranging from 500 to 1.2 million Tcf (trillion cubic feet). Oceanic deposits at the edge of the continental rise, the steep slope that connects the submerged continental shelf with a deep ocean basin, speculatively range from 110 thousand to 270 million Tcf (T. S. Collett. Natural-Gas Hydrates, Resource of the Twenty-First Century? In M. W. Downey et al., eds. Petroleum Provinces of the Twenty-First Century. *American Association of Petroleum Geologists Memoir* 2001;74:85–108).

28. The figures represent a 5% chance of finding 2.7 billion barrels and a 95% chance of 1.5 billion barrels (U.S. Geological Survey National Oil and Gas Resource Team. *1995 National Assessment of United States Oil and Gas Resources* [U.S. Geological Survey Circular 1118, 1995]).

29. In 1993, the Alberta Chamber of Resources selected "oil sands" as the new brand-name euphemism for dirty-sounding "tar sands" (Dan Woynillowicz et al. *Oil Sands Fever: The Environmental Implications of Canada's Oil Sands Rush* [Pembina Institute, November 2005]; this is an excellent comprehensive report on the magnitude of Canadian oil sand resources and the environmental impacts of production).

30. J. W. Smith. Synfuels, Oil Shale, and Tar Sands. In L. C. Ruedisili and M. W. Firebaugh, eds. *Perspectives on Energy*, 3rd ed. (New York: Oxford University Press, 1982), 225–249; B. W. Pipkin et al. *Geology and the Environment*, 4th ed. (Belmont, California: Brooks/Cole-Thompson Learning, 2005), 398–399; C. W. Montgomery. *Environmental Geology*, 5th ed. (New York: WCB McGraw-Hill, 1997), 318–320.

31. D. C. Duncan and V. E. Swanson. *Organic-Rich Shale of the United States and World Land Areas* (U.S. Geological Survey Circular 523, 1965); J. R. Donnell. Storehouse of Energy Minerals in the Piceance Basin. In O. J. Taylor, Comp. *Oil Shale, Water Resources, and Valuable Minerals of the Piceance Basin, Colorado: The Challenges and Choices of Development* (U.S. Geological Survey Professional Paper 1310, 1987), 21–28.

32. D. A. Rickert et al., eds. *Synthetic Fuels Development, Earth Science Considerations* (U.S. Geological Survey, 1979), 10; Walter Youngquist. 1998. Shale Oil: The Elusive Energy. *M. King Hubbert Center for Petroleum Supply Studies Newsletter* 98(4); P. A. Fischer. Hopes for Shale Oil Are Revived. *World Oil* 2005;226(8); J. T. Bartis et al. *Oil Shale Development in the United States: Prospects and Policy Issues, Summary* (RAND Corporation, 2005), ix; L. G. Weeks. The Next Hundred Years Energy Demand and Sources of Supply. *Geotimes* 1960;5:18–21, 51–55.

33. Bengt Söderbergh et al. A Crash Program Scenario for the Canadian Oil Sands Industry. *Energy Policy* 2006. Available: www.elsevier.com/wps/find/journaldescription.cws_home/30414/description#description (accessed June 2006).

34. Anthony Andrews. *Oil Shale: History, Incentives, and Policy* (Congressional Research Service RL33359, 13 April 2006).

35. Mining and processing tar sand takes about 700 cubic feet of gas for each barrel of bitumen, and in-place production requires 1,200 cubic feet of gas per barrel. Increased coal-bed methane production will partly offset demand for conventional gas, but natural gas also is needed to generate electricity. Nuclear power, promoted in early 2006 as a substitute for dwindling gas supplies in the Alberta tar sands area, would generate electricity to separate hydrogen from water by electrolysis. But bringing the plant on line would take at least 10 years, with substantial cost increases.

36. Diagrammatic representations show producing beds interlayered with beds of high fracture porosity (allowing free movement of liquified kerogen toward wells), capped both top and bottom by impermeable layers. This extraction method has close to zero likelihood in the real world. Time will tell if Shell can make it work on a large scale (H. R. Johnson et al. *Strategic Significance of America's Oil Shale Resource*: Vol. 2. *Oil Shale Resources, Technology and Economics* (U.S. Department of Energy 2004), 17).

37. Bartis et al. Oil Shale Development in the United States; Randy Udall and Steve Andrews. The Illusive Bonanza: Oil Shale in Colorado (Association for Study of Peak Oil-USA). Available: www.aspo-usa.com (accessed 2 November 2005). Other, very dubious, in situ technologies are described by the U.S. Office of Technology Assessment (*An Assessment of Oil Shale Technologies* [New York: McGraw-Hill, 1980], chapter 5).

38. P. J. McCabe et al. The Future of Energy Gases. *U.S. Geological Survey* Circular 1115 (1993), 1.

39. U.S. Geological Survey National Oil and Gas Resource Team. *1995 National Assessment of United States Oil and Gas Resources*. The credibility of the U.S. Geological Survey adding 322 trillion cubic feet "reserve growth" due to technological improvements is challenged by J. H. Laherrère (*Reserve Growth: Technological Progress, or Bad Reporting and Bad Arithmetic?* [2001]. Available: www.greatchange.org/ov-laherrere,reserve_growth_techno-logical_Progress.html; see also Jean Laherrère. *Estimates of Oil Reserves* (Laxenburg, Austria: International Institute for Applied Systems Analysis, 2001), 47–60. Available: www.greatchange.org/ov-laherrere.June_10_01.pdf). Technological oil recovery improvements are of two types—those that can extract previously unrecoverable oil and increase reserves, and those that recover oil more rapidly but do not increase reserves. The last 20 years or so have seen no significant improvements to additional recovery techniques.

40. The mean U.S. Geological Survey estimate of approximately 885 trillion cubic feet (Tcf) combines estimates for the continental United States plus state offshore areas (U.S. Geological Survey National Oil and Gas Resource Team. *1995 National Assessment*, table 1) and the Minerals Management Service estimate for federal offshore areas (U.S. Department of the Interior, Minerals Management Service. *An Assessment of the Undiscovered Hydrocarbon Potential of the Nation's Outer Continental Shelf* [OCS Report MMS 96–0034, 1996], table 1). Some experts predict that world gas production will peak around 2030 at a production rate of about 135 Tcf/year (*Association for Study of Peak Oil Newsletter* August 2004;No. 45:387.

41. In October 2004, the U.S. senate approved a 36-inch pipeline from northern Alaska, supported by more than $14 billion in U.S. construction loan guarantees, which would deliver only 0.5 trillion cubic feet of gas per year, barely 2% of current consumption and much less at the projected growth in demand.

42. The U.S. Geological Survey estimates that six major basins in the five states have an even chance of producing enough coal-bed methane for 1.9 years supply at current U.S. consumption levels, and a 95% chance of 1.4 years supply (U.S. Geological Survey. *Coal-Bed Gas Resources of the Rocky Mountain Region* [Fact Sheet FS-110–01, 2001]).

43. The U.S. Geological Survey figure of 1,530,000 gas wells is small compared to the National Petroleum Council's projection that four million new gas wells and 14 million new oil wells would be necessary to meet U.S. demand by 2030 (U.S. Geological Survey National Oil and Gas Resource Team. *1995 National Assessment*). The National Petroleum Council is an industry advisory board to the Department of Energy, formerly led by George W. Bush's vice president, Richard Cheney, while he was employed by Halliburton Energy Services, Inc.

44. The Bureau of Land Management estimates coal-bed methane development will disturb 7.2 acres per well site, including ancillary facilities (U.S. Department of the Interior, Bureau of Land Management. *Powder River Basin EIS Newsletter* 2001; No. 1). The four million new gas wells to be drilled by 2030 would directly disturb 28.8 million surface acres, more than half the size of Utah. The supposedly necessary 14 million new oil wells would carve up an area of at least 100 million acres, nearly the size of California.

45. T. S. Collet. Gas Hydrates as a Future Energy Resource. *Geotimes* November 2004;24–27.

46. B. U. Haq. Methane in the Deep Blue Sea. *Science* 1999;285:543–544; J. H. Laherrère. Uncertain Resource Size, Enigma of Oceanic Methane Hydrates. *Offshore* 1999;59:140–141, 160; J. H. Laherrère. Data Show Methane Hydrate Resource Overestimated. *Offshore* 1999;59:156–158; G. R. Dickens et al. Direct Measurement of In Situ Methane Quantities in a Large Gas-Hydrate Reservoir. *Nature* 1997;385:426–428; see also National Research Council. *Charting the Future of Methane Hydrate in the United States* (Washington, DC: National Academies Press, 2004). A recent report claims commercial hydrate-producing potential for deep-water fluid discharge areas, based on estimates of gas reserves in 40 structures spread over 200 km^2 (H. Shoji et al. Hydrate-Bearing Structures in the Sea of Okhotsk. *EOS, Transactions, American Geophysical Union* 86 No. 2005;2:13, 18). Estimates put the amount of gas at 0.5% annual U.S. consumption. The presence of marine hydrates is linked to a particular seismic reflection, the bottom-simulating reflector (BSR)—the boundary between hydrate-bearing sediment and underlying strata containing free gas and water. The BSR reflector can be an indicator of the base of gas hydrate deposits, but in places may correspond to a submarine temperature boundary (C. J. Campbell. *The Availability of Non-conventional Oil and Gas* [prepared for the Office of Science and Innovation, Department of Trade and Industry; London: Association for the Study of Peak Oil, 2006]). Drill cores and remotely operated submersibles have recovered methane hydrates at a number of sites (R. Chapman et al. Thermogenic Gas Hydrates in the Northern Cascadia Margin. *EOS, Transactions of the American Geophysical Union* 2004;85[38]:361, 365). Whatever the real magnitude of methane hydrate deposits, recoverable amounts probably are small, and no accurate estimates are available (Keith Kvenvolden, personal communication, September 2006).

47. Hurricane Ivan amply demonstrated the vulnerability of Gulf of Mexico drilling platforms to submarine mud slides in 2004. Recent studies examined drill cores from the northern Gulf of Mexico to determine if deep sediments contain significant gas hydrate deposits and whether gas hydrates may have contributed to submarine landslides near Mississippi Canyon, but did not come up with answers (T. D. Lorenson et al. *EOS, Transactions of the American Geophysical Union* 2002;83:601, 607). One study found that solid hydrate-cemented sediments overlie highly fluidized sands or muds—a setting similar to avalanche-prone

snow-covered mountains. These unstable conditions probably are responsible for massive sub-marine slides on Blake Ridge, east of Florida (R. E. Kayen and H. J. Lee. Slope Stability in Regions of Sea-Floor Gas Hydrate, Beaufort Sea Continental Slope. In W. C. Schwab et al., eds. *Submarine Landslides, Selected Studies in the U.S. Exclusive Economic Zone* [U.S. Geological Survey Bulletin 2002, 1993], 97–103; see also R. L. Kleinberg and P. G. Brewer. Probing Gas Hydrate Deposits. *American Scientist* 2001;89:244–251).

48. U.S. Geological Survey. *Natural Gas Hydrates—Vast Resource, Uncertain Future*, USGS Fact Sheet FS-021–01, 2001. Other problems and hazards could arise when hydrates reform and plug pipelines, a dangerous and expensive condition. Solid hydrate plugs melt at the pipe walls, allowing the remaining plug to move down the pipeline at measured velocities of 185 miles per hour, compressing the gas ahead of it until the pipe ruptures from high pressure. Heating the hydrate plug from the outside to melt it can evolve gas at the plug ends, which can increase pressure and burst the pipeline (Sloan. Fundamental Principles and Applications of Natural Gas Hydrates, 353–359).

49. C. J. Cleveland et al. Energy Returns on Ethanol Production. *Science* 2006;312:1746; C. J. Cleveland. *Energy Quality, Net Energy and the Coming Energy Transition.* Paper presented at the Association for Study of Peak Oil-USA conference, Boston, 25–27 October 2006; Charles Hall et al. Hydrocarbons and the Evolution of Human Culture. *Nature* 2003;426:318–322; C. A. S. Hall and C. J. Cleveland. *EROI: Definition, History, and Future Implications.* Paper presented at the Association for Study of Peak Oil conference, Denver, Colo., 10 November 2005; C. S. A. Hall, David Murphy, and Nate Gagnon. *Order from Chaos (Hopefully): A Preliminary Protocol for Determining EROI for Fuels.* Paper presented at the Association for Study of Peak Oil-USA conference, Boston, 25–27 October 2006; Cutler Cleveland and Ida Kubiszewski. Energy Return on Investment (EROI) for Wind Energy. In Peter Saundry, ed. *Encyclopedia of Earth* (Washington, DC: Environmental Information Coalition, National Council for Science and the Environment, 2006). Other causes of reduced net energy yield, called energy return on investment, or energy return on energy invested, from nonrenewable resources include higher energy costs for finding smaller and smaller deposits, and production from unconventional deposits, such as tight formations and those located in deep-water and remote areas.

50. The George W. Bush administration is still plugging and funding the idea that clean-coal technologies are up to the task of creating zero-emissions coal-fired power plants (U.S. Department of Energy. *Clean Coal Today Newsletter* (March 2005); see also J. J. Heinrich et al. *Environmental Assessment of Geologic Storage of CO_2.* Paper presented at the Second National Conference on Carbon Sequestration, Washington, DC, 2003; H. J. Herzog and D. Golomb. Carbon Capture and Storage from Fossil Fuel Use. In C. J. Cleveland, ed. *Encyclopedia of Energy* [New York: Elsevier Science, 2004], 277–287). Coal gasification can produce liquid and gas fuels, including methanol, diesel, hydrogen, and methane. But production and use yield huge amounts of greenhouse CO_2 along with other greenhouse and smog-forming gases. When CO_2 can be captured before release, it may be sequestered by injecting it into abandoned oil fields or deep saline aquifers, for example. But sequestration is expensive, the integrity of long-term containment is unproven, and there are no regulations in place to control procedures. At each step, CO_2 capture and disposal processes are laden with potentially huge environmental effects (G. S. Bodvarsson et al. Initiative Addresses Subsurface Energy and Environment Problems. *EOS, Transactions of the American Geophysical Union* 2006;82:18, 20).

51. The calculation uses A. A. Bartlett's exponential expiration time (EET) equation (see table 12.3), British thermal units (Btu) of total fossil energy consumption for the year 2000 (83.9×10^{15} Btu), and the Btu content of recoverable coal resources, in proportions of 46% anthracite plus bituminous coal (25×10^{6} Btu/ton) and 54% lignite plus subbituminous coal (20×10^{6} Btu/ton), and assumes a 10-year average annual consumption increase of 1.5%.

52. Illum. *Oil-Based Technology and Economy*, 34–35.

53. Brice Smith. *Insurmountable Risks* (Takoma Park, Md.: IEER Press, and Berkeley, California: RDR Books, 2006), 86–89; see also R. C. Armstrong and E. J. Moniz. *Report of the Energy Research Council* (Cambridge: Massachusetts Institute of Technology, 2006), 20–21.

54. John Gever et al. *Beyond Oil: The Threat to Food and Fuel in the Coming Decades* (Boulder: University Press Colorado, 1991), 65–68. Thinner coal beds, a shift to more energy-intensive surface mining, and a 1% per year decline in average heat content of the bituminous coal mined from 1955 to the mid-1970s account for the decline in net energy from coal (C. J. Cleveland et al. Energy and the U.S. Economy: A Biophysical Perspective. *Science* 1984;225:890–897).

55. T. D. Beamish. *Silent Spill, The Organization of an Industrial Crisis* (Cambridge: MIT Press, 2002).

56. Larry Bleiberg. Oil From Sands of Alberta. *Dallas Morning News* 13 February 2000.

57. Principal U.S. oil shale deposits are located in the already overallocated upper Colorado River Basin (see chapter 9), where water supplies are not parceled out casually (Johnson et al. *Strategic Significance of America's Oil Shale Resource*). Production of 400,000 barrels of oil per day would require about 70,000 acre-feet of water per year (K. L. Lindskov and B. A. Kimball. *Water Resources and Potential Hydrologic Effects of Oil-Shale Development in the Southeastern Uinta Basin, Utah and Colorado* [U.S. Geological Survey Professional Paper 1307, 1984]).

58. For example, mining sufficient oil shale to produce 500,000 barrels of oil per day would involve daily processing and disposing of more than three times the daily wastes from Kennecott's Bingham Canyon copper mine. Even the smallest feasible oil shale production of 50,000 barrels per day would dig open pits comparable in size to the world's largest iron and copper mines. A highly unlikely scenario of open-pit mining for deeply buried oil shale deposits would dig rock from 2,500 foot depths (1,000 feet of overburden, 1,500 feet of shale) (Johnson et al. *Strategic Significance of America's Oil Shale Resource*, 5).

59. Water produced with oil, gas, and coal-bed methane can have total dissolved solid values as much as 350,000 ppm, about one-third the salinity of equatorial sea water. The salts include dissolved sodium, chlorine, calcium, bicarbonate, and other inorganic solids (U.S. Geological Survey. *Water Produced with Coal-Bed Methane* [USGS Fact Sheet FS-156–00, 2000]; C. A. Rice et al. *Water Co-produced with Coalbed Methane in the Powder River Basin, Wyoming: Preliminary Compositional Data* [U.S. Geological Survey Open-File Report 00–37, 2000], table 2). Produced waters are health risks because of their high radium contents (L. M. H. Carter, ed. *Application of Geoscience to Decision-Making* [U.S. Geological Survey Circular 1108, 1995], 22–27, 111–113).

60. The Bureau of Land Management created a National Energy Office with the sole mission of implementing the new energy policy, issuing "41 Tasks," a broad internal directive to expedite review of "impediments to oil and gas development in the Intermountain West; review lease stipulations that protect wildlife and other natural values for revision; expedite

requests for permission to drill wells; and fast-track the granting of energy-related rights-of-way, covering roads, pipelines, and other structures." State Bureau of Land Management directors now have to require their state offices to justify any oil or gas or coal permit denial in writing, specifying possible adverse impacts from lost production, missed exploration opportunities, and other economic issues (U.S. Department of the Interior Bureau of Land Management. *National Energy Policy Tasks* (24 August 2001)).

61. The Bureau of Land Management issued 1,803 drilling permits in 1999 and 6,399 in 2004, more than 95% in the five Rocky Mountain states.

62. The Bureau of Land Management has approved invasive exploration activities within Utah's Lockhart Basin, Hatch Point, Dome Plateau, and Goldbar Rim—all magnificent scenic areas as well as important wildlife habitat and proposed wildernesses. National wildlife refuges are not protected from oil and gas exploitation. Twenty-seven percent of the nation's 575 refuges have been either explored, drilled and exploited, or laced with pipelines, and 105 refuges contain a total of 4,406 oil and gas wells, 1,806 currently active. Thirty-five refuges contain only pipelines, which are damaging enough (see chapter 5) (U.S. General Accounting Office. *National Wildlife Refuges: Opportunities to Improve the Management and Oversight of Oil and Gas Activities on Federal Lands* (GAO-03-517, 2003); U.S. General Accounting Office. *National Wildlife Refuges: Improvement Needed in the Management and Oversight of Oil and Gas Activities on Federal Lands* [GAO-04-192T, 2003]).

63. U.S. Geological Survey Central Region Assessment Team. *Qualitative Appraisal of Oil, Gas, Coal, and Coal-Bed Methane in 21 U.S. National Monuments Established or Expanded Between 1996–2001* (March 2001). A 2003 National Petroleum Council report to the Secretary of Energy estimated that 40% of natural gas resources in the Rocky Mountain region is inaccessible to exploration. But while the new national monuments and expansion of existing ones made significant oil and gas reserves inaccessible, the Clinton administration also opened about two million other acres *per year* to oil and gas drilling—a higher rate than either the Reagan or first Bush administration. The U.S. Geological Survey assigned only three newly protected areas a high likelihood of having *moderate* amounts of oil, only one a moderate likelihood of a *modest* amount of oil, and rated the remaining 15 at low to zero chance of having *any* oil. For natural gas, the U.S. Geological Survey assessed only one site with high and another with moderate likelihood for large amounts, and gave a third moderate likelihood of small gas amounts, and the remaining 16 sites low to no chance of finding gas. The prospects for coal and coal-bed methane rated even lower: A single site has high probability for coal deposits or coal-bed methane, while the remaining 18 have low to no probability for either.

64. G. A. Petzet. Monument Jeopardizes Area of Utah's Resources. *Oil and Gas Journal* 24 March 1997;69–72.

65. Twenty-seven dry holes have been drilled at 12 of the favorable locations, and 24 other holes yielded only "shows," meaning "oil stains or bitumen" in drill cuttings.

66. The National Petroleum Council's "Rocky Mountains" resource area comprises five basins assessed by U.S. Geological Survey for undiscovered oil and gas—the Paradox, Powder River, San Juan, Southwestern Wyoming, and Uinta-Piceance provinces (U.S. Geological Survey. *Assessment of Undiscovered Oil and Gas Resources in Selected Rocky Mountain Provinces for the Energy Policy and Conservation Act of 2000 (EPCA)* [Fact Sheet FS-149-02, 2002]).

67. U.S. Department of the Interior Bureau of Land Management. *Final Environmental Impact Statement and Proposed Plan Amendment for the Powder River Basin Oil and Gas*

Project (2002), table S-2. Approvals for the 39,000–65,000 wells were granted in January 2003. The environmental impact statement's preferred alternative envisions 39,367 new coal-bed methane wells, 25,997 drill pads, 17,276 miles of roads, 14,127 miles of two- to three-inch pipeline, 6,347 miles of 12-inch pipeline, 5,311 miles of overhead electric lines, 1,281 produced water infiltration facilities, 37 contaminant impoundments, 285 produced water injection wells, and 476,216 acre-feet (160 billion gallons) of produced water coming to the surface.

68. For example, from 1982 through 2004, as U.S. dependence on foreign oil increased from 28% to nearly 58%, the government opened 229 million acres of federal lands in 12 western states for oil and gas drilling. During the 15 years from 1989 to 2003, oil production from western public lands amounted to 53 days supply at current consumption levels (Environmental Working Group. *Big Access, Little Energy: The Oil and Gas Industry's Hold on Western Lands* [2004]. Available: www.ewg.org/oil_and_gas; Environmental Working Group. *15 Years of Wide-Open Access to Western Public Lands Produced Just 58 Days of Oil* [2005]. Available: www.ewg.org/).

69. U.S. Government Accountability Office. *Securing U.S. Nuclear Materials: Poor Planning Has Complicated DOE's Plutonium Consolidation Efforts* (GAO-06–164T, 2005).

70. Even if global demand for electricity were to increase at the same rate as a very optimistic nuclear production, bringing one nuclear power plant online every 15 days to 2050, carbon emissions from the electricity sector would continue to rise at the same rate as today (Smith. *Insurmountable Risks*, 26–27).

71. Natural uranium is 99.3% uranium-238 with only 0.7% uranium-235, the easily split bomb-fuel isotope. Highly enriched bomb-grade uranium is about 20% uranium-235. Uranium-238 is relatively more stable than uranium-235 and much less radioactive. To sustain heat-generating nuclear fission reactions, natural uranium must be enriched with uranium-235. The optimum low-enriched uranium mixture cannot be used for atomic bomb fuel (Institute for Energy and Environmental Research. *Fact Sheet, Fissile Material Basics* (20 March 1996). Available: www.ieer.org/fctsheet/fm_basic.html; Institute for Energy and Environmental Research. *Fact Sheet, Uranium: Its Uses and Hazards* (July 2005). Available: www.ieer.org/fctsheet/uranium.html).

72. A modern light-water power reactor contains "several hundred" fuel channels, each holding 12 fuel bundles weighing 50 pounds. (J. A. L. Robertson. *Decide the Nuclear Issues for Yourself*, chapter 3: Do-It-Yourself Reactor Design (2002). Available: www.magma. ca/~jalrober/Publications.htm).

73. Major accidents have included the 1957 Windscale plant uranium fire in England, 1979 Three Mile Island loss of coolant meltdown in Pennsylvania, and the 1986 Chernobyl, Ukraine, meltdown and fire. Although far from a maximum credible accident, the Windscale nuclear power plant blanketed English countryside, farms, and towns with hot radiation, contaminating locally produced milk with radioactive iodine-131, especially threatening to infants and children (see chapter 7, box 7.1). At the accident's peak, a radioactive cloud passed over London, 300 miles away. Like Chernobyl 29 years later, tricky graphite heating characteristics sparked the fire.

74. Health statistics, averaging the population all around the plant, seemed to show no increase in cancer rates for the Three Mile Island region after the meltdown. But when the data are selected for populations living downwind from the plant at the time of the Three Mile Island accident, the epidemiological picture changes radically toward increased cancer

rates (Steven Wing et al. A Reevaluation of Cancer Incidence near the Three Mile Island Nuclear Plant, the Collision of Evidence and Assumptions. *Environmental Health Perspectives* 1997;105:52–57).

75. The Nuclear Regulatory Commission (NRC) and First Energy, the Davis-Besse plant operator, shared responsibility for the pressurized water reactor vessel's dangerous condition. The NRC had two full-time inspectors at the plant but considered First Energy a "good performer" and so required few inspections or questions about conditions. This lax oversight led to improper shutdown procedures even after problems were reported. First Energy noticed corrosion in April 2000 photographs, which showed rust streaming from the vessel head and about 900 pounds of crystallized boric acid caked on the reactor's corroded lid. First Energy admitted it did not turn the photos over to the NRC until later, because the NRC had not asked for them. NRC staff recommended immediate shutdown when alerted, but NRC commissioners overruled them (U.S. General Accounting Office. *Nuclear Regulation: NRC Needs to More Aggressively and Comprehensively Resolve Issues Related to the Davis-Besse Nuclear Power Plant's Shutdown* [GAO-04–415, 2004]; *NOW with Bill Moyers* [Public Broadcasting System, 24 January 2003]). In 2005, the NRC proposed a $5.45 million fine against First Energy.

76. The apparent sabotage at the small (200-kilowatt) SL-1 power reactor near Idaho Falls, Idaho, occurred during routine servicing on the night of 3 January 1961. Three technicians had shut it down to check sticking control rods. To extract control rods with machinery, the technicians had to hand-lift some a short distance out of the reactor core. Although all of the technicians knew that the central rod would start nuclear reactions if raised more than four inches, apparently one of them did just that. After investigation, an internal Atomic Energy Commission memo stated that "the…accident is now known to have been initiated on purpose by one of the operators," suspicious of an affair between his wife and a shift partner (Daniel Ford. *The Cult of the Atom: The Secret Papers of the Atomic Energy Commission* [New York: Simon and Schuster Touchstone Books, 1984] 203–204).

77. In response to critics of its mock-attack "Operational Safeguards Response Evaluation" test program, which tested nuclear power plant defense arrangements against terrorist invasions with a six-month advance warning, the Nuclear Regulatory Commission discontinued the program (Daniel Hirsch. The NRC: What, Me Worry? *Bulletin of the Atomic Scientists* 2002;58:38–44). The Nuclear Regulatory Commission then developed and classified new team protocols to avoid publicity about the makeup of mock terrorist teams. Prominent media reports revealed that the new teams comprised a number of members "less than twice" the old number, however, now known to be five external members with two insiders.

78. In the western United States, uranium was just a waste product of radium mining between 1912 and 1924, and then from vanadium mining up to 1945. Uranium mining for military purposes began only after 1947 (W. I. Finch. *Uranium, Its Impact on the National and Global Energy Mix* [U.S. Geological Survey Circular 1141, 1997]).

79. World Nuclear Association. *Supply of Uranium* (World Nuclear Association, September 2005). Estimates of proven world reserves total about 3.9 million short tons, with current consumption at about 75,000 short tons per year. Extravagant claims, such as "Since uranium is ubiquitous and plentiful in the earth's crust, its availability is determined almost entirely by the willingness to find it," assumes enough energy can be supplied for mining average crustal rocks and the oceans for uranium (Jeremy Whitlock. *Canadian Nuclear FAQ. How Much Longer Will the World's Uranium Reserves Last?* Section G: Uranium. Available: www. nuclearfaq.ca/cnf_sectionG.htm#ujranium_supply [accessed April 2006]).

80. Deffeyes. *Beyond Oil*, chapter 8. This silly idea is based on the average 2.8 ppm uranium abundance in Earth's crust and 0.003 ppm average ocean concentration.

81. L. J. Carter and T. H. Pigford. 2000. Confronting the Paradox in Plutonium Policies. *Issues in Science and Technology* Winter 1999–2000;29–36. In the 1970s, Presidents Gerald Ford and then Jimmy Carter withdrew government support for commercial reprocessing and plutonium recycling because of the proliferation risks. After 2000, the George W. Bush administration effectively abandoned the established U.S. nonproliferation policy without input from the state, energy, or defense departments. The Department of Energy began promoting new research programs for advanced commercial nuclear power reactors and fuel cycle technologies, through the Generation IV (Gen IV) and Advanced Fuel Cycle (AFCI) Initiatives in South Africa, Argentina, Brazil, South Korea, Switzerland, Japan, Canada, the United Kingdom, and France. Five of those countries formerly ran clandestine nuclear weapons programs. Gen IV is expected to develop a reactor type competitive with fossil fuel power plants, and AFCI (formerly called the Spent Fuel Pyroprocessing and Transmutation Program) calls for building a massive commercial reprocessing plant by 2015, with a longer term goal to develop new reprocessing technologies and fast reactors for transmuting nuclear waste. Clearly, these programs would revive the breeder reactor program (Natural Resources Defense Council. *DOE's Nuclear Energy Research Programs Threaten National Security* [2003]. Available: nrdc.org/nuclear/bush/freprocessing.asp; Natural Resources Defense Council. *Memo of Dissent Regarding Generation IV Technology Planning* [Natural Resources Defense Council, October 2002]. Available: nrdc.org/nuclear/bush/bushinx.asp).

82. Breeder reactors use the neutrons not needed to sustain uranium-235 fission to bombard uranium-238, producing fissionable plutonium-239 for additional fuel. The process yields plutonium very slowly, however. Reaching energy break-even, when more fuel has been produced than consumed, could take several decades (Montgomery. *Environmental Geology*, 329).

83. U.S. Government Accountability Office. *Nuclear Weapons: Views on Proposals to Transform the Nuclear Weapons Complex* (GAO-06–606T, 2006).

84. Daniel Metlay, staff member for the Nuclear Waste Technical Review Board, cited in R. C. Ewing and Allison Macfarlane. Yucca Mountain.

85. Including (but not limited to) parts of Wyoming, Montana, Arizona, and southeastern California.

86. Ewing and Macfarlane. Yucca Mountain.

87. The great heat and pressure at the sun's core fuse hydrogen nuclei to make other chemical elements. Hydrogen bombs initiate the fusion process with a fission bomb explosion. Physicists have not found a way to control even tiny fusion reactions, let alone develop practical applications. For a clear discussion of fusion processes, see David Goodstein. *Out of Gas, the End of the Age of Oil* (New York: W.W. Norton, 2004), 107–110.

88. Michael Collier et al. *Dams and Rivers* (U.S. Geological Survey Circular 1126, 1996), 2; this is an excellent, very readable account of the downstream effects of dams.

89. R. K. Mark and D. E. Stuart-Alexander. Disasters as a Necessary Part of Benefit-Cost Analysis. *Science* 1977;197:1160–1162. Federal agencies generally ignore this significant element in dam cost–benefit assessments. A hand-written notation on an internal Bureau of Reclamation memorandum called the Mark and Stuart-Alexander report "a dangerous idea." The 1976 collapse of the Teton Dam in Idaho illustrates the peril. The resulting flood took

11 lives, but the toll could have been much worse if the dam had failed at night instead of during the day (Marc Reisner. *Cadillac Desert* [London: Penguin Books, 1986], chapter 11).

90. At least 2,000 irrigation dams in the United States have filled with sediment already. See chapter 9 for estimates of filling rates of larger dams.

91. Fraser Goff and Cathy Janik. Geothermal Systems. In *Encyclopedia of Volcanoes* (San Diego, California: Academic Press, 2000), 817–834; W. A. Duffield and J. H. Sass. *Geothermal Energy, Clean Power from the Earth's Heat* (U.S. Geological Survey Circular 1249, 2003).

92. Goff and Janik. Geothermal Systems.

93. Growing the equivalent of global energy use would use more than 10% of Earth's total land mass (M. I. Hoffert et al. Advanced Technology Paths to Global Climate Stability, Energy for a Greenhouse Planet. *Science* 2002;298:981–987). Total annual global energy consumption is about 13 trillion watts, of which 85% (11 trillion watts) comes from fossil fuels (R. F. Service. Is It Time to Shoot for the Sun? *Science* 2005;309:548–551). With projected global energy demands calling for another 30 trillion watts by 2050, the amount of global agricultural land available to grow nothing but energy is hopelessly inadequate.

94. W. E. Rees. Footprint, Our Impact on Earth Is Getting Heavier. *Nature* 2002;420:267–268; see also Dukes. Burning Buried Sunshine.

95. To electrify the city of 100,000 using wood chips, the 500,000 woodland acres would have to produce about 1.2 tons of wood per acre every year indefinitely. Average electrical demand for a city this size is about a billion kilowatt-hours (David Pimentel et al. Renewable Energy, Current and Potential Issues. *BioScience* 2002;52:1111–1120).

96. John Perlin. *A Forest Journey, the Role of Wood in the Development of Civilization* (New York: W.W. Norton, 1989), 15–16.

97. Pimentel et al. Renewable Energy, 1113.

98. "The International Institute for Sustainable Development's Global Subsidies Initiative (GSI) is a project designed to put the spotlight on subsidies and the corrosive effects they can have on environmental quality, economic development and governance" (Global Subsidies Initiative. Available: www.globalsubsidies.org).

99. Six gallons of ethanol fuel has the energy content of about four gallons of gasoline, but making the ethanol requires three gallons of gasoline or the energy equivalent. So making six gallons of ethanol offsets one gallon of gasoline (T. W. Patzek. The Real Corn-Ethanol Transportation System [20 June, 2006]. Available: petroleum.berkeley.edu/patzek/BiofuelOA/Materials/TrueCostofEtOH.pdf). The same result is obtained from the energy output:input ratio of 1.34 for ethanol production claimed by Hosein Shapouri et al. *The Energy Balance of Corn Ethanol: An Update* (Agricultural Economic Report No. 814, U.S. Department of Agriculture, 2002). A slightly rosier scenario uses the maximum theoretical ethanol yield of 2.64 gallons per bushel of corn and gives energy credit for feed byproducts requiring just 3.9 gallons of ethanol to offset one gallon of gasoline. But the area required to offset current gasoline consumption still is about 75% of the contiguous United States (T. W. Patzek. The Real Biofuel Cycles [online supporting material for letter]. *Science* 2006;312:1747. Available: www.petroleum.berkeley.edu/patzek/BiofuelQA/Materials/RealFuelCycles-Web.pdf [accessed July 2006]).

100. Corn yields about 2.5 gallons of ethanol per bushel. The average national corn harvest (2000–2005) yielded 10.2 billion bushels from 71.7 million acres, averaging 142.5 bushels

per acre. The yield is reported as bushels per harvested acre; the yield per planted acre (average 79.2 million acres) as opposed to 71.1 million harvested is only 129 bushels, due to crop failure or other cause.

101. The complex question of how much energy it takes to convert corn to ethanol, compared to the energy content *of* the ethanol, has a wide range of answers, from a net energy loss of 20–29% to a net gain of 67%. The differences are due to different energy input values, changes in ethanol production processes, and accrediting the energy value of ethanol production byproducts, mainly animal feed (see n. 99). Recent useful sources of information on the production and use of ethanol include David Pimentel (Limits of Biomass Utilization. In *Encyclopedia of Physical Science and Technology*, 3rd ed. [San Diego, California: Academic Press, 2000], 159–171), Shapouri et al. (*The Energy Balance of Corn Ethanol: An Update*), T. W. Patzek. (Thermodynamics of the Corn-Ethanol Biofuel Cycle. *Critical Reviews in Plant Sciences* 2004;23:519–567), David Pimentel and T. W. Patzek (Ethanol Production Using Corn, Switchgrass, and Wood; Biodiesel Production Using Soybean and Sunflower. *Natural Resources Research* 2005;14:65–76), A. E. Farrell et al. (Ethanol Can Contribute to Energy and Environmental Goals. *Science* 2006;311:506–508), and Patzek (The Real Biofuel Cycles).

102. Nathanael Greene. *Growing Energy: How Biofuels Can Help End America's Oil Dependence* (Natural Resources Defense Council, 2004); R. D. Perlack et al. *Biomass as Feedstock for a Bioenergy and Bioproducts Industry: The Technical Feasibility of a Billion-Ton Annual Supply* (U.S. Department of Energy and U.S. Department of Agriculture, DOE/GO-102005–2135, 2005).

103. Patzek. The Real Biofuel Cycles; T. W. Patzek and David Pimentel. Thermodynamics of Energy Production from Biomass. *Critical Reviews in Plant Sciences* 2006;24:327–364. See also the *Science* Letters section (2006;312:1743–1748) for critiques of a number of biofuels issues in 2006 *Science* articles. A large Department of Energy effort is under way to address the cellulose breakdown problem (U.S. Department of Energy. *Breaking the Biological Barriers to Cellulosic Ethanol: A Joint Research Agenda, a Research Roadmap Resulting from the Biomass to Biofuels, Proceedings of Workshop held 7–9 December 2005, Rockville, Maryland* [2006]).

104. A bushel of soybeans produces about 1.4 gallons of biodiesel oil. Soybean crops yields are between 87 and 44 bushels per acre, with lower yields in arid western states (U.S. Department of Agriculture. *Ethanol Conversion Factors* [2003]. Available: www.fsa.usda.gov/daco/bioenergy/2002/2002FactorsNFormulas.pdf; Teresa Halvorsen. *Grower's Demands Increase Soy Biodiesel Usage* [Iowa Farm Bureau Federation, 8 February 2003]. Available: www.agan-denvironment.com/news/news_20030208e.htm).

105. D. A. Pfeiffer. *The Dirty Truth About Biofuels*. Available: www.oilcrash.com/articles/pf_biolhtm (accessed June 2006). The University of Massachusetts Political Economy Research Institute ranks Archer Daniels Midland Corporation, chief cheerleader and beneficiary of corn ethanol, 10th among the 100 top corporate air polluters in the United States. The rank is indexed to the number of pounds of toxics released, multiplied by toxicity and extent of public exposure.

106. Patzek. The Real Biofuel Cycles, 1.

107. M. A. Delucchi. *Conceptual and Methodological Issues in Lifecycle Analyses of Transportation Fuels* (Institute of Transportation Studies, University of California Davis, UCD-ITS-RR-04–45, 2004).

108. Pimentel and Patzek. Ethanol Production Using Corn, Switchgrass, and Wood; Patzek and Pimentel. Thermodynamics of Energy Production from Biomass.

109. These genetic modifications reduce the energy a plant uses for making reproductive structures (A. J. Ragauskas et al. The Path Forward for Biofuels and Biomaterials. *Science* 2006;311:484–489).

110. Chris Henry and Rick Koelsch. *What Is an Anaerobic Digester?* (University of Nebraska, Lincoln). Available: manure.unl.edu/adobe/v7n10_01.pdf.

111. The motionless stalks of one such development near Pacheco Pass, California, are still visible from California Highway 152. Others, including small wind farms near Highway 58 north of Mojave, California, were abandoned after short periods of use.

112. One kilowatt-hour will light a 100-watt light bulb all night or run a typical hair dryer for one hour.

113. Western U.S. wind farms employ a wide variety of wind generator types and sizes. In 1995, the 6,500 Altamont Pass, California, turbines included 26 different types. Rotor diameters of the turbines varied from 33 to 149 feet, but most had 35- to 60-foot diameters. Towers varied from 80 to 140 feet high (Predatory Bird Research Group. *A Pilot Golden Eagle Population Study in the Altamont Pass Wind Resource Area, California* [National Renewable Energy Laboratory, 1995], 136–138). Some older Altamont Pass generators were replaced by turbines with rotor diameters of about 200 feet and heights of nearly 300 feet. Latest generation turbines have rotor diameters between about 164 and 262 feet. The amount of captured wind energy increases with the area swept by the rotors, so increasing rotor diameter by a factor of 2 increases the amount of captured wind energy by a factor of 4.

114. Marc Fioravanti. *Wind Power Versus Plutonium, an Examination of Wind Energy Potential and a Comparison of Offshore Wind Energy to Plutonium Use in Japan* (Takoma Park, Maryland: Institute for Energy and Environmental Research, 1999), 13, 18. Array spacings are calculated from tables given by Fioravanti, assuming flat land and uniform wind conditions. Applying the tables to an array of 100 turbines, in 10 rows of 10 turbines each, with an average rotor diameter of 50 feet, the total space requirement at the lowest efficiency (five rotor diameter spacing) calculates to 143.5 acres; spacing for highest efficiency (nine rotor diameters) requires 465 acres. The tables suggest actual spacings for 100 turbines of little more than 9,000 acres for five-rotor-spacing arrays, and about 30,000 acres for nine-rotor-spacing arrays. Actual spacings may be very different from these theoretical values, however, due to the large range of rotor diameters actually used on turbines. At Altamont Pass and other windy areas, hilly terrain both limits turbine placement and violates the tables' flat land and uniform wind assumptions.

115. Pacific Northwest Laboratory. *An Assessment of the Available Windy Land Area and Wind Energy Potential in the Contiguous United States* (Pacific Northwest Laboratory, 1991); Science for Democratic Action. *Large-Scale Wind Energy Development in the United States* 9(4) (August, 2001), 9.

116. Experimental Mojave Desert solar electricity-generating plants Solar One and Solar Two at Daggett, California, consisted of hundreds of reflectors (heliostats) that tracked the sun, focusing the rays on a liquid-filled tank at the central tower. Solar One used water. At Solar Two liquid nitrate salt in the tank rose to about 1,650°F, and heated water for a steam turbine. Both designs instantly vaporized any bird flying through the focal area of reflected sun rays.

117. Howard Wilshire and Douglas Prose. Wind Energy Development in California, USA. *Environmental Management* 1987;11:13–20; H. G. Wilshire et al. *Geologic Processes at the Land Surface* (U.S. Geological Survey Bulletin 2149, 1996).

118. The tally of killed raptors totaled 881–1,300: 75–116 golden eagles, 209–300 red-tailed hawks, 73–333 American kestrels, and 99–380 burrowing owls (K. S. Smallwood and C. G. Thelander. *Developing Methods to Reduce Bird Mortality in the Altamont Pass Wind Resource Area* [BioResource Consultants, Final Report to the California Energy Commission, Contract No. 500–01-019, 2005], 3).

119. Interlaboratory Working Group on Energy-Efficient and Clean-Energy Technologies. *Scenarios for a Clean Energy Future* (prepared for U.S. Department of Energy, Office of Energy Efficiency and Renewable Energy, 2000). This report was reviewed by outside experts from industry, government, and universities.

120. J. A. Johnson. *Benchmark 2005, Energy Benchmark for High Performance Buildings*, version 1.1 (New Buildings Institute, 2005); U.S. Green Building Council. *LEED, Leadership in Energy and Environmental Design*, reference package version 2.0 (U.S. Green Building Council, 2001). Improving building design also will have significant benefits in reducing greenhouse gas emissions: In 2000 buildings accounted for 48% of CO_2 emissions, larger than both transportation and industrial CO_2 emissions (Gil Friend. The 2030 Climate Challenge and West Coast Green [2 October, 2006]. Available: www.worldchanging.com/archives/005005.html).

121. In 2001, the United States consumed a total of 3.3 terawatts of power (Service. Is It Time to Shoot for the Sun?).

122. Youngquist, GeoDestinies, 252.

123. M. T. Klare. *Blood and Oil: The Dangers and Consequences of America's Growing Dependency on Imported Petroleum* (New York: Metropolitan Books, 2004).

124. R. L. Hirsch et al. *Peaking of World Oil Production: Impacts, Mitigation, and Risk Management* (U.S. Department of Energy, 2005), 4. A combination of factors lay behind rapidly escalating 2006 oil prices: rapidly increasing world oil demand, fierce competition for energy resources between importing nations, political instabilities in exporting nations potentially threatening access to their oil, and rising energy demand in the exporting nations (J. J. Brown. What the Mainstream Media Are Not Telling You About the Run Up in Oil Prices. *Energy Bulletin* 20 April 2006). Available: www.energybulletin.net.

Chapter 13

1. C. O. Sauer. The Agency of Man on the Earth. In W. L. Thomas, Jr., ed. *Man's Role in Changing the Face of the Earth* (Chicago: University of Chicago Press, 1956), 49–69; John Perlin. A *Forest Journey: The Role of Wood in the Development of Civilization* (New York: W.W. Norton, 1989), 35–101; Tim Flannery. *The Eternal Frontier* (New York: Atlantic Monthly Press, 2001), chapter 13.

2. The Enrich-Holdren equation, $I = PAT$, represents the proposition that the intensity I of human Impact on the Earth is proportional to population (P), affluence (A), and technology (T) (J. P. Holdren and P. R. Erlich. Human Population and the Global Environment. *American Scientist* 1974;62:282–292; see P. K. Haff. Neogeomorphology. *EOS, Transactions of the American Geophysical Union* 2002;83:310, 317).

3. The world's rivers erode and carry away 14–23 billion tons of soil and rock-derived sediment every year, glaciers move about four billion tons, and wind erosion shifts about one billion. Humans move about 35 billion tons per year (Haff. Neogeomorphology; Roger L. Hooke. On the Efficiency of Humans as Geomorphic Agents. *Geological Society of America, GSA Today* 1994;4:217, 224–225, 310).

4. The Earth's tectonic plate boundaries generally mark the center lines of ocean basins and coincide with major mountain belts, but some lie along continental margins. The plates grow continuously at lines of mid-oceanic volcanoes. At other boundaries the plates either collide with or push past each other. At some collision zones, tectonic forces shove ocean floor rocks and sediments into deeper Earth regions. When forced deep enough, earth materials become transformed by metamorphic processes and even melt, producing volcanoes. Earthquakes in continental interiors mark either collisions between continent fragments at the edges of tectonic plates or adjustment to collisions at or beyond a continent's margin (U.S. Geological Survey. *Major Tectonic Plates of the World* [27 June 2001]. Available: geology.er.usgs.gov/eastern/plates.html).

5. Geologists define three universal rock types. *Igneous rocks* ("born in fire") start as melts deep inside Earth. *Sedimentary rocks* start as eroded rock debris (sediment) or chemical deposits, which become rocks through burial, compression, and natural cementation. Both rock types can transform into *metamorphic rocks* when exposed to marked changes in heat and pressure, aided by fluid interactions. Metamorphic rocks can be converted to other metamorphic types under a range of different pressure–temperature conditions. All types may be eroded at the Earth's surface or forced deep underground and melted or partially melted.

6. Where ice sheets exist, in high mountains or polar regions, they also transport sediment rubble. Global warming is rapidly eliminating most valley glaciers and polar ice caps, however (S. A. Schumm. *The Fluvial System* [New York: John Wiley and Sons, 1977]; A. G. Brown and T. A. Quine, eds. *Fluvial Processes and Environmental Change* (New York: John Wiley and Sons, 1999); A. S. Goudie et al. 1999. *Aeolian Environments, Sediments, and Landforms* (New York: John Wiley and Sons, 1999)).

7. See n. 6.

8. S. H. Cannon. Regional Rainfall-Threshold Conditions for Abundant Debris-Flow Activity. In S. D. Ellen and G. F. Wieczorek, eds. *Landslides, Floods, and Marine Effects of the Storm of January 3–5, 1982, in the San Francisco Bay Region, California* (U.S. Geological Survey Professional Paper 1434, 1988), 35–42; U.S. Geological Survey. *Debris-Flow Hazards in the United States* (U.S. Geological Survey Fact Sheet 176–97, 1997).

9. U.S. Geological Survey. *Videotaped Interviews of Debris-Flow Victims, Santa Cruz Mountains* (February–March 1998).

10. A gully has a cross-sectional area greater than one square foot (P. J. Whiting et al. Depth and Areal Extent of Sheet and Rill Erosion Based on Radionuclides in Soils and Suspended Sediment. *Geology* 2001;29:1131–1134).

11. Crumbling at the upper end of gullies is mainly due to high pressures in water-filled soil pores at the soil's surface.

12. W. H. Wischmeier and D. D. Smith. *Predicting Rainfall Erosion Losses, a Guide to Conservation Planning* (U.S. Department of Agriculture Handbook 537, 1978). Note that this handbook has been replaced (K. G. Renard et al. *Predicting Soil Erosion by Water, a Guide to Conservation Planning with the Revised Soil Loss Equation (RUSLE)* [U.S. Department

of Agriculture Handbook 703, 1997]). The Revised Universal Soil Loss Equation (RUSLE) uses the same approach used in the USLE (for details, see D. S. Jones et al. *Calculating Revised Soil Loss Equation (RUSLE) Estimates on Department of Defense Lands—a Review of RUSLE Factors and U.S. Army Land-Condition Trend Analysis (LCTA) Data Gaps* [Fort Collins, Colorado: Center for Ecological Management of Military Lands, 1995]).

13. W. S. Chepil and N. P. Woodruff. The Physics of Wind Erosion and Its Control. *Advances in Agronomy* 1963; 15; see also D. A. Gillette. On the Production of Soil Wind Erosion Aerosols Having the Potential for Long Range Transport. *Journal de Recherches Atmosphériques* 1974;8:735–744.

14. Goudie et al. *Aeolian Environments.*

15. T. L. Péwé, ed. *Desert Dust, Origin, Characteristics, and Effect on Man* (Geological Society of America Special Paper 186, 1981); see also M. C. Reheis et al. Geochemical Evidence for Diversity of Dust Sources in the Southwestern United States. *Geochimica et Cosmochimica Acta* 2002;66:1569–1587; M. C. Reheis. A 16-Year Record of Eolian Dust in Southern Nevada and California, USA: Controls on Dust Generation and Accumulation. *Journal of Arid Environments* 2006;67:487–520.

16. W. E. Larson et al. The Threat of Soil Erosion to Long-Term Crop Production. *Science* 1983;219:458–465. See the glossary for description of a soil profile. Excellent textbooks that deal with soil genesis, properties, and stability are P. W. Birkeland (*Soil and Geomorphology*, 3rd ed. [New York: Oxford University Press, 1999]) and William Dubbin [*Soils* (Ames, Iowa: Iowa State University Press, 2001)]. *Science* magazine [vol. 304, 2004] published an important series of articles on soil [Andrew Sugden et al., Ecology in the Underworld, 1613–1615; Jocelyn Kaiser, Wounding Earth's Fragile Skin, 1616–1618; Erik Stokstad, Defrosting the Carbon Freezer of the North, 1618–1620; Elizabeth Pennisi, The Secret Life of Fungi, 1620–1622; R. Lal, Soil Carbon Sequestration Impacts on Global Climate Change and Food Security, 1623–1627; J. R. McNeill and Verena Winiwarter, Breaking the Sod: Humankind, History, and Soil, 1627–1629; D. A. Wardle et al., Ecological Linkages Between Aboveground and Belowground Biota, 1629–1633; and I. M. Young and J. W. Crawford, Interactions and Self-Organization in the Soil-Microbe Complex, 1634–1637]). See also Suzanne Panderson et al. (Proposed Initiative Would Study Earth's Weathering Engine. *EOS, Transactions of the American Geophysical Union* 2004;85:265, 269).

17. A 10-square-foot section of good soil can support populations of 200,000 arthropods and enchytraeids (small worms lacking enzymes for digesting organic matter) and billions of microbes (Martin Wood. *Soil Biology* [New York: Chapman and Hall, 1989], 20, 10 [table 1.7]; E. Lee and R. C. Foster. *Australian Journal of Soil Resources* 1991;29:745). High-quality soil contains an average of more than a ton each of earthworms and arthropods, more than 300 pounds each of protozoa and algae, nearly two tons of bacteria, and three tons of fungi in each 2.5 acres. Residing in the topmost layer of the soil, these critical elements of soil health become victims of erosion (David Pimentel et al. Environmental and Economic Costs of Soil Erosion and Conservation Benefits. *Science* 1995;267:1117–1123).

18. Jayne Belnap. Surface Disturbances, Their Role in Accelerating Desertification. *Environmental Monitoring and Assessment* 1985;37:39–57.

19. Preston Sullivan. *Sustainable Soil Management* (ATTRA [Appropriate Technology Transfer for Rural Areas], September 2001), 6–8. Available: www.attra.org/attra-pub/PDF/soilmgmt.pdf.

20. H. G. Wilshire. The Impact of Vehicles on Desert Soil Stabilizers. In R. H. Webb and H. G. Wilshire, eds. *Environmental Effects of Off-Road Vehicles, Impacts and Management in Arid Regions* (New York: Springer-Verlag, 1983), 31–50; Jayne Belnap and D. A. Gillette. Vulnerability of Desert Biological Soil Crusts to Wind Erosion, the Influences of Crust Development, Soil Texture, and Disturbance. *Journal of Arid Environments* 1998;39:133–142; M. J. Singer and Isaac Shainberg. Mineral Soil Surface Crusts and Wind and Water Erosion. *Earth Surface Processes and Landforms* 2004;29:1065–1075.

21. Daniel Hillel. *Out of the Earth, Civilization and the Life of the Soil* (Berkeley: University of California Press, 1991), 160. A commonly cited estimate of the mean rate of soil formation is one foot in 10,000 years, but it varies considerably, depending on climate, soil parent material, shape of the land, and the sequences of climatic conditions affecting soil at a particular site.

22. D. D. Richter, Jr., and Daniel Markewitz. *Understanding Soil Change, Soil Sustainability over Millennia, Centuries, and Decades* (Cambridge: Cambridge University Press, 2001); see also Hillel. *Out of the Earth*, 160–161.

23. H. G. Wilshire et al. *Geologic Processes at the Land Surface* (U.S. Geological Survey Bulletin 2149, 1996). Winds blew fine dust off the Pleistocene glaciers and deposited it in a narrow belt across the lower Midwest, beyond the glacier's farthest southern stand. This deposit, called "loess," became one of nation's richest soils.

24. Jayne Belnap and Steve Warren. Patton's Tracks in the Mojave Desert, USA, an Ecological Legacy. *Arid Land Research and Management* 2002;16:245–259; this study showed more rapid recovery of biological crusts (about 100 years) beneath perennial plant cover, but much slower recovery (more than 1,000 years and perhaps as much as 2,000 years) in areas between shrubs. The rates of recovery vary with lichen species, but a biological crust community is not considered to have recovered until the most sensitive species has recovered under the harshest conditions in the habitat.

25. L. B. Leopold et al. *Channel and Hillslope Processes in a Semiarid Area, New Mexico* (U.S. Geological Survey Professional Paper 352-G, 1967).

26. M. G. Anderson et al., eds. *Floodplain Processes* (New York: John Wiley and Sons, 1996).

27. James Buchanan Eads's work influenced the development of Mississippi flood control works. Quotation from J. M. Barry. *Rising Tide, the Great Mississippi Flood of 1927 and How It Changed America* (New York: Simon and Schuster, 1997), 26. This remarkable book outlines the evolution of flood-control policy along American rivers, and the origin of federal disaster relief. It also shows the need for caution in building houses on floodplain lands.

28. Anderson et al. *Floodplain Processes.*

29. J. F. Mount. *California Rivers and Streams, the Conflict Between Fluvial Processes and Land Use* (Berkeley: University of California Press, 1995); David Knighton. *Fluvial Forms and Processes, a New Perspective* (New York: John Wiley and Sons, 1998).

30. The groundwater zone also is called the zone of saturation or "phreatic" zone (from the Greek word for well or spring) because the water level in a well represents the top of the water table, and a spring represents places where the phreatic zone emerges at the surface.

31. The water table's elevation above sea level at many different points (measured by drilling wells) defines the "potentiometric surface." Groundwater flows from the highest to lowest areas on the potentiometric surface. Darcy's Law relates the groundwater flow rate between

two locations to the elevation difference between the water at each position and the horizontal distance between them. (Rigorously, the groundwater velocity is directly proportional to the elevation difference between the water at each location and inversely proportional to the horizontal distance between them.)

32. K. Pruess. On Water Seepage and Fast Preferential Flow in Heterogeneous, Unsaturated Rock Fractures. *Journal of Contaminant Hydrology* 1998;30:333–362.

33. Atmospheric nuclear bomb tests between 1944 and 1963 blasted large quantities of tritium, a radioactive isotope of hydrogen, into the air, which rain washed into soils all over the world. Tritium forms a radioactive water molecule useful for tracing subsurface water. The depth in the ground that the tritium has reached since 1963 yields an average rate of water movement for predicting how long contaminated water will take to go deeper or farther downslope (J. A. Izbicki et al. Movement of Water Through the Thick Unsaturated Zone Underlying Oro Grande and Sheep Creek Washes in the Western Mojave Desert, USA. *Journal of Hydrology* 2002;10:409–427; J. A. Izbicki. Geologic and Hydrologic Controls on the Movement of Water Through a Thick, Heterogeneous Unsaturated Zone Underlying an Intermittent Stream in the Western Mojave Desert, Southern California. *American Geophysical Union, Water Resources Research* 2002;38[3]:DOI 10.1029/2000WR000197). Experiments show that fissured and fractured desert soils transmit water up to 350 times faster than do unfissured soils, with infiltration rates ranging from 0.33 to 2.75 inches per year (B. R. Scanlon. Moisture and Solute Flux Along Preferred Pathways Characterized by Fissured Sediments in Desert Soils. *Journal of Contaminant Hydrology* 1991;10:19–46).

34. National Research Council. *Rock Fractures and Fluid Flow, Contemporary Understanding and Applications* (Washington, DC: National Academies Press, 1996).

35. T. C. Winter et al. *Ground Water and Surface Water: A Single Resource* (U.S. Geological Survey Circular 1139, 1999), . Quotation of hydrologist H. E. Thomas from R. C. Heath. Basic Ground-Water Hydrology (*U.S. Geological Survey* Water-Supply Paper 2220, 1983), 136. J. N. Moore and S. N. Luoma. Hazardous Wastes from Large-Scale Metal Extraction. *Environmental Science and Technology* 1990;24:1279–1285. A brief report giving an excellent overview of many of the problems associated with toxic element releases from metal mining and smelting; E. V. Axtmann and S. N. Luoma. Large-Scale Distribution of Metal Contamination in the Fine-Grained Sediments of the Clark Fork River, Montana, U.S.A. *Applied Geochemistry* 1991;6:75–88.

37. "Chaotic" unsaturated zone water pathway patterns do not take straight line or simply curved paths (Naomi Oreskes et al. Verification, Validation, and Confirmation of Numerical Models in the Earth Sciences. *Science* 1994;263:641–646).

38. J. A. Fallin. *Hydrogeology, Environmental Contamination, Assessment, Remediation, and Effects of Superfund (CERCLA), and Other Hazardous Waste Sites in Western North America* (Boulder, Colorado: Janus Press International, 1994).

39. J. A. Davis et al. Assessing Conceptual Models for Subsurface Reactive Transport of Inorganic Contaminants. *EOS, Transactions of the American Geophysical Union* 2004;85:449, 455.

40. Conceptual models of expected contaminant behavior are used to envision and remediate groundwater contamination in complex systems. As new data accumulate for testing conceptual models, as many as 20–30% of the models applied to specific problems prove to be invalid. The usefulness of any model depends on rigorous data collection and the willingness to abandon models that don't work (John Bredehoeft. The Conceptualization

Model Problem—Surprise. *Journal of Hydrogeology* 2005;13:37–46; National Research Council. *Conceptual Models of flow and Transport in the Fractured Vadose Zone* [Washington, DC: National Academies Press, 2001]).

41. A National Research Council panel is examining "natural scrubbing" processes of contaminant plumes, called attenuation, in the hopes of constructing future artificial attenuation schemes that might use biological or inorganic agents to protect threatened aquifers (see B. Bekins et al. Natural Attenuation Strategy for Groundwater Cleanup Focuses on Demonstrating Cause and Effect. *EOS, Transactions of the American Geophysical Union* 2001;82:53–58).

42. Colloids are typically about 40 billionths to 40 millionths of an inch in diameter and invisible to the unaided eye. Cloudy water is charged with colloids. See glossary.

43. G. B. Allison et al. A Review of Vadose-Zone Techniques for Estimating Groundwater Recharge in Arid and Semiarid Regions. *Journal Soil Science Society of America* 1994;58:6–14; Daniel Hillel. Unstable Flow in Layered Soils, a Review. *Hydrological Processes* 1987;1:143–147; D. B. Stephens. A Perspective on Diffuse Natural Recharge Mechanisms in Areas of Low Precipitation. *Journal Soil Science Society of America* 1994;58:40–48; S. P. Neuman. Trends, Prospects and Challenges in Quantifying Flow and Transport Through Fractured Rocks. *Journal of Hydrogeology* 2005;13:124–147; G. S. Bodvarsson et al. Development of Discrete Flow Paths in Unsaturated Fractures at Yucca Mountain. *Journal of Contaminant Hydrology* 2003;62–63:23–42.

44. For example, see R. E. Jackson and Varadarajan Dwarakanath. Chlorinated Degreasing Solvents, Physical-Chemical Properties Affecting Aquifer Contamination and Remediation. *Groundwater Monitoring and Remediation* 1999;19:102–110.

45. R. E. Jackson. The Migration, Dissolution, and Fate of Chlorinated Solvents in the Urbanized Alluvial Valleys of the Southwestern USA. *Hydrogeology Journal* 1998;6:144–155. For example, acid mine drainage from late 1880s copper mining in the Pinal Creek Basin of central Arizona created a plume of acidic water, enriched in iron, manganese, cobalt, copper, zinc, aluminum, and nickel, which extends some 15 miles from the source. Interaction between the plume and carbonate minerals in the ground moderates the water's acidity, changing it from a highly acid pHs of 2–3 to much less acid pH 5–6 values. As the acidity changes, copper sorbs onto iron oxides, and cobalt, zinc, and nickel contaminants spread less rapidly. These interactions have lowered metallic concentrations in the plume by about 60%.

46. The colloid particles sorb ions in the salt water, which neutralizes the colloid's charges, promoting coagulation. Chalcedony and agate form from silica colloids that congeal out of volcanic water, and cherts form in lakes or shallow ocean water.

47. "Sufficient amounts of many elements, such as copper, selenium, arsenic, and lead, have potentially been supplied by weathering and erosion during geological time to cause serious poisoning of the ocean had not some process of elimination of these substances been active" (Brian Mason. *Principles of Geochemistry* [New York: John Wiley and Sons, 1966], 177).

48. R. W. Buddemeier and J. R. Hunt. Transport of Colloidal Contaminants in Groundwater, Radionuclide Migration at the Nevada Test Site. *Applied Geochemistry* 1988;3:535–548.

49. Interagency Floodplain Management Review Committee. *Sharing the Challenge, Floodplain Management into the 21st Century*, Part I. Administration Floodplain Management Task Force (Washington, DC, 1994); Wilshire et al. *Geologic Processes at the Land Surface*, 2–13.

50. U.S. General Accounting Office. *Water Quality: Identification and Remediation of Polluted Waters Impeded by Data Gaps* (GAO/T-RCED-00–88, 2000).

51. G. V. Jacks and R. O. Whyte. *Vanishing Lands, a World Survey of Soil Erosion* (New York: Arno Press, 1939), 4; see also W. S. Fyfe. Soil and Global Change. *Episodes* 1989;12:249–254; V. G. Carter and T. Dale. *Topsoil and Civilization*, rev. ed. (Norman, Okla.: University of Oklahoma Press, 1974).

Conclusions

1. Admiral Hyman G. Rickover delivered the speech titled "Energy Resources and Our Future" at a banquet of the Annual Scientific Assembly of the Minnesota State Medical Association, St. Paul, Minnesota, 14 May 1957. Available: www.energybulletin.net/newswire. php?id=23151.

2. Every year, the worlds rivers erode and carry away 14–23 billion tons, glaciers move about four billion tons, and wind erosion shifts about one billion (P. K. Haff. Neogeomorphology. *EOS, Transactions of the American Geophysical Union* 2002;83:310, 317; Roger L. Hooke. On the Efficiency of Humans as Geomorphic Agents. *GSA Today* 1994;4:217, 224–225).

3. A pile of the total earth materials humans have moved over the past 5,000 years (see chapter 13, figure 13.2), both intentionally and unintentionally, would make a mountain 12,000 feet high, 25 miles wide, and more than 60 miles long, larger than Oregon's Mt. Hood. If population pressures increase earth-moving activities, "we could double the length of our mountain range in the next 100 years" (Roger L. Hooke. On the History of Humans as Geomorphic Agents. *Geology* 2000;28:843–846).

4. Roger L. Hooke (Spatial Distribution of Human Geomorphic Activity in the United States, Comparison with Rivers. *Earth Surface Processes and Landforms* 1999;24:687–692) notes that dumping all the earth moved by humans in the United States each year into the Grand Canyon would fill it in less than 400 years. The Colorado River carved the canyon over a period of four million to eight million years.

5. Other causes of habitat destruction (and proportion of species affected) are reservoirs and other water developments (30%), outdoor recreation (27%), livestock grazing (22%), infrastructure, including roads (17%), modified fire ecology (14%), logging (12%), mining and oil/gas development (11%). The percentages add to more than 100 because different activities may threaten more than one species (D. S. Wilcove et al. Quantifying Threats to Imperiled Species in the United States. *BioScience* 1998;48:607–615; see also R. F. Noss and R. L. Peters. Endangered Ecosystems: A Status Report on America's Vanishing Habitat and Wildlife. *Defenders of Wildlife* December 1995). Assessing interactions between the various causes of habitat loss yields the same results (Brian Czech et al. Economic Associations Among Causes of Species Endangerment in the United States. *BioScience* 2000;50:593–601).

6. Paul Hawken, Amory Lovins, and L. Hunter Lovins. *Natural Capitalism: Creating the Next Industrial Revolution* (Boston: Little, Brown, 1999).

7. Gretchen C. Dailey, ed. *Nature's Services: Societal Dependence on Natural Ecosystems* (Covelo, Calif.: Island Press, 1997), 3–4.

8. Daniel Quinn. *Ishmael* (paperback edition; New York: Bantam Books, 1995; first published 1992), 33–75.

9. Chip Ward. *Canaries on the Rim: Living Downwind in the West* (New York: Verso Press, 1999), 147.

10. Garrett Hardin. The Tragedy of the Commons. *Science* 1968;162:1243–1248. Available: www.garretthardinsociety.org/articles/art_tragedy_of_the_commons.html.

11. Some of the most vicious dictatorships in modern history were (and many of today's are) tightly allied with capitalism. For instance, capitalists Joseph P. Kennedy and Prescott Bush, progenitors of three United States presidents, supported Hitler and Mussolini in the 1930s. On the socialist side, communist dictatorships did not exploit their resources as rapaciously as they may have intended, so their inheritors retain a higher proportion of mineral wealth than many western democracies. But the old Soviet Union implemented the world's most environmentally degrading and destroying policies right up until China surpassed them.

12. Hawken et al. Natural Capitalism, 6–9.

13. World Commission on Environment and Development. *Our Common Future* (New York: Oxford Paperbacks, 1987).

14. Jared Diamond. *Collapse: How Societies Choose to Fail or Succeed* (New York: Viking Press, 2005), 514–515.

15. Reisner. *Cadillac Desert*, 115 (Penguin ed.).

16. Information about management-intensive grazing practices can be found on many Web sites, including Alice Beetz (*Rotational Grazing: Livestock Systems Guide* [ATTRA Publication IP086, National Sustainable Agriculture Information Service, November 2004]. Available: attra.ncat.org/attra-pub/rotategr.html) and Henry M. Bartholomew, ed. (*Getting Started Grazing* (Ohio State University Extension Information). Available: ohioline.osu.edu/gsg/index.html; see also Rick Duff. *What Other Farmers Think About Management Intensive Grazing* (Ohio State University Extension Information). Available: ohioline.osu.edu/gsg/gsg_1.html; Ed Rayburn, Principles of Management-Intensive Grazing. *Forage Management* (West Virginia University Extension Service Newsletter) January 2003. Available: www.wvu.edu/~agexten/pubnwsltr/TRIM/5828.pdf). A very accessible account of exemplary management-intensive grazing practices focuses on Virginia management-intensive grazier and farmer Joel Salatin (Michael Pollan. *The Omnivore's Dilemma: A Natural History of Four Meals* [New York: Penguin Press, 2006], 49–53).

17. Post Carbon Institute. *The End of Suburbia* [DVD] (Electric Wallpaper Co., 2004); James Howard Kunstler. *The Long Emergency: Surviving the Converging Catastrophes of the Twenty-First Century* (New York: Grove Atlantic Press, 2005).

18. K. S. Deffeyes. *Beyond Oil: The View from Hubbert's Peak* (New York: Hill and Wang, 2005); Richard Heinberg. *Power Down: Options and Actions for a Post-carbon World* (Gabriola Island, Canada: New Society Publishers, 2004); see also monthly newsletters of the Association for the Study of Peak Oil and Gas. Available: www.peakoil.net; *Energy Bulletin*. Available: www.energybulletin.net; *Oil Drum*. Available: www.theoildrum.com.

19. Kunstler. *The Long Emergency*.

20. Heinberg. *Power Down*.

Appendix 1

1. Modified and expanded from the National Wilderness Preservation System. *Timeline for Management of Public Lands in the United States*. Available: www.wilderness.net/nwps/nwps_timeline.cfm. A comprehensive account of national environmental laws is presented by Zygmunt

J. B. Plater, Robert H. Abrams, and William Goldfarb. *Environmental Law and Policy, Nature, Law, and Society* (St. Paul, Minn.: West Publishing Company, 1992). We highly recommend this book for its clear writing and examples that illustrate the workings of each law.

2. U.S. Department of Agriculture Forest Service. *The Permanent Good of the Whole People* (1965); see also D. R. Whitnah, ed. *Greenwood Encyclopedia of American Institutions, Government Agencies* (Westport, Conn.: Greenwood Press, 1983).

3. U.S. Department of Agriculture Forest Service. A *Description of Forest Service Programs and Responsibilities* (Rocky Mountain Forest and Range Experiment Station, 1989).

Appendix 2

1. Information mainly from Zygmunt J. B. Plater et al. *Environmental Law and Policy*, 243–255.

2. Federal agencies responsible for permitting proposed projects typically contract with private companies for Environmental Impact Statement (EIS) preparation, but the project proposers pay for them. Those EISs have tended to be general and formulaic, rather than site specific, with carefully considered descriptions of potential impacts and their mitigations. The National Environmental Policy Act (NEPA) contains provisions aiming to prevent biased EISs—through requirements that responsible regulatory agencies evaluate EISs, but they generally lack the expertise. NEPA also mandates reviews by other agencies with appropriate expertise, but these are rarely obtained.

3. Richard Caplan. *Permit to Pollute, How the Government's Lax Enforcement of the Clean Water Act Is Poisoning Our Waters* (U.S. Public Interest Research Group, August 2002), 1.

4. Caplan, Permit to Pollute.

5. "Cradle to grave" is inaccurate because so many "disposal" methods cannot provide a final resting place for the wastes, and waste problems continue beyond the grave, including pollution coming from cemeteries.

6. Under the Resource Conservation and Recovery Act, "solid" wastes can include "solid, liquid, semisolid, or contained gaseous material resulting from industrial, commercial, mining, and agricultural operations" if the liquids and gases can be held in containers.

7. U.S. General Accounting Office. *Hazardous Waste: Effect of Proposed Rule's Extra Cleanup Requirements Is Uncertain* (GAO-01-57, 2001), 1.

8. Stated the National Environmental Policy Act review committee chairwoman, "What started as an overly vague single-paragraph statute is now 25 pages of regulations, 1,500 court cases, and hundreds of pending lawsuits that are blocking important projects and economic growth. Too often we are hearing horror stories about endless reams of paper needed to complete the environmental impact statements." Quoted in Rachel's Precaution Reporter No. 14 (29 November, 2005). Available: www.precaution.org/lib/05/ht051130.htm.

Appendix 5

1. U.S. Department of Energy. *Closing the Circle on the Splitting of the Atom* (1996), 21.

2. U.S. Department of Energy. *Linking Legacies* (1997), 29; Peter Kuran. *Trinity and Beyond* (Documentary Film Works Production, Goldhill DVD, 1999).

3. Peter Kuran. *Trinity and Beyond*; quotation is part of narration; P. P. Parekh et al. Radioactivity in Trinitite [fused soil at Trinity site] Six Decades Later. Journal of Environmental Radioactivity 2006;85:103–120.

4. A 15 June 2000 news item notes that in March 2000, U.S. Navy tests in the Bahamas, which generated intense underwater noise or explosions, may have caused a mass stranding of healthy beaked whales. The whales had signs of hemorrhaging in or around their ears, consistent with the effects of distant explosions or intense acoustic effects—so we can only imagine the sea life holocaust that may have followed oceanic nuclear bomb blasts. More beaked whale strandings occurred in 2002, possibly due to naval exercises using powerful underwater air guns in the Gulf of California and off the Canary Islands in the eastern Atlantic. Researchers claim there is no evidence that the air guns harm whales, but they have not sought any evidence for testing the claims (David Malakoff. Suit Ties Whale Deaths to Research Cruise. *Science* 2002;298:722–723).

5. John Eichelberger et al. Nuclear Stewardship, Lessons from a Not-So-Remote Island. *Geotimes* 2002;20–23.

Appendix 6

1. The test area can be found on a U.S. Geological Survey topographic map quadrangle named Groom Mine SW 7.5′, but three different documents assigned ground zero for the experiment to different coordinates. The State of Nevada and U.S. Department of Defense place ground zero in a different quadrangle than the U.S. Geological Survey site location and are almost certainly in substantial error. Another document, also clearly in error, places the center of Area 13 more than 22 miles to the southeast, within the Desert Wildlife Range (U.S. Department of Defense, Defense Environmental Restoration Program. *Annual Report to Congress for Fiscal Year 1995* [1996]; U.S. Department of Energy. *CERCLA Preliminary Assessment of DOE's Nevada Operations Office Nuclear Weapons Testing Areas* [prepared for DOE by the Desert Research Institute II, Off-Site Areas, 1988], figure 3.10.3).

2. P. W. Merlin. *Project 57, Plutonium Dispersal Took Place near Groom Lake* (unpublished manuscript, 1995); the warhead was a XW-25 MB-1 "Genie" air-to-air missile, with a design yield of 1.5 kilotons.

3. The contaminated area shows plutonium concentrations above 10–40 picocuries per gram (pCi/g). Air-suspended plutonium particulates spread over much greater areas. High-volume air monitors at Tempiute, about 28 miles north-northeast of ground zero, recorded alpha-particle emissions of six disintegrations per minute per cubic meter ($d/m/m^3$), and monitors at Caliente, about 150 km (80 miles) east of ground zero, showed 1.0 $d/m/m^3$ (O. R. Placak et al. *Report of Off-Site Radiological Safety Activities for Project 57 Nevada Test Site* (M-7003; Albuquerque, N.M.: U.S. Atomic Energy Commission, Office of Test Operations, 1957)). The high-concentration plume orientation near ground zero suggests that the Caliente measurements sampled only the airborne plume edges.

4. U.S. Defense Nuclear Agency. *Plumbob Series 1957, United States Atmospheric Nuclear Weapons Tests* (Nuclear Test Personnel Review 6005F, 1981 [the 1981 publication date is not verified]).

5. R. O. Gilbert et al. Radionuclide Transport from Soil to Air, Native Vegetation, Kangaroo Rats and Grazing Cattle on the Nevada Test Site. *Health Physics* 1988;55:869–887.

6. Prominent perennial shrub and grass species include four-winged saltbush (*Atriplex canescens*), shadscale (*Atriplex confertifolia*), winterfat (*Eurotia lanata*), bud sagebrush (*Artemisia spinescens*), spiny hop-sage (*Grayia spinosa*), green-molly (*Kochia americana*), wolfberry (*Lycium andersonii*), and Indian ricegrass (*Oryzopsis hymenoides*). Nine additional perennial shrub and grass species and 43 species of annuals and herbaceous perennials were identified on the site (E. M. Romney et al. Some Ecological Attributes and Plutonium Contents of Perennial Vegetation in Area 13. In P. B. Dunaway and M. G. White, eds. *The Dynamics of Plutonium in Desert Environments* [NVO-142, U.S. Atomic Energy Commission, Nevada Operations Office, 1974], 91–106; A. Wallace et al. The Challenge of a Desert: Revegetation of Disturbed Desert Lands. In M. G. White et al., eds. *Transuranics in Desert Ecosystems* [NVO-181; U.S. Department of Energy, Nevada Operations Office, 1977], 17–40; J. Barth et al. Solubility of Plutonium and Americium-241 from Rumen Contents of Cattle Grazing on Plutonium-Contaminated Desert Vegetation in *in vitrio* Bovine Gastrointenstinal Fluids, August 1975 to January 1977. In W. A. Howard et al., eds. *The Radioecology of Transuranics and Other Radionuclides in Desert Ecosystems* [NVO-224; U.S. Department of Energy, Nevada Operations Office, 1985), 243–281]).

7. These sandy loam soils have well-developed horizons to depths of about five feet, typically consisting of a surface layer of vesicular silt, an underlying oxidized layer in which clay is relatively concentrated, and a lower layer enriched in calcium carbonate. Old soils commonly have a lower zone, primarily of calcium carbonate (caliche) (V. D. Leavitt. Soil Surveys of Five Plutonium-Contaminated Areas on the Test Range Complex in Nevada. In P. B. Dunaway and M. G. White, eds. *The Dynamics of Plutonium in Desert Environments* (NVO-142, U.S. Atomic Energy Commission, Nevada Operations Office, 1974), 21–27).

8. R. O. Gilbert and E. H. Essington. Estimating Total $^{239+240}$Pu in Blow-Sand Mounds of Two Safety-Shot Sites. In M. G. White et al., eds. *Transuranics in Desert Ecosystems* (NVO-181, U.S. Department of Energy, Nevada Operations Office, 1977), 367–408.

9. Gilbert and Essington (Estimating Total $^{239+240}$Pu) assign a very small percentage of mounds to animal activities and in tabulations call the noncoppice dunes around the bases of shrubs "complex mounds," implying in text that they are "blow sand." Mounds formed by colonies of burrowing animals in the Mojave Desert commonly are much larger than coppice dunes.

10. The National Research Council (*Management and Disposition of Excess Weapons Plutonium* [Washington, DC: National Academies Press, 1994], 19) uses a figure of 4 kilograms of plutonium per weapon; actual figures are classified. Nearly 9 pounds of plutonium-239 emits 252 curies, so the figure of 46 curies probably represents only about one-fifth of the actual plutonium this experiment originally dispersed.

11. R. O. Gilbert. Revised Total Amounts of ^{239,240}Pu in Surface Soil at Safety-Test Sites. In M. G. White et al., eds. *Transuranics in Desert Ecosystems* (NVO-181, U.S. Department of Energy, Nevada Operations Office 1977), 423–429. The average plutonium inventory at nine other safety test sites is 11 curies per site.

12. Includes plutonium-239 and plutonium-240 isotopes; these isotopes are normally not separated.

13. E. M. Romney et al. ^{239,240}Pu and ^{241}Am Contamination of Vegetation in Aged Plutonium Fallout Areas. In M. G. White and P. B. Dunaway, eds. *The Radioecology of*

Plutonium and Other Transuranics in Desert Environments (NVO-153, U.S. Department of Energy, Nevada Operations Office, 1975), 43–88; E. M. Romney et al. Plant Uptake of [239,240]Pu and [241]Am Through Roots from Soils Containing Aged Fallout Materials. In M. G. White et al., eds. *Environmental Plutonium on the Nevada Test Site and Environs, Nevada Applied Ecology Group* (NVO-171, U.S. Energy Research and Development Administration, Nevada Operations Office, 1977), 53–63.

14. Diethylene triamine pentaacetic acid (DTPA).

15. F. H. F. Au and W. F. Beckert. Influence of Microbial Activities on Availability and Biotransport of Plutonium. In M. G. White et al., eds. *Environmental Plutonium on the Nevada Test Site and Environs, Nevada Applied Ecology Group* (NVO-171, U.S. Energy Research and Development Administration, Nevada Operations Office 1977), 219–226.

16. D. D. Smith and D. E. Bernhardt. Actinide Concentrations in Tissues from Cattle Grazing a Contaminated Range. In M. G. White et al., eds. *Transuranics in Desert Ecosystems* (NVO-181, U.S. Department of Energy, Nevada Operations Office 1977), 281–303.

17. C. Blincoe et al. Studies of Transuranic Element Ingestion by Fistulated Steers Grazing Area 13 of the Nevada Test Site. In W. A. Howard et al., eds. *The Radioecology of Transuranics and Other Radionuclides in Desert Environments* (NVO-224, U.S. Department of Energy, Nevada Operations Office, 1985), 289–301; R. G. Patzer et al. Passage of Sand Particles Through the Gastrointestinal Tract of Dairy Cows. In M. G. White et al., eds. *Environmental Plutonium on the Nevada Test Site and Environs, Nevada Applied Ecology Group* (NVO-171, U.S. Energy Research and Development Administration, Nevada Operations Office, 1977), 151–165. The ingested material contained 3,600–11,100 pCi of plutonium-238, 85,000–400,000 pCi of plutonium-239, and 11,000–56,000 pCi of americium-241 (D. D. Smith. Grazing Studies on a Contaminated Range of the Nevada Test Site. In M. G. White et al., eds. *Environmental Plutonium on the Nevada Test Site and Environs, Nevada Applied Ecology Group* (NVO-171, U.S. Energy Research and Development Administration, Nevada Operations Office 1977), 139–149).

18. Smith and Bernhardt. Actinide Concentrations in Tissues from Cattle, 281–303; cattle ingested about 2 pounds of forage per 100 pounds of body weight.

19. Blincoe et al. Studies of Transuranic Element Ingestion, 289–301.

20. Barth et al. Solubility of Plutonium and Americium-241, 139–149.

21. Smith and Bernhardt. Actinide Concentrations in Tissues from Cattle, 281–303.

22. Barth et al. Solubility of Plutonium and Americium-241, 139–149.

23. Most of the radioactivity at Area 13 is in the 20–53 micrometer soil size fraction, dimension, meaning 20–53 millionths of an inch. The most damaging respirable fraction (<5 micrometers) contains about 5% of the total soil radioactivity. The mean percentage of total-soil plutonium-239 in the resuspendable (<100 micrometers) soil fraction was 88% (range, 73–99% for 10 samples) (E. M. Romney et al. Plant Root Uptake of Pu and Am. In W. A. Howard et al., eds. *The Radioecology of Transuranics and Other Radionuclides in Desert Environments* (NVO-224, U.S. Department of Energy, Nevada Operations Office, 1985), 185–199).

24. Concentrations of plutonium-239 measured in kangaroo rats (*Dipodomys microps*): 2,400 pCi in pelts, 4,700 pCi in gastrointestinal tracts, and 8.6 pCi in carcasses. Kangaroo rats move substantial amounts of soil in their burrowing activities, exposing themselves to

radionuclide contaminants, and may ingest contaminants on or in leaves and seeds (K. S. Moor and W. G. Bradley. Ecological Studies of Vertebrates in Plutonium-Contaminated Areas of the Nevada Test Site. In P. B. Dunaway and M. G. White, eds. *The Dynamics of Plutonium in Desert Environments* (NVO-142, U.S. Atomic Energy Commission, Nevada Operations Office, 1974), 187–212.

25. Richard Stone. New Effort to Thwart Dirty Bombers. *Science* 2002;296:2117–2118.

26. Richard Stone. Plutonium Fields Forever. *Science* 2003;300:1220–1224.

Appendix 7

1. The Soviet Union detonated 124 nuclear devices between 1965 and 1988 as part of their peaceful nuclear explosives program, 81 on Russian territory: 33 for seismological research, 21 for oil and gas extraction, 19 to create underground storage chambers for oil and gas, 2 for underground disposal of toxic liquid wastes, 1 for sealing a natural gas well, and 5 miscellaneous. Environmental consequences included widespread contamination in the depths and on the surface of the Earth, especially air pollution, contaminated groundwater, and enhanced geologic hazards (triggering landslides and earthquakes) (Vladislav Larin and Eugeny Tar. A Legacy of Contamination. *Bulletin of the Atomic Scientists* May-June 1999;18–20).

2. Trevor Findlay. *Nuclear Dynamite, the Peaceful Nuclear Explosion Fiasco* (Sydney: Brassey's Australia, 1990), 1–2.

3. Although the nuclear dynamite schemes described in this appendix seem bizarre, they only emulated the equally crackpot schemes for radiation after X-rays and natural radioactivity were discovered in the mid-1890s. Radiation was used to treat minor ailments such as ringworm and acne, and women's depression—by irradiating their ovaries. Discovered in 1898, radium was prescribed for heart trouble, impotence, ulcers, depression, arthritis, cancer, high blood pressure, blindness, and tuberculosis. Radioactive toothpaste and skin cream were marketed. Monsanto Research Corporation later proposed a plutonium-powered coffee pot, which would boil water for 100 years without a refueling, and a Boston company proposed radioactive uranium cufflinks because uranium is heavier than lead and the weight would prevent the cuffs from riding up (Catherine Caulfield. *Multiple Exposures, Chronicles of the Radiation Age* [New York: Harper and Row, 1989]; *Rachel's Environment and Health Weekly*. The Major Cause of Cancer—Part 1. 2000;No. 691). In 2001, schemes were announced that would recycle radioactive waste into steel for household implements.

4. Project Chariot, a proposal to blast out a deep harbor near Point Hope, Alaska, is one of the more fascinating Plowshare program stories (Dan O'Neill. *The Firecracker Boys* [New York: St. Martin's Press, 1994]; see also Six Questions for Australians About the Proposal to Use Nuclear Explosives to Dig a Harbor in Northwest Australia [editorial]. *Environment* 1969;11:16–19).

5. R. G. West and R. C. Kelly. A *Selected, Annotated Bibliography of the Civil, Industrial, and Scientific Uses for Nuclear Explosions* (TID-4500; Oak Ridge, Tenn.: U.S. Atomic Energy Commission, Division of Technical Information Extension, 1971).

6. J. B. Krygier. Project Ketch, Project Plowshare in Pennsylvania. *Ecumene* 1998;5:311–322.

7. Findlay. *Nuclear Dynamite,* 159.

8. Findlay. *Nuclear Dynamite,* 175.

9. A 1946 *Science Digest* article proposed this project years before the Plowshare Program's birth (Ron Kroese. Industrial Agriculture's War Against Nature. In Andrew Kimbrell, ed. *Fatal Harvest, the Tragedy of Industrial Agriculture* [Covelo, California: Island Press, 2002], 26).

10. O. W. Perry et al. *Project Carryall, Feasibility Study. California State Division of Highways, the Atchison, Topeka and Santa Fe Railway Co.* (U.S. Atomic Energy Commission, 1963).

11. National Research Council. *Application of the Plowshare Program of Nuclear Excavation Experimentation to Highway Construction* (Circular No. 20, 1966); B. C. Hughes. The Corps of Engineers' Nuclear Construction Research Program. In L. E. Weaver, ed. *Education for Peaceful Uses of Nuclear Explosives* (Tucson: University of Arizona Press, 1970), 81–103; F. N. Houser and E. B. Eckel. *Possible Engineering Uses of Subsidence Induced by Contained Underground Nuclear Explosions* (U.S. Geological Survey Professional Paper 450-C, 1962), C17–C18; see also West and Kelly. *A Selected, Annotated Bibliography.*

12. West and Kelly. *A Selected, Annotated Bibliography.*

13. D. J. Supkow et al. *Potential Site Investigation for Nuclear Energy Crater Experiment and Water Management in Arizona* (Tucson: University of Arizona, Hydrology and Water Resources Office and Engineering Experiment Station, 1967); G. D. Cohen and F. M. Sand. *Water Resource Applications, Underground Storage of Natural Gas, and Waste Disposal Using Underground Nuclear Explosions* (PNE-3008; Princeton, N.J.: Mathematica, 1967), 24–30.

14. A. M. Piper. *Potential Applications of Nuclear Explosives in Development and Management of Water Resources, Preliminary Canvass of the Ground-Water Environment* (U.S. Geological Survey Report TEI-873, 1968), 67.

15. Cohen and Sand. *Water Resource Applications,* 33.

16. F. E. Williams et al. *Potential Applications for Nuclear Explosives in a Shale-Oil Industry* (U.S. Department of the Interior Bureau of Mines Information Circular 8425, 1969).

17. M. A. Lekas. Economics of Producing Shale Oil, the Nuclear In-Situ Retorting Method. *Quarterly Journal of the Colorado School of Mines* 1966;61:91–107.

18. Harlan Zodtner, ed. *Industrial Uses of Nuclear Explosives, Plowshare Series* (Report No. 1, UCRL-5253; Livermore: University of California Radiation Laboratory, 1958).

19. J. W. Reed. *A Nuclear Explosion to Determine the Effects on Hurricanes During the International Geophysical Year* (Contract AT(29–1)-789; Albuquerque, New Mexico: Sandia Corporation, 1956).

20. Findlay. *Nuclear Dynamite,* 293–294; U.S. Atomic Energy Commission. *Significant Peaceful Nuclear Explosion Events and Related Activities* (NVO-134; Las Vegas, Nevada; U.S. Atomic Energy Commission, Nevada Operations Office, 1973).

21. The water came out of rock salt, which had melted and formed a pool about 176,000 cubic feet in volume at the bottom of the cavity, enough to overfill 16 Olympic swimming pools. Some of the water also came out of about 282,000 cubic feet of rock salt that collapsed into the pool. Smaller amounts of water (about 116 gallons per day) may have leaked from the overlying Culebra aquifer through a shaft seal (Vernon Brechin, personal communication, 1999).

22. Findlay. *Nuclear Dynamite,* 165 (emphasis added).

23. Inventories of radionuclides in the area around the Buggy crater included 18.5 curies of plutonium, 3.2 curies of americium-241, and 1.4–1.7 curies each of cobalt-60, cesium-137, and strontium-90 (R. D. McArthur and S. W. Mead. *Nevada Test Site Radionuclide Inventory and Distribution Program, Report #5. Areas 5, 11, 12, 15, 17, 18, 19, 25, 26, and 30* (NV10384-26; U.S. Department of Energy Nevada Operations Office, 1989).

24. In 1941, the lifetime limit of allowable radiation exposure for a human body was set at a tenth of a microcurie of radium (0.0000001 curie), or the equivalent amount of another radioactive material. Limits today generally are lower, but not given in terms of curies. *See* chapter 7 n. 35.

25. The thermonuclear bombs slated for most excavation projects required a fission triggering device to detonate them. Fission bombs create longer lived radionuclides, the "dirty" part of the explosives (see Arjun Makhijani and Scott Saleska. The Production of Nuclear Weapons and Environmental Hazards. In A. Makhijani et al., eds. *Nuclear Wastelands, a Global Guide to Nuclear Weapons Production and Its Health and Environmental Effects* [Cambridge, Mass.: MIT Press, 1995], 23–63).

26. H. P. Metzger. *The Atomic Establishment* (New York: Simon and Schuster, 1972).

27. A. W. Klement, Jr. *Nuclear Cratering Explosion Effects for Interoceanic Canal Feasibility Studies* (NVO-67, Revision 1; U.S. Atomic Energy Commission Nevada Operations Office, 1971).

28. Radioactive levels were 30–40 times normal background in Idaho Falls, Idaho; Spokane, Washington; Salt Lake City, Utah; and Denver, Colorado (Findlay. *Nuclear Dynamite*, 167–168).

29. Findlay. *Nuclear Dynamite*, 174.

30. Radioactivity showed up in three separate flaring tests, which burned gas at the well head. The total radioactivity released included 1,064 curies of krypton-85, 2,824 curies of radioactive hydrogen-3 (tritium), and 2.4 curies of carbon-14. Hydrogen is used for H-bomb (fusion) fuel. Tritium helps hydrogen atoms fuse together, creating the thermonuclear explosion.

31. Gerald Johnson, quoted in Findlay. *Nuclear Dynamite*, 190–191.

32. C. E. Arthur and F. E. Armstrong. *A Study of Radioactive Contamination Resulting from the Use of Nuclear Explosives for Stimulating Petroleum Production* (Report of Investigations 6684, U.S. Department of the Interior Bureau of Mines, 1965).

33. Larin and Tar. A Legacy of Contamination.

34. *Rachel's Environment and Health Weekly*. The Major Cause of Cancer.

35. Quoted in R. Hewlett and F. Duncan. *Atoms for Peace and War* (Berkeley: University of California Press, 1989), 529.

36. Held hostage by some senators demanding support for unrelated legislation, the U.S. Senate refused to ratify the Comprehensive Test Ban Treaty in August 1999. In 2001, the George W. Bush administration abandoned attempts to ratify the treaty.

37. *Rachel's Environment and Health Weekly*. The Major Cause of Cancer.

Appendix 8

1. Dr. Eric Grosfils's research into the issue for his geophysics course in the Department of Geology, Pomona College, Claremont, California, inspired and informed this account.

2. R. W. Nelson. Low-Yield Earth-Penetrating Nuclear Weapons. *Journal of the Federation of American Scientists* 2001;54:1–8; Stephen Young. Penetrating the EPW Myth, Why Bunker Busters Are a Bad Idea. *Catalyst* 2002;1:8–9 [EPW means earth-penetrating weapon]; Union of Concerned Scientists. *The Troubling Science of Bunker-Busting Nuclear Weapons* (Fact Sheet, 2003); C. E. Paine et al. *Countering Proliferation, or Compounding It?* (Natural Resources Defense Council, 2003). Available: nrdc.org.

3. In April 2003, U.S. warplanes dropped nonnuclear high-explosive "bunker-busters," consisting of four 2,000-pound bombs, on a building in Baghdad, Iraq, from an elevation of 20,000 feet. High-explosive bombs do not present the same need for containment, although their penetrating nose cones may be made from radioactive depleted uranium.

4. In misleading weaponeers' language, mini-nukes are "reduced collateral damage" weapons. The rationales for how cleanly bombs will be used, and how well the radiation will be contained, are nearly identical to the ones that supported "Plowshare" nuclear bomb tests, ostensibly for peaceful projects (see appendix 7).

5. Mark Bromley et al. *Bunker Busters, Washington's Drive for New Nuclear Weapons* (British American Security Information Council, 2002). To assure more-or-less containment of the B61-Mod 11 would require detonation at a depth greater than 2,600 feet (Natural Resources Defense Council. *The Bush Administration's Misguided Quest for Low-Yield Nuclear Bunker Busters* [2003]).

6. A poorly contained nuclear explosion simultaneously ejects a vertical column of radioactive material and a ringlike ground-level debris cloud, called a base surge, which speeds away from ground zero in all directions. The radius (R) in feet of the circular base surge area is approximately $4,000(Kt^{0.333})$. After the underground debris is ejected, the surface above the blast collapses into a crater with a radius on the order of $50(Kt^{0.333})$. For a 5-Kt bomb, R equals about 6,800 feet, translating to about 5.3 square miles covered with radioactive debris, and a crater about 170 feet across (Nelson. Low-Yield Earth-Penetrating Nuclear Weapons, 1).

7. Expressions for calculating the more conservative penetration depths include $D = \geq 450(Kt^{0.294})$, $D = \geq 300(Kt^{0.333})$, and $D = \geq 400(Kt^{0.333})$ (Christopher Paine, Natural Resources Defense Council, personal communication, 2 July 2003).

8. Nelson. Low-Yield Earth-Penetrating Nuclear Weapons, 5; R. W. Nelson. Low-Yield Earth-Penetrating Nuclear Weapons. *Science and Global Security* 2002;10:8.

9. Nelson, Low-Yield Earth-Penetrating Nuclear Weapons (2001), 5; Eric Grosfils, personal communication, 19 February 2003.

10. U.S. Congress Office of Technology Assessment. *The Containment of Underground Nuclear Explosions* (OTA-ISC-414; Washington, DC: U.S. Government Printing Office, October 1989), 54. To "improve" the situation, the Defense Science Board recommended developing payloads "capable of delivering and emplacing highly intrusive sensors and sensor arrays in support of adaptive strike planning." The sensors would be delivered in a salvo of "interrogation rounds," with the science fiction aims of obtaining deeply buried target images and assessing the target area's penetrability. After such an interrogation, however, the actual bomb attack might not be a surprise. (Defense Science Board. *Report of Task Force on Future Strategic Strike Forces* [2004], 6–5–6–6. Available: www.fas.org/irp/agency/dod/dsb/fssf.pdf).

11. Nelson. Low-Yield Earth-Penetrating Nuclear Weapons (2002), 8.

12. R. D. Jacobi et al. Methodology for Remote Characterization of Fracture Systems in Bedrock of Enemy Underground Facilities. *Reviews in Engineering Geology* 2001;14:27–59.

13. Nelson. Low-Yield Earth-Penetrating Nuclear Weapons (2002), 8. Sandia National Laboratories in New Mexico has patented a penetrator bomb, claiming that the weapon can dive up to 35 feet into reinforced concrete. But it penetrated only 12 feet in initial tests (Michael Scherer. Building a Better Bomb. *Mother Jones* May/June 2002;15–16). Neither depth could contain even the smallest of the contemplated nuclear explosions. The Defense Science Board asserted that penetration to 330 feet is sufficient to contain a 3-Kt bomb (*Report of Task Force on Future Strategic Strike Forces*, 6–13), but conservative calculations indicate a required depth of 575 feet.

14. Nelson, Low-Yield Earth-Penetrating Nuclear Weapons (2002), 2; see also M. A. Levi. Dreaming of Clean Nukes. *Nature* 428:892. Recommendations that multiple penetrators should be fired down the same hole, and penetration enhanced with "a sequence of optimally timed chemical explosions to blast a very deep crater, or to use massive shaped charges," either make the problem of containment more acute or pretty much eliminate the possibility (Defense Science Board. *Report of Task Force on Future Strategic Strike Forces*, 6–7, 6–13). Proposals to reduced the angle of impact sensitivity with precursor blasts to "clear surface obstacles and fracture surface layers" (6–13) only add to penetration problems. Again, such explosive preambles would be something of a warning.

15. Quotation in Nelson, Low-Yield Earth-Penetrating Nuclear Weapons (2002), 2.

16. A 1-Kt explosion will produce 41 billion curies of radiation one minute after detonation, falling to 10 million curies in 12 hours. Venting of a 1-Kt explosion may release 500,000 to one million curies of radiation. See n. 24 in appendix 7. (IRIS Consortium. *Nuclear Testing and Nonproliferation* [prepared at the Request of the Senate Committee on Governmental Affairs and the House Committee on Foreign Affairs of the United States Congress, February 1994], II-17).

17. Young. Penetrating the EPW Myth, 8–9.

18. Natural Resources Defense Council. *The Bush Administration's Campaign for a Return to Nuclear Testing* (2003). Available: nrdc.org.

19. Nelson. Low-Yield Earth-Penetrating Nuclear Weapons (2001), 1–8.

20. Attempts to repeal the Spratt-Furse Amendment failed in 2000, but the U.S. Senate approved repeal on 21 May 2003, along with $15.5 million for research on "a large hydrogen 'bunker buster' bomb with yields of tens of kilotons to a megaton." The Defense authorization bill permits only research on design and costs of both low-yield nuclear weapons and a "Robust Nuclear Earth Penetrator" (bunker buster); development and production require specific congressional authorization. Bunker buster funding was cut in half to $7.5 million (SecurityNet. *The "Mini-Nukes" Provision* (Union of Concerned Scientists, 19 November 2003). All funding for the Robust Nuclear Earth Penetrator, and for research on other nuclear bomb designs, was eliminated from the 2005 federal budget. Early 2006, however, brought the news of plans for redesigning the nuclear bomb stockpile in a head-to-head competition between the Lawrence Livermore and Los Alamos labs. Of course, the new designs will not need to be tested as the "plan is to develop a design that lies well within the experience—and within what we call the 'sweet spot'—of our historical test base" (lab official quoted in Ian Hoffman. Lab Officials Excited by New H-Bomb Project. *Oakland Tribune* 7 February 2006).

Appendix 9

1. Depletion of an oil field starts when the first barrel comes out of the ground. The answer to how much recoverable oil remains in the ground at any given time involves politics, economics, and technology, not just the amounts, and is not easy to discover. But production profiles show that many of the world's major producing fields have peaked and are in long-term decline. Close to home, examples include both of the U.S. supergiant fields, in east Texas and Prudhoe Bay, Alaska, which peaked many years ago. Neither improved technology nor price increases have stopped their steep production declines (C. J. Campbell. The Oil Peak, a Turning Point. *Solar Today* July/August 2001:40–43; M. R. Simmons. The World's Giant Oilfields. *M. King Hubbert Center for Petroleum Supply Studies Newsletter* 2002(1)).

2. C. J. Campbell and J. H. Laherrère. The End of Cheap Oil. *Scientific American* 1998;278:79. An educated approach to predicting future petroleum discoveries could be based on discovery trends, graphing cumulative oil discoveries against cumulative exploratory wells drilled ("creaming curve"), and factoring in statistical analyses of field sizes. This approach produces relatively credible numbers for producing fields, holding most of the small puddles that can be found in the future and leaving the few untested potential fields open for guesswork (C. J. Campbell. The Assessment and Importance of Oil Depletion. In K. Aleklett and C. Campbell, eds. *Proceedings of the First International Workshop on Oil Depletion* [Uppsala, Sweden, 2002]. Available: www.isv.uu.se). The U.S. Geological Survey estimates involve substantially more guesswork, however (Rembrandt Koppelaar. USGS WPA [World Petroleum Assessment] 2000 part 1: A Look at Expected Oil Discoveries. *The Oil Drum* 27 November 2006; Association for the Study of Peak Oil and Gas. *ASPO Newsletter* March 2006; No. 63:10–11).

3. Noncredible oil production figures abound. Trade journals, including *Oil and Gas Journal* and *World Oil*, generally accept "reserve" reports from companies or government agencies at face value, even if skewed or deliberately misrepresented. For example, *World Oil* reported 1,160 billion barrels in reserves in 1998, followed by 979 billion barrels in 1999. If both numbers were correct, they imply that the world consumed 181 billion barrels of oil in one year—seven times the known 1999 consumption figure. In 2000, the U.S. Geological Survey put cumulative world production at 710 billion barrels, whereas an estimate of 699 billion barrels was given for 1993 (effective date of estimate) (U.S. Geological Survey World Energy Assessment Team. *World Petroleum Assessment 2000* [4 CD-ROM disks; Digital Data Series-60, 2000]). In 2000, world production averaged on the order of 22 billion barrels per year, so the U.S. Geological Survey report implied a cumulative production increase of only 11 billion barrels from 1993 to 2000, 7% of the actual figure (U.S. Geological Survey. *Changing Perceptions of World Oil and Gas Resources as Shown by Recent USGS Petroleum Estimates* [Fact Sheet-145–97, 1997]).

4. Past production of conventional ("regular") oil from *Association for the Study of Peak Oil and Gas Newsletter* November 2006;No. 71. Available: www.peakoil.

5. "Reserves" also are called "proved reserves" (Campbell. The Assessment and Importance of Oil Depletion, 5). The Securities and Exchange Commission promulgated the regulations to prevent reserve estimate fraud, but their requirement ignores probable reserves, resulting in recoverable resources underestimates. In practice, reserves in old fields were estimated mainly by extrapolating the decline rates of all the wells to zero. Whole-field reserve estimates were impossible because of highly fragmented well ownership. Operators nevertheless recognized

probable and possible amounts of producible oil beyond what qualified as "proved," and expe-rience has shown that ignoring these quantities resulted in significant underestimation of resources present. Increased production beyond Securities and Exchange Commission–based proved reserves has been ascribed to improved technology, but actually is an artifact of the estimation method.

6. K. S. Deffeyes. *Hubbert's Peak, the Impending World Oil Shortage* (Princeton, New Jersey: Princeton University Press, 2001), 147; Campbell and Laherrère. The End of Cheap Oil, 80–81; Jean Laherrère. Published Figures and Political Reserves. *World Oil* January 1994:33; for other examples of reserve manipulation, see C. J. Campbell. *Middle East Oil: Reality and Illusion* (Uppsala Hydrocarbon Depletion Study Group, Uppsala University, Sweden, 2004). Available: www.peakoil.net/; Klaus Illum. *Oil-Based Technology and Economy: A Short Introduction to Basic Issues and a Review of Oil Depletion Projections Derived from Different Theories and Methods* (Copenhagen, The Netherlands, 2004), 14.

7. Illum. *Oil-Based Technology and Economy*, 14; M. R. Simmons, *Twilight in the Desert* (Hoboken, N.J.: John Wiley and Sons, 2005), chapter 12; Deffeyes. *Hubbert's Peak*, 6.

8. Campbell. Middle East Oil: Reality and Illusion.

9. U.S. Geological Survey World Energy Assessment Team. *World Petroleum Assessment 2000.*

10. In the 1995 U.S. Geological Survey report, scientist C. D. Masters wrote,

There is a lot of oil and gas in the world. Numerically, we can consider that we have already discovered sufficient resources (that is, we have *Reserves*, 1,103 billion barrels of oil and 5,136 trillion cubic feet of gas) for more than 50 years of continuing sub-stantial production. Of course, demand will increase to challenge any arithmetic cal-culation, but we have a lot of *Undiscovered Resource* potential....[W]e believe that the *Ultimate Resources* of recoverable conventional oil in the world are slightly in excess of two trillion barrels of which we have consumed, to date, about 700 BB (billion bar-rels). And, for gas, in a BTU sense, about an equal amount of resource energy exists, of which we have only used less than half as much as oil, or some 300 billion barrels of oil equivalent. In a resource sense, then, we have remaining about three-quarters of our oil and gas endowment....(C. D. Masters. Energy Realities of the World, a Projected Role for Fossil Fuels. In L. M. H. Carter, ed. *Energy and the Environment, Application of Geoscience to Decision-Making* (U.S. Geological Survey Circular 1108, 1995), 7–8; emphasis original)

In 2000, the U.S. Geological Survey projected a world oil endowment fully one trillion barrels higher than Masters's estimate (U.S. Geological Survey World Energy Assessment Team. *World Petroleum Assessment 2000*; see also Deffeyes. *Hubbert's Peak*, 155, 157).11. U.S. Geological Survey World Energy Assessment Team. World Petroleum Assessment 2000.

12. U.S. Geological Survey. *Reserve Growth Effects on Estimates of Oil and Natural Gas Resources* (Fact Sheet FS-119–00, 2000); U.S. Geological Survey. *The Significance of Field Growth and the Role of Enhanced Oil Recovery* (Fact Sheet FS-115–00, 2000); see also Leonardo Maugeri. Oil: Never Cry Wolf: Why the Petroleum Age Is Far from Over. *Science* 2004;304:1114–1115.

13. Campbell contends that

over all these years the major companies simply reported as much as they needed...to deliver a satisfactory financial result, and...that was normal commercial prudence....As the result of underreporting what they found…the reserves naturally grew over time…in reality technology did not add much to the reserves themselves, although it did keep production higher for longer and delivered more profit. (Campbell. The Assessment and Importance of Oil Depletion, 8)

Geologist Jean Laherrère considers "reserve growth" to be in large part an artifact of the Securities and Exchange Commission–based proved reserve estimates. He showed that technological advances enhanced conventional oil production from the old giant fields but did not significantly affect the amount of oil ultimately recovered (J. H. Laherrère. *Reserve Growth: Technological Progress, or Bad Reporting and Bad Arithmetic?* [2001]. Available: www.greatchange.org/ov-laherrere,reserve_growth_technological_Progress.html; see also Jean Laherrère. *Estimates of Oil Reserves* ([Laxenburg, Austria: International Institute for Applied Systems Analysis, 2001], 47–60. Available: www.greatchange.org/ov-laherrere.June_10_01.pdf). Laherrère's interesting account of oil and gas reserves terminology (a tangled web) (Jean Laherrère. *Uncertainty On Data and Forecasts* [San Rossone Italy: Association for the Study of Peak Oil, 18–19 July 2006]) also explains, "Reserve growth occurs when reserves are reported as the minimum (proved), but does not occur statistically when reported as mean (expected) value." The U.S. Geological Survey's belief in huge U.S. field growth, which apparently yielded its extreme world reserve growth estimates, is a phenomenon arising from the Securities and Exchange Commission requirement that only proved reserves be reported. An in-depth examination of reserve growth by Koeppelaar reveals the complexity in reserve growth numbers games (Rembrandt Koeppelaar. A Primer on Reserve Growth, Parts 1–3. *The Oil Drum* 16 December 2006 [pt. 1]; 25 December 2006 [pt. 2]; 8 January 2007 [pt. 3]. Available: www.theoildrum.com).

14. Campbell. The Assessment and Importance of Oil Depletion, 11–13; L. B. Magoon. *Are We Running Out of Oil?* (poster; U.S. Geological Survey Open-File Report 00–320, 2000). For a lucid discussion of the shape of the world production curve, see Richard Heinberg. *Power Down: Options and Actions for a Post-carbon World* (Gabriola Island, Canada: New Society Publishers, 2004), 34–45; K. S. Deffeyes. *Beyond Oil: The View from Hubbert's Peak* (New York: Hill and Wang, 2005), chapter 3. Three different forecasters expect the world oil production peak between 2003 and 2016 (R. C. Duncan. Three World Oil Forecasts Predict Peak Oil Production. *Oil and Gas Journal* 26 May 2003;18–21). In addition to geological exploration and engineering efforts, the rate of postpeak decline depends on many social and economic factors (see also L. F. Ivanhoe. Future World Oil Supplies: There Is a Finite Limit. *World Oil* October 1995;77–78, 80, 82, 86, 88; M. R. Simmons. Global Crude Supply: Is the Oil Peak Near? *World Energy* 2004;7(1); R. L. Hirsch et al. *Peaking of World Oil Production: Impacts, Mitigation, and Risk Management* [U.S. Department of Energy, February 2005]).

15. R. A. Kerr. USGS Optimistic on World Oil Prospects. *Science* 2000;289:237. One respected U.S. Geological Survey petroleum expert has expressed a much less rosy view (Magoon. *Are We Running Out of Oil?*). Fanciful predictions of 2112 as the peak year for world oil production are based on U.S. Geological Survey's 5% chance of new oil discoveries from 1995 through 2025, assuming 0% growth in demand from 2000 on. Note that the 3,000 billion barrel mean estimate of ultimately recoverable oil extends peak production by less than eight years over more judicious estimates of 2,400 billion barrels, given a plausible 1.6% increase in demand (Illum. *Oil-Based Technology and Economy*, 9). The Association for Study of Peak Oil's updated depletion model indicates peak production of conventional oil in 2005 and for all liquids by 2010. (*ASPO Newsletter 64* April 2006, item 692).

16. The Association for the Study of Peak Oil and Gas Newsletter presents a monthly assessment of predicted depletion profiles of various classes of oil and gas (Available: peakoil.net).

17. A thorough summary of technical and political aspects of world oil and gas peak estimates is given by J. H. Laherrère. *Oil and Gas: What Future?* (2006). Available: www.oilcrisis.com/laherrere/groningen.pdf.

18. The U.S. Geological Survey World Petroleum Assessment 2000 projected potential discoveries from 1995 to 2025, totaling a mean of 649 billion barrels of oil, plus 612 billion barrels of reserve growth outside the United States. Discoveries of natural gas liquids for that period are projected to total 207 billion barrels plus 42 billion barrels of reserve growth. For the United States, the U.S. Geological Survey National Oil and Gas Resource Team (*National Assessment of United States Oil and Gas Resources* (U.S. Geological Survey Circular 1118, 1995)) estimated mean discoveries of 83 billion barrels, including natural gas liquids plus 76 billion barrels of reserve growth. The total mean prognostication for the world in the 30 year period is 1,420 billion barrels of conventional oil, including reserve growth, but excluding natural gas liquids, implying average discoveries of 47.3 billion barrels/year (Association for Study of Peak Oil and Gas. *ASPO Newsletter* March 2006;63. Available: www.peakoil.net). The actual performance of discovery over the first eight years of the 30-year period falls far short of projections: The U.S. Geological Survey attributed the actual discovery of only 69 billion barrels in the eight-year period to some areas being closed to exploration (T. R. Klett et al. An Evaluation of the U.S. Geological Survey World Petroleum Assessment 2000. *American Association of Petroleum Geologists* August 2005;89:1033–1042).

Appendix 10

1. Tom Biegler (Fuel Cells: A Perspective. *ATSE Focus* (Australian Academy of Technological Sciences and Engineering) 1 January 2005) describes types of fuel cells and their attendant problems very well. The 2004 federal budget proposed $1.2 billion for hydrogen research and development, largely transferred from renewable energy and conservation research and development funding. Only $17 million is allocated to renewable resources, an $86 million cut. More than half of the hydrogen support will go to automakers and the energy industry, and another $22 million is earmarked for coal, nuclear power, and natural gas (B. C. Lynn. Outfront, Hydrogen's Dirty Secret. *Mother Jones* May/June 2003;15–17).

2. Hydrogen is only 0.0084 of the Earth's total composition (Walter Youngquist. *GeoDestinies* [Portland, Oregon: National Book Company, 1997], 257). It is, however, concentrated at the Earth's surface in the form of water and so is readily accessible.

3. Electrolysis of water to hydrogen plus oxygen first requires distilling the water, which yields mineral wastes. Electrolysis of seawater produces brines that must be disposed of. The energy-eating aspects of producing, transporting, and using hydrogen in fuel cells include converting high-voltage AC power to low-voltage DC power, distilling water, compressing or liquefying H_2 gas for transport, and then transporting and storing it. Together, those processes absorb 50% of the electric energy used for electrolysis. Reconversion of the hydrogen to electricity plus water in 50%-efficient hydrogen fuel cells, and subsequent DC/AC conversion, together absorb another 25%, leaving only 25% of the original electrical energy for consumer use (Ulf Bossel et al. The Future of the Hydrogen Economy: Bright or Bleak?

Fuel Cell Seminar 3–7 November 2003; Ulf Bossel. We Need a Renewable Energy Economy, Not a Hydrogen Economy. *Hydrogen and Fuel Cell Letter* 2003;18[9]).

4. A. B. Lovins. Twenty Hydrogen Myths. *Rocky Mountain Institute* 20 June 2003; 2(5). Available: www.rmi.org. Photochemical processes are being explored by S. U. Khan et al. (Efficient Photochemical Water Splitting by a Chemically Modified n-TiO_2. *Science* 2002;297:2243–2245; see also R. F. Service. Catalyst Boosts Hopes for Hydrogen Bonanza. *Science* 2002;297:2189–2190; Scott Fields. Making the Best of Biomass Hydrogen for Fuel Cells. *Environmental Health Perspectives* 2003;111:A38–A41). Other hydrogen-producing schemes include reforming biomass sugar and alcohol to extract H_2. This requires a platinum catalyst and high energy inputs to reach temperatures of 435–510°F and pressures of 390–785 pounds per square inch (Esteban Chornet and Stefan Czernik. Harnessing Hydrogen. *Nature* 2002;418:928–929; R. D. Cortright et al. Hydrogen from Catalytic Reforming of Biomass-Derived Hydrocarbons in Liquid Water. *Nature* 2002;418:964–967).

5. National Research Council. *The Hydrogen-Economy, Opportunities, Costs, Barriers, and R&D Needs* (Washington, DC: National Academies Press, 2004), ES-5.

6. P. M. Grant. Hydrogen Lifts Off, with a Heavy Load. *Nature* 2003;424:129–130.

7. Internal combustion gasoline engines are only 20–30% efficient, so theoretically the United States actually should need three million barrels of petroleum per day for transportation. We actually use 12 million barrels per day just for transportation, but engine inefficiencies waste nine million. To replace this usage with equivalent energy from hydrogen demands about 250,000 tons daily, twice the current global daily hydrogen production and 10 times the U.S. production (Lovins. Twenty Hydrogen Myths, 22, 39). To generate the required hydrogen by electrolysis requires adding 400 billion watts of continuously available electric power to the national electric grid, nearly doubling present U.S. average power capacity (Grant. Hydrogen Lifts Off, 129–130). According to Lovins (Twenty Hydrogen Myths, 24), "new nuclear power plants would deliver electricity at about two to three times the cost of new wind power, 5–10 times the cost of new gas-fired cogeneration in industry and buildings, and 10 to 30+ times that of efficient use, so they won't be built, with or without a hydrogen transition....Under no conceivable circumstances would a market economy choose nuclear power."

8. Both Lovins (Twenty Hydrogen Myths, 13) and Jeremy Rifkin (*The Hydrogen Economy* [New York: Jeremy P. Tarcher/Putnam, 2002], 193, 209) envision a "transition" period of unspecified length when hydrogen will be produced mostly in numerous small reformers at gas stations and in buildings running devices powered by hydrogen fuel cell electricity. Collecting polluting byproducts from all those disseminated sources will not be feasible. Centralizing the processes of reforming natural gas or gasifying coal for hydrogen production makes CO_2 capture simpler. But there is as yet no way to dispose of the huge CO_2 amounts that they would produce. Neither process addresses the limited supply of nonrenewable fossil fuels (David Goodstein. *Out of Gas, the End of the Age of Oil* [New York: W. W. Norton, 2004], 39; see also David Goodstein. Energy, Technology and Climate. In *Running Out of Gas, New Dimensions in Bioethics*. Available: www.its.caltech.edu/~dg/Essay2.pdf).

9. S. C. Davis and S. W. Diegel. *Transportation Energy Data Book*, edition 23 (U.S. Department of Energy, 2003), table 1.14. Available: www-cta.ornl.gov/data. Before oil prices started rising substantially in 2004, transportation was expected to account for 72% of oil consumption by 2020.

10. Producing a 2001 U.S. domestic car required an average of 3,309 pounds of metals and 256 pounds of plastics (derived from petroleum), plus glass, fluids, lubricants, and other materials (Davis and Diegel. *Transportation Energy Data Book*, table 4.15).

11. Figures based on an average medium-sized German car, driven 8,000 miles per year for 10 years, show that an average car generates only 40% of its total pollution and waste during its driving life. The remaining 60% comes from mining and transporting the raw materials, building the cars, and later disposing of them—generating 58% of car-associated pollution and waste before it even hits the road. For *each* car made in Germany, where environmental standards are higher than in the United States, just extracting the raw materials produces 55,000 pounds of waste and nearly 15 billion cubic feet of polluted air; transporting the materials to factories yields another 15 billion cubic feet of polluted air and three gallons of crude oil pollutants; producing the car itself adds 3,300 pounds of waste and 2.6 billion cubic feet of polluted air; operating the car for 10 years at 24 miles per gallon of gas contributes 49 tons of CO_2, 10.6 pounds of sulfur dioxide, 103 pounds of nitrogen dioxide, 717 pounds of carbon monoxide, and 79 pounds of hydrocarbons to air and water, along with 40 pounds of road, tire, and brake abrasion products to water and soil (*Oeko-bilanz Eines Autolebens* [Heidelberg, Germany: Umwelt-und Prognose-Institut Heidelberg, 1993], reported in John Whitelegg. *Dirty from Cradle to Grave* [Eco-Logica Ltd., Transport and Environment Consultancy, 1993]). Operation of all vehicles on hydrogen fuel would reduce the pollutants from operating cars while only partially eliminating pollutants from producing the hydrogen.

12. The energy content in 2.2 pounds of hydrogen is the same as in one gallon of gasoline, weighing 6.2 pounds (Lovins. Twenty Hydrogen Myths, 5.)

13. Youngquist. *GeoDestinies*, 257–258; Bjørnar Kruse et al. *Hydrogen, Status and Possibilities* (Bellona Foundation Oslo, Report 6–2002, 2002). Available: www.bellona.no.en/energy/hydrogen/report_6–2002/index.html.

14. Compression to 800 bars of pressure absorbs 13% of the energy carried by hydrogen, and liquifaction absorbs 30–50% (Ulf Bossel. *Does a Hydrogen Economy Make Sense?* (European Fuel Cell Forum, Oberrohrdorf, Switzerland, 2005), 11. Available: www.efcf.com).

15. Kruse et al. (*Hydrogen, Status and Possibilities*) and Davis and Diegel (*Transportation Energy Data Book*, table 6.9) compare the advantages and disadvantages of hydrogen storage methods.

16. Richard Stone and Phil Szuromi. Powering the Next Century. *Science* 1999;285:677.

17. Hydrogen gas stored as metal hydride must be 99.999% pure because even in very small quantities, other gases reduce the binding forces between hydrogen and metal powders. But for some fuel cell applications, cheaper 99.99% or 99.9% purity may suffice (Lovins. Twenty Hydrogen Myths, 21). The powders are held in pressurized containers capable of maintaining at least 150 Psi pressure.

18. To store 4.4 pounds of hydrogen, equivalent to only two gallons of gasoline, a metal hydride cartridge might weigh as much as 400 pounds. The equivalent of a 20-gallon tank of gasoline would add more than 4,000 pounds to the weight of a car (Bossel et al. The Future of the Hydrogen Economy, 12).

19. Lovins. Twenty Hydrogen Myths, 16. Filament-wound carbon-fiber tanks add "only several hundred dollars" to the price of a hydrogen fuel cell car, but only the rich likely can

afford the total cost for some time (D. W. Keith and A. E. Farrell. Rethinking Hydrogen Cars. *Science* 2003;301:315–316). Safety is predicated on leaked hydrogen dispersing rapidly, its low radiant heat when burning, and smokeless burning (Lovins. Twenty Hydrogen Myths, 9). Higher pressure tanks can improve the refueling range, tanks able to contain pressures of 10,000 Psi are under development, but a puncture that allows rapid hydrogen release could turn a light-weight car into a misguided missile.

20. Initial cost estimates for creating a hydrogen refueling infrastructure exceed $5,000 per vehicle, even assuming savings from large-scale development (Keith and Farrell. Rethinking Hydrogen Cars, 315–316). The cost of a stationary fuel cell, supplying 200 kilowatts of electricity to a small commercial building or 14 average homes, currently is $900,000. The cells have a 20-year life expectancy and require a $300,000 overhaul every five years. Transporting hydrogen by road to refueling stations shares the impracticalities of hydride transportation: Carrying hydrogen able to generate the energy contained in a single gasoline tanker would require 22 pressurized hydrogen gas tankers of the same gross weight, or 4.5 tankers carrying liquid hydrogen (Bossel. *Does a Hydrogen Economy Make Sense?* 10).

21. Invisible and odorless hydrogen fuel has certain problematic properties. The flame also is invisible and odorless, and not very hot, but the fire potential is high because a broader range of H_2 concentrations is flammable in air compared to any other fuels in use or under consideration. Hydrogen cannot be odorized like natural gas for leak detection because it has the lowest molecular weight of any material and so is more mobile than any other gas. Also, many materials, including the sulfur compounds that odorize natural gas, would irreversibly poison fuel cell catalysts (Russell Moy. Liability and the Hydrogen Economy. *Science* 2003;301:47). These properties contribute to the significant proportion of industrial accidents from undetected hydrogen leaks (National Aeronautics and Space Administration. *Safety Standard for Hydrogen and Hydrogen Systems, Guidelines for Hydrogen System Design, Materials Selection, Operations, Storage, and Transportation* [A-109, 1997]).

22. Lovins believes that "industry norms for hydrogen leak detection and safety interlocks are convincingly effective" (National Academy of Sciences. *The Hydrogen Economy*, ES-9).

23. M. J. Prather. An Environmental Experiment with H_2? *Science* 2003;302:581–582.

24. About half of atmospheric hydrogen gas comes from oxidizing methane and other hydrocarbons, and the rest from biogenic processes and combustion, including biomass (Prather. An Environmental Experiment, 582; see also Thom Rahn et al. Extreme Deuterium Enrichment in Stratospheric Hydrogen and the Global Atmospheric Budget of H_2. *Nature* 2003;424:918–921). Replacing 50% of current fossil fuels with hydrogen might increase atmospheric H_2 an average 3–10% (M. G. Schultz et al. Air Pollution and Climate-Forcing Impacts of a Global Hydrogen Economy. *Science* 2003;302:624–627). Unintended H_2 releases to the atmosphere also might increase water vapor abundance in the stratosphere, aiding the destruction of stratospheric ozone and changing atmosphere–biosphere interactions in unpredictable ways (T. K. Tromp et al. Potential Environmental Impact of a Hydrogen Economy on the Stratosphere. *Science* 2003;300:1740–1742). A Letters section in the journal *Science* contained an illuminating discussion on the possible consequences of molecular hydrogen releases to the atmosphere from expanded hydrogen fuel cell use (*Science* 2003; 302:1329–1333).

25. Photochemical H_2 removal process takes up one OH^- and releases one HO_2^{-3} plus water vapor. Eliminating H_2 from the atmosphere reduces OH^-, the primary methane sink, increasing the abundance of methane (M. J. Prather and D. H. Enhalt. Atmospheric

Chemistry and Greenhouse Gases. In J. T. Houghton et al., eds. *Climate Change 2001, the Scientific Basis* [Cambridge: Cambridge University Press, 2001], 239–287).

26. Schultz et al. Air Pollution and Climate-Forcing, 624–627.

27. Rifkin. *The Hydrogen Economy*, 215.

28. R. F. Service. The Hydrogen Backlash. *Science* 2004;305:958–961; Nurettin Demirdövenj and John Deutch. Hybrid Cars Now, Fuel Cell Cars Later. *Science* 2004;305:974–976; S. Pacala and R. Socolow. Stabilization Wedges: Solving the Climate Problem for the Next 50 Years with Current Technologies. *Science* 2004;305:968–972.

INDEX